Lecture Notes in Production Engineering

Lecture Notes in Production Engineering (LNPE) is a new book series that reports the latest research and developments in Production Engineering, comprising:

- Biomanufacturing
- Control and Management of Processes
- Cutting and Forming
- Design
- Life Cycle Engineering
- Machines and Systems
- Optimization
- Precision Engineering and Metrology
- Surfaces

LNPE publishes authored conference proceedings, contributed volumes and authored monographs that present cutting-edge research information as well as new perspectives on classical fields, while maintaining Springer's high standards of excellence. Also considered for publication are lecture notes and other related material of exceptionally high quality and interest. The subject matter should be original and timely, reporting the latest research and developments in all areas of production engineering. The target audience of LNPE consists of advanced level students, researchers, as well as industry professionals working at the forefront of their fields. Much like Springer's other Lecture Notes series, LNPE will be distributed through Springer's print and electronic publishing channels. To submit a proposal or request further information please contact Anthony Doyle, Executive Editor, Springer (anthony.doyle@springer.com).

Konrad von Leipzig · Natasha Sacks · Michelle Mc Clelland
Editors

Smart, Sustainable Manufacturing in an Ever-Changing World

Proceedings of International Conference on Competitive Manufacturing (COMA '22)

Springer

Editors
Konrad von Leipzig
Department of Industrial Engineering
Stellenbosch University
Stellenbosch, South Africa

Natasha Sacks
Department of Industrial Engineering
Stellenbosch University
Stellenbosch, South Africa

Michelle Mc Clelland
Department of Industrial Engineering.
Stellenbosch University
Stellenbosch, South Africa

ISSN 2194-0525 ISSN 2194-0533 (electronic)
Lecture Notes in Production Engineering
ISBN 978-3-031-15601-4 ISBN 978-3-031-15602-1 (eBook)
https://doi.org/10.1007/978-3-031-15602-1

This Springer imprint is published by the registered company Springer Nature Switzerland AG
The registered company address is: Gewerbestrasse 11, 6330 Cham, Switzerland

International Program Committee

Conference Chair

K. Von Leipzig, Stellenbosch University, South Africa (kvl@sun.ac.za)

Co-chairs

V. Hummel, Reutlingen University, Germany (Vera.Hummel@Reutlingen-University.DE)
N. Sacks, Stellenbosch University, South Africa (natashasacks@sun.ac.za)

International Program Committee

F. Ansari, Technical University of Vienna, Austria
L. Bam, Stellenbosch University, South Africa
A. H. Basson, Stellenbosch University, South Africa
T. Becker, Stellenbosch University, South Africa
D. Blaine, Stellenbosch University, South Africa
A. Braun, Reutlingen University, Germany
J. B. Byiringiro, Dedan Kimathi University of Technology, Kenya
I. Campbell, Loughborough, England
P. Conradie, Stellenbosch University, South Africa
O. Damm, LIIA Management Consultants, South Africa
C. J. du Plessis, Spatial Edge, South Africa
N. du Preez, Stellenbosch University, South Africa
C. J. Fourie, Stellenbosch University, South Africa
R. Genga, University of the Witwatersrand, South Africa

About CIRP

CIRP was founded in 1951 with the aim to address scientifically, through international cooperation, issues related to modern production science and technology. The International Academy of Production Engineering takes its abbreviated name from the French acronym of College International pour la Recherche en Productique (CIRP) and includes some 600 members from 50 countries. The number of members is intentionally kept limited, so as to facilitate informal scientific information exchange and personal contacts.

CIRP aims, in general, at:

- Promoting scientific research, related to
  - manufacturing processes,
  - production equipment and automation,
  - manufacturing systems, and
  - product design and manufacturing
- Promoting cooperative research among the members of the academy and creating opportunities for informal contacts among CIRP members at large
- Promoting the industrial application of the fundamental research work and simultaneously receiving feedback from industry, related to industrial needs and their evolution.

CIRP has its headquarters in Paris, staffed by permanent personnel, and welcomes potential corporate members and interested parties in CIRP publication and activities in general.

CIRP Office, 9 rue Mayran, 75009 Paris, France. Web: http://www.cirp.net.

Foreword

Welcome to the eighth International Conference on Competitive Manufacturing hosted by the University of Stellenbosch and organized by the Department of Industrial Engineering.

The recent COVID pandemic yet again illustrated the interconnectedness of countries, companies, and individuals and also highlighted the dependencies on one another. In a small world where global trade is a given, international competitiveness stays a challenge. It requires high-quality products manufactured with state-of-the-art technologies at low cost under the assumption of highly efficient operations management as well as clear corporate goals and strategies. This in turn is facilitated by and dependent on improved engineering training, education, and relevant applied research, fueled by active interaction between academia and industry.

The main objective of COMA '22, the International Conference on Competitive Manufacturing, is to present recent developments, research results, and industrial experience accelerating improvement of competitiveness in the field of manufacturing. The close to 80 papers and presentations invited or selected to be delivered at the Conference deal with wide aspects related to product design and realization, production technologies and systems, operations management as well as enterprise design and integration. The worldwide participation and range of topics covered indicate that the Conference is truly a significant meeting of people striving for similar aims. The event is an additional opportunity for communication between paper authors and attendees, which undoubtedly will serve as a further step toward exciting developments in the future. It also provides ample opportunities to further exploit international research collaboration and collaboration between academia and practice. As in the past, we hope that the event will lead to tangible new outreach endeavors not only between existing collaborators, but also opening new opportunities to stimulate increased productivity and entrepreneurial ideas, so vital for an economy challenged by the COVID pandemic.

The chairmen and the organizing committee express heartfelt thanks and gratitude to the members of the international program committee, who have given their help and expertise in refereeing the papers and will chair the plenary and technical sessions during the Conference, as well as to the authors for participating and ensuring that the high standards required on an International Conference were maintained. These thanks and gratitude are extended to our highly regarded plenary speakers.

The chairmen convey sincere thanks to the conference sponsors for their generous support, which made this event possible.

The International Academy of Production Engineering (CIRP) is gratefully acknowledged for the scientific sponsorship given to the Conference.

Finally, the tremendous effort of the organizing committee is appreciated. Grateful thanks are due particularly to the Conference secretariat for ensuring the success of COMA '22.

We hope that you will find the Conference interesting and stimulating!

Stellenbosch, South Africa

Mr. K. H. von Leipzig
Conference Chair

Submission Review Process

A formal "Call for papers" for the 8th International Conference on Competitive Manufacturing (COMA '22) was issued in May 2021 to submit an 'Abstract' within the identified tracks/themes. Abstract submissions were subjected to an internal reviewing process, whereby successful submissions were notified and invited for presentation to the conference. Authors were subsequently invited to submit the 'Full Paper', which was published as a conference proceeding. Both the Abstracts and Full Papers were submitted online through the EasyChair submission page https://easychair.org/my/conference?conf=coma22 where acknowledgement of receipt was sent to authors. Authors were informed that a double-blind review process is applied to Full Paper submissions.

The following dates were set by the organising committee:

- Call for papers (1st May 2021)
- Submission of abstracts (12th July 2021)
- Notification of acceptance of abstracts (16th July 2021)
- Submission of full papers (28th January 2022)
- Feedback on paper reviews (8th February 2022)
- Revised paper submissions (15th February 2022)

Abstracts were required to be a maximum length of 400 words. Full Papers were required to be a maximum length of 6 pages, but leniency was given for the Author biographies and references. Full Paper submissions were required to adhere to a specific template and format which was placed on the conference site here: https://blogs.sun.ac.za/coma/callforpapers/.

A double-blind reviewing process was used for the Full Paper submissions. As such, both the reviewer and author identities are concealed from the reviewers, and vice versa, throughout the review process. Each Full Paper submission was sent to a minimum of two reviewers, with a third reviewer being requested in case of non-consensus between the first two reviewers. The reviews were completed by national and international academics, and experts in the respective field, listed on the International Programme Committee page.

A total of 45 reviewers participated in the review process, each reviewing between two and five papers. Reviewers were asked to review submissions according to the following criteria and were encouraged to provide recommendations and suggestions

- Does the title reflect the contents of the paper?
- Does the paper relate to what has already been written in the field?
- Do you deem the paper to be proof of thorough research and knowledge of the most recent literature in the field of study?
- Is the paper clearly structured, easy to read and with a logical flow of thought?
- Are the arguments employed valid and supported by the evidence presented?
- Are the conclusions clear and valid?
- Does the paper conform to accepted standards of language and style?
- Any other recommendation(s)?
- Select reviewer recommendation: 'Accept Submission', 'Revision Required', or 'Decline Submission'

Reviewer feedback was saved on the submission system, where acceptance emails together with review comments were sent to the authors, allowing them to revise the submission. The authors were given between 2 and 4 weeks to incorporate changes, after which the final document was submitted for approval and publication as a conference proceeding.

Topics

Papers were invited in the following areas relevant to the conference themes:

Product Design and Realisation:
Design for manufacturing and assembly, reverse engineering, CAD/CAE, concurrent engineering, design for additive manufacturing, biologically inspired design approaches, virtual prototyping, networks in product development, open design.

Production Technologies:
Expert systems in manufacturing, CAD/CAM Systems, HSC, EDM, forming, additive manufacturing, casting, metrology, mechatronics, precision manufacturing, biomanufacturing, robotics, sensing, assembly, automation, intelligent manufacturing, biologically inspired manufacturing processes, non-conventional machining, environmental aspects, machining of materials, abrasive processes, hybrid processes, laser-based manufacturing, green manufacturing, coating technology.

Production Systems and Organisations:
Production planning and control, logistics, modelling and simulation, SW-applications, communication networks, 5G network applications, social manufacturing, learning factory, digital factory, biological transformation in production systems, cyber-physical approaches, big data, predictive maintenance,

asset management, human-machine collaboration, employee qualification, human resource management, IoT in manufacturing, manufacturing digitization challenges, augmented and virtual reality, lean manufacturing, sustainable manufacturing.

Enterprise Design and Integration:
Knowledge management, product life cycle, human interface, integrated design and manufacturing, technology and innovation management, total quality management, distributed control systems, socio- economic and environmental issues, artificial intelligence and machine learning, digitals twins, virtual setup, subscription vs selling.

Supply Chain Management:
Supply chain track and tracing; digital supply networks, blockchain in supply chains, circular economy, artificial intelligence for supply chains, biological transformation in supply chains.

COVID-19: Manufacturing and Supply Chain:
Post-pandemic business models, Supply chain localisation, manufacturing as a service, Rapid medical device manufacturing, Distributed manufacturing, Constrained supply chains, Resilient supply chains.

Materials and Manufacturing:
Smart materials, Recycling, Remanufacturing, Future materials, Biomaterials, Sustainable materials, Nanomaterials, Coatings, Metal matrix composites.

Acknowledgements

Sincere thanks to our distinguished supporters and sponsors, whose generosity made possible the success of this Conference.

Stellenbosch
UNIVERSITY
IYUNIVESITHI
UNIVERSITEIT
forward together
sonke siya phambili
saam vorentoe

CIRP

ENGINEERING
EZOBUNJINELI
INGENIEURSWESE

Contents

Novel Engineering Concept and Innovation

Human Integration

COVID-19 Manufacturing, Services, Business Models

Production Systems and Organizations

Identification of Residual Development Efforts in Agile Ramp-Up Production

Thomas Bergs, Sebastian Apelt, Malte Becker, Alexander Beckers, and Sebastian Barth

Abstract Agile product development is increasingly finding its way into the development of physical products. The subsequent transfer of a planned and still unstable manufacturing process into stable series production after the design freeze is the goal of ramp-up production, but confronts manufacturing companies with different challenges. A currently high level of changes to the product geometry and the planned manufacturing sequence due to not achieved requirements in late phases of the ramp-up production (Residual Development Efforts—RDE) results in time-consuming and cost-intensive changes to the product and manufacturing sequence, which leads to failure to achieve ramp-up targets. The goal of current research is therefore to increase the agility of ramp-ups and to integrate the ramp-up production into the phase of agile product development. This offers the potential to use the increased dynamics of the product development process and the knowledge already generated for the validation and stabilization of the manufacturing process. However, due to the integration of ramp-up production into product development, there are additional far-reaching effects of product and technology uncertainties prevalent in agile product development on the design of agile ramp-up production. Additional uncertainties regarding the product geometry due to non-finalized designs and the resulting

T. Bergs · S. Apelt (✉) · M. Becker · A. Beckers · S. Barth
Laboratory for Machine Tools and Production Engineering WZL of RWTH Aachen University, Chair of Manufacturing Technology, Aachen, Germany
e-mail: s.apelt@wzl.rwth-aachen.de

T. Bergs
e-mail: t.bergs@wzl.rwth-aachen.de

M. Becker
e-mail: malte.becker@rwth-aachen.de

A. Beckers
e-mail: a.beckers@wzl.rwth-aachen.de

S. Barth
e-mail: s.barth@wzl.rwth-aachen.de

T. Bergs
Fraunhofer Institute for Production Technology IPT, Aachen, Germany

K. von Leipzig et al. (eds.), *Smart, Sustainable Manufacturing in an Ever-Changing World*, Lecture Notes in Production Engineering,
https://doi.org/10.1007/978-3-031-15602-1_1

uncertainties regarding the probability of use and achievement of the requirements of the manufacturing technologies initially result in additional residual development efforts. Furthermore, the interactions between the manufacturing technologies in the manufacturing sequence are thus subject to additional uncertainties, which also leads to increased RDEs. To meet this challenge, it is necessary to analyze prevailing uncertainties and predict their impact on potential changes in agile ramp-up production. Therefore, a methodology is presented, which enables the analysis of product and technology uncertainties and thus the identification of product and process-related changes (RDE) in agile ramp-up production.

Keywords Agile ramp-up production · Technology planning · Uncertainties · Residual development efforts · Manufacturing sequence

1 Introduction

In the course of globalization, manufacturing companies are confronted with various challenges in a dynamic competitive environment. These challenges include shorter product life cycles, changing customer requirements and increasing product variety [1]. In order to meet these challenges, an optimized product development process and a controlled transition from product development to series production, which is referred to as ramp-up production, are necessary [2]. The design freeze describes the point at which no more changes to the developed product are allowed and at which the release for the ramp-up production is set [3]. In conventional ramp-up production, which follows the design freeze, ramp-up targets are often not achieved. This means, for example, that the time-to-market and time-to-volume as well as the ramp-up budget are exceeded [4]. The non-achievement is due to instabilities in the manufacturing sequence, which are caused by fluctuating employee and technology capabilities on the one hand [5]. On the other hand, the instabilities are caused by residual development efforts in the ramp-up production. Residual development efforts are necessary subsequent developments to the product or to the technologies of a manufacturing sequence, which are based on insufficient product and technology maturity [2] as well as on an insufficient coordination between product development, technology planning and ramp-up management [3]. Manufacturing sequences describe the combination of value-adding process steps and handling technologies for the manufacture of a component [6]. It is difficult to address these challenges in the planning phase because manufacturing technologies are not physically connected to generate a manufacturing sequence until ramp-up production. Only through this connection previously unknown interactions between technology and product, but also between the technologies themselves, become visible [4].

Residual development efforts (RDEs) in ramp-up production are similar to constantly changing customer requirements in product development. Since the concept of agile product development is finding its way into product development to meet this challenge, current research focuses on the adaptation of agile methods to the ramp-up production [4]. In agile product development, the scrum approach is widespread, which divides a development project into short development cycles, referred to as sprints [7]. The increase in agility and the integration of ramp-up production into agile product development is referred to as agile ramp-up production. On the one hand, the integration aims to enable improved coordination between product development, technology planning and ramp-up management. On the other hand, the ramp-up production should become more agile through agile product development, so that problems (like RDEs or instabilities of the manufacturing sequence) can be identified earlier to be counteracted. Due to the integration, the ramp-up production no longer starts after the design freeze. Therefore, uncertainties regarding the product are present in the agile ramp-up production [4]. Since the technologies for the manufacturing sequence are planned in parallel with product development, there are technology uncertainties as well. The challenge for technology planning in agile ramp-up production is to validate a manufacturing sequence planned under uncertainty based on uncertain information. As a consequence of agile ramp-up production, the existing uncertainties have to be analyzed to avoid additional residual development efforts.

To address this challenge, the state of the art regarding existing methods for ramp-up production and product development is analyzed and the objective of the developed methodology is presented. Subsequently, the methodology is explained in detail and validated in a case study.

2 State of the Art

For agile ramp-up production, it is necessary that both product development and ramp-up production are considered. In addition, the manufacturing sequence and residual development efforts must be taken into account. Due to the integration of ramp-up production into product development, uncertainties are present in agile ramp-up production, which must also be taken into account. In the scientific literature, a variety of approaches exist which address either the ramp-up production or the product development. Most of the analyzed approaches addressing ramp-up production describe it from an organizational and socio-technical point of view and neglect the manufacturing sequence and residual development efforts, such as the approaches of Laick [8], Winkler [9], Dyckhoff et al. [10]. The approaches of Lanza and Stauder consider manufacturing technologies as a part of a manufacturing sequences but handling technologies and residual development efforts are not sufficiently addressed [3, 5]. It is concluded that existing approaches regarding ramp-up production do not allow a comprehensive consideration of residual development efforts.

The following section analyzes approaches from product development and their transferability to agile ramp-up production. Examples of such approaches are Cooper et al. and Sommer et al. However, these approaches describe product development from an organizational and socio-technical perspective [11, 12]. Rey's dissertation deals with the combination of technology planning and product development. Handling technologies and instabilities during ramp-up production are not considered [13]. Summed up, the ramp-up production is not considered in any of these approaches.

In addition to the described approaches, first approaches which consider both product development and ramp-up production exist, e.g. from Basse and De Lange. However, manufacturing sequences, modeling of uncertainties, and the identification of residual development efforts are not or insufficiently discussed [14, 15].

Conclusively, existing approaches do not allow a cross-phase consideration of product and technology uncertainties as well as a consideration of residual development efforts in agile ramp-up production. Furthermore, existing approaches neglect handling technologies in the manufacturing sequence. Therefore, a methodology is needed that considers the manufacturing sequence, product and technology uncertainties and the residual development efforts in the ramp-up production. This requires modeling product and technology uncertainties as well as predicting and evaluating residual development efforts in ramp-up production. The developed methodology is described in detail below.

3 Objective

The objectives of the methodology presented are to increase the agility of ramp-up productions and to enable users to systematically model existing uncertainties for the identification of residual development efforts under consideration of planned manufacturing sequences. For this purpose, uncertainties resulting from the integration of ramp-up production into product development must be taken into account in order not to additionally threaten ramp-up targets. The methodology enables the identification of problems at an early stage in the ramp-up production and the initiation of targeted measures to eliminate the problems. This improves target achievement during the ramp-up production and reduces the development time to series maturity.

4 Case Study

For a better understanding of the methodology, the details are given by means of a case study from industrial practice. As a consequence of the increasing demand for vehicles with low CO_2 emissions, legal regulations for the reduction of CO_2 emissions as well the high importance of a successful ramp-up production in the automotive industry, the validation of the developed methodology is carried out on the basis of a cylinder crankcase for an engine. One solution to the problem of reducing exhaust emissions and fuel consumption is to reduce friction. There is potential in optimizing the tribological system between the cylinder bore and the piston ring. The friction

Fig. 1 Case study [16]

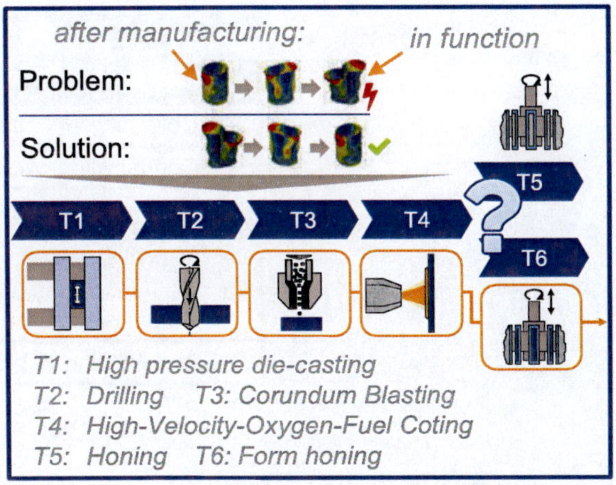

is largely dependent on the contour of the cylinder bore surface [16]. The aim is to manufacture the cylinder bore surface of the cylinder crankcase with a containment contour, see Fig. 1.

The development of the cylinder crankcase is conducted agile. Uncertainties exist with regard to the general use of a containment contour, the exact contour dimensions and shape, position and dimensional tolerances. In the case study, the containment contour could be realized by form honing (T6). However, there are uncertainties with regard to the tool life and the handling step that transfers the cylinder crankcases to the form honing technology. In order to achieve the shortest possible time to market, the ramp-up production is carried out in parallel with agile product development.

5 Methodology

In this section, the developed methodology is described. Uncertain manufacturing sequences and product characteristics represent the input of the methodology. The conceptual design of the methodology consists of two steps (see Fig. 2). In the first step, the uncertainty situation is modeled by collecting single information, which are afterwards combined to aggregated information. In the second step, RDEs are identified based on the modeled uncertainties, which are finally evaluated for the prioritization for the execution of prototype tests and validation in the ramp-up production. The output of the methodology are prioritized potential residual development efforts for which countermeasures have to be determined. The determination of countermeasures is not part of this paper. In the following, both steps are presented in detail.

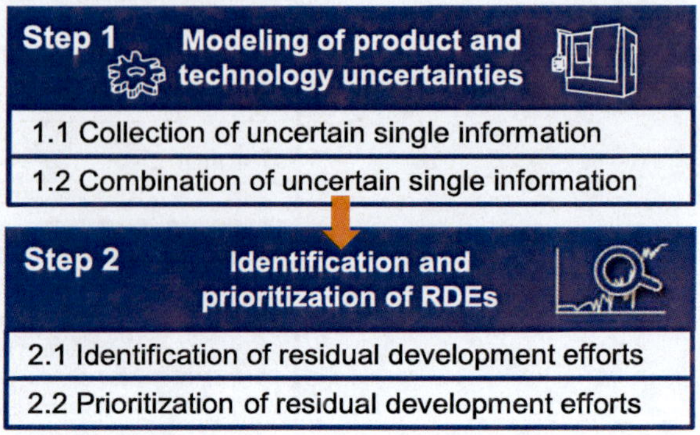

Fig. 2 Conceptual design of the methodology

5.1 Modeling of Uncertainties

The understanding of uncertainty in this approach is based on the generalized information theory, in which the terms information and uncertainty are linked with each other. According to this theory, uncertainties are information deficits, which can be reduced by appropriate additional information [17]. Furthermore, in this approach, uncertainties are divided into product and technology uncertainties. Product uncertainties result from uncertain product characteristics. Technology uncertainties are information gaps for the manufacturing of product characteristics as well as for handling. This step serves to quantify existing uncertainties regarding the developing product and the manufacturing sequence. This step is modeled based on the methodology of Rey, which allows the combination of various individual pieces of information into product requirement and technology capability profiles [13].

5.1.1 Collection of Uncertain Single Information

The first step of the methodology is to collect information regarding product requirements and technological capabilities of the planned manufacturing and handling technologies. A single Information (A) can be acquired from various sources such as standards, journals, technical books or expert statements. The certainty of single information (CI) describes how certain an information source is when providing a single information. A certainty of 100% corresponds to the highest certainty of an information source regarding a single information, whereas a certainty of 0% corresponds to the lowest certainty. Furthermore, the user of the methodology determines the reliability of a single information (RI), see Fig. 3.

The reliability of a single information describes the percentage of trustworthiness of an information source. This is relevant, because different sources of information have different credibility [13]. Subsequently, the collected information is modeled using the evidence theory from Dempster [18] and Shafer [19].

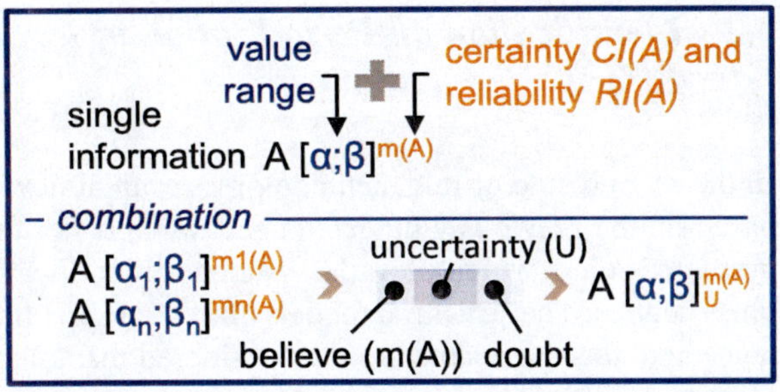

Fig. 3 Modeling on Uncertainties

In this theory of plausible reasoning, evidence describes an immediate insightfulness of findings and unprovable statements, for which the correctness can only be determined by their occurrence, or non-occurrence [20]. Evidence (also called base dimension) results from the certainty and reliability of the occurrence of a single piece of information (A). To do this, the certainty and reliability of a single information are multiplied, see formula (1) [13].

$$m(A) = CI(A) \cdot RI(A) \tag{1}$$

For example, an evidence of 100% results if an expert provides a single information with a certainty of 100% and the user of the methodology fully trusts this expert. In the case study introduced, one of the uncertainties is the tool life of the honing stones used in the manufacture of the cylinder crankcase. For this purpose, information is obtained from the design and technology planning departments, which differ in their credibility and in the specified interval range.

5.1.2 Combination of Uncertain Single Information

By combining different information from different sources, it is possible to generate an aggregate information considering the reliability of each source of information [20]. Thereby, fixed single values as well as value ranges can be specified by the information sources. By combining this single information (in the form of value ranges or single information), aggregated value ranges are evaluated with an uncertainty (based on the evidence). The uncertainty thus provides information about the certainty with which the final expression lies within the aggregated value (Interval). For this purpose, the combination rule of Yager [21] is used. This combination rule is a derivative of the combination rule according to Dempster [18], which is also suitable for processing contradictory information [21]. The result of the step is a combined information with a total certainty and uncertainty (U) [13]. Based on the case study, the information regarding the tool life from design and technology planning are combined. The result is a value range with a specified uncertainty that the actual tool life is assigned to. By acquiring more information, the range of values can be narrowed or the uncertainty can be reduced.

5.2 *Identification an Periodization of Residual Development Efforts*

In the second step of this methodology, residual development efforts are identified based on the previously modeled uncertainties. For this step, structural models are presented below, which allow the systematic identification of RDEs based on modeled uncertainties. The structural models offer the user a first point of reference for assistance and must be adapted to the considered manufacturing task. Subsequently, the identified residual development efforts are prioritized.

5.2.1 Identification of Residual Development Efforts

To identify RDEs, a fundamental distinction is made between product- and technology-driven uncertainties. Both structural models (product and technology) are analogous to a tree structure. The first level represents uncertainty classes to which the present uncertainties are assigned. Uncertainty classes based on product uncertainties are, for example, uncertainties regarding functionality fulfillment or geometric uncertainties. Regarding the use case, product uncertainty classes are differentiated into mechanical, hydraulic, pneumatic, electrical and magnetic, optical-physical, medical-biological, acoustic, optical-physical [22] and tribological [23] functionalities as well as geometrical uncertainties. Examples for classes of technology uncertainties are uncertain manufacturability, uncertain handling and uncertain process design (see Fig. 4).

The uncertainty classes are subdivided into subgroups (uncertain elementary function) on a second level, if possible. An elementary function is the smallest unit of a

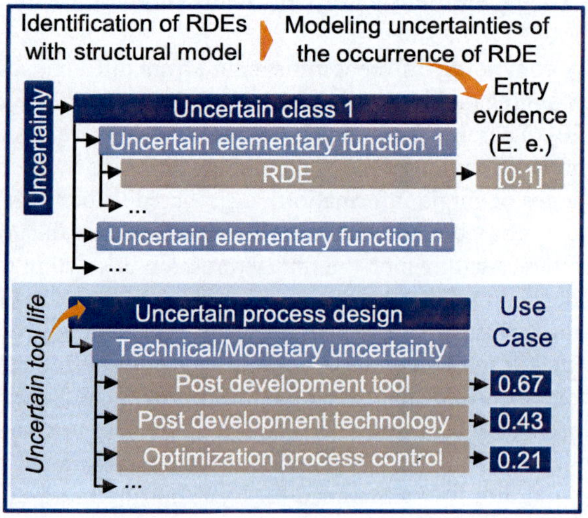

Fig. 4 Structural models for the identification of RDEs

function (or an uncertainty) that cannot be further subdivided. For example, Koller has already described elementary functions for energy and material converting technical systems. These are “transforming”, “enlarging or reducing”, “changing direction”, “conducting or isolating”, “dividing or collecting” and “mixing or separating” [24]. The elementary functions of uncertain manufacturability (technology uncertainties) are based on the classification of the performance of manufacturing processes for the manufacturability of component functionalities into the macroscopic, microscopic, mesoscopic and nanoscopic levels according to Klocke et al. [23]. Uncertainties regarding process design are divided in the elementary functions technical and monetary uncertainties as well as unintended effects. The next level is the derivation of RDEs from uncertainties within the elementary functions. An example of a residual development effort for unintended effects is damage to the component. An exemplary residual development effort for a conversion uncertainty is a design change of a macro-characteristic.

In the introduced use case, there is uncertainty regarding the tool life of the honing stones. These are technological uncertainties, which are assigned to the uncertainty class “uncertainties in process design”. With regard to the technological/monetary uncertainties, it is possible that the tool or the technology must be post-developed or the process control must be optimized.

Subsequently, the entry evidence is determined for each potential RDE. Following the procedure of a failure mode and effect analysis, various information are collected. The acquired information about the occurrence of the RDE are modeled based on the existing uncertainty analogously to Sect. 5.1 using the evidence theory of Dempster and Shafer. The modeled information are aggregated to an entry evidence (E. e.) using Yager’s combination rule. The entry evidence defines the certainty with which the residual development effort will occur.

5.2.2 Prioritization of Residual Development Efforts

The final step of the methodology is the prioritization of the residual development efforts (RDEs). The prioritization serves as a basis for deciding which RDE should be validated in the next sprint in order to be able to initiate appropriate countermeasures. For this purpose, the RDEs are weighted. To weight the residual development efforts the entry evidence (E. e) of Sect. 5.2.1 is multiplied by the uncertainty (U) from Sect. 5.1.

$$prio = U \cdot E.e. \tag{2}$$

For example, the resulting uncertainty for post-development on the tool was evaluated with an uncertainty of $U = 0.35$. The entry evidence for the post-development on the tool was evaluated with $E.\ e. = 0.67$. Thus, the prioritization value is $prio = 0.23$. The available residual development efforts are then sorted in a table according to descending priority (prio) (Fig. 5). Finally, the RDEs are clustered. For this purpose, the ABC analysis is used [25]. For clustering, the prioritization values of the RDEs

Fig. 5 Prioritization of RDE

RDE	U	E. e.	prio
...	[0;1]	[0;1]	U · E. e.
Post development tool	0.35	0.67	0.23
...	...	...	...
Design change macro-characteristic	0.15	0.21	0.12
...	...	...	...

Cluster A Cluster B

are cumulated and the RDEs are assigned to the classes. Class A (80%—very important RDEs) contains the RDEs which must be analyzed in the next sprints. Class B (15%—moderately important RDEs) contains RDEs that have a certain urgency but whose occurrence probability is relatively low. Class C (5%—relatively unimportant RDEs) contains RDEs that do not need to be analyzed immediately because their occurrence probability is considered to be very low [26]. Based on the prioritized residual development efforts, countermeasures for their elimination must be determined in the next step.

6 Conclusions

The presented methodology allows technology planners to model uncertainties as well as to identify and prioritize residual development efforts. Uncertainties are modeled using the evidence theory of Dempster-Shafer [18, 19] and the combination rule of Yager [21]. Uncertainty modeling considers the technological level of the manufacturing sequence. In addition, the methodology enables the identification and prioritization of residual development efforts through the application of two newly developed structural models. A case study was used to demonstrate the practical application of the methodology.

The methodology forms the basis for analyzing and reducing RDEs in agile production ramp-up. In future research it will be necessary to develop models and methods that support users in the analysis of RDEs through the targeted derivation of test plans for prototype production. In addition, an evaluation methodology must be developed, which, in addition to the probability of occurrence, considers the effects of the RDEs of agile ramp-up production as well as the validation time and the validation options in the agile ramp-up production.

Acknowledgements "Funded by the Deutsche Forschungsgemeinschaft (DFG, German Research Foundation) under Germany's Excellence Strategy—EXC-2023 Internet of Production—390621612".

References

1. Rey, J., Apelt, S., Trauth, D., Mattfeld, P., Bergs, T., Klocke, F.: Highly iterative technology planning: processing of information uncertainties in the planning of manufacturing technologies. Prod. Eng. Res. Devel. **13**(3–4), 361–371 (2019)
2. Kukulies, J., Schmitt, R.: Stabilizing production ramp-up by modeling uncertainty for product design verification using Dempster-Shafer theory. CIRP J. Manuf. Sci. Technol. **23**, 187–196 (2018)
3. Lanza, G.: Simulationsbasierte Anlaufunter-stützung auf Basis der Qualitätsfähigkeiten von Produktionsprozessen. Dissertation (2005)
4. Bergs, T., Apelt, S., Beckers, A., Barth, S.: Agile ramp-up production as an advantage of highly iterative product development. Manuf. Lett. **27**, 4–7 (2021)
5. Stauder, J.: Anlauforientierte Gestaltung von Fertigungssystemen. Dissertation, Aachen (2017)
6. Bergs, T., Apelt, S., Rey, J., Barth, S.: Modelling and analysis of manufacturing tasks for combined phase-out and ramp-up production. 30th CIRP Des. Conf. Procedia CIRP 2020(91), 164–169 (2020)
7. Rey, J., Grünebaum, T., Trauth, D., Bergs, T.: Highly iterative planning of manufacturing technologies: evaluation of manufacturing technology capabilities considering information uncertainties. In: Print Proceedings of the ASME 2019 14th International Manufacturing Science and Engineering Conference (MSEC), Vol. 1 (2019)
8. Laick, T.: Hochlaufmanagement—Sicherer Produktionshochlauf durch zielorientierte Gestaltung und Lenkung des Produktionsprozesssystems. Dissertation (2003)
9. Winkler, H.: Modellierung vernetzter Wirkbeziehungen im Produktionsanlauf, pp. 151. Dissertation, Hannover, (2007)
10. Dyckhoff, H., Müser, M., Renner, T.: Ansätze einer Produktionstheorie des Serienanlaufs. Z. Betriebswirt. **82**(12), 1427–1456 (2012)
11. Cooper, R.G., Sommer, A.F.: The agile-stage-gate hybrid model: a promising new approach and a new research opportunity. J. Prod. Innov. Manag. **33**(5), 513–526 (2016)
12. Sommer, A.F., Hedegaard, C., Dukovska-Popovska, I., Steger-Jensen, K.: Improved product development performance through agile/stage-gate hybrids: the next-generation stage-gate process? Res. Technol. Manag. **58**(1), 34–45 (2015)
13. Rey, J.: Methodik zur hochiterativen Gestaltung von Fertigungsprozessfolgen. Dissertation, Aachen (2020)
14. Basse, F.: Gestaltung eines adaptiven Änderungssystems für einen beherrschten Serienhochlauf (Design of an adaptive engineering change system for a stable series ramp-up). Dissertation (2018)
15. DeLange, J.D.M.: Prozessmodell für den Werkzeugbau im iterativen Produktentwicklungsprozess. Diss. RWTH Aachen (2020)
16. Flores, G., Wiens, A., Lahres, M., Hoffmeister, H.-W.: New optimization potential for the reciprocating engine. ATZ Produktion **4**(01/100), 38–43 (2011)
17. Barhoumi, S., Kallel, I.K., Bouhamed, S.A., Bosse, E., Solaiman, B.: Generation of fuzzy evidence numbers for the evaluation of uncertainty measures. In: 2020 5th International Conference on Advanced Technologies for Signal and Image Processing (ATSIP). 2020 5th International Conference on Advanced Technologies for Signal and Image Processing (ATSIP), pp. 1–6. IEEE, Sousse, Tunisia, 9/2/2020–9/5/2020 (2020)
18. Dempster, A.P.: Upper and lower probability inferences based on a sample from a finite univariate population. Biometrika **54**(3/4), 515 (1967)
19. Shafer, G.: A Mathematical Theory of Evidence, p. 314. Princeton University Press, Princeton, NJ (1976)
20. Deng, Y.: Uncertainty measure in evidence theory. Sci. China Inf. Sci. **63**(11) (2020)
21. Yager, R.R.: On the dempster-shafer framework and new combination rules. Inf. Sci. **41**(2), 93–137 (1987)
22. Brecher, C.: Integrative Produktionstechnik für Hochlohnländer. Springer, Berlin, Heidelberg (2011)

23. Klocke, F., Brinksmeier, E., Weinert, K.: Capability profile of hard cutting and grinding processes. CIRP Ann. **54**(2), 22–45 (2005)
24. Koller, R.: Konstruktionslehre für den Maschinenbau: Grundlagen zur Neu- und Weiterentwicklung technischer Produkte mit Beispielen, 4, Auflage Springer, Berlin (1998)
25. Dunsmuir, W.T.M., Snyder, R.D.: ABC analysis in inventory control—the issue of stability. OR Insight **3**(3), 24 (1989)
26. Uppin, E.P., Kelkar, A.A.: Application of ABC analysis for material management and planning and scheduling using MSP. Int. Res. J. Eng. Technol. (IRJET) **06**(07), 1196–1200 (2019)

Prof. Dr.-Ing. Thomas Bergs studied mechanical engineering at the University of Duisburg GH and design engineering at the RWTH Aachen University. In 2018 he was called to the RWTH Aachen University as Professor of the Chair of Manufacturing Technology at the Laboratory for Machine Tools and Production Engineering (WZL).

Sebastian Apelt studied business administration and engineering at the RWTH Aachen University. Since 2018 he works as research assistant in the workgroup Technology Planning at Laboratory of Machine Tools and Production Engineering (WZL) of RWTH Aachen University.

Malte Becker studied product development at RWTH Aachen University. Since 2020 he works as student research assistant in the workgroup Technology Planning at Laboratory of Machine Tools and Production Engineering (WZL) of RWTH Aachen University.

Alexander Beckers studied business administration and engineering at the RWTH Aachen University. Since 2017 he works as research assistant and took over the lead of the workgroup Technology Planning at Laboratory of Machine Tools and Production Engineering (WZL) of RWTH Aachen University in 2020.

Dr.-Ing. Sebastian Barth studied business administration and engineering at the RWTH Aachen University. Since 2019 he works as Division Director Technology Planning and Grinding Technology at Laboratory of Machine Tools and Production Engineering (WZL) of RWTH Aachen University.

Cross-Process Modeling of Manufacturing Process Sequences with Consideration of Model Uncertainties

Thomas Bergs, Alexander Beckers, Sebastian Apelt, Tim Hommen, and Sebastian Barth

Abstract The cost optimized manufacturing of components represents a central success factor for manufacturing companies. For this purpose, the required final state characteristics of components must be manufactured at the lowest possible efforts. These efforts depend on the companies' targets and may include the costs of the whole manufacture as well as environmental impacts caused by the manufacturing processes. Each manufacturing process of a process sequence has an influence on both, the final state characteristics of the component and the efforts, so that an improvement of individual processes in a process sequence does not have to lead to the technical and economical (global) optimum. To take this challenge (reaching global optimum) into account, a systematic procedure to combine different models of single manufacturing processes to cross-process models under consideration of the individual model uncertainties and their effects on cross-process models is presented. These cross-process models predict the final state characteristics of a component considering all manufacturing processes of the process sequence. For validation, the systematic approach is applied to a manufacturing process sequence for the manufacturing of indexable inserts.

T. Bergs · A. Beckers (✉) · S. Apelt · T. Hommen · S. Barth
Laboratory for Machine Tools and Production Engineering WZL of RWTH Aachen University, Chair of Manufacturing Technology, 30 Campus-Boulevard, 52074 Aachen, Germany
e-mail: a.beckers@wzl.rwth-aachen.de

T. Bergs
e-mail: t.bergs@wzl.rwth-aachen.de

S. Apelt
e-mail: s.apelt@wzl.rwth-aachen.de

T. Hommen
e-mail: t.hommen@wzl.rwth-aachen.de

S. Barth
e-mail: s.barth@wzl.rwth-aachen.de

T. Bergs
Fraunhofer Institute for Production Technology, 17 Steinbachstr, 52074 Aachen, Germany

K. von Leipzig et al. (eds.), *Smart, Sustainable Manufacturing in an Ever-Changing World*, Lecture Notes in Production Engineering,
https://doi.org/10.1007/978-3-031-15602-1_2

Keywords Technology planning · Manufacturing process sequences · Cross-process modelling

1 Introduction

In a competitive environment, the economically-optimized manufacturing of components is a success factor for companies. Here, the component characteristics (like macro or micro geometry) required by the customers must be manufactured at minimum cost [1]. Sequentially linked manufacturing processes are used in manufacturing process sequences [2] to manufacture the components with the required characteristics [3]. In principle, the final component characteristics are not only generated by the last manufacturing process of the sequence, but also by different processes of the process sequence. Each manufacturing process influences component states, and thus individual component characteristics, which have an influence on the following processes [4].

This requires methodical support to identify relevant process parameters in the process sequence. Subsequently, existing correlations and process models must be combined to cross-process models for the prediction of the final component state characteristics. Here, it is challenging to evaluate the uncertainty of these combined models resulting from each process model uncertainty. Moreover, the effects of different process designs on the economic profit of the process sequence must be evaluated. An optimization of the process parameters follows on the basis of the technological and economic evaluation [5].

Up to now, manufacturing processes are often designed individually but are only linked in later planning phases. For a holistic economical optimization, the entire process sequence and the relationships between the processes must be considered in process design [6].

The focus in this paper is on the prediction of component state at the end of a manufacturing process sequence depending on different process parameter designs and especially on the evaluation of uncertainties in cross-process models (see Fig. 1).

2 State of Research

According to the focus of this paper on the prediction of component characteristics by cross-process models and the evaluation of uncertainties in process models contained in manufacturing process sequences, the state of research is divided into two parts. First, approaches in which process sequences are designed holistically are examined in terms of how the cross-process models are generated and how uncertainties of individual process models are taken into account. Subsequently, approaches to evaluate the influence of uncertainties by single models on cross-process models are analyzed.

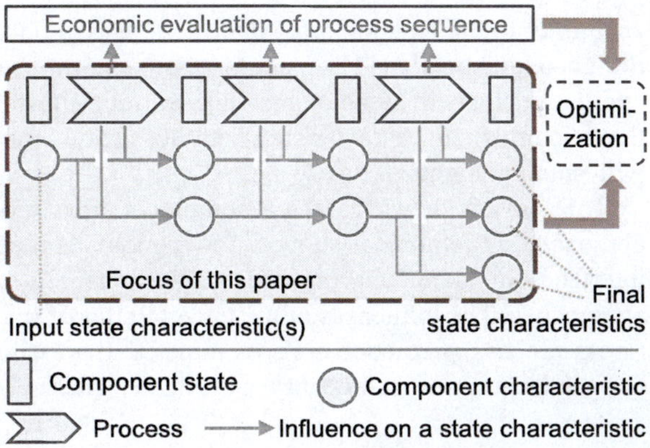

Fig. 1 Focus of this paper

2.1 *Approaches to Design Manufacturing Process Sequences*

Many research approaches address the generation of manufacturing process sequences. These approaches support the selection of suitable manufacturing technologies, the determination of process parameters (process design) and the determination of optimized schedules and batch sizes. Beckers et al. provide an overview over these different approaches [7]. In the following, approaches for the design of manufacturing process sequences are analyzed.

A method supporting the economical evaluation of process sequences and for linking technological with economical models as indicated in Fig. 1 are explained in [7]. However, it does not include methodological support for generating cross-process models or for dealing with uncertainties. In their approach, Mukherjee and Ray describe the relevance of the cross-process design and demonstrate it with a two-stage honing process. The authors analyzed alternative metaheuristic approaches for optimization of component quality. For the modeling of the processes, Mukherjee and Ray refer to empirical response surface models and multivariate regressions of the individual processes as well as to the identification of transfer values. The authors determined the quality of the regression of each process with different test statistics. An evaluation of the impact of the uncertainty by a single model on the overall (cross-process) model is not contained [8]. Denkena et al. assign great importance to the cross-process design and the technological interfaces between processes (transfer values), too. The authors describe that individual processes and interfaces must first be analyzed and simplified together with experts. In the next step, the process model is created and the output values are interpreted as transfer values. How the individual process models are generated and how relevant interfaces are identified is not described. By integrating the models into simulation software, a holistic process sequence analysis can be done [9]. Klocke et al. used technological interfaces and generated process models as well as systematic parameter variation to optimize process sequences [10]. In their approach, Bera and Mukherjee generated regression models for each process and they combined them with each other according to the technological interfaces. Based on this, the process sequence was designed holistically [11]. A similar approach was used by Hejazi et al. [12]. Further approaches that are based on machine learning show promising results in the prediction of component characteristics and subsequent process design but are often limited with small amounts of data.

The analysis shows that the generation of cross-process models is part of different approaches. Methodical support for modeling processes and identifying relevant interfaces in manufacturing process sequences is of lower importance in these approaches. The influences of the uncertainties of the process models on the overall (cross-process) prediction are not evaluated. It is assumed that comprehensive expert knowledge and sufficient data are available. These deficits are addressed in Sect. 3.

2.2 *Approaches to Deal with Uncertainties for the Design of Process Sequences*

As Rey et al. showed, technology planning is often subject to information uncertainties. These uncertainties relate e.g. to the individual processes and interactions between the processes and thus have an impact on the final decision regarding the design of process sequences. Therefore, Rey et al. developed an approach to evaluate uncertainties based on the Dempster Shafer theory. Information collected by expert interviews was evaluated and aggregated so that the impact of individual uncertainties on the overall decision quality could be analyzed. However, the focus of the approach is on processing information from experts rather than examining process models [13].

Suri and Otto describe the importance as well as the propagation of process uncertainties (output variations) in process sequences. The authors clarified that a solo reduction of a single process' variation is not always the best for the final output variation of the whole manufacturing process sequence. Figure 2 shows that although an operating point (a) with a low variation was chosen in the first process, the variation after process 2 is higher than for an operating point (b) with a high variation in the first process [14].

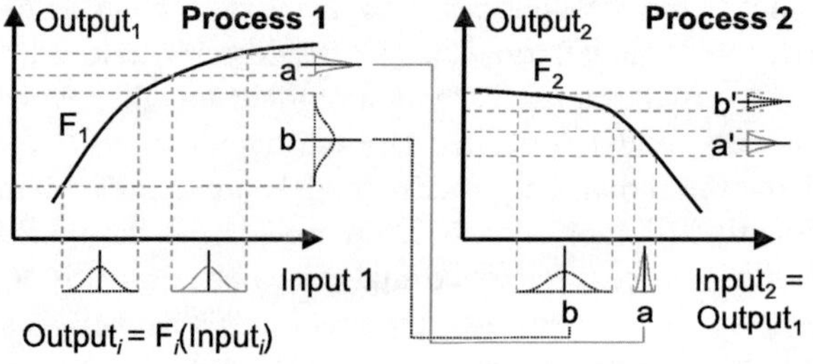

Fig. 2 Variation propagation of two processes referring to Suri and Otto [14]

Suri and Otto developed a method to determine robust operating points for the entire manufacturing process sequence and to reduce final component variation. To model the variations, the authors used a simplified root-sum squares analytic. The variation propagation was modeled by mathematical relations [14]. It should be emphasized that the approach is designed for one (linearized) operating point per process. Thus, the approach is not directly suitable for the design of manufacturing process sequences with multiple parameter combinations and non-linear correlations. A literature analysis on comparable approaches is included in [15].

Overall, there is a lack of methodological support to link process model uncertainties along a process sequence so that the impact of each model on the quality of a cross-process model can be evaluated. This deficit is addressed in the following section.

3 Methodology for Cross-Process Modeling Including Uncertainty Evaluation

Based on the deficits of the existing approaches, a methodology for the cross-process modeling and the evaluation of model uncertainties regarding their impact on the cross-process model was developed. To counteract the analyzed deficits, the methodology is also suitable when only limited amounts of data (and no comprehensive models) are available. Therefore, the evaluation of uncertainties is of high relevance.

The methodology contains three steps. In the first step, the manufacturing processes are modeled and the process models are evaluated regarding their prediction quality. Subsequently, the models are linked and the effects of the individual process model uncertainties on the prediction quality of the cross-process model are evaluated. In Step 3 it is determined which models need to be improved for the design of optimized manufacturing process sequences.

To demonstrate the practical applicability as well as to validate the methodology in a case study, data from an industrial project are used, in which 200 indexable inserts were manufactured by using sintering, grinding, brushing and coating. For reasons of confidentiality, individual data were slightly adjusted, whereby the goal of the application example (demonstration of applicability and validation of the methodology) are not affected.

3.1 Step 1—Process Models and Evaluation

There are many alternative techniques for modeling manufacturing processes, which are e.g. based on high data volumes, known physical relationships or expert knowledge. For the scope of the present approach, it is assumed that process models can even be generated on the basis of small amounts of data and without comprehensive knowledge, whereby the different types of modeling can be integrated. In the following, modelling techniques are used, which describe unknown relations between input variables and response values of a process by an approximation of known system states. The resulting models (e.g. generated by classical regression or kriging modeling) have an analytical form and are called metamodels in reference to [16, 17]. Challenges in modelling such as overfitting, which occur in the context of small amounts of data, must be taken into account, especially when transferring the results.

For the generation of metamodels, practitioners of the methodology can use existing data of manufacturing processes, whereby several modeling approaches can be applied by computer support. If not enough data are available, the practitioner has to execute experiments, since metamodels can be improved by adding more data. Here it is recommended to use design of experiments methods as well as expert knowledge (if available) to reduce the experimental effort. For example, the Latin Hypercube method offers the possibility to cover many areas of the system space by comparatively small numbers of experiments referring to Myers et al. [18]. Further, more comprehensive experiments are not recommended until the uncertainties of the individual process models and their effect on the cross-process model as well as the relevance of individual response variables on the final component characteristics have been evaluated. If a large amount of data is available, alternatives for predicting component characteristics, e.g. by means of machine learning or other approaches, should be tested.

For the case study, nine metamodels for the prediction of component characteristics in different manufacturing processes were generated on the basis of existing data (form 200 indexable inserts) and a stepwise regression in Matlab. Figure 3 shows the section of the metamodel used to predict the edge radius after brushing (b_3). It shows e.g. the dependence of the response variable on the process time (t_b) and on component characteristics before the process.

As shown in the state of research, the determination of individual prediction qualities (of process models) is relevant to evaluate the influence on the cross-process models of the process sequence. For the evaluation of the process models values like maximum absolute error (MAE), root mean squared error (RMSE) or the coefficient of determination (R^2) can be used. For details on the named values above and as on other values, it is referred to Ryberg et al. [19].

Fig. 3 Metamodels in the case study

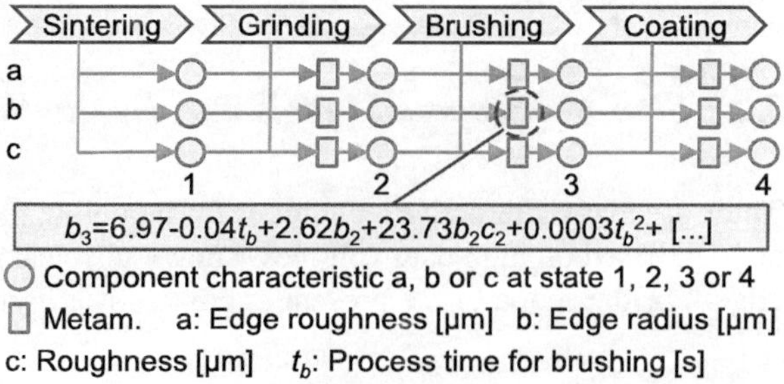

Fig. 4 Coefficient of determination

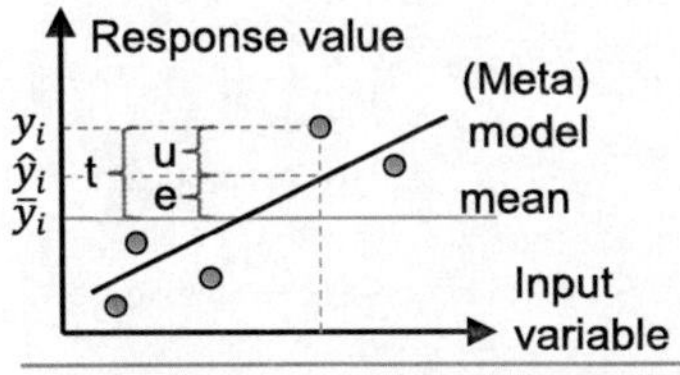

$$R^2 = \frac{e}{t} = \frac{\sum_{i=1}^{n}(\hat{y}_i - \bar{y}_i)^2}{\sum_{i=1}^{n}(y_i - \bar{y}_i)^2} = 1 - \frac{\sum_{i=1}^{n}(y_i - \hat{y}_i)^2}{\sum_{i=1}^{n}(y_i - \bar{y}_i)^2} \quad (1)$$

y_i: real value $\bar{y}_i$: mean of real values $\hat{y}_i$: predicted value
t: total/true variation (v.) u: unexplained v. e: explained v.

In the present approach, R^2 is used to evaluate the uncertainties of models. This value indicates to what extent the total variation of response values can be explained by the model used, cf. Figure 4, Eq. (1). For example, the model for the prediction of b_3 partially shown in Fig. 3 has an R^2 of 0.865.

3.2 *Step 2—Generation of Cross-Process Models and Uncertainty Evaluation*

For the design of optimized manufacturing process sequences, a holistic design approach is necessary. Therefore it must be ensured that the final component characteristics meet the requirements. Therefore, the individual process models must be transferred into cross-process models for the prediction of the final component characteristics. For this purpose, the models (starting with the last process step) must be integrated into each other according the transfer values. At the same time, models that describe a single process but are irrelevant for final component characteristics are removed.

If, as in the case study, data from the entire process sequence are available, the R^2 of the cross-process models (e.g. 0.773 for the prediction of b_4—edge radius after coating) can be determined directly. However, on the basis of this procedure, the evaluation of the effects of individual process model uncertainties on the cross-process models is not possible. It does not become clear which process models should be improved in order to improve the cross-process models. Therefore, in the following approach it is presented how the individual model uncertainties can be linked to the cross-process uncertainty to enable suggestions for improvement (see Sect. 3.3).

For this it is introduced, how variances are passed on along a process sequence. Referring to the approach of Arras [20] as well as Suri and Otto [14], the first-order of the Taylor series expansion at the operating point x is used for the approximation of f(x), see Eq. (2) in Fig. 5 for the case with one variable. Assumptions in this context are that the variance propagation is only valid in the linear range (close to the operating point) and for normally distributed variations. The variance of a process depending on several variables with their own variances can be determined using Eq. (4). This shows that the variance of a response value is determined from the

Fig. 5 Variance propagation by Arras [20]

$$y \approx f(\mu_x) + \frac{\partial f}{\partial x}\Big|_{x=\mu_x} (x - \mu_x) \quad (2)$$

$$\mu_y = f(\mu_1, \mu_2, \ldots, \mu_n) \quad (3)$$

$$\sigma_y^2 = \sum_{i=1}^{n} \left(\frac{\partial f}{\partial x_i}\right)^2 * \sigma_i^2 + 2 * \sum_{i=1}^{n-1} \sum_{k=i+1}^{n} \left(\frac{\partial f}{\partial x_i}\right) * \left(\frac{\partial f}{\partial x_k}\right) * \sigma_{ik} \quad (4)$$

σ_y^2: Variance of response variable y μ_i: Operating point i
n: Number of input variables f: Function for process model
σ_{ik}: Covariance for variables i and k

variances of the variables weighted based on their influence on the response value as well as the covariances of the variables (also weighted). For the detailed derivation of Eq. (4), it is referred to Arras [20].

In extension to the approaches of Arras as well as Suri and Otto, the variances resulting from the previous process are to be multiplicated with the R^2 of the corresponding model (see Eq. (5) in Fig. 6), so that the explainable variance is reduced. In Fig. 6, this extension is shown for an example from the case study without covariance and with a linear relationship based on Eqs. (1) and (4). The variances as well as R^2 for the individual processes (process models) were determined from the available industry data (200 indexable inserts). The edge roughness after brushing (a_3) depends in particular on the edge radius after grinding (b_2). Thus, the variance of the edge roughness also depends on the variance of the input edge radius. Since the radius can only be explained with an R^2 of 0.2 ($R^2_{b3,I}$), see Eq. (5), it leads to a low explainable variance for the edge radius ($\sigma^2_{a3,pe}$) compared to the true variance ($\sigma^2_{a3,t}$) considering (model) uncertainty propagation, see Eq. (6). Thus, the combined R^2 ($R^2_{a4,I}$), which represents the input for the following process, is a low value. At this point, it should be noted that the individual process' (meta)models have been standardized to allow easier comparability with regard to the effects of individual variables.

This case shows how model uncertainties affect the explainable variances along a process sequence. Since Fig. 6 shows an example without covariance, the equation Eq. (8) for the generally valid (linear) case is shown below in Fig. 7, which was also used for the other relationships in the case study.

The formulas presented above allow to pass variances as a function of R^2 for linear relationships. Since in process sequences there are often (as in the case study, cf. Figure 3) non-linear correlations, further methodological support is necessary, as explained in the state of research and in Fig. 2. Therefore, in the presented approach, the individual models are partially linearized. This is shown in Fig. 8 for the case in which the response value depends on one variable, although the application to multi-dimensional models is possible. Here, the total function is decomposed with regard to the terms of the individual variables. Starting at a point (e.g. a boundary of the parameter range) a linear function is applied according to the gradient of the non-linear function. Since this linear function deviates from the non-linear function with increasing x (see Fig. 8), a maximum acceptable deviation/threshold (evaluated

Fig. 6 Example from case study for the propagation of model uncertainties

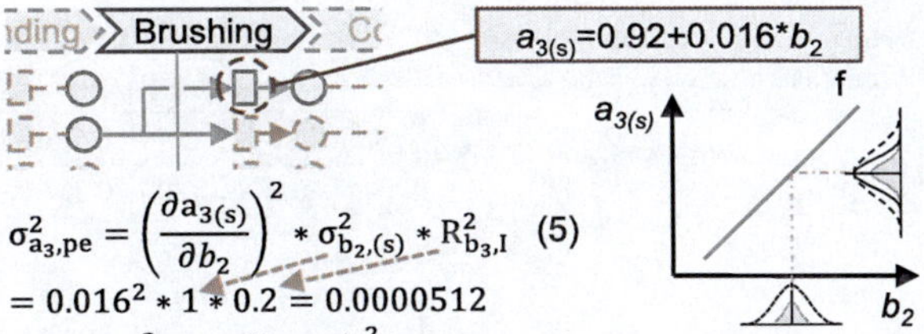

$$\sigma^2_{a_3,pe} = \left(\frac{\partial a_{3(s)}}{\partial b_2}\right)^2 * \sigma^2_{b_2,(s)} * R^2_{b_3,I} \quad (5)$$
$$= 0.016^2 * 1 * 0.2 = 0.0000512$$

$$\sigma^2_{a_3,t} = \frac{\sigma^2_{a_3,e}}{R^2_{a_3}} = \left(\frac{\partial a_{3(s)}}{\partial b_2}\right)^2 * \sigma^2_{b_2,(s)} * \frac{1}{R^2_{a_3}} \quad (6)$$
$$= \frac{0.016^2 * 1}{0.64} = 0.0004$$

$$R^2_{a_4,I} = \frac{\sigma^2_{a_3,pe}}{\sigma^2_{a_3,t}} = \frac{0.0000512}{0.0004} = 0.128 \quad (7)$$

$a_{i(s)}$: Edge roughness after process *i* (standardized)
b_i: Edge radius after process *i*
$\sigma^2_{x_i,z}$: Variance for characteristic *x* after process i, with *z*: (s) (standardized), e (explained), pe (explained including variance propagation and model uncertainty (R^2) from process *i*-1)
$R^2_{x_i}$: Coefficient of determination from process model to predict characteristic *x* after process *i*
$R^2_{x_i,I}$: Input coefficient of determination for process model (process *i*) regarding characteristic *x*

Fig. 7 General propagation of model uncertainties (linear) along process sequences (8)

Cf. Eq. (5) in Figure 6 with added covariance

$$R^2_{x_i,pe,n} = \frac{\sigma^2_{pe}}{\sigma^2_t} = \frac{\sum_{i=1}^{n}\left(\frac{\partial f}{\partial x_i}\right)^2 * \sigma^2_{x_i,t} * R^2_{x_i,I,n} + 2 * \sum_{i=1}^{n-1}\sum_{k=i+1}^{n}\left(\frac{\partial f}{\partial x_i}\right)\left(\frac{\partial f}{\partial x_k}\right)\sigma_{ik}}{\frac{\sum_{i=1}^{n} * \sigma^2_{x_i,t} + 2 * \sum_{i=1}^{n-1}\sum_{k=i+1}^{n}\left(\frac{\partial f}{\partial x_i}\right)\left(\frac{\partial f}{\partial x_k}\right)\sigma_{ik}}{R^2_{x_i,n}}}$$

Cf. Eq. (7) in Figure 6 | Cf. Eq. (6) in Figure 6 with added covariance

by root mean squared error) is defined. This determines the range in which the linear function fits in the part of the non-linear function sufficiently. To reduce the error, the corresponding range is linearized in the middle in a second step, see Fig. 8 right. This procedure is repeated until the entire function is linearized. For linearization, Eq. (2) was used.

Fig. 8 Linearization of functions

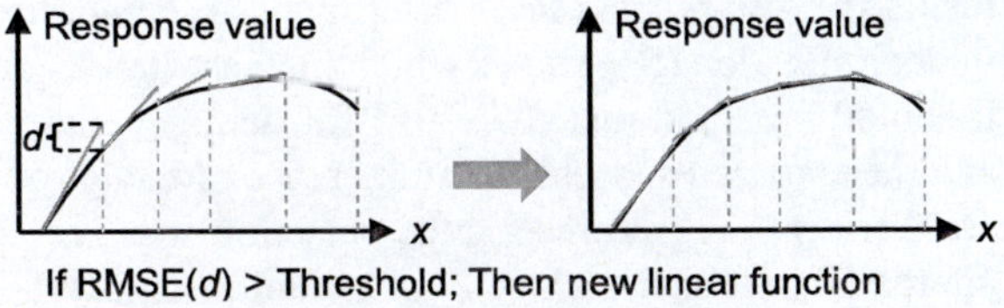

Fig. 9 Extract of the results from case study

	Grinding			Brushing			Coating		
	calc	true	diff	calc	true	diff	calc	true	diff
R_a^2	0.144	0.144	0 %	0.005	0.005	0 %	0.552	0.558	1 %
R_b^2	0.200	0.200	0 %	0.837	0.848	1.3 %	0.757	0.773	2.1 %

Now variances and R^2 can be calculated for the individual linear ranges of the process models. Afterwards, a weighted average of the resulting R^2 for the separated cross-process models is calculated according to the width of the individual ranges. A linearization with a threshold of 0.1 as well as the formulas shown above (concerning the propagation of variances considering the R^2) were applied to the case study. This resulted in the calculated (calc) values in Fig. 9. The true values are those derived directly from the data of the indexable inserts manufacture. The difference (diff) shows the percentage deviation between the calculated and true values. In the case study, there was a maximum deviation of 2.1% for the R^2 of the edge roughnesses and the edge radius. By reducing the threshold, this difference can be reduced by increasing calculation effort.

It also becomes clear that despite the partially low R^2, the deviations in the following processes can be explained significantly better (see $R^2{}_a$, difference between brushing and coating). This underlines the motivation described above, that individual models are to be identified, which have a high influence on the quality of the cross-process models. The case study clarified that all combined R^2 (cross-process models) are lower than the R^2 of the individual models. The partially low R^2 (e.g. after grinding) result from the fact that the available data were not collected specifically for predictions and do not represent a complete representation of the process. Nevertheless, they are applicable for the validation of the method.

3.3 *Step 3—Evaluation Method for Improvement Options of Cross-Process Models*

Now, it is known how the R^2 of the cross-process models can be determined based on, among other things, the individual R^2 of the process models. So, it is possible to evaluate the influence of the process models on the cross-process model. For this purpose, a sensitivity analysis is first performed in which the R^2 of the process models is increased individually (as an assumption) by the same absolute value and the resulting cross-process R^2 is determined. A comparison of the resulting coefficients of determination (R^2) shows which increase had the most positive effects. However, it should be noted that the effort for data generation through further experiments and the expected success for an increase of the R^2 of the individual process models are different. Therefore, for the selection it is recommended to choose a relative approach, in which each $R^2{}_x$ (of model x) is increased by the value $(1-R^2{}_x) * Y$, while Y is indicating the increase factor (e.g. 0.1).

4 Conclusions

In addition to the economic evaluation of the single processes, the fulfillment of the required final component characteristics is of central relevance for the optimization of manufacturing process sequences. For the design of these sequences, it is therefore necessary to determine the effects of individual process parameters on the final component characteristics and to evaluate them with regard to the prediction quality. By applying the presented approach, practitioners are supported in the generation of cross-process models for the prediction of component characteristics. By extending the approaches of Arras, Suri and Otto, practitioners are enabled to evaluate the effects of the R^2 of the process models on the cross-process model and, based on this, to derive promising possibilities for model improvement. The validation of the approach by means of a case study has shown the applicability of the methods and models. Further research is needed in the detailing of the method for the cost and expected benefit evaluation of model improvements as well as in the integration of optimization methods when state characteristics can be predicted and the economic efficiency can be evaluated.

Acknowledgements The authors would like to thank the Deutsche Forschungsgemeinschaft (DFG, German Research Foundation) for funding this research within the project "Methodology for generating cross-technology metamodels" Project-ID: 441745638.

References

1. Chakraborti, A., Nagarajan, H.P.N., Panicker, S., Mokhtarian, H., Coatanéa, E., Koskinen, K.T.: A dimension reduction method for efficient optimization of manufacturing performance. Procedia Manuf. **38**, 556–563 (2019)
2. Klocke, F., Fallböhmer, M., Kopner, A., Trommer, G.: Methods and tools supporting modular process design. Robot. Comput. Integr. Manuf. **16**(6), 411–423 (2000)
3. Afazov, S.M.: Modelling and simulation of manufacturing process chains. CIRP J. Manuf. Sci. Technol. **6**(1), 70–77 (2013)
4. Wuest, T., Klein, D., Seifert, M., Thoben, K.-D.: Method to describe interdependencies of state characteristics related to distortion. Mat. wiss. u. Werkstofftech. **43**(1–2), 186–191 (2012)
5. Yu, C., Dong, S., Yang, Z., Sun, J., Wang, G.: Optimization of multi-stage production operations based on genetic algorithm **17**, 18.1–18.8 (2016)
6. Denkena, B., Behrens, B.-A., Charlin, F., Dannenberg, M.: Integrative process chain optimization using a Genetic Algorithm. Prod. Eng. Res. Devel. **6**(1), 29–37 (2012)
7. Beckers, A., Stauder, L., Grünebaum, T., Barth, S., Bergs, T.: Design of economically-optimized manufacturing process sequences using cross-process models. CIRP J. Manuf. Sci. Technol. **33**, 15–29 (2021)
8. Mukherjee, I., Ray, P.K.: Optimal process design of two-stage multiple responses grinding processes using desirability functions and metaheuristic technique. Appl. Soft Comput. **8**(1), 402–421 (2008)
9. Denkena, B., Henjes, J., Henning, H.: Simulation-based dimensioning of manufacturing process chains. CIRP J. Manuf. Sci. Technol. **4**(1), 9–14 (2011)
10. Klocke, F., Buchholz, S., Stauder, J.: Technology chain optimization: a systematic approach considering the manufacturing history. Prod. Eng. Res. Devel. **8**(5), 669–678 (2014)

11. Bera, S., Mukherjee, I.: A multistage and multiple response optimization approach for serial manufacturing system. Eur. J. Oper. Res. **248**(2), 444–452 (2016)
12. Hejazi, T.H., Seyyed-Esfahani, M., Mahootchi, M.: Optimization of degree of conformance in multiresponse-multistage systems with a simulation-based metaheuristic. Qual. Reliab. Engng. Int. **31**(4), 645–658 (2015)
13. Rey, J., Grünebaum, T., Trauth, D., Bergs, T.: Highly iterative planning of manufacturing technologies: evaluation of manufacturing technology capabilities considering information uncertainties. In: Proceedings of the ASME 14th (2019)
14. Suri, R., Otto, K.: Manufacturing system robustness through integrated modeling. J. Mech. Des. **123**(4), 630–636 (2001)
15. Mondal, S.C., Ray, P.K., Maiti, J.: Modelling robustness for manufacturing processes: a critical review. Int. J. Prod. Res. **52**(2), 521–538 (2014)
16. Jurecka, F.: Robust design optimization based on metamodeling techniques. Zugl.: München, Techn. Univ., Diss., 2007. Shaker, Aachen
17. Hitz, R., Manopulo, N., Hora, P., Barlat, F., Moon, Y.H., Lee, M.G.: Simulation and knowledge based process planning through the use of metamodels. In: NUMIFORM 2010: Proceedings of the 10th International Conference on Numerical Methods in Industrial Forming Processes Dedicated to Professor O. C. Zienkiewicz (1921–2009), Pohang (Republic of Korea), pp. 888–895. 13–17 June 2010. AIP (2010)
18. Myers, R.H., Anderson-Cook, C., Montgomery, D.C.: Response surface methodology: process and product optimization using designed experiments, Fourth, Edition Wiley, Hoboken, New Jersey (2016)
19. Ryberg, A.-B., Domeij Bäckryd, R., Nilsson, L.: Metamodel-Based Multidiscipli-nary Design Optimization for Automotive Applications (2012)
20. Arras, K.O.: An introduction to error propagation: derivation, meaning and examples: Technical Report (1998)

Prof. Dr.-Ing. Thomas Bergs studied mechanical engineering at the University of Duisburg GH and design engineering at the RWTH Aachen University. In 2018 he was called as Professor of the Chair of Manufacturing Technology at the Laboratory for Machine Tools and Production Engineering (WZL).

Alexander Beckers studied business administration and engineering at the RWTH Aachen University. Since 2017 he works as research assistant and took over the lead of the workgroup Technology Planning at Laboratory of Machine Tools and Production Engineering (WZL) of RWTH Aachen University in 2020.

Sebastian Apelt studied business administration and engineering at the RWTH Aachen University. Since 2018 he works as research assistant in the workgroup Technology Planning at Laboratory of Machine Tools and Production Engineering (WZL) of RWTH Aachen University.

Tim Hommen studied business administration and engineering at the RWTH Aachen University and industrial engineering at Tsinghua University. Since 2021 he works as research assistant in the workgroup Technology Planning at Laboratory of Machine Tools and Production Engineering (WZL).

Dr.-Ing. Sebastian Barth studied business administration and engineering at the RWTH Aachen University. Since 2019 he works as Division Director Technology Planning and Grinding Technology at Laboratory of Machine Tools and Production Engineering (WZL) of RWTH Aachen University.

Modeling Interactions and Dependencies in Production Planning and Control

An Approach for a Systematic Description

Alexander Mütze, Simon Lebbing, Simon Hillnhagen, Matthias Schmidt, and Peter Nyhuis

Abstract In PPC, there are many interactions and dependencies that are difficult for the practical user to keep track of. These insecurities often lead to a configuration of PPC, which is locally optimized but not optimally carried out in the overall context of a company's target system. With a specially developed description and modeling approach, the interactions between PPC tasks, procedures, and logistical objectives are shown systematically and transparently to support the holistic and target-oriented design of PPC in perspective. The interdependencies presented in this paper have been compiled by contributions of various authors and transformed into a unified descriptive logic. To provide a comprehensive understanding of the interactions and interdependencies within PPC configuration, three levels of detail have been defined: First, the logistical objectives have to be examined in isolation. In this process, the central interdependencies at the level of the logistical objectives must be identified and presented in causal diagrams. The created impact network of interdependencies then serves as a basis for investigating the impact and interaction of PPC tasks. For each of the tasks then, primary and secondary interdependencies must be identified and visualized transparently with the help of isolated causal diagrams. On the final level of detail, the object of investigation has to change from task-specific to procedure-specific interdependencies, presenting (specific) input and output variables and impacts. Within the scope of this paper, an overview of the identified levels is given. For this purpose, the focus is exemplary on in-house production planning and control and the PPC task of order release.

A. Mütze (✉) · S. Lebbing · P. Nyhuis
Institute of Production Systems and Logistics, Leibniz University Hannover, Hannover, Germany
e-mail: muetze@ifa.uni-hannover.de

P. Nyhuis
e-mail: nyhuis@ifa.uni-hannover.de

S. Hillnhagen · M. Schmidt
Institute of Product and Process Innovation, Leuphana University of Lüneburg, Lüneburg, Germany
e-mail: simon.hillnhagen@leuphana.de

M. Schmidt
e-mail: mattschm@leuphana.de

K. von Leipzig et al. (eds.), *Smart, Sustainable Manufacturing in an Ever-Changing World*, Lecture Notes in Production Engineering,
https://doi.org/10.1007/978-3-031-15602-1_3

Keywords Production planning and control · PPC · Production configuration · Modeling · Procedures · Objectives

1 Introduction

Companies are held back in configuring their production as well as their Production Planning and Control (PPC) system by a widespread lack of transparency regarding the interaction of different PPC tasks, actuating and control variables, and logistical objectives (cf. [1]). As a result, decision-makers face considerable challenges in selecting and parameterizing appropriate procedures for the PPC tasks within PPC configuration to ensure the best possible accomplishment of the company's objectives [2]. Nevertheless, although various empirical studies confirm that companies are aware of the central importance of logistical parameters (e.g. [3]) and thus are interested in designing their production/PPC system in the best possible way, in practice, there is still a lack of a valid system understanding by the decision-makers [4, 5].

To counteract this condition and promote the target-oriented design of PPC, a variety of models can be found in literature describing tasks and processes qualitatively or determining the quantitative effects of specific PPC procedures/procedure classes on the achievement of logistical objectives (cf. [1, 6–9]). However, it should be stressed that none of the existing models holistically promotes a system understanding of PPC, which ranges from the tasks to be performed to the procedures to be selected, including their quantitative and qualitative effect on the logistical objectives. So far, neither the complete interdependencies between the logistic objectives nor their influenceability by the interaction of various PPC tasks and procedures have been comprehensively described. Existing models, such as the Manufacturing Control Model according to Lödding [6] and the Hanoverian Supply Chain Model by Schmidt and Schäfers [7], are thus mostly limited to the description of primary input & output variables respectively to the observation level of PPC tasks.

Therefore, a comprehensive model is needed, which supports the practical user in PPC configuration, pointing out mutual dependencies and linking these with specific models, e.g., to quantify the effect of a particular procedure. In order to provide companies with transparent and systematic decision support when configuring their PPC, the comparability of PPC tasks is to be improved with the help of a systematic investigation and the linkage with logistic models. The description of the interdependencies should thus explicitly not end at the system boundaries of the directly affected logistical objectives. Instead, a comprehensive and valid systems understanding is to be generated by the presentation of complete chains of effects in the logistic target influence.

2 Fundamentals

The field of production planning and control has been a focus of various research projects and contributions for decades. As a result, a variety of frameworks have been developed to describe and analyze PPC.

The Manufacturing Control Model of Lödding [6] has to be highlighted as a significant contribution to the description of PPC fundamentals, providing a unique and often adapted modeling approach. The manufacturing control model divides the major PPC task of production control into the four tasks of order generation, order release, capacity control, and sequencing. In addition to a general description of these main tasks, the model also aims to visualize interdependencies between the individual tasks and the major logistical objectives. For this purpose, Lödding uses a modeling scheme that defines the effects of PPC tasks via actuating and control variables and objectives. This modeling approach served as the basis for several other PPC models, including the impact model of in-house Production Planning and Control of Schäfers [10].

Another essential PPC framework is the Hanoverian Supply Chain Model (HaSupMo) (see Fig. 1) [7]. It combines the modeling approach of Lödding, with its limitation to the subarea of production control, with the understanding of the tasks and processes running in the PPC by the extended Aachen PPC Model [8, 9]. Thus, HaSupMo provides a comprehensive overall view on PPC by describing all PPC tasks in general and linking them. Further, the model is subdivided into a PPC- and a supply-chain-part, whereby the tasks of PPC are brought into a logical sequence and are connected with the logistical objectives of the core processes of a company's internal supply chain using actuating and control variables. It therefore provides an integrated perspective on the complex *PPC*.

In addition to the presented model-based approaches that qualitatively describe the interdependencies between PPC tasks and the logistical objectives, there are also control loop-based approaches. For example, Wiendahl and Breithaupt presented a

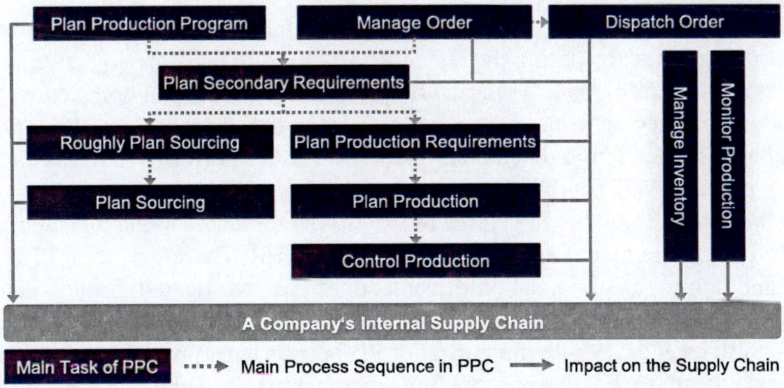

Fig. 1 A comprehensive view on the described PPC tasks within HaSupMo [7]

concept for an automatic production control system based on controllers for WIP and backlog [11].

Although there is a generally valid description of the PPC tasks, there is no holistic view, including all specific procedures for fulfilling these tasks. While the existing models visualize the interdependencies within PPC, their scope is strictly limited to the task level; so there is no or at most limited consideration of the procedure-specific interdependencies. (cf. [1]).

3 Research Question and Design

The focus of this paper and the underlying research project is to systematically examine the impact of PPC procedures on the logistical objectives of a company's internal supply chain. In this context, the approach can be considered as an extension to existing frameworks, since so far, these stop at the level of the impact of PPC tasks and only provide a brief overview on relationships and interactions.

The modeling approach presented is based on interdependencies that have been compiled from studies of various authors. A central part of the research work was to establish a unified description and modeling logic and to develop a level-based view on the interactions of decisions within PPC configuration. For this purpose, the research activity as a whole is based on the methodology of reference modeling, which is supported by both argumentative-deductive knowledge generation and bibliometric studies. (cf. [12]).

4 Structuring Levels to Describe Interactions within PPC

Within the model-based approach, various levels to examine relations and interdependencies within PPC are observed (see Fig. 2).

The first level, which is the basis for any configuration decision, is the level of logistical objectives, including the elementary corporate logistics objectives and the core processes of a company's internal supply chain (cf. [7, 13]). For the core process of production, serving in the following as role model, these are specified as short throughput times and high schedule reliability (logistics performance) as well as low work in process (WIP) and high capacity utilization (logistics costs).

On the second level, the PPC tasks of HaSupMo are linked to the logistical objectives following the modeling approach of Lödding [6]. Therefore, each PPC task's direct and indirect interactions on the objectives are investigated through actuating and control variables.

The third level deals with the impact of PPC procedures/procedure classes on the logistical objectives. For this purpose, the relationship between the single PPC procedures, actuating/control variables, and other influencing factors are further broken down, examined, and modeled.

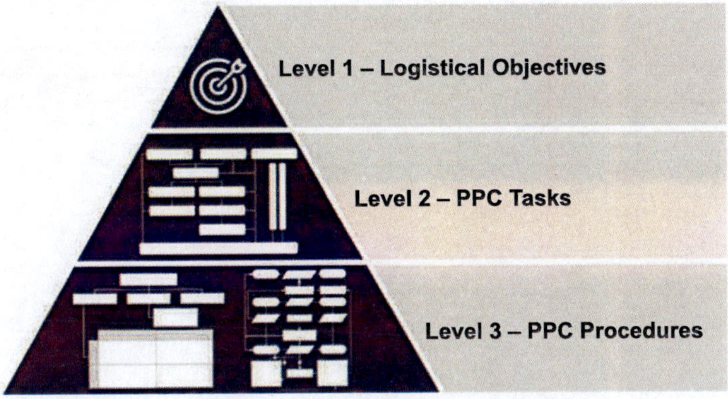

Fig. 2 Overview of the levels of consideration within this contribution

This paper gives extracts of a widely entangled network of impacts. For this purpose, we focus on describing the effect of an inventory-oriented order release (production control) in Level 3, which is the overall objective of the following sections, each describing the individual level in general and giving an example strongly related to inventory control.

4.1 Level 1: Logistical Objectives

4.1.1 General Approach

In order to define a systematic approach towards the description of interactions in PPC, the logistical objectives are first examined in isolation. Here, both qualitative and quantitative dependencies are identified and transparently prepared in the form of separately developed models. In case of the core process of production, the result is four separately developed causal diagrams, presenting the central input and output variables for the respective scope of the underlying logistical objective. Based on these isolated dependencies, a summarizing causal diagram was composed combining the essential findings of the individual models.

4.1.2 Example: Objective "Work in Process"

As an example, the modeling logic on the level of objectives is described for the logistical objective work in process. The simplified corresponding causal diagram is shown in Fig. 3.

At the center of each causal diagram, the possible forms of expression (mean value, spread, standard deviation, etc.) of the respective logistical objective, and if applicable, a further breakdown are displayed.

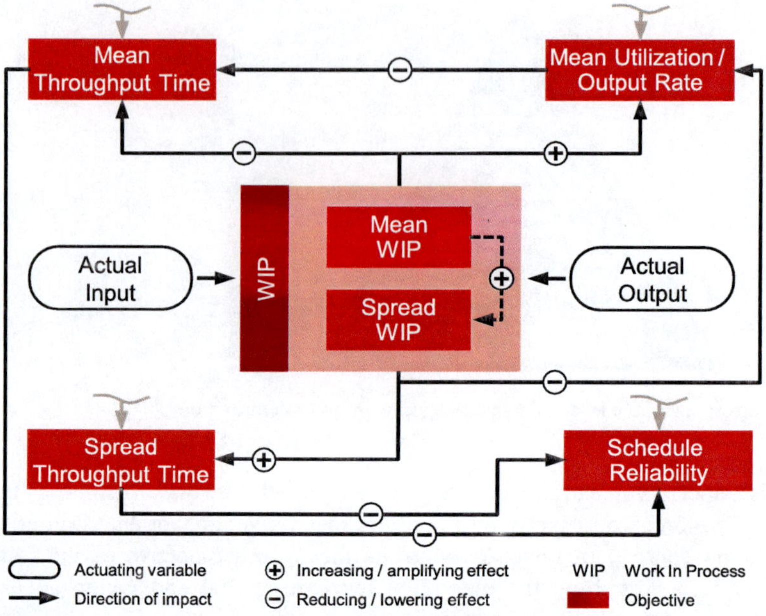

Fig. 3 Interactions and interdependencies among production objectives focusing WIP

In case of the WIP, a distinction was made between mean and spread. Unlike other objectives, the WIP is also a control variable and thus directly dependent on actuating variables, which is shown by the incoming arrows. The interdependencies between the specific logistical objective and the corresponding input and output variables/objectives are also shown in form of incoming and outgoing arrows, which are marked with signs indicating the relationship. Here, it is assumed that the starting variable increases and that in the system, no events occur which are not related to this change. In the case of Fig. 3, the understanding would be that as the mean WIP increases, the mean throughput time also increases. This and the other qualitative impact relationships are taken from the works of Schmidt and Nyhuis [14], Lödding [6] and Schäfers [10] to which is made reference at this point.

In addition to qualitative modeling, the single cause-effect relationships can be quantified with the help of (logistic) models. In Fig. 4, the conflicting logistical objectives are visualized with logistic operating curves as decisively postulated by Nyhuis and Wiendahl [13]. Here, the quantitative effect of the mean WIP on the mean output rate is shown by the output rate operating curve. In addition to the output rate, within logistic operating curve theory models have been explored describing the WIP-related behavior of the schedule reliability and the average throughput time, with their exact characteristics being determined with the aid of, e.g., the first funnel formula [15] and Little's Law [16], which provide mathematical functions for these dependencies.

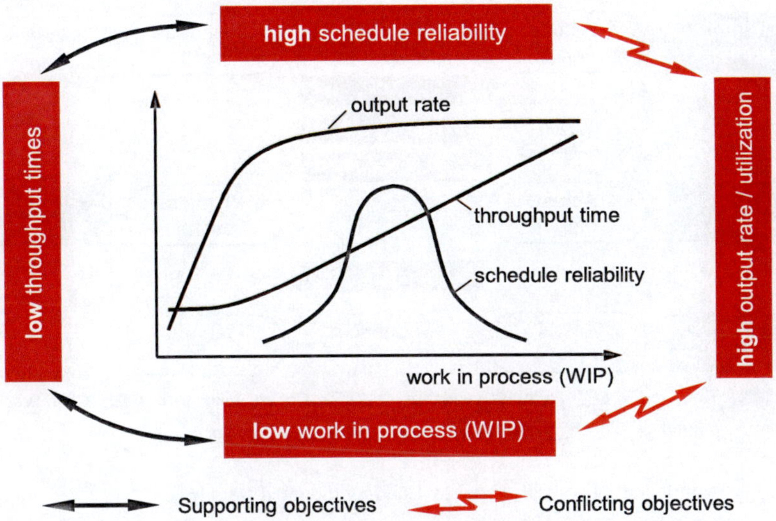

Fig. 4 Visualization of conflicting objectives with logistic operating curves [17]

4.2 Level 2: PPC Tasks

4.2.1 General Approach

The developed causal diagrams on the level of logistical objectives serve as a basis for establishing a link between the PPC tasks and the logistical objectives. As previous studies have shown, the in-house production planning and control tasks are particularly suitable for modeling, as logistic models can capture their quantitative influence effectively (cf. [6, 10]). In this contribution, the PPC tasks of in-house production control are thus focused.

For each PPC task, dependencies between the tasks and actuating variables are identified. For example, sequencing influences the actual sequence in the input of a production area and ultimately affects the objective of schedule reliability. At the same time, however, the value of this objective also depends on the control variable backlog, which in turn is influenced on the one hand by production planning and on the other hand by the PPC task capacity control (see Fig. 5).

As in the Hanoverian Supply Chain Model, the presented approach also follows the modeling logic of Lödding [6] and subsequent developments [10]. Thereby existing models are extended by secondary effects and functional chains in order to give a holistic view. Due to the uniform and structured representation, the resulting models for different PPC tasks are highly comparable with each other so that the identification of linking elements and interactions is promoted.

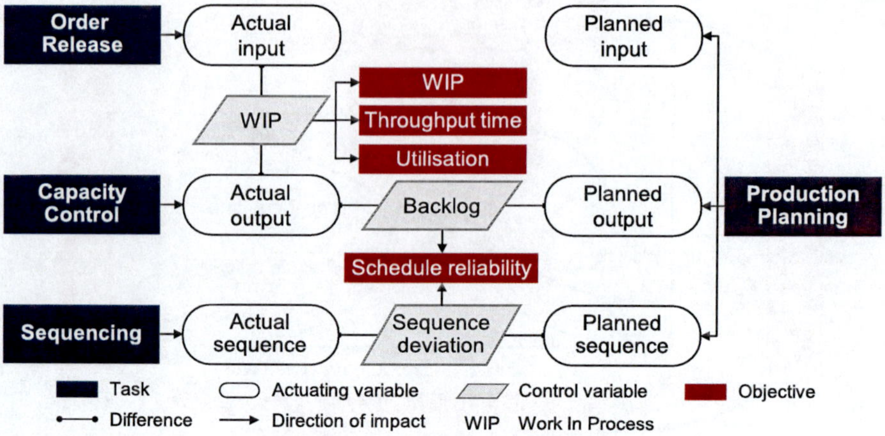

Fig. 5 The manufacturing control model by Lödding (adopted from [6])

4.2.2 Example: PPC Task "Order Release"

Within the PPC task of order release, production orders are released for processing and fed into production. The central dependencies to the logistical objectives are shown in Fig. 5.

The WIP is the central control variable influenced by order release. In this context, the WIP is also influenced by capacity control defining the output of production. The determination of the incoming quantity, the timing, and the sequence of the orders have a decisive influence on a large number of logistical objectives. Within the framework of the created causal diagram on the level of objectives (Fig. 3), the mean value and the spread of the WIP, the utilization, and the schedule reliability are highlighted. Order Release actively influences the incoming work content and the load on the associated work systems through the physical scheduling of production orders [6]. Both, the influence of order release on the mean value of the WIP and the associated spread of the WIP must thus be taken into account.

As a result of longer queues in front of the work systems, both performance parameters and throughput times increase [13]; the longer the throughput times are, the more they scatter (see also Fig. 3). Ultimately, long and highly scattered throughput times have a negative effect on schedule deviation behavior, schedule adherence, and schedule reliability. The spread of the throughput time also reduces the planning basis and therefore has a negative impact on the scheduling situation. If production orders tend to be released late or early, this results in a positive or negative backlog in the receipt of production; this influences the mean value of the lateness distribution. The backlog-related lateness in the inward movement of goods immediately influences the outward lateness and thus has a central influence on its scheduling behavior [10].

4.3 Level 3: PPC Procedures/Procedure Classes

4.3.1 General Approach

In a final step, the depth of observation is changed from the task level to the level of PPC procedures.

In contrast to dependencies within PPC tasks, the impact of PPC procedures can be explicitly calculated. In general, additional input variables must be taken into account for this purpose. For example, the impact intensity of a sequencing rule is highly dependent on the length of the queue of a working system.

In order to have an easily understandable and systematic modeling approach for this purpose, it is necessary to systematize the PPC procedures and procedure classes available for fulfilling a PPC task. With this, the final complexity can be significantly reduced. In addition, this subdivision allows a modular way of reasoning to be integrated into modeling, which brings immense advantages, especially when considering cross-task or cross-objective considerations over several partial causal diagrams.

Following the modeling approach for a procedure class of the PPC task, Order Release is presented.

4.3.2 Example: Procedure Class "Inventory-Oriented Order Release"

PPC procedures, in general, can be classified in different ways (e.g. same input variables). To categorize procedures of the PPC task Order Release that show similar behavior concerning given interdependencies, a classification was made extending the existing subdivision provided by Lödding [6] (see Fig. 6). Despite, e.g., criteria towards the trigger logic, a more profound division of Load/WIP regulation Order Release procedures was made comprising a subdivision into card- and non-card-based procedures and into centralized/decentralized and with/without work system-specific load balancing.

The (simple) inventory-oriented order release (central; without work system-specific load balancing) is generally characterized by the fact that the physical dispatching of an order only takes place if a previously defined inventory level within production is not exceeded by the release of the corresponding order [6]. The inventory-oriented order release aims thus to keep the inventory level of the entire production constant [10, 14].

Figure 7 illustrates the derived impact network for the procedure class of inventory-oriented order release. The overall goal here is to reduce the spread of the WIP in order to archive a comparable high output rate with comparable low WIP [6, 13]. Due to the reduced spread of the WIP, the probability of inventory-related capacity utilization losses decreases so that the output rate rises [6, 14]. Furthermore, the throughput times are kept constantly low so that the predictability of the order

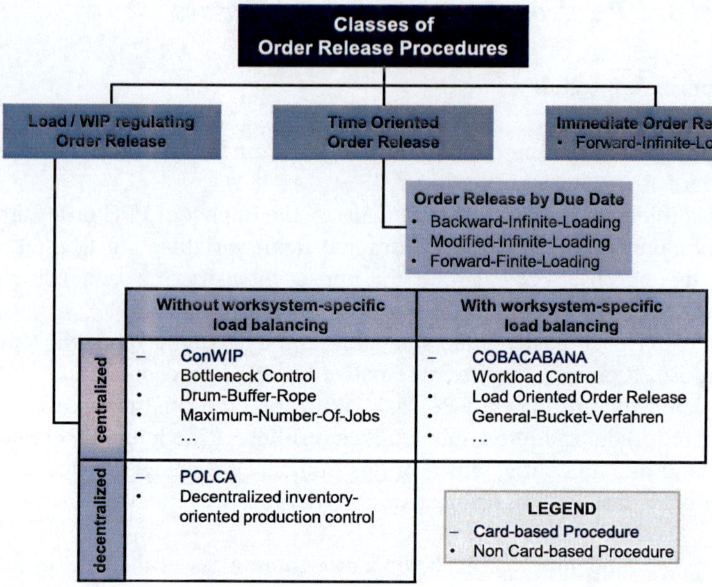

Fig. 6 Classes of order release procedures (extended, following [6])

throughput and the planning quality increase, giving a chance that the entire due-date behavior is positively influenced.

The realization of load balancing requires a targeted blocking of orders, which can extend beyond the planned start date of the orders. If necessary, there is the possibility of dispatching orders ahead of time so that the inventory-oriented order release generally does not lead to an exact implementation of the planned specifications [6]. Using the inventory-oriented procedure logic, the actual and planned dates of order receipt regularly diverge; this causes a backlog-related schedule deviation in the input of production; both positive and negative backlogs in the receipt of the work system are possible [10].

In addition, Schäfers [10] points out the influence of order release on the actual sequence in the production input, impacting the logistical objective of schedule reliability and interacting with the backlog-induced part.

Assuming a high level of planning quality, the increased inbound schedule deviation tends to have a negative impact on the outbound schedule deviation of the work system and thus jeopardizes the realization of a high level of on-time delivery and adherence to schedules.

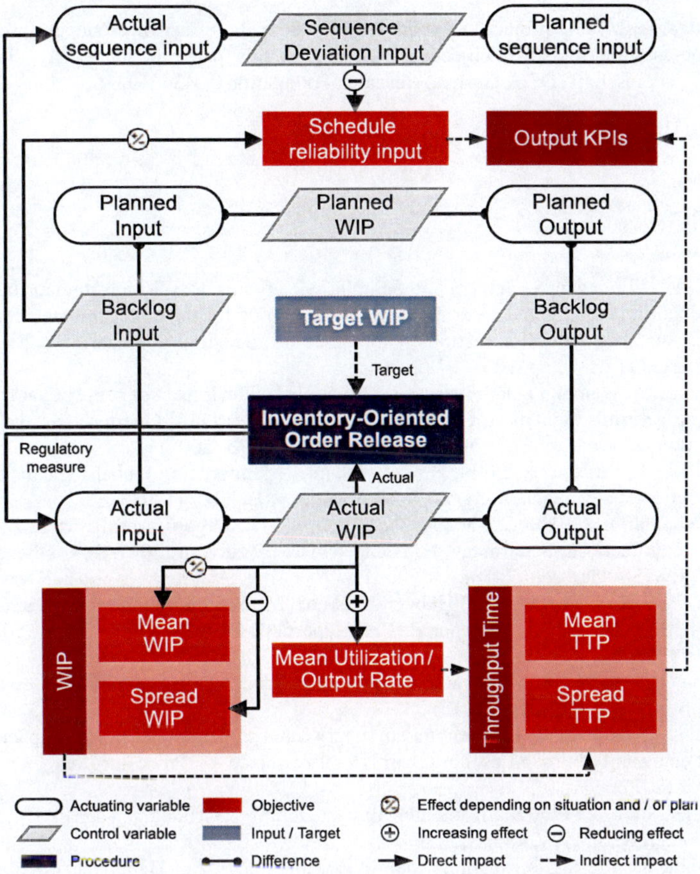

Fig. 7 Example of an impact network: class *inventory-oriented order release*

5 Conclusions

The presented approach and method regarding the systematic description of interdependencies and interactions within production planning and control lay the basis for a cross-level assessment. It is likely that the system's understanding of decision-makers will be significantly enhanced, resulting in better positioning of PPC in the trade-off between logistical objectives and overall system design.

In subsequent steps, the collected findings and the modeling approach will be integrated into the Hanoverian Supply Chain Model (www.hasupmo.education). Based on a previous delimitation of the PPC procedure classes under consideration, these are explained using profiles and the elaborated modeling approach presented.

Acknowledgements The project "Systematic analysis of the effect of production planning and control processes on logistical objectives" leading to this paper is funded by the Deutsche Forschungsgemeinschaft (DFG, German Research Foundation)—434659386

References

1. Mütze, A., Hillnhagen, S., Schäfers, P., Schmidt, M., Nyhuis, P.: Why a systematic investigation of production planning and control procedures is needed for the target-oriented configuration of PPC. In: 2020 IEEE international conference on industrial engineering and engineering management (IEEM), pp. 103–107 (2020)
2. Bozarth, C.C., Warsing, D.P., Flynn, B.B., Flynn, E.J.: The impact of supply chain complexity on manufacturing plant performance. J. Oper. Manage. Publ. Quarterly by the American Production & Inventory Control Society, Inc 27(1), 78–93 (2009)
3. Handfield, R., Straube, F., Pfohl, H.-C., Wieland, A.: Embracing global logistics complexity to drive market advantage, p. 81. DVV Media Group, Hamburg (2013)
4. Seitz, M., Härtel, L., Hübner, M., Merkel, L., be Isa, J., Engehausen, F., Meluzov, N., Rost, R., Schmidhuber, M., Sauermann, F., Hünnekes, P.: PPS-Report 2017/18: Studienergebnisse, p. 42. TEWISS, Garbsen (2018)
5. Mundt, C., Winter, M., Heuer, T., Hübner, M., Seitz, M., Schmidhuber, M., Maibaum, J., Bank, L., Roth, S., Scherwitz, P., Theumer, P.: PPS-Report 2019: Studienergebnisse, p. 50. TEWISS, Garbsen (2020)
6. Lödding, H.: Handbook of manufacturing control: fundamentals, description, configuration, p. 577. Springer, Heidelberg (2013)
7. Schmidt, M., Schäfers, P.: The Hanoverian supply chain model: modelling the impact of production planning and control on a supply chain's logistic objectives. Prod. Eng. Res. Devel. **11**(4–5), 487–493 (2017)
8. Schuh, G., Stich, V.: Produktionsplanung und -steuerung 1: Grundlagen der PPS, 4. Aufl. 2012 ed. Springer, Berlin, XV, 485. (2012)
9. Hackstein, R.: Produktionsplanung und -steuerung (PPS): Ein Handbuch für die Betriebspraxis, 2nd edn., p. 372. VDI-Verl, Düsseldorf (1989)
10. Schäfers, P.: Modellbasierte Untersuchung der Wirkung von Planungs- und Steuerungsverfahren auf die Termintreue einer Produktion, p. 189. TEWISS, Garbsen (2020)
11. Wiendahl, H.-P., Breithaupt, J.-W.: Production planning and control on the basis of control theory, In: Okino, N., Tamura, H., Fujii, S. (eds.) Advances in Production Management Systems. Perspectives and future challenges, pp. 351–362. Springer, Boston, MA (1998)
12. Wilde, T., Hess, T.: Forschungsmethoden der Wirtschaftsinformatik: Eine empirische Untersuchung. Wirtschaftsinf. **49**(4), 280–287 (2007)
13. Nyhuis, P., Wiendahl, H.-P.: Fundamentals of production logistics: theory, tools and applications, p. 320. Springer, Berlin (2009)
14. Schmidt, M., Nyhuis, P.: Produktionsplanung und -steuerung im Hannoveraner Lieferkettenmodell: Innerbetrieb-licher Abgleich logistischer Zielgrößen, 1st edn., p. 217. Springer, Berlin (2021)
15. Bechte, W.: Theory and practice of load-oriented manufacturing control. Int. J. Prod. Res. **26**(3), 375–395 (1988)
16. Little, J.D.C.: A Proof for the Queuing Formula: $L = \lambda W$. Oper. Res. **9**(3), 383–387 (1961)
17. Nyhuis, P.: Produktionskennlinien: Grundlagen und Anwendungsmöglichkeiten. In: Nyhuis, P. (ed.) Beiträge zu einer Theorie der Logistik, pp. 185–218. Springer, Berlin (2008)

Alexander Mütze (M.Sc.) studied industrial engineering at Leibniz University Hannover and worked as a research associate at the Institute of Production Systems and Logistics (IFA) since 2018, focusing on production planning and control and logistic modeling.

Simon Lebbing (M.Sc.) studied industrial engineering at Leibniz University Hannover and has been working as a student assistant at the Institute of Production Systems and Logistics (IFA) since 2019.

Simon Hillnhagen (M.Sc.) studied industrial engineering at Leuphana University of Lüneburg and has been working as a research associate at the Institute for Product and Process Innovation (PPI) in the group production management since 2020.

Prof. Dr.-Ing. habil. Matthias Schmidt holds the chair of production management at the Institute for Product and Process Innovation (PPI) at the Leuphana University of Lüneburg since 2018. In addition, he became the head of the PPI in 2019.

Prof. Dr.-Ing. habil. Peter Nyhuis has been head of IFA at Leibniz University Hannover since 2003. After obtaining his doctorate in engineering (Dr.-Ing.) and subsequent habilitation, he worked as an executive in the field of supply chain management.

ARTI-Based Holonic Manufacturing Execution System Using the BASE Architecture: A Case Study Implementation

A. Wasserman, K. Kruger, and A. H. Basson

Abstract With industry's drive to adopt Industry 4.0 technologies in manufacturing processes, intelligent automated manufacturing has become largely sought after. With defining features such as robustness, reconfigurability and scalability, the Holonic Manufacturing Execution System (HMES) approach shows great potential to satisfy Industry 4.0 requirements. Implementations of these systems has been aided by the development of holonic reference architectures, such as the Product-Resource-Order-Staff-Architecture (PROSA) and its recent revision the Activity-Resource-Type-Instance (ARTI) architecture. This paper presents an ARTI-based HMES, implemented using the Biography-Attributes-Schedule-Execution (BASE) architecture. The BASE architecture guides the development of digital administration shells for resources, and it is deemed suitable for implementing the different holon types of the ARTI architecture. The case study system is a Fischertechnik Industry 4.0 Training Factory—a small-scale manufacturing system. The complexity of the case study, which comprises several interacting subsystems, provides a good basis to evaluate the ARTI and BASE architectures for HMES development. The paper reports that the ARTI architecture provides a well-defined structure for the conceptual design of HMESs, while the BASE architecture effectively supports the implementation of ARTI-based HMESs.

Keywords Holonic manufacturing execution system · ARTI · BASE

A. Wasserman · K. Kruger (✉) · A. H. Basson
Department of Mechanical and Mechatronic Engineering, Stellenbosch University, Stellenbosch, South Africa
e-mail: kkruger@sun.ac.za

A. Wasserman
e-mail: awasserman@sun.ac.za

A. H. Basson
e-mail: ahb@sun.ac.za

K. von Leipzig et al. (eds.), *Smart, Sustainable Manufacturing in an Ever-Changing World*, Lecture Notes in Production Engineering,
https://doi.org/10.1007/978-3-031-15602-1_4

1 Introduction

The move towards digitization and automation of manufacturing systems as part of the trailing end of the 'Digital Revolution' and the beginning of the 'Fourth Industrial Revolution' (I4.0), intends to improve, amongst others, the manufacturing industry, by augmenting (and designing from the ground up) manufacturing systems with digital technologies. This process has introduced a multitude of manufacturing execution system concepts.

Holonic Manufacturing Execution Systems (HMESs) have seen increasing deployment over the years. A HMES is a "highly distributed control paradigm" [1] that consists of a system of holons that work together to accomplish a common goal/task within a manufacturing system. Although the holons work together, they are individually autonomous and possess the ability to carry out multiple functions or behaviours. Each holon in the HMES cooperates with others in order to form a network of functionalities and behaviours that can carry out complex manufacturing and control tasks [2].

The biggest advantages of HMESs are that they: are relatively easily reconfigurable; are robust and fault tolerant; offer reduced software development costs (as system complexity is reduced); and are generally scalable.

Research in this field has made the implementation of HMESs more accessible due to the creation of Holonic reference architectures for manufacturing systems. Some of the most influential are the Adaptive Holonic Control Architecture (ADACOR) [3], Product-Resource-Order-Staff Architecture (PROSA) [1] and its revision, the Activity-Resource-Type-Instance (ARTI) architecture [4]. These architectures were designed as references for the conceptual development of HMESs, but provide limited guidance for the actual implementation of such systems.

Although HMESs have been considered in academia for more than two decades, the ARTI architecture is still relatively new and very few implementations thereof on complex manufacturing systems have been reported. As such, there is a shortage of case studies to inform further implementations of the ARTI architecture in academic studies or industry applications.

Strongly influenced by the above-mentioned holonic reference architectures is the Biography-Attributes-Schedule-Execution (BASE) architecture [5]. The BASE architecture was originally developed to guide the implementation of digital administration shells for human workers. The architecture design was guided by the requirements for Resource Holons (RHs) (as also defined within the ARTI architecture). Upon inspection, close relations can be seen between the BASE architecture and the ARTI reference architecture—indicating that the BASE architecture could potentially be used in an implementation of the ARTI reference architecture for an HMES.

This paper presents a case study implementation of an ARTI-based HMES using the BASE architecture. The presented case study aims to demonstrate the ARTI architecture's suitability as a tool to facilitate holonic system design, and the BASE architecture's suitability as a platform for HMES implementation.

The paper is structured as follows: Sect. 2 provides the necessary background on the ARTI and BASE architectures, Sect. 3 introduces the case study system and requirements, Sect. 4 describes the HMES implementation, Sect. 5 presents a preliminary evaluation of the HMES, and Sect. 6 provides a conclusion and discussion of planned future work.

2 Architecture Overviews

2.1 ARTI Architecture

The ARTI reference architecture was developed as update to the PROSA reference architecture, addressing some required refinements and the terminology used in the PROSA architecture, as its terminology was manufacturing-specific, and it was difficult to apply PROSA to neighbouring fields [4]. The ARTI reference architecture was thus designed to have universal terminology and functionality.

ARTI consists of four different kinds of holons: Resource Type and Instance holons, as well as Activity Type and Instance holons. These four holons are then each sub-divided into decision making and reality reflection components: Intelligent Agents (IAs) and Intelligent Beings (IBs) respectively.

Activity Type (AT) holons refer to holons that represent a class of an activity. The holon has the activity-related knowledge that would be common amongst all instances of that activity, such as process plans, material requirements, expected activity time, etc., but it does not contain instance specific information, such as the instance's current state values or its activity progress. Each different activity within a holonic system, including process activities, maintenance, transport, worker activities, etc., would have its own AT holon, and this holon would communicate with the Activity Instance holons to complete an activity.

Activity Instance (AI) holons are then holons that represent the real-world activity taking place. These AIs contain the current state information specific to that instance of the activity's execution only, as well as the activity's history, and rely on their ATs for the information that is general to the activity. AIs also manage the execution of their real-world activity, which includes resource selection and management of schedule bookings with the resource. These AIs would exist for every instance of a real-world activity, whether it is production related, maintenance related, or worker related. AIs are known to be decision-intensive and have significant responsibilities.

Resource Type (RT) holons are similar to AT holons in that they represent a class of a certain type of resource. The holon has the resource-related knowledge that would be common amongst all instances of that resource, such as capabilities, dimensions, maintenance needs, etc., but it does not contain instance specific information, such as the resource instance's current state values or schedule. In practice, all different types of physical resources, such transport, materials, machinery, workers, infrastructure, etc., would need RT holons.

Resource Instance (RI) holons represent real-world instances of resources. As with AI holons, they contain the current state information specific to that instance of the resource only, as well as the resource's history. They rely on their RTs for the information that is general to the resource. RIs also manage the operation of their real-world resource, which includes, but is not limited to, state tracking, management of schedule bookings with the activity, process execution, etc. Similar to AIs, the RIs are also decision-intensive components with several responsibilities.

All the above-mentioned holons are separated into two different components: Intelligent Agents (IAs) and Intelligent Beings (IBs). IBs are seen as the reality-reflecting components of a holon, using either sensors, models or other types of inputs to mirror the reality of their World of Interest (WOI). IAs are seen as the components of a holon that are responsible for decision-making using a variety of methods, including computationally simple principles, or more computationally complex methods.

2.2 BASE Architecture

The BASE architecture was developed to guide the implementation of administration shells for a human RHs. Comparisons can be drawn between the BASE architecture and RAMI4.0 [6], however BASE is a "separation of concerns based on time" [5] for functionalities provided to the human worker, whereas RAMI4.0 separates concerns of I4.0 systems based on physical layers. An outline of the BASE structure is provided in Fig. 1.

The architecture specifies generic core components, with additional plugin components to add application-specific functionalities. The Biography, Attributes,

Fig. 1 BASE architecture showing the core and plugin components [3]

Schedule, and Execution components (indicated as "B", "A", "S" and "E" in Fig. 1), which together with the Communication Manager component, constitute the core components of the architecture. The Biography deals with past events that have transpired within the holon and its surroundings. Biography data may include logs from completed activities, maintenance logs, errors, etc. Attributes deals with the properties of the holon that are considered stable and that do not change (or change slowly). This would include information such as a holon's specifications, capabilities, age, type, etc. The Schedule deals with future events that are scheduled for execution by the holon (using the three-stage activity lifecycle structure [5]). Execution deals with the holon's present state and interfaces with sensors to acquire state information. Lastly, the Communication Manager facilitates the communication between internal components, and between the holon and external holons/systems.

In Fig. 1, the yellow-coloured corner blocks are the plugin components. These plugins consist of the Analysis, Scheduling, Reflection and Execution plugins. These plugins interface mainly with their adjacent core components (as depicted in Fig. 1). Scheduling Plugins (SPs) represent algorithms, software systems and/or decision-maker interfaces, which create, manage, and optimise the scheduled activities of the HRH. The activities that are to be scheduled can originate from external holons as service requests or internal (administration shell component) requests. Execution Plugins (Eps) are responsible for managing the execution of scheduled activities. EPs instantiate the execution of scheduled activities by monitoring and communicating with the resource through the Execution component. Reflection Plugins (RPs) create and maintain biographic entries of completed activities or events—entering data into the Biography after activity completion. RPs can also gather data on an activity post execution, such as through reviews, quality checks, etc. Analysis Plugins (Aps) generate value from the data recorded in the Biography with the aim of updating and amending the Attributes.

The BASE architecture was implemented in the Erlang/OTP programming language. Erlang enables scalable soft real-time systems with high availability and natively supports concurrency, distribution and fault tolerance [7]. As such, the BASE architecture is an easily scalable and fault-tolerant platform on which to build holonic systems.

The original implementation of the BASE architecture was extended in [8]—allowing for multiple BASE instances to be managed from a single platform and the addition of 'basic resources' to the system, which allows external services to communicate with the BASE holons.

3 Case Study

3.1 Hardware

The case study was implemented on a Fischertechnik Industry 4.0 Training Factory, seen below in Fig. 2, which is a small-scale manufacturing system. It has the capability to produce red, blue, and white workpieces, each with NFC tags embedded to allow for product tracking within the factory. The manufacturing cell consists of six different automated stations: a high-bay warehouse (HBW) for storing unprocessed workpieces; a multi-processing station (MPO) where various processes are executed on the workpieces; a sorting line (SLD) that sorts workpieces based on colour; a vacuum gripping robot (VGR) that transports the workpieces between different stations; a delivery station (DSO) where the completed workpieces are taken to be dispatched from the factory; and an input station (DSI) where raw workpieces are delivered to the factory.

The factory is controlled, on the lowest level, by six Fischertechnik TXT controllers, which interact directly with the hardware and communicate with a central controller and cloud dashboard via Message Queuing Telemetry Transport (MQTT) messaging. The control routines on these controllers were updated to remove much of the automated functionality and to allow activities/processes to be initiated through commands sent from the high-level controller via MQTT messages.

Fig. 2 Fischertechnik industry 4.0 training factory

3.2 Requirements

The requirements for the case study system were predominantly qualitative and subjective. The requirements that were identified as being of the most value are that: the system should be reconfigurable, in order to allow for the addition or substitution of new resources with minimal programming/development effort; the system should be easily scalable, so that numerous instances of resources and activities could be created; and the system needed to be robust and reactive, so that if one of the resources in the production line went offline, the system would be able to adapt and select an alternative production path (if available).

4 HMES Design and Implementation

The implementation of the case study HMES builds on a hierarchical architecture (shown in Fig. 3), with Type holons at the top of the hierarchy, followed by Instance holons at the level below. The architecture is also fractal in nature, as each Instance holon could contain an internal architecture of the same structure.

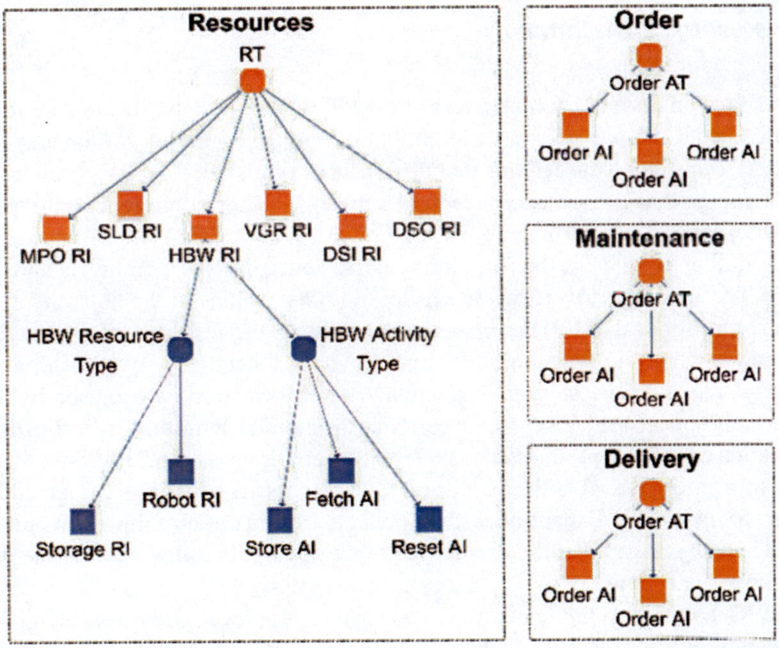

Fig. 3 Case study system architecture

4.1 Resource Type

The system has only one RT that represents the six different resource stations in the system, since all the resources are similar in operation and execution (differing only in MQTT topics and attributes). The RT provides the same services to all the RIs in this case study. An alternative approach would be to create a RT for each different station; however, this approach would have led to unnecessary code repetition, with no added advantages, in this case.

The RT has the role of being the major decision maker and manager for the RIs. RIs request the RT to perform complex computations on their behalf, as a service. This frees the RIs from computationally heavy tasks, to be available to handle and execute their tasks and communications with hardware, which may be sensitive to delays. An advantage of using Erlang for the implementation of the system is that processes can be spawned as needed, which is beneficial for the RT, since a handler process can be spawned for each request that is received from RIs. Thereby, bottlenecks can be avoided. RTs are also responsible for managing their RIs, e.g. providing their attributes after being spawned (as a part of attribute inheritance), and also to monitor the state of their tasks and provide this information to ATs as a service.

4.2 Resource Instance

The system comprises of six instances of the RT, one for each hardware station in the system. Each RI is spawned with a slightly different configuration, which is specified in the start-up file of the RI and includes information such as configuration name, MQTT topics that the instance needs to subscribe to for communication with the hardware, and default attributes for the RI.

Scheduling of tasks for the RI holon is carried out using the Contract Net Protocol (CNP). The RI holon advertises its services to other holons in the holarchy. Once a Request for Proposal (RFP) has been received, the RI requests a service from the RT to calculate when the first available timeslot in its schedule is (by considering the RI's schedule and process time attributes). This information is then used by the RI to send a proposal back to the service requesting holon with the earliest estimated completion time for the requested task. Upon receiving an 'accept' response from the requesting holon, the RI adds the requested task to its schedule, to start execution at the agreed upon time. Execution within the RI is carried out by a finite state machine (FSM), which ensures that the correct sequence of tasks is followed for that specific resource to execute the requested task successfully.

The RI holon communicates to its respective low-level controller using JSON messages sent via MQTT to that resource's topic. MQTT messaging is offered to the HMES as a service by a 'basic resource'. Having one basic resource handling all MQTT messaging could potentially create a bottleneck. While problem was not encountered in the small-scale case study, it can be alleviated by simply spawning another instance of the MQTT service basic resource.

The RI holon stores data from its executed tasks, such as the scheduled start time, actual start time, completion time and completion state. This information can later be used to update the attributes of the RI through various means of analysis—the updated attributes can then be used for improved scheduling of tasks.

These execution and analysis processes can be seen as the Intelligent Agent (IA) part of the holon, as these components encapsulate the decision-making functionality. Correspondingly, the scheduling and reflection components can be seen as the Intelligent Being (IB) part of the holon, as these components handle reality reflection of the holon's world of interest.

4.3 Activity Type

Three AT holons were used in the system, namely: Order, Delivery and Maintenance ATs. The Order AT managed orders that were placed by the user via a custom web dashboard, which allows for the ordering of red, white, and blue workpieces. Once an order has been placed, the AT spawns an AI holon, as well as an FSM with which it handles the order. Once the AI has initialized, the AT sends the first process step in the order process, as listed in the AT's configuration file, to the AI. The process steps are provided according to the Next-Execute-Update (NEU) protocol [9]. In a similar manner, the Delivery and Maintenance AT holons manage the execution of maintenance and delivery activities.

The AT offers services to AIs, just as the RT does to RIs, while also performing additional managerial functions. The AT is responsible for monitoring the state of activities that the AI is busy handling, attribute inheritance, and initiating and terminating AIs. The AT also communicates and interacts with the dashboard service to receive user input, as well as to display information to the user. As a part of the dashboard interfacing, the AT also collects the state of all resources in the system to display this information to the user.

4.4 Activity Instance

AIs are dynamically spawned as activities are started by the AT. For each activity that the AT initiates, an AI is spawned. When spawned, the AI registers for service handling of its specific type, thereby notifying its AT that it is initialized and ready for activity handling. As mentioned in Sect. 4.3, AIs receive process steps to execute to complete the activity, one at a time, and notify the AT once each process step is completed.

Production process steps are carried out by the RIs that advertise service handling for that specific process step. The coordination of the production process steps is coordinated by the AIs, through service requests (using the CNP) to the various RIs. For Activities, the AIs are responsible for service handling instead of the AT, since the AT must manage and perform computations for multiple AIs.

Once an AI has completed its list of required process steps and the activity has been completed, it sends a message to its AT containing an information packet with all relevant information from the completed activity (e.g. activity state, start time, predicted completion time and actual completion time). The AT then terminates the AI and analyses the collected activity information to update the attributes for future instances of the activity.

4.5 Scheduling of Operations

This case study used a simple method of scheduling. Scheduling of operations was performed by AIs contracting the services of RIs through the CNP for each process step that was carried out. Once the first process step was completed, the CNP would be followed again to secure a service provider for the second step and so on.

Since the system state was constantly monitored, this method allowed for the case where if one resource went offline, that the system would be able to detect the offline resource and carry on with the activity, provided that there was another resource of the same type in the system.

4.6 Implementation with the BASE Architecture

For the HMES case study implementation, the BASE architecture was used to create a digital administration shell for each holon (RTs, ATs, RIs and AIs) within the HMES. These administration shells all made use of the same core components, but specific plugins (of all the plugin types)—implementing the functions of the various holon types—were developed and deployed. A detailed description of the implementation using the BASE architecture is presented in [10].

5 Evaluation

The HMES was deployed to the Training Factory and, in testing, successfully coordinated the production operations. A key benefit of HMES is the inherent support that is provided for reconfigurability—the ability to change production capability (e.g. product type, mix, volume, etc.) by changing, adding or removing resources. As a preliminary evaluation of the implemented HMES, a reconfigurability experiment was performed. The results, and broader implications of the results, are discussed briefly here—further details are presented in [10].

5.1 Reconfigurability Experiment

In this experiment, a new type of resource was added in series to the existing production line. The new resource was a Packaging Station (PKG), which would simulate the packaging of workpieces after they came off the sorting line. With the addition, workpieces were moved from the SLD to the PKG, and from the PKG to the DSO, by the VGR. To utilize the added PKG, new process steps were then added to the order process.

5.2 Results and Discussion

The reconfigurability experiment required the addition of 616 lines of code to the implementation. The added code reused a substantial portion of existing code (from the implementation of other resources), with a code reuse rate (percentage of total lines of new code that was reused from existing code) of 72%. Furthermore, the added code required 7 h of development time. To incorporate the new sequence of production steps, only the configuration file of the AT had to be updated.

The result show that the implementation simplifies the reconfiguration of the Training Factory by allowing for code reuse and by limiting changes required in the greater HMES. This result is a consequence of the modularity of the HMES, as supported by the ARTI design and BASE architecture implementation.

In general, the HMES is considered highly modular and scalable, since numerous RIs of existing RTs can be spawned with no added lines of code, with minor changes to the configuration file to reflect the correct MQTT topics of each resource. A new resource configuration of the same existing RT could easily be added by spawning another instance with a new FSM Erlang module, a modified configuration file and less than 10 lines of new code in the BASE plugins for interfacing. Additionally, a completely new resource (Type and Instance) could be added to the system with the addition of only eight BASE plugins, with each plugin consisting of a large percentage of reused code. This pertains to the addition of Activities as well.

The process of compartmentalising and splitting functions and behaviours into different respective holons was appreciably aided by the ARTI architecture. Separating generic functions from instance-specific functions and placing them into a separate holon considerably reduced system complexity and system development time, as this reduced the percentage of code duplication that was needed in the implementation. Another benefit of using the ARTI architecture was that communication protocols are well defined in the architecture, which simplified the development of communication functions. The BASE architecture is the backbone of the HMES implementation and reduced development time and difficulty through eliminating the need to develop an administration backbone for each holon. The BASE architecture provides for the communication, biography and execution methods, with development only needed for the inner workings of these functions for application-specific functionalities.

6 Conclusion

This case study managed to demonstrate that ARTI is a valuable framework for use in the implementation of HMESs, as it assists in functionally and logically splitting the system into its respective holons, and it guides the developer in the implementation thereof. The secondary goal of the case study was also partly achieved, in that it was demonstrated that the BASE architecture is suitable for the manufacturing environment, and it decreases the development effort of implementing holonic systems. It also has the potential to improve the ease with which holonic systems can be scaled and reconfigured.

The presented case study can be extended in future work—the most notable being the implementation of a Delegate Multi-Agent System (DMAS) to improve scheduling of tasks. DMAS can be used to implement a complex scheduling system that uses micro-processes to explore, schedule and coordinate alternative process routes [4].

Acknowledgements The financial assistance of the National Research Foundation (NRF) towards this research is hereby acknowledged. Opinions expressed and conclusions arrived at, are those of the authors and are not necessarily to be attributed to the NRF.

References

1. Van Brussel, H., Wyns, J., Valckenaers, P., Bongaerts, L., Peeters, P.: Reference architecture for holonic manufacturing systems: PROSA. Comput. Ind. **37**, 255–274 (1998)
2. Kruger, K., Basson, A.: JADE multi-agent system holonic control implementation for a manufacturing cell (2018)
3. Leitão, P., Restivo, F.: ADACOR: a holonic architecture for agile and adaptive manufacturing control. Comput. Ind. **57**, 121–130 (2006)
4. Valckenaers, P., Van Brussel, H.: Design for the unexpected: from holonic manufacturing systems towards a humane mechatronics society. Butterworth-Heinmann (2015)
5. Sparrow, D.E., Kruger, K., Basson, A.H.: An architecture to facilitate the integration of human workers in Industry 4.0 environments. Int. J. Prod. Res. 1–19 (2021)
6. Plattform Industrie 4.0 2015, Status Report - RAMI4.0. ZVEI—German electrical and electronic manufacturers 0:28
7. Ericsson Computer Science Laboratory, Erlang. https://www.erlang.org/ (2021) Accessed 13 Sep 2021
8. Van Niekerk, D.J.: Extending the BASE architecture for complex and reconfigurable cyber-physical systems using holonic principles. Masters thesis, Department of Mechanical and Mechatronic Engineering, Stellenbosch University (2021)

9. Valckenaers, P., De Mazière, P.A.: Interacting holons in evolvable execution systems: the NEU protocol. In: Mařík, V., Schirrmann, A., Trentesaux, D., Vrba, P. (eds.) Proceedings of 7th International Conference, HoloMAS 2015, pp. 120–129. Springer International Publishing, Cham (2015)
10. Wasserman, A.: ARTI-based holonic control implementation for a manufacturing system using the BASE architecture. Masters thesis, Department of Mechanical and Mechatronic Engineering, Stellenbosch University (2022)

Alexander Wasserman has completed his M.Eng. at Stellenbosch University as a member of the Mechatronics, Automation and Design Research Group. His research focussed on HMES design and implementation.

Karel Kruger obtained his Ph.D. from Stellenbosch University, South Africa. He is a senior lecturer in the Department of Mechanical and Mechatronic Engineering and is co-leader of the Mechatronics, Automation and Design Research Group at Stellenbosch University, South Africa.

Anton H. Basson obtained his Ph.D. in Aerospace Engineering at Penn State University. In 1997, he was appointed as Professor in Mechanical Engineering at Stellenbosch University and is co-leader of the Mechatronics, Automation and Design Research Group.

Production Systems and Organizations II

Bridging the Gap Between Digital Human Simulation to Standard Performance with Human Work Design

Peter Kuhlang, Martin Benter, and Maria Neumann

Abstract One major task in industrial companies is designing human work processes. Productivity and ergonomic factors are key performance indicators for this task. To measure and improve these indicators, practitioners use different methods to analyse and design work processes. Most methods either focus on productive or ergonomic aspects, but not both. Furthermore, they require a lot of manual effort regarding data collection and interpretation. These factors limit the number of industrial workplaces that are designed in a productive and ergonomic way. The process building block system MTM-HWD® is one method that was developed specifically to allow an integrated analysis of both productive and ergonomic aspects. However, applying MTM-HWD® still requires considerable effort. The application effort for methods such as MTM-HWD® can be reduced by automatically interpreting digitized human motion data. Motion data digitally depicts human movements. The data includes, for instance, joint positions, covered distances or object interactions. Particularly the technologies human simulation and virtual reality are suitable for generating or interpreting motion data. This article shows how motion data that originates from human simulations can be used to derive a partially automated MTM-HWD® analysis. The technologies thus reduce the required manual effort for MTM-HWD®. This way, human work processes can be designed to be productive and ergonomically sound, without increasing the required design effort. To clarify the process the article primarily focusses on the simulation software imk ema. First, the article shows the necessary information for a creating a MTM-HWD® analysis. It then dissects, which information can be generated automatically by human simulations tools and which data has to be added manually. By manually adding the information that cannot be determined automatically, it is then possible to derive a MTM-HWD® analysis. To clarify this process, an example is shown using the software imk ema.

Keywords Human motion data · Human simulation · MTM-HWD®

P. Kuhlang (✉) · M. Benter · M. Neumann
MTM ASSOCIATION e. V., 22609 Hamburg, Germany
e-mail: peter.kuhlang@mtm.org

K. von Leipzig et al. (eds.), *Smart, Sustainable Manufacturing in an Ever-Changing World*, Lecture Notes in Production Engineering,
https://doi.org/10.1007/978-3-031-15602-1_5

1 The MTM-HWD® Process Building Block System

The aim of the MTM process building block system MTM-HWD® (Methods-Time Measurement-Human Work Design, short: HWD) is the combined productive and ergonomic evaluation of human work [1]. To achieve this aim, the performed actions and their productive and ergonomic influencing factors must be recorded. To simplify and systematize the recording, a classification of these variables according to human body parts is needed [1].

Due to its objective description of human movements, it can also be used as a basis for the digitization of the planning and design of human work [2, 3]. Feasible times can be assigned to digitally recorded work sequences [4, 5]. Furthermore, it can be ensured that these sequences do not harm the ergonomics of the involved employees [1]. Therefore, it is a key for the successful application of digital planning tools [3], such as the simulation software imk ema [6].

2 Application of MTM-HWD®

To succeed in the application of HWD, it is necessary to correctly assess the HWD building blocks and their influencing factors. The central components are the building block elements that describe the work sequence. They consist of the following components:

- object
- action
- active limb
- passive limb [7]

The influencing variables must be determined for each building block to evaluate the sequence in terms of time and ergonomics.

The process building block system MTM-HWD® comprises of 25 influencing factors (Fig. 1), which are classified into the areas:

- lower limbs
- trunk
- head/neck
- upper limbs

2.1 Building Block Elements

The most important building block element is the action that the worker performs, i.e., whether he picks an object up or moves it. The type of object with which the worker performs his actions (e.g., actuator or tool) is also part of the building block

description segment	name
lower limbs	path
	floor conditions
	stability
	basic position
	leg posture
trunk	trunk flexion
	trunk rotation
	trunk inclination
head/ neck	head posture
	eye travel

description segment	name
upper limbs	upper arm posture
	hand position
	arm extension
	wrist posture
	weight
	force
	direction of force
	distance class
	supply
	place accuracy
	assembly position
	positioning conditions
	grasp motion
	type of grasp
	vibration

Fig. 1 HWD influencing factors

description. Furthermore, it is important to record which limb is actively performing the action and what the passive limb is doing during the action [1].

In addition to the building block elements, the user records the relevant influencing factors for the action. These factors are described briefly in the following sections [1, 6, 8].

2.2 *Influencing Factors—Lower Limbs*

The influencing factors of the lower limbs are the first segment that is assessed. The distance an employee covers during an action is described by the factor path. HWD differentiates between walking, climbing steps/ladders and crawling. The floor conditions indicate whether the floor is uneven or whether obstacles may block the path and thus, a greater level of control is required while walking. The position of the body at the end of the action is described by the basic position. A distinction is made between standing, sitting, kneeling, crouching, or lying down. The influencing factor leg posture shows if the worker extends or bends the legs, and the factor stability depicts if the posture of the lower extremities allows a secure stand.

2.3 *Influencing Factors—Trunk and Head/Neck*

The next influencing variables describe the posture of the workers trunk and head. The posture of the upper body is reflected by the factors trunk flexion, rotation and inclination. Whether the torso is bend forward or backward is shown in the flexion, the deflection to the side is reflected by the inclination and the rotation describes how much the shoulders are twisted in relation to the hips. Analogously, the head posture describes whether the head is turned, tilted forward, backward, or sideways or if it has a normal posture. Finally, time relevant eye movements are represented by the factor eye travel.

2.4 *Influencing Factors—Upper Limbs*

The upper limbs include a series of influencing variables, such as the posture of the arms and hands, the necessary forces and their directions and elements that reflect the required control effort of the actions. The upper arm posture describes whether the arms are deflected to the front, back or side. The hand position shows the height of the hand in relation to the shoulder joints. The distance between hand and shoulder is reflected by the arm extension. An influencing variable which is similar to trunk posture is wrist posture. It represents the bending, inclination, or rotation of the wrist.

The user can also specify the weight of the object as well as the force necessary to handle the object. Furthermore, the direction of the force (up/down, forward/backward, sideways) can be recorded.

Now the influencing variables that describe the control effort are explained. They have a high time relevance and include distance class, supply, place accuracy, assembly position, positioning conditions, grasp motion, type of grasp and vibration.

The distance class describes the distance covered by the hand during an action. The influencing factor supply shows the arrangement of the objects that have to be grasped. Place accuracy, assembly position and positioning conditions describe the way an object has to be placed. Grasp motion is the motion performed to gain control over an object or to shift control. Type of grasp describes the required posture of the fingers to gain or relinquish control of the object. The final factor is vibration, which indicates a sudden impulse or vibration affecting the hand and the arm during an action.

Once all the building block elements and influencing variables have been assessed, the required time and the ergonomic load of the work task can be determined. This provides the basis for a target-oriented design of the workplace.

3 Categorization of the Influencing Factors

To obtain an HWD analysis from a human simulation tool, all building block elements and influencing factors must be assessed correctly. Therefore, it is essential to know which elements can be determined automatically and which elements must be put in manually.

To answer this question systematically, a classification of the HWD influencing factors into categories has been made [7]. Figure 2 shows a possible categorization of the factors.

An influencing variable is building block specific if it can be assigned to an action. A good example is the path, because it is only relevant for one of the possible actions and must be evaluated for each of these actions.

description segment	influencing factor	category			
		building block specific		overarching	
		accuracies	distances	postures	forces
lower limbs	path		x		
	floor conditions	x			
	stability			x	
	basic position			x	
	leg posture			x	
trunk	trunk flexion			x	
	trunk rotation			x	
	trunk inclination			x	
head/ neck	head posture			x	
	eye travel		x		
upper limbs	upper arm posture			x	
	hand position			x	
	arm extension			x	
	wrist posture			x	
	weight				x
	force				x
	direction of force				x
	distance class		X		
	supply	x			
	place accuracy	x			
	assembly position	x			
	positioning conditions	x			
	grasp motion	x			
	type of grasp	x			
	vibration				x

Fig. 2 Categorization of HWD influencing factors

On the other hand, overarching influencing factors are not limited to specific actions. Furthermore, they can remain valid for several actions in succession. Additionally, they are not unique to HWD, but can also be found in other methods. The basic position is a good example for that. It can remain the same for several actions or entire work sequences and is also relevant for ergonomic methods such as EAWS (Ergonomic Assessment Worksheet) [9].

Accuracies and distances form the two subcategories of the specific influencing factors. Accuracies show the required control effort and include several influencing factors, such as grasp motion and type of grasp. The object being used affects these variables the most. For example, a screw, which is small, and lays jumbled with others in a container, requires a greater accuracy while grasping than a screwdriver.

The necessary motion length of the actions is recorded under distances. The layout of the workstation mainly determines the distances. In conclusion, a better layout leads to shorter distances.

The overarching influencing factors are divided into postures and forces. Postures describe the position of the head, torso, arms, hands, and legs. Only by recording them correctly, a successful ergonomic evaluation of the work process is possible. Like distances, the layout of the workstation influences them the most. Thus, a poorly designed workplace leads to an increased ergonomic load.

Finally, the applied forces are addressed. The load weights of the objects and the required force necessary to move or position an object are part of this. The applied forces are, like postures, important for the assessment of the ergonomics. The difference is, they are mainly affected by the used objects. The heavier the used tools, the higher the required forces.

This classification of the HWD influencing factors into categories can also be used by other human work assessing methods. For example, the MTM-UAS (Universal Analysing System) method [10] primarily uses accuracies and distances for the time evaluation of work processes. In contrast, the EAWS method [9] is based on postures and forces for ergonomic evaluation.

4 Determining the Information Through Human Simulations

The focus now is the examination of the building block elements as well as the categories of the influencing factors. As a result, the extent on how human simulations are capable of providing the required information for an HWD analysis can be determined. This article takes a closer look at the tool imk ema [6] and its possibilities. In the case of missing information, the article elaborates on how this information can be filled in. Figure 3 shows the result of the examination.

HWD information		determination of information	necessity of additional user activity
building block elements	object	necessary input	no
	action	assignment of action sequences	yes (verification)
	limbs	necessary input	no
influencing factors	accuracies	additional input	yes (manual input)
	distances	automatic determination	no
	postures	automatic determination	no
	forces	necessary/ additional input	partial (manual input)

Fig. 3 Determination of HWD information using human simulations

4.1 Building Block Elements

Simulation tools such as imk ema show human work processes digitally. To create these processes correctly, a lot of information and parameters are needed which are similar to the required data for creating an HWD analysis. For example, the used object including the object information (e.g., weight) and the performing extremity are necessities in both cases. Thus, an important part of the building block elements is already present in the simulation. However, the actions which represent an essential part of a correct HWD application cannot be easily obtained from the simulation input. This problem can be solved by identifying frequently used objects and defining their corresponding relevant action sequences.

This approach ensures that building block elements can be determined for a large number of use cases. However, since an accurate result cannot be guaranteed in all cases, a trained HWD user needs to verify the results and correct them if necessary.

4.2 Influencing Factors—Accuracies

Similar to the actions, accuracies cannot be derived directly from the simulation input. Thus, assessing them correctly is rather difficult. Therefore, the accuracies are integrated as additional manual inputs during the creation of the modelling processes. For example, the influencing variable supply is indicated as additional information when the simulation objects are entered. It is recommended that an experienced HWD user enters these inputs or verifies the simulation results.

4.3 *Influencing Factors—Distances*

By using a digital human model, simulation tools such as imk ema calculate and visualize human movements. As a result, the generated motion data contains, among other things, the coordinates of the body parts during the work process. This data is then used to derive the distances. By doing so, distances can be determined very accurate, probably even more accurate than the manual recording by an HWD user because estimates and measurement errors are avoided. But the determined distances are only as good as the simulation itself. An incorrect model of the workplace or unrealistic simulated movements can lead to unrealistic calculated distances.

4.4 *Influencing Factors—Postures*

The postures can be derived from the motion data of the human simulation as well as the distances. Because the currently existing posture definitions were originally only intended for manual assessment, explicit values for determining the postures were not always available. Therefore, the existing definitions have been specified for the digital application. For example, it was not clearly defined at which body angles an employee lies or sits.

4.5 *Influencing Factors—Forces*

Similar to the accuracies, forces cannot be derived directly from the data of the digital human model but are integrated as additional information during the modelling process. For example, the weights of the used objects can be added. Other occurring forces have to be entered manually for a correct HWD analysis. Thus, the input or verification by an experienced user is necessary here as well.

5 Implementation With the Imk Ema Simulation Tool

Together with the company imk automotive GmbH, the MTM ASSOCIATION e. V. has launched a project for the automated derivation of HWD analyses from human simulation tools. The results of this project are the described necessary additions and adaptations that must be made.

After the simulation model has been created with the additional data and the work process has been simulated, the generated data is transferred to the MTM software TiCon via a jointly developed interface. Then the transferred data and the HWD rules are checked by TiCon and appropriate MTM times are assigned. Moreover, the work

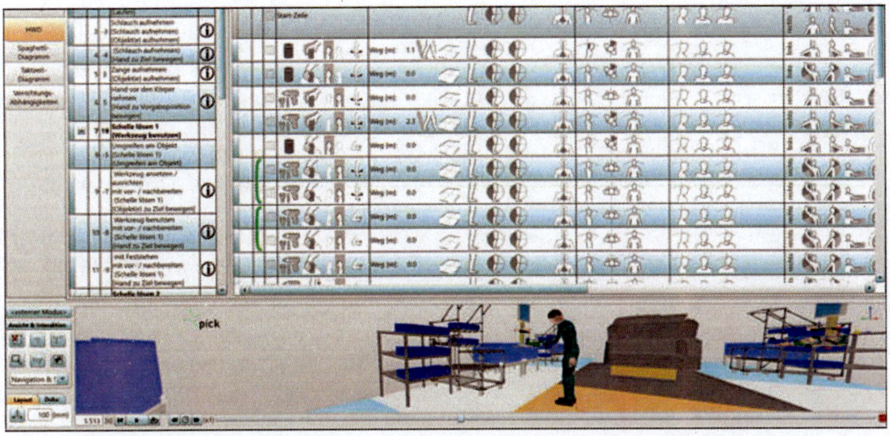

Fig. 4 Illustration of a derived HWD analysis in imk ema

task is ergonomically evaluated. As a result, the user gets a simulation analysis that provides a reliable planning time and enables a meaningful ergonomic evaluation.

Because the work process that took place in the simulation tool is reflected by the simulation analysis, incorrect or humanly impossible workflows would be visible. This also makes faulty simulation visible, so that the user can take immediate action.

This solution is currently being tested extensively in a joint validation project that focuses on the automated derivation of an HWD analysis. An example of this validation can be seen in Fig. 4. It shows the user interface of the imk ema software. The workstation layout as well as the simulated human performing the work task are shown on the bottom of the figure. In this case, the task is the assembly of a hose on a passenger car.

A section of the corresponding correct HWD analysis can be seen in the upper area. The ema functionalities allow the user to jump to certain actions in the sequence and see the corresponding part of the simulation.

6 Conclusion

This paper described how MTM-HWD® analyses can be derived from digital human motion data. It was revealed that a large part of the necessary information is already entered or generated when simulation work processes. Only the missing information needs to be added by the HWD user. At the end, the user gets a simulation analysis that contains both, time, and ergonomic evaluation of the simulated work process.

The article also explains how and why using human simulation tools for creating an MTM-HWD® analysis is possible and useful. This is of high importance for a productive and ergonomic design of human work in the increasingly digitalized production.

References

1. Finsterbusch, T.: Entwicklung einer Methodik zur Bildung von Bausteinsystemen für die Gestaltung menschlicher Arbeit, Technische Universität Dresden (2016)
2. Finsterbusch, T., Kuhlang, P.: A new methodology for modelling human work—evolution of the process language MTM towards the description and evaluation of productive and ergonomic work processes. In: Proceedings of 19th Triennial Congress of the IEA (2015)
3. Kuhlang P.: Positionen der Deutschen MTM-Vereinigung e. V. zu Assistenzsystemen und zur Verarbeitung von digitalen Bewegungsdaten. In: MTM-Schriftenreihe Industrial Engineering, Vol 12 (2019)
4. Faber, M., Przybysz, P., Latos, B.A., Mertens, A., Brandl, C., Finsterbusch, T., Härtel, J., Kuhlang, P., Nitsch, V.: Empirical validation of the time accuracy of the novel process language Human Work Design (MTM-HWD®). Prod. Manuf. Res. **7**(1), 350–363 (2019)
5. Finsterbusch, T., Petz, A., Faber, M., Härtel, J., Kuhlang, P., Schlick, C.M.: Empirical evaluation of the novel process language MTM-human work design for modeling human tasks. In: 7th International Conference on Applied Human Factors and Ergonomics (2016)
6. Leidholdt, W., Fritzsche, L., Bauer, S.: Editor menschlicher Arbeit (ema), Homo Sapiens Digitalis—Virtuelle Ergonomie und digitale Menschmodelle. Springer Vieweg, Berlin (2016)
7. Benter, M., Kuhlang, P.: Kategorisierung der MTM-HWD®-Einflussgrößen zur Bewertung der Ableitbarkeit aus digitalen Bewegungsdaten, GfA Herbstkonferenz (2019)
8. Kuhlang P.: Produktive und ergonomiegerechte Arbeit—Von Grundsätzlichem zur Prozesssprache MTM über die Ergonomiebewertung zu Human Work Design (MTM-HWD®), Leistung und Entgelt, Vol 2/2018 (2018)
9. Schaub, K., Caragnano, G., Britzke, B., Bruder, R.: The European assembly worksheet. Theor. Issues Ergon. Sci. **14**(6), 616–639 (2012)
10. Bokranz, R., Landau, K.: Produktivitätsmanagement von Arbeitssystemen (2006)

Prof. Dr. Peter Kuhlang, is the CEO of the MTM ASSOCIATION e. V. and the head of the MTM Institute. He is responsible for the prosperous development of all worldwide research, development, consulting, software and network activities.

Dr.-Ing. Martin Benter, works at the MTM Institute and is responsible for the digitization and automation of MTM methods.

Maria Neumann, works at the MTM Institute and specializes in the digitization & automation of MTM methods.

Interface Holons in the BASE Architecture for Human-System Integration in Cyber-Physical Systems

D. J. van Niekerk, K. Kruger, and A. H. Basson

Abstract The Biography-Attributes-Schedule-Execution (BASE) architecture is a holonic reference architecture that was developed to integrate human workers into Industry 4.0 environments. The architecture guides the development of Human Resource Holon Administration Shells (BASE shells), which raise humans to a Cyber-Physical System (CPS) level so that they can interact with other CPSs. Existing research on the BASE architecture has already shown how interface services can be used to enable the cyber-physical interfacing between humans and their associated BASE shells. This paper extends this research by introducing dedicated Interface Holons in the BASE architecture. Interface Holons encapsulate interface services provided by cyber-physical interfacing resources, such as sensors, loudspeakers, augmented/virtual reality systems, etc. The details of the BASE architecture plugins required in Interface Holons are discussed. Furthermore, standard inter- and intra-holon communication protocols between the BASE shells of Interface Holons and Human Resource Holons are explained.

Keywords Industry 4.0 · Human-system integration · Cyber-physical systems · Holonic manufacturing systems

1 Introduction

Tay et al. [1] defined Industry 4.0 (I4.0) as an aggregation of existing ideas and technologies into a new value chain. This involves connecting systems in a self-organising manner that enables dynamic control within an organisation. Human workers form a big part of the I4.0 revolution because of their unmatched flexibility [2].

Derigent et al. [3] observed that in the past two decades, the holonic systems paradigm served as major driver of Intelligent Manufacturing Systems. They further show how Holonic Control Architectures (HCAs), which are built on the concept of

D. J. van Niekerk · K. Kruger (✉) · A. H. Basson
Department of Mechanical and Mechatronic Engineering, Stellenbosch University, Stellenbosch, South Africa
e-mail: kkruger@sun.ac.za

K. von Leipzig et al. (eds.), *Smart, Sustainable Manufacturing in an Ever-Changing World*, Lecture Notes in Production Engineering,
https://doi.org/10.1007/978-3-031-15602-1_6

holons, have evolved and can address the needs of I4.0. Koestler [4] defined a holon as an entity with communication and decision-making capabilities that is composed of a set of sub-level holons, yet at the same time is part of a holarchy of higher-level holons. HCAs are composed of multiple autonomous and cooperating holons that are ordered in a holarchy [5].

Derigent et al. [3] stated that service orientation is a key enabler for Industry 4.0 and that the distributed and autonomous behaviour of holons enables the use of holonic services. The service-oriented and holonic paradigms both employ decentralised control architectures, which are decomposed into services and holons, respectively [6]. The combination of holonic systems and Service-oriented Architectures (SoAs) in the manufacturing context facilitate the development of Service-oriented Holonic Manufacturing Systems (SoHMSs), which are known for their flexibility and reactiveness [6].

While HCA and SoHMS research have considered the integration of humans as holons, it was only explored in depth with the development of the Biography-Attributes-Schedule-Execution (BASE) architecture. The BASE architecture uses administration shells (referred to as BASE shells) to raise humans to a Cyber-Physical System (CPS) level, supporting interaction with other CPSs [7].

Sparrow et al. [8] showed how interface services, performing observation and informative actions, can be used to enable cyber-physical interfacing between humans and their BASE shells. This paper will present how the BASE architecture can be applied to non-human holons and, thereby, be used to develop Interface Holons that support cyber-physical interfacing services.

2 The Base Architecture

2.1 Background

The development of the BASE architecture was originally driven by the need to integrate human workers into I4.0 environments. This development was strongly influenced by the requirements and responsibilities of Resource Holons in Holonic Manufacturing Systems (HMSs) [7].

Sparrow et al. [8] proposed the BASE shell, a Human Resource Holon Administration Shell (HRH-AS), to raise humans to a CPS level, which would enable them to interact with other CPSs. This BASE shell adds digital storage, processing, and communication abilities to the human's natural abilities by making use of human–machine interfaces and information and communication technology.

2.2 Core Components and Plugins

The BASE architecture consists of five core components and four types of plugins, as shown in Fig. 1. The BASE architecture's five core components are the Schedule, Execution, Biography, Attributes and Communication Manager. Schedule stores all activities that must still be executed, Execution stores all activities that are being executed and Biography stores all activities that have been completed.

In addition to its storage component, Execution consists of a State Blackboard, an Observer and an Informer. The State Blackboard stores the human's current state (physical, mental and biological) as updated by the Observer. The Observer gathers information about the human from observation services, like sensors and cameras, and delivers the information in *Value-Confidence-Timestamp* format. The Informer delivers information to the human and uses the Attributes to tailor this information delivery.

The Attributes component stores a holon's attributes, i.e. the properties of the holon that do not change during activities. Sparrow et al. [7] created two categories of attributes, namely Personal Attributes and Contextual Attributes. Personal Attributes are persistent data about a resource and its BASE shell and forms the digital model of the resource. Contextual Attributes are application specific and defines the resource within the context it is in.

The Communication Manager is the communication gateway of the BASE shells and coordinates the flow of information in the shell. Inter-holon communication is also coordinated by the Communication Managers of holons' BASE shells.

The five core components are not application specific and, as such, these components are not expected to add any value on their own. To add application-specific value to a BASE shell, [7] created BASE plugins which can interact with the BASE

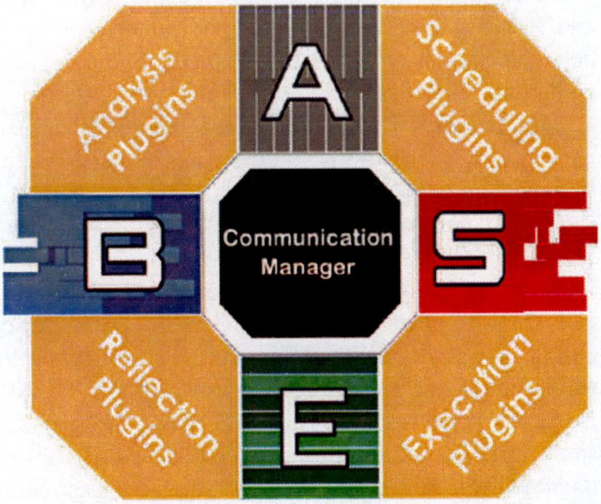

Fig. 1 The BASE architecture [7]

core components. There are four types of BASE plugins, namely: Scheduling Plugins (SPs), Execution Plugins (Eps), Reflection Plugins (RPs) and Analysis Plugins (Aps).

A BASE shell's SPs schedule new activities in its Schedule using smart algorithms, consideration of attributes, and interactions with other holons. Scheduled activities are initiated, monitored and driven by EPs, which update the activities' execution progress in the Execution. When activities are completed, RPs are used to save the activity data in the Biography and add any post-execution data to the activities. APs analyse activity data stored in the Biography in order to update the BASE shell's attributes in the Attributes component.

2.3 The Three-Stage Activity Lifecycle

The BASE architecture uses the Three-Stage Activity Lifecycle (3SAL) model, proposed by [9], for the structuring of its activity data. In the 3SAL model, an activity progresses through three temporal stages: *scheduled*, *in execution* or *completed*. The activity data is structured according to the stage at which the data was created. Schedule data is edited before an activity is started, execution data is edited while an activity is being executed and post-execution data is added after an activity has been completed.

3 Interface Holons

Devices like temperature sensors, accelerometers, location sensors, loudspeakers, computer monitors, augmented or virtual reality systems, touch screen interfaces, and services like WhatsApp, provide interfaces by which data can be exchanged between humans and their BASE shells. These devices can be represented as holons and are henceforth referred to as Interface Holons.

The extended BASE architecture has two standard types of Interface Holons, namely Observer Holons and Informer Holons. When a human's BASE shell needs to exchange data with its physical part, rather than managing the cyber-physical interfacing connections and communications itself, it can make use of an available Interface Holon.

Observer Holons enable data flow from the physical to the cyber layer and typically involve sensor devices. BASE shells can create Service Contracts with Observer Holons to obtain observed data on the state of their associated human. Conversely, Informer Holons enable data flow from the cyber to the physical layer—allowing BASE shells to communicate with their associated humans—and typically involve some form of actuating device. Examples of Informer Holons are loudspeakers, displays, warning lights and vibrating devices. Some Interface Holons are both Observer and Informer Holons, e.g. touch screen devices and Virtual Reality (VR) and Augmented Reality (AR) headsets.

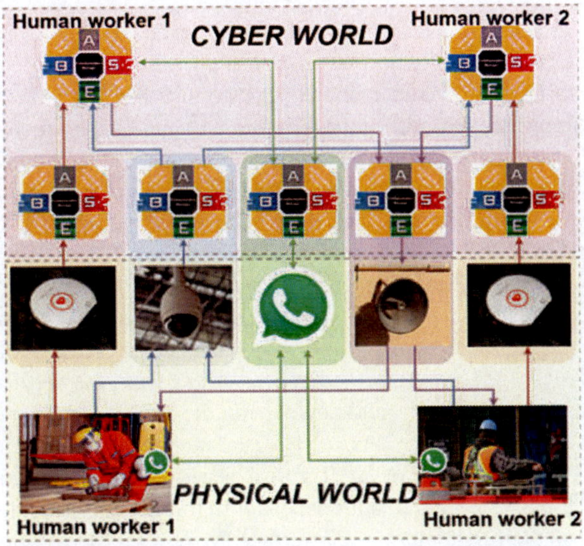

Fig. 2 Interface holon examples

The use of Interface Holons to enable data flow between the physical and cyber worlds is illustrated in Fig. 2. This figure shows two human workers and five cyber-physical interfaces, each represented by their own BASE shell.

Sparrow et al. [8] categorized cyber-physical interfaces into two classes: *personal* and *environmental*; where a personal interface is dedicated to a single human and an environmental interface can be used by multiple people. The scenario shown in Fig. 2 has three environmental interfaces (camera, speaker and WhatsApp service) and each worker has a personal interface (wearable heart rate monitor). The camera and two heart rate monitors are examples of Observer Holons, while the speaker is an example of an Informer Holon. The WhatsApp interface is an example of an Observer and Informer Holon combined, since it allows information to flow in both directions. An interesting aspect to note about the WhatsApp interface is that although workers have their own phones, the physical part of the WhatsApp interface is shown as one resource. This is because the BASE shell of the WhatsApp interface does not individually communicate with each phone, but uses a third-party service, like Twilio [10], to communicate with the different phones.

4 Base Architecture Plugins For Interface Holons

Every Interface Holon needs two standard plugins, namely an Observer/Informer SP and an Observer/Informer EP (depending on the interface type). These plugins are generic for all Interface Holon BASE shells, but call interface-specific code contained in the Interface Module. The Interface Module contains the code required to start the custom Interface Process and is also used by the Observer EP to verify Identifiers (discussed in Sect. 5.2). When an Interface Holon is initialised, its Observer/Informer EP calls a function in its holon's Interface Module to start the Interface Process, which starts and manages the connection with the holon's physical device. This Interface Process must be able to send and/or receive messages to and from the Observer/Informer EP that started it. Each Observer/Informer EP needs to monitor its holon's Interface Process and restart it if it fails.

When new data has been observed about another holon by an Observer Holon's physical device, this physical device sends the data to its BASE shell's Interface Process. The Interface Process forwards this new data to its Observer EP, which shares the new data with the BASE shells that requested it. When another holon requests an "Inform" service from an Informer Holon, the Informer EP sends a message to its Interface Process to execute the service. This Interface Process then messages its physical device with the custom details required to present the specified information.

An example of an "Inform" service would be a human's BASE shell requesting the BASE shell of a loudspeaker to inform the associated human about a new activity on their Schedule. The Interface Process sets up a connection with the PC controlling the speaker (e.g. via TCP or HTTP) and sends the "inform" message to this PC. The PC translates the message using some text-to-speech software and broadcasts the speech over the loudspeaker.

5 Service Descriptions

5.1 Overview

Observer and Informer Holons have the same Service Description structure, except that Informer Holons do not have a "Topics" field. The highest level of the Service Description has two fields: "Resources" and "Interface description". The "Interface description" field can be used by developers to add any descriptive information about the interface. The "Resources" field contains all the resource types for which interfacing is possible, and each resource type field contains three fields, namely: "Available identifiers", "Identifier type" and "Topics". Identifiers and Topics are discussed in more detail in Sects. 5.2 and 5.3, respectively.

5.2 Identifiers

Identifiers are used by Interface Holons to distinguish between the different holons with which they interact. When using Informer Holons, human holons do not need to specify an identifier—their unique holon identity (ID) will always be used. One very important functionality of Informer Holons is that they prevent one holon's BASE shell to communicate directly with another holon's physical part. This functionality promotes modularity, ensuring that holons remain independent entities that cannot directly control each other and need to request each other's services.

Like Informer Holons, Observer Holons can use a holon's ID as Identifier, but Observer Holons can also use one of the holon's attributes. For example, if all human holons in a factory have a QR code printed on their clothing and this is used by cameras in the factory to distinguish humans, this QR code will be the Identifier used by the camera holons. The type of Identifier that an Observer Holon is able to use is specified in its "Identifier type" field. Note that the identifier type can differ for the different type of holons that an Observer Holon can interface with.

There are two ways in which a holon can specify for which Identifier they are making a service request to an Observer Holon:

- The holon can add an "Identifier" field to the request arguments of the Service Contract with one of its attributes used as the value of this field.
- The holon can specify no Identifier in the Request Arguments, in which case the Interface Holon will use the holon's ID.

In both scenarios, the Observer Holon will verify the Identifier by calling the *valid_identifier* function in its Interface Module. Observer Holons allow more than one holon to use the same Identifier, e.g. all the human holons in a factory can request to observe the weather conditions from a weather Observer Holon. The Identifier used might then be something similar to "Location", which must be an attribute in all the humans' BASE shells. However, in most cases every Identifier is meant for one holon only.

When the "Available identifiers" field is equal to "all", the Interface Holon can (but will not necessarily) enable cyber-physical interfacing for all holons of some resource type, as long as they provide a valid Identifier. When an "Available identifiers" field is not equal to "all", but is a list of identifiers, the Interface Holon can only enable interfacing for the Identifiers specified, and will reject requests from holons with Identifiers that are not in this list.

5.3 Topics

Topics are the names/descriptions of observed data, e.g. "Temperature", and each Topic specifies the estimated accuracy and unit of the observed data. When making a request to an Observer Holon, a holon can specify which Topics' data it wants to receive or not specify anything, in which case all of the Topics' data will be shared with the holon.

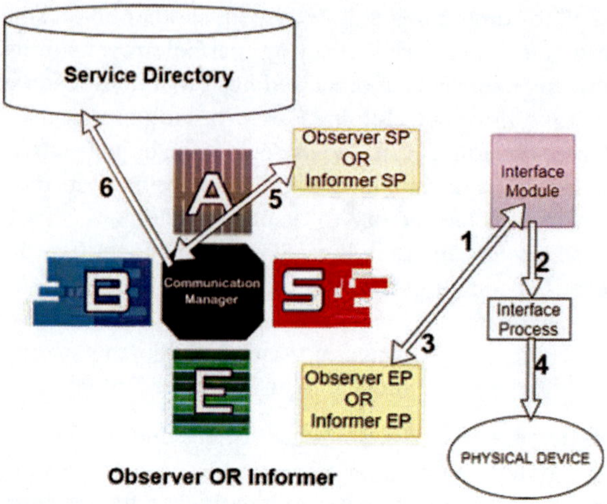

Fig. 3 Interface holon initialisation

6 Inter- and Intra-Holon Communications

6.1 Initialisation of Interface Holons

Figure 3 shows the interactions necessary to add a new Interface Holon to a system. The initialisation of all Interface Holons follows the same procedure. The Interface EP (Observer EP or Informer EP) calls the *start_interface* function in its holon's Interface Module (arrow 1). The Interface Module starts the Interface Process (arrow 2) and returns the address of this process to the Interface EP (arrow 3), since this EP needs to monitor the Interface Process to restart it if necessary. As mentioned in Sect. 4, the Interface Process is different for each Interface Holon and will manage the connection with the holon's physical device or with a gateway to the device (arrow 4). The Interface Holon's Communication Manager obtains the holon's services ("Observe" and/or "Inform" service description—as discussed in Sect. 5.1) from its Observer/Informer SP (arrow 5). It then adds the Interface Holon's ID, with its services, to a service directory (arrow 6) so that other holons in the system can find and use it.

6.2 Observer Holons

A holon's BASE shell can determine if an Observer Holon can provide data to it by inspecting the Observer Holon's service description. Figure 4 shows the BASE shell of a human holon requesting an Observer Holon's "Observe" service after it determined that the Observer can provide data to it. The Observer Holon's Communication Manager forwards this request to its Observer SP (arrow 1), which checks two

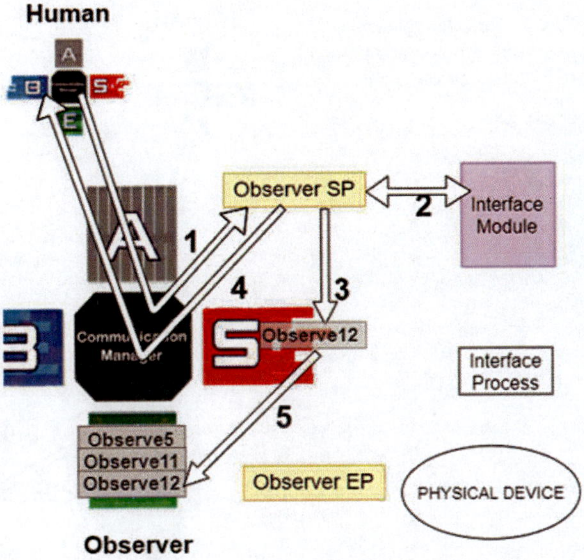

Fig. 4 Request to observer holon

requirements before accepting the service request and scheduling a new observation activity. These requirements are to check (in order):

1. If the requesting holon's resource type is one of the resource types in this Observer Holon's service description.
2. If the Identifier provided by requesting holon is valid (arrow 2) and included in the "Available identifier" list (discussed in Sect. 5.2).

If both requirements are met, the Observer SP schedules a new "observe" activity ("Observe 12" in Fig. 4) immediately (arrow 3), and accepts the service request via its Communication Manager (arrow 4). Observer EP starts this activity (arrow 5) and adds the requesting holon to its list of clients.

Similarly, Fig. 5 shows the Observer Holon's Interface Process receiving and responding to new data from its physical device.

6.3 *Informer Holons*

Informer Holons handle "Inform" requests similar to how Observer Holons handle "Observe" requests, except that when the inform activity is started, it is executed and finished and not kept in Execution as with Observer Holons. Figure 6 shows the interactions between the Informer EP and its Interface Process to execute an "Inform" service requested by a human holon. The Informer EP messages its Interface Process (arrow 1), requesting it to execute the "Inform" service request encapsulated in the inform activity ("Inform 23" in Fig. 6). The Interface Process communicates with

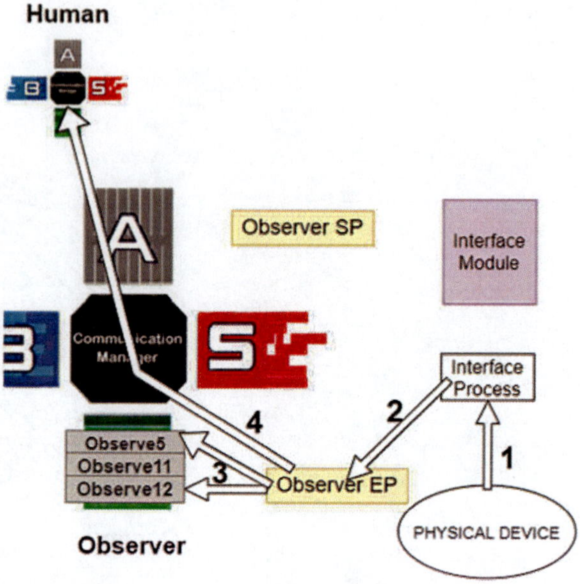

Fig. 5 New data observed by observer holon and passed to a human holon client

the Informer Holon's physical device (arrow 2), to deliver a message to the human of the BASE shell that made the "Inform" request. The way in which the physical device delivers the message is dependent on the device (e.g. screen or speaker. The message delivery result is returned to Informer EP (arrow 3), which informs the human's BASE shell whether the message delivery was successful or not (arrow 4). Informer EP writes this result to the stage 2 data of the "Inform23 (SPA)" activity and then finishes this activity (arrow 5 and arrow 6).

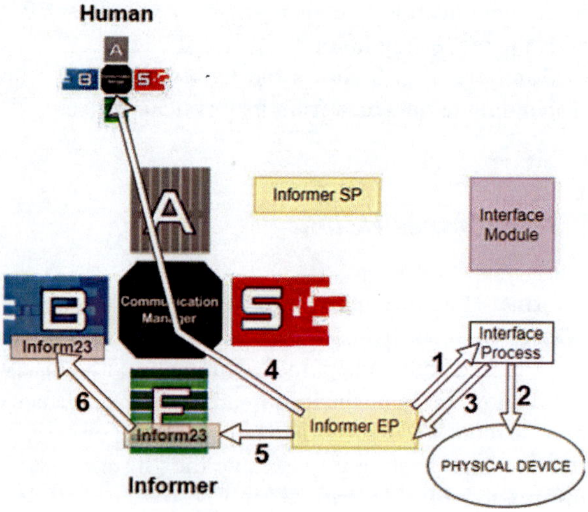

Fig. 6 Informer holon executing "inform" service

7 Interface Holon Activities and Attributes

As described in Sect. 6, each type of Interface Holon has its own interface activity type: "Observe" or "Inform", which is used to encapsulate the interfacing services provided by the Interface Holon. In addition, all Interface Holons have three common types of activities: "Device State Change", "Interface State Change" and "Restart Interface".

Activities of type "Device State Change" are created when the Interface Process informs its EP that the physical device's state has changed from working to down or vice versa. This provides developers with a mechanism to track the state of an Interface Holon's physical device. However, this functionality requires that the physical device be programmed to communicate its state to the Interface Process.

Activities of type "Interface State Change" are created when the Interface Process fails or has been restarted successfully. An Interface Process's failure is detected by the EP that started it (Observer EP or Informer EP), since this EP monitors the Interface Process after starting it via the Interface Module. It is the responsibility of this EP to start and finish an activity of type "Interface State Change" and to try to restart the Interface Process. If the restart succeeds, the plugin will have to log another activity of type "Interface State Change", with the Result set to "Success". Activities of type "Restart interface" can be scheduled if an Interface Holon's Interface Process becomes unresponsive.

8 Discussion

Interfacing between humans and their BASE shells is a critical functionality of a CPS implemented with the BASE architecture. By representing interfaces as Interface Holons, cyber-physical interfacing errors can be isolated and contained, improving the CPS's fault-tolerance. Furthermore, Interface Holons improve the diagnosability of cyber-physical interfacing errors, since they capture the details of these errors in their respective Biographies. Human holons' BASE shells can use their CPS's service directory to dynamically find and utilise new Interface Holons. Consequently, human holons' BASE shells do not need to be reconfigured to use new interfacing methods, since Interface Holons can be added and removed independently from the human holons. These benefits of Interface Holons further improve the BASE architecture's ability to integrate humans into I4.0 environments, by accurately reflecting their state (through Observer Holons) and allowing their BASE shells to communicate important information to them (through Informer Holons).

9 Conclusion

The BASE architecture facilitates the integration of human workers into I4.0 environments. In order for a BASE shell to communicate with and reflect the state of the associated human, cyber-physical interfacing devices/services, like sensors and human machine interfaces (HMIs), are required. This paper introduced Interface Holons to encapsulate these devices/services and explained the plugins, service descriptions and interaction protocols required for integrating these holons effectively into a CPS. Future work will extend the BASE architecture as a reference architecture for non-human entities as well, providing all resource holons with advanced functionalities through effective interfacing between the physical and cyber layers.

Acknowledgements The financial assistance of the National Research Foundation (NRF) towards this research is hereby acknowledged. Opinions expressed and conclusions arrived at, are those of the authors and are not necessarily to be attributed to the NRF.

References

1. Tay, S.I., Lee, T.C., Hamid, N.A., Ahmad, A.N.: An overview of industry 4.0: definition, components, and government initiatives. J. Adv. Res. Dyn. Control Syst. **10**, 1379–1387 (2018)
2. Özkiziltan, D., Hassel, A.: Humans versus machines: an overview of research on the effects of automation of work. SSRN Electron. J. (2020). https://doi.org/10.2139/ssrn.3789992
3. Derigent, W., Cardin, O., Trentesaux, D.: Industry 4.0: contributions of holonic manufacturing control architectures and future challenges. J. Intell. Manuf. (2020)
4. Koestler, T.G.: The Ghost in the Machine. Arkana Books, London (1967)
5. Bussmann, S.: An agent-oriented architecture for holonic manufacturing control. In 1st International Workshop on IMS, Lausanne (1998)
6. Gamboa Quitanilla, F., Cardin, O., L'Anton, A., Castagna, P.: A modelling framework for manufacturing services in service-oriented holonic manufacturing systems. Eng. Appl. Artif. Intell. 55, 26–36 (2016)
7. Sparrow, D., Kruger, K., Basson A.H.: An architecture to facilitate the integration of human workers in Industry 4.0 environments. Int. J. Prod. Res. (2021)
8. Sparrow, D., Taylor, N., Kruger, K., Basson, A.H., Bekker, A.: Interfacing with humans in factories of the future: holonic interface services for ambient intelligence environments. In: Service Oriented, Holonic and Multi-Agent Manufacturing Systems for Industry of the Future: Proceedings of SOHOMA 2020, pp. 299–312 (2021)
9. Sparrow, D., Kruger, K., Basson, A.H.: An abstract activity life-cycle description for communication in cyber-physical systems, in service oriented, holonic and multi-agent manufacturing systems for industry of the future. In: Proceedings of SOHOMA 2019, pp. 85–97 (2019)
10. Twilio, "Twillio—Communication APIs for SMS, Voice, Video and Authentication. [Online]. Available: https://www.twilio.com/ (2021). Accessed: 1 Jun 2021

Daniel van Niekerk is an M.Eng. (Mechatronics) candidate at Stellenbosch University. He is currently doing research on holonic control architectures as part of the Mechatronics, Automation and Design Research Group.

Karel Kruger obtained his Ph.D. from Stellenbosch University, South Africa. He is a senior lecturer at the Department of Mechanical and Mechatronic Engineering and is a co-leader of the Mechatronics, Automation and Design Research Group at Stellenbosch University, South Africa.

Anton Basson obtained his Ph.D. in Aerospace Engineering at Penn State University. He was appointed in 1997 as Professor in Mechanical Engineering at Stellenbosch University and is co-leader of the Mechatronics, Automation and Design Research Group.

Design and Construction Framework to Enable the Modular Block Building Methodology to Broaden South African Oceans Economy

H. Theunissen, T. Van Niekerk, and J. H. C. Pretorius

Abstract From the design to final completion of sea trials, the complexity of a ship is built into every aspect of the build. From planning the purpose of the ship and its economic viability, to the level of technology used in the design and build, where it will be built and the production flow of the build, is all linked and restricted by the capability and capacity of the shipyard. The modern era of technology, design methods and computer capabilities (CAD/CAM), has enabled multiple sites to contribute towards the total build. This has facilitated for modular sections (blocks) of ships being built in different locations, with the blocks fitting together when shipped to one location for final assembly. This type of cooperative methodology is often preferred by large government contracts where it is desired that the build benefit as many in the industry as possible, not just one shipyard. Additionally, each shipyard may have a speciality area of production that will lead to a more cost effective "block" being built, saving on overall costs. This ultimately leads to a cheaper, better quality and faster construction of the ship. A qualitative and quantitative, mixed methods research methodology approach was used to identify the South African industry capabilities using a questionnaire that was developed to engage the maritime industry and identify capabilities, needs and future requirements. This paper will; Review the knowledge design and construction areas required for ship building and apply the knowledge areas to propose a multi-site modular block building framework for South African shipbuilding.

H. Theunissen (✉) · T. Van Niekerk
Department of Marine Engineering, Faculty of Engineering, Built Environment and Technology, Nelson Mandela University, Gqeberha, South Africa
e-mail: howard.theunissen@mandela.ac.za

T. Van Niekerk
e-mail: Theo.vanNiekerk@mandela.ac.za

J. H. C. Pretorius
School of Engineering Management, Faculty of Engineering and Built Environment, University of Johannesburg, Johannesburg, South Africa
e-mail: jhcpretorius@uj.ac.za

K. von Leipzig et al. (eds.), *Smart, Sustainable Manufacturing in an Ever-Changing World*, Lecture Notes in Production Engineering,
https://doi.org/10.1007/978-3-031-15602-1_7

Keywords Modular flexible manufacturing methodology · Block building · Oceans economy · Design and construction · Multi-site · Ship building · Project management · Maritime sustainability · Maritime project management

1 Introduction

There is a scarcity of experts (approximately 10 Naval Architects) [1] and role-players (ship/boat builders) in the South African industry. Most of these ship and boat builders are operating individually and in an un-coordinated manner. It is also noted that majority of the work that the South African industry does is for international, Northern Hemisphere, clients [2]. South Africa has increasingly relied on Northern Hemisphere guidance and expertise in selecting ship designs and builders. As a result, much of the South African large vessel fleet, including SA Agulhas II and SA Navy Frigates were designed and built overseas. In South Africa, the most significant recent builds have been the seven Voith Schneider 70 tonne bollard pull tugs built by the Sandock Austral Shipyard [3] that were built between 2007 and 2011 as part of a contract to supply Transnet. Additional tugs, ATD Tug 2909 "Inyathi" and ATD Tug 2909 "Imvubu" were supplied to the SA Navy in 2016 and 2015, respectively. The Fast Ferry 3209 "Sikhululekile" was supplied to the Robben Island Museum in 2008. These were supplied by Damen Shipyard South Africa [4]. Most other vessels built in South Africa fall into the leisure and luxury industry with South Africa being an exporter of yachts, catamarans and smaller powered vessels, typically up to a length of 40 m, but not included in this research.

Currently, on the national (government) order book [2], the only significant current build, at approximately 100 m, is the SA Navy's hydrographic survey vessel being built by Sandock Austral Shipyards in Durban. This vessel has been designed in Finland and is based on the SA Agulhas II, also designed and built in Finland. However, according to the Urban Soul Group [2] study, the opportunity and potential future order book for vessels to support the South African national requirements are summarised in Table 1, which illustrates the expected demand for the future ship builds for South Africa at 64 vessels.

Using qualitative and quantitative, mixed methods research methodology to identify the South African industry capabilities, a questionnaire was developed to engage the SA maritime industry and evaluate the construction techniques used locally and compare to the modern best practices globally for government vessels of similar size. This paper will benchmark the knowledge design and construction techniques required for ship building and apply the knowledge areas to propose a multi-site modular block building sustainable solution for South African shipbuilding.

Table 1 Opportunity matrix: A potential future order book [2]

Vessel type	Organisation	Size
Scientific research Vessels (2) OPV (2) (Offshore patrol vessels) IPV (Inshore patrol vessels)	Department of agriculture, forestry and fisheries (DAFF) DAFF/Department of environmental affairs (DEA)	1 × 18 m 1 × 60 m 2 × 80 m 3 × 40 m
Navy vessels: IPV (3) OPV (3) Frigates (4) Hydrographic Research vessel (1)	Department of defence (DOD)	3 × 40 m 3 × 80 m 4 × 100 m 1 × 100 m
Fishing vessels: Recapitalisation (20)	Fishing industry	20 × 15 m
Cable laying vessels (3)	Department of transport	3 × 110 m
Oil/Gas: OSV (4) (Offshore supply vessel) PSV (4) (passenger supply vessel)	Oil and gas industry	4 × 80 m 4 × 60 m
Harbours tugs (10)	Transnet port authority (TNPA)	10 × 40 m
Pollution control vessels (4)	DEA	4 × 40 m
Search and rescue tugs (2)	South African maritime safety authority (SAMSA)	2 × 80 m
Total planned future builds		64

2 Modern Ship Building Techniques

Durban's Sandock Austral Shipyard's, in Durban, seven tugs built for Transnet South Africa [3] were built using the building technique where the hull was built and welded upside down. Once the hull was completed, it was right-sided and the superstructure assembled on top. Once painted, it was launched and outfitting was completed quay-side. This method of building is adequate for smaller vessels like tugs. For larger ships, the industry in South Africa needs to consider using a flexible and modular manufacturing methodology. This is referred to as "block building" in the maritime manufacturing industry. This methodology is used due to the size and complexities of the modern ship and its technology. The ship is built in smaller sections or "blocks", including pre-outfitting, painting, piping, wiring and even, in some cases, full fitment of cabins. These are eventually joined together to form the whole ship. This method saves time and money, as the smaller sections can be planned and simultaneously built, even at different locations. These sections are then sent to the final location and assembled to the final structure. "A very large crude oil carrier with deadweight of 350,000 tons comprises of approximately 120 blocks. Blocks generally weigh between 200 and 300 tons, even up to 500 tons depending on the available facilities [5].

The last ship over 100 m to be built in South Africa was the SAS Drakensberg, launched in 1987 [6], for the SA Navy. In 2019, Sandock Austral Shipyard started building the 95 m hydrographic survey vessel. To build in accordance with international methodology, they are using the block building method.

3 Ship Building/Construction Tenders and Contracts

The current tenders for ship building in South Africa generally follows a linear process, illustrated in Fig. 1. Where the client (government) broadly defines the technical and operation requirements of the vessel. After the tender process, the tender is awarded to a single shipyard. Ultimately, the shipyard manages the programme and controls the build process to completion. The negative consequence of this is that the yard has control and could manipulate the design, build, finances and results. There is no, or limited, oversight other than what might have been the laid out in the original contract.

Ship building starts with a contract. Fisher [7], states the purpose of a contract "is to define unique as well as some non-unique features which the contractor is to provide the purchaser". Eyres and Bruce [8] state that the legal document between the tendering ship builder (contractor) and the owners (customer) normally include information in the form of technical specifications. Most shipbuilding contracts are based on a uniform, or standard contract between the purchaser and the builder. For Eyres and Bruce, unless the details are incorporated into the specifications and plans, generally the purchaser will have to accept the ship builder's solution(s) to the issue. Most often these are the least-cost solutions in a fixed cost contract [7].

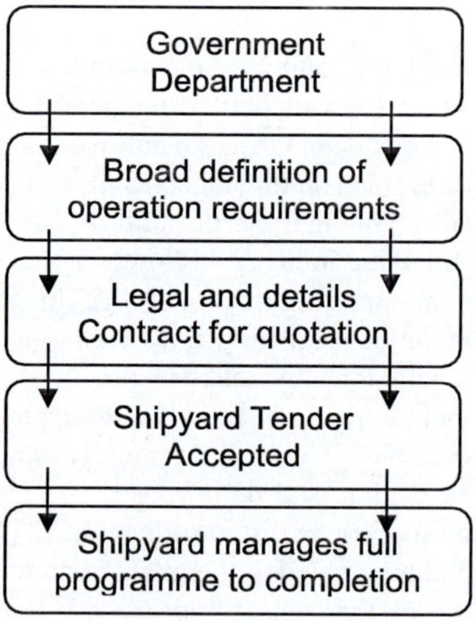

Fig. 1 Basic flow of SA ship build process

4 Ship Design Office, Drawing Office and Production

Modern ship building, as with all modern design and production, is deeply-integrated with software from initial concept to sea trials and sign-off as well as life-cycle management. Each stage uses software, but, significantly, all the stages are linked with software packages that bring together all individual components to achieve a whole. It is important that the design office considers the capacity and capability of the shipyard that will be manufacturing the ship. This is so that when designing the production process, all parts and/or assemblies will be able to be moved and lifted into position with the available facilities at the yard. Otherwise, the yard will need to upgrade their capabilities to support the build.

From the 3D ship product model developed by the naval architect, the mould lines, structural sections and block sections are all developed and defined and captured into the Production Control System (PCS). This system manages the whole production of the ship, from ordering to cutting and processing to block assembly and the final assembly.

5 International Benchmarking of Block Building Flexible Manufacturing Technology

In countries with many shipyards, modern shipbuilding methodology, standardisation and data sharing has further enabled these shipyards to be able to cooperate with each other. This allows individual parts and block assemblies to be manufactured by the different shipyards. These blocks are then transported to one primary yard where they are assembled to the final ship [9].

The drive to reduce costs and improve productivity facilitated parallel sections/blocks being built, whether at the same yard or at a different location. Standardisation and software commonisation means that sufficient accuracy allows the sections to be fitted and welded together without any, or minimal adjustment.

This type of cooperative methodology is often preferred by large government contracts where it is desired that the build benefit as many in the industry as possible, not just one shipyard. A recent significant example of this cooperation is the Aircraft Carrier Alliance, in the United Kingdom, that has built the two aircraft carriers for the British Royal Navy. Shown in Fig. 2 are the various sections built by the different companies and yards in England and Scotland [10].

6 Maritime Build Strategy For Economic Growth In SA

For South Africa to be competitive in a global ship building economy, it will need shipyards that have the correct equipment and facilities to be able to build according

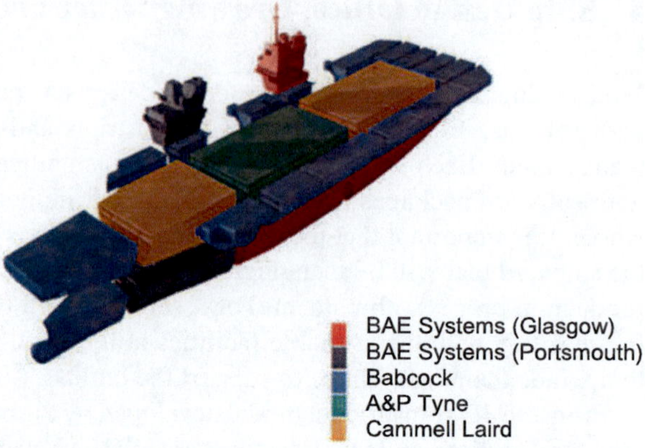

Fig. 2 Aircraft carrier alliance cooperative build [10]

to modern techniques. This might require investment, including cranes and land to set-up an efficient operation.

Questions that need to be investigated:

1. what investment is required?
2. are the larger shipyards using their monopoly due to size to gain the larger contracts?
3. are there other options or solutions for ship building to be more inclusive and sustainable that requires minimal investment?

To analyse the South African maritime industry, a questionnaire was developed [11] and coded into the online survey website QuestionPro.com [12]. It was sent to the approximately 500 maritime contacts during the height of the Covid-19 pandemic in 2020. The number of completed surveys was 28 and 63 partially completed questionnaires. However, the responses that were received provided significant information and insight into the maritime industry in South Africa, as they were completed by key industry personal.

The results of the questionnaire, that were relevant to this paper, were imported and analysed using the qualitative analysis software, Atlas-Ti [13]. The Atlas-Ti generated network in Fig. 3 shows, from the respondents, that the maritime industry in South Africa is capable of building diverse sizes and types of marine craft, and, if planned and managed, can incorporate block building flexible processes into production. There are shipyards that can manufacture/build ships up to 150 m in length, and even larger if the yard focused on block building technology.

Table 1 noted that the largest vessels planned for construction were up to 110 m. When considering the economy and development of the maritime blue economy in South Africa, the data shows that there was only one yard that could manufacture up to 150 m. Another yard could build up to 85 m. This was restrictive on the growth of the sector as only two yards would receive the bulk of the future construction. This would prevent growth in the industry and only keep two yards in business.

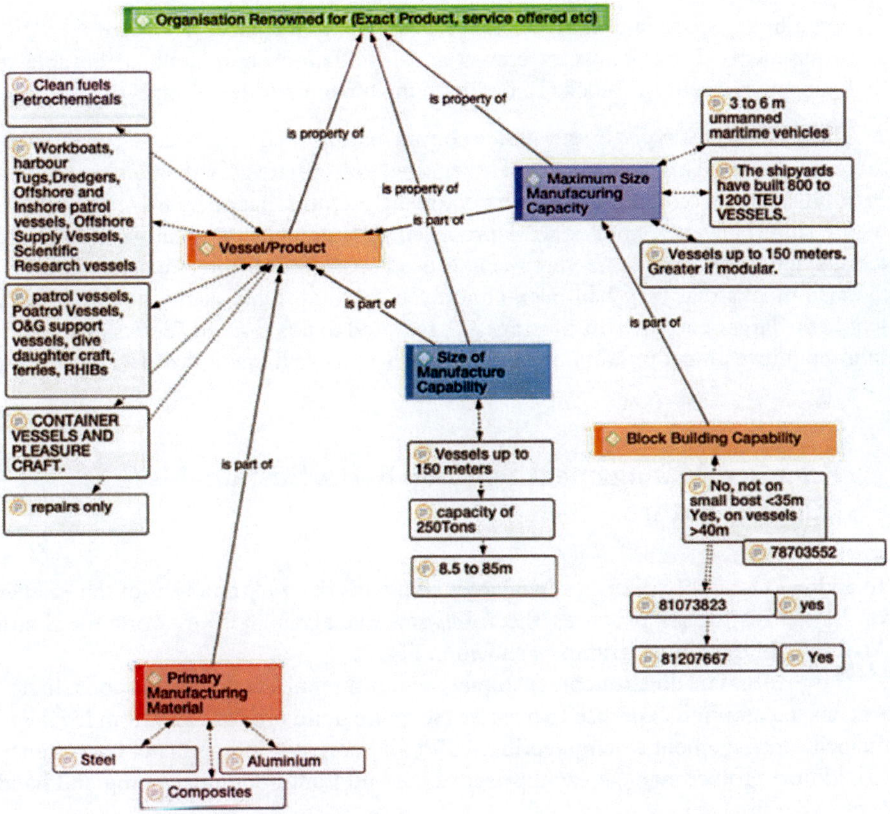

Fig. 3 Network of product relationship

The network in Fig. 4, from the questionnaire results, highlights the areas of investment required to increase the block building capabilities.

Analysing the requirements from the respondents highlighted that the following were required for moving the ship building industry in South Africa forward:

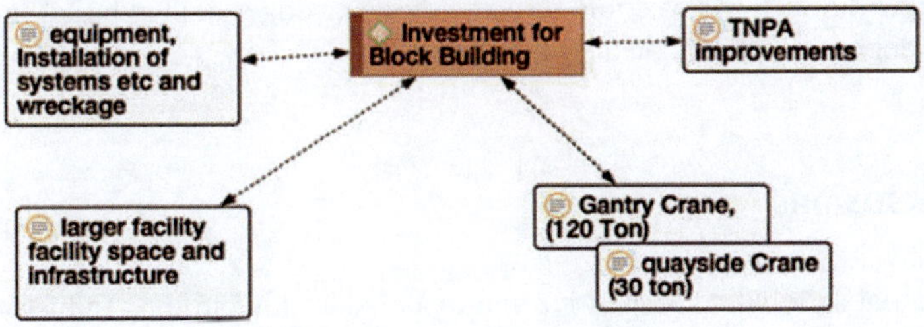

Fig. 4 Industry requirement for block building development

1. Facilities—Space and infrastructure investment in the harbours
2. Equipment—Large capacity cranes and installation equipment, to be able to manoeuvre the large blocks to facilitate the block building assembly.

The marine industry in South Africa comments, as noted in the questionnaire, that there is significant financial investment required to facilitate growth in block building capability. This would certainly be based on the premise that one yard will build the whole ship. However, applying the proposed flexible building plan would require less capital investment. The yards would need space and lifting capacity based on the size blocks that they had been contracted to build. Only at the final assembly would the larger capacity lifting cranes be required to position the blocks. Therefore minimum investment required for larger cranes at more than one yard.

7 Proposed Management Solution For A Sustainable Ship Building Future

To address monopoly of one or two larger shipyards due to limitations of the smaller yards, the researcher proposes the following management framework for South African public build programmes shown in Fig. 5.

The government department (customer) would appoint a management consultancy that has the required expertise to oversee the entire build programme, from legal and financial management to engineering, build and commissioning. This consultancy would then project manage each aspect of the build until commissioning and hand over of the final vessel.

Financially, the consultancy requires a management fee based on achieving the agreed objectives. It would not receive the entire cost to build, and the finances would stay at the appointing department and that department would pay the various contractors when the consultant approved the milestones. In this way, there could be many recipients to the build and the skills developed, which would be broadly beneficial. This would also mitigate the risk of a single shipyard going bankrupt or failing owing to poor performance. If this occurred, that block could be transferred to another yard to complete.

In this way, South Africa could be on the 'cutting edge' of ship building in Africa and even begin to compete on a global scale.

8 Conclusions

To ensure that a globally competitive and sustainable ship building industry in South Africa grows, more of the yards will need to ensure that they have the capability to build using flexible block technology. This will enable the smaller yards to be

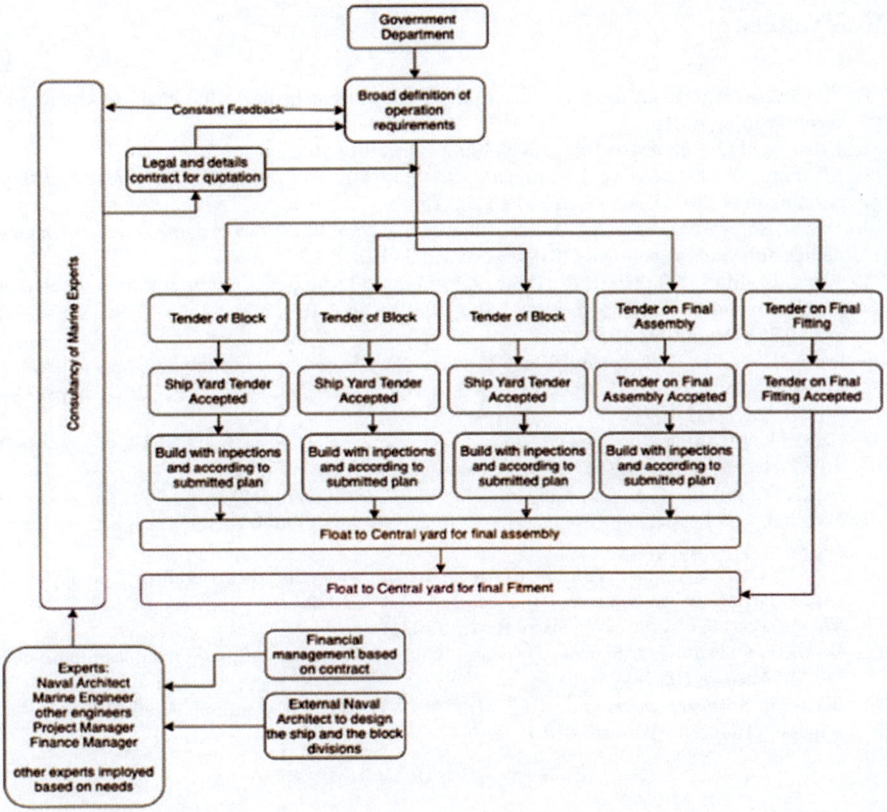

Fig. 5 Proposed management framework for South African public build programmes

economically-competitive for the larger builds. By manufacturing a block that is within its capacity, then the completed block could be shipped on a barge to be assembled at a larger yard. This would be similar to the building of the British Aircraft Carriers by the Aircraft Alliance Cooperative Build (see Fig. 2). Ultimately, this cooperative approach could enable smaller yards to contribute towards, and benefit from building a much larger vessel, and not just one large capacity yard. Management of this future and planned approach will enable a globally-competitive, quality engineered and built product from South Africa.

The additional benefit of this methodology is that more jobs can be created at different centres of manufacturing (shipyards). Multiple shipyards can benefit from the big project, rather than just one yard.

References

1. Mukandila, A.: vessel design, a determinant component in the value chain of the marine manufacturing (2018)
2. Urban Soul Group: Algoa Bay Shipbuilding Feasibility Study. Cape Town (2017)
3. SAShipyard: Shipbuilding SA Shipyard. SA-Shipyards.co.za. http://sa-shipyards.co.za/shipbuilding.html (2019). Accessed 24 Feb 2020
4. Damen Shipyard: Damen SA Builds. www.damen.com, https://www.damen.com/companies/damen-shipyards-cape-town (2019). Accessed 24 Feb 2020
5. Wang, C., Mao, Y.S., Hu, B.Q., Deng, Z.J., Shin, J.G.: Ship block transportation scheduling problem based on greedy algorithm. J. Eng. Sci. Technol. Rev. **9**(2), 93–98 (2016). https://doi.org/10.25103/jestr.092.15
6. Wessels, A.: Sas Drakensberg'S first 25 years: the life and times of the SA Navy's foremost grey diplomat, 1987–2012. Sci. Mil. S. Afr. J. Mil. Stud. **4**(12), 1987–2012 (2013). https://doi.org/10.5787/41-2-1071
7. Fisher, K.W.: Shipbuilding specifications: best practice guidelines. Int. J. Marit. Eng. **146**(a4), 10 (2004). https://doi.org/10.3940/rina.ijme.2004.a4.1004
8. Eyres, D.J., Bruce, G.J.: Ship Construction, 7th ed. Elsevier (2012)
9. Molland, A.: The Maritime Engineering Reference Book: A Guide to Ship Design, Contruction and Operation (2008)
10. Aircraft Carrier Alliance. Aircraft Carrier Alliance. www.aircraftcarrieralliance.co.uk, www.aircraftcarrieralliance.co.uk (2020). Accessed 04 Apr 2020
11. Wilson, P.A.: SA Maritime Industry Report (2021)
12. © 2021 QuestionPro Survey Software. QuestionPro.com. https://www.questionpro.com (2020). Accessed 07 May 2020
13. Scientific Software Development GmbH, "Atlas-Ti." (2020), [Online]. Available: https://atlasti.com/product/what-is-atlas-ti/

Howard Theunissen has a M.Tech. (Mechanical Engineering) from Nelson Mandela University, a Certificate in Naval Architecture from Lloyds Maritime Academy and is a Member of the Royal Institute of Naval Architects. He is the Head of Department of Marine Engineering at the Nelson Mandela University.

Prof. Theo van Niekerk research interests include Mobile Industrial Robotics, Automation of Manufacturing Systems, Process Control and Instrumentation. He teaches in the Department of Mechatronics.

Prof. Jan Harm C Pretorius has B.Sc. Hons, M.Ing and D.Ing degrees in Electrical and Electronic Engineering from Rand Afrikaans University and an MSc (Laser Eng & Pulse Power) at the University of St Andrews in Scotland. He is the head of the Postgraduate School of Engineering Management at the Faculty of Engineering and Built Environment at the University of Johannesburg.

Production Controlling Governance to Ensure Homogenous Information Systems and Targeted Decision-Making Processes

T. M. Demke, A. Mütze, and P. Nyhuis

Abstract In order to increase the effectiveness and efficiency of (digital) production controlling, information systems are useful tools to provide and display required information. Regarding condensing information related to production logistics, logistic models and frameworks, like the Supply Chain Operations Reference (SCOR) model, provide a collection of Key Performance Indicators (KPIs). The transfer of KPIs from logistic models to effective information systems faces the challenge of providing user-specific and appropriate information through individual key figures on strategic, tactical and operational levels. Additionally, a missing structural and procedural organization of (digital) production controlling prevents effective information management and consequently homogenous and targeted decisions within production controlling. To exploit the potential of data-driven production based on the transformation of raw data into useful information with the aim of effective information management, an approach for a production controlling governance is presented below. This approach systematizes corporate strategics, organizational structures, the production controlling process, KPIs resulting from logistic models, and information needs to ensure homogenous information systems and targeted decision-making processes in the context of production controlling.

Keywords Production controlling · Information systems · Structural and procedural organization · Governance · Production logistics

T. M. Demke (✉) · A. Mütze · P. Nyhuis
Institute of Production Systems and Logistics, Leibniz University Hannover, Hannover, Germany
e-mail: demke@ifa.uni-hannover.de

A. Mütze
e-mail: muetze@ifa.uni-hannover.de

P. Nyhuis
e-mail: nyhuis@ifa.uni-hannover.de

K. von Leipzig et al. (eds.), *Smart, Sustainable Manufacturing in an Ever-Changing World*, Lecture Notes in Production Engineering,
https://doi.org/10.1007/978-3-031-15602-1_8

1 Introduction

Current market competition demands adaptability, connectivity, digitalization, learning ability and sustainability as corporate attributes from competing companies [1, 2]. In particular, high logistic performance and low logistic costs as competitive factors are focused by organizational improvement efforts, requiring order fulfillment processes that are transparent, stable, efficient and aligned with corporate objectives [3, 4]. Therefore, all decisions within a company's supply chain, starting with the structural design up to the recurring planning and controlling of processes, have to be coordinated and aligned. Thus, companies can be characterized as decision factories [5] since decision-making processes are an omnipresent component of all business processes.

In order to support decision-making and the coordination of processes within the internal supply chain, control loops and KPIs as well as visual instruments are generally used in production controlling [3, 6]. However, cross-departmental controlling rarely works in practice due to target systems as isolated sources of information not being defined end-to-end. Thus, partial optimization often occurs in internal supply chains instead of achieving holistic optimization resulting in decisions that harm the company's overall performance [7].

The preparation, assistance and support of target-oriented decision-making across departments and different hierarchical levels require homogeneous information systems for all actors and a holistic target system based on superordinate objectives providing department-specific targets [3, 7]. Consequently, inhomogeneous information systems cause divergent and, from a holistic corporate perspective, non-targeted decision-making processes. Besides homogeneous information systems, it is also necessary to determine at which hierarchical level which decision-making authority exists and what information basis is required. Particularly, in times of progressing digitalization, this is becoming increasingly important, as possibilities of data-based decision support increase with increasing data storage and data availability. At the same time, the complexity associated with decision-making processes rises if an underlying governance is missing.

This paper addresses the correlation between inhomogeneous information systems and divergent decision-making processes. An approach is presented that ensures homogeneous information systems and consequently consistent and target-oriented decision-making processes by means of a production controlling governance. The focus of this approach is on providing adequate information so that decision-makers get target-oriented support. One example would be the support of production management in dispatching orders to production based on current and forecast inventory data, which is represented by key figures and visualized through logistics models. This enables better control of inventory and keeps inventory levels and lead times constant.

The development of this generic approach is based on existing governance concepts, considers possible organizational structures and integrates the production controlling process as well as models and key figures of production logistics.

After an introduction to governance, this paper presents production controlling and outlines current challenges related to information management in the context of production controlling. Subsequently, the approach for a production controlling governance is described, and its elements are discussed.

2 Theoretical Background and Literature Review

2.1 The Idea of Governance

2.1.1 Definition and Reasons for Application

"Governance" is a term used in connection with various scientific fields, such as social and political science, law, and economics, although there is no uniform understanding of it. An exemplary usage occurs in "corporate governance", covering structures and processes related to production, decision-making, controlling, etc. within an organization [8]. Concerning corporate governance, the literature describes governance models to be implemented in company structures, e.g. for management, information technology (IT), information flow and data management [8–12]. In the context of target-oriented (digital) production controlling based on a production controlling governance, the definitions of "governance" regarding IT and information flow are considered in particular.

Logan [11] provides a detailed definition of "information governance":

> Information governance is the specification of decision rights and an accountability framework to encourage desirable behavior in the valuation, creation, storage, use, archival and deletion of information. It includes the processes, roles, standards and metrics that ensure the effective and efficient use of information in enabling an organization to achieve its goals.

Additionally, in an informational context, Dinter et al. associate the term "governance" with a structural and procedural organization to support decision-making processes [9].

The approach for a production controlling governance presented in Sect. 3 is based on the definition of "information governance". This transition results from the common context of sharing data and information to support decision-making processes.

2.1.2 Components of an Information Governance

Based on the previous definition by Logan [11], processes, roles, metrics, and standards can be identified as components of an information governance. Furthermore, based on a literature review, Tallon et al. [12] structure the composition of information governance by structural, procedural and relational aspects. Structural aspects include the locus of information-based decision making as well as steering committees, boards and oversight. Processual aspects imply data principles, data quality, metadata, data access and data life cycle. The education of users as well as business and IT partnership complete the relational aspects. Tallon et al. [12] add further antecedents that include performance strategy, competitive strategy, IT strategy, organizational structure, IT structure, diversification breadth, product relatedness/mix, IT process harmonization, IT architecture, IT culture, industry regulation, and data volume or data growth rate.

2.2 *Production Controlling Within Companies*

The term "production controlling" implies the planning, controlling and monitoring of all production-related activities, considering the underlying corporate objectives. Production controlling requires the provision of relevant information using information systems [13].

According to Gottmann [3], production controlling comprises the tasks of linking objectives resulting from production and controlling, increasing productivity, identifying suitable key figures for production coordination and controlling, and collecting, analyzing and interpreting them.

In this context, various correlations between influencing, manipulated, controlled and target variables [13] that support decision-making processes in controlling tasks have to be considered. In addition to key figures, specific decision models have been developed to support decision-making processes in this context [14].

To execute the tasks of production controlling within today's digital factories, in addition to defined responsibilities, the use of information technology is required. In this context, the term "digital production controlling" is applied. Digital (also: real-time) production controlling is supported by instruments of Industry 4.0, such as cyber-physical systems (CPS)- and digital twins, which contribute to the increasing data availability and utilization [15].

2.2.1 Production Controlling Process

The production controlling process (see Fig. 1) is based on a loop consisting of six steps [14]. First, it is necessary to quantify the various target variables, taking into account their interdependencies. Following the quantification of target variables, the task is to determine target values serving as reference variables. In order to assess the current process behavior, corresponding actual values have to be recorded using feedback data. The following target-actual comparison identifies deviations. In the case of deviations, the next step is to analyze their causes. Finally, based on this analysis, measures have to be derived, e.g. by identifying root causes [7], to correct identified deviations. [14]

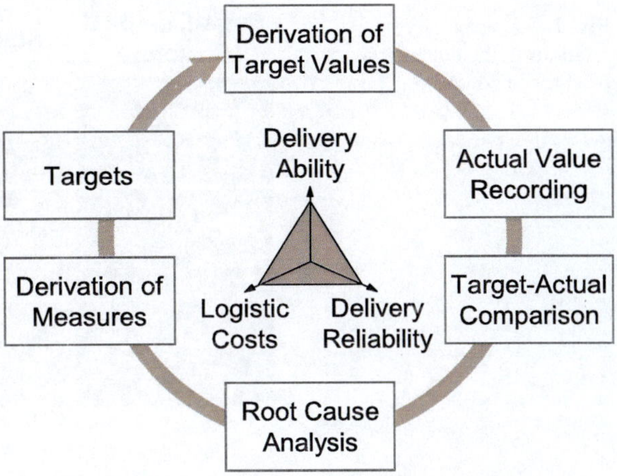

Fig. 1 Process loop of the production controlling process (adapted from [14])

2.2.2 Problem Definition and Research Question

With regard to production controlling, the process loop (see Fig. 1) provides an orientation for a procedural organization. However, there are no comprehensive approaches in literature providing a holistic structural and procedural organization according to a governance (see Sect. 2.1). Existing approaches to systematically support production controlling are primarily dealing with developing KPI systems (cf. [3]), do not map the entire supply chain (cf. [16]) or lack methods for a comprehensive root cause analysis and derivation of measures (cf. [16]). Thus, the development of model-based procedures to identify root causes of unsatisfactory performance (cf. [7]), or the link between KPI systems and the organization of production controlling is not entirely established (cf. [13]).

As aspects concerning the operation and the continuous development of information systems become more important than their initial development, a holistic perspective on all activities is necessary [9]. For this reason, a structural and proce dural organization in the form of a governance is required with regard to production controlling. Through a production controlling governance, the effectiveness and efficiency of the use of information within production controlling processes can be ensured following the previous definition of information governance (see Sect. 2.1). Consequently, the achievement of the underlying corporate objectives can be facilitated. Following Hahn and Lassmann [13] and Ullmann [16], the definition of a structural and procedural organization for production controlling requires the following questions to be answered:

- Who receives which information when, how, regarding what and on which data basis?
- Who takes which decisions when, how, regarding what and on which information/data basis?

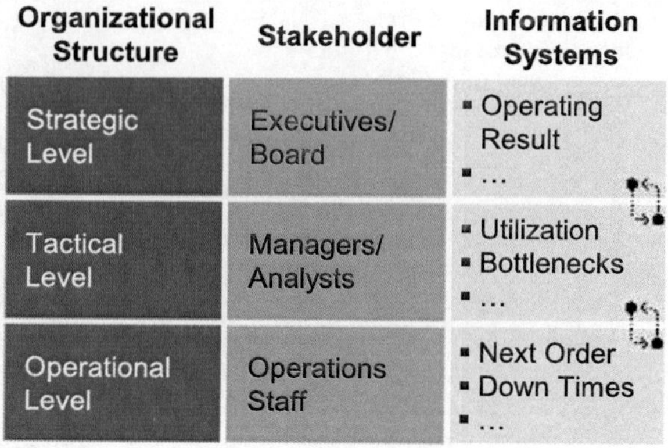

Fig. 2 Representative relations in the context of production controlling

The following section presents a production controlling governance approach that addresses the previously described challenges in connection with the development of homogeneous information systems to support homogeneous and goal-oriented decision-making processes within production controlling through a structural and procedural organization. Figure 2 shows representative relations between the organizational structure, various stakeholders within production controlling, and information needs to be considered within appropriate information systems. In order to achieve homogeneous information management to support targeted decision-making processes, it is necessary to consider these requirements within a production controlling governance.

3 Approach for a Production Controlling Governance

The approach for a production controlling governance presented in the following is based on the composition of an information governance described in Sect. 2.1.2 following Logan [11] and Tallon et al. [12]. The approach includes the core elements *strategy, organization and roles, processes, key figures and display elements, information technology*, and *standards and requirements* (see Fig. 3). These core elements are presented in the next sections in greater detail.

3.1 Strategy

To ensure target-oriented decision-making processes according to the corporate strategy and derived objectives within production controlling, it is necessary to align production controlling and controlling objectives with the corporate strategy. For this reason, the aspect of *strategy* forms the first element of a production controlling governance. With reference to the components of an information governance

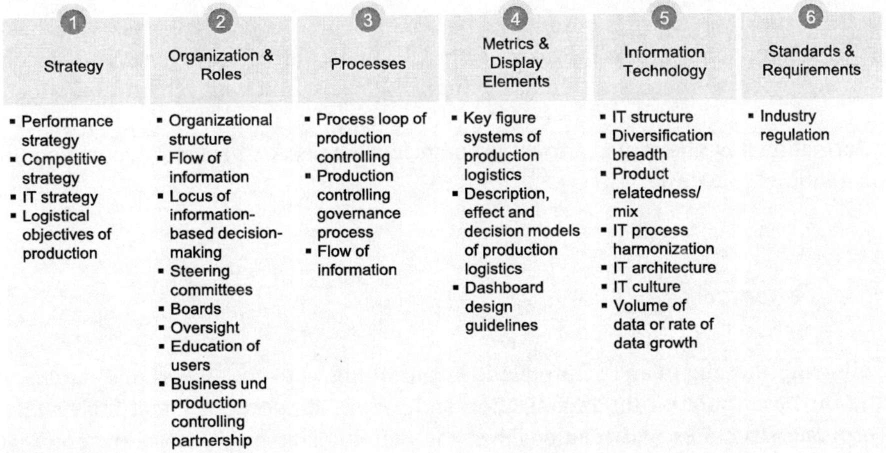

Fig. 3 Core elements of the production controlling governance

(see Sect. 2.1.2), the aspect of *strategy* in the context of production controlling governance considers particularly the performance and competitive strategy, which supports coordinated decisions within production controlling. Due to the relation of production controlling to the entire production logistics (from procurement to delivery), the target system of production logistics (cf. [14]) and related correlations have to be taken into account. Based on the defined reference of production controlling governance to digital production controlling (see Sect. 2.2), it is also necessary to anchor the IT strategy within this strategic element.

3.2 Organization and Roles

Considering the dependency of corporate strategy and structure, whereby structure results from strategy [17], structural and relational aspects follow the strategic element in line with the described composition of an information governance (see Sect. 2.1.2). In connection with the defined information governance element roles, the structural and relational aspects, as well as roles in the context of production controlling governance, are summarized within the element *organization and roles*. In connection with this element, it is necessary to organize digital production controlling based on the existing organizational structure, and in particular on the basis of defined roles, such as plant controller, functional controller and project controller [13]. With the aim of homogeneous information management to support targeted decision-making processes within production controlling, these roles need to be defined particularly based on individual decision-making competencies, information requirements and flows. Therefore, aligned with the composition of an information governance, this governance element also integrates the locus of information-based decision making, steering committees, boards, oversight, education of users and business and production controlling partnership. With the aim of homogeneous decision-making processes across different hierarchical levels, and with regard to the locus of information-based decision-making processes, it is essential to identify relevant information flows between various corporate levels and take them into account within the required role description.

3.3 Processes

Following the alignment of production controlling with the underlying strategies and the description of the organization and roles, taking into account higher-level corporate structures and relationships, the definition of processes corresponds to the described composition of information governance (see Sect. 2.1.2). On the one hand, the definition of processes serves to establish the procedural organization in production controlling. On the other hand, it is used to determine the contribution of production controlling to the operating result beyond the scope of production controlling. The procedural organization within production controlling needs to be designed following the process loop of production controlling presented in Sect. 2.2.1. Concerning the process-related integration of production controlling beyond the scope of controlling, a generally applicable production controlling governance process needs to be developed based on the IT governance process, e.g. by the Control Objectives for Information and Related Technology (COBIT) reference model (cf. [10]). To effectively define processes within the production controlling governance, it is also essential to consider relevant information flows.

3.4 Key Figures and Display Elements

Following the composition of information governance described above, the definition of corresponding key figures is still required for the targeted and homogeneous execution of the processes defined above by specified roles. Due to the use of the production controlling governance in the context of production logistics, this element focuses in particular on key figure systems with reference to correlations related to production logistics (cf. [3, 18]). These correlations can be identified through existing description, effect and decision models of production logistics (cf. [7, 14]). In addition to the definition of key figures, the information systems to be used within production controlling (see Sect. 2.2) also require corresponding display elements. On the one hand, existing logistic models can be used in this context due to their graphical properties (cf. [14]). On the other hand, guidelines for dashboard design can be applied to the development of new display elements. In this context, the work by Few [19] should be named as an example.

3.5 *Information Technology*

Due to the intended integration of information systems into digital production controlling to support targeted and homogeneous decision-making processes (see Sect. 2.2), it is necessary to consider information technology aspects. Therefore, the following element includes appropriate aspects. In order to organize information systems for targeted and homogeneous information management within the context of digital production controlling and beyond, the IT structure, diversification breadth, product relatedness/mix, IT process harmonization, IT architecture, IT culture, and volume of data or rate of data growth have to be taken into account within this element following the described composition of an information governance. The integration of these aspects ensures the alignment of the necessary IT with the underlying IT strategy on the one hand and the consideration of information technology requirements on the other.

3.6 *Standards and Requirements*

Finally, according to the described components of an information governance, the implementation of an element for the consideration of standards within the production controlling governance is required. In order to take the above-mentioned aspect of industry regulation (see Sect. 2.1.2) into account, the production controlling governance includes an element that contains standards and requirements. By summarizing standards and requirements intended in this element, their fulfillment can be ensured in the context of digital production controlling.

4 Conclusion

This paper presents an approach for a production controlling governance. This approach is based on the described need for target-oriented and homogeneous information management to promote aligned decision-making processes within digital production controlling using information systems. The presented approach includes six core elements: *strategy, organization and roles, processes, key figures and display elements, information technology* as well as *standards and requirements*. On the one hand, the consideration of the core elements *strategy* as well as *organization and roles* leads to a structural and procedural organization of digital production controlling that is aligned with corporate strategies, organizational structure as well as information needs. The core elements *processes*, as well as *key figures and display elements*, support the processes within digital production controlling and their integration into higher-level corporate processes in terms of homogeneity and target orientation. The implementation of the core elements *information technology* as well as *standards and requirements* within the production controlling governance ensures the consideration of relevant internal and external conditions within the digital production controlling.

5 Outlook

In the context of information governance, existing literature argues that there is no one-size-fits-all solution [20]. Applied to the field of production controlling, this means that there is no need for a general production controlling governance in corporate practice, but rather for a guideline for developing company-specific production controlling governances. In addition, there is a need for further research, especially in the context of the identification and description of role- and process-dependent information needs within production controlling. Finally, these identified and described information needs result in a significant potential regarding the collection of key figures as well as the design of corresponding display elements for the implementation within aligned information systems to support homogeneous and goal-oriented decision-making processes within digital production controlling.

References

1. Westkämper, E., Löffler, C.: Strategien der Produktion: Technologien, Konzepte und Wege in die Praxis, pp. 309. Springer Vieweg, Berlin, Heidelberg (2016)
2. Farhikhteh, S., Kazemi, A., Shahin, A., Mohammad Shafiee, M.: How competitiveness factors propel SMEs to achieve competitive advantage? CR **30**(3), 315–338 (2020)
3. Gottmann, J.: Produktionscontrolling, p. 233. Springer Fachmedien, Wiesbaden (2019)
4. Mangan, J., Lalwani, C.S.: Global logistics and supply chain management, 3rd ed, pp. 393. Wiley, Chichester, West Sussex (2016)
5. Martin, R.L.: Rethinking the decision factory. Havard Bus. Rev. **91**(10), 96–104 (2013)
6. Schäfers, P., Mütze, A., Nyhuis, P.: Integrated concept for acquisition and utilization of production feedback data to support production planning and control in the age of digitalization. Procedia Manuf. **31**, 225–231 (2019)
7. Härtel, L., Nyhuis, P.: Systematic data analysis in production controlling systems to increase logistics performance. In Advances in production research. proceedings of the 8th congress of the german academic association for production technology (WGP), pp. 3–13, Nov 19–20. Springer, Aachen (2018)
8. Keasey, K., Wright, M.: Issues in corporate accountability and governance: an editorial. Account. Bus. Res. **23**(sup1), 291–303 (1993)
9. Dinter, B., Lahrmann, G., Winter, R.: Information logistics as a conceptual foundation for enterprise-wide decision support. J. Decis. Syst. **19**(2), 175–200 (2010)
10. Johannsen, W.: Information-Governance—Herausforderungen in verteilten Umgebungen. In: Fähnrich, K.-P., Franczyk, B. (eds.) Informatik 2010. Service Science—neue Perspektiven für die Informatik. Band 2. Gesell-schaft für Informatik e.V, Bonn, pp. 311–316 (2010)
11. Logan, D.: What is information governance? And why is it so hard? Gartner Blog. https://blogs.gartner.com/debra_logan/2010/01/11/what-is-information-governance-and-why-is-it-so-hard/ (2010). Accessed Sept 18 2021
12. Tallon, P.P., Ramirez, R.V., Short, J.E.: The information artifact in IT governance: toward a theory of information governance. J. Manag. Inf. Syst. **30**(3), 141–178 (2013)

13. Mütze, A., Hillnhagen, S., Schäfers, P., Schmidt, M., Nyhuis, P.: Why a systematic investigation of production planning and control procedures is needed for the target-oriented configuration of PPC. In: 2020 IEEE International Conference on Industrial Engineering and Engineering Management (IEEM), pp. 103–107 (2020)
14. Nyhuis, P., Wiendahl, H.-P.: Fundamentals of Production Logistics: Theory, Tools and Applications, pp. 320. Springer, Berlin (2009)
15. Leng, J., Zhang, H., Yan, D., Liu, Q., Chen, X., Zhang, D.: Digital twin-driven manufacturing cyber-physical system for parallel controlling of smart workshop. J Ambient Intell. Hum. Comput. **10**(3), 1155–1166 (2019)
16. Ullmann, W.: Controlling logistischer Produktionsabläufe am Beispiel des Fertigungsbereichs, pp. 86. Dissertation, Hannover (1994)
17. Chandler, A.D.: Strategy and structure: Chapters in the history of the industrial enterprise, 22, print, p. 463. MIT Press, Cambridge, Mass. (2001)
18. Supply Chain Council: SCOR: Supply Chain Operations Reference Model: Revision 11.0, pp. 976 (2012)
19. Few, S.: Information dashboard design: displaying data for at-a-glance monitoring, 2nd edn., p. 246. Analytics Press, Burlingame, CA (2013)
20. Weber, K., Otto, B., Österle, H.: One size does not fit all: a contingency approach to data governance. J. Data Inf. Qual **1**(1), 4:1–4:27 (2009)

Tabea Marie Demke (M. Sc.) studied production and logistics at Leibniz University Hannover and has been working as a research assistant at the Institute of Production Systems and Logistics (IFA) since 2021 in the field of production management.

Alexander Mütze (M. Sc.) studied industrial engineering at Leibniz University Hannover and has been working as a research assistant at the IFA since 2018 in the field of production management focusing on production planning and control and logistics modeling.

Prof. Dr.-Ing. habil. Peter Nyhuis studied mechanical engineering at Leibniz University Hannover and subsequently worked as a research associate at the IFA. He has been heading the IFA since 2003. In 2008 he became managing partner of the IPH—Institut für Integrierte Produktion Hannover gGmbH.

Automation, Human–Machine Interaction, Interfaces

Sustainable Utilization of Industrial Robotic Systems by Facilitating Programming Through a Human and Process Centred Declarative Approach

Titanilla Komenda, Jorge Blesa Garcia, Maximilian Schelle, Felix Leber, and Mathias Brandstötter

Abstract This paper presents an approach of how declarative programming can be applied for sustainable industrial robotic system integration by semi-automatically creating and modifying a robotic system's control programme—not only focusing on the manipulator but rather taking into account the end-effector, sensors as well as peripheral equipment. The developed method is illustrated on two industrial use-cases while also giving a critical evaluation on the benefits and drawbacks of the presented approach.

Keywords Robot programming · Declarative programming · Industrial robotic system

T. Komenda (✉) · M. Schelle
Fraunhoter Austria Research GmbH, Theresianumgasse 7, 1040 Vienna, Austria
e-mail: titanilla.komenda@fraunhofer.at

M. Schelle
e-mail: maximilian.schelle@fraunhofer.at

J. B. Garcia · F. Leber
University of Applied Sciences Technikum Wien, Hochstädtplatz 6, 1200 Vienna, Austria
e-mail: blesagr@technikum-wien.at

F. Leber
e-mail: leber@technikum-wien.at

M. Brandstötter
JOANNEUM Research Forschungsgesellschaft mbH, ROBOTICS Institute, Lakeside B13b, 9020 Klagenfurt am Wörthersee, Austria
e-mail: mathias.brandstoetter@joanneum.at

K. von Leipzig et al. (eds.), *Smart, Sustainable Manufacturing in an Ever-Changing World*, Lecture Notes in Production Engineering,
https://doi.org/10.1007/978-3-031-15602-1_9

1 Introduction

Even though robotic systems are representatives for flexible tools in automated production environments, market driven changes mostly require manual modifications of control programmes. The lack of practicable flexibility in a necessary market-driven need for modifying production systems inhibits resilient application of industrial robots but also of collaborative human–robot work systems. Changes in product design, number of units to be produced or the economic degree of automation and thus deployment of personnel, require changes which are mostly carried out by qualified personnel.

This increases dependency and contradicts concepts of flexibility [1]. Moreover, modifications result in increased error potential in terms of standstills, hazards or wrong process sequences due to system's complexity as a result of mutually interacting peripheral units [2]. Further, the manual effort does not end with re-programming but with re-certifying any automated system after changes have been made—as required by current safety guidelines [3]. This has led to a number of research activities developing automated or semi-automated programming approaches for robotic systems.

In comparison to other technical machines, the possibilities for programming a robotic system are enormous. Almost every manufacturer of hardware—be it robotic manipulators or peripheral equipment such as conveyors, end effectors or sensors—provides its own programming environment with various interfaces enabling e. g. machine-to-machine communication, the transfer of programme code from independent programming platforms or direct control on joint level. These manifold possibilities increase the susceptibility to errors in programme code generation as well as the dependence on different experts. In addition, control sequences for different types of hardware must be orchestrated which makes programming of industrial robotic systems a multi-disciplinary task.

Furthermore, robotic system programming is shifting from system integrators towards end-users which also changes the required programming approach [4]. End-users require a problem- or process-oriented programming approach rather than a robot-oriented one. As their expertise lies in the implementation on processes and not on executing robot motions or the motions of any other hardware, programming needs to be conducted on a higher level of abstraction. However, this also means that different information and computational pipelines in terms of process modelling and code computation are required to allow machine-readable computation of end-user-specific process representations.

2 State of the Art of Industrial Robotic System Programming

Industrial processes benefit from their structured interaction between manufacturing systems and well-defined process steps. For this reason, almost all industrial robots are still programmed directly by a programmer. While most of the modern programming approaches only focus on the manipulator itself—in terms of trajectory planning by applying deep reinforcement learning or learning by or from demonstration, i.e., teach-in or playback, [5]—some of them try to include peripheral equipment, such as end-effectors, sensors, conveyors, or other machines and consider changing environmental conditions [6]. However, learning based methods are still restricted due to limited explainability [6] which must be based on fundamental understanding of AI environments [7].

Beside programming approaches considering the creation of robot control programmes in a robot-oriented way, meaning the creation of explicit motion commands for the robotic system, other approaches focus on different types of process models, such as task-, problem- or data-orientation, resulting in, for example, declarative programming languages [8] or the development of different data exchange formats such as AutomationML [9] and corresponding ontological semantics for interpreting those data formats [10].

A promising approach to automatically program robots has been the use of so-called skills [11]. The application of skills increases the reusability and modularity of programs, as skills can be combined to generate more complex ones. Besides, skills can be seen as services in Service Oriented Architectures (SOAs) for production [12] while increasing the level of abstraction in robot programming and easing application of artificial intelligence methods [13]. However, one of the main drawbacks of using skills is the lack of well-established standards for organizing, describing, characterising, and presenting skills to clients [14]. Another major obstacle is the actual implementation of skills for specific hardware. Usually, similar devices produced by different manufacturers require ad-hoc programming of desired skills for each device.

A hardware-agnostic Plug & Produce approach was introduced by Profanter [15]. They developed different concepts based on OPC UA communication and skill representation of tasks enabling different devices to be interchangeable and thus increasing system flexibility. The proposed approach of human and process centred declarative robotic system programming is based on that concept.

3 Evaluating Robotic System Programming Approaches

Despite of different approaches targeting the objective of simplifying industrial robotic system programming, a recent review on industrial robot programming methods showed that there is no common benchmark to evaluate these developments [16]. This leads to the question of whether a lack of evaluation criteria has hindered implementability of those programming methods in an industrial context, or whether a definition of industrially relevant criteria for the application of intuitive programming techniques is key to their applicability.

In addition, a general challenge in robotic system programming is the growing number of possible system states depending on the process complexity. Often, robotic actions depend on a variety of process and sensor data that can quickly create unconsidered system states, if handled incorrectly. This, in turn, can cause downtimes or incorrect system behaviour such as deadlocks or livelocks, if logical contradictions occur. For example, the theoretical number of system states increases by 2^n only by adding *n* sensors with Boolean states to a system. Thus, by adding a sensor with Boolean outputs, 2 additional system states need to be considered in the logical representation. In addition, a discretisation of a continuous robot trajectory might lead to an exponential increase of system states dependent on the interval of discretisation.

Table 1 gives an overview of a qualitative comparison of different programming approaches from a user's point of view. Teach-in, playback and 3D-based graphical approaches are considered imperative while block-based graphical as well as text-based approaches can be considered as both—dependent on their implementation.

The programming methods are compared in terms of the possibility in identifying and/or preventing errors, the independence of manufacturers, the modifiability of the robotic system control programmes, the traceability of the implemented control programmes, the offline compatibility, the independence of expertise as well as the reference to the reality and the overall robotic system.

While error prevention is partially applicable with almost every approach as errors in the control programme can directly be seen in an 3D-environment, this

Table 1 Qualitative comparison of robot programming approaches from a programmer's point of view ○ not applicable ◑ partially applicable ● applicable

	Teach-In	Playback	3D-based	Text-based	Block-based
Error prevention	◑	◑	◑	○	◑
Independence of manufacturers	○	●	◑	○	●
Modifiability	◑	◑	◑	◑	●
Traceability	○	○	◑	◑	●
Offline compatibility	○	○	●	●	●
Independence of expertise	○	●	◑	◑	●
Reality reference	●	●	◑	○	◑
System reference	○	○	◑	●	●

might be more difficult with text-based approaches. However, formalised block-based approaches have the possibility of verification which enhances error prevention. The independence of manufacturers is given with playback as well as block-based approaches as they are hardware-agnostic. Approaches using graphical 3D-programming interfaces are partially independent as some tools use real hardware representations, but others use specifically designed hardware translators.

Even though modifiability is given for every approach, the traceability of the implemented modifications is not given for the teach-in as well as the playback approach and only partially given for 3D- and text-based approaches. Further, independence of expertise is only given by applying playback as well as block-based methods. While the reference to real robots is fully given in the most imperative approaches, the system reference including peripheral equipment might only be considered in text- and block-based approaches.

In contrast to a qualitative approach in evaluating robotic system programming approaches, quantitative indicators with reference to evaluation approaches in the fields of embedded systems development might be applicable for robotic systems as well. Indicators might be divided into indicators evaluating the programming effort as well as indicators evaluating the quality of the control programme. The programming effort can be calculated based on the hourly rate of programming, testing, modifying, and optimizing the control programme. On the contrary, the programme quality might be formulated based on different indicators shown in Table 2.

An analysis of five different industrial use cases showed a high variation in the presented indicators for evaluating the programme quality. Even though all use cases were pick-and-place applications, the different application types required different specificities leading to more or less complex robotic system control programmes. The use cases included sorting, palletising, packaging, commissioning as well as depalletising tasks respectively. Table 2 gives an overview of the indicators collected. The high variance in the indicators presented, makes the need for a uniform or standardised programming approach necessary that allows less variability from the user's point of view due to a higher level of abstraction but a formalised translation towards machine-executable code.

4 Proposed Approach

In the programming approach proposed, the system is defined through YAML configuration files. These configuration files describe information flows available from sources such as sensors, hardware components themselves such as robots or actuators, as well as skills of individual hardware components. In this context, skills refer to aggregated, process-centric action descriptions. Skills are modelled in a similar way to the work presented in [17].

The Canonical Robot Command Language (CRCL) is used to define specific action blocks [18], while an OPC UA server is used for communication and interaction with different hardware components, i.e., agents. This allows for accessing

Table 2 Quantified comparison of robotic control programmes of different types of pick-and-place applications

Indicator	Description	Sorting	Palletising	Packaging	Commissioning	Depalletising
Degree of programme complexity	Ratio of programme size to number of programme lines for functional programme parts	2:1	13:1	17:1	2000:1	40:1
Degree of code anomaly	Ratio of error states to programme size	1:200	1:180	1:25	1:5000	1:33
Degree of reliability	Ratio of errors to the shift duration	1:8	1:1.6	1:23.25	1:0.016	1:24
Degree of programme criticality	Ratio of safety-relevant programme lines to the total number of programme lines	1:75	1:225	1:25	1:500	1:133
Degree of programme quality assurance	Ratio of programming effort to the testing effort	2:1	50:1	5:1	0.8:1	0.75:1

detailed representations of agent status and allowing requesting actions to be performed by the agents. The OPC UA standard is used for information modelling and communication. OPC UA was selected in order to ensure hardware agnosticism. In addition, OPC UA offered several additional advantages as it supports encryption, provides semantic modelling capabilities, and has a better performance with its open-source implementation open62541 [19] compared to ROS [20]. Besides, OPC UA is a well-established technology in industry, which facilitates integration and has been proven to enable the implementation of deterministic systems when combined with Time Sensitive Networking (TSN) [21].

Following on from this, a skill in the OPC UA server is an object node that represents a finite state machine with different possible states, such as Ready, Running, Halted or Suspended. The transitions between states are represented as events in the OPC UA server. These events are triggered by calling the methods Start, Halt, Suspend or Reset contained within the skill object node. The skill state transition events, combined with changes in variable nodes representing sources of information (sensors, alarms, etc.), are the events that inform about the transition of the system into a new state. Every skill object also contains a set of variable nodes that hold the values of the skill parameters and are read by the corresponding agent executing the skill at runtime. While every primitive (CRCL command) used for defining a skill has an associated parameter in the OPC UA server, not every parameter is meant to be modified in each instantiation of the skill. For example, in a pick-and-place operation, the pick position and the place position might be exposed to the end user through a middleware, nevertheless other parameters available in the OPC UA server, such as speed, acceleration or the open/close setting of the gripper remain constant and not available in the middleware.

Finally, the combination of different mechanisms implemented in OPC UA, such as the use of variable and object nodes, methods, events, subscriptions or the inclusion of guards in the modelled CRCL commands, allows for generating flexible machine code that follows a complex flow in an automated way. At the same time, the generated code for a configured system can be easily modified through an OPC UA server by using an OPC UA client. The OPC UA client can have a GUI and be used manually by a human but can also be part of a more complex software, for example a controller implementing a certain policy or any kind of middleware.

5 Use Case Description and Evaluation

In this section, an exemplary implementation of the proposed declarative programming approach is compared to the imperative description. Table 3 summarizes the textual and visual comparison between the two approaches. The process block "Pick & Place" is extended and generalised compared to the usual pure pick-and-place activity, as can be seen in the examples.

Thus, for the first use case, palletising is to be handled by a cobot mounted on a mobile platform, implementing a sorting application. After the arrival of the mobile robot, the declarative process module "Pick & Place" starts the palletising of carriers in a defined configuration onto a conveyor belt. For this process, it is important to consider the changing pick and place parameters, since the carriers are stored in stacks and are similarly to be palletised in stacks onto the conveyor carrier. The important parameters that can be modified in this use case are the start and end positions of the carriers. The combination of different movement sequences and the gripping activity proves to be advantageous for the fast process mapping. Furthermore, due to the previous initialisation of the process with a gripping point, only the target position and orientation need to be described.

In the second use case, the "Pick & Place" process is dealing with the transport of goods via a conveyor belt system in logistics applications. In contrast to the previous use case, the object is not manipulated by a cobot, but instead by the coupling of differently oriented conveyor belts connected by switches. The declarative approach in combination with the system interface allows a simple modelling and modification of the route target after initialisation and definition of the target position. With the newly defined approach, the route decisions are mapped by an internal logic.

Table 3 Comparison of two use cases using declarative and imperative programming approaches

	Depalletizing use case	Commissioning use case
Declarative model		
Declarative description	*Depalletize workpiece carriers*	*Move container from the warehouse to the commissioning station*
Imperative model (*BPMN*)	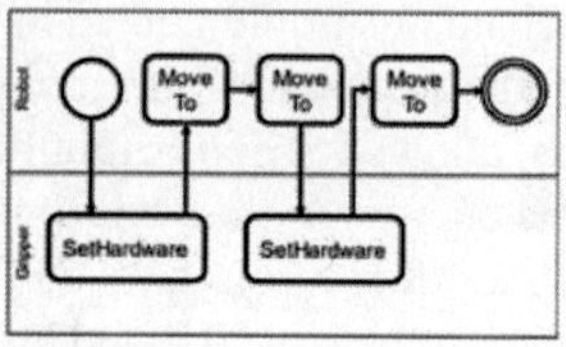	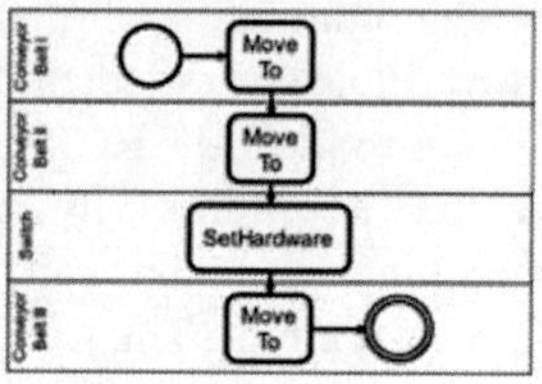
Skill representation (*YAML*)	*Depalletizing_Skill{* *SetHardware* *MoveTo* *MoveTo* *SetHardware* *MoveTo}*	*Commissioning_Skill{* *MoveTo* *MoveTo* *SetHardware* *MoveTo}*
Imperative description (*CRCL/OPC UA*)	*Set do_gripper, 1;* *MoveJ Offs(pObject,0,0,100), v200, z20, t_gripper;* *MoveL pObject, v100, fine, t_gripper;* *Set do_gripper, 0;* *Wait 0.3;* *MoveL Offs(pObject,0,0,100), v200, z20, t_gripper;*	*Move conveyor_part1 with v250;* *Move conveyor_part2 with v250;* *Set do_switch, 1;* *Move conveyor_part3 with v250;*

For the use cases described, the required skills were defined in the following manner: *MoveTo* skill composed of two CRCL commands, i.e., *SetTransSpeed* and *MoveTo*. The skill requires three parameters, i.e., a Boolean indicating whether the movement is linear or not, a double setting the linear motion speed and a pose. *SetHardware* requires an input for setting a digital output such as opening and closing the gripper or a switch including an input for a necessary waiting time due to physical communication principles.

6 Discussion and Conclusions

This paper presents a continuous model computation pipeline for the purpose of semi-automated robotic system programming. The approach is presented on two fragmental industrial use cases. This research has shown, that as long as programming of robotic systems is done from a robot's perspective, programming experts incorporate their expert knowledge on robot capabilities, behaviour as well as application-specific factors of the entire system. However, if the programming paradigm changes from mapping these explicit robot-specific actions, i.e. motion commands, to a higher level of abstraction, the underlying software basis changes decisively. Processes are then modelled and represented by a chain of robotic skills or task descriptions satisfy given specifications while at the same time allowing verifications of modification against specifications. Future work will focus on evaluating the process abstraction process using the proposed declarative programming approach on four more complex industrial use cases while at the same time quantifying the expected reduction of programme complexity by keeping the required process quality.

Acknowledgements This project is funded by the Federal Ministry for Climate Protection, Environment and Energy, Innovation and Technology, BMK, and carried out within the framework of the programme Production of Future under the grant agreement number 877362 within the project SAMY—Semi-Automated Modification in Control Programmes of Industrial Collaborative Robotic Systems.

References

1. Gaede, C., Ranz, F., Hummel, V., Echelmeyer, W.: A study on challenges in the implementation of human-robot collaboration. J. Eng. Manage. Oper. **1**, 29–39 (2019)
2. Ingrand, F.: Recent trends in formal validation and verification of autonomous robots software. In: Proceedings of the Third IEEE International Conference on Robotic Computing, pp. 321–328 (2019)
3. Fisher, M., Collins, E., Dennis, L., Luckcuck, M., Webster, M., Jump, M., Page, V., Patchett, C., Dinmohammadi, F., Flynn, D., Robu, V., Zhao, X.: Verifiable self-certifying autonomous systems. In: Proceedings of the IEEE International Symposium on Software Reliability Engineering Workshops, pp. 341–348 (2018)
4. Schmidbauer, C., Komenda, T., Schlund, S.: Teaching cobots in learning factories—user and usability-driven implications. Procedia Manuf. **45**, 398–404 (2020)
5. Ravichandar, H., Polydoros, A.S., Chernova, S., Billard, A.: Recent advances in robot learning from demonstration. Annu. Rev. Control Rob. Auton. Syst. **3**, 297–330 (2020)
6. Kardos, C., Kovács, A., Váncza, J.: Towards feature-based human-robot assembly process planning. Procedia CIRP **57**, 516–521 (2016)
7. Bengio, Y., Courville, A., Vincent, P.: Representation learning: a review and new perspectives. IEEE Trans. Pattern Anal. Mach. Intell. **35**(8), 1798–1828 (2013)
8. Peterson, J., Hager, G.D., Hudak, P.: A language for declarative robotic programming. Proc. IEEE Int. Conf. Rob. Autom. **2**, 1144–1151
9. Drath, R., Luder, A., Peschke, J., Hundt, L.: AutomationML—The glue for seamless automation engineering. In: Proceedings of the IEEE International Conference on Emerging Technologies and Factory Automation, pp. 616–623 (2008)
10. Hua, Y., Mende, M., Hein, B.: Modulare und wandlungsfähige Robotersysteme, at-Automatisierungstechnik, 63/10, 33–37 (2015)
11. Bøgh, S., Nielsen, O., Pedersen, M., Krüger, V., Madsen, O.: Does your robot have skills?. In: Proceedings of the 43rd International Symposium on Robotics, pp. 1–6 (2012)
12. Perzylo, A., Grothoff, J., Lucio, L., Weser, M., Malakuti, S., Venet, P., Aravantinos, V., Deppe, T.: Capability-based semantic interoperability of manufacturing resources: A BaSys 4.0 perspective, IFAC-Papers OnLine, 52/13:1590–1596 (2019)

13. Konidaris, G., Kaelbling, L.P., Lozano-Perez, T.: From skills to symbols: Learning symbolic representations for abstract high-level planning. J. Artif. Intell. Res. **61**, 215–289 (2018)
14. Schlenoff, C., Prestes, E., Madhavan, R., Goncalves, P., Li, H., Balakirsky, S., Kramer, T., Miguelánez, E.: An IEEE standard ontology for robotics and automation. In: Proceedings of the IEEE/RSJ International Conference on Intelligent Robots and Systems, pp. 1337–1342 (2012)
15. Profanter, S., Perzylo, A., Rickert, M., Knoll, A.: A generic plug and produce system composed of semantic OPC UA skills. IEEE Open J Ind. Electron. Soc. 128–141 (2021)
16. Heimann O., Guhl, J.: Industrial robot programming methods: a scoping review. In: Proceedings of the 25th IEEE International Conference on Emerging Technologies and Factory Automation **1**, 696–703 (2020)
17. Profanter, S., Breitkreuz, A., Rickert, M., Knoll, A.: A Hardware-Agnostic OPC UA Skill Model for Robot Manipulators and Tools. In: Proceedings of the IEEE 24th International Conference on Emerging Technologies and Factory Automation, pp. 1–8 (2019)
18. Proctor, F., Balakirsky, S., Kootbally, Z., Kramer, T., Schlenoff, C., Shackleford, W.: The canonical robot command language (CRCL). Ind. Rob. Int. J. **43**, 495–502 (2016)
19. https://github.com/open62541
20. Profanter, S., Tekat, A., Dorofeev, K., Rickert, M., Knoll, A.: OPC UA versus ROS, DDS, and MQTT: Performance evaluation of industry 4.0 protocols. In: Proceedings of the IEEE International Conference on Industrial Technology, pp. 1–8 (2019)
21. Pfrommer, J., Ebner, A., Ravikumar, S., Karunakaran, B.: Open source OPC UA PubSub over TSN for real-time industrial communication. In: Proceedings of the IEEE 23rd International Conference on Emerging Technologies and Factory Automation, pp. 1–4 (2018)

Titanilla Komenda obtained her M.Sc. degree in Mechatronics/Robotics from the University of Applied Sciences Technikum Wien. Since 2017 she has worked as project manager at Fraunhofer Austria in the field of robotic system integration.

Jorge Blesa Garcia is a junior researchers at the University of Applied Sciences Technikum Wien in Vienna, where he studies Mechatronics and Robotics.

Maximilian Schelle received his M.Sc. degree in Computational Methods in Mechanical Engineering from the UAS Hamburg, Germany in 2020. He currently works at Fraunhofer Austria on optimizing robotic systems.

Felix Leber is a junior researchers at the University of Applied Sciences Technikum Wien in Vienna, where he studies Mechatronics and Robotics.

Mathias Brandstötter received his Ph.D. in Robotics and Control at UMIT in 2016. He has been deputy director at ROBOTICS - Institute of Robotics and Mechatronics at JOANNEUM RESEARCH, since 2015.

Productivity Driven Dynamic Task Allocation in Human–Robot-Collaboration for Assembly Processes

M. Euchner and V. Hummel

Abstract The numerous challenges, such as mass customisation, globalisation and digitalisation, pose major challenges for Industry 4.0 in the industrial environment. Smart collaborative robotics, which has been identified as a key technology, still lags behind expectations. The synergies that arise from combining strengths of humans and robots offer many possibilities for adapting work systems to the new challenges. An ergonomic and economic consideration of HRC still reveals potential for improvement in order to promote the widespread use of collaborative robots. While the technology is primarily intended to support humans ergonomically, new stresses arise, such as a forced posture, monotonous work or the feeling of machine-determined work. In addition to the ergonomic burdens, the currently applicable safety standards still severely restrict the economic use of collaborative robots at the present time. This paper describes an approach that enhances developed methods considering ergonomic optimisation by combining it with economic improvement of human–robot collaborations by means of semi-autonomous group work. Based on the ergonomic and economic criterias within a work system, the task allocation of a workplace will be adjusted. Productivity describes the relationship between output and input. Productivity could be increased by reducing the input or by increasing the output. This method aims to increase output by dynamically allocating work tasks based on productivity. By comparing actual and target times, individual assembly processes are taken over by the robot or handed over to the human. The method is intended to help cushion the impact on productivity in the event of a resource failure through dynamic task allocation. In addition to optimising output, the aspects of semi-autonomous group work should also increase employee motivation. In summary, it can be said that the developed method should contribute to the ergonomic and economic use (e^{2}use) of collaborative robots in assembly.

Keywords Human–robot-collaboration · Dynamic task allocation · Productivity

M. Euchner · V. Hummel (✉)
ESB-Business School, Reutlingen University, Reutlingen, Germany
e-mail: vera.hummel@reutlingen-university.de

M. Euchner
e-mail: marc.euchner@reutlingen-university.de

K. von Leipzig et al. (eds.), *Smart, Sustainable Manufacturing in an Ever-Changing World*, Lecture Notes in Production Engineering,
https://doi.org/10.1007/978-3-031-15602-1_10

1 Introduction

The entire economy is facing more and more challenges. While the shift to mass production released synergies in terms of cost reduction, today companies are again faced with the difficulty of small batch sizes and customised products up to batch size 1 [1]. In addition the so-called environmental turbulences, such as currency crises, climate change and rising raw material prices, are leading to increased complexity in processes [2]. Human–robot collaboration (HRC) has been identified as a key technology to handle these requirements in the future [3]. Furthermore, the term Industry 5.0 defines the three premises such as human-centricity, resilience and sustainability that are necessary to sucessfully cope with the new challenges [4].

This paper puts a spotlight on the resilience of hybrid work systems. With the help of dynamic work system adaptations in the form of dynamic task allocation, the resilience is to be increased and the productivity to be stabilised. Also in view of demographic change, the resilience of a work system comes into focus as productivity should be maintained regardless of the age of the employee and his performance.

This paper's aims to provide a method that uses the advantages of HRC in order to achieve a resilient production and maintain productivity, even in the face of environmental turbulences that lead to short-term changes in capacity requirements. In the spirit of Industry 5.0, it is also a human-centric approach and meant to lead to greater acceptance and penetration of human–robot collaboration in production in the long term.

2 State Of The Art

In the following chapter the state of the art, which is the basis for the solution approach of a resilient production environment through a resilient, human-centered HRC, is outlined.

2.1 *Reasons for More Flexible and Responsive Production*

Since the manufacturing industry is facing multiple challenges nowadays, a flexible and responsive production should serve as a solution for future challenges [5]. Due to growing turbulences the demands on production planning and control are increasing and therefore a necessity for an increase in responsiveness and flexibility in production arises [6].

Reasons for needing a more flexible design of the production are according to [7] following turbulences:

- *in procurement*: turbulence due to material deviations and unreliable suppliers.

- *in production:* turbulence caused by unexpected quality deviations and process uncertainties due to machine and personnel failure.
- *by the customer*: turbulence due to changes in the end product or changes in the order.

While certain turbulences only have a medium impact on production, short-term turbulences pose major challenges. Time for an adequate reaction is usually too short, which is why short-term turbulences can lead to production targets not being met. In the future, volatile markets will mainly lead to short-term fluctuations within a company. Especially strong fluctuations within one day will pose great challenges to companies [5].

2.2 Resilient Manufacturing

The term resilience describes the ability to cope with turbulence [9] One major challenge of nowadays production is to meet the customers expectations in terms of timely delivery of personalized products [8]. Resilient production also should be able to react to different uncertainties, such as material, component and energy shortages or human-induced disturbances, in order to lead the system from an undesired back to a desired state [8]. Due to global factors such as climate change and pollution the interest in resilient systems is increasing [2, 8]. Characteristics of resilience and implications for manufacturing include productivity, processes, capacity, reliability, sustainability, quality, etc. [8]. Resilient systems are aiming to avoid or at least reduce losses in productivity as well as efficiency [9]. An approach to implement resilient production systems lies in the area of production planning. Independent of turbulent influences, production goals are to be achieved [8, 9]. This paper considers the resilience of HRC workplaces in the context of an overall resilient production system.

2.3 HRC and Dynamic Allocation

Collaborative robotics, as well as HRC, count as a key technology to meet the increased requirements of, for example, small batch sizes [10]. HRC is understood as the hand-in-hand cooperation of a collaborative robot with a human. This means that human an robot work at the same time in a common workspace on the same workpiece. By combining the 'advantages' of human and robot, synergies are to be created in order to implement [11] ergonomic and economic optimisations. Above all, the increased system flexibility that can be implemented through HRC serves as an important characteristic for establishing a resilient system that simultaneously pursues a human-centered approach [12]. An approach for an appropriate use of HRC and automation in which the human is the system's center is described as human centered Lean automation [13] The presented approach builds on the described Lean Automation approach. It aims to use the right amount of automation in order to create robust, reliable and not overly complicated solutions [14]. There are various approaches to task allocation in HRC. [15–17] follow a capability-oriented approach. The result of these methods is a suitability assessment of the resources with regard to the individual assembly steps. Based on this assessment, an allocation is selected. The result of these methods is one allocation which is considered the best based on defined parameters for example of the workpiece. With regard to a dynamic allocation, several allocations can be considered. An approach of a dynamic adjustment of the allocation was worked out by [18]. The individual interaction scenarios are compared on the basis of a utility value and the allocation of the tasks is changed in case of an unexpected event. This paper explores how system flexibility can contribute to resilient manufacturing.

2.4 *Productivity in HRC Through Human-Centred Automation*

Productivity describes the ratio of output to input. It represents the relationship between the results of an economic or operational activity (output) and the resources used for this activity (input). Productivity is considered to be an overall measure of efficiency [19]. Already [13] consider a human-centered Lean automation approach, which not only leads to increased productivity but also to an improvement of safety, ergonomics and human well-being. The approach of [13] shows that by means of an HRC the productivity of the work system can be increased.

Komenda et al. [20] investigated the effects of different task allocation on productivity. A wide variety of interaction scenarios were considered, differing in their allocation of tasks between humans and robots. The experiments showed that the total cycle time varies according to certain patterns of task allocation. Decisive for this are precedence conditions and resulting waiting times. The complex dynamic problem of task allocation can be solved by simulation and an optimiser.

2.5 *Semi-Autonomous Group Work in HRC*

A semi-autonomous group work is defined by a common work order that can be executed in an artifactual manner, joint organisation of actions to accomplish the order, jointly negotiated decisions, and responsibility for the use of resources and work results [21]. Different goals can be pursued by the implementation of a group work. These include for example a higher job satisfaction as well as an increase in performance and motivation [22]. In the sense of Industry 5.0, group work can play a central role in the implementation of people-centered workplaces.

3 Methodical Approach

The methodical approach is based on the design science research process according to [23]. This approach includes a total of six steps up to the communication of the result. The first step describes the identification of the problem. Hereby the increasingly frequent turbulences that lead to short-term changes in capacity requirements were identified [5]. This requires resilient production systems that can achieve the production target despite external influences [9]. As collaborative robots are considered as a key technology to meet the increased demands on production, the objective of this paper was to realise a resilient production by using HRC. In the following steps a concept is developed (step 3) which is implemented in a prototype (step 4). The evaluation (step 5) is done by calculating the limits of the established method and a critical review. Step 6 of the design science research process is the communication of the result.

4 Solution Approach

This chapter describes a solution approach, how a dynamic allocation of tasks within a human–robot collaboration based on productivity can lead to a more resilient production. The necessary steps are shown in Fig. 1. Dynamic allocation is the changing allocation of work content to a resource based on changing process parameters.

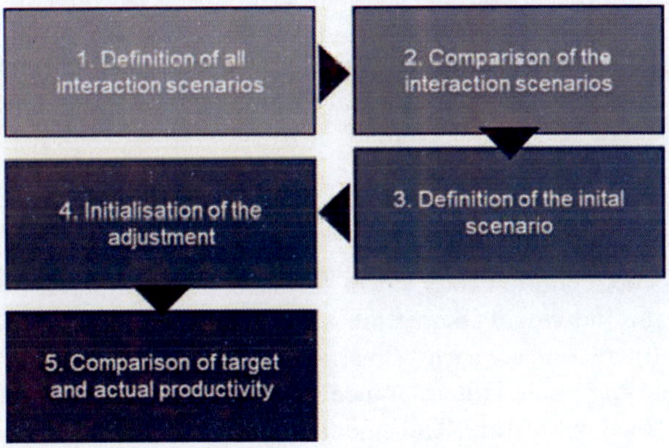

Fig. 1 Steps of the solution approach

Table 1 Examplary list of interactions scenarios

IS	Assembly steps (MS)			Processing time (TIS)
	MS 1	MS 2	MS 3	
1	H	R	R	37
2	H	H	H	42
3	R	H	H	38

4.1 Resilience Through Group Work in HRC

4.1.1 Define All Possible Interaction Scenarios

A dynamic allocation requires a pool of all possible interaction scenarios (IS). The idea of these interaction scenarios originates from the described method of [18]. An interaction is understood as a defined allocation of a resource to an assembly step. In this solution approach, an IS is understood to be a defined sequence of individual process steps as well as the allocation of the process steps to a resource. In this approach, it is not about finding an allocation that is best suited for a work center. Rather, it is about identifying multiple ways in which a human and a robot can work together within a work system. A prerequisite for a dynamic allocation is that there are several interaction scenarios. Meant, multiple ways in which the human and the robot can work together in a workplace. This implies that the human can also take over assembly steps for which the robot would actually be better suited and vice versa. The only prerequisite is that the human is capable of performing the task within a human-friendly framework.

Table 1 shows an exemplary list of interaction scenarios that differ in their allocation of the individual assembly steps and the processing time.

4.1.2 Comparison of the Interaction Scenarios

The identified interaction scenarios are compared using a tolerance range (TbIS), the individual interaction scenarios are compared based on their processing time. Interaction scenarios that have a processing time that is far too high are not included in the pool. This tolerance ensures that orders can be processed within the predefined order time. The calculation of the tolerance range (TbIS) is based on a quality management approach for process capability calculation [24].

$$Tb_{IS} = \frac{1}{n}\sum T_{IS} \pm x \times \sigma \quad (1)$$

TbIS describes the tolerance limit time of the interaction scenarios and is calculated from the mean value of the individual processing times of the interaction scenarios (TIS), whereby the x-fold value of the standard deviation (σ) is added or subtracted. The value x depends on economic and strategic aspects of the company, since a larger x leads to the situation when more interaction scenarios with a higher processing time remain part of the pool. All interaction scenarios that have a processing time within these tolerance limits are still suitable for the interaction scenario pool. An interaction scenario whose time exceeds the defined limit of TbIS is not included in the pool, since processing of the order in the predefined order time is at risk.

4.1.3 Definition of the Incipient Interaction Scenario

Based on the identified interaction scenarios that lie within the tolerance range, the scenario that has a medium processing time is selected at the beginning. A medium processing time ensures that the work system can react in both directions. If unforeseen turbulences occur in production, an interaction scenario with a shorter or longer processing time can be selected in order to avoid efficiency losses.

4.1.4 Initialisation of the Adjustment on Basis of Productivity

Based on the mean interaction scenario, the actual productivity of the work system is calculated according to formula 2.

$$As - is - Productivity = \frac{Output(IS_{current})}{Input} \tag{2}$$

$$Target - Productivity = \frac{Output_{target}}{Input} \tag{3}$$

The target productivity describes the necessary productivity to achieve the production target. If the target productivity changes due to turbulences and results in an increase in the target productivity, a suitable interaction scenario is selected.

This results in three scenarios:

1. Target productivity = Actual productivity
2. Target productivity > Actual productivity
3. Target productivity < Actual productivity

A change in target productivity can be caused by the environmental turbulences described in Sect. 2.1. The initialisation, i.e. the adaptation of the work system, is based on the comparison of target and actual productivity. Depending on the execution time of the defined interaction scenarios and their associated setup time, a suitable IS is selected. In case the target productivity exceeds the actual productivity, an IS with a lower execution time is selected. This gives the possibility to reach the production target despite turbulences.

4.1.5 Comparison of Target and Actual Productivity

The adjustment of the target and actual productivity is necessary as soon as parameters, which alter one of the two productivity values and bring the system out of equilibrium, change. Such a parameter change can occur due to the environmental turbulences mentioned. Not only a change in the target productivity due to an order change, but also a change in the actual productivity due to a machine or personnel failure can be a reason for an adjustment of the productivity parameters. The adjustment of the interaction scenario is done in agreement with the employee and by providing the necessary information that requires an adjustment of the system. It is not in the spirit of semi-autonomous group work and human-centered workplace design that the work system is automatically adjusted based on the target productivity. The adjustment is made for example on the basis of a jointly negotiated decision of the semi-autonomous group. The worker decides to adapt the interaction scenario based on the information provided [21].

5 Flexible Production Through Semi-Autonomous Group Work At Werk150

The presented solution approach was built in the Werk150, the learning and research factory of the ESB Business School at Reutlingen University [25].

The solution approach of a resilient production through the implementation of a dynamic allocation in the sense of a semi-autonomous group work was carried out within the scooter production line of Werk150. Figure 2 shows the focus aspects considered.

In this production line, the collaborative robot will be used for bolting and pick-and-place processes in the future. During the assembly of the running board, there

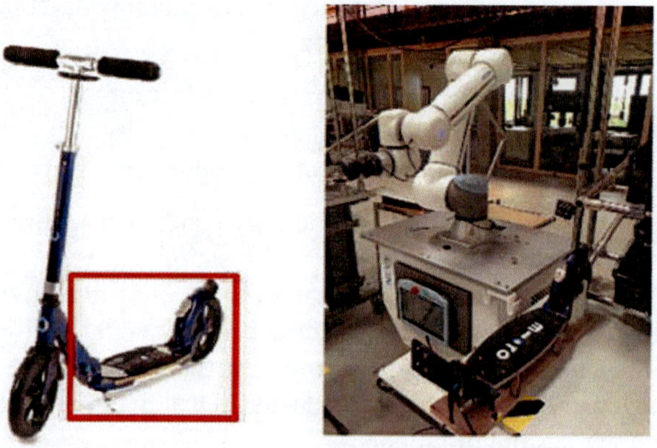

Fig. 2 Focused assembly group and cobot

Table 2 Identified interaction scenario

IS	Assembly steps (MS)			Processing time (TIS) in [sec]
	MS 1	MS 2	MS 3	
1	H	H	H	100
2	K	K	H	124
3	R	K	H	100
4	H	R	H	108
5	K	R	H	120
6	R	H	H	88

are currently six interaction scenarios that differ in the allocation of work content. Table 2 shows the allocation of tasks between human (H), robot (R) and both (K).

The processing time per unit TIS results from the execution times of the individual assembly steps and a process knowledge of which assembly steps can be executed simultaneously. There are currently six feasible scenarios for the selected assembly step. By adapting the end effector of the collaborative robot, further interaction scenarios could be defined. Based on formula 1 and a simple σ, the following tolerance limit times result:

- upper tolerance limit: 119 s
- lower tolerance limit: 94 s

Interaction scenarios that are above the tolerance limit time are not further considered. All interaction scenarios that are below the upper tolerance limit time can be considered for an adjustment in order to realise a resilient production. The lower tolerance limit can be used in control mode to guarantee stable processes. In the case of strong turbulence, IS that lie below the lower tolerance limit can also be taken into account in order to realise a very high target productivity.

Based on the IS within the tolerance limit, the productivity for each IS is calculated. Table 3 shows the IS that are not considered further due to the upper tolerance limit time in red colored lines. In addition, the productivity was calculated in pieces per hour. IS3 is selected as the middle interaction scenario because it is a human–robot collaboration compared to IS1. This IS offers the possibility to react in both directions in case of a change of the target productivity. If the target productivity is greater than the actual productivity, IS6 is selected to increase the actual productivity.

6 Limits and Challenges

Based on the values determined for the footboard production in Werk150, there is a margin of 53 assemblies produced, which could be realised via a 7 h production. In case of a short-term change in the target productivity before the start of production, + 34 required assemblies can be produced over a whole day through an adjustment

Table 3 Comparison based on tolerance limit

Interaction scenario	Processing time	Productivity in [pieces/h]
1	100	36
2	124	29
3	100	36
4	108	33
5	120	30
6	88	41

to interaction scenario 6. This is the maximum possible adjustment. Depending on change time, the number of assemblies to be produced decreases. An increase in the target productivity by six assemblies in the last hour to be produced can no longer be implemented with the current interaction scenarios. An extension of the automation in order to include the bolting process could lead to faster interaction scenarios, which could further increase the resilience of the system.

Basically, collaborative robots can be used for a variety of tasks. In addition to pick-and-place activities, they can also be used for screwdriving and gluing processes. The more manufacturing processes can be taken over, the more interaction scenarios are available to adapt the system. However, this creates an economic discrepancy, as a change of the end effector would lead to increasing set-up times, which therefore would reduce the productivity and thus also the economic efficiency of a dynamic adaptation. In the context of semi-autonomous group work, however, it would be possible to use.

7 Critical Questioning

In the implementation of dynamic allocation, basic requirements for a successful and effective use of dynamic allocation were defined. The current safety requirements would require extensive safety time certification for each adjustment. Certification of the overall system would be required to avoid recertification due to an adjustment of the allocation. Therefore an economical deployment depends on the certification of the entire system. Because a recertification would take too much time. Basically, the workplace should be suitable for an HRC application. An elementary prerequisite is that the individual interaction scenarios can be exchanged without major adjustments, such as a necessary exchange of the end effector. Large adjustments due to dynamic allocation hinder the economic use of collaborative robots. In addition, in the spirit of Industry 5.0 and human-centered workplaces, humans should be involved in the design process and agree that the task content is dynamically adapted.

8 Conclusion and Outlook

The solution approach describes a possibility for implementing resilient production, which can also take into account the aspects of semi-autonomous group work and thus enables human-centered hybrid work systems. However, the focus is on the resilience of an overall system, which is increased by a dynamic adaptation of the interaction scenarios. To a certain extent, short-term environmental turbulence can be cushioned by the method described. This depends mainly on the number of interaction scenarios and the time difference between the mean interaction scenario and the shortest interaction scenario. In addition, it must be noted that a permanent use of the shortest interaction scenario is the most favoured option from a business point of view, however, due to the human-centered approach, not only business factors are at stake. The goal is to realise a human-centered production that is capable of enduring turbulences through its resilience.

Outlook: The method presented for the realisation of a resilient production unfolds its full potential once it has been implemented in an automated way in production planning and control. An implementation in the production, planning and control system enables an immediate reaction to occurring turbulences that affect productivity. In addition, the use of multi-deffectors plays a crucial role in making a dynamic adjustment of the interaction scenario economically reasonable. Set-up times between adjustments would contribute to reducing the productivity of the system.

References

1. Wallaschkowski, S.: Die Entstehung des modernen Konsums. Springer Fachmedien Wiesbaden, Wiesbaden (2019)
2. Bauernhansl, T.: Fabrikbetriebslehre 1. Springer, Berlin (2020)
3. Buxbaum, H.-J. (ed.): Mensch-Roboter-Kollaboration. Springer Fachmedien Wiesbaden (2020)
4. Breque, M., De Nul, L., Petridis, A.: Industry 5.0: Towards a sustainable, human-centric and resilient European industry (2021)
5. Spath, D., Ganschar, O., Gerlach, S., Hämmerle, M., Krause, T., Schlund, S.: Produktionsarbeit der Zukunft - Industrie 4.0, Stuttgart
6. Wiendahl, H.-H.: Auftragsmanagement der industriellen Produktion: Grundlagen, Konfiguration, Einführung, 2011th edn. Springer, Berlin (2011)
7. Wiendahl, H.-P.: Betriebsorganisation für Ingenieure. Carl Hanser Verlag GmbH & Co, KG, München (2019)
8. Kusiak, A.: Resilient manufacturing **31**, 269 (2020)
9. Heinicke, M.: Framework for resilient production systems. In: Advances in Production Management Systems. Innovative and Knowledge-Based Production Management in a Global-Local World, p. 200. Springer Berlin (2014)
10. Schleicher, T.: Kollaborierende Roboter anweisen: Gestaltungsempfehlungen für ergonomische Mensch-Roboter-Schnittstellen, 1st edn. Springer Fachmedien Wiesbaden GmbH; Springer Vieweg, Wiesbaden (2020)

11. Bauer, W., Bender, M., Braun, M., Rally, P., Scholz, O.: Leichtbauroboter in der manuellen montage—einfach einfach anfangen: Erste Erfahrungen von Anwenderunternehmen, Stuttgart (2016)
12. Kuhlenkötter, B., Hypki, A.: Wo kann Teamwork mit Mensch und Roboter funktionieren?. In Mensch-Roboter-Kollaboration, p. 69. Springer Fachmedien Wiesbaden, Wiesbaden (2020)
13. Malik, A.A., Bilberg, A.: Human centered Lean automation in assembly, Procedia CIRP **81**, 659–664
14. Jackson, M., Hedelind, M., Hellström, E., Granlund, A., Friedler, N.: Lean Automation: Requirements and Solutions for Efficient use of Robot Automation in the swedish Manufacturing Industry, 3rd edn. (2011)
15. Malik, A.A., Bilberg, A.: Complexity-based task allocation in human-robot collaborative assembly **46**, 471 (2019)
16. Linsinger, M., Sudhoff, M., Lemmerz, K., Glogowski, P., et al.: Task-based potential analysis for human-robot collaboration within assembly system. In: Tagungsband des 3, p. 1. Kongresses Montage Handhabung Industrieroboter, Springer Vieweg, Berlin, Germany (2018)
17. Ranz, F., Hummel, V., Sihn, W.: Capability-based task allocation in human-robot collaboration (2017)
18. Nikolakis, N., Sipsas, K., Tsarouchi, P., Makris, S.: On a shared human-robot task scheduling and online re-scheduling **78**, 237 (2018)
19. Schneider, H.: Das Rätsel der Produktivität. Springer Fachmedien Wiesbaden, Wiesbaden (2020)
20. Komenda, T., Ranz, F., Sihn, W.: Influence of Task Allocation Patterns on Safety and Productivity in Human-Robot-Collaboration, Lisbon (2019)
21. Hacker, W.: Arbeitsanalyse zur prospektiven Gestaltung der Gruppenarbeit. In: Gruppenarbeit in Unternehmen: Konzepte, Erfahrungen, Perspektiven, p. 49. Beltz Psychologie-Verl.-Union, Weinheim (1994)
22. Wall, T.D., Clegg, C.W.: A longitudinal field study of group work redesig **2**, 31 (1981)
23. Peffers, K., Tuunanen, T., Rothenberger, M.A., Chatterjee, S.: A design science research methodology for information systems research **24**, 45 (2007)
24. Jakoby, W.: Qualitätsmanagement für Ingenieure: Ein praxisnahes Lehrbuch für die Planung und Steuerung von Qualitätsprozessen. Springer Fachmedien Wiesbaden, Wiesbaden (2019)
25. ESB Business School, ESB Logistik-Lernfabrik wird zum Werk150 ESB Business School. https://www.esb-business-school.de/de/fakultaet/aktuelles/meldungen/2019/esb-logistik-lernfabrik-wird-zum-werk150/. Accessed 9 Nov 2020

Marc Euchner (M.SC.) (University of applied sciences Reutlingen, Germany)

He holds a M.Sc. degree in Operations Management and is a member of Werk150, focusing on Human–Robot-Collaboration, Areas: research and teaching, collaborative robots, semi-autonomous group work in HRC, digital twins.

Prof. Dr.-Ing. Vera Hummel (University of applied sciences Reutlingen, Germany)

Her research focus areas include: Procurement, Production and Transportation Logistics, Industrial Engineering. Roles: Vice-Dean Research, Head of Logistics Department, Initiator and Project Manager, Establishment of "Werk150" the Learning and Research Factory on the campus of Reutlingen University.

Artificial Intelligence Based Robotic Automation of Manual Assembly Tasks for Intelligent Manufacturing

Alexej Simeth and Peter Plapper

Abstract Increasing product customization and shortening product life cycles in an ever-changing world is challenging for automation. This is especially true for assembly tasks, requiring a high level of perception, skill, and adaptability. With the rise of smart manufacturing, intelligent manufacturing, and other aspects related to Industry 4.0, the hurdles for automation of the aforementioned tasks are getting reduced. Especially Artificial Intelligence (AI) is expected to enable smart and flexible automation since it is possible to deduct decisions from unknown multi-dimensional correlations in sensor data, which is critical for the assembly of highly customized products. In this research paper, three different conventional and AI-based glue detection models are proposed with the target to automate a gluing process in a manual assembly of highly customized products in a batch size one production scenario. A conventional, one-dimensional rule-based model, and two hybrid models using a support vector machine image classifier (SVM) and either Tamura features or convolutional neural network (CNN) feature extraction are presented and compared. The obtained results demonstrate the efficiency and robustness of AI-based algorithms, as the CNN and SVM hybrid model outperforms the other two approaches achieving a prediction accuracy of >99% at the fastest classification speed.

Keywords Liquid detection · Convolutional neural network · Artificial intelligence · High mix low volume automation

1 Introduction

As one of the last processes in production operation, assembly plays a significant role in manufacturing systems [1]. A highly efficient and optimized assembly process is

A. Simeth (✉) · P. Plapper
High Performance Manufacturing, Department of Engineering, University of Luxembourg, Esch-sur-Alzette, Luxembourg
e-mail: alexej.simeth@uni.lu

P. Plapper
e-mail: peter.plapper@uni.lu

K. von Leipzig et al. (eds.), *Smart, Sustainable Manufacturing in an Ever-Changing World*, Lecture Notes in Production Engineering,
https://doi.org/10.1007/978-3-031-15602-1_11

key for a competent manufacturing line [2]. Assembly processes however are still the part of manufacturing with a low overall degree of automation [3]. On the one hand, this is due to the nature of assembly tasks which often require a high level of perception, skill, and logical thinking [4]. On the other hand, the flexibility and adaptability of manual assembly required is opposite to the increased output of automated assembly systems [5]. Especially in present boundary conditions characterized by an increasingly changing demand (mass customization) manufacturing lines face the trade-off between automated and manual systems. With the rise of smart manufacturing, intelligent manufacturing, and other aspects related to advanced manufacturing paradigms, the hurdles for automation in assembly are getting reduced [6]. Regarding replacing human intelligence in production, the authors of [7] state that intelligent manufacturing utilizes Artificial Intelligence (AI) techniques to minimize human involvement, which is key for the automation in assembly.

The author's research focusses on the automation of manual assembly processes in high variance low volume production scenarios up to batch size one via the application of AI. The selected assembly consists of a pick-and-place and gluing process. A varying number of different parts is placed into corresponding bores in a carrier workpiece and then bonded with glue. The required amount of glue is unknown due to the specific properties of the carrier workpiece and inserted parts. Further the amount may change with every inserted part. Thus, volumetric process control is not applicable and alternative solutions for the automation are necessary.

As one step in achieving the automation of the full manual assembly process, this paper presents different image classification methods to detect the fluid level of glue in workpieces so that the mentioned gluing process can be started, monitored, and stopped. The different methods, namely a one-dimensional conventional approach, a multi-dimensional, and an AI-based approach, are compared against each other. All methods are working with an optical sensor signal (RGB-Camera), which can be used as well for other sensing task in the described assembly.

The rest of the paper is structured as follows: Sect. 2 provides a brief overview on state-of-the-art liquid detection methods. In Sect. 3 the selected approaches for this paper are introduced and explained. The experimental setup and obtained results of each approach are presented in Sect. 4. The last sections of this paper summarize the results and list the future work to be done to perform the final task, i.e., the glue level detection.

2 State of the Art

A brief overview of related work is given in this chapter. An elaborated version of this was presented on APMS conference 2021 [8].

In manufacturing automation, existing detection problems to identify specific features, objects, parts, etc. are solved differently depending on the exact task. In the area of glue, liquid, and fluid detection in production environment, most solutions are proposed in the context of bottle filling and electronics manufacturing [9–15]. The majority of reviewed publications applies static edge detection algorithms in order to identify the surface or silhouette of the liquid [9–11]. The border of the liquid is detected and compared to a predefined position or threshold to measure the width of an applied glue line [9], or to determine the upper filling level of a bottle [10, 11]. In [12] the dark liquid inside the monitored bottle is separated from the image background solely based on colour values and the contour of the identified area is taken to estimate the filling level. A less complex algorithm based on changes in histogram is applied in [13] to measure the volume of a liquid and a bubble phase in translucent cylindric vessels. Key of this method is the experimental setup which amplifies the liquid and bubble phase. In [14], the authors compare a conventional detection approach to identify liquid and bubble phase in bottles via mean filters with Convolutional Neural Network (CNN) approach. The classification results are slightly improved with the CNN despite its very simple structure of three layers. To detect the variation in a dispensed glue drop on a workpiece, the authors of [15] selected principal component analysis. Target is to identify whether a fault in the dispensing system exists.

Most of the introduced methods by other researchers apply conventional models to identify the surface, the silhouette or the colour of a liquid and compare it against predefined thresholds or references. Further they are depending on a constant environment with specific settings for each feature to be detected. To start, monitor, and stop a gluing process in a constantly changing production environment with always changing products, e.g., high difference between products or robot mounted system, more robust and flexible detection algorithms are necessary.

3 Case Study: Vision-Based Gluing

For this research a gluing process in a final assembly in a batch size one production scenario is selected. Every carrier workpiece has a different number of inserted workpieces which must be bonded with glue. The position of the glue inlet and the amount of glue differs between every single gluing activity. Currently, this process is conducted manually. To automate this process, a robust decision rule to automatically detect the filling level of glue is required. Three different approaches are explained in the following.

3.1 Patch Brightness

The first approach is based on a single dimension, namely the brightness of the image patch. Similarly to the approaches presented in Sect. 2, this is a simple conventional detection rule. The region of interest (ROI), i.e., the outlet hole, is cropped out of the camera image and monitored (cf. 4.2). The colour of the glue is white and thus it is expected that the overall brightness of the image patch of the ROI increases with the glue level increasing in and exiting the outlet hole. At a certain brightness level, the process is stopped. The average brightness is calculated by converting the image patch from RGB to grayscale and taking the average of the intensity values of the pixels. The threshold is defined based on gluing trials explained in Sect. 4.

3.2 Tamura Features

Secondly, a multidimensional detection rule based on several image features is applied. Image features can be classified based on pixel properties as brightness, gradient, colour, or texture [16]. Texture feature descriptors play an important role and are widely applied in computer vision classification tasks. Tamura's texture features [17] have high correlation to human visual perception and are proven to describe texture human-like most accurately and efficiently [18]. Since human vision shall be replaced by an automated system for glue detection, Tamura features are selected for the second glue detection method. Tamura's texture descriptors are based on six textural features: coarseness, contrast, directionality, line-likeness, regularity, and roughness [17]. Coarseness represents the image granularity and is a measure of the micro-texture of an image. Contrast measures the variation of gray-level intensity of the pixels and is a measure of image quality. The existence of directional patterns like horizontal, vertical, diagonal, etc., is characterized by the feature Directionality. Line-likeness is the characteristic of texture that is composed of lines. A group of edge pixels is considered as a line when the pixels have the same edge direction. The sum of texture variation in an image is characterized by Regularity. It is measured as the sum of the variation of the four previous listed features. Roughness is used to emphasize the effects of coarseness and contrast and approximated as the sum of both features.

The extracted features are forwarded to a Support Vector Machine (SVM) image classifier. The SVM is trained on images and a detection rule is defined automatically.

3.3 CNN AlexNet Feature Extraction

The third proposed method to robustly detect the glue is a hybrid model based on a pretrained CNN and an SVM, which was presented in [8]. The CNN is applied to extract image features, which are used for classification. Implementing a pretrained CNN is usually significantly faster and simpler than designing a new network [19]. In contrast to our initial publication [8], the less complex eight layer deep CNN AlexNet is applied, which is trained on the ImageNet database with 1000 object categories [20]. The reason for the selection of a CNN is the outstanding performance in several classification challenges, e.g., [21, 22]. Similar to the second approach, the features are forwarded to train a SVM image classifier, which achieves a similar performance and accuracy as deep learning classifiers, i.e., classification by the CNN itself, while reducing complexity and computational effort [23].

4 Experimental Setup and Results

4.1 Experimental Setup

The selected algorithms need to be exposed to actual data so that they can be trained and optimized. To generate the initial data set a test stand is designed to imitate the real industrial gluing process as depicted in Fig. 1. A smartphone camera is mounted parallel to the surface on a frame built of aluminium profiles. In the field of view the carrier workpiece with inserted parts is placed. A gluing tool mounted on the frame is arranged so that the nozzle covers the inlet hole. While glue is fed into the inlet hole, the outlet hole is monitored by the camera and videos are taken. From the videos single frames are extracted and the region of interest (ROI), i.e., the outlet hole, is cropped out of the frames. The resulting images of 41 $\times$ 41 $\times$ 3 pixels are labelled into the two categories “empty” and “full”. The two classes represent the states where either insufficient glue is filled into the workpiece and the process shall be continued or sufficient glue is filled, and the gluing process shall be ended. In total 605 images are obtained following this procedure with 380 of the label empty and 225 of the label full. To improve the data basis for training and testing, the initial data set is augmented to increase the size. Augmentation of the frames has been conducted via reflection and rotation following the method of importance sampling [24]. Thus, only the data which is expected to be most influential is augmented. In this use case, images which are labelled as empty but close to label full and vice versa are considered for augmentation. Examples for obtained images are given in Table 1. With augmentation, the total data set size is increased to 1790 images.

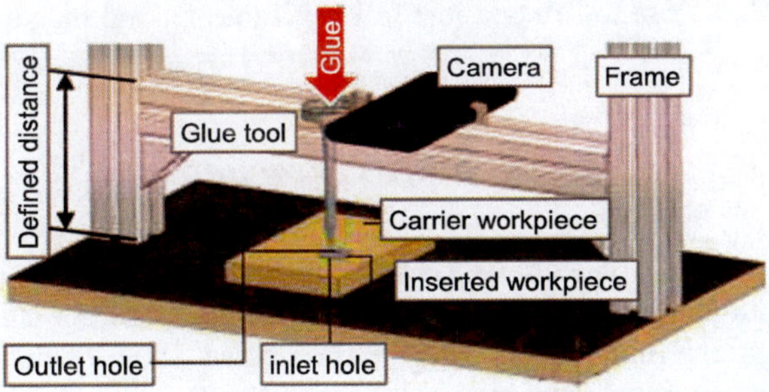

Fig. 1 Setup for data set generation [8]

Table 1 Examples of generated data set images

Label "EMPTY"	
Label "FULL"	

4.2 Conventional Approach: Patch Brightness

For the conventional approach based on the average brightness of the ROI, a fixed or adaptive threshold must be defined. This threshold indicates that enough glue is filled into the workpiece. While monitoring the outlet hole, it is assumed that the average brightness increases with a strong gradient once the glue gets visible. A chronological sequence of the filling of the outlet hole with glue until exiting of the same is given in Table 1. However, not all gluing processes are as smooth as the given images indicate. In Fig. 2 the average brightness of the ROI is given for three different gluing sequences. The first sequence represents a standard sequence from which images as in Table 1 are obtained. Once the glue gets visible in the ROI the brightness increases continuously. At some point the exiting glue forms a sphere above the outlet. The shadow of that sphere reduces the brightness until the shadow exits the ROI. This is indicated by the dip in average brightness at frame position 300 in the left graph in Fig. 2. In some cases, bubbles emerge on the surface of the glue significantly impacting the brightness. In the middle graph the dip in brightness is strongly visible. At this point however, the gluing process needs to be continued since the targeted level is not reached yet. The third graph on the right shows a gluing process in a brighter environment (sunny day). The overall level of brightness is much higher, although a constant light source is applied. To avoid such light disturbances an encapsulation would be required. Lastly, the surface of the outlet hole material has high differences in reflectivity so that it appears in different shades from dark gray to nearly white. Overall, no threshold applicable for all trials could be identified. The existence of glue can be detected but it is not possible to distinguish between empty but close to full and just full (cf. Table 1), and thus to robustly control the process.

4.3 Tamura Feature Extraction and SVM

In the second approach a machine learning SVM image classifier is trained on Tamura image features. Therefore, the generated data set is split into training and test data in the ratio 80:20. With the training set (80% of data) the SVM is trained. Tamura features are extracted and forwarded to the classifier. After training, the classification model is tested on the test set, i.e., the remaining 20% of the data set. An example of extracted Tamura features is given in Table 2 for a completely empty and a full image.

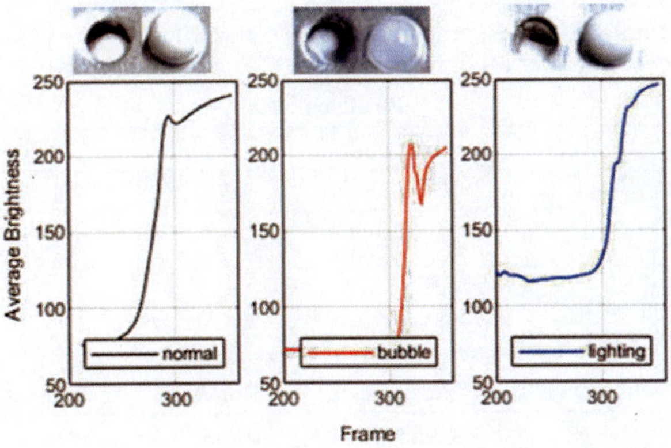

Fig. 2 Plot of average brightness in different scenarios: Left: Regular gluing behaviour. Middle: Unregular behaviour with formation of bubbles. Right: Bright scenario

Table 2 Tamura features of depicted image patches

	Coa	Con	Lin	Dir	Reg	Rou
	6.215	91.06	29.32	0.328	1.0	97.27
	7.868	59.64	28.80	0.747	1.0	67.50

Features are coarseness, contrast, line-likeness, directionality, regularity, and roughness

Besides regularity, all other five features of the two images vary. Most significant changes are in features contrast and roughness, which is based on contrast. The high difference is explained by the change from dark outlet hole to white glue. The overall prediction results are given in the confusion matrix in Table 3. The confusion matrix contains the values for true positive (TP), true negative (TN), false positive (FP), and false negative (FN) predictions. TP and TN are counted if a 'full' image is classified as 'full' or an 'empty' image is classified as 'empty' respectively. The case FP describes the classification of an 'empty' image as 'full' and FN the classification of a 'full' image as 'empty'. From the data it is visible, that the classification model classifies nearly all 'empty' images correctly but has a higher share of false predictions in the label 'full'. Every fourth full image is classified falsely. Thus, the classification model tends to stop the process too late, since the chance of a classification of an actual full image into label empty is given.

Based on the confusion matrix, the performance metrics accuracy, precision, and recall are calculated. Furthermore, the time to train the model, i.e., training time, and the time to classify a test set image is recorded. All results are given in Table 4. The overall prediction accuracy is 85.8%. i.e., out of 100 images, approx. 86 images are classified correctly. The high precision indicates, that if an image is labelled as 'full', the probability that it is actually 'full' is very high. In contrast, recall is significantly lower. Thus, a high share of 'full' images is not identified and classified as 'empty'.

Table 3 Confusion matrix of classification with Tamura features and SVM with true positives (TP), true negatives (TN), false positives (FP), and false negatives (FN)

	Predicted class			
Actual class	Label	Full	Empty	TP = 133 TN = 174 FP = 5 FN = 46
	Full	133	46	
	Empty	5	174	

Table 4 Evaluation metrics of Tamura features and SVM classifier

Accuracy	Precision	Recall	Training time	Testing time
0.858	0.963	0.743	58.04 s	2.54 s

The training of the model is rather fast and takes less than a minute. After training, approx. 2–3 s are necessary to classify a new image. In an industrial process, this period is quite long and critical to monitor the gluing process. In Fig. 2 it is visible, that the duration from first appearance of the glue to full is about 50 frames. Considering the recording frame rate of 30 frames per second, the timespan from empty to full is less than two seconds. However, the processing time highly depends on the used system and implementation. In a different setup, the testing time might be faster and sufficient.

4.4 *CNN AlexNet and SVM*

Like the second approach, the machine learning SVM image classifier is trained on extracted features. This time the features are extracted by the pretrained CNN AlexNet. Thus, the CNN itself is not changed and used for feature extraction only. The procedure of training and testing of approach two and three are same. The augmented data set is split in the ratio 80:20 into training and test data. Based on the training set the SVM classifier is trained and then tested and validated with the remaining test set. In Table 5, the prediction results on the test set for this model in form of a confusion matrix are presented. On the given test set, the model works almost perfectly. All positive images have been classified correctly. One false prediction occurred, and an 'empty' image was classified as 'full', thus a false positive prediction.

The evaluation metrics of the third approach are given in Table 6. All positive, i.e., 'full' images have been identified and the recall is one. Also, the precision is close to one since nearly all positive predictions have been correct. Overall, the presented third approach achieves a very high accuracy of 99.7% correct predictions. Training the model requires approx. a minute. More important however is the prediction time of 0.37 s.

Table 5 Confusion matrix of classification with CNN and SVM with true positives (TP), true negatives (TN), false positives (FP), and false negatives (FN)

	Predicted class			
Actual class	Label	Full	Empty	TP = 179 TN = 178 FP = 1 FN = 0
	Full	179	0	
	Empty	1	178	

Table 6 Evaluation metrics of CNN and SVM classifier

Accuracy	Precision	Recall	Training time	Testing time
0.997	0.994	1.000	61.53 s	

5 Discussion of Results

The presented approaches in this paper significantly differ in their ability to robustly detect the fluid level in bores. With the first conventional, rule-based approach it was not possible to define a threshold which works for the hole data set. The simple detection based on the average brightness of ROI was always indicating a strong increase in brightness when glue started to fill the workpiece and exit the outlet hole. However, the average brightness was significantly different between the test set images at a comparable gluing level. Thus, it was not possible to determine from the graph at a point in time, whether a hole is half full, full, or overfull, which refers to the images at end of the first row and the second row in Table 1. The appearance of bubbles on the glue, the change in reflectivity, and the overall lighting situation (daylight, evening, night) have further a high impact on the outcome, which can not be compensated with the simple rule (cf. Fig. 2).

The second and third approach apply the same machine learning SVM image classifier but use different image features. For approach two Tamura features and for approach three features extracted by the pre-trained CNN AlexNet are used. The Tamura features are based on six calculated image characteristics. From AlexNet, a fully connected layer of 1000 image classes is extracted, which contains a much higher number of features used to predict the label. Comparing the prediction quality of images of the class 'empty' both classifiers perform well although the AlexNet and SVM hybrid model performs better with only one false prediction. For the other label 'full' the result of the Tamura features-based model is not satisfactory since approx. 25% of the images are falsely classified. Overall, the Tamura features-based model is clearly outperformed by CNN-based model indicated by the far higher accuracy of 99.7% of CNN versus 85.8% of Tamura in terms of prediction accuracy. Comparing the timing aspect of the approach two and three the training time for Tamura is a few seconds shorter. More important for the overall evaluation is the actual testing time since this time is affecting the ability to monitor a real, live gluing process. The training of the model itself can be done offline not at the production line. The CNN based model classifies new images more than six times faster the Tamura model. However, another implementation of the Tamura model may significantly reduce process time. All tests have been run on the same machine.

Overall, the hybrid approach three outperforms the other presented approaches in all aspects. Prediction quality and prediction time is best with approach three. Only the required time for training is longer compared to approach two, but this is not critical since it has no impact on a latter industrial gluing process.

6 Conclusion and Future Work

In this paper different conventional and AI-based methods to robustly detect the glue level in bores are presented and compared. The target is to automate an industrial gluing process in a robot-based automation for a manual assembly line. The three different approaches are the following: A conventional, rule-based algorithm and two hybrid models using Tamura features or image features extracted by the convolutional neural network (CNN) AlexNet and a support vector machine (SVM) image classifier. The aim of the image classification is the determination whether enough glue is filled into a workpiece by classifying the glue outlet hole as 'full' or 'empty'. With the rule-based method glue can be detected, but the glue level differs significantly when using a predefined threshold. Based on this output, the process cannot be controlled. The hybrid models are trained and tested on the same data set. In all evaluation metrics beside training time, the CNN-based model outperforms the other and achieves overall the best results. Over 99% of all test images are classified correctly. In the test set, also different special cases are considered. Classification speed is highest with this model, which is important for a later industrial process. Overall, a robust glue detection model is presented applying both deep learning (DL) and machine learning (ML) techniques.

In a next step, the presented hybrid DL and ML model shall be validated on additional, new data sets. Therefore, it is planned to apply the model on a real technology demonstrator to monitor a live gluing process. With this procedure the functionality of the model when exposed to new data and further the sufficiency of classification time can be evaluated.

References

1. Sanders, D., Gegov, A.: AI tools for use in assembly automation and some examples of recent applications. Assem. Autom. (2013)
2. Outón, J.L., Villaverde, I., Herrero, H., Esnaola, U., Sierra, B.: Innovative mobile manipulator solution for modern flexible manufacturing processes. Sensors **19**(24), 5414 (2019)
3. Kleindienst, M., Ramsauer, C.: Der Beitrag von Lernfabriken zu Industrie 4.0-Ein Baustein zur vierten industriellen Revolution bei kleinen und mittelständischen Unternehmen. Industrie-Management **3**, 41–44 (2015)

4. Scholer, M.: Wandlungsfähige und angepasste Automation in der Automobilmontage mittels durchgängigem modularem Engineering - Am Beispiel der Mensch-Roboter-Kooperation in der Unterbodenmontage. Universität des Saarlandes (2018)
5. Lotter, E.: Hybride Montagesysteme. In: Lotter, B., Wiendahl, H.-P. (eds.) Montage in der industriellen Produktion, 2nd edn., pp. 167–194. Springer Verlag Berlin Heidelberg, Berlin, Heidelberg (2012)
6. Wang, L.: From intelligence science to intelligent manufacturing. Engineering **5**(4), 615–618 (2019). https://doi.org/10.1016/j.eng.2019.04.011
7. Wang, B., Tao, F., Fang, X., Liu, C., Liu, Y., Freiheit, T.: Smart manufacturing and intelligent manufacturing: a comparative review. Engineering **7**(6), 738–757 (2021). https://doi.org/10.1016/j.eng.2020.07.017
8. Simeth, A., Plaßmann, J., Plapper, P.: Detection of fluid level in bores for batch size one assembly automation using convolutional neural network. In: Dolgu, A., et al. (eds) Advances in Production Management Systems. Artificial Intelligence for Sustainable and Resilient Production Systems. IFIP International Federation for Information Processing. APMS 2021, IFIP AICT 632, pp. 86–93. Springer International Publishing, Cham (2021)
9. Huang, Y., Wu, C., Chang, C.-Y.: An application of image processing in flat panels. In: 2012 International Conference on Wavelet Active Media Technology and Information Processing (ICWAMTIP), pp. 8–11 (2012). https://doi.org/10.1109/ICWAMTIP.2012.6413427
10. Felipe, M.A.A., Olegario, T.V., Bugtai, N.T., Baldovino, R.G.: Vision-based liquid level detection in amber glass bottles using OpenCV. In: 2019 7th International Conference on Robot Intelligence Technology and Applications (RiTA), pp. 148–152 (2019). https://doi.org/10.1109/RITAPP.2019.8932807
11. Pithadiya, K.J., Modi, C.K., Chauhan, J.D.: Machine vision based liquid level inspection system using ISEF edge detection technique. In: Proceedings of the International Conference and Workshop on Emerging Trends in Technology, pp. 601–605 (2010). https://doi.org/10.1145/1741906.1742044
12. Gonzalez Ramirez, M.M., Villamizar Rincon, J.C., Lopez Parada, J. F.: Liquid level control of Coca-Cola bottles using an automated system. In: 2014 International Conference on Electronics, Communications and Computers (CONIELECOMP), pp. 148–154 (2014). https://doi.org/10.1109/CONIELECOMP.2014.6808582
13. Ma, H., Peng, L.: Vision based liquid level detection and bubble area segmentation in liquor distillation. In: 2019 IEEE International Conference on Imaging Systems and Techniques (IST), pp. 1–6 (2019). https://doi.org/10.1109/IST48021.2019.9010097
14. Beck, T., Gatternig, B., Delgado, A.: Schaum-und Füllstanderkennung mittels optischer Systeme mit neuronalen Algorithmen. In: Fachtagung „Experimentelle Strömungsmechanik, pp. 17.1–17.7 (2019)
15. Huang, Z., Angadi, V.C., Danishvar, M., Mousavi, A., Li, M.: Zero defect manufacturing of microsemiconductors—an application of machine learning and artificial intelligence. In: 2018 5th International Conference on Systems and Informatics (ICSAI), pp. 449–454 (2018). https://doi.org/10.1109/ICSAI.2018.8599292
16. Karmakar, P., Teng, S., Zhang, D., Liu, Y., Lu, G.: Improved tamura features for image classification using kernel based descriptors. In: International Conference on Digital Image Computing: Techniques and Applications (DICTA), pp. 1–7 (2017)
17. Tamura, H., Mori, S., Yamawaki, T.: Textural features corresponding to visual perception. IEEE Trans. Syst. Man. Cybern. **8**(6), 460–473 (1978). https://doi.org/10.1109/TSMC.1978.4309999
18. Chi, J., Yu, X., Zhang, Y., Wang, H.: A novel local human visual perceptual texture description with key feature selection for texture classification. Math. Probl. Eng. **1**, 2019 (2019). https://doi.org/10.1155/2019/3756048
19. MathWorks: Pretrained Deep Neural Networks (2021). https://www.mathworks.com/help/deeplearning/ug/pretrained-convolutional-neural-networks.html. Accessed 8 April 2021
20. ImageNet. http://www.image-net.org. Accessed 21 Aug 2021

21. Russakovsky, O., et al.: Imagenet large scale visual recognition challenge. Int. J. Comput. Vis. **115**(3), 211–252 (2015). [Online]. Available: http://arxiv.org/abs/1409.0575
22. Bochkovskiy, A., Wang, C.-Y., Liao, H.-Y.M.: YOLOv4: optimal speed and accuracy of object detection (2020). [Online]. Available: http://arxiv.org/abs/2004.10934
23. Arun Kumar, T.K., Vinayakumar, R., Sajith Variyar, V.V., Sowmya, V., Soman, K.P.: Convolutional neural networks for fingerprint liveness detection system. In: 2019 International Conference on Intelligent Computing and Control Systems (ICCS), pp. 243–246 (2019). https://doi.org/10.1109/ICCS45141.2019.9065713
24. Arouna, B.: Adaptative Monte Carlo method, a variance reduction technique. Monte Carlo Methods Appl. **10**(1), 1–24 (2004). https://doi.org/10.1515/156939604323091180

Alexej Simeth holds M.Sc. degree in Mechanical Engineering and Business Administration from the RWTH Aachen. He is currently a doctoral researcher at the Department of Engineering. His area of research is the development and implementation of AI-based smart automation for high mix low volume processes.

Peter Plapper obtained his Ph.D. degree in Machine tools and Production Engineering (WZL) from the Technical University of Aachen (RWTH). Since 2010 he is Professor for Tool Machines and Production Technologies at the University of Luxembourg.

Development of an AI-Based Method for Dynamic Affinity-Based Warehouse Slotting Using Indoor Localisation Data

Jan Schuhmacher and Vera Hummel

Abstract In industrial warehouse environments, the allocation of storage locations to goods, also known as slotting, is often based on static historical information (turnover frequency, quantities to be stored, etc.) of the corresponding stored goods. These static slotting methods relying on historical planning data often lead to deficits such as storage capacity bottlenecks or long process times for logistics staff in the warehouse to pick the required assortments of goods. These shortcomings often appear only subliminally in warehouse environments, while the causes of these performance losses cannot be (quantitatively) proven. Therefore, a method for dynamic affinity-based slotting of shelf racks has been developed, which uses artificial intelligence algorithms and enables continuous monitoring of performance-relevant parameters and influencing factors. An indoor localization system is used to generate close-to-real-time location and movement data of logistic staff during order picking processes. By applying k-Means cluster analysis methods, the localization data from the indoor localization system can be used in combination with other data sources (e.g. customer orders to be fulfilled, historical performance data etc.) for picking pattern recognition to achieve an optimized slotting in terms of process times and storage capacity. In addition, a graphical user interface for monitoring target variables as well as heatmaps for visualizing the frequency or duration of stay of employees in the corresponding warehouse areas have been implemented. The developed method specifically uses the potentials of AI in the area of data processing and preparation of possible decision alternatives based on close-to-real-time data for human decisions in the sense of hybrid (human/machine) decision-making. This can facilitate implementation in industrial warehouse systems by increasing acceptance. For a practice-oriented development and demonstration of the method, a demonstrator has been developed and set up at Werk150, the factory of the ESB Business School on the campus of Reutlingen University.

J. Schuhmacher (✉) · V. Hummel
ESB Business School, Reutlingen University, Reutlingen, Germany
e-mail: jan.schuhmacher@reutlingen-university.de

V. Hummel
e-mail: vera.hummel@reutlingen-university.de

K. von Leipzig et al. (eds.), *Smart, Sustainable Manufacturing in an Ever-Changing World*, Lecture Notes in Production Engineering,
https://doi.org/10.1007/978-3-031-15602-1_12

Keywords Warehouse · Storage · AI · Cluster analysis · Dynamic slotting · K-means

1 Introduction

The goal of an optimized storage location assignment, also referred to as slotting, is to find the best (performance optimized) storage location to store goods or stock keeping units (SKUs) in the warehouse [1]. Due to the rise of lot-size-one production of personalized products in line with Industry 4.0 static slotting strategies, which rely on a fixed storage locations in the warehouse for each good, are pushed to their limits [2, 3]. Optimized dynamic slotting strategies can help reduce picking process times (and thus costs) and likewise can support in using the available storage spaces as effectively and efficiently as possible [1]. Dynamic slotting strategies allow dynamic influencing factors and variables such as turnover rates, new product configurations and sales forecasts of the sales department to be taken into account when allocating storage locations to stored goods in order to reduce process times and increase the availability of storage locations that can be assigned [4]. The use of processes and methods of artificial intelligence (AI) offers far-reaching possibilities for a time efficient data processing and clustering to determine and analyse these dynamic influencing factors and variables and to integrate them into the optimized warehouse slotting by continuously including the relevant influencing factors and variables in the decision-making process [5]. However, many of the machine learning (ML) and AI methods or applications have a "black-box" character, so that the interrelationships are so complex that humans or even AI experts can no longer fully comprehend them and the level of resistance of humans towards decisions made by the AI is rising [6, 7]. This level of resistance can be lowered significantly by including the human in the decision-making and execution process [7]. Therefore, the developed method specifically uses the potentials of AI in the area of data processing and analytics for the preparation of possible decision alternatives based on close-to-real-time data to provide the logistic decision-maker with a sound basis for decision-making, which s/he can enrich with his/her implicit (experience) knowledge.

2 Warehouse Slotting Strategies

In companies, static slotting strategies often lead to storage capacity bottlenecks and long walking distances of logistics employees, which result in long process times [8, 9]. Dynamic slotting strategies can lead in particular to an increase in the availability of free storage locations as well as to an increase in the throughput performance in the warehouse by reducing the required process times [1, 10, 11]. The improvement of the availability of free storage locations is a common advantage of dynamic slotting, as there is no fixed allocation (or reservation) of storage locations to stored goods

[4, 9]. Considering the process time, the travel time of the logistics employee is of particular interest since the travel time is usually the most significant part of the overall picking time [12].

2.1 Common Dynamic Slotting Strategies

In order to reduce process times in the warehouse and to increase the availability of storage spaces that can be occupied, random, turnover-based and affinity-based strategies are used in particular in the field of dynamic slotting strategies. To find an appropriate slotting strategy for the development of an AI-based method for dynamic slotting using close-to-real-time location data, the following strategies have been investigated:

- **Random slotting**: By following a random slotting strategy, incoming SKUs are randomly assigned to an available storage location without any constraints [1, 9, 13]. This is a very simple and frequently used method, which also helps to distribute the picker traffic in the warehouse. However, this strategy can potentially lead to long walking distances during picking in the warehouse due to a complete lack of constraints [1].
- **Turnover-based slotting**: The goods are arranged in the warehouse depending on their picking frequency/demand quantity in a manner that "fast-moving" goods are positioned at handle height and as far to the front of the rack aisles as possible [9, 13]. This can significantly reduce walking distances and thus process times for picking frequently requested goods [9, 14]. In addition, this strategy delivers particularly good results if there are no common dependencies in the goods to be picked [1].
- **Affinity-based slotting (Clustering)**: With this strategy, goods that are frequently ordered/picked together are placed close to each other, which can reduce walking distances of pickers and corresponding process times [1, 14]. However, this allocation procedure can lead to disadvantages if fast-moving goods are also involved, which can cause congestion from several pickers in the corresponding storage areas [1].

2.2 Limitation of Existing Methods

Existing dynamic slotting methods using the above mentioned strategies are mostly based on complex (AI) algorithms such as genetic algorithms for affinity-based [10] and turnover-based [11] slotting and (meta)heuristics for combined turnover- and affinity-based [1] and random [8] slotting. These methods do not integrate the logistic decision-maker in the decision-making process and do not use his knowhow and implicit (experience) knowledge for optimizations of the processes related to warehouse slotting. In addition, these methods are not using real-time position data and corresponding process time data for the optimization of storage location allocation. The use of this real-time position data thus enables, among other things, a particularly precise performance-based analysis of human picking processes to improve warehouse slotting without manual data acquisition processes or complex WMS or ERP systems as a data source.

2.3 Selection of Slotting Strategy

The literature review of slotting strategies (see Sect. 2.1) showed that a random slotting strategy does not provide sufficient reliability to improve process times in the warehouse. Zoning the warehouse based on the picking frequency (turnover-based) of products is already widely used in practice and can be implemented without applying AI methods (e.g., by ranking items by picking frequency and then assigning storage locations starting with the most common item to the storage location with the least walking distance). To demonstrate the potential of AI with warehouse slotting optimization, the decision was made to apply an affinity-based bin allocation strategy for the developed method and demonstrator, as the potentials of using AI-based methods (especially cluster analyses to identify common pick order combinations) with subsequent optimization can be shown particularly clearly here and offers great potentials for industry.

3 Method for Affinity-Based Slotting Using Indoor Localization Technology

The developed method is based on the acquisition of close-to-real-time position data of the picker via an indoor localization system, the enrichment of these data with external data sources and the involvement of human decision-makers via corresponding data analysis and optimization interfaces and dashboards (see Fig. 1). The individual steps of the method can be run through continuously by the logistic decision-maker in order to iteratively achieve ever better degrees of target achievement. The considered targets are the reduction of process times (travel time and times for goods picking), improvement of storage space utilization and availability of free storage spaces through the application of dynamic, affinity-based slotting. The individual steps of the developed method and the involvement of the logistic decision-maker are explained in more detail in Fig. 1.

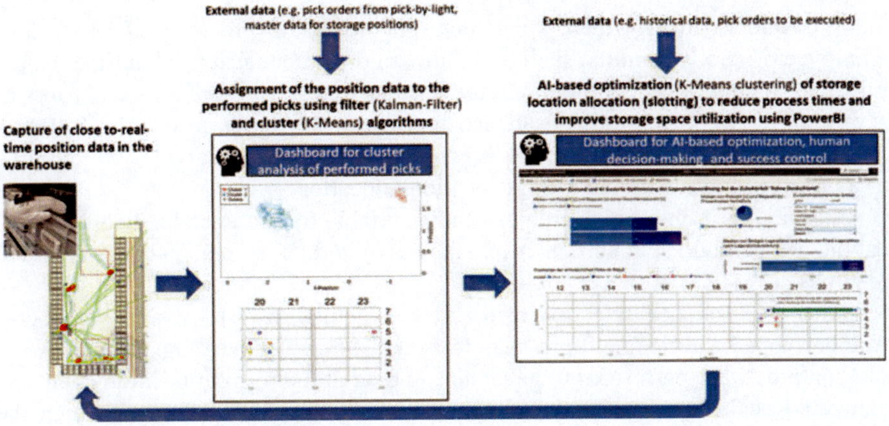

Fig. 1 Overview of developed method

3.1 Capturing of Position Data

An ultrasound-based indoor localization system is used to capture the necessary position and time information during the execution of the pick orders. For this purpose, localization receivers were installed in the warehouse area and localization tags (transmitters) were used to be attached to a work glove or via a wristband on the picker's wrist. The position and time information obtained by the indoor localization system is provided via a data acquisition and analysis portal of the localization system, which calculates the travel and pick times for the picking orders performed and generates a heatmap of the warehouse environment based on the duration of the picker's stay in the respective warehouse area within a certain period of time. A more in-depth analysis and AI-supported storage location optimization is performed by exporting the generated position and time information of the picking orders from the data acquisition portal and automated import into a PowerBI report with corresponding AI-supported analysis and optimization functions for the logistic decision-maker.

3.2 Assignment of to Position Data to Performed Picks

Before further analysis of the indoor localization system measurement data, Kalman filter-based filtering of the measurement data was performed to reduce the influence of measurement errors due to disturbances such as signal echoes on the further analyses. In order to cluster the position data generated by the indoor localization system to the respective picking positions in the shelf and to optimize the storage location assignment, it was necessary to select a suitable method. By clustering the pre-filtered position data (only positions in a defined interval at the front of the shelf are relevant for the picking process), groups (clusters) of similar (spatially close) positions can be formed, which represent the picking processes of the logistics employee of goods from the shelf and allow to draw conclusions about the picked goods (or their storage location) and the process times. The determined clusters of several different picking orders can also be used to identify corresponding hotspots of frequently frequented shelf areas (high picking frequency of goods), which serve as a basis for optimizing the storage location allocation.

Cluster algorithms, which are potentially suitable for the intended field of application, can be divided in particular into partitioning, hierarchical and density-based methods [15, 16].

For a well-founded selection of a suitable method, a test scenario was therefore developed for picking 4 products from the warehouse (see Fig. 2) and several picking runs were performed to generate test data in the form of position data. The generated position data was then imported into a PowerBI project in which the partitioning methods K-Means, Affinity Propagation and Means Shift, the hierarchical BIRCH and Agglomerative Clustering methods as well as the density-based methods DBSCAN and OPTICS for clustering the pre-filtered position data were investigated for comparison purposes. The cluster analysis methods were implemented in PowerBI through phython script. Except for the 2 cluster algorithms "k-Means" and "Agglomerative Clustering", where the number of clusters can be specified in advance, none of the investigated methods provided (without any further manual manipulation of the data) the correct number of 4 clusters and thus 4 picking positions in the warehouse.

Fig. 2 Test scenario with 4 pick positions for the selection of a suitable cluster algorithm

Since the cluster algorithm k-Means, as a partition- or center-point-based method, directly provides the determined centers of the identified clusters without any further calculation steps, the decision was made to use the k-Means algorithm for the demonstrator.

3.3 AI-Based Optimization of Storage Location Allocation

By determining the centers of the identified clusters, the pick positions of the executed pick orders can be determined in connection with inventory data of the warehouse. Likewise, an optimized affinity-based slotting can be realized with the help of minimizing the cluster distances to each other by dynamically changing the bin locations in the warehouse using the location data (see Fig. 3).

This AI-based storage location optimization is performed by importing the generated position measurement data into a PowerBI report with AI-supported analysis and optimization functions. In addition to the position and time information, external data sources such as the material stock lists with the corresponding article, quantity and storage location information (also involving X, Y and Z center coordinates of the storage locations) as well as executed and open pick orders with article data and quantities are included in the analyses and optimizations. In this way, the logistics decision-maker can perform more detailed analyses of selected pick orders in PowerBI with regard to the goods to be picked, the pick and travel times, the pick positions in the warehouse and the storage location utilization, and initiate optimization runs.

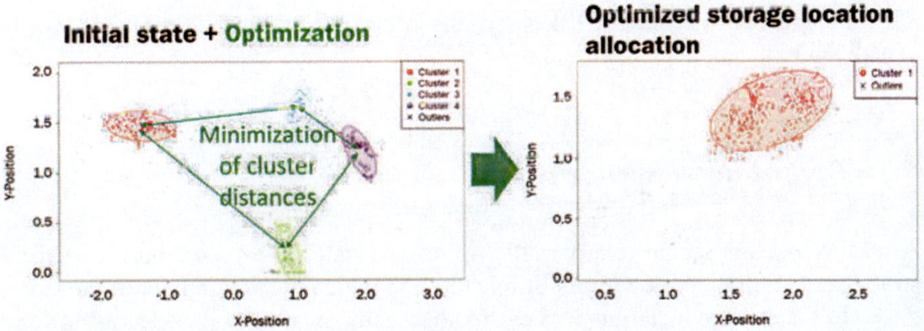

Fig. 3 Optimized affinity-based slotting using the k-Means algorithm

4 Demonstrator at Werk150

To support the industry transfer of the developed method, a digital and physical demonstrator for companies was developed and integrated into the warehouse environment of Werk150—The factory of the ESB Business School (Reutlingen University). The realized demonstrator consists of a physical part, which contains the hardware for data acquisition and pick process execution, and a digital part, which includes the measurement data processing as well as the developed AI-supported analysis and optimization solutions, including visualizations and dashboards for the logistics decision-maker. Different scenarios were developed and implemented to validate and demonstrate the developed method, starting with a static slotting strategy as a baseline scenario and ending with a dynamic affinity-based target scenario using the developed method. The picking process in the demonstrator can be carried out by the instructor or by the seminar participant her-/himself.

4.1 Baseline Scenario

The baseline scenario reflects the typical condition of a static storage location assignment based on historical planning data with long walking distances (and thus long process times) as well as a low availability of free storage locations (impending storage capacity bottleneck due to high storage location utilization rate). The key figures of the total process times (incl. the time shares of picking times and walking times) as well as the high warehouse occupancy rate due to a fixed assignment of storage bins to storage goods can thus be taken from the generated PowerBI report after the practical execution. Likewise, the routes taken by the employee can be visualized in a generated heatmap.

4.2 Optimization Scenario

After the baseline scenario, a partially optimized state in the warehouse is demonstrated as a starting point before optimization, in which the assignment of the stored goods to the storage locations was carried out using an affinity-based slotting some time ago. This is to illustrate that even when using dynamic storage location allocation strategies, continuous monitoring of the target parameters or constraints through the use of AI methods (here: k-Means cluster analysis method) is required to always maintain a target-optimized state in the warehouse. The order to be picked is an order for the pre-picking of components of a silver-colored product variant, which is a standard product. Accordingly, this represents a very common combination that was accounted for in the affinity-based slotting by placing the corresponding bins a short distance apart on the shelf using the k-Means cluster algorithm enriched with the (experience) knowledge (e.g. product and market trends, component handling characteristics) of the logistic decision-maker.

However, this order now increasingly contains an additional accessory part that was not yet a "common component" for these parts combination in the originally performed affinity-based slotting run, and thus is located a long distance from the standard components. The long travel distance can be seen on the one hand in the generated heatmap (see Fig. 4) and on the other hand by analyzing the process times. In Fig. 4, the real-time position of the picker can be seen as a blue circle in the heatmap and the pick positions in the warehouse at which the picker had previously been for the execution of the picks are shown in light green to red (long stay) depending on the time of stay. Through the analysis and optimization solution implemented in PowerBI using the k-Means algorithm for cluster analysis, the optimization of the bin allocation can now be performed according to the strategy of affinity-based slotting. The k-Means cluster analysis now shows the decision-maker two clusters for the initial situation with a large distance between the cluster of standard parts and the new accessory part before optimization (see Fig. 5).

By using the k-Means clustering with a minimization of the cluster distances and including external data sources such as master data on available storage locations as well as historic and future pick orders, an optimization of the storage bin allocation can thus be triggered and the optimization result can be viewed in the PowerBI dashboard (see Fig. 6). Based on the X, Y and Z center coordinates of the storage

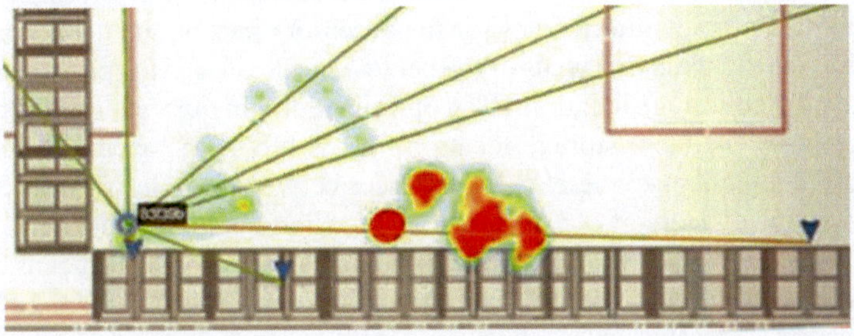

Fig. 4 Heatmap before optimization

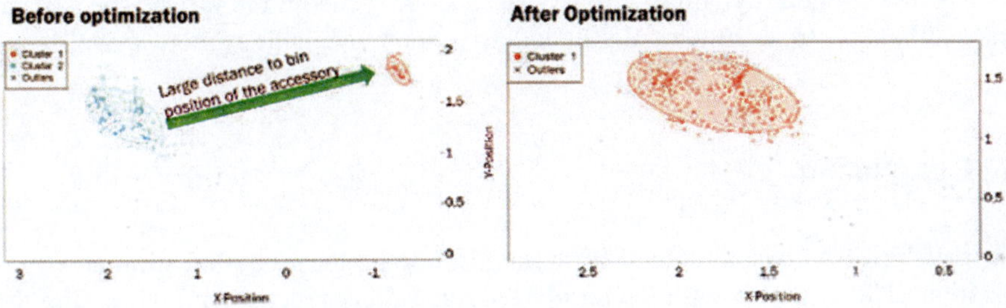

Fig. 5 K-Means cluster analysis before and after optimization

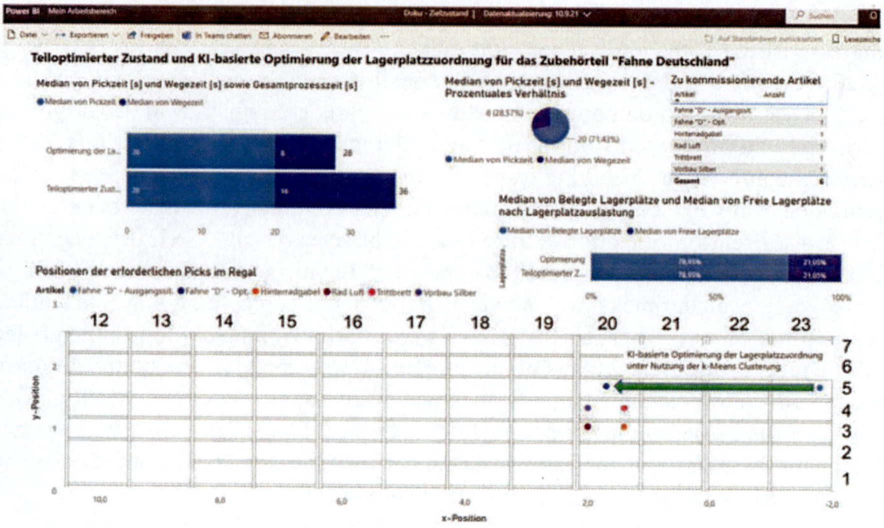

Fig. 6 PowerBI dashboard

locations, the distances between the goods to be picked can thus be reduced through several alternative calculation runs for slotting the SKUs on the shelf also considering the required sequence of picking. In this way, the logistical decision-maker receives various storage location alternatives for the accessory part on the dashboard, taking into account the summed distances of the center coordinates of the pick positions and availability of free storage locations. The optimization run thus leads to the allocation of an alternative available storage location for the bin of the accessory part above the standard components, whereby the distances between the pick positions and thus also between the clusters of bin positions on the shelf could be reduced.

The optimization is likewise reflected in the visualization of the k-means clustering in the form of a common cluster of the components concerned (see Fig. 5 right). The affinity-based assignment of an alternative storage location for the accessory thus leads, in the example explained, to a reduction in the average total process time from 36 to 28 s after optimization. Likewise, compared to the baseline scenario with static slotting, the applied (dynamic) affinity-based slotting method improved the storage space utilization from 96.5 to 78.9% leading to more available storage space.

5 Conclusion

The developed AI-based method for dynamic storage location allocation and optimization of warehouses with the aim of reducing process times and increasing the availability of free storage locations has been successfully validated. The developed solution uses the potentials of AI in the field of k-Means cluster analysis for the discovery of similarity structures of pick orders, the assignment of the determined position data to the performed picks as well as for the optimization of the storage location allocation in the warehouse. Following the principle of hybrid (human/machine) decision-making, the decision-making power remains with the logistic decision-maker, who receives a well-founded basis for decision-making through the developed solution. In this way, the acceptance of the developed solution in industrial practice can be increased and the implicit (experience) knowledge of the logistician can also be included in the optimizations. Nevertheless, further comparisons of the different methods of random, turnover-based and affinity-based slotting apart from the chosen approach based on k-Means clustering are to be aimed at. For a more accurate determination of the picking times, the development of a glove with built-in sensors for the detection of the gripping processes by sensing the movement of the fingers is also planned with an industrial partner. Furthermore, a closer (quantitative) analysis of the generated position data with respect to ergonomic aspects offers still another potential to improve the storage location allocation (e.g. for heavy goods) with the goal of reducing process times and decreasing the ergonomic load for manual warehouses.

Acknowledgements Research supporting these results have been funded by the Steinbeis Innovationszentrum Transferplattform BW Industrie 4.0.

References

1. Kofler, M., et al.: Affinity based slotting in warehouses with dynamic order patterns. In: Klempous, R., et al. (eds.) Advanced Methods and Applications in Computational Intelligence, pp. 123–143. Springer International, Heidelberg (2014)
2. Papcun, P., et al.: Augmented reality for humans-robots interaction in dynamic slotting "chaotic storage" smart warehouses. In: Ameri, F., et al. (eds.) Advances in Production Management Systems. Production Management for the Factory of the Future, pp. 633–641. Springer International (2019)
3. Bauernhansl, T.: Die Vierte Industrielle Revolution – Der Weg in ein wertschaffendes Produktionsparadigma. In: Vogel-Heuser, B., et al. (eds.), Handbuch Industrie 4.0 Bd.4, pp. 1–31. Springer Berlin Heidelberg (2017)
4. MicroChannel. Information on https://www.microchannel.com.au/company-blog/march-2015/warehouse-management-static-vs-dynamic-slotting (2015)
5. Zunic, E., et al.: Smart warehouse management system concept with implementation. In: 2018 14th Symposium on Neural Networks and Applications (NEUREL), pp. 1–5. IEEE (2018)
6. Schaaf, N., Wiedenroth, S.J., Wagner, P.: Information on http://publica.fraunhofer.de/dokumente/N-630667.html (2021)
7. Klumpp, M.: Automation and artificial intelligence in business logistics systems: human reactions and collaboration requirements. Int. J. Log. Res. Appl. **21**(2018), 224–242 (2018)
8. Quintanilla, S., et al.: Heuristic algorithms for a storage location assignment problem in a chaotic warehouse. Eng. Optim. **47**(2015), 1405–1422 (2015)
9. Sponheimer, A.: Lagerkonzepte: Vorteile und Anforderungen einer chaotischen Lagerung in der pharmazeutischen Industrie. TechnoPharm **2**(2012), 94–99 (2012)
10. Li, J., Moghaddam, M., Nof, S.Y.: Dynamic storage assignment with product affinity and ABC classification—a case study. Int. J. Adv. Manuf. Technol. **84**(2016), 2179–2194 (2016)

11. Bottani, E., et al.: Optimisation of storage allocation in order picking operations through a genetic algorithm. Int. J. Log. Res. Appl. **15**(2012), 127–146 (2012)
12. Rushton, A., Croucher, P., Baker, P.: The Handbook of Logistics and Distribution Management: Understanding the Supply Chain, sixth edn. Kogan Page, London (2017)
13. Viveros, P., et al.: Slotting optimization model for a warehouse with divisible first-level accommodation locations. Appl. Sci. **11**, 936 (2021)
14. ten Hompel, M., Schmidt, T.: Management of warehouse systems. In: ten Hompel, M., Schmidt, T. (eds.) Warehouse Management, pp. 13–62. Springer, Berlin Heidelberg, Berlin, Heidelberg (2007)
15. Ali, A. et al.: Systematic review: a state of art ML based clustering algorithms for data mining. In: 2020 IEEE 23rd International Multitopic Conference (INMIC), pp. 1–6. IEEE (2020)
16. Hass, G., Simon, P., Kashef, R.: Business applications for current developments in big data clustering: an overview. In: 2020 IEEE International Conference on Industrial Engineering and Engineering Management (IEEM), pp. 195–199 (2020)

Jan Schuhmacher, M.Sc. MBE is a research associate at ESB Business School and part of the research, training and consultant group of "Werk150—The Factory on Campus of ESB Business School". His major field of research is the design and optimization of changeable, autonomously controlled intralogistics systems.

Vera Hummel, Prof. Dr.-Ing., Dipl.-Ing., has been a professor at the ESB Business School, Reutlingen University since 2010. She is a founding member of the "Initiative of European Learning Factories" and the initiator and head of the "Werk150—The Factory on Campus of ESB Business School" for research, education and industry training on the campus of Reutlingen University.

Enterprise Design and Digital Twins

A Framework for Leveraging Twin Transition in the Manufacturing Industry

Lukas Stratmann, Volker Stich, Ruben Conrad, Gerrit Hoeborn, Felix Optehostert, and Minh Phuc Phong

Abstract The manufacturing industry consumes 54% of global energy and attributes for 20% of global CO_2 emissions, demonstrating the industry's role as global driver of climate change. Therefore, reducing its carbon footprint has become a major challenge as its current energy and resource consumption are not sustainable. Industry 4.0 presents a chance to transform the prevailing paradigms of industrial value creation and advance sustainable developments. By using information and communication technologies for the intelligent networking of machines and processes, it has the potential to reduce energy and material consumption and is considered a key contributor to sustainable manufacturing as proclaimed by the European Commission in the term "twin transition". As organizations still struggle to utilize the potential of Industry 4.0 for a sustainable transformation, this paper presents a framework to successfully align their own twin transition. The framework is built upon three key design principles (micro level: leverage eco-efficient operations, meso level: facilitate circularity and macro level: foster value co-creation) derived using case study research by Eisenhardt, and four structural dimensions (resources, information systems, organizational structure and culture) based on the acatech Industry 4.0 Maturity Index. Eleven interconnected areas of action are defined within the

L. Stratmann (✉) · V. Stich · R. Conrad · G. Hoeborn · F. Optehostert · M. P. Phong
Institute for Industrial Management FIR at the RWTH Aachen University, Aachen, Germany
e-mail: Lukas.stratmann@fir.rwth-aachen.de

V. Stich
e-mail: volker.stich@fir.rwth-aachen.de

R. Conrad
e-mail: ruben.conrad@fir.rwth-aachen.de

G. Hoeborn
e-mail: gerrit.hoeborn@fir.rwth-aachen.de

F. Optehostert
e-mail: felix.optehostert@i40mc.de

M. P. Phong
e-mail: minh.phuc.phong@rwth-aachen.de

K. von Leipzig et al. (eds.), *Smart, Sustainable Manufacturing in an Ever-Changing World*, Lecture Notes in Production Engineering,
https://doi.org/10.1007/978-3-031-15602-1_13

framework and offer a holistic and practical approach on how to leverage an organization's twin transition. Within the conducted research, the framework was applied to the challenge of information quality and transparency required for high-value secondary plastics in the manufacturing industry. The result is a digital platform design that enables information transactions for secondary plastics and establishes a circular ecosystem. This shows the applicability of the framework and its potential to facilitate a structured approach for designing twin transitions in the manufacturing industry.

Keywords Industrial sustainability · Manufacturing · Twin transition · Industry 4.0 · Sustainability · Circular economy · Plastics industry · Business ecosystems · Digital platform design

1 Introduction and Objective

While the manufacturing sector provides 36 million jobs in Europe, the industry is the third largest contributor to greenhouse gas emissions and the main cause of resource scarcity and loss of biodiversity [1]. Worldwide, 54% of global energy consumption and 20% of CO_2 emissions are attributable to the manufacturing industry [2]. As such, the current consumption patterns of production are not sustainable [3]. In the last five years an unprecedented dynamic of sustainable development can be seen globally. The 2016 Paris Climate Agreement [4], the climate report of the Intergovernmental Panel on Climate Change (IPCC) in 2018 [5] and the European Green Deal in 2019 [6] signal the urgency for action. The 2016 Paris Climate Agreement [4], the climate report of the Intergovernmental Panel on Climate Change (IPCC) in 2018 [5] and the European Green Deal in 2019 [6] signal the urgency for action. Twin Transition as term to describe achieving sustainability through measures of Industry 4.0, the European Commission's Horizone Europe research program signals a strong reliance on this combination. Greater understanding and awareness of the consequences of environmental degradation among society, the public sector and nongovernmental organizations lead to urgent demands for social responsibility and environmentally sustainable action from the industry [3, 7].

Simultaneously, Industry 4.0 transforms the prevailing paradigms of industrial value creation. Rigid and pre-determined value chains are replaced by globally connected, flexible value networks and new forms of collaboration. Data-driven business models shift the focus from products to customers and solutions. Availability, transparency and access to data are the key success factors of Industry 4.0 [8]. By 2025, the economic potential of Industry 4.0 is estimated to be between €70–140 billion [9]. Supported by a comprehensive database, Industry 4.0 is expected to improve environmental issues and social benefits through more efficient energy and material consumption as well as adaptive working environments [10]. Experts in the industry agree that Industry 4.0 is a key contributing factor towards achieving sustainability in manufacturing [11]. Defined as industrial sustainability, the required transformation targets a state in which industry is part of an overall socially, environmentally and economically sustainable system and actively contributes to a healthy planet. Corporate strategies embed themselves in the overarching system responsibility in this target state.

The objective of this paper is to derive leverage points of Industry 4.0 for sustainability and provide a holistic framework with concrete areas of action, that help manufacturing companies rationalize their Industry 4.0 strategy and sustainability objectives.

2 Study Background and Perspective

There are several opportunities in the context of Industry 4.0 to facilitate sustainability throughout the value chain. Many manufacturing companies employ digital technologies, services and solutions to drive their sustainability strategies. Even though synergies between these two industry trends are apparent and discussed in practice and among scholars, only few frameworks for the systematic implementation of Industry 4.0 to leverage sustainability exist in literature.

The relevant frameworks derive either from scientific articles conducted through literature reviews or white papers obtained through studies. As for the scientific articles, Kamble et al. [12] offer a sustainable Industry 4.0 framework and Parida et al. [13] provide frameworks regarding digitalization, Business Model Innovation, and sustainable industry. The white papers present two frameworks: One showing a Target Vision for Sustainable Tool and Die Making [14] and the other visualizing Development Paths for Industry 4.0 and Ecological Sustainability [11].

Due to the nature of literature reviews, Kamble et al. and Parida et al. investigate the convergence of the two topics sustainability and Industry 4.0 from a conceptual perspective. As such, they do not provide tangible courses of action for the implementation of sustainability with Industry 4.0. Furthermore, Kamble's et al. Sustainable Industry 4.0 Framework has a predominantly internal perspective on companies. The framework concentrates on processes within a smart factory and their potential for sustainable outcomes. The aspects of networking and collaboration in value creation networks, which are crucial for the transition to a circular economy, are disregarded. Since the frameworks of Boos, Kelzenberg et al. [14] and the Plattform Industry 4.0 [11] are the results of studies published in white papers, they are inherently biased regarding their conclusion. Neither of the white papers deals with the negative consequences of Industry 4.0 on sustainability, nor do they critically assess the proposed solutions regarding their actual sustainable impact. Although digital technologies are considered key enablers for most action fields, the opportunities of Industry 4.0 for the employee dimensions are omitted and the transformational implications of Industry 4.0 on the social aspect of sustainability are not specified.

In summary, the reviewed frameworks either do not consider practical applications or do not critically evaluate the opportunities of Industry 4.0 for sustainability. As a result of these research gaps, this paper aims to establish a comprehensive and practice-oriented model built upon a theoretically sound foundation, that helps manufacturing companies to rationalize the implementation of sustainable manufacturing through Industry 4.0.

3 Methodology

Eisenhardt's case study research approach [15] is an appropiate methodology to understand Twin Transition as an emergent field from a practice-oriented view. Case study research aims at building scientific theories from valid empirical data sets. Through iteratively cross examining of case studies and overlapping qualitative data with quantitative data, the underlying mechanisms and principles of a problem are derived. Case study research ensures the practicality of a resulting theory due to its setup based on multiple cases and as such, the emergent propositions are deeply grounded, accurate and generalizable [15, 16]. Furthermore, the case study research ensures the practicality of the resulting theory because its data is based on differing empirical evidence [15, 16].

3.1 Eisenhardt's Case Study Research Methodology

Eisenhardt's case study research methodology is applied to develop a theoretical framework for the Twin Transition in the manufacturing industry. The approach consists of several steps that aim at building a theory from real case studies [15]. A total of eight manufacturing companies are selected as case studies. Selecting relevant cases, special attention is paid to ensure that the selected companies show a strong commitment to environmental and social issues. This is ensured by screening reports on quantified goals and the compliance to global sustainability standards such as the United Nations' 17 Sustaniable Development Goals.

The selection is focused on companies with an extensive portfolio of digital products, services, and solutions. Information from case studies is collected, including press releases, interviews with managers, annual reports, sustainability reports, and other public documents from the companies' websites. Based on the aggregated information, the case studies are analyzed. Through in-depth examination of impactful use cases and projects, the foundation for cross-case comparisons to identify overall design principles is established. Additional literature from journal articles, whitepapers and studies is used to validate and confirm the identified design principles and further enrich its content.

Table 1 Selected case studies

		Company details		
Case study	Industry	Business segments	Revenue [billions]	Number of employees
Robert Bosch GmbH	Manufacturing	Automotive, consumer goods, energy technology	€77.7	398,000
Schneider Electric SE	Manufacturing	Industrial automation, electric energy supply	€27.2	135,000+
Continental AG	Manufacturing	Automotive, tires, powertrain	€37.7	236,386
DMG Mori AG	Manufacturing	Machine tools, industrial services	€1.83	6672
Hilti AG	Manufacturing	Power tools, measuring systems, fastening solutions	~€4.98	29,549
Siemens AG	Multi-industry company	Automation, medical technology, power generation	€57.1	293,000
ABB Ltd	Manufacturing	Electrification, process automation, motion, robotics & discrete automation	~€21.85	105,000
Ørsted A/S	Energy	Offshore wind farms, Onshore wind and solar PV, Bioenergy	~€7.07	6179

3.2 Case Studies in the Manufacturing Industry

As selected case studies, the companies in Table 1 present how organizations from the manufacturing sector use Industry 4.0 technologies to facilitate their sustainability strategy. The eight case studies include companies from different sizes which are part of different business segments. The difference in these categories allow an observation from multiple perspectives resulting in a holistic analysis.

4 A Framework for Leveraging the Twin Transition

4.1 Design Principles for Leveraging the Twin Transition

Based on the case studies, three design principles were derived. The design principles with the corresponding solutions and exemplary use cases are shown in Fig. 1.

Design principle	Solutions	Exemplary use cases
Micro level: Leverage eco-efficient operations	• Smart connected products • Eco design • Energy management system • Smart grids	• Schneider Electric Green Premium Program • ABB Ability Suite • Bosch Local Power • Siemens Isabela Island Hybrid Power Plant
Meso level: Facilitate circularity	• Servitization of products • Digital availability of sustainability information • Recycling, Reusing	• Hilti Fleet Management • Schneider Electric EcoFit Programm • ABB in the Circular Initiative
Macro level: Foster value co-creation and flexible collaboration	• Industrial Symbiosis • Corporate Venture Services • Supplier assessment platforms	• Ørsted in the Kalundborg Symbiosis • ABB and TaKaDu • Hilti, Continental and DMG Mori with EcoVadis and IntegritiyNext

Fig. 1 Design principles of leveraging the Twin Transition

Leveraging sustainability through Industry 4.0 in manufacturing companies, the initiatives must be considered on three different levels: the micro, meso and macro level. The micro level observes specific operations and activities within the factory. Stock and Seliger view the micro perspective as the horizontal and vertical integration of value creation modules within a smart factory, which encompass equipment, humans, organization, processes, products and their interactions and interrelationships [17]. On the micro level, manufactures run resource- and emission-efficient operations that are enabled by smart and connected machines, data-based decision-making and flexible processes. The meso level functions as a bridge between the micro and macro level. On the meso level, manufacturing companies facilitate circular economy with the technological opportunities of Industry 4.0. The transition to a circular economy enables manufacturing companies to decouple their economic growth from their resource consumption [18]. This allows companies to exploit the eco-efficiency benefits on the micro level and enhanced collaboration on the macro level. The macro level covers interactions and relationships between stakeholders, products and equipment along the value chain [17]. The third design principle calls for fostering value co-creation and flexible collaboration with relevant stakeholders. Figure 2 illustrates the different levels of analysis in relation to the design principles. In the following chapters, the design principles are outlined in detail.

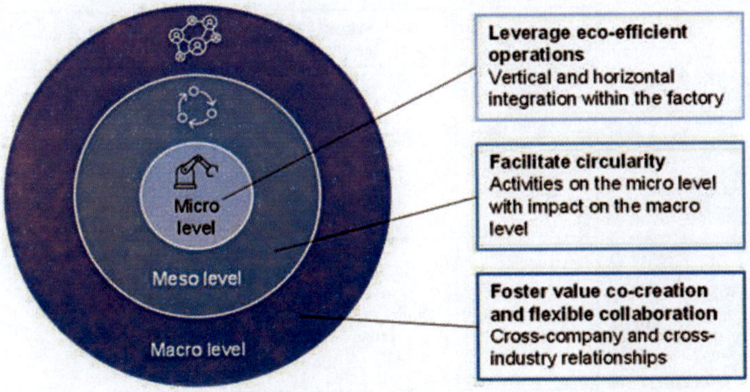

Fig. 2 The design principles cover the entire scope of a company's activities

4.2 *Areas of Action for Leveraging the Twin Transition*

In the following, concrete areas of action are highlighted to specify how manufacturing companies can tackle the Twin Transition along the three design principles. As reference model for continuous transformation within the Industry 4.0, the acatech Industry 4.0 Maturity Model is applied [19]. The four structural dimensions *resources*, *information systems*, *organizational structure* and *culture* represent the entirety of a manufacturing organization [20]. The reference model ensures that the implementation of the design principles for leveraging sustainability will cohere to the holistic digital transformation of a manufacturing company. In conjunction with the design principles, the four structural dimensions form a framework representing Twin Transition initiatives in a manufacturing company at the micro, meso, and macro level. The framework functions as a general structure to systematically categorize industry-proven areas of action for the Twin Transition. The identified eleven areas of action are depicted in Fig. 3.

4.2.1 Resources

In the acatech Industry 4.0 Maturity Index, resources refer to physical, tangible resources [20]. These consist of human resources, machinery, equipment, tools, materials, and the final product. The appropriate design and configuration of technical resources and competencies enable companies to reduce data and action latency, thus maximizing value creation from data [20]. To leverage sustainability in context of the manufacturers' resources, three industrial practices are identified:

- Micro level: Connected products and data-driven services
- Meso level: Product-as a Service
- Macro level: Industrial symbiosis.

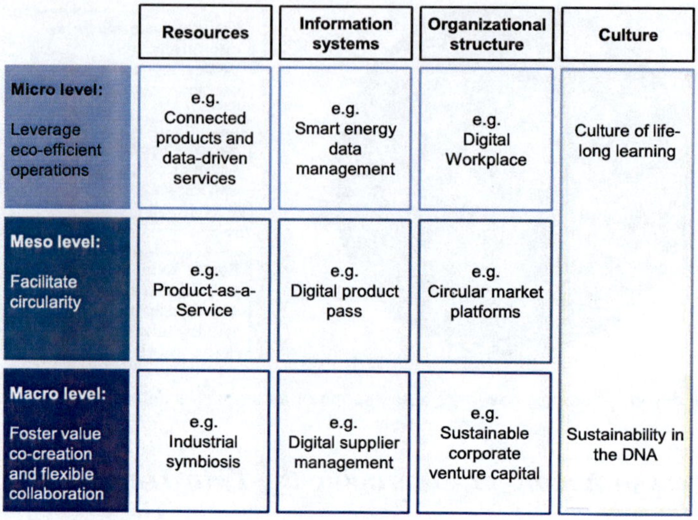

Fig. 3 Framework for leveraging the Twin Transition

Manufacturers achieve eco-efficient operations through the employment of smart, connected products and the complementary data-driven services for their customers. By implementing a Product-as-a-Service business model, manufacturers can increase the utilization and extend the lifetime of their machines, therefore facilitating circularity [21]. Industrial symbiosis enables companies to retain the value of resources by reusing waste and by-products from other companies as raw materials. Since industrial symbiosis enables companies to acquire raw materials from other companies through a conscious collaboration effort, this industrial practice belongs to the structural dimension resources on the macro level [22].

4.2.2 Information Systems

Information systems encompass socio-technical systems that are responsible for the preparation, processing, storage, and transfer of data and the subsequent context-based provision of information. They include information and communication technologies that are critical for companies to ensure availability of information supporting decision-making. The creation of a comprehensive information system requires standardized interfaces, flexibility, openness, comprehensive IT security and appropriate data quality [20, 23]. In the context of sustainability, information systems provide real-time transparency and insights into products, production systems and processes in a supply chain. The following industrial practices enable manufacturers to aggregate the necessary data for sustainable operations throughout the value chain.

- Micro level: Smart energy data management
- Meso level: Digital product pass

- Macro level: Digital supplier management.

Smart energy data management eases companies' transition to renewable energy sources in their factories [24]. The introduction of a digital product passport solves information disparities that otherwise inhibit operations in a circular economy. On one hand, it benefits the manufacturer by enabling tracking and tracing of the products' information throughout its lifecycle. On the other hand, it creates the possibility to share information with relevant stakeholders to facilitate circular strategies such as re-manufacturing, retrofitting or recycling. Digital supplier management allows manufacturers to monitor and influence their suppliers' sustainability practices more efficiently. Therefore, it can directly affect a company's partnership with others [25].

4.2.3 Organizational Structure

While resources and information system represent technological enablers of a digital transformation, the implementation requires a corresponding organizational structure. The organizational structure encompasses an external and internal view on the organization. It refers to the internal structure and operational processes as well as an organization's position in the value creation network. With the right organizational structure, mandatory rules are determined to orchestrate internal as well as external collaboration [20]. As such, the industrial practices corresponding to organizational structure are:

- Micro level: Digital Workplace
- Meso level: Circular market platforms
- Macro level: Sustainable corporate venture capital.

A digital workplace and flexible working models show the ability to reduce carbon emissions by e.g. decreasing travel and commute [26]. Setting up an organizational structure that fosters digital tools and workplaces enables more efficient, digital processes as well as attracting a wider area of workers. On the meso level, circular market platforms allow manufacturers to effectively organize reverse logistics for their end-of-life components and machines. While it enables companies to increase the integration of second-use components and materials, it also opens opportunities to directly cooperate with other actors from the value creation network [27]. Sustainable corporate venture capital is located at the macro level and forms symbiotic partnerships between established companies and external start-ups. While the start-ups benefit from opportunities to scale-up and further develop their solutions, companies gain a new source for sustainable innovation and technologies [28].

4.2.4 Culture

Organizational culture is "*defined as a set of beliefs and values shared by members of the same organization, which influence their behaviors*" [29, 30]. Schein differentiates between three levels of culture: Artifacts, espoused beliefs and values and, underlying assumptions [29]. At the surface level, artifacts are observable and visible features of a company's culture and include e.g., products, architecture, published reports or press releases. Yet, artifacts are ambiguous and hard to decipher. Their actual meaning can only be understood at a deeper level of culture—a company's espoused beliefs and values. Shared beliefs and values include what leaders and employees proclaim on how and why things are visible through the company's strategy, goals, and philosophies. If the beliefs and values are implemented in a way that an organization takes these concepts for granted, they become underlying assumptions. These underlying assumptions form the ultimate source of values and action and dictate how a culture is experienced and lived by employees [29].

Complementary to the organizational structure which covers the hard factors of collaboration, a coherent corporate culture, encompassing the soft factors of collaboration, is needed to align the behavior of a company's employees. An environment to foster collaboration and trust between employees ensures the unrestricted sharing and exchange of knowledge within the company [20].

For manufacturing companies, two components of a sustainable corporate culture in the Industry 4.0 are identified: a culture of life-long learning and sustainability in the DNA. First, a culture of life-long learning ensures the long-term employability of workers in the dynamic paradigm shift of Industry 4.0 [31]. Secondly, companies should embed sustainability in their corporate DNA to align their employee's behavior with the company's sustainability strategy [32]. Both components are relevant at the micro, meso, and macro level, therefore neither component is assigned to a specific design principle. Rather, the culture components originate and permeate throughout the company's organizational structure bottom-up and top-down.

5 Application in the Plastics Industry

The plastics industry presents a valid example to apply the developed framework. The plastics manufacturing industry alone generates large quantities of secondary plastics of various types. With only 13.7% of the plastic used coming from recycled materials Germany, the reuse as secondary plastics shows enormous potential [33].

Although over time, the processual solutions and technologies of the recycling process have been optimized, the reputation of recycled materials remains the same as before: low quality, contaminated and not suitable for reuse in new products. This is largely due to a lack of transparency and information regarding the composition and origin of recycled materials. In addition to missing data in traceability, there is a lack of trust between the various actors along the plastic recycling ecosystem.

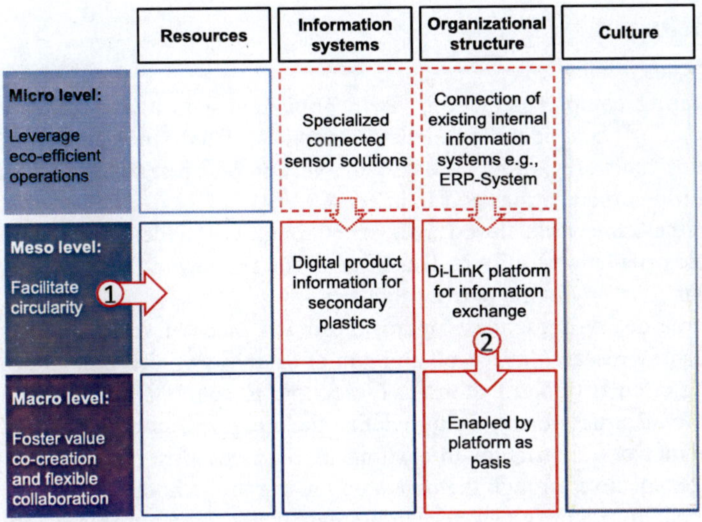

Fig. 4 Applied framework for leveraging the Twin Transition at Di-Link

The question on how to leverage sustainability through Industry 4.0 in the plastic industry can be answered using the framework. To facilitate circularity of industrial plastic waste, a solution at the meso-level is needed (cf. 1 in Fig. 4).

Considering these obstacles, Di-Link, a research project at the FIR at RWTH Aachen University developing an information platform, offers a solution through the following value propositions: providing information on secondary plastic materials, controlling the quality of the materials before moving them along the circular economy, playing the consultant role and matchmaking the different materials between actors along the platform.

Applying the developed framework, Di-Link involves not only an organizational structure, but also the structural dimension of information systems. The platform not only collects information about second life materials in a product passport but provides an infrastructure for comprehensive data exchange. As shown in Fig. 4 to facilitate a circular economy, it is necessary to work with information systems and structural organizations at the micro level. The use of the micro level enables benefits at the meso level.

In this way, using specialized connected sensor solutions, it is possible to increase the internal efficiency of data collection. On the micro level, process improvements can be achieved by e.g., the use of internal Enterprise-Resource-Planning systems.

With Di-Link as information platform, which is available for all actors of the circular ecosystem through a specific software, it is possible to reach the macro level in the Framework (Fig. 4, Number 2). At this level, the formerly sole companies that have been part of a linear value chain are now connected within a circular ecosystem. Here, different value chains coexist and coopetition is enabled with Di-Link acting as one of the central digital platforms.

6 Conclusion

Manufacturing companies tackling Twin Transition gain a framework to streamline their efforts and identify relevant strategies. This paper creates clarity and comprehensiveness regarding the emergent, yet complex research area of sustainable manufacturing organizations and Industry 4.0.

Within the framework, eleven areas of action have been identified and can be used as strategic orientation for Twin Transitions. The framework has been validated on the circular economy in the plastics industry.

The three design principles (micro, meso and macro level) have been derived in a case study research approach and aim at establishing eco-efficient operations, facilitating circularity and fostering co-creation. Resources, Information Systems, Organizational structure and Culture define the areas of action within the levels.

The framework contributes to existing literature by observing the topic from a practical perspective through the case study research methodology. The framework helps scholars to identify new research areas, while it encourages practitioners as a decision support for the systematic and holistic implementation of sustainability with Industry 4.0. Nevertheless, there are limitations to the capability of Industry 4.0 to foster industrial sustainability.

Industry 4.0 and sustainable development are two extensive transformation processes that beyond the technical implementation, require socio-economic considerations and integrative change processes. While this framework offers an orientation for Twin Transition initiatives, it does not support the implementation of comprehensive transformations or industrial practices. Also, the framework is aimed at the manufacturing industry and therefore offers limited support for other sectors. Future research should focus on defining each area of action in greater detail by estimating its potential impact on sustainable development and derive approaches to implement industrial practices.

Acknowledgements The presented research is part of the results of the research project "Di-Link: Digitale Lösungen für industrielle Kunststoffkreisläufe" funded by the Bundesministerium für Bildung und Forschung (BMBF), under the funding reference number 033R235C. The responsibility for the content of this publication lies with the authors. The authors would like the BMBF for the kind support within this research project.

References

1. Juschkat, K.: Klimaschutz Und Industrie – Wie Geht Das? [Online]. Available: https://www.konstruktionspraxis.vogel.de/Klimaschutz-Und-Industrie-Wie-Geht-Das-A-848631/. Accessed: 6 June 2021
2. World Economic Forum (Wef): [White Paper] Global Lighthouse Network: Insights from the Forefront of the Fourth Industrial Revolution. Mckinsey & Company; World Economic Forum (Wef), Geneva, Switzerland, 2020. Accessed: 10 June 2021. [Online]. Available: http://www3.weforum.org/Docs/Wef_Global_Lighthouse_Network.pdf

3. Bonilla, S., Silva, H., Da Terra Silva, M., Franco Gonçalves,R., Sacomano, J.: Industry 4.0 and sustainability implications: a scenario-based analysis of the impacts and challenges. Sustainability **10**(10), 3740–3764 (2018). https://doi.org/10.3390/Su10103740
4. Portier, C., Frey, B., Midgley, L.: Business reporting on the SDGs: integrating the SDGs into corporate reporting: a practical guide. Global Reporting Initiative (Gri); United Nations Global Compact (Un Global Impact); PWC (2018)
5. IPCC: Climate Change 2021: The Physical Science Basis. Intergovernmental Panel On Climate Change (2021). Accessed: 30 Sept 2021. [Online]. Available: https://www.ipcc.ch/Report/Ar6/Wg1/Downloads/Report/Ipcc_Ar6_Wgi_Full_Report_Smaller.pdf
6. European Commission: The European Green Deal (2019). [Online]. Available: https://ec.europa.eu/Info/Sites/Default/Files/European-Green-Deal-Communication_En.pdf
7. Mcwilliams, A., Parhankangas, A., Coupet, J., Welch, E., Barnum, D.T.: Strategic decision making for the triple bottom line. Bus. Strat. Env. **25**(3), 193–204 (2016). https://doi.org/10.1002/Bse.1867
8. Bundesministerium Für Wirtschaft Und Energie (Bmwi): [Progress Report] Shaping Industrie 4.0: Autonomous, Interoperable and Sustainable. In: 2019 Progress Report (2019). Accessed: 14 May 2021. [Online]. Available: https://www.plattform-I40.de/Pi40/Redaktion/En/Downloads/Publikation/2019-Progress-Report.Pdf?__Blob=Publicationfile&V=7
9. Auer, J.: [Report] Industry 4.0—Digitalisation to Mitigate Demographic Pressure. Deutsche Bank Research, Frankfurt Am Main (2018)
10. Voigt, K.-I., Kiel, D., Müller, J.M., Arnold, C.: Industrie 4.0 Aus Perspektive Der Nachhaltigen Industriellen Wertschöpfung. In: Bär, C., Grädler, T., Mayr, R. (eds.) Digitalisierung Im Spannungsfeld Von Politik, Wirtschaft, Wissenschaft Und Recht, pp. 331–343. Springer Berlin Heidelberg, Berlin, Heidelberg (2018)
11. Plattform Industrie 4.0: Sustainable production: actively shaping the ecological transformation with Industrie 4.0: Impulse Paper Task Force Nachhaltigkeit. Berlin (2020). Accessed: 1 June 2021. [Online]. Available: https://www.plattform-I40.de/Pi40/Redaktion/En/Downloads/Publikation/Sustainable-Production.Pdf?__Blob=Publicationfile&V=2
12. Kamble, S.S., Gunasekaran, A., Gawankar, S.A.: Sustainable Industry 4.0 framework: a systematic literature review identifying the current trends and future perspectives. Process Saf Environ Prot **117**, 408–425 (2018). https://doi.org/10.1016/J.Psep.2018.05.009
13. Parida, V., Sjödin, D., Reim, W.: Reviewing literature on digitalization, business model innovation, and sustainable industry: past achievements and future promises. Sustainability **11**(2), 391–409 (2019). https://doi.org/10.3390/Su11020391
14. Boos, W., Kelzenberg, C., Helbig, J., Busch, M., Graberg, T., Schweins, J.: [White Paper] Wettbewerbsfaktor Nachhaltigkeit - Ein Differenzierungsmerkmal Für Den Werkzeugbau. Wba Aachener Werkzeugbau Akademie Gmbh; Werkzeugmaschinenlabor Wzl Der Rwth Aachen, Aachen (2020)
15. Eisenhardt, K.M.: Building theories from case study research. Acad. Manag. Rev. **14**(4), 532 (1989). https://doi.org/10.2307/258557
16. Eisenhardt, K.M., Graebner, M.E.: Theory building from cases: opportunities and challenges. Amj **50**(1), 25–32 (2007). https://doi.org/10.5465/Amj.2007.24160888
17. Stock, T., Seliger, G.: Opportunities of sustainable manufacturing in Industry 4.0. Procedia CIRP **40**, 536–541 (2016). https://doi.org/10.1016/J.Procir.2016.01.129
18. Ellen Macarthur Foundation: [Report] Towards The Circular Economy: Economic And Business Rationale For An Accelerated Transition. Ellen Macarthur Foundation 1, 2013. Accessed: 1 June 2021. [Online]. Available: https://www.ellenmacarthurfoundation.org/Assets/Downloads/Publications/Ellen-Macarthur-Foundation-Towards-The-Circular-Economy-Vol.1.pdf
19. Schuh, G., Anderl, R., Dumitrescu, R., Krüger, A., Ten Hompel, M.: Industrie 4.0 Maturity Index: Managing the Digital Transformation of Companies—Update 2020. Acatech – Deutsche Akademie Der Technikwissenschaften E.V., Acatech Study (2020)
20. Schuh, G., Anderl, R., Dumitrescu, R., Krüger, A., Ten Hompel, M.: Using the Industrie 4.0 Maturity Index in Industry: Current Challenges, Case Studies and Trends. Acatech – Deutsche Akademie Der Technikwissenschaften E.V., Acatech Cooperation (2020)

21. Han, J., Heshmati, A., Rashidghalam, M.: Circular economy business models with a focus on servitization. Sustainability **12**(21), 8799 (2020). https://doi.org/10.3390/Su12218799
22. Domenech, T., Bleischwitz, R., Doranova, A., Panayotopoulos, D., Roman, L.: Mapping industrial symbiosis development in Europe_typologies of networks, characteristics, performance and contribution to the circular economy. Resour. Conserv. Recycl. **141**(7), 76–98 (2019). https://doi.org/10.1016/J.Resconrec.2018.09.016
23. Schuh, G., Frank, J., Jussen, P., Rix, C., Harland, T.: Monetizing Industry 4.0: design principles for subscription business in the manufacturing industry. In: Co-creating Our Future: Scaling-Up Innovation Capacities Through the Desgin and Engineering of Immersive, Collaborative, Empathic and Cognitive Systems: 2019 IEEE International Conference on Engineering, Technology and Innovation (ICE/ITMC) : Sophia Antipolis Innovation Park, France, 17–19 June 2019, Valbonne Sophia-Antipolis, France, pp. 1–9 (2019)
24. Marinakis, V., et al.: From big data to smart energy services: an application for intelligent energy management. Futur. Gener. Comput. Syst. **110**(3), 572–586 (2020). https://doi.org/10.1016/J.Future.2018.04.062
25. Mastos, T.D., et al.: Industry 4.0 sustainable supply chains: an application of an iot enabled scrap metal management solution. J. Clean. Prod. **269**(4), 122377 (2020). https://doi.org/10.1016/J.Jclepro.2020.122377
26. Yalina, N., Rozas, I.S.: Digital workplace: digital transformation for environmental sustainability. IOP Conf. Ser.: Earth Environ. Sci. **456**, 12022 (2020). https://doi.org/10.1088/1755-1315/456/1/012022
27. Berg, H., Wilts, H.: Digital platforms as market places for the circular economy—requirements and challenges. Nachhaltigkeitsmanagementforum **27**(1), 1–9 (2019). https://doi.org/10.1007/S00550-018-0468-9
28. Hegeman, P.D., Sørheim, R.: Why do they do it? Corporate venture capital investments in cleantech startups. J. Clean. Prod. **294**, 126315 (2021). https://doi.org/10.1016/J.Jclepro.2021.126315
29. Schein, E.H.: Organizational Culture and Leadership, 2nd edn. Jossey-Bass, San Francisco (1992)
30. Erez, M., Gati, E.: A dynamic, multi-level model of culture: from the micro level of the individual to the macro level of a global culture. Appl. Psychol. **53**(4), 583–598 (2004). https://doi.org/10.1111/J.1464.-0597.2004.00190.X
31. Schuh, G., Frank, J.: Maturity-based design of corporate culture in the context of Industrie 4.0. In: 2020 International Conference on Technology and Entrepreneurship—Virtual (ICTE-V), 1–8 (2020). https://doi.org/10.1109/Icte-V50708.2020.9113784
32. Walkiewicz, J., Lay-Kumar, J., Herzig, C.: The integration of sustainability and externalities into the "corporate DNA": a practice-oriented approach. Cg, Ahead-Of-Print, Ahead-Of-Print, P. 333 (2021). https://doi.org/10.1108/Cg-06-2020-0244
33. Lindner, C., Schmitt, J., Hein, J.: Stoffstrombild Kunststoffe In Deutschland 2019: Kurzfassung Der Conversio Studie (2020). [Online]. Available: https://www.plasticseurope.org/de/Resources/Publications/3377-Stoffstrombild-Kunststoffe-Deutschland-2019-Kurzfassung

Lukas Stratmann holds an M.Sc. RWTH degree in Mechanical Engineering and Business Administration from the RWTH Aachen University. He is currently doctoral candidate at the FIR Institute for Industrial Management. His specific focus is in the area of the twin transition regarding digitization and sustainability for the manufacturing industry.

Prof. Stich is head of the Institute for Industrial Management (FIR) at the RWTH Aachen University since 1997. Prof. Dr.-Ing. Volker Stich worked for the St. Gobain Automotive Group for 10 years and led the management of European plant logistics. In addition, he was responsible for the worldwide coordination of vehicle development projects.

Ruben Conrad is head of the business transformation department at RWTH Aachen University. He joined the department after his studies in mechanical engineering at the RWTH Aachen University. His work experience spans three continents, Asia, North America, and Europe, where he worked with Bayer AG and the University of St. Gallen. He has worked on several research and industry projects with a focus on the design of new value creation models and management systems in various industries.

Gerrit Hoeborn, M.Sc. M.Sc. is head of the ecosystem Design section at RWTH Aachen University. He joined the department after his studies in industrial engineering at the RWTH Aachen University and Tsinghua University. His research focus is business transformation and digital business strategies.

Felix Optehostert, Dr.-Ing. is Senior Manager at the Industry 4.0 Maturity Center. He supports manufacturing companies in developing their digital transformation strategy and roadmap. Together with his team, he brings in deep expertise in the areas of digital transformation and data-driven business development. Before working at the I4.0MC, Felix studied mechanical engineering and business administration at the RWTH Aachen University. He worked as a consultant for data-driven service development and led the service engineering group at the Institute for Industrial Management FIR.

Minh Phuc Phong holds an M.Sc. RWTH degree in Mechanical Engineering and Business Administration from the RWTH Aachen University. Previously, he worked as an analyst at the Industry 4.0 Maturity Center and developed expertise in the areas of digital transformation strategy and roadmap development. He is an expert on Industry 4.0 and currently consulting in this field.

Virtual Reality for Interacting with a Manufacturing System Digital Twin—A Case Study

Yafet Haile-Melekot, Karel Kruger, and Jörg Niemann

Abstract Globally, and across several sectors, the Industry 4.0 vision of smart, highly connected systems is starting to become reality. Amongst other key enabling technologies, Industry 4.0 relies on the advancement of Digital Twins (DTs) and Virtual Reality (VR). DTs are accurate virtual representations of the structure, behaviour, and state of physical systems. In recent years, the DT concept has received notable research attention in the field of manufacturing systems as a means to support data-led decision making, optimization and enhanced control. VR entails the creation of immersive virtual, simulated environments, which has great potential for visualization of—and interaction with—DT models and data. The paper describes the development of a VR experience that is integrated with the DT of a physical manufacturing system. The VR experience is developed using HTC Vive hardware and Unity3D software, and integrated with a Fischertechnik Learning Factory 4.0 miniature manufacturing system. The presented VR experience allows users to view, navigate and interact with the model of the physical manufacturing system. Furthermore, real time production data of the physical system can be visualized in the virtual environment and the user can affect the operation of the physical system through control commands. Lastly, the paper discusses the possibilities and challenges for deploying such VR experiences in real manufacturing environments.

Keywords Virtual reality · Internet of things · Digital twin · Manufacturing · Industry 4.0

Y. Haile-Melekot · J. Niemann
Faculty of Mechanical and Process Engineering, University of Applied Sciences Düsseldorf, Düsseldorf, Germany
e-mail: yafet.haile-melekot@study.hs-duesseldorf.de

J. Niemann
e-mail: joerg.niemann@hs-duesseldorf.de

K. Kruger (✉)
Department of Mechatronic Engineering, Stellenbosch University, Stellenbosch, South Africa
e-mail: kkruger@sun.ac.za

K. von Leipzig et al. (eds.), *Smart, Sustainable Manufacturing in an Ever-Changing World*, Lecture Notes in Production Engineering,
https://doi.org/10.1007/978-3-031-15602-1_14

1 Introduction

The networking of smart systems is becoming increasingly important for companies worldwide and across all industries. Towards Industry 4.0, supporting key technologies such as Virtual Reality (VR), Digital Twin (DT), and Internet of Things (IoT) offer the potential to gain decisive knowledge in planning and simulation of processes and operations in the manufacturing environment and thus increase the willingness of industrial companies to invest. Along with important connected industry building blocks such as hardware equipment or connectivity, supporting Industry 4.0 technologies are predicted to have a compound annual growth rate of 37% (see Fig. 1).

One of the reasons for categorizing VR as an Industry 4.0 supporting technology is the ability of VR to create a virtual representation of physical objects. The virtual experience and the functions of VR are predominantly characterized by means of the three I's, which stand for the Immersion, Interactivity, and the Intensity of information [2, 3]. Immersivity describes the degree to which the user is immersed in the virtual world. VR applications, such as the Head Mounted Display (HMD), can be used to view recreated digital representations in a fully immersive environment. The interactivity was mainly shaped by technologically more advanced input/output mechanisms, which increased the degree of interactivity in contrast to standard CAD systems [4]. Information intensity describes the fact that VR systems use information that is human-sense related.

In the manufacturing environment in particular, VR is taking on functions and roles that rightly characterise it as an Industry 4.0 supporting technology. VR can be used as an analysis tool in the early design phase, to simulate interaction with the final product, to monitor and control processes remotely and many other applications [5].

Another important supporting Industry 4.0 technology is that of DT's. DTs are highly accurate digital replicas of physical elements that typically exist in the real

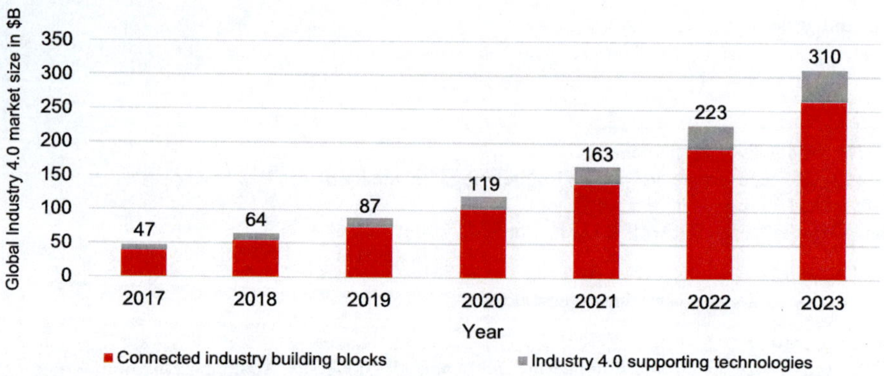

Fig. 1 Global Industry 4.0 market size 2017–2023 according to [1]

world [6]. The DT existing in the virtual environment should be able to reveal the greatest possible behavioural information [7]. In this way DT's can support the design and optimisation in digital factory planning [8]. A special feature of the DT is the possibility of providing it with real-time data from ongoing operations based on an interface connection. Here, a distinction can be made between static and dynamic sources of information. Statistic information includes geometric dimensions, bill of materials or processes, whereas dynamic information can change with the ongoing product life cycle [9]. The extent of the potential is illustrated above all by the increase in effectiveness achieved, which can be attributed to linking DTs with the IoT. In this way, in addition to optimising operational efficiency, maintenance or improved repair measures, the monitoring and analysis of real-time data allows control, strategy and design improvements [10].

The reason for classifying IoT as the most important supporting and key technology in Industry 4.0 is due to its role in networking physical objects in the real world and digital objects in the virtual world [11]. IoT enables the collection of sensor data using smart devices that quickly and accurately transmit information about the physical object that exists in the real world [12]. The communication takes place via wireless connections such as Zigbee, Bluetooth, and Wifi [12]. The most widely used protocol for machine-to-machine communication is the Message Queuing Telemetry Transport (MQTT) protocol, which enables the collection of data from heterogeneous devices and simplifies the remote monitoring of these devices [13]. More than almost any other technology, IoT characterises smart manufacturing and the Industry 4.0 vision.

In this paper, the implementation of integrated IoT, DT, and VR experience will be presented. The focus is on the visualisation of real-time data in a virtual environment. The implementation uses Unity3D as VR software and the HTC Vive as VR hardware, and the Fischertechnik Training Factory Industry 4.0 is used as a case study. The paper is structured as follows: Sect. 2 describes the VR experience. Section 3 explains the real time data integration based on the connection with the local MQTT broker. Section 4 describes the real time data visualisation in the virtual environment and Sect. 5 discusses opportunities and challenges. Finally, Sect. 6 summarizes and evaluates future work possibilities.

2 Virtual Reality Experience

For the implementation of a satisfying virtual experience, a replica of the physical object that is as close to reality conditions as possible is a mandatory prerequisite. Here, not only the resolution but also the size ratio of the DT existing in the virtual world should be considered. Combined with the assurance of a reliable software and hardware environment, this can provide users with the basis for a flawless experience.

Unity3D is a development platform that is primarily used in the gaming sector for game development. However, Unity3D can also be used for other interactive 3D graphics applications and supports augmented reality developments in addition to

VR. Unity3D uses Scenes to realise the development of the virtual environment. Scenes can contain many GameObjects that serve as placeholders for content and can be linked to the objects located in the virtual environment. The GameObjects can be accessed and manipulated by means of so-called scripting. The programming language used here is C#.

The CAD files of the Fischertechnik Training Factory were imported using the Unity plugin Pixyz and scaled. Due to Pixyz's compatibility with Unity, imports of high resolution and large CAD files can be carried out without any problems.

The creation of the virtual environment is also strongly influenced by SteamVR. The SteamVR software development kit (SDK) is a library provided for free by the software developer Valve. SteamVR SDK is also a Unity plugin and can be accessed and imported at any time from the Unity asset store alongside Pixyz. As one of the few HMDs, the HTC Vive is supported by the SteamVR SDK and provides the Player GameObject. The Player GameObject is necessary to detect, place, track, and navigate the HMD and the controllers in the virtual environment. In addition to HMD detection, the VR experience is to be enriched by the most flexible locomotion types possible. The VR-locomotion types are divided into motion-based, roomscale-based, controller-based, and teleportation-based [14].

In the context of the virtual experience implemented here, a mix of roomscaled-based and teleportation-based locomotion is used, which includes locomotion by walking and teleportation by pressing the controller button. The SteamVR plugin provides a Teleportation script, which like the Player script can be placed in the Scene as a GameObject (see Figs. 2 and 3). The HTC Vive controllers have a total of 4 actuation buttons (1, 3, 7, 8), a tracking sensor (6) and a trackpad (2) (see Fig. 4). The button necessary for teleportation in the virtual environment is limited to button number 2.

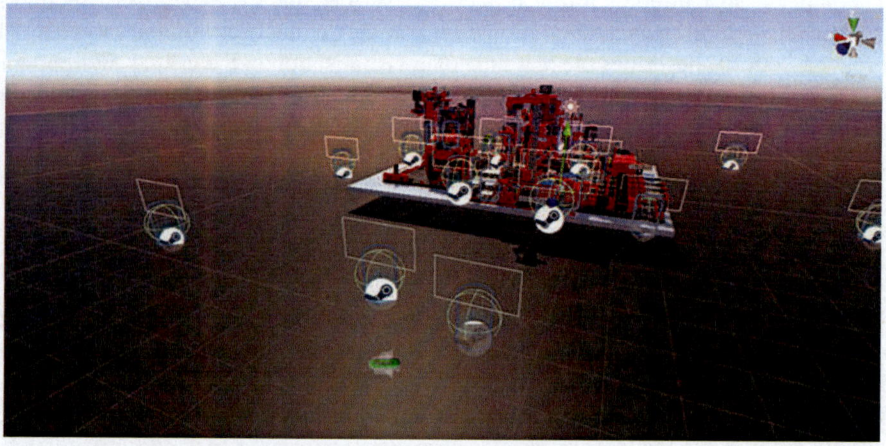

Fig. 2 GameObject placements in the scene

Fig. 3 Teleporting during simulation

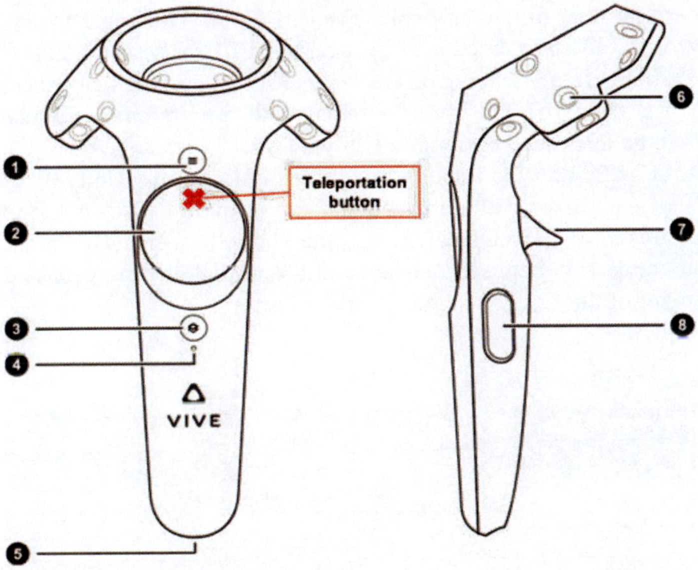

Fig. 4 Teleporting button on HTC Vive controller [15]

3 Real-Time Data Integration

The Fischertechnik Training Factory Industry 4.0 can be used to simulate smart manufacturing processes. The factory can produce nine products, divided into three products each in the colours red, white, and blue. In addition, a built-in environmental sensor enables the measurement of air temperature, humidity, air pressure, air quality and brightness. The operations of the Fischertechnik Training Factory can be controlled and displayed in the Fischertechnik Cloud provided for this purpose. The operations include starting the order process or displaying the measured data from the environmental sensor.

The Fischertechnik Training Factory consists of five subsystems: the high-bay warehouse (HBW), the vacuum gripper (VGR), the sorting system (SLD) and the multi-processing station (MPO). Each of these components has a TXT Controller.

The respective TXT Controllers are networked with each other and communicate using the MQTT message protocol. A central role is taken by the main TXT Controller, which simultaneously assumes the function of MQTT Cloud Client and MQTT Broker (see Fig. 5). Its role as an MQTT broker is made clear by the fact that the data of all TXT Controllers of the Fischertechnik subsystems are sent to the main TXT Controller. The MQTT Cloud Client role of the main TXT Controller on the other hand is characterised by sending the data to a remote MQTT Broker. The data is made available to the Fischertechnik Cloud via the remote MQTT Broker and can be visualised using the Fischertechnik Cloud Dashboard.

To access the real-time data of the Fischertechnik Training Factory, it is first necessary to establish a connection to the local MQTT Broker. The M2MQTT library, which is available free of charge on Github, is used to establish a connection to the local MQTT Broker [16]. The library is based on the programming language C# and can therefore be integrated easily into Unity3D.

The M2MQTT library is imported by inserting it into the Unity3D asset folder and can then be accessed within the Unity3D application. It also contains C# scripts that can be linked to GameObjects within the Unity3D scene. This is important to establish a connection between the scene in the virtual environment and the local MQTT Broker of the Fischertechnik Training Factory.

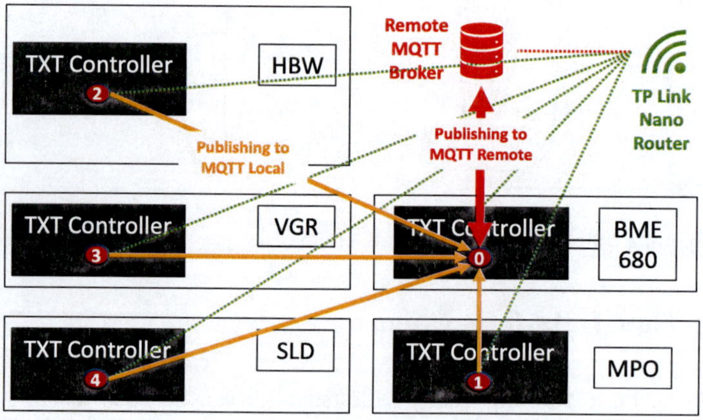

Fig. 5 Fischertechnik block diagram according to [17]

The communication takes place via the local MQTT Broker, which manages the subscribers and receives the messages from the publishers. The messages from the publishers are filtered, stored, and forwarded by the broker to the corresponding subscribers. Therefore, the first step is to subscribe to the published data sets of the environmental sensor within the M2MQTT script, which can also be determined by means of the use of the MQTT Explorer. The prerequisite here is the prior dialing into the IP address of the local MQTT broker (TXT Controller 0). This can be entered within Unity3D in the user interface of the M2MQTT GameObject.

In order to establish a connection from the Unity3D Scene to the Broker, it is also necessary to dial into the Fischertechnik TP Link Nano Router from the computer to be used. This can be done via the computer's common Wi-Fi connection. After the successful connection, the Unity3D environment or the computer to be used takes on the client role.

In addition to the scripts presented so far, such as for teleporting, the SteamVR plugin also includes interaction scripts. By using some of the interaction scripts, it is possible to grab, throw or move objects in the virtual environment. These can be combined with the XR label available in the M2MQTT library. The XR label is a user interface (UI) label and is linked to the M2MQTT script. The UI is specially tailored to VR applications and provides the basis for the successful import of real-time data in the virtual environment. It consists of different layers that interact to enable the visualisation of the subscribed sensor data. The most important layers are the Console Input Field and the Status Panel, which together divide the UI into two sections. The sensor data subscribed by the local MQTT broker are displayed in the Console Input Field. The Status Panel in turn allows conclusions to be drawn about the IP address to be used, the connection status with the local MQTT Broker, the execution of a test publish or the clearing of the Console Input Field.

To make the virtual real-time data experience realistic and rich in interaction, the visualisation of the sensor data should only be triggered when a button is pressed. To implement this, the scripts of the SteamVR SDK can be used and assigned with the M2MQTT script in the compound. These are assigned to the button box. The interactable and hover button scripts link the button to the button box associated with it and identify the button box and button as objects. In addition, the scripts allow the assignment of functionalities depending on the button status (see Fig. 7).

4 Real Time Data Visualization in the Virtual Environment

The real-time data visualisation is predominantly characterised by button actuation in the virtual environment. By pressing the red button (shown in Fig. 6), which is linked to a grey button box, the data collected by the Fischertechnik environmental sensor can be displayed. Pressing and then releasing the button triggers several actions simultaneously (see Fig. 6). By pressing the button, the connection to the local MQTT broker is started, the visualisation of the real-time data records is executed, and the data visualisation window (Console Input Field) is cleared. If the button is released the connection to the local MQTT broker is interrupted. The connection status can be tracked using the status panel. The status panel contains the fields Connect Fischertechnik (green), Disconnect Fischertechnik (red), Text Publish (yellow) and Clear (grey) (see Fig. 8). The status fields light up constantly but turn grey when the respective status becomes true. Next to the real-time data label, the key is shown, which complements the real-time data displayed. This is to ensure that all users to whom the real-time data is displayed can understand exactly what data is being displayed to them. The data displayed includes temperature (t), humidity (h), air pressure (p), air quality (aq), gas resistance (gr) and a time stamp (ts) according to ISO8601. If the button is held down continuously, the captured real-time data from the environmental sensor updates every three seconds. The new measured sensor data sets are listed below the previously measured sensor data set. This allows a direct and clear comparison so that measurement deviations can be easily noticed (Fig. 8).

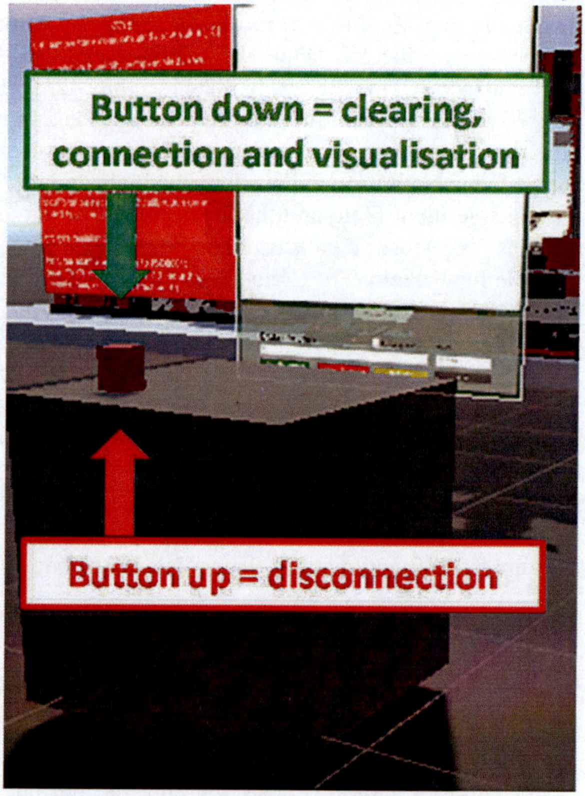

Fig. 6 Functionalities with button usage

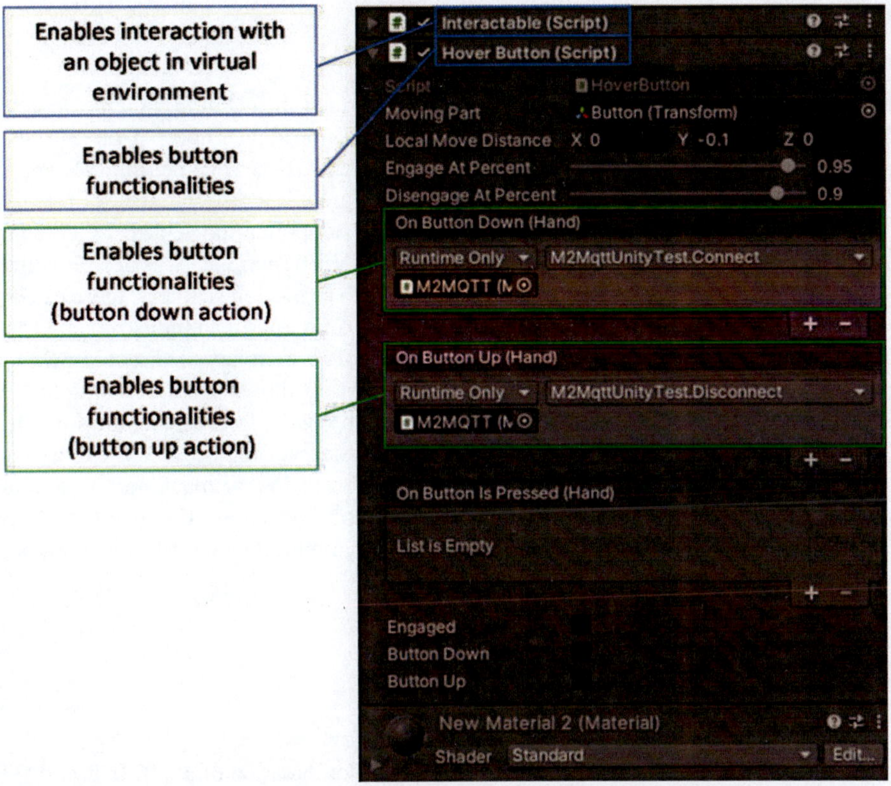

Fig. 7 Overview of interaction and button scripts as well as button assignments

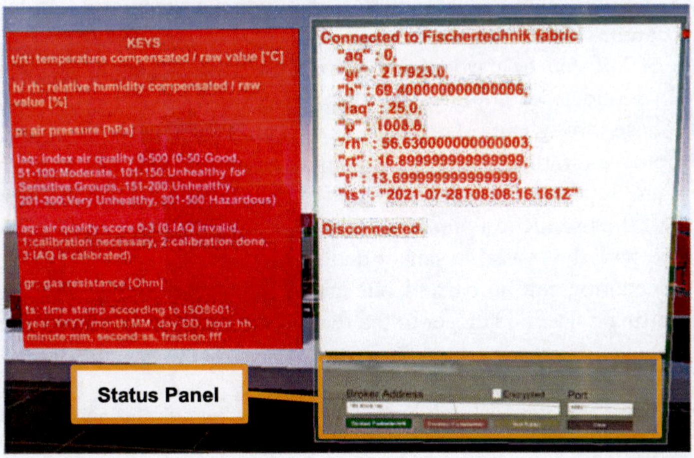

Fig. 8 Real time data label

5 Possibilities and Challenges

SteamVR's interaction scripts can be easily integrated into Unity3D GameObjects, which has helped to put a greater focus on real time data import. Unity3D also impresses with its popularity in the use and implementation of games and VR applications. Due to the large number of users, there is great interest on the part of developers in providing a remedy for the different areas of application. This is reflected in the free availability of the M2MQTT library. The VR experience was accompanied by side effects such as dizziness and nausea. Known as motion sickness, these effects are caused by immersion in a fully immersive environment. In an industrial environment, these can be crucial as health and time factors which have a direct impact on employees and profits. The real-time data visualisation shows that the linking of VR and IoT can already be successfully implemented by using freely available software packages. This is illustrated by the message transfer via the local MQTT broker, which is of great importance in the implementation of IoT or machine-to-machine communications in the manufacturing environment. By scaling the Fischertechnik Learning Factory Industry 4.0 in the virtual environment, the factory can be aligned to manufacturing-like dimensions.

6 Conclusion

This paper outlines the experience around the implementation of a VR, IoT, and DT experience through the usage of the Fischertechnik Learning Factory Industry 4.0. The implementation is mainly driven by the interaction of the SteamVR SDK and the M2MQTT library. The implementation proves that a successful VR, IoT, and DT experience can be realised with few resources.

The use of VR can be particularly helpful for simulations in production environments. In product development, more precise knowledge can be gained during the design phase through simulations close to reality. Another way to use VR is for training employees in the manufacturing environment. A DT of a production plant can prepare employees for potential dangers for safety training even before the actual completion of the production plant.

Combined with the available sensor data of a cyber physical system, many other simulation scenarios can be carried out in the industrial production environment, which are getting closer and closer to the real production process due to the IoT data flow.

Acknowledgements The financial assistance of the National Research Foundation (NRF) towards this research is hereby acknowledged. Opinions expressed and conclusions arrived at, are those of the authors and are not necessarily to be attributed to the NRF.

References

1. IOT Analytics. Available from: https://iot-analytics.com/industry-4-0-and-smart-manufacturing/. Accessed 27 Aug 2021
2. Tsao, Y., Shu, C., Lan, T.: Development of a reminiscence therapy system for the elderly using the integration of virtual reality and augmented reality. Sustainability **11**(17), 4792, 1–10 (2019)
3. Vince, J.: Virtual Reality System. Addison-Wesley, Wokingham (1995)
4. Peng, Q.: Virtual reality technology in product design and manufacturing—the design and implementation of a course for the graduate study. The Canadian Design Engineering Network (CDEN) and the Canadian Congress on Engineering Education (CCEE), Winnipeg (2007)
5. Liagkou, V., Salmas, D., Stylos, C.: Realizing Virtual Reality Learning Environment for Industry 4.0. Procedia CIRP **79**, 712–717 (2019)
6. Redelinghuys, A.J.H., Basson, A.H., Kruger, K.: A six-layer architecture for the digital twin: a manufacturing case study implementation. J. Intell. Manuf. **31**(6), 1383–1402 (2020)
7. Kuhn, T.: Digitaler Zwilling. Informatik Spektrum **40**(5), 440–444 (2017)
8. Niemann, J., Westkämper, E.: Digitale Produktion – Herausforderung und Nutzen. In: Spath, D., Westkämper, E., Bullinger, H.J., Warnecke, H.J. (eds) Neue Entwicklungen in der Unternehmensorganisation, VDI-Buch. Springer Vieweg, Berlin/Heidelberg (2017)
9. Schroeder, G., Steinmetz, C., Pereira, C., Espíndola, D.: Digital twin data Modeling with AutomationML and a communication methodology for data exchange. IFAC-PapersOnLine **49**(30), 12–17 (2016)
10. Kuehn, W.: Digital twins for decision making in complex production and logistic enterprises. Int. J. Des. Nat. Ecodyn. **13**(3), 260–271 (2018)
11. Pistorius, J.: Industrie 4.0—Schlüsseltechnologien. Springer, Berlin/Heidelberg (2020)
12. Souza, V., Cruz, R., Silva, W., Lins, S., Lucena, V.: A digital twin architecture based on the industrial internet of things technologies. In: 2019 IEEE International Conference on Consumer Electronics (ICCE), pp. 1–2 (2019)
13. Jain, V., Chatterjee, J., Kumar, A., Rathore, P.: Internet of Things and Machine Learning in Agriculture. De Gruyter, Berlin/Boston (2021)
14. Boletsis, C.: The new era of virtual reality locomotion: a systematic literature review of techniques and a proposed typology. Multimodal Technol. Interac. **1**(4), 1–17 (2017)
15. Circuitstream. Available from: https://cdn.circuitstream.com/uploads/2019/09/OculusControllersViveControllers-min.png. Accessed 4 Sept 2021
16. Patierno, P.: M2Mqtt. Available from: https://m2mqtt.wordpress.com. Accessed 5 Sept 2021
17. Fischertechnik. Available from: https://www.fischertechnik.de/-/media/fischertechnik/fite/service/elearning/lehren/lernfabrik/fabrik_2019_englisch_neu.ashx. Accessed 5 Sept 2021

Yafet Haile-Melekot holds a B.E. in Industrial Engineering from the Cologne University of Applied Sciences. Currently he is studying International Industrial Engineering (M.Sc.) at the University of Applied Sciences Düsseldorf.

Karel Kruger obtained his Ph.D. from Stellenbosch University, South Africa. He is a lecturer at the Department of Mechanical and Mechatronic Engineering and is part of the Mechatronics, Automation and Design research group.

Jörg Niemann obtained his Ph.D. degree in Manufacturing Engineering from the University of Stuttgart. He is Professor in Business Management and Mechanical Engineering at the Düsseldorf University of Applied Sciences, Germany.

Accuracy in Digital Twinning; An Exploration Based on Asset Location

Eric Lutters and Roy Damgrave

Abstract Digital twins usually aim to develop a correct, encompassing, and well-aligned representation of any reality under consideration. However, digital twins can impossibly mimic actuality exactly; hence, it is essential to identify and find ways to deal with the inaccuracies involved in delineating an object/aspect system in the real world for the different perspectives involved. From a digital twinning approach, it is explored what the notion accuracy entails in asset loca(lisa)tion and in the digital models involved. Here, the overarching approach is to re-interpret the role that data and accuracy play—from a pervasive need for (more) detail, to instruments in purposeful decision making. This implies that the focus moves from achievable accuracy to required accuracy, allowing for different means to deal with accuracy and data resolution. An approach based on octrees demonstrates this, to stress that 'more data is not always better'.

Keywords Digital twinning · Accuracy · Manufacturing environment · Asset location

1 Introduction

Engineers traditionally rely on models and modelling efforts in well-nigh every aspect of their work. And always, they have been very aware of the fact that any model is a mere partial and imperfect representation of reality. In fact, the deficiencies of models and model-making have become ingrained in the experience and education of engineers in a wide variety of disciplines. However, still engineers are regularly

E. Lutters (✉) · R. Damgrave
Design Engineering, University of Twente, Enschede, The Netherlands
e-mail: e.lutters@utwente.nl

R. Damgrave
e-mail: R.G.J.Damgrave@utwente.nl

E. Lutters
Industrial Engineering, Stellenbosch University, Stellenbosch, South Africa

K. von Leipzig et al. (eds.), *Smart, Sustainable Manufacturing in an Ever-Changing World*, Lecture Notes in Production Engineering,
https://doi.org/10.1007/978-3-031-15602-1_15

confronted with the observation that reality does not adhere to their model(s). Such perceptions are instrumental, as they help engineers to be and remain aware of the limitations of their models, but also to advance their models and to meliorate their modelling skills.

Nowadays, engineering models are often labelled as digital twins. In many cases, however, a digital twin merely represents the structured data realm that can underlie one or more models. Even in abroad view, a digital twin is, essentially, an advanced engineering model. Nevertheless, it seems that digital twins are more and more treated as potentially ultimate and conclusive representations of reality. There are two main reasons for this. Firstly, digital twins are rooted in data, where often no a-priori relations between the data content, types and sources are available. Hence, digital twins tend to emerge from data, rather than having an envisaged structure as their basis. This may significantly reduce any engineer's ability to overview and purposefully employ a digital twin. Secondly, the staggering digitalisation (potential) observed in industry and academia may (implicitly) feed a tendency to believe that harvesting more data yields a better digital twin. Be it a fear-of-missing-out, or a belief in progress, companies run a severe risk of becoming 'data rich, but insight poor' (DRIP) [1]. Moreover, many companies accumulate vast stores of data they have no idea what to do with, and no hope of learning anything useful from [2]. Fundamentally, however, digital twins are not necessarily more resistant to a profusion of data than the 'traditional' engineering models are. In the race for digitalisation, the essence and existence of inaccuracy is often compromised in how digital twins are established and used. Here, the sheer availability of data is all too often seen as the panacea for all abeyant factors.

This publication focuses on the accuracy of digital twins, and on how uncertainties can be addressed. Given the extremely broad application area of digital twins, the publication will not depict an overall answer or approach. Rather, it focuses on elementary aspects of uncertainty, and exemplifies how these play a role in asset loca(lisa)tion in factory environments. This case study leads to reflections that may enable an alternate way of addressing accuracy in digital twins. In this, a 'more data is better' paradigm will certainly not be leading.

2 Digital Twinning

Establishing a digital twin (DT) usually aims to develop a twin into a correct, encompassing, and well-aligned representation of the reality under consideration. This may be partially possible for systems that are quite limited in scale, scope, number of perspectives and complexity -quite like 'traditional' modelling. Here, simulations, examinations, or disquisitions based on the digital twin may be adequately transferable to reality. However, even for the simplest system, it is impractical to prove that a digital twin fully coincides with reality. The endeavours involved in realising digital twins for complex, dynamic, and emergent systems are intrinsically open-ended. This essentially results in deficient and unfinished approximations of reality,

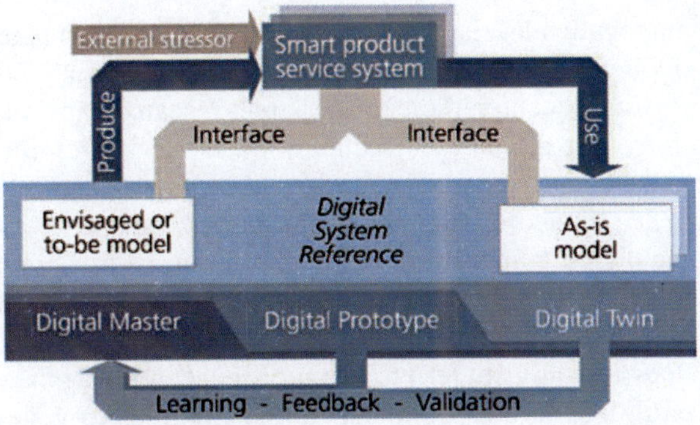

Fig. 1 Digital system reference [3, 4]

rendering the digital twin an interface to the real world that is meaningful and useful, yet imperfect.

Setting aside the many definitions of the notion Digital Twin that are available, here, the focus is more on the intent of the digital twinning concept and the decision making it can support. Hence, a digital twinning framework is used: the digital system reference [3] that simultaneously focuses on the as-is situation ('digital twin'), on the to-be situation ('digital master'), but certainly also on the could-be situation ('digital prototype'). Figure 1 depicts this digital system reference and illustrates how captured reality and designed future are related—connected by a means to explore possible futures [5].

Where the digital master represents the nominal design of the object under consideration, there will always be discrepancies with the real-world data in the digital twin. The digital system reference does not aim to mitigate such discrepancies, but rather to make them explicitly instrumental. Disparities help to evolve currently deficient instantiations of the design, to guide the production process, but also to understand repercussions of design decisions (master) in reality (twin). Even more, the digital prototype enables exploring and exploiting the gap between master and twin to simulate, validate, and verify potential design directions. However, also the digital twin itself contains inaccurate representations of reality. This makes the reliability of the digital twin (e.g., in simulations) less distinct, and hence also obscures the relation between digital twin and digital master. This immediately makes decision making based on the digital twin less effective and efficient.

Accuracy challenges in digital twins stem from incorrect input data (e.g. incorrect sensor data, untimely sampling, algorithmic inconsistencies and inadequate modelling). Simultaneously, however, the need to deal with accuracy challenges, in the context of real-time geometry assurance, are driven and amplified by faster optimisation algorithms, more computer power, and increased amounts of available data [6]. Oftentimes, mitigating strategies aim at 'correcting' the digital twin. For example, multilevel calibration methods in a robot-cell may yield sufficient accuracy to allow for offline planning [7]. However, it can be argued that such approaches aim at resolving symptoms rather than at solving the root cause, especially at higher aggregation levels in the entire system. Overall, it is reasonable to assume that faults in digital twins can compromise intelligent cyber-physical production processes [8].

Consequently. there are apparent research gaps related to (i) real-time identification of the digital twin accuracy status, (ii) self-correcting digital twins with respect to the data feed and its models, and (iii) mechanisms to self-adapt the digital twin [4, 9, 10]. In addressing these gaps, there seems to be an inherent tendency to provide the digital twin with more data, to improve the similarity between the digital twin model and reality. Whenever the digital twin is indeed incomplete or does not cover all aspects or perspectives, this may make sense. However, in many cases, more data does certainly not lead to that more exact replica of reality, nor does more data inherently lead to better understanding. Contradictory, more data may come with more modelling and processing efforts, potentially hampering decision making. Additionally, if more data types are integrated to establish the validity of already existing data, such additional data (sources) come with their own inaccuracies and risks. For example, consequences of inaccurate digital twins in a robot cell may be counteracted by adding more data for operators, providing data-driven security features, or means to prevent or stop propagating/cascading faults. Such measures, however, do not address root causes.

The digital system reference in Fig. 1 focuses on adequately retrieving the data that is required and does contribute to the decision making that is required. Hence, it is based on data-pull rather than on data-push. This extends to the notion inaccuracy: a digital twin should aim to provide the accuracy that is required, not the highest accuracy possible. After all, if the accuracy of the digital twin is higher than what a decision model can use, the quality of the decision will not increase. As an illustration: in industrial practice, myriad examples exist of sensor data that by far exceed (in terms of accuracy, significant numbers or frequency) the requirements in decision making. Where this may provide a false sense of security, it may also obfuscate, and complicate decision making or reduce the effectiveness or efficiency thereof. Hence, rather than pursuing the most and most accurate data, the digital system reference aims to build the digital twin based on adequate data for all perspectives involved, with an accuracy that meets the requirements of the decision making at hand.

3 Asset Loca(lisa)tion

In digital twinning, any measurement (like a sensor on a machine) or estimate (like remaining processing time, or location of an asset) leads to a non-exact representation of values. For some sensors, the effect is one-dimensional, in the sense that the inaccuracy impacts only one value (e.g., temperature). For other situations, like in establishing the location of an asset, the situation is more intricate. Hence, this section illustrates the uncertainty and dependencies that occur in asset location.

Within the scope of a manufacturing environment, the location of assets (be it a product, a machine, a tool, ...) is decisive for the technical and logistic execution of production with adequate quality. Whereas in the physical reality the location of any asset may seem obvious, in digital twinning the location of assets is less equivocal. For example, when materials are on their way to the manufacturing environment, a GPS signal of a truck might periodically represent the material location with an accuracy of a few meters. At arrival at a docking bay, an accuracy of a few centimetres is required, within a certain time window. During internal transport of the material (or any asset), the location data may stem from, e.g., a forklift truck, an automated guided vehicle (AGV), or a real-time positioning system (RTLS) [11]. The accuracy and timeliness of the location vary with the type of transport, but even with the coordinate system in which the location is expressed. At handing-over a component between AGV/robot/.../machine, again a different accuracy is required. In logistics, a transaction-based indication (timestamped placing in a warehouse/bay/slot) or passing a gate at a certain time (e.g., by RFID) may yield sufficient data. Contrarily, if a CNC production machine is involved, location and time need to be profoundly aligned, for example in positioning a tool with respect to a component.

In physical reality, observations may be leading. In a digital twin, however, it is never certain if an asset is exactly at the stipulated location. Even within the limited scope of one production machine this is impossible, due to e.g., setup errors or geometric deviations caused by tolerances or process inaccuracies resulting from earlier process steps. However, especially 'handing-over' from one location reference to another may cause/introduce inaccuracies. Hence, at the production floor, there are multiple frameworks of reference (see Fig. 2), each with their inherent accuracy, but also with overlaps, misalignments, and errors between them. This hampers the use of the digital twin for e.g., tolerance chain analysis or accuracy-related what-if analyses in digital prototypes. During production, twins can be 're-aligned' by measuring stages; it, however, goes without saying that this leads to additional work and even introduces inaccuracies related to the measuring itself.

Moreover, digital twins simultaneously serve different perspectives: accuracy requirements for a CNC-controlled process are certainly different than for logistic decisions. Traditionally, perspectives relied on their own observations, leading to mostly concomitant, parallel, and disconnected systems. Within the context of a digital system reference, it becomes counterintuitive to not integrate such systems.

Fig. 2 Different frames of reference, with inexactness within and between them

This, however, does have repercussions on how different sensoring methods merge to arrive at one unequivocal way of representing the location of an asset at a certain point in time.

4 Accuracy

With asset location as an illustration, a digital twin cannot be expected to be exhaustively complete, encompassing, exact, and unequivocal. Sections 2 and 3 already used notions like accuracy, (un)certainty and precision without specifying their exact meaning and impact. To fathom what makes a digital twin 'accurate', it is relevant to focus on the notions involved.

4.1 Nomenclature

The treatise here is based on ISO 5725-1 [12]. Intended to depict the accuracy of a measurement method, the nomenclature can provide insights for the accuracy in digital twinning. The standard uses the terms 'trueness' and 'precision' to describe the accuracy of a measurement method. 'Trueness' refers to the closeness of agreement between the arithmetic mean of a large number of test results and the true or accepted reference value. 'Precision' refers to the closeness of agreement between test results (see Fig. 3). As derivatives, 'precision error' represents the distribution of random errors; 'bias' is the total systematic error. In terms of asset location, the precision error in a real-time location system is illustrated by a varying readout of the location for a non-moving asset. Bias can be illustrated by a slightly misplaced docking station for an AGV, or by a calibration error in a tool. Both trueness and precision are inherently linked to the notion resolution: the minimum increment a measurement/sensor is capable of. Although obvious, it is quite important to stress that trying to obtain trueness for one observation beyond the available resolution is generally pointless.

Some values suffer from volatility, e.g., by drift: the trend signal may change over time without an impetus in the physical asset. An 'outlier' is a measurement or value that can be disregarded as improbable for the quantity observed. Whereas an outlier is usually easily recognisable for humans, in digital twins that can be more difficult, especially for measurements that happen irregularly or with low frequency. Next to the sheer geometric value, the timeliness of measurements is a separate dimension in accuracy: the frequency or interval, the regularity, but also the latency in a value becoming available each cause misalignments between the digital twin and reality.

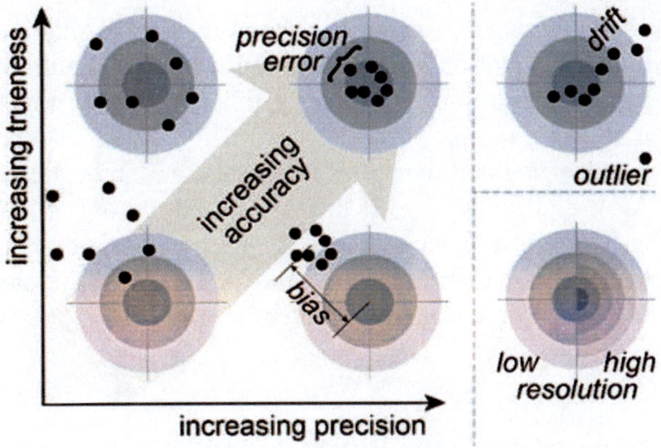

Fig. 3 Accuracy nomenclature (after [12])

4.2 *Accuracy in Loca(lisa)tion*

From a theoretical stance, the nomenclature for accuracy is clear. However, there are some phenomena in real world physical systems that bring along consequential repercussions. Firstly, the definition of trueness refers to a 'reference value', which, in asset location, may be an inoperable intellection. Oftentimes, no 'nominal position' is available that could serve as that reference. Moreover, in considering the many frames of references that play a role (see Fig. 2), a nominal position would depend on which frame of reference (or co-ordinate system) is applied. After all, it is cumbersome to work with an overarching co-ordinate system that is reliable for an entire production facility. The reasons for this encompass, e.g., (i) frames of reference of assets dynamically transform and rotate with respect to each other, (ii) there is a mix of a.o. Cartesian (e.g. milling machine) and polar (e.g. robot) co-ordinate systems, (iii) different frames of reference may rely on different timings, and iv) the accuracy within one co-ordinate system may rely on the location in that system.

Different perspectives and applications impose different restrictions on frequency, intervals, timeliness and latency. These restrictions are often quite intricate, as they can be directly related to safety [8] or product quality [13]. With that, the absence of an overarching co-ordinate system renders all frames of reference relative to each other. Nonetheless, between groups of frames of reference, hierarchical relations can exist—either permanently or temporarily. For example, the reference for a robot gripper will depend on the reference of the robot pedestal; a pallet may provide a temporal parent reference for a product that is clamped on it. However, especially in handing-over between different references that are not hierarchically dependent (i.e., AGV to robot to … to machine to … to warehouse), there will always be overlap and various types of misalignments. On such occasions, it is certainly not always clear which frame of reference is dominant. Moreover, the handing-over is actually characterised by uncertainty, ambiguity and volatility at the same time. At such moments, a digital twin is confronted with at least two aberrant 'truths', potentially leading to inconsistent or incorrect decision making. Such situations are often addressed by making the hand-over more robust, e.g., by using mechanical stops or guides. Not only do such solutions have inherent inaccuracies, but they are also foremost a solution for larger batches or mass production. A more resilient solution is to, for example, use vision systems. These also come with inherent inaccuracy but may be instrumental in facilitating and interpreting the transfer between, say, an AGV and a machine. If the vision system, over time, can become instrumental in learning from inaccuracies (e.g., in compensating bias over multiple observations), handovers may even become antifragile [14]. Still, also a vision system adds its own frame of reference, with inherent inexactness.

5 Accuracy in Digital Twinning

Based on the depictions in Sects. 2–4, digital twinning is about dealing with open-ended systems or environments. Moreover, the level of coincidence between twin and reality is not necessarily increased by adding more data. This is especially true if the added data does not align with the perspective (logistic, quality control, …) that requires more alignment, or the data accuracy exceeds the resolution available. More fundamentally, however, is the fact that a digital twin only needs to provide data at the level where it still improves decision making. For a one-dimensional physical quantity, this implies the distinction between merely signalling a threshold value versus high frequency capturing sensor data with high trueness. For asset location it implies that, rather than impetuously capturing high accuracy locations of all assets with high frequency, it is better to retrieve/produce the location data that is required by different perspectives with the accuracy and timeliness required.

This is where a paradigm shift in thinking about accuracy in digital twinning may emerge: rather than aiming for statistical significance and averaging many values (as mentioned in the definition of trueness), the probability of data being conclusive can drive decision making. This means that the digital twin should provide stakeholders or perspectives with data that does not exceed the accuracy they need or can use in decision making. With this, a production planner, a quality manager, and a warehouse manager may receive differently expressed location data on the same asset at the same time. This disentangles the interfering frames of reference (Fig. 2) for most perspectives and allows for (temporarily) assigning dominant co-ordinate systems for relevant perspectives.

The essence of this approach is that the accuracy required by the location request is converted into a probabilistic occupancy estimate. This estimate considers if the geometric entity under consideration is at a certain location at a certain time, with a specified tolerance. Mathematically speaking, a 4D space (3 spatial dimensions and time) is subdivided in sections that meet the accuracy of the request. Stated differently, rather than providing a location with the highest possible accuracy, the 4D space answers whether the geometry entity is encapsulated in a specific region. Where the size of the regions can be adjusted to the type of request, the required accuracy of the request can be converted to the resolution of regions in the 4D space.

For a logistic planner the resolution of the 4D space can be significantly lower than for a quality engineer—thus inherently blurring accuracy issues at higher trueness levels.

5.1 Octrees for Asset Loca(lisa)tion

Probabilistic occupancy estimates have already proven quite useful in a wide variety of fields, ranging from obstacle detection in robotics [15] and game engines to rendering in computer graphics. There is quite a variety of mathematical approaches available, but to illustrate the principle, here octrees are applied. Basically, an octree is a tree data structure in which each internal node has exactly eight children. Octrees are most often used to partition a three-dimensional space by recursively subdividing it into eight octants (see Fig. 4). At the lowest level represented, that octant is the resolution of the 4D space mentioned above. In e.g., obstacle avoidance in robotics, the recursive subdivision is done instantaneous, in near-real-time. However, the octree approach is also instrumental for more long-term capturing of geometry. A simplified example is shown in Fig. 5. The advantages of octrees are that they are fast, can function at different scales and resolutions, can adaptively change the accuracy/resolution, and can efficiently represent sparse spaces. Moreover, octants can selectively be subdivided to locally achieve the accuracy/resolution required. An octree starts from one reference point, making it rather effortless to combine them in a way that matches the relative frames of reference in Fig. 2. Moreover, an entire octree can transform/rotate by merely moving its reference point. Consequently, octrees can quickly represent (changing) positions with the level of accuracy required. Using octrees does not imply that the entire production facility with all (moving) assets requires continuous mapping, rather an on-demand (or pull) approach can provide decision makers with the appropriate data. The appropriateness lies in the subset of data, the accuracy and the timeliness of the data. Here, the digital prototype (see Fig. 1) can be instrumental in determining what data would be purposeful and accurate enough. Moreover, it is not even required that all the location data stems from sensoring in the real world. For locations that are less sensitive, less dynamic or sufficiently predictable, octrees can easily assimilate artefacts from the real, modelled and simulated world. This facilitates decision making; it is also valuable in comparing models, extrapolations, and simulations with the real world.

Currently, a setup (see Fig. 6) is established as a proof-of-concept for accuracy research. In this, the 'real world' and the digital twin/prototypes are explicitly used in conjunction. This allows for a probabilistic approach based on octrees. In this setup, different location systems (internal/inherent, RTLS, vision) are combined in a peer-to-peer setup; later, e.g., fencing and RFID can be included as well. Simultaneously, a realistic assembly line is under development that facilitates the interaction of human users with robots, machines, and other assets. In these setups, accuracy is indeed addressed as a constructive element in decision making. The setup in Fig. 6 aims to develop a method to determine the resolution the octree-representation needs

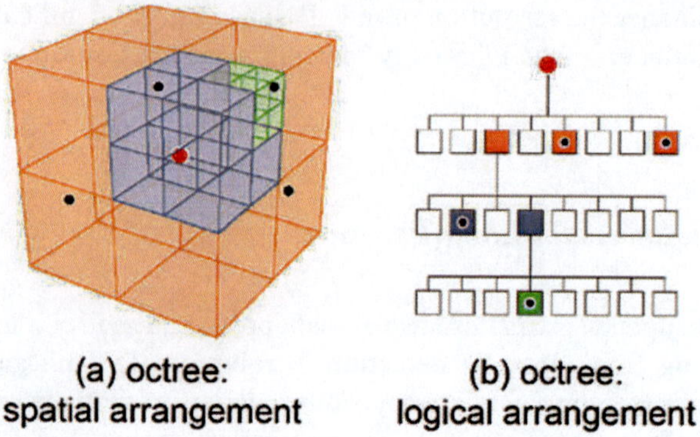

Fig. 4 Octree principle

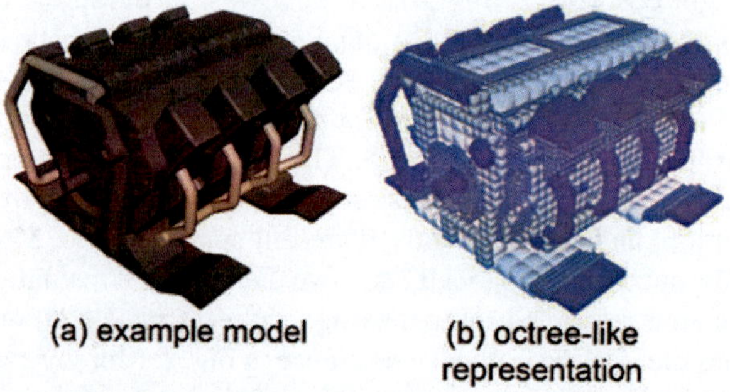

Fig. 5 Perspective dependent and probability driven, octree-like representation

to provide to facilitate decision making for different perspectives. In this, focus is foremost on the hand-overs between e.g., the 'turtlebot' (AGV) and the 3D-printer, to elaborate how the octrees, the digital twin and the actual frames of reference can conjointly lead to resilient and potentially antifragile approaches.

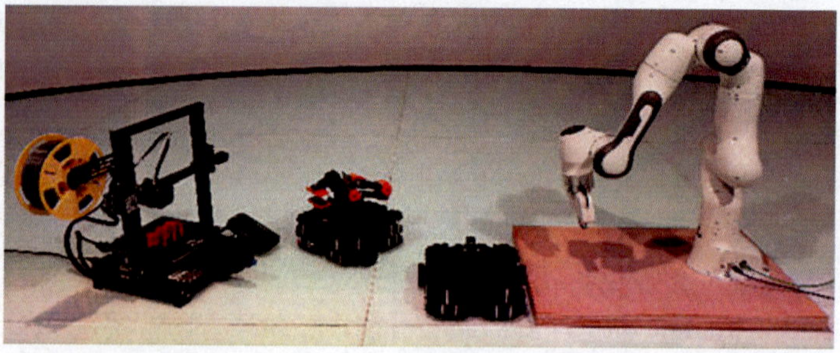

Fig. 6 Initial test setup for octree-based probablilistic accuracy

6 Concluding Remarks

Exploring the uncertainties in asset loca(lisa)tion, both in the real world and in its digital counterparts, straightforwardly expresses that accuracy is not an absolute notion. It vastly differs for different perspectives, but foremost with different decisions. It is argued that the accuracy that is required in decision making should drive the data acquisition strategy in digital twinning. Where the digital twin can never be an all-encompassing and exact replica of reality, the digital twin is essentially open-ended. With that, what priorities and efforts are involved in data acquisition should be driven by the role that data will play in decision making. That does stress that any digital twin indeed yields a de-facto indeterminate model. This, however, underlines that uncertainty is not a limitation of a digital twin, it is a mere characteristic of a digital twin.

Current research aims to elaborate the relation between decision making and required/provided accuracy in the digital twinning approach. Different approaches (like octree) will be explored and combined, with the intent to make the notion inaccuracy an inherent and instrumental characteristic of decision making, rather than something that must be mitigated.

References

1. Peters, T.J., Waterman, R.H.: In Search of Excellence: Lessons from America's Best-Run Companies. Collins Business Essentials. Warner books, New York (1982)
2. Marr, B.: Big Data in Practice. Wiley, Chichester (2016)
3. Lutters, E., de Lange, J., Damgrave, R.G.J.: Virtual dashboards in pilot production environments. In: COMA'19; International Conference on Competitive Manufacturing: Stellenbosch (SA), pp. 22–27 (2019)
4. Tomiyama, T., Lutters, E., Stark, R., Abramovici, M.: Development capabilities for smart products. CIRP Ann. **68**(2), 727–750 (2019)
5. Slot, M., Lutters, E.: Digital twinning for purpose-driven information management in production. Procedia CIRP **100**, 666–671 (2021)
6. Söderberg, R., Wärmefjord, K., Carlson, J.S., Lindkvist, L.: Toward a Digital Twin for real-time geometry assurance in individualized production. CIRP Ann. **66**(1), 137–140 (2017)
7. Erdős, G., Paniti, I., Tipary, B.: Transformation of robotic workcells to digital twins. CIRP Ann. **69**(1), 149–152 (2020)
8. Preuveneers, D., Joosen, W., Ilie-Zudor, E.: Robust digital twin compositions for Industry 4.0 smart manufacturing systems. In: 2018 IEEE 22nd EDOCW, pp. 69–78 (2018)
9. Vrabič, R., Erkoyuncu, J.A., Farsi, M., Ariansyah, D.: An intelligent agent-based architecture for resilient digital twins in manufacturing. CIRP Ann. **70**(1), 349–352 (2021)
10. Abramovici, M., Göbel, J.C., Dang, H.B.: Semantic data management for the development and continuous reconfiguration of smart products and systems. CIRP Ann. **65**(1), 185–188 (2016)
11. Thiede, S., Sullivan, B., Damgrave, R.G.J., Lutters, E.: Real-time locating systems (RTLS) in future factories. In: 54th CIRP Conference on Manufacturing Systems: Athens (2021)
12. ISO: Accuracy (trueness and precision) of measurement methods and results—Part 1: General principles and definitions (1994)
13. Kiesel, R., van Roessel, J., Schmitt, R.H.: Quantification of economic potential of 5G for latency critical applications in production. Procedia Manuf. **52,** 113–120 (2020)

14. Kennon, D., Schutte, C.S.L., Lutters, E.: An alternative view to assessing antifragility in an organisation: a case study in a manufacturing SME. CIRP Ann. **64**(1), 177–180 (2015)
15. Hornung, A., Wurm, K.M., Bennewitz, M., Stachniss, C., Burgard, W.: OctoMap: an efficient probabilistic 3D mapping framework based on octrees. Auton. Robot. **34**(3), 189–206 (2013)

Eric Lutters obtained his Ph.D. degree in Mechanical Engineering at the University of Twente, The Netherlands. At that university, he currently holds a position as associate professor in the faculty Engineering Technology. He also is a professor extraordinary at the Industrial Engineering department at Stellenbosch University.

Roy Damgrave obtained his Ph.D. degree in Industrial Design Engineering at the University of Twente. Currently, he is assistant professor in the Department of Design, Production and Management (faculty Engineering Technology) at that university.

Simultaneous Multi-stakeholder Digital Twinning for Anticipated Production Environments

R. G. J. Damgrave and E. Lutters

Abstract Establishing, monitoring, or changing production environments involves significant time, efforts and investments. Digital Twins, Digital Prototypes and Digital Masters, provide different perspectives on these production environments. To utilise these perspectives effectively and efficiently as tooling in the development trajectory and operation of production environments, adequate employment is required. This activity, referred to as 'digital twinning', enables anticipating and experiencing the behaviour of envisaged production environments, while simultaneously reducing technical, logistic, and financial risks. Digital twinning also addresses the alignment between these digital support systems and the production environment, in terms of effectiveness, quality, and configurability. This paper focusses on providing a framework that oversees the configuration possibilities that different configurations of a digital twin and digital prototype have. The framework allows for a functional and structured configuration of these tools, depending on the use condition and demands of the stakeholders.

Keywords Digital twinning · Synthetic environments · Manufacturing environment · Decision making

1 Introduction

The complexity of production environments is growing substantially, due to the vast increase of relations and interdependencies between real assets (e.g., tangible machines) and virtual assets (e.g., simulations of process optimisation). The fast,

---

R. G. J. Damgrave (✉) · E. Lutters
Design Engineering, University of Twente, Enschede, The Netherlands
e-mail: r.g.j.damgrave@utwente.nl

E. Lutters
e-mail: e.lutters@utwente.nl

E. Lutters
Industrial Engineering, Stellenbosch University, Stellenbosch, South Africa

K. von Leipzig et al. (eds.), *Smart, Sustainable Manufacturing in an Ever-Changing World*, Lecture Notes in Production Engineering,
https://doi.org/10.1007/978-3-031-15602-1_16

ample, and multifarious technological developments in the field present manufacturing companies with significant challenges in strategically and tactically outlining the effects and results of anticipated production environments.

Establishing or modifying production environments that benefit from new technological possibilities involves significant time, efforts and investments [1]. Any ability to anticipate or even experience the behaviour of envisaged environments can be instrumental in substantially reducing technical, logistic, and financial risks [2, 3]. In the context of Industry 4.0, this interweaving, or digital transformation, is infused by the technological and connected world through e.g. the research and development in the area of cyber physical systems. To deliberately compare current conditions to potential realities three perspectives on the data are discerned in a so-called Digital System Reference (DSR) [4]:

- digital master (DM) ('to-be'): the definition of the envisaged entity,
- digital twin (DT) ('as-is'): the current/previous condition of an entity,
- digital prototype (DP) ('could-be'): simulations between the design intent and the actual conditions.

The way a DT or DP is visualised or experienced can vary from a complete digitalised environment, up to a complete physical model (Fig. 1), with a gliding scale between these extremes. Not only the level of digitalisation in digital twins and prototypes is variable, also other aspects like the scale of physical assets, the level of interaction, the resolution of simulation or the balance between representation and simulation can vary.

To better understand the consequences of proposed changes or variants, and to obtain a better grip on the uncertainties involved, the use of so-called 'digital prototypes' is proposed. A DP allows the development, testing, improvement, simulation, and upscaling of (parts of) a product or production environment while not hampering primary processes. Additionally, such a systems avoids avoiding investments whenever possible. A DP allows for the evaluation of e.g., alternative process chains, hardware selection and layouts with respect to, among others, quality, reliability, lead time and manufacturing costs in anticipated production environments. Therefore, a DP can be virtual where possible, and physical where required. With that, it can be considered a Synthetic Environment (SE) [5].

There is no predetermined or single form in which a DT or a DP should be presented or used. For every intended purpose, the most appropriate configuration should be considered. The same accounts for the presented state of the production environment: whether this is the current state, an expected state, a future state or a historical state.

Whereas a DT/DP can be a powerful tool in effectively and efficiently prognosticating anticipated solutions, there is a clear risk that the development of DT/DP themselves become a high-risk, complex, lengthy, and expensive endeavour. At the same time, DT/DP development needs to be fast, lean, and effective to allow for tailored (instantiated) solutions. However, it requires expertise from many perspectives, thus quickly becoming an intricate pursuit for something that companies essentially see as a one-off activity. Nonetheless, in explicitly not regarding the development of a

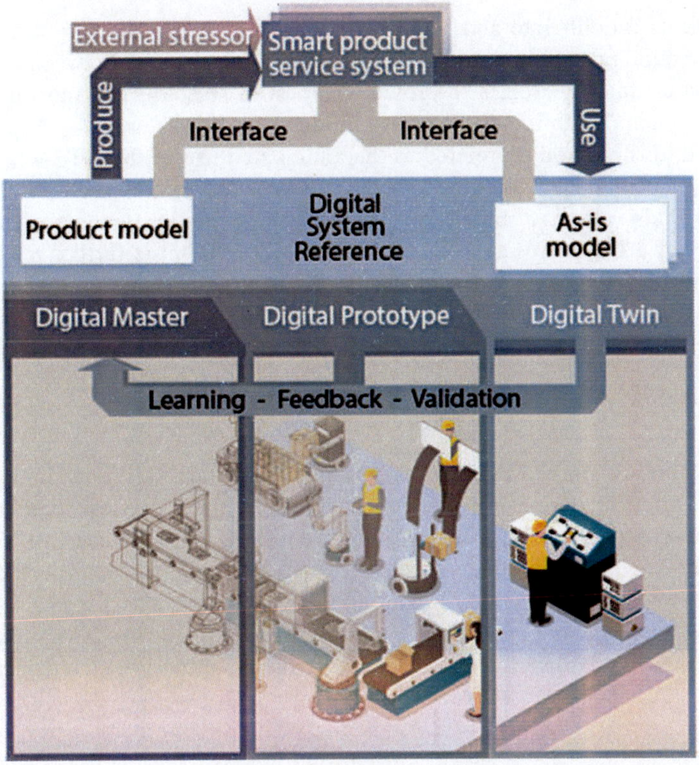

Fig. 1 The digital system reference and its different perspectives

DT/DP as an isolated and single-purpose activity, 'reinventing the wheel' can be prevented and DT/DP development can benefit from cross-fertilisation and re-using elements from other DT/DPs. This implies that a backbone or framework is required, against which DT/DP design and development can be executed in a structured, fast, flexible, and purposeful manner.

Currently, such an underlying approach for DT/DP development is not available, thus hindering agile DT/DP development. This publication aims to present a framework that supports companies in swiftly designing and evolving ideas for digital twinning into adequate and useful tooling. The design approach facilitates structured and goal-oriented configuration and development of DT/DPs that, from the start, support the development of production environments by combining real and virtual entities, without being forced into premature investments and commitments.

Within these environments, a DT, DP or DM, is represented in a hybrid experience with physical and digital (AR/VR) elements. Here, a combination of real and virtual entities addresses all stakeholder perspectives, data sources and assets involved, while using both measured and simulated data sources. This allows for assessing alternatives, explorations, and what-if analyses, rendering the development of production environments a deliberate and informed decision-making trajectory. The physical size, interaction possibilities and available actuators may differ between the options. The focus of this research is to provide support in selecting the most appropriate configuration and functionality of the solution.

Overall, this is strongly related to the anticipated use of the DT/DP and should therefore be considered as tailored tooling for the stakeholder. The virtual elements of a DT/DP also allow to change any of these characteristics on the go. This type of utilisation of a DT or DP as a functional and tailored tool is further referred to as 'digital twinning'.

2 Digital Twinning

Digital Twinning aims to represent a certain mix of measured and simulated data, while also offering tailored experiences based on e.g., the active stakeholder, context, environment, goal, etc. Since this is always a filtered representation of data, every use-situation requires an adequate configuration of tools to visualise, interact and experience the solution. The challenge in effective digital twinning is configuring the most suitable and relevant representation of the data in the synthetic environment. This requires a well-considered match between the available (measured or simulated) data, tools, and the anticipated effect.

2.1 User Support

The digital system reference aims to support companies in anticipating, or prototyping, potential futures of production environments. In this, the many different stakeholders and perspectives are considered, as are the tools and techniques (either existing, new or to be developed) that will capacitate and empower the environment. Here, especially the opportunity to make early and underpinned comparisons of possible alternative solutions for (part of) the production environment is instrumental. It is principally the flexibility that is required in assessing possible alternatives that makes it impossible to plainly bring together physical assets to evaluate any alternative. Hence, digital twinning heavily relies on bringing together real and virtual elements. The resulting synthetic environment is composed of a wide variety of tools, techniques, information, hardware and software components that are configured to allow for insightful decision making. Here, the increasing availability of cyber-physical systems acts as a catalyst [3]. The modular approach in a SE immediately benefits the potential solution; with that, establishing a digital twinning tool can be seen as a configuration problem rather than as a unique design endeavour. This implies that different instantiated DPs within the DSR can often be constructed by exchanging functional modules; additionally, the performance of the solution can be assessed in different environments by exchanging modular contextualisations [6].

To prevent 'reinventing the wheel' and to allow for continuous improvement in DT/DP development, a need occurs to document the rational and the configuration of developed DT/DPs in such a way that they are comparable and sharable. Eventually this could even lead to modular setups that stimulate reusability of parts of the solutions. The configuration approach stresses the need for a framework for digital twinning development; after all, the more structured and aligned a DT/DP development is, the more individual development trajectories will benefit from the existing modules. This anticipates a framework that combines the expertise, technology and experience available in all realised DT/DPs, thus being able to effectively and efficiently actualise new and tailored DT/DPs for specific companies and circumstances.

2.2 *Requirements*

In the development of the framework, a research-by-design approach is applied. This implies that the requirements on the framework are established while employing initial versions of the framework. In this way, knowledge on and experience with existing DT/DP can be captured and use to evolve the framework definition. Consequently, the DT/DP will also immediately benefit from the evolving framework.

This bi-directional learning approach prevents that all requirements for the framework should be determined before starting the development of the framework, or assuming that the requirements cannot change over time. Furthermore, it distinguishes the requirements for digital twinning and the ones for the framework. Additionally, access to an environment with (real-time) relevant data from production systems and experiences makes digital twinning usable for simulations, reviewing, validation and what-if analyses [7].

3 Collaborative Interaction

3.1 *Context*

In digital twinning, all the current, potential, and anticipated realities can be experienced interchangeably or even simultaneously. A SE makes this possible, while also facilitating different stakeholders/perspective concurrently. Collaborating with multiple perspectives simultaneously, in a personalised and tailored DT/DP, provides more insight in the relation between expertise of different stakeholders. A direct connection between multiple DT/DPs is essential, as is relying on a shared data repository. This allows many new collaboration and twinning potentials, such as:

- Remote and local collaboration
- Multiple and single user experiences

- Multiple variants of the anticipated environment can be compared to each other simultaneously
- A direct switch can be made between different realities: as-is, to-be, could-be and has-been
- Worst-case and best-case scenarios
- Simulation validation
- …

The amalgamation of real and virtual elements in digital twinning allows for the development of a solution in which a sub-set of the assets is not physically present (at the moment) but is rather taken into account by means of simulation. The far-reaching integration of online and offline elements lead to an environment in which physical and virtual assets are comprehensively intertwined. With that, digital twinning renders aspects like quality, time and cost insightful in the context of the envisaged environment, but also in comparison to alternative manifestations of that same environment. This allows the stakeholders to already experience an environment that does not yet exist, e.g. by means of virtual dashboarding [8] and VR/AR approaches. This includes interfaces at different levels of detail and aggregation, anticipating the dynamics and growth of the type and number of stakeholders involved.

3.2 Approach for Configurating DT/DPs

The approach should allow for fast and efficient building of a DT/DP in an easy to adjust (e.g. modular) way, while simultaneously incorporating a structured insistence on effective, reliable, real-time, and secure data collection and utilisation. With that, the development approach should not aim to instantiate 'yet another SE', but should rather offer a systematic yet agile, but foremost (cost-)efficient guidance in (envisaged) production environments. This requires a purposeful combination of cyber-physical system, human-technology integration and use of digital twins [9]. Reasoning from this guidance approach, every DT/DP is therefore treated as a manifestation of the underlying experience and expertise available in its development environment.

Generally, the goal of a DT/DP is to increase autonomy, improve communication and monitoring, and facilitate self-diagnosis at multiple perspectives on the of analysis between machines, humans and systems. In this, using a DT/DP will lower the risk of (extended) downtime, of technical and logistic incongruences and of many other phenomena with unexpected consequences. Additionally, the integration of sensoring (e.g. IoT) and information sources (from ERP to PLM) is a significant key focus. The company, therefore, wants to have a DT/DP that, with as little effort as possible, renders an adequate depiction of the current or a potential future environment. This not only allows the company to make underpinned decisions on investments, capabilities, capacities, and suitability of the environment, it additionally allows for training of staff members in an environment that does not yet (fully) exist. This implies that a DP can also be a learning factory [10], where staff can practice, but can also be exposed to (extreme) scenarios and what-if situations. Consequently, once the components of the DP become part of the primary process of a company in a DT or DM there will hardly be a learning curve for e.g. operators [11]. At the same time, however, the DT/DP can be used to observe the (learning) behaviour to improve the DT/DP or the system design. Especially the simulation of extreme cases and what-if scenarios can lead to more insight in the solution, its constituents and its behaviour. With that, digital twinning becomes a means to facilitate a bi-directional learning factory.

4 Configuring Digital Twinning

Observations in industry and experience with pilot production environments indicate that many functions re-appear in practically the same structure in different environments. Hence, a modular approach [12, 13] not only avoids re-inventing the wheel, it also allows for purposeful re-usability and configuration of assets, logistic and control entities. Whereas from a single perspective the advantages of modularity or reusability are not always clear, both strongly influence the speed with which solutions can be established. It, for example, implies that a rough-cut yet effective DT/DP can be available promptly, and time-consuming detailing only takes place once the evolvement of the solution has been appropriate thus far. From a framework perspective, every instantiated DT/DP should be a value-adding compromise between, among others, speed, quality, accuracy, uncertainty, comprehensiveness, flexibility, agility, realism, and goal orientation.

4.1 Framework

The framework supports the concurrent establishment of multiple DT/DPs, that each fit to different contexts, or provide variants of solution for the same context. The principle of the framework is shown in Fig. 2; it is based on a 'docking station' that spins out multiple DT/DPs based on a shared approach and an overarching knowledge/content base. The framework enables the configuration of a DT/DP and helps in determining what content is needed. Additionally, it supports in structuring the modules and the information content that together constitute the DT/DP.

In the center of the framework the building block for the configuration of a SE, as introduced in [14] is used. This is a functional representation of common denominators as the interface of a SE to the DSR and simulation realm. The left side of the framework provides the selection of the content that, in terms of common denominators, represent the functionality and positioning of the solution. This part determines the content of a DT/DP and captures the overall rationale and experience

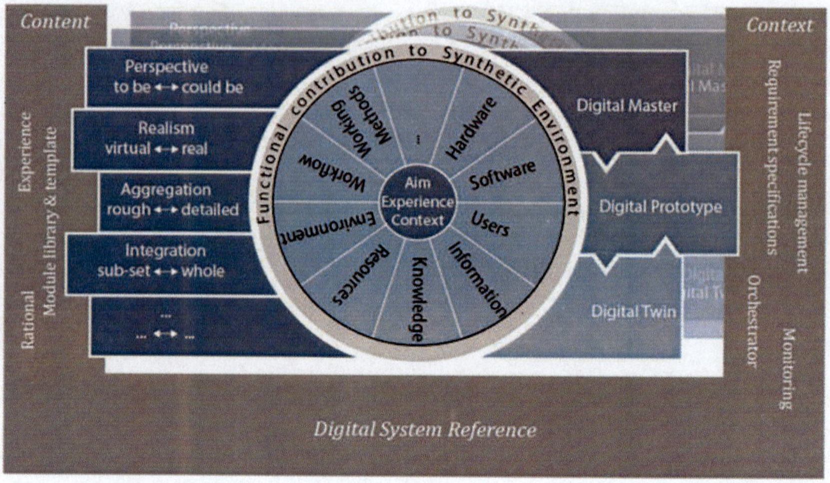

Fig. 2 Framework for digital twinning development

of the solutions that have already been instantiated. The right side provides the anticipated functionality that allow for purposeful organization of a DT/DP. This includes, for example, managing the requirement specification, monitoring the status, orchestrating the DT/DP and managing the life cycle of the DT/DP. The solution can be configured to structure a 'to-be' environment (digital master), 'as-is' (digital twin) or 'could be' (digital protype).

Both sides of the framework conjointly provide the ingredients that can be tailored to the requirements for a specific solution. The overall overview of the DT/DP resides in the Digital System Reference, that acts both as a backbone for the individual solution and as a bus architecture for interconnection the instantiated solutions. Moreover, the framework interrelates all DT/DP developments for the benefit of the individual DT/DP(s), as for the evolvement of the framework and its expertise/modules.

4.2 *Instantiated Digital Twinning Solution*

The framework contains all the information regarding the possibilities, preferences, and requirements of possible configurations. With every update or review of a DT/DP (comparable to 'docking' the DT/DP to the framework), the performance (diagnostics and use) data of that specific DT/DP is analysed to further improve the understanding of the consequences of certain DT/DP configurations. The demarcation between the docking station and the (docked) solution are the interfaces between both and are determined by the data transfer. For this there is no intend to set a standard for every interface, but to allow the use of (customized) connectors to facilitate this communication. This bi-directional learning approach facilitates an environment in which each DT/DP instantiation can benefit from the experiences and expertise obtained from other DT/DPs. Every connected SE can be optimised based on the data analysis of other DT/DP, or errors can be corrected when recognised in another DT/DPs with a comparable configuration. The effects recognised in one DT/DP can solve issues in another (unrelated) solution. With this, a well-founded reconfiguration of a DT/DP can be made, based on the relations set in the DSR. After establishing the proposed digital twinning activity, the actual DT/DP can be spun out (or 'undocked') and put to practice.

5 Practical Application

To validate the usability and modular approach of the framework, and to assess the applicability and performance of the instantiated DT/DPs, an example project established. This envisaged production environment aims to combine different production and assembly assets to construct a customisable product, here described anonymously as a 'phone stand'.

In this production environment, a 3D printer, a CNC engraver, multiple robotic arms and multiple AGVs are deployed to produce the product. To allow for quick adjustments for different scenarios and low impact of errors, the environment consists of scaled versions of common production machines (Fig. 3). These scaled assets intentionally use the same instruction set and communication protocols as industrial versions. For example, the robotic arms use the same operating system and language (ROS) as industrial size robots, while the system relies on the OPC UA protocol for communication. The dimension and configuration of the desired 'phone stand' is based on the 'phone of the visitor'. With this information, a selection is made of semi-manufactured parts. These parts are available in a small warehouse and are transported by means of AGVs to the 3D printer, the CNC engraver, or the robotic assembly station. In this respect, the scaled assets conjointly act and function as a real(istic) production environment. DT/DPs are constructed based on available templates and modules in the framework. The case study is used to gain understanding on the effects of changes to modules, and to get insight in missing modules or modules requiring improvement. To keep track of the status of the production environment, different types of IoT sensors and virtual dashboards (VR/AR) are integrated in the DT and DP. Multiple parameters, such as the position of the AGVs, are monitored and stored continuously or on request. This information leads to, for example, heatmaps of the transportation activities, which are used to rearrange assets to improve e.g., logistics. These changes are first made to the DP, so that multiple variants can be evaluated simultaneously. This information is used to facilitate the decision making on a digital master configuration by providing more data on the captured effect of realised configurations.

Whereas instantiating one DT and DP has been rather straightforward and unambiguous, using the framework approach has also immediately demonstrated instrumental, as, according to expectations, a clear and constructive connection between

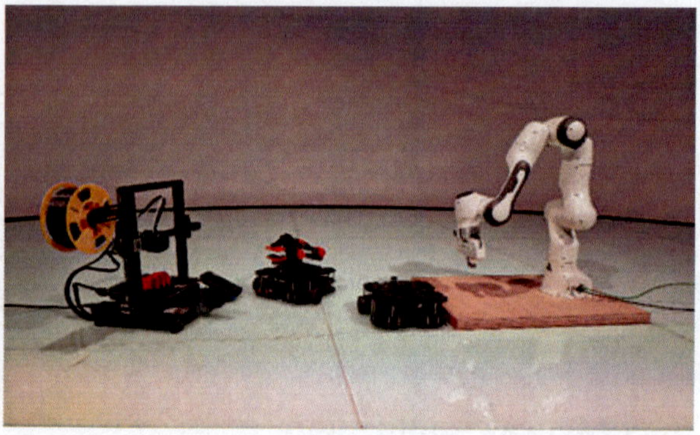

Fig. 3 A digital prototype in the form of a scaled version of a production environment

the various stakeholders appeared almost inherently and implicitly. Not only did the modules and knowledge in the framework benefit from this, but especially the DT/DPs themselves have seen a more effective evolvement. The major benefit has, however, been in the efficiency of the framework usage: extemporaneously, different digital twinning solution came to rely on the same functional and physical modules, contributing to the continued development of these modules and thus strengthening other DT/DPs.

6 Conclusion

Actual projects for the development of digital twinning solutions led to the observation that many companies are reinventing the wheel. This is mainly caused by the fact that an overarching, generic backbone is lacking for the purposeful support in the development of DT/DPs. To make optimal use of digital twinning, a tailored configuration is needed. The developed framework facilities this digital twinning development by using a modular approach to quickly develop DT/DP proposals in a comparable manner. The data resulting from multiple DT/DPs is collected, sorted, and analysed to support decision making for all stakeholders involved.

The information content can be used to alter the configuration of the DT/DP to align to stakeholders' expertise and to optimally aim for different objectives, such as context influences, data accuracy or asset availability. Simultaneous use of multiple stakeholder-dependent perspectives increases the insight in the current and upcoming activities in an (anticipated) production environment. To further optimise the use of a DT/DP, the solution should be able to deal with varying levels of accuracy, possibilities, users, and interaction. Here, the modular approach increases the effectiveness and efficiency of configuring and adjusting the DT/DP. Decision making processes are further facilitated by providing tailored simulations and visualisations. Existing production environments can be compared to designed 'to-be' and simulated 'could-be' situations, including virtual extensions with assets that are not (yet) available.

The framework offers support in the development of digital twinning solutions, that provides the stakeholders with the means of identifying the most appropriate tooling. This will support decision making, risk management and maintenance processes, while functioning as a flexible, bi-directional learning environment.

References

1. Papakostas, N., O'Connor, M.J., Hargaden, V.: Integrated simulation-based facility layout and complex production line design under uncertainty. CIRP Ann. **67–1**, 451–454 (2018)
2. Damgrave, R.G.J., Lutters, E.: Smart industry testbed. Procedia CIRP **84**, 387–392 (2019)
3. Salunkhe, O., Gopalakrishnan, M., Skoogh, A., Fasth-Berglund, Å.: Cyber-physical production testbed: literature review and concept development. Procedia Manuf. **25**, 2–9 (2018)
4. Lutters, E.: Pilot production environments driven by digital twins. S. Afr. J. Ind. Eng. **29**, 14 (2018)
5. Damgrave, R.G.J., Lutters, D.: Enhancing development trajectories of synthetic environments. CIRP Ann. **67–1**, 137–140 (2018)
6. Stark, R., Kind, S., Neumeyer, S.: Innovations in digital modelling for next generation manufacturing system design. CIRP Ann. **66–1**, 169–172 (2017)
7. Tomiyama, T., Lutters, E., Stark, R., Abramovici, M.: Development capabilities for smart products. CIRP Ann. **68–2**, 727–750 (2019)
8. Lutters, E., de Lange, J., Damgrave, R.G.J.: Virtual dashboards in pilot production environments. In: COMA'19; International Conference on Competitive Manufacturing: Stellenbosch (SA), pp. 22–27 (2019)
9. Stark, R., Fresemann, C., Lindow, K.: Development and operation of Digital Twins for technical systems and services. CIRP Ann. **68–1**, 129–132 (2019)
10. Abele, E., Metternich, J., Tisch, M., Chryssolouris, G., Sihn, W., ElMaraghy, H., et al.: Learning factories for research, education, and training. Procedia CIRP **32**, 1–6 (2015)
11. Schallock, B., Rybski, C., Jochem, R., Kohl, H.: Learning factory for Industry 4.0 to provide future skills beyond technical training. Procedia Manuf. **23**, 27–32 (2018)
12. Li, X., Bayrak, A.E., Epureanu, B.I., Koren, Y.: Real-time teaming of multiple reconfigurable manufacturing systems. CIRP Ann. **67–1**, 437–440 (2018)
13. Tomiyama, T., Moyen, F.: Resilient architecture for cyber-physical production systems. CIRP Ann. **67–1**, 161–164 (2018)
14. Damgrave, R.G.J., Slot, M., Thiede, S., Lutters, E.: Reality-infused simulations for dashboarding potential realities. Procedia CIRP **100**, 882–887 (2021)

Roy G. J. Damgrave obtained his Ph.D. degree in Industrial Design Engineering at the University of Twente. Currently, he is assistant professor in the Department of Design, Production and Management, University of Twente.

Eric Lutters obtained his Ph.D. degree in Mechanical Engineering at the University of Twente, The Netherlands. At that university, he currently holds a position as associate professor in the Faculty of Engineering Technology. He is also an extraordinary professor at the Department of Industrial Engineering, Stellenbosch University.

Novel Engineering Concept and Innovation

A Digital Service Engineering Training Course to Raise Competitiveness in Advanced Service Industries

Jörg Niemann, Claudia Fussenecker, Martin Schlösser, and Dominik Kretschmar

Abstract Modern business models are increasingly seen as a source of outstanding organizational performance and competitive advantage that either synergizes with the previous business model or completely replaces the previous strategy. For enabling this transition the paper describes an advanced model to master the digital business transformation and ends-up with a proposal for an academic digital service engineering training course. Based on a large literature review the paper will identify useful methods and tools which are used in modern industrial companies. The findings serve as blueprint for the academic education and training of future service engineers.

Keywords Service · Education · Service engineering · Online training

1 Transition from Traditional to Modern Business Models

Today's trends such as lean supply chain, smart manufacturing, cloud platforms, big data management, artificial intelligence, augmented and virtual reality, mobility, smart e2e transparency, additive manufacturing, customization, service-orientated business models and outsourcing etc. are all based on the changes of customer mentality and technological advancements.

J. Niemann (✉) · C. Fussenecker · M. Schlösser · D. Kretschmar
Department of Mechanical and Process Engineering, University of Applied Sciences Düsseldorf, Düsseldorf, Germany
e-mail: joerg.niemann@hs-duesseldorf.de

C. Fussenecker
e-mail: claudia.fussenecker@hs-duesseldorf.de

M. Schlösser
e-mail: martin.schloesser@hs-duesseldorf.de

D. Kretschmar
e-mail: dominik.kretschmar@hs-duesseldorf.de

K. von Leipzig et al. (eds.), *Smart, Sustainable Manufacturing in an Ever-Changing World*, Lecture Notes in Production Engineering,
https://doi.org/10.1007/978-3-031-15602-1_17

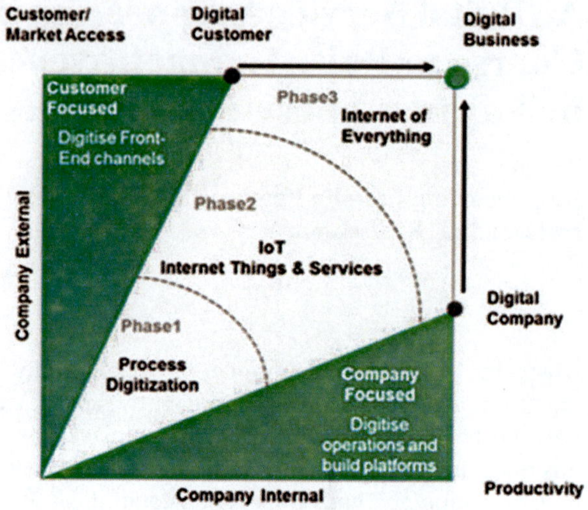

Fig. 1 Digital transformation (modified and enhanced according to [2])

This is only possible, if companies can digitalise information and data about products, customers, processes and services and thereby digitally transform their business model. With this new working method, a high amount of data is collected about business procedures and production processes, customer demands, as well as data about internal and external communication, requiring a high amount of management and data analysis.

Digital transformation means a re-orientation of products, services, processes and business models towards the continuously digitalised world and results in faster transactions and more reliability through quality and security and therefore leads to higher customer satisfaction [1]. The digital transformation of business models can be implemented in three general phases (see also Fig. 1):

- Phase 1: Digitise the current business and build a platform for digital processes.
- Phase 2: Integrate Internet of Things (IoT) functionalities into the platform and develop digital services.
- Phase 3: Close e2e loop of the entire business operations and modularize platform services [2].

Figure 1 depicts the journey from the traditional to the digital business. The model is divided into company internal and external elements, as the digitalisation of a business model can only work, if both the customer side and the own company can be "digitalised".

This begins through the digitalisation of the channels and processes used to create or provide value. Afterwards, the digitalisation of products, services and other objects are included in the value chain. Finally, a full digitalisation of all transactions and procedures with a high automation level leads to a fully digitalised business model [3].

Modern business models are increasingly seen as a source of outstanding organizational performance and competitive advantage that either synergizes with the previous business model or completely replaces the previous strategy.

New business models such as pay-per-use (usage-based payment e.g.: Car2go), peer-to-peer (trade between private individuals e.g.: Airbnb) or performance-based contracting (payment for the final performance e.g.: Rolls Royce) have revolutionised entire industries. Therefore, many companies have changed their model to move from pure product sales to the sale of problem solutions and services. When servitization moves a manufacturer all the way to becoming a solution provider there are major changes on the business model.

For enabling this transition, several frameworks are described in the literature [4, 5]. Figure 2 shows a modified and advanced model based on Bucherer [6] which is applied for the development of a new business model on the basis of an existing model. It consists of several phases in which different activities are proposed. After each phase there is a gate which requires a verification, if the planned solutions and the meaningfulness of the concepts are given. When this is fulfilled the next phase starts, otherwise there is a need to start from scratch with the previous phase. This model is to be understood as a cycle and should serve to question and optimize the business model during the entire life cycle.

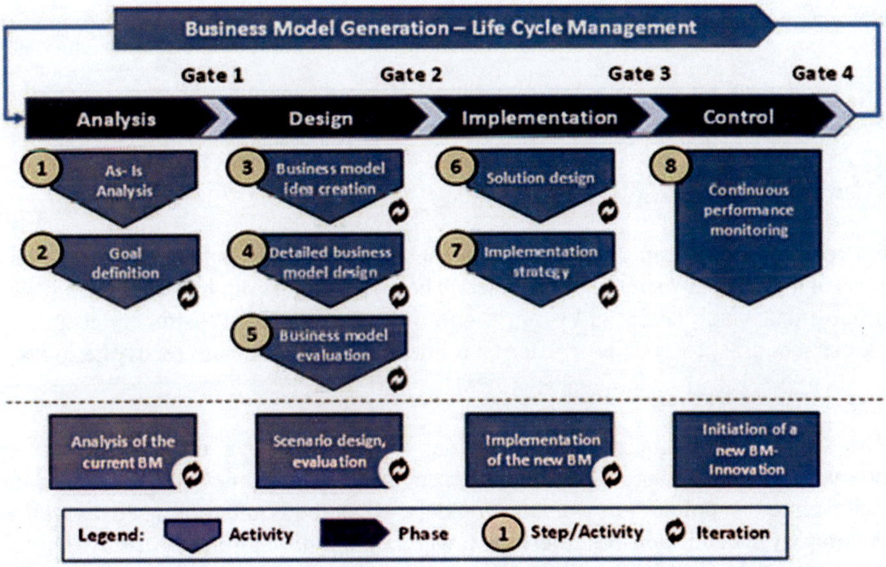

Fig. 2 Process model for business model innovation (modified and enhanced according to [6])

2 Methodology for Transformation

To develop a new business model and to transform the current business model, numerous methods exist to support the process. To provide extensive applicability to any kind of business sector, the stepwise approach presented was based on a combi-nation of methodologies from various authors.

A methodology is defined as "a system of broad principles or rules from which specific methods or procedures may be derived to interpret or solve different problems within the scope of a particular discipline. Unlike an algorithm, a methodology is not a formula but a set of practices" [7]. The following chapters will explain the created methodology and shows the recommended methods that can be used. The set of methods in use have been selected based on a literature research and interviews with practitioners from industrial companies [8–11].

However, this process and its tools must be followed and used by a competent team with a suitable set of skills. The selection of the team members is highly important, as the business model they will find could be crucial to success. The aim of the business model transformation is to improve the existing business or change it with the result of success and higher profitability. One of the keys to success is not only the team, but also the dedication of the top management to increase innovation and change [12, 13].

3 Steps and Methods

3.1 Phase of Business Analysis

Companies are a complex of various elements and interdependencies. Three main parts of the company should be analysed: The own business model (customers, value proposition, value chain and profit model), the stakeholder (customer incentives, partners, competitors) and the external influences on the business (ecosystem) [14].

Step 1—As-Is Analysis

The As-Is-analysis includes all functions and departments of a company, the product spectrum, technology, production depth, quantity framework, financial data, customer and supplier data, organization data, all methods and tools used, as well as the employees and their relationships toward each other and the company. During the As-Is-analysis deficits will be found, which are a result of various reasons. For example, wastage of resources or potential of the employees or not seizing opportunities by acting in a non-future orientated way [15]. A literature review shows that the methods have been listed to be the most suitable and applicable in operational business:

- SWOT [4, 5, 8]
- Benchmarking [9, 15]
- Ishikawa diagram [4, 5, 16].

Step 2—Goal definition

The highest level of a company goal is the vision. The vision is the long-term goal of a company that describes the general purpose of the company [8] An example for a vision from Procter & Gamble is the following:

"We will provide branded products and services of superior quality and value that improve the lives of the world's consumers, now and for generations to come. As a result, consumers will reward us with leadership sales, profit and value creation, allowing our people, our shareholders and the communities in which we live and work to prosper" [17].

When developing goals the company should decide what they want to achieve, where they want to stand in the future and in which amount of time the goal should be achieved. According to a literature review the following tools are essential in practical use:

- Goal pyramid [4, 8]
- Gap analysis [4, 8]
- Scenario analysis [4, 5, 8].

3.2 *Phase of Design*

In this phase ideas for the new business model will be systematically developed. This is mainly done by work groups applying creativity techniques.

The identified business models will then be designed and evaluated according to the company goals and capabilities.

Step 3—Business model idea creation

The idea finding phase marks the beginning of the business model development process. The goals have been defined and the development direction set. Either an existing product or service is to be improved or a completely new idea is to be developed. In either case a key aspect is to take the customers into account. It is crucial to understand what the customer needs and where his inconveniences lie, as well as what the customer expects from the company. Another question to be answered is, how the company can position itself in a way to satisfy the customers need in the best manner in comparison to the competition. All methods to develop creative ideas and to emphasise creativity underlie the same principles: understanding of the challenge, loosening of transfixed stereotypes and assumptions, recombining existing approaches and solutions and refining of ideas through criticism and improvement [18].

This phase helps to create new ideas through creative thinking, without being influenced by existing business models and ideas, as well as the current business model in place [19]. The following sections will explain some methods to develop creative, innovative ideas. The most commonly used idea generation methods are mind mapping and brainstorming. However, these will not be looked at further, as they are very simple and well known. For operational business applications the literature references the following tools:

- Destroy your business [11]
- The empathy map [13]
- St. Galler business model navigator [14].

Step 4—Detailed business model design

After some ideas have been gathered, they should be thought through systematically. All aspects should be considered, to enable a holistic perspective of the business model and to ensure its functionality. For this the most proven tools used in practice are

- Business Canvas [1, 13]
- SIPOC [9, 20].

Step 5—Business model evaluation

Having described the business models through the canvas and SIPOC method, the business models can now be evaluated. However, this can be a difficult task due to incomparability or through incomplete perspectives on the business models. Therefore, first of all an environmental analysis for each business model that is estimated to be promising should be made and then an objective evaluation with a systematic procedure and reasonable evaluation criteria should be followed. Literature reports to apply the following tools:

- PESTEL [19]
- Porters five forces [19]
- Value benefit analysis [21].

3.3 Phase of Implementation

In this phase the solution starts the design for the new found business model. All relevant aspects and details have to be considered and included into the new solution.

After the final approval of the solution design the business model has to be implemented into the daily operation and business routines. For this a detailed implementation strategy has to be elaborated to master this transition with regard to all kinds of management aspects.

Step 6—Solution design

This step focuses on the business model that is planned to be used and helps to prepare and consider all aspects that are relevant for the implementation of the new business model. According to several authors this step is being performed by the application of the following tools.

- Detailed process design [4, 5, 9]
- Resource and investment plan [4, 5, 9, 22]
- Business case [9, 22–24]
- Performance management [12, 17, 18].

Step 7—Implementation strategy

When all the framework has been set, the implementation strategy can be developed. The goal of the implementation strategy, is that the current processes can be transformed without major delays and downtimes. Furthermore, the resource availability is to be considered and the influence on the running operations or departments before setting up a project. It is recommended to start a pilot first, to stabilise the processes and find gaps and potentials for improvement. The old and the new process should be operated in parallel, so that the new process can gain maturity and stability [9]. The implementation of a new business model within a firm should be done through a project. Beforehand, some planning must be done to define the framework. Therefore, project management elements will be used to support the transformation [19]. Beginning with the project charter, that defines all important elements of the project a second important document will be presented, that is used throughout the entire transformation phase: the project plan. Furthermore, a RACI plan is put together, to define the roles and responsibilities during the transformation project. Finally, a budget plan for the project is set up. Applicable tools to finish this step are:

- Project charter [9]
- Action plan [9]
- RACI matrix [9]
- Project plan [9]
- Transformation plan [11, 12].

3.4 Phase of Control

The final phase aims at the continuous evaluation of the newly implemented busi-ness model. For this the operational performance figures have to continuously moni-tored and controlled in a structured manner.

Step 8—Continuous performance monitoring

The objective of this final step is to ensure a durable and sustainable development of the new business model. Therefore the current operational performance is permanently monitored and benchmarked against previously defined key performance figures. By this deviations and according counter measures can be taken at an early stage. Various authors recommend the following tools to master this step:

- Balance scorecard [12, 14]
- Break even analysis [12, 14]
- Rolling forecast [12, 14].

4 Development of a Training Program

The challenges outlined above place enormous demands on the developers of services in the future. Therefore, the goal was to develop a consistent methodology for the development of services, which takes up and implements these requirements. In this context, a training course was developed at the Flix Research Centre for Life Cycle Excellence at the University of Applied Sciences Duesseldorf. The target group of the course are enginneering students and professionals (see Fig. 3). The course teaches the basics of designing and developing modern services using "hands-on" training modules.

Figure 4 shows a rough overview of the topics, methods and practical parts of the course contents.

The course participants develop and design a service step by step using methods and tools that are learned and tested within the course. In this way, the course participants simultaneously learn the methodological tools in a creative course atmosphere according to the needs identified from the literature and the industry survey (cf. Sect. 4). The course lasts approximately five days and can be extended or shortened on a modular basis depending on the group of participants and the objectives. At the

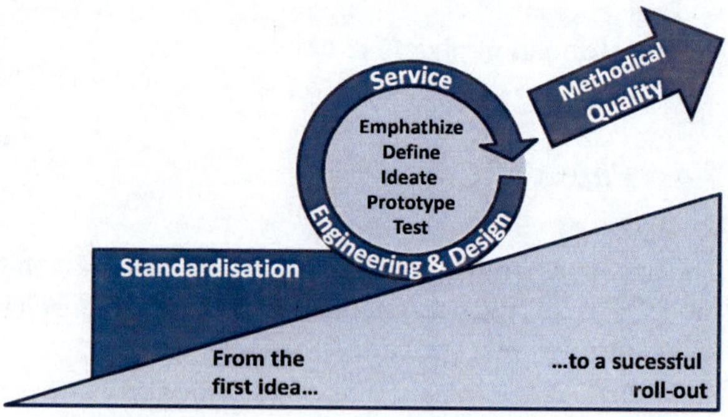

Fig. 3 Service Engineering training course [*source* own graph]

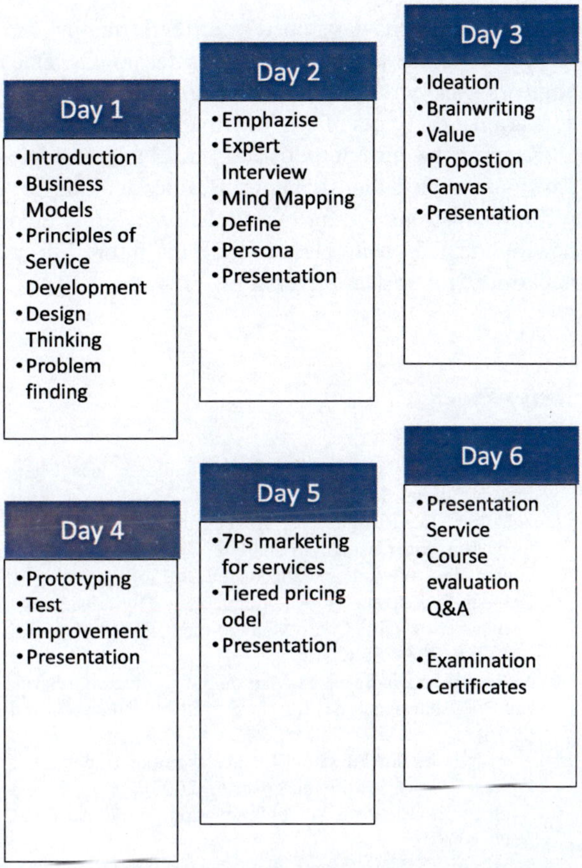

Fig. 4 Course content [*source* own graph]

end of the course, in addition to the presentations of the developed servcies, there is also an examination of the theoretical basics (online examination). The course is also suitable for delivery in an online format. However, experience has shown that a face-to-face format is advantageous due to the numerous group work and practical interactions. The course language and the teaching materials are in English.

5 Summary and Outlook

The paper describes a process and a toolbox of methods for the digital business transformation. The single phases have been subdivided into smaller single steps. To master the steps a literature review has been performed to identify useful methods for the practical execution. This toolbox of methods can be used as a blueprint for the education and training of future service engineers. According to the findings FliX research centre has developed a certified trianing course in the field of service engineering for students and business professionals. Due to the callenges of tomorrows engieers it is recommended to integrate the suvject of service engineering into the educational curricula of engineering studies.

Keeping the current industrial, social and ecological changes in mind the content development is rather dynamic. Besides looking for a suitable funding possibility, recommendations for further adjustments are welcome. A proposal is to set up a new online and onsite class to train the future European service engineers within a network of interested universities.

References

1. Kreutzer, R.T.: Konzeption und Grundlagen des Change- Managements. Wirtschaftswissenschaftliches Studium **1**(46), S. 10–17 (2017)
2. Schallmo, D., Rusnjak, A., Anzengruber, J., Werani, T., Jünger, M. (Hg.): Digitale Transformation von Geschäftsmodellen. Grundlagen, Instru-mente und Best Practices. Springer Fachmedien Wiesbaden, Wiesbaden (2017)
3. Dehmer, J., Kutzera, A.-A., Niemann, J.: Digitalisierung von Geschäfts-modellen durch plattformbasiertes Value Chain Manage-ment. ZWF - Zeitschrift für wirtschaftlichen Fabrikbetrieb **112**(4), S. 253–256 (2017)
4. Niemann, J.: Die Services-Manufaktur, Industrielle Services planen –entwickeln – einführen. Ein Praxishandbuch Schritt für Schritt mit Übungen und Lösungen. Aachen, Shaker Verlag (2016)
5. Niemann, J., Tichkiewitch, S.: Westkämper Engelbert: Design of Sustainable Product Life Cycles. Springer, Heidelberg Berlin (2009)
6. Bucherer, E.: Business Model Innovation-Guidelines for a Structured Approach. Shaker Verlag, Aachen (2010)
7. Schallmo, D.R.A.: Design Thinking erfolgreich anwenden. So entwickeln Sie in 7 Phasen kundenorientierte Produkte und Dienstleistungen. Springer Fach-medien Wiesbaden, Wiesbaden (2017)
8. Kohne, A.: Business Development. Kundenorientierte Geschäftsfeldentwicklung für erfolgreiche Unternehmen. Springer Vieweg, Wiesbaden. Online available: https://doi.org/10.1007/978-3-658-13683-3. Last access 4 April 2020
9. Niemann, J., Reich, B., Stöhr, C.: Lean Six Sigma - Methoden zur Produktionsoptimierung. Springer Viehweg, Heidelberg, Berlin (2021)
10. Pattern Cards: St. Gallen Business Model Navigator. BMI Lab AG, Schweiz (2013)
11. Gassmann, O., Frankenberger, K., Csik, M.: Geschäftsmodelle entwickeln. 55 innovative Konzepte mit dem St. Galler Business Model Navigator. Hanser, München. Online available: http://www.hanser-elibrary.com/action/showBook?doi=10.3139/9783446437654. Last access 4 April 2020
12. Doleski, O.D.: Integrated Business Model: Applying the St. Gallen Management Concept to Business Models. Springer Fachmedien Wiesbaden, Deutschland (2015)
13. Chizzoli, C., Raccagni, D., Busacca, B.: Customer Value Analysis: Problems and Applications. EGEA, Italien (2020)
14. Brugger, R.: Der IT Business Case. Kosten erfassen und analysieren, Nutzen erken-nen und quantifizieren, Wirtschaftlichkeit nachweisen und realisieren. 2. Springer-Verlag, Aufl. s.l. (2009) (Xpert.press). Online available: http://site.ebrary.com/lib/alltitles/docDetail.action?docID=10297037. Last access 4 April 2020
15. Niemann, J., Pisla, A.: Life Cycle Management of Machines and Mechanisms. Switzerland, Springer Nature (2021)

16. Fussenecker, C., Niemann, J., Paul, A., Schlösser, M., Schöning, M.: Educating future engineers—meeting challenges at secondary school and university. In: Yilmaz, M., Niemann, J., Clarke, P., Messnarz, R. (eds.) Systems, Software and Services Process Improvement, 27th European Conference, EuroSPI 2020, Düsseldorf, Germany, September 9–11, 2020, Proceedings. CCIS 1251, pp. 742–753, Springer Nature Switzerland AG (2020). https://doi.org/10.1007/978-3-030-56441-4_56
17. Stöhr, C., Janssen, M., Niemann, J., Reich, B.: Smart services. Procedia Soc. Behav Sci **238**, 192–198 (2018)
18. Bergmann, R., Garrecht, M.: Organisation und Projektmanagement. 2. Aufl. 2016. Springer Gabler (BA KOMPAKT), Berlin, Heidelberg (2016). Online available: https://doi.org/10.1007/978-3-642-32250-1. Last access 4 April 2020
19. Barsh, J., Capozzi, M.M., Davidson, J.: Leadership in innovation. McKinsey Quarterly. Online available: http://www.mckinsey.com/business-functions/strategy-and-corporate-finance/our-insights/leadership-and-innovation. Last access 4 April 2020
20. Niemann, J.: Ökonomische Bewertung von Produktlebensläufen- Life Cycle Controlling. In: Spath, D. (Hrsg.) u.a.: Neue Organisationsformen im Unternehmen In: Spath, D., Westkämper, E., Bullinger, H.-J.; Warnecke, H.-J. (Hrsg.) Neue Entwicklungen in der Unternehmensorganisation. Springer-Vieweg, VDI Buch, Berlin (2017)
21. Niemann, J.: Life Cycle Management- das Paradigma der ganzheitlichen Produktlebenslaufbetrachtung. In: Spath, D., Westkämper, E., Bullinger, H.-J., Warnecke, H.-J. (Hrsg.) Neue Entwicklungen in der Unternehmensorganisa-tion. Springer-Vieweg, VDI Buch, Berlin (2017)
22. Niemann, J., Fussenecker, C., Schlösser, M., Ahrens, T.: ELIC—Teacher as a medium to built a new generation of skilled engineers, In: Proceedings of the International Conference on Competitive Manufacturing (COMA'19), 30 January–1 February 2019, Stellenbosch, South Africa, S. 234–238
23. Gassmann, O., Frankenberger, K., Csik, M.: The Business Model Navigator. 55 Models that Will Revolutionise Your Business. Pearson, Harlow (2014)
24. Glauner, F.: Zukunftsfähige Geschäftsmodelle und Werte. Strategieentwicklung und Unternehmensführung in disruptiven Märkten. Springer Gabler, Berlin, Heidelberg (2016). Online available: https://doi.org/10.1007/978-3-662-49242-0. Last access 4 April 2020

Jörg Niemann obtained his Ph.D. degree in Manufacturing Engineering from the University of Stuttgart. In 2012 he was appointed Professor in Business Management and Mechanical Engineering at the Düsseldorf University of Applied Sciences, Germany.

Claudia Fussenecker holds a M.A. in European Management from the FH Bund in Brühl and a Diploma in Industrial Management from EUFH in Brühl. She is currently a scientific assistant in the Department of Mechanical and Process Engineering, focusing on Marketing and Services and Life Cycle Management.

Martin Schlöesser holds a B.E. with specialization in Product Development and Production from the University of Applied Sciences Duesseldorf. He is currently a scientific assistant in the Department of Mechanical and Process Engineering, focusing on production simulation and Life Cycle Management.

Dominik Kretschmar holds a M. Sc. in industrial engineering. He is a research assistant in the Department of Mechanical and Process Engineering at Duesseldorf University of Applied Sciences with a focus on innovation, entrepreneurship, and product service systems.

Modern Project Approaches in Shortening the Lead Time in Innovation for Young Emerging Companies Based on the Experienced Seniors Knowledge

Adrian Pîslă, L. Nae, Calin Vaida, Eduard Oprea, Bogdan Gherman, Michel Deriaz, and Doina Pisla

Abstract The paper represents the preliminary findings in developing a multi-role, innovative digital platform entitled "WisdomOfAge", as a Learning Management System (LMS) addressing to the economic and technological dimension of a company (especially start-ups) in reaching competitive manufacturing advantage. The platform is developed within the frame of an EU project [1], offering finance to support retired seniors' active living. The project capitalizes the existing trend of having an aging population within the EU, in parallel with a gap in number and quality of skilled workers. The multirole digital platform creates a unique synergic solution that copes with the coexistence of different technological cultures using the resourceful retired seniors' expertise to compensate the lack of knowledge and experience required by a lot of newly established companies. "WisdomOfAge", is meant to create an efficient but pleasant environment for seniors and companies to easily

A. Pîslă (✉) · C. Vaida · B. Gherman · D. Pisla
Technical University of Cluj-Napoca, Memorandumului 28, 400114 Cluj-Napoca, Romania
e-mail: Adrian.Pisla@muri.utcluj.ro

C. Vaida
e-mail: Calin.Vaida@mep.utcluj.ro

B. Gherman
e-mail: Bogdan.GHERMAN@mep.utcluj.ro

D. Pisla
e-mail: Doina.Pisla@mep.utcluj.ro

L. Nae · E. Oprea
Digital Twin, Bd. Mircea Voda 24, 030667 Bucharest, Romania
e-mail: laurentiu.nae@digitaltwin.ro

E. Oprea
e-mail: eduard.oprea@digitaltwin.ro

M. Deriaz
Yumytech, 38 route d'Ambilly, 1226 Thônex, Switzerland
e-mail: Michel.Deriaz@unige.ch

K. von Leipzig et al. (eds.), *Smart, Sustainable Manufacturing in an Ever-Changing World*, Lecture Notes in Production Engineering,
https://doi.org/10.1007/978-3-031-15602-1_18

overcome the retirement moment, when the professional activity is abruptly interrupted, leading to feelings like: out of business, lack of purpose, lack of achievements and a lot of unallocated time.

Keywords Seniors customized knowledge transfer · Lead time to innovation · Sustainability · Risk management

1 Introduction

The paper presents initial results from a research project [2, 1] considering the digitalization potential with focus on companies in their early stage of development, knowing that within emerging industries companies must overcome many entry barriers to become profitable.

The early-stage companies are looking for new businesses to ensure the sustainability [3] while the start-ups are projected to develop and validate a scalable business model, for a new product/solution/technology etc. that aims to grow large beyond the solo founder [4–6].

At the beginning every start-up faces high uncertainty [3] and have high rates of failure, but a minority of the start-ups will go on to be successful, influential [7], or even become unicorns as privately companies valued at over 1 billion US$.

Considering the successful ones, it is already known that 90% of the start-ups end up in bankruptcy, mostly determined by the generations' gap, unsuitable technological capabilities approach, or due to the lack of experience in generating a correct market projection and a poor customer profile assessment [8].

Innovation increases the start-up chances to react to changes, and to discover new opportunities helping in fostering competitive advantage, leading to better products and services. The innovation could be a success key for any company [9], but the poor management and the lack of experience can make even brilliant innovations to fail.

ISO TC 279 in the standard ISO 56000:2020 [10] for "Innovation Management", defines innovation as "*a new or changed entity realizing or redistributing value*". Together with the innovations implementation the start-ups, are developing new businesses with relevant impact in gaining stronger positions. For any business, the challenge of sustainable transformation is balancing the need to keep business in motion while making changes to reach development ambitions. Having a competitive manufacturing or providing competitive services depends on the approach in taking small steps into the unknown as the "Lead Time in Innovation".

In accordance with the new learning paradigm supported by OECD (Organisation for Economic Co-operation and Development) [11], the research results are meant to reduce the lead time in innovation, by integrating a combination of economic, social and technological aspects to compensate some of the lack of knowledge, experience and managerial skills.

The regular solution is to address to specialized companies with regular instructions that may advise on company lifecycle. Applying 2 of the 3 Iron triangle rule, Cost-Time-Quality; it may be noticed that all three characteristics seem to be critical in the case of start-ups that are using regular training: is costly, requires a lot of time and the quality is mostly related to the personal development and not on the company development Therefore is needed to consider different approaches and solutions.

The solution, presented in the paper, starts from the capacity to use the potential of experienced seniors (in an early retirement stage) to provide valuable input towards YEC (Young Emerging Companies) mediated by a customized knowledge transfer digital platform. The opportunity came from the industrial evidences that although the technology is ever changing and tools are permanently improving, product processing is fundamentally constant for most of the industrial domains. Starting from that, the research was conducted by a company that wants not only to develop but also to operate an adequate knowledge transfer digital platform in the form of a Learning Management System (LMS).

The paper is structured in 8 chapters. After the introduction (Sect. 1), consist in presentation of conditions and facilities that governs entrepreneurial activities; in Sect. 2 is revealed the way a digital platform becomes a solution for a complex industrial-social problem, highlighting the necessity of such a digital platform. Section 3 offers insights that would make the proposed LMS a successful digital platform, presenting elements regarding the conceptual approach, followed by the stakeholder representation in Sect. 4. Section 5 presents the design implementation. Section 6 offers a brief description of the envisaged Business plan. Section 7 offers a description of the technology behind the digital platform and in Sect. 8 are presented some conclusions regarding the project development.

2 Necessity of a Digital Platform

Companies are investing a lot in human resources, in average 4000 US$ per person [12]. The selection is first oriented on "hiring culture" fit for the company, and secondly on "hiring skills". Skills can be trained, and 84% of the executives know that their future depends on innovation, understanding that training must be provided regularly to sharpen the employees' skills, absolutely necessary within the company lifecycle.

The regular training doubles the hiring investment leading also to issues like adequate scheduling, employees' availability, and activities continuity. Therefore, a regular schedule of training is valid for larger companies with medium and long term investments plans, with training dedicated especially for the "green" new hired people that actually need the right tools to be progressive and efficient.

On other side, the communities are striving to use their accumulated knowledge, and to valorise the expertise of their members offering longer active and self-sustained living to senior members [4]. At least in the early retirement stage; the meaningful occupation is a desired intermediate activity between work and free time over retirement years.

The current pandemic situation generated by COVID-19 has generated further restrictions regarding distancing, which pointed towards a digital solution, which is remotely able to contribute to the training of the YEC employees.

Several attempts to capitalize the existing potential have been achieved, like ProMe [13], which provides multiple facilities for informal communication, or the T-Echo (developed in Japan) [14]. Attempts to access any of these platforms shows have been unsuccessful, which indicates that these have not reached the commercial status, as expected. Analysing the achievements described in [13, 14] and the actual situation it has been concluded that the facts that have led to their failure fall within the following: that these projects have been too general, without a specific subject for the training experience, are addressing rather the social aspects than the technical ones (where the competition is already much stronger), as well as some lack of background and experience.

All mentioned drawbacks are eliminated from the start by the project consortium, as it will be showed within the next chapters, enabling a successful solution as an integrated part of the already existing services delivered by the Digital Twin Company. The digital platform "WisdomOfAge" is supporting both the trainers [15] and trainees (companies) in a specific professional field of "engineering" [16], having as starting point mechanical engineering (CAD-CAE-CAM), which sets all the premises for a successful endeavour, in a customized, adaptable and affordable solution.

3 The Concept Behind the Platform

A modern approach is proposed by generating a digital platform consisting of an adapted industrial software that perfectly fits with the tutoring activity and industrial training, based on knowledge assistance and transfer from experienced seniors, in a customized digital environment.

For most of the people, the actual formal education systems in technical higher education starts from 18 or 19, followed by a 3–6 years of studies and with about other 10 years of exploratory activity. At the age of 35, or later, people may start to become specialists and eventually create innovative developments.

At the age of 65, with over 30 years of experience in a field, they have reached retirement, which provides more personal time. But this is almost instantly doubled by a feeling of usefulness, lack of purpose and incapability of creating added value for themselves and the society [17]. The current learning strategy aims to develop new competences which come as a response to the existing uncertainties. These are holistic, such as, curiosity, imagination, resilience and self-regulation. The competences, inherited by the new generations enable to adapt it, generating also natural gaps in specific topics, strategy displayed in Fig. 1.

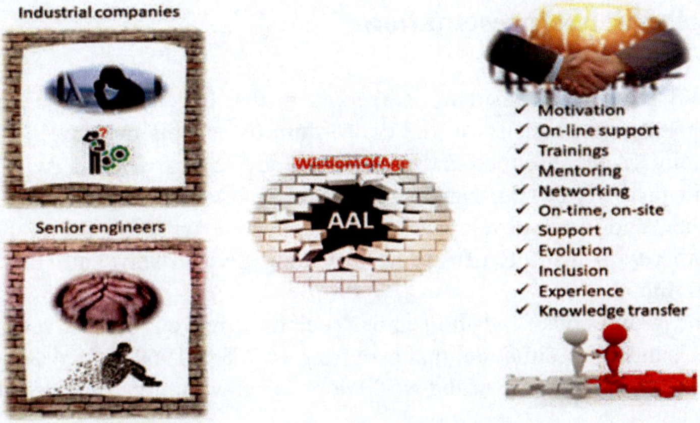

Fig. 1 Knowledge means power with "WisdomOfAge"

Here is where the senior experts, with extensive knowledge and specific experience, can help to ensure a faster transition and adaptation of the younger generation towards specific problem solving.

The "valorisation" of this situation comes from the fact that external information sourcing may offer lead-time advantage in product innovation [17].

Digital Twin as a Siemens PLM partner uses the Siemens professional software modules and gets access to Siemens facilities like the Siemens future pathfinder, the answers of 775 senior executives concerning the decade transformation objectives [18]. It leads to four transformation objectives with the measured importance within three regional insights: Asia, Europe, US. The four transformation objectives are: Decarbonisation, Digitalization, Holistic Impact and Resource efficiency. For Europe, Decarbonization and Holistic Impact which have got the same importance score (73.8%), while the Digitalization with Resource Efficiency have received the same importance score (74.6%). That totally confirms the correct approach and the necessity of the digital platform "WisdomOfAge", as Europe is concerned a lot about the Resources efficiency and the valorisation of the unused human capital in terms of knowledge and experience. This is even more true with the large scale digitalization process within Industry 4.0, which includes the concerns for the development and training of the human resources.

Considering Resource Efficiency, this relates to innovation and lead time to innovation: design products and processes that use sustainable resources; the capacity to incorporate used material from other processes; minimize the generation; enhancing customer satisfaction while optimizing commercial operations [19]. Principles used in the developed content of the "WisdomOfAge".

4 Stakeholder Representation

There are 3 groups of considered stakeholders that benefit from the digital platform. The 1st group consists in the consortium that forms the development team for "WisdomOfAge", grouped in the "technical developers" Digital Twin Company, Technical University of Cluj-Napoca, University of Genève, Fig. 2.

And the "end-users developers" Yumytech and ArxiT as Swiss IT companies, together with Institute of Ageing Research (Switzerland) and Happy Ageing (Belgium), Fig. 3.

The 2nd group consist of the senior specialists, retired, but active. Loneliness became a "pandemic" situation, that rose from 16.7% in 1969 to 28.4% in 2019 [6], in parallel with the increase of the world population with more than double (212%) [20].

The daily interaction in the working place was an important form of connection, suddenly disrupted by retirement [21]. In their case, "WisdomOfAge" platform answers these challenges in an organic way, as a dedicated instrument through

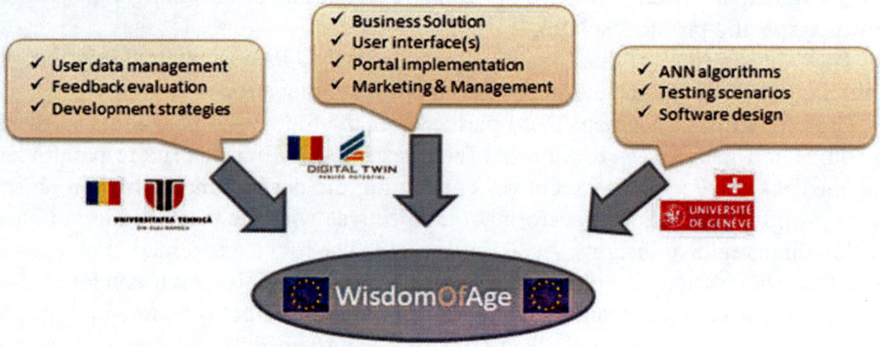

Fig. 2 WisdomOfAge Consortium

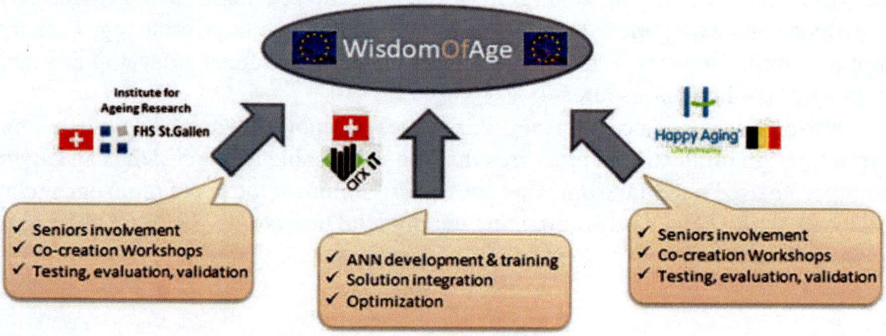

Fig. 3 WisdomOfAge Consortium

adequate tools design, friendly interfaces and tools functionalities for: counselling, mentoring and tutoring, considering their willingness have a comfortable balance between free time and professional activities.

The 3rd group is the motivated young emerging companies, with basic domain skills, but high digital skills, suffering from lack of experience to develop their businesses in a competitive environment.

5 The LMS Implementation

The "WisdomOfAge" starting webpage is presented in Fig. 4, focusing on building confidence for seniors and companies, advocating the advantages they will obtain by using it.

The training and coaching abilities together with the power and customization capabilities of the involved IT modules are the key factors for promoting the senior specialists.

The seniors must be aware about the existence and capabilities of "WisdomOfAge", while the platform should provide the means of an easy familiarization with the tools and the digital environment.

Actually, the entire design must lead to a friendly user interface, enabling the seniors to feel comfortable in sharing their knowledge, encouraging adequate partner discussions, implementing the required measures with regard to the data protection.

The companies that need to have access to the training and knowledge transfer are stepping in a new, unknown world defined and described by the seniors. Their employees will land in a friendly environment, where the young person feels comfortable, where the trainers are coming out from their desk and greet the newly entered, inviting them to "sit together on the nearby sofa", make him or her to feel cosy (like a coffee chat) to facilitate in the most efficient way the communication.

Fig. 4 "WisdomOfAge" platform starting page

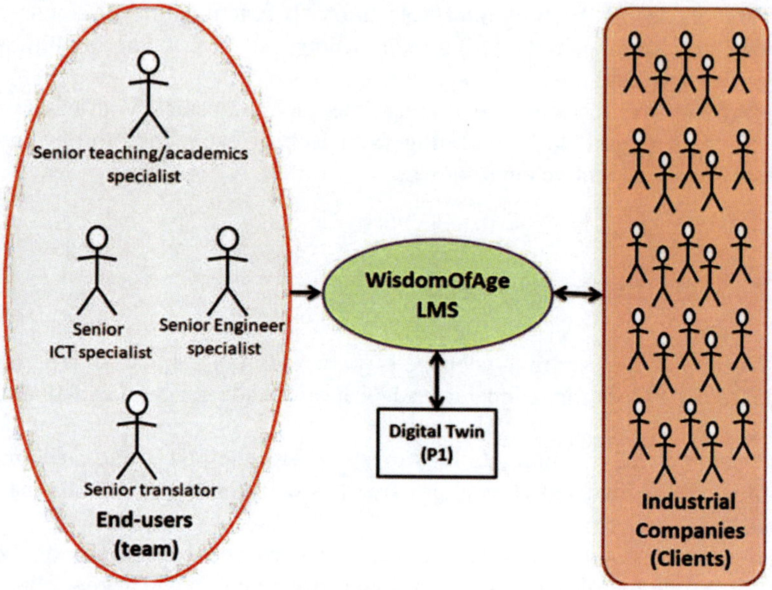

Fig. 5 First step implementation of the Learning Management System

Following this perception, the objectives implementation is planned in two steps. First to initiate the program and customize the digital platform to interface the two main groups of end-users, Fig. 5.

The second step consists in a continuous service for the digital platform maintenance, performed by Digital Twin, setting up a call-centre able to find and guide beneficiaries within the LMS in all phases: Seniors and industrial companies, signup, achieve content and training activities. In the same time, Digital Twin has to maintain the pool of professionals, as is depicted in Fig. 6.

All stakeholders are involved in the implementation process. The knowledge transfer and the direct on topic assistance, in gaining experience, are two major aspects that must be implemented and transmitted through the "WisdomOfAge" digital platform.

6 Business Development Plan

Nowadays everything goes to be trendy, to be modern, to be digitalised and only somewhere in the background seem to exist sustainability and functionality, as pale features in face of profit. These things will change provided that a long-term vision is arising, due to the change in the price of energy, resources availability and due to a weak spot reported at European level: the increasing lack of proficient specialists (technicians, skilled manpower, experienced engineers, to cover specific tasks

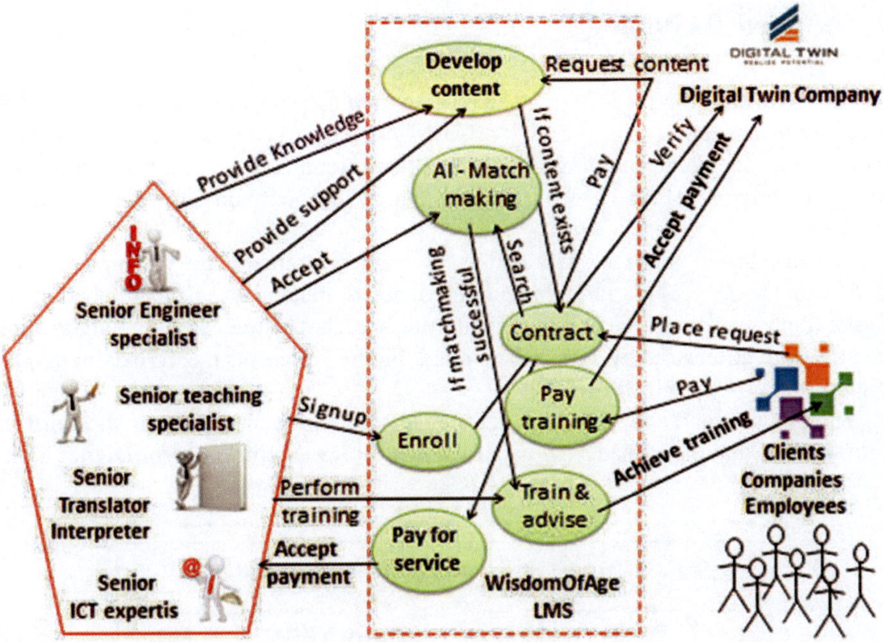

Fig. 6 Structural implementation of the Learning Management System

within the company: functional design, assembly, equipment maintenance, or parts manufacturing).

The crises we face today actually started about 2 decades ago, by treating the industrial activities like shopping: shopping products, shopping services, shopping energy, shopping time or the right to pollute. The large amount of affordable data takes the humans out of equation, just by combining the products, logistics and customer profile data. The result: the customer is a "customized box", the flexibility to adapt to every new challenge being determined by the existence of "old school" people.

This affordability slowly disappears and "WisdomOfAge" offers a solution, by using updated digital technology and the people formed in the "old school" to transform the "shoppers" (theoretical active users) that plays a pattern integration role in "active users", answering to the need for expertise and experience.

The business plan is leading buy a subscription model to a continuous generation of "customized shopping store" for each "shopper company", having in the same place every domain and every kind of accumulated experience (brand & style).

This approach is inducing the traditional discovery process for the knowledge transfer, where every company has to search for exactly what it wants and generate trained people and data that offers exactly what they need.

7 Applied Technology

Digital Twin has already developed an on-line training platform in CAD/CAM/CAE tools, with a general non-industry-oriented content. WisdomOfAge comes as a natural enhancement, enabling personalized solutions for customers ensuring the lead time shortening for new technologies and high competitiveness on the concurrent markets.

The structured databases with end-users: trainers; trainees, the teaching module with offers and the client module with the demand, the end-users customized interfaces: the trainer interface and the client interface, the AI matchmaking algorithms and the evaluation strategy will be developed. Figure 7 presents the "WisdomOfAge" digital platform architecture.

The modular and adapted user interface, integrates functionalities in the spirit of Industry 4.0, having already the expertise of training engineers form Digital Twin

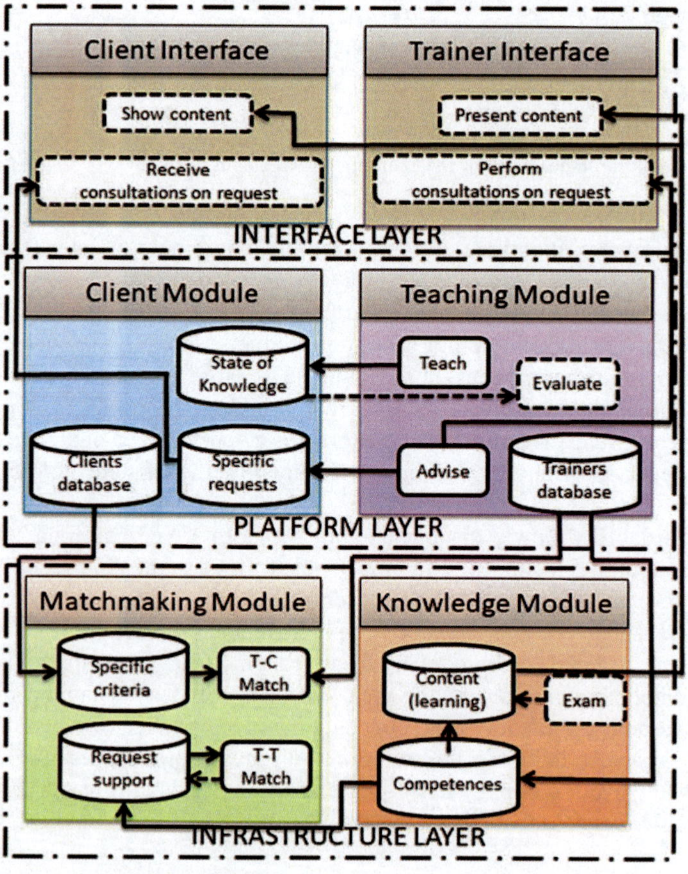

Fig. 7 WisdomOfAge platform architecture

in Siemens PLM product development. Their expertise is built on training activities with over 1500 end-users in the first four years after market launch. The result is that the development team around "WisdomOfAge" may alleviate the disadvantages observed in real life practice: low interaction, low specificity, lack of motivation, difficulties in managing the system losing the purpose of the platform, the knowledge transfer.

The potential VR representation and the cloud-based experience must be scaled in two levels: for professionals (companies) considering Oculus VR headsets, but also for the initiated that are only using regular smartphones or desktop browsers. That is possible, similar with the case of interconnection platforms: Zoom, Skype or Teams; because the software relies on open standards graphics API, the WebGL, available across devices and browsers. The virtual reality (VR) augmented reality (AV) and enhanced reality (EA) can be combined with 3D technologies in a general concept of mixt reality (MR) enabling mentors to create a digital environment based on their knowledge and for trainee to benefit from the digital production using it directly in their everyday activity or replicate it for the new type of applications.

8 Conclusions

The aim of the presented research is to develop a business oriented digital LMS platform, that reduces the lead-time to innovation for young emerging companies, as a direct answer to the companies' demand of having handy and customized tutoring services to reduce the lack of experience for innovative products and processes. Beside the software interface, facilities and functionalities, the platform intends to exploit the expertise of senior professionals by answering in the same time to a societal demand for having a longer active living of the professionals after retirement, once with the increase of the population average lifespan.

For this purpose an international development team has designed a performant digital connection between the skilled professionals and young companies for reducing the lead time to innovation, considering that a dedicated expertise in a project is possible in matter of weeks instead of months. Being focused on a specific topic the risk failure and the costs are substantially reduced.

The lead-time to innovation decrease is facilitated from the pilot stage. To efficiently assess the quality of the LMS, the development team has defined quantifiable goals that will be used to evaluate the platform 24 months after the market launch. Covering multiple aspects, these KPIs also enable the fast identification of any faults and uses the clients' feedback to find solutions to eliminate them.

Digital Twin is the main beneficiary of: ► the project Intellectual Property (IP) results, ► a comprehensive business plan and a business model, ► direct access of the trained pool of seniors, ► advertising facilities, ► direct contact with the involved companies and ► European scalability of the "WisdomOfAge" platform.

Acknowledgements This work was supported by a grant of the Romanian Ministry of Research and Innovation, CCCDI-UEFISCDI and AAL Programme, co-funded by the European Union's Horizon 2020 research and innovation programme project AAL-CP-AAL-2020-7-83-CP-WisdomOfAge within PNCDI III.

References

1. AAL-Ageing Well in the Digital World. http://www.aal-europe.eu/
2. Gherman, B., et al.: WisdomOfAge: designing a platform for active and healthy ageing of senior experts in engineering. In: IHAW2021. Cyprus, in press (2021)
3. Schmitt, A.: A dynamic model of entrepreneurial uncertainty and business opportunity identification: exploration as a mediator and entrepreneurial self-efficacy as a moderator. Entrep. Theory Pract. **42**(6), 835–859 (2018). https://doi.org/10.1177/1042258717721482.S2CID148840401
4. Linkedin: The differences between entrepreneurs and startup founders. www.linkedin.com. Retrieved 30 May (2019)
5. Katila, R., Chen, E.L., Piezunka, H.: All the right moves: How entrepreneurial firms compete effectively. Strategic Entrepreneurship JNL **6**(2), 116–132 (2012). https://doi.org/10.1002/sej.1130. Retrieved 18 May 2017
6. Robehmed, N.: What is a startup? Forbes. Retrieved 30 April 2016
7. Kane, L.: Usability for Seniors: Challenges and Changes. UX Design for Seniors (Ages 65 and older), 3rd ed., p. 214. Nielsen Norman Gr. (2019)
8. Erin, G.: Why startups fail, according to their founders, Fortune.com, 25 Sept 2014
9. VIIMA, www.viima.com/importanceofinnovation
10. www.iso.org
11. OECD: Future of education and skills 2030, OECD Learning Compass 2030 (2019)
12. Entrepreneur Europe, 21 February 2021
13. Kostopoulos, G., et al.: ProMe: a mentoring platform for older adults using machine learning techniques for supporting the "live and learn" concept. Mobile Inf. Syst. **2018**, 8 pp (2018)
14. Nagai, Y., et al.: T-echo: promoting intergenerational communication through gamified social mentoring. UAHCI/HCII, Part IV, LNCS 8516, pp. 582–589 (2014)
15. Nedopil, C., et al.: The art and joy of user integration in AAL projects. AAL Association (2013)
16. Gwen, M.: Combating loneliness when you live alone and work from home. Fast Company Journal, 5 Dec 2020
17. Van Criekingen, K.: External information sourcing and lead-time advantage in product innovation. J. Intellect. Capital **21**(5), 709–726 (2020)
18. https://pathfinder.dc.siemens.com/ [22]
19. World Economic Forum: The Future of Jobs: Employment, Skills and Workforce Strategy for the Fourth Ind. Revolution, 176 pp. (2016)
20. Nullmeier, F.: The Failure of a Welfare Market: World Population by Year—Worldmet, www.worldmeters.info
21. State-Subsidized Private Pensions Between Economic Developments and Media Discourses, The Dynamics of Welfare Markets, Work and Welfare in Europe. P. Macmillan, Cham, pp. 131–159 (2021)

Adrian Pîslă Professor of engineering and management focused on machine-tools and robots numerical control, Head of Siemens PLM Training Center, TUCN, Romania.

Calin Vaida Professor at TUCN, specialist in Robotics, conceptual and structural design, robots simulation and programming, robotic systems manufacturing and implementation.

Eduard Oprea Co-founder and CEO at Digital Twin, a highly valued expert in PLM and Manufacturing Systems, with and a portfolio that includes the largest successful implementation PLM projects in Romania.

Bogdan Gherman Associate professor at TUCN, specialist in Robotics, structural design, mathematical modelling, robots simulation, robotic systems manufacturing and implementation.

Michel Deriaz PhD University of Geneva, specialist in Artificial Intelligenceindustrial expertise with achievements like FoxyTag camera warning system, and FoxyTour a museum and city guide.

Doina Pisla Profesor for Robotics, Vice-rector for doctoral studies within the Technical University of Cluj-Napoca (TUCN), Head of CESTER—Research Center for Industrial Robots Simulation and Testing.

Opportunities Presented by Industrial 4.0 Revolution to Revitalize the Railway Sector: A Review

Sinegugu Tshabalala and Khumubulani Mpofu

Abstract The Industry 4.0 paradigm presents a significant transformation in the way products are produced or how supply chains are managed. This is seen through the rapid growth and expansion of the Internet, information exchange and autonomous systems for the resolve of accomplishing collective goals. These opportunities can revitalize the railway sector specifically, in passenger and freight transportation. The paper is a review of how Industry 4.0 concepts such as the internet of things (IoT), big data analytics, artificial intelligence and block chain can do for the following areas of the railway ecosystem: Smart signalling and operations automation, automatic fare collection systems. The paper will review the technologies and opportunities presented by the technologies in the three described areas of the railway ecosystem. The review will intensely focus on how the technology will assist in revitalizing the rail sector. This section will also touch socio-economic issues such as local manufacturing (localization), job creation and skills development (I4.0 related skills).

Keywords IoT · Block chain · Big data · Artificial intelligence

1 Introduction

The rail sector has the potential to be the backbone of the economy as it provides accessibility for passenger and freight activities. Rail provides an environmentally friendly solution and considered a cost effective freight transport option [1]. Economic efficiency is what rail has to offer based on the capacity of moving people and a variety of commodities ranging from automotive vehicles, coal, and maize. Despite its vital contribution to the economy and the transportation sector,

S. Tshabalala (✉) · K. Mpofu
Tshwane University of Technology, Pretoria, South Africa
e-mail: sntshabalala@outlook.com

K. Mpofu
e-mail: MpofuK@tut.ac.za

K. von Leipzig et al. (eds.), *Smart, Sustainable Manufacturing in an Ever-Changing World*, Lecture Notes in Production Engineering,
https://doi.org/10.1007/978-3-031-15602-1_19

rail transportation in South Africa has been declining since 2016, due to dilapidation of rail infrastructure, thus, the transportation sector has seen inclining volumes to road freight. The reason for the decline is that the rail system has been plagued with vandalism and theft of critical infrastructure, causing rail transportation to be unreliable.

While many initiatives and studies have been put in place by the Department of Transportation to revitalize the rail sector, incorporating industry 4.0 can be an added advantage. These technologies offer security, accuracy and management of data as an asset to enable decision-making. The aim of the paper is to integrate the Industry 4.0 paradigm into rail infrastructure systems to increase and maintain the operational efficiency of current manufacturing processes and service delivery operations, and advance current automation methodologies to new heights.

The paper presents a review of what Industry 4.0 concepts using internet of things (IoT), big data analytics, artificial intelligence and block chain can do for the following areas of the railway ecosystem: smart signalling and operations automation, automatic fare collection systems and secondary data from PRASA's annual report to highlight problems and source opportunities.

1.1 Background

The railway sector is plagued with deterioration of key operations infrastructure, which has caused a steady decline in rail activities over the past years since 2008 [2]. Rail systems are considered to be complex systems as these systems involve the integration of system technologies and their sub-systems based on the functionality. These systems are intended to increase rail capacity, speed and mobility [1]. The systems we are going to discuss in this paper are rail signalling, operation automation and automatic fare collection.

1.1.1 Problems Affecting Rail Transportation

The theft and vandalism of signalling and overhead traction equipment is the root cause of late and cancelled trains, which directly affects the fare revenue stream. The theft has been so unprecedented that major rail routes have been closed, forcing commuters to use alternative means of transport [2]. Data quantifying contributing factors within the signalling component has not been made available in the annual report (Fig. 1).

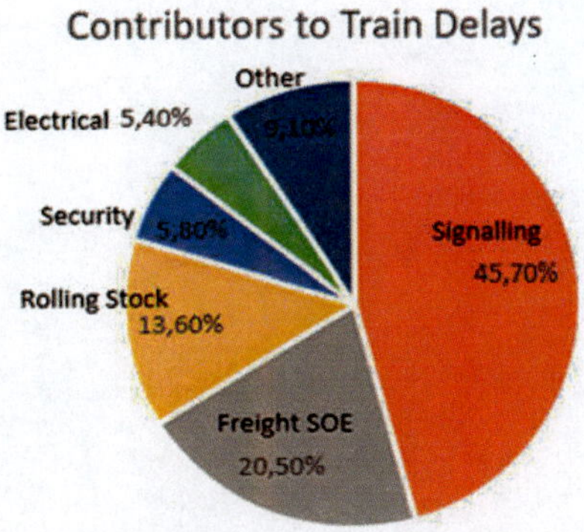

Fig. 1 Train delay pie graph [2]

2 Smart Signalling Using IoT and Data Analytics

2.1 Systematic Approach

2.1.1 Intelligent Rail Framework Design

China Railway built a railway intelligent transportation system (RITS) using artificial intelligence and related technologies to manage economic development through the rail sector. In Fig. 2, is the RITS architecture, which has outlined the structure systematically. This architecture is clear and attainable. Locally, adaption of this system would be feasible with Rail Safety of rail infrastructure as the priority, followed by Rail Efficiency and later Rail Quality of Services. The Rail Safety component is implemented first to collect data, create reliable train schedules and reduce signalling root causes with the current installed equipment. The national strategy was to bring back security guards and continue with the fencing. However, security camera's with night vision even facial recognition can be more effective. These days, the technology is able to send signals even to the police station. Modernization would be futile if the rail infrastructure is not secured first. The goal is to reduce the theft and vandalism occurrences as much as possible. This is so that the current routes are effective and efficient. A snow balling effect will occur to support opening closed routes. The rail passengers and local businesses will regain trust in rail transportation once again.

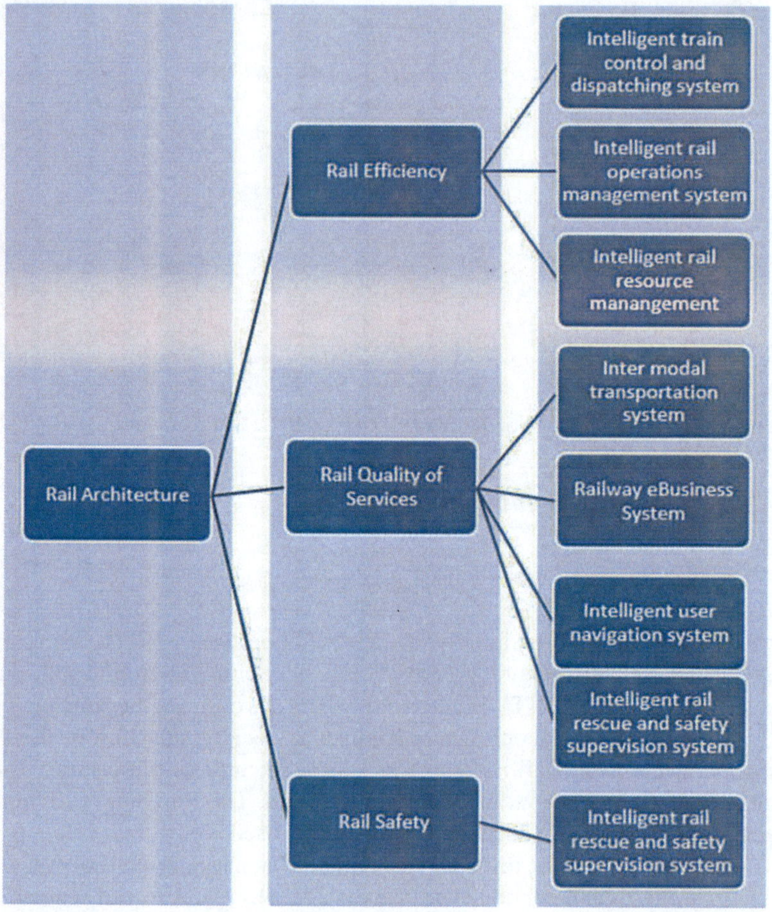

Fig. 2 Intelligent rail framework [3]

3 Data Analytics to Optimize Rail Operations

3.1 Proposed Framework

3.1.1 *Data Analytics Framework Using Deep Learning Algorithms*

Data science is the key companies can use to improve efficiencies, it is trending and analytically deals to solve complex problems. The architecture of artificial intelligence must emphasize evaluation and redesign the nature of the design process [4].

Currently, scheduling of trains is a big operational challenge and this is known by the data showing train are delayed as a contributing factor. Protected signalling and overhead traction equipment will result in accurate data coming through from the installed IoT sensors and equipment sensors. Big data analytics provides important information to enable the rail subsections to become optimized. Rail operations such as planning, can now be placed as firm train schedules, increasing the efficiency of the rail operations.

An MIS-IoT (Modular Intelligent Server Based Internet of Things Framework including Big Data and Machine Learning) framework, done by Aras Can Onal was reviewed. It is structured to allow future extensions to the programme to occur. It is intelligent by providing machine learning and deep learning methods on big data that is coming from IoT devices [5].

The proposed framework was built on a three-tiered framework:

- Machine learning and analytics layer,
- Data acquisition layer and
- ETL (extract, transform and load) layer.

The purpose of the framework was used to find sensor faults and find anomalies on historic sensor data. The proposed IoT framework has an advantage over the other IoT frameworks by implementing modular machine learning and deep learning methods. Modular developed architecture is created by retrieving data from a database using functions that are used as application programming interface (API) [5].

This type of framework would work best on the rail system due to large volumes of data coming from IoT devices. The system requires a learning module that will learn tasks and cluster data in models that can be used to simplify the data because of the complexity in rail systems. This deep learning module allows the mining of data to focus on specific data sets when required. This is because deep learning is considered to be more advantageous than conventional machine learning principles in producing accurate results. Also, there is a requirement for the system to predict analytics. In addition, it allows additional layers to be built on it. Complex systems require the flexibility offered by this framework. The end goal is to have efficient train schedules, however, the framework needs to systematically manage operating the signalling function for the accumulation of data to occur and the management of the volume of data coming through. The deep learning module will first improve the current system and will provide data for operational planning.

One of the problems that has been presented with AI and IoT is security [4]. Having connected devices over the internet may leave systems vulnerable to hackers. Blockchain was identified as the right technology to help buffer security. Blockchain plays a vital role by providing security to the system as each process is locked in blocked cell structures. The encryption on the programming does not allow any access to hackers and malicious viruses that are intended to corrupt the systems software. While blockchain can provide smart contracts, it also creates a layer of security in the system. The rail system is a complex system and these technologies can provide the necessary support that will not destabilize the existing system if well implemented. These technologies have the potential to meet and exceed the expectations to support digitalization needs.

4 Automated Fare Collection

4.1 Fare Collection Using Cloud Computing

Rail quality of services include the automation of fare collection methods as the current method is manual and tedious. Commuters spend time in queues also not all train depots sell tickets, so the flexibility of boarding at convenient stations is not possible. Currently, tickets are sold at designated rail stops and train conductors that operate from inside the railcar train, ensure everyone on board has a paid ticket. Tickets are only checked once passengers are in the train. This process is also labour intensive, with an automated fare collection system in place these manual operations are eliminated. The labour can be utilized in other operations with the business.

An automated fare collection system to benchmark from, currently exists with the Gautrain and operates efficiently as passengers without the smart card cannot go beyond the paid areas. The paid area and the unpaid area are demarcated and fenced. However, this process can become a wireless activity using cloud computing.

In the Automatic Fare Collection System (AFCS) passengers proceed to one of the gates that separate the paid area from the unpaid area [6].

In this scenario, the transporters have different manual and automatic ticketing methodologies for fare collection. The ticket is in the form of a smart card, printed-paper ticket, and both purchased from the ticket counter [6].

To remove the constraints experienced with the current fare collection model, the proposed model is an advanced mobile ticketing system (MTS) that utilizes mobile communication, cloud computing technology and RF detection. The ticket is procured using the commuter's mobile number, as one would do to buy airtime for instance. Thus, the system will recognise the commuter through a unique subscriber number. The model is designed to eliminate the smart card, paper ticket, manpower reduction at terminals, manage commuter data as an asset for decision making, easy and automated billing.

This system would need to be adapted to fit the local market because mobile service providers in South Africa operate differently and would require an authenticator at the train terminals. However, the fare collection can also be authenticated through a mobile application that can retrieve the payment directly from the commuter's bank account and charge it to a card or unique code that can be scanned when entering the paid area.

This system may require a thorough investigation by a population that has access to smart phones and build a flexible collection model that will be inclusive.

4.2 Local Manufacturing Opportunities

The forth industrial revolution should be seen as a catalyst for job creation than it is for job loss and that is to make South Africa the manufacturing hub of I4.0 equipment and devices. Jobs that are replaced by robots can be transferred into the manufacturing sector. Manufacturing opportunities that are in the rail sector that be assembly of IoT sensors, electronic components of signalling equipment or overhead traction equipment. 3D printing can be used to print spare parts of components that are old or those that have long lead times.

Local manufacturing of automated drones that can be used as surveying guards to secure the rail lines from theft and vandalism.

5 Conclusions

The South African rail system is faced with different challenges than other rail systems around the globe. It is plagued with theft and vandalism of expensive equipment. The data shows that signalling is indeed the biggest contributor to train delays. While, the data was mainly for the passenger rail sector, the system can be duplicated for freight rail systems.

The fourth industrial revolution presents technologies that can facilitate the revitalization of the rail sector. The main action takes place in the IoT architecture. Deep learning and blockchain are programming type modules that are embedded in the architecture to bring out the desired outcomes. Rail Systems incur large amounts of data and would require deep learning modules to handle the data coming from the connected sensors to produce good predictive analytics.

The paper looked at what has been done previously and how the adoptions of those proposed methodologies can aid the revitalization process. I4.0 is also a catalyst to increasing manufacturing activities with the country to service global industrialization. Indeed, there is great potential in commercializing these ideologies for a better Africa.

Acknowledgements I would like to thank the Department of Industrial Engineering at TUT. I acknowledge my sponsors, Gibela, they have awarded me this opportunity to study further. I would like to acknowledge my supervisor Prof Mpofu for his motivation and valuable contribution. I would like to thank my mother, always standing by me. I cannot thank the Almighty God enough for His divine strength and wisdom.

References

1. Railways Explained, (date posted 5 Sep 2020), How Railway Signalling Evolved from flags to 4G Network, [Online Video], URL: https://www.youtube.com/watch?v=ABVT8MOYb1g, Accessed: Date [14 Sep 2021]
2. Krishna, A.S.H., Mahalakshmi, D.M., Jose, P.S.: Automatic fare collection system for public transport corporation using fingerprint recognition with help of UIDAI. In: 2019 IEEE International Conference on Electrical, Computer and Communication Technologies (ICECCT), pp. 1–6 (2019). https://doi.org/10.1109/ICECCT.2019.8869312

3. Li, P., Jia, L.-M., Nie, A.-X.: Study on railway intelligent transportation system architecture. In: Proceedings of the 2003 IEEE International Conference on Intelligent Transportation Systems, vol. 2, pp. 1478–1481. https://doi.org/10.1109/ITSC.2003.1252729
4. Krishna, C.V., Rohit, H.R., Mohana: A review of artificial intelligence methods for data science and data analytics: applications and research challenges. 2018 2nd International Conference on I-SMAC (IoT in Social, Mobile, Analytics and Cloud) (I-SMAC)I-SMAC (IoT in Social, Mobile, Analytics and Cloud) (I-SMAC), pp. 591–594 (2018). https://doi.org/10.1109/I-SMAC.2018.8653670
5. Onal, A.C., Berat Sezer, O., Ozbayoglu, M., Dogdu, E.: MIS-IoT: modular intelligent server based internet of things framework with big data and machine learning. In: 2018 IEEE International Conference on Big Data (Big Data), pp. 2270–2279 (2018). https://doi.org/10.1109/BigData.2018.8622247
6. Chandra, D.G., Prakash, R., Lamdharia, S.: Mobile ticketing system for automatic fare collection model for public transport. In: 2013 5th International Conference and Computational Intelligence and Communication Networks, pp. 600–603. https://doi.org/10.1109/CICN.2013.131
7. Ghimire, A., Thapa, S., Jha, A.K., Adhikari, S., Kumar, A.: Accelerating business growth with big data and artificial intelligence. In: 2020 Fourth International Conference on I-SMAC (IoT in Social, Mobile, Analytics and Cloud) (I-SMAC), pp. 441–448 (2020). https://doi.org/10.1109/I-SMAC49090.2020.9243318
8. Ramamoorthi, S., Muthu Kumar, B., Mohamed Sithik, M., Thinesh Kumar, T., Ragaventhiran, J., Islabudeen, M.: Enhanced security in IOT environment using Blockchain: a survey. Mater. Today: Proc. (2021)
9. Saxena, S., Bhushan, B., Ahad, M.A.: Blockchain based solutions to secure IoT: background, integration trends and a way forward. J. Netw. Comput. Appl. **181**, 103050 (2021)

Hybrid Approach on the Project Development Within Automotive Industry

Ovidiu Popa, Cristian Făgărăşan, Adrian Pîslă, Cristina Mihele, and Rus Felician

Abstract The automotive industry is one of the most spread and synergic industries in dynamic markets, gathering around many different sectors from material extractions to robotics and cloud computing. The current research analyses the influences and particularities of project management and identifies the challenges and advantages of the two principal methodologies used in the software development projects within the automotive industry, the Agile and the Waterfall ones. Once with I4.0 and worldwide digitalization, the global software development industry becomes a very significant and dynamic pillar in the contemporary environment. It is more and more evident that almost all devices and machines in our lives become interconnected and more intelligent (software-driven). From ordinary smartphones to complex, innovative airplanes, the automotive industry is part of global digitalization. The software development industry evolution has two essential characteristics: agility and speed. These aspects have a real impact in all other industries, including software development, practically in any product or process development. In the last 30 years, automotive development started to focus significantly on vehicle's "brain" capabilities. In the automotive world, automation and controls represented the forerunner of the software development industry, but today, logic and the car's computers started to work with artificial intelligence. The complexity and the development dimensions are many times increased, and the industry took significant steps forward. The safety aspects of the vehicle are crucial for the entire product development. For this reason, the industry has many regulations and standards that ensure the product's quality and safety. Therefore, in automotive, the software development is not aligned with the general software development from the speed and agility point of view and the paper shows a possible harmonization of the methodology in automotive software Project development through a hybrid solution.

Keywords Automotive · Software project management · Hybrid methodologies

O. Popa (✉) · C. Făgărăşan · A. Pîslă · C. Mihele · R. Felician
Technical University of Cluj-Napoca, Memorandumului 28, 400114 Cluj-Napoca, Romania
e-mail: popaovidiu92@yahoo.com

A. Pîslă
e-mail: Adrian.Pisla@muri.utcluj.ro

K. von Leipzig et al. (eds.), *Smart, Sustainable Manufacturing in an Ever-Changing World*, Lecture Notes in Production Engineering,
https://doi.org/10.1007/978-3-031-15602-1_20

1 Introduction

Nowadays it is impossible to ignore the global value of the automotive industry that employs directly around five million people and indirectly many times more. Only in South Korea, this industry employs over 1.4 million people. Besides that, the automotive sector is one of the most dynamic fields reporting continuous and sustainable growth. It gathers around different other industries from raw materials extraction, materials development, design, machine-tools and robotics, manufacturing process, technologies development, cabling, testing, maintenance, ensuring, dealership, planning, programming, AI, and cloud computing, bringing to the market more than 60 million cars and trucks in a single year.

In the paper it is considered the evolution and the approach of the project management within the automotive industry based on efficient processes, focused on some aspects of reducing methodological disadvantages, aiming for a sustainable solution identification that may ensure the delivery of reliable and flexible software products. The impact of the envisaged methodology on the quality standards and project documentation is considered a vital part of automotive software development. By influencing the product development lifecycle, there is an opportunity to develop a structured framework that can be adopted to build software products more flexibly and agilely.

2 Automotive Industry

2.1 Industry Characteristics

The Automotive industry is dynamic and in continuous development. If at the begging of the XX century, in the United States, there was a ratio of 4.87 people at one car and today the ratio is approximately 1.3. The production volumes are significantly increased. From the complexity point of view, the development and car manufacturing include a series of industries like (Fig. 1):

- Machine building
- Electronic and Electrotechnics
- Mechatronic
- Software development.

In the XVIII century, the focus in the car's production was oriented on component's production, mechanics, and hydraulic system. Taking that into consideration, waterfall project management was a good solution for development and production. Today, many companies use pure waterfall development, especially in hardware development projects, but it is a challenge to continue just with the waterfall approach. The car's comfort and performance become a priority for the market. Major

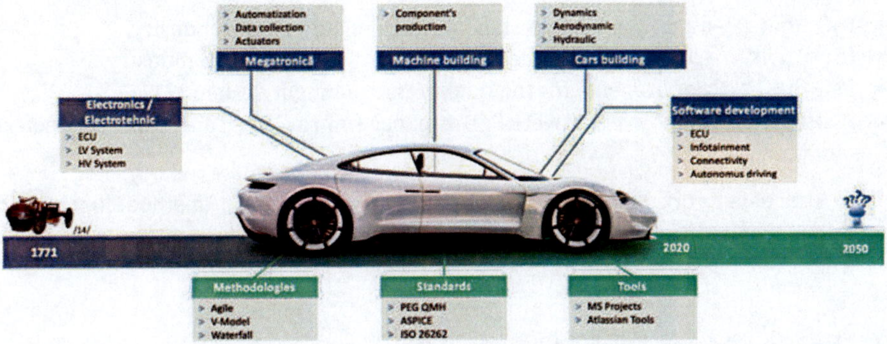

Fig. 1 Industries and key tools involved in the automotive industry

contributors are defined by the progress in the Software Development industry, electronics, mechatronics, aerodynamics, dynamics, and materials development industry. Looking at all areas, one of them (software development) has a critically needed characteristic: agility. The agility is specific for this industry, and because the industry had the main development process as a standalone industry, it could develop its own culture and particularities. The production and development differences are visible in all areas: simplified processes, minimal documentation requirement, fewer bounders, continuous improvement, etc. All these characteristics proved a very high efficiency in software development, but it can also face challenges. People assimilate the characteristics and the values that the industry promoted [1]. When the industry is integrated into other more waterfall industries the organizations can encounter some challenges, as resistance to change. The agile FRAMEWORK is indeed very agile, but the humans behind the process are still humans [2]. When people understand the benefits of an iterative project management methodology, it becomes very difficult to readopt a waterfall approach, even the reasons to do that are justified. That is a big challenge for the automotive industry, and some aspects of this challenge are presented in the paper.

2.2 *Standards in Automotive*

In 1771 when the first car was developed (Nicolas-Joseph Cugnot), the quality standards weren't a priority for developers and the reasons are more than obvious. Today, at the current production scale, it could be impossible to produce reliable and safe products without quality standards. For the actual automotive industry development, sustainable and safe products represent an important success market factor. The quality standards represent today a big part of the car's development and production. In favour of what was mentioned, specific for the automotive industry there are important standards such as:

- ISO 26262—Functional safety standard for the automotive industry.
- ISO/16949—Quality management system for the automotive industry.
- ISO 9001—Requirements for the quality management system [3].
- ASPICE—Automotive Software Performance Improvement and Capability determination [4].

In the paperwork, the focus is on project management methodologies and processes standards.

3 Project Management Methodologies

3.1 *Waterfall Methodology*

In the mechanical industries, it is proved that the waterfall approach (Fig. 2) has high advantages. The hardware productions and releases require a high level and detailed planning, and during the history, this approach was validated. For the manufacturing sector, iterative research could increase significantly costs [5].

3.2 *Iterative Methodology*

An iterative methodology (Fig. 3) came with a significant advantage from an agility and adaptability point of view. Nowadays the iterative approach is a commonly adopted solution in the software development industry [6].

Fig. 2 Waterfall approach in PM [5]

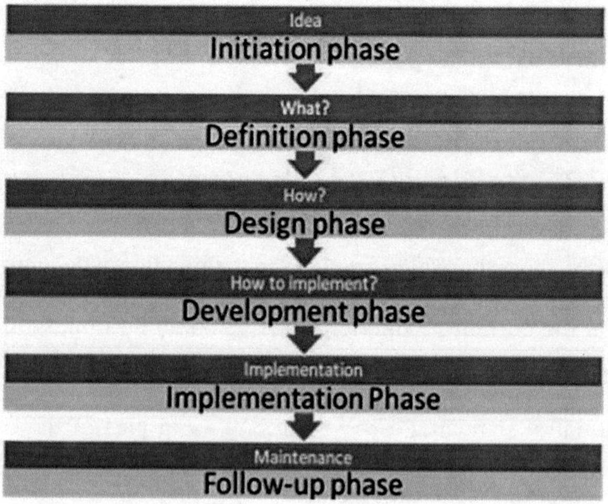

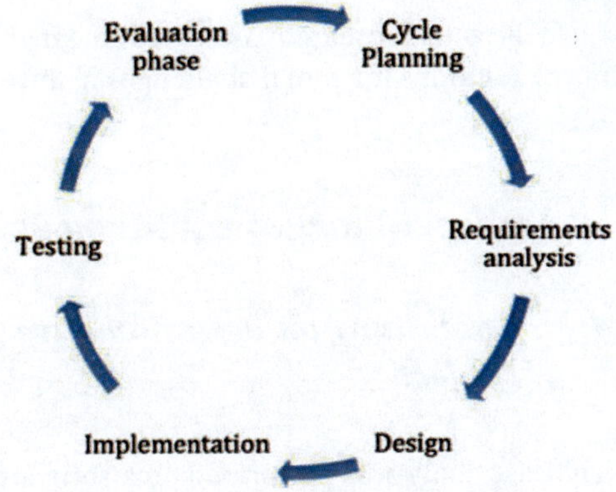

Fig. 3 Iterative approach in PM

3.3 *Hybrid Methodology*

Since a while ago, the industries understood the advantages and the disadvantages of the first two important methodologies. The industries managed to combine them in hybrid methodologies (Fig. 4). In essence, the principle is working, but the methodology is concentrated in the development processes. The required documentation level and the required quality processes become a challenge for the automotive software development world [2].

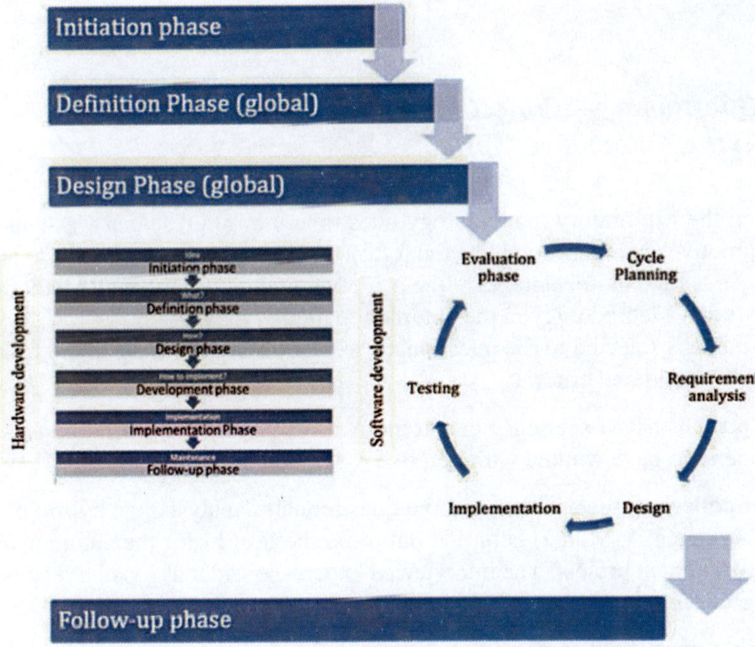

Fig. 4 Iterative approach in PM [5]

In the actual research, presented also in the paper, the focus is on the identification of the solutions for actual challenges in the automotive software development world.

4 Project Management Methodologies Under the Magnifier

4.1 Case Study on an Automotive Software Development Project

The case study was performed on a software development project, which has as the main scope to develop a web-based application for the series cars infotainment. The applied project management methodology was Agile, an iterative methodology.

From the standards point of view, the requirement was to have a complaint project with ISO9001 and ASPICE level 2. After a detailed analysis and many trials to find the best way to be agile in the automotive industry, there were identified the next challenges:

- The required documentation level affects the team processing and development velocity.
- The required quality management system increases significantly the project costs;
- The project team has difficulties following the waterfall approach in documentation, considered as the basis for the required standards implementation approach.

4.2 Questionary—Project Management Methodologies in the Automotive World

Applying the exploratory methodology questionary-based research was conducted (35 automotive engineers complete the form). The questionary scope was to analyze the challenges and the advantages of the Agile Methodology, Waterfall Methodology, and V-Modell Methodology in the automotive industry.

The persons selected to the questionary were chosen according to the following two major considered criteria:

- The persons have experience in automotive software development projects
- The persons have worked with iterative & waterfall methodologies.

After collecting the answers and the questionnaire analysis, the hybrid methodology (Agile and V-Model) is turned out to be the best fit for the automotive software development project. The interviewed experts consider as top challenges in the software development projects within the automotive industry:

- Achievement of the required documentation levels

- Achievement of the required quality standards
- Ensuring the interconnectivity and the dependencies between the hardware and the software production.

The specialist's answers and recommendations result in flexibility in the automotive software development projects and a faster adaption capability of the development team to the project processes.

5 Another Approach

In the first part of the study case, the challenges were identified and analyzed. In the second phase, an iterative V-model was mapped for the studied project. The software development workflow was applied to the software development project dedicated to the infotainment system, centered on the central media system from a car console. In the example presented below (Fig. 5), the project aim is to deliver the product just with the unit and the integration tests.

As it is shown in Fig. 5, the customer requirements play an essential role at the beginning of the project. It is necessary to elicit clear and precise customer requirements. The client Practical Application Guidelines (PAG) raises the business requirements with the company's internal or external implementation partner.

The Agile implementation methodology teams need to be cross-functional; they need to have all the skills necessary to deliver increments of working software. Usually, the development team has a business analyst who takes the client requirements and evaluates them together with the team. In this phase, the team collaborates with the client to refine and clarify the business requirements and transform them into functional requirements.

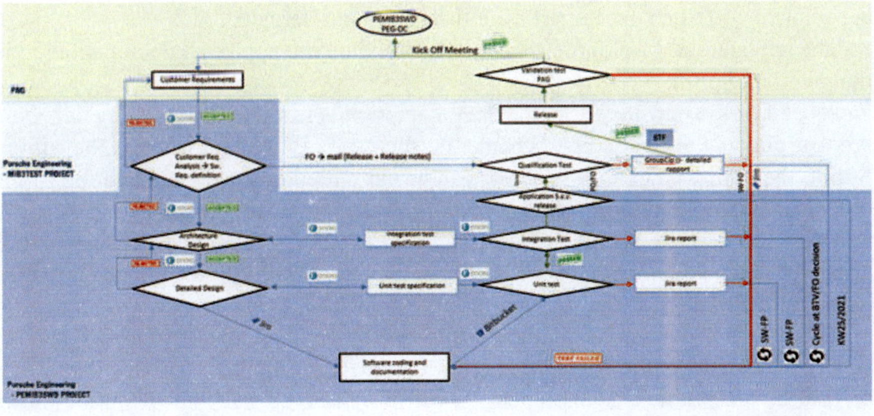

Fig. 5 Development workflow for a software component used in the infotainment system

After the client has approved the specifications, a more detailed analysis occurs, and the final software requirements are defined. After they are revised and validated, the application architecture is developed, which must be designed, reviewed, and approved.

The development team's next step is to create a detailed design of the application. Only after this phase is complete, the unit test specifications description follows, based on the design specifications.

In correlation with the defined architecture, the integration tests specifications are prepared. If during the testing phase, the acceptance criteria are validated, the final elaborated version can be delivered to the client.

In parallel, the customer may have his processes to validate the acceptance criteria. For big projects, the companies have an acceptance team that verifies that the functionality provided is according to the business requirements and the implementation satisfies the business need.

After this validation stage, the customer is doing qualification and validation tests followed by the approval forms.

As it may be observed, a V-Model methodology is used for the development of the software component. However, the application can be developed in much more iterations by leveraging the implementation methods of the Agile methodology. In this sense, the most popular ones are Scrum and Kanban, which are well-known for their success rates and easiness of usage. Project iterations are called "Sprints" in the Scrum framework.

The Agile development process in the documentation and processes was also adjusted to create a new working methodology, as shown in Fig. 6.

All projects have kick-offs meetings during the initiating phase. After this formal meeting with the project team, as Fig. 6 shows, "Sprint X" is started, as the first sprint of the development phase. In this iteration, the first version of the project's documentation is produced, and any technical project setup can be done.

Once the documentation is revised and approved during the sprint review meeting, the priorities for the next iteration are discussed and defined.

The application development can begin based on the approved documentation in the previous Sprint.

As part of the new iteration, the documentation is further developed by adjusting it or introducing any missing elements. Furthermore, in the first week of the current Sprint, the approved change requests can be submitted in the documentation so the client can adapt the requirements according to the dynamic changes or needs.

Once the second Sprint is finalized, a new version of the documentation is approved. This documentation will become the base of the application development in the third Sprint.

Through this interactive method, the project can permanently be flexible related to new requirements of the client through a Change Request process, taking the overall picture of the product development. It can be observed that the V-Model allows systematic documentation and planning for applications in the automotive industry.

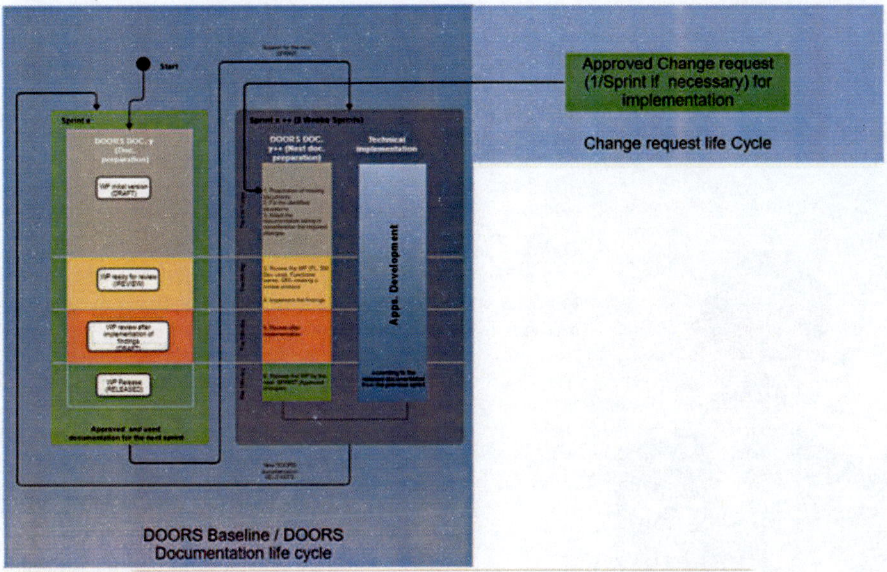

Fig. 6 Iterative methodology integrated into V-Model development workflow

6 Conclusions

The Project management methodologies have as main scope to help to correctly finish the projects, regardless of the market on which must perform. The automotive industry has procedures and standards, which are well-defined. Still, the sector should inspect and continually adapt the development procedure to be faster, more flexible, and incorporate the Agile mindset. Even though the hardware development and the standard's requirements come with many limitations, the results can be significantly increased if the Agile spirit is adopted in all industry branches. In the third phase of the study case, the feasibility of the developed methodology was analyzed. The project team started to implement the processes and the documentation needed for the quality standard without major blocker. Even that, the team should be planted taking into consideration additional effort between 0.2 and 0.4. For the automotive software development area, a combination of the Agile software development lifecycle and a V-Model documentation process adapted to work in short, incremental releases can be a suitable solution in this dynamic and rigorous context.

References

1. Richard, L.H.: Leadership: enhancing the lessons of experience. Ginnett, & Curphy, McGraw-Hill/Irwin (2010)
2. Kelley, O.: Transitioning from Waterfall to Agile Project Management (2015)

3. International Standard ISO 9001: Quality Management System-Requirements (2021)
4. Author VDA QMC Working Group 13/Automotive SIG, 2021Automotive SPICE, Process Reference Model, Process Assessment Model, Version 3.1
5. Baars, W.: Project Management Handbook (2006)
6. Narcis, P.: Project Management MZT815 Materiale Auxiliare (2021)

Ovidiu Popa is Engineering and Management Ph.D. student, at the Technical University of Cluj-Napoca, Romania with a focus on introducing the adaptive software project management methods into the automotive industry.

Cristian Făgărăşan is an Engineering and Management Ph.D. student at the Technical University of Cluj-Napoca, Romania, with a focus on software project management area.

Adrian Pîslă, Director of Siemens PLM Training Center at Industrial Engineering, Robotics and Production Management Faculty, Technical University of Cluj-Napoca, Romania.

Cristina Mihele is Engineering and Management Ph.D. student, at the Technical University of Cluj-Napoca, Romania with a focus on the differences between management and leadership, with an emphasis on the present leadership principles.

Rus Felician is Engineering and Management Ph.D. student, at the Technical University of Cluj Napoca, România with a focus on the studies on sustainable energy alternatives.

Human Integration

The Application of ArchiMate for Planning the Implementation of Manufacturing Management Systems

D. W. Gibbons and A. F. van der Merwe

Abstract This paper investigates the use of a model-based approach for planning the implementation of manufacturing management systems. Manufacturing management systems such as ERPs, MESs and PLMs are defined by various standards and frameworks and are developed and implemented based on the specific industry they are to be operated within. Such implementations require conformance to large document-centric requirement sets and are depicted in cumbersome vendor-specific specifications. This paper utilizes the ArchiMate language and notation for planning such implementations and engaging stakeholders prior to domain-specific implementation.

Keywords Manufacturing systems · Model-based systems engineering · ArchiMate · Architecture frameworks · Industrie 4.0

1 Introduction

Modern manufacturing is becoming more automated and complex with the increase of smart and advanced technologies and systems. For companies to stay competitive in these modern manufacturing environments they have to be able to be competitive on the four fundamental production objectives, namely cost, quality, flexibility and time [1]. To improve on these metrics, the manufacturing and production systems need to be engineered and effectively managed with these metrics in mind. Implementing Manufacturing Management Systems (MMS) is by no means a trivial task and requires a host of "hard" and "soft" skills to perform successfully. This paper addressed the "hard" factors of MMS implementation and integration. "Soft" factors such as social, change, environmental and organizational factors were excluded from

D. W. Gibbons (✉) · A. F. van der Merwe
Department of Industrial Engineering, Stellenbosch University, Stellenbosch, South Africa
e-mail: duncang@sun.ac.za

A. F. van der Merwe
e-mail: andrevdm@sun.ac.za

K. von Leipzig et al. (eds.), *Smart, Sustainable Manufacturing in an Ever-Changing World*, Lecture Notes in Production Engineering,
https://doi.org/10.1007/978-3-031-15602-1_21

the scope of this research. Implementing MMSs require significant investments to purchase software licenses, improve infrastructure and train personnel. Such undertakings additionally require considerations for organisational change and can take years before their benefits are realized. These risks are exacerbated for small and medium-sized enterprises [2].

The risks associated with implementing MMSs have been identified in previous literature [2–4]. Some of the causes for MMS implementation failures include:

- Not involving system users in the planning phase.
- Poor communication amongst stakeholders.
- Poor alignment between process and technology and resource domains.
- Lack of an effective implementation methodology.
- Poor specification of goals and requirements.

This paper aims at investigating the use of a model-based approach to planning the implementation of MMSs and address these implementation risks.

2 Literature Review

2.1 Manufacturing and Production Systems

The term *manufacturing system* is often used interchangeably with the term *production system* and there exist many definitions for both [5–8]. The term *manufacturing* is inherently related to transforming materials into products, whereas *production* is a more generic term used to define a process of producing or provisioning an item. Hitomi [7] defines manufacturing as the production of *tangible* goods, whereas production is defined as the making of something new, tangible or intangible. A manufacturing system is a specific type of production system, but a production system is not always a manufacturing system. Hitomi [7] notes that manufacturing is recognised as a function of production, and the manufacturing system is, therefore, a subsystem or subset as Suh, Cochran and Lima [8] define it, of the wider production system. Groover [5] supports this notion and defines manufacturing systems as "components of a larger production system, which is defined as the people, equipment, and procedures that are organized for the combination of materials and processes that comprise a company's manufacturing operations". De Weck et al. [9] define manufacturing systems similarly as "the equipment, processes, people, organization, and knowledge, as well as the interactions of these that are involved in the manufacturing of a given end product". Both the production and manufacturing systems contain subsystems and functions, Hitomi [7] defines manufacturing systems in terms of three fundamental aspects:

1. Structural, being the facility layout and assemblage of physical components. The static view of the system.

2. Transformational, concerning the conversion and logistics processes that transform resources and materials into products and transport materials within the facility.
3. Procedural, being the management aspect of the system to meet objectives. Further defined in terms of planning, implementation and control.

Expanding on the procedural aspect, the management system, Hitomi [7] defines the function of an MMS as "plan and implement productive activities to convert raw materials into products to meet production objectives, and control this process to reduce or eliminate the degree of deviation of actual performance from the plan". From this definition of MMSs, it is evident that the effective design, engineering and implementation thereof is a key aspect in ensuring production objectives, such as cost, quality, productivity and time, are met. This ultimately leads to improved manufacturing competitiveness.

Groover [5] conceptually breaks down production systems into facilities, largely physical, and manufacturing support systems, largely functional and containing the managerial aspect of production. Figure 1 presents the technical aspects of a production system according to Groover [5].

Similarly, Hitomi [7] presents the key functions of a manufacturing firm as presented in Fig. 2. Along with these functions, Hitomi [7] identifies material and information flow as the core flows that flow between these functions. Noting information can be further defined in terms of managerial, technological and strategic management information flows. At a high level, the function of management,

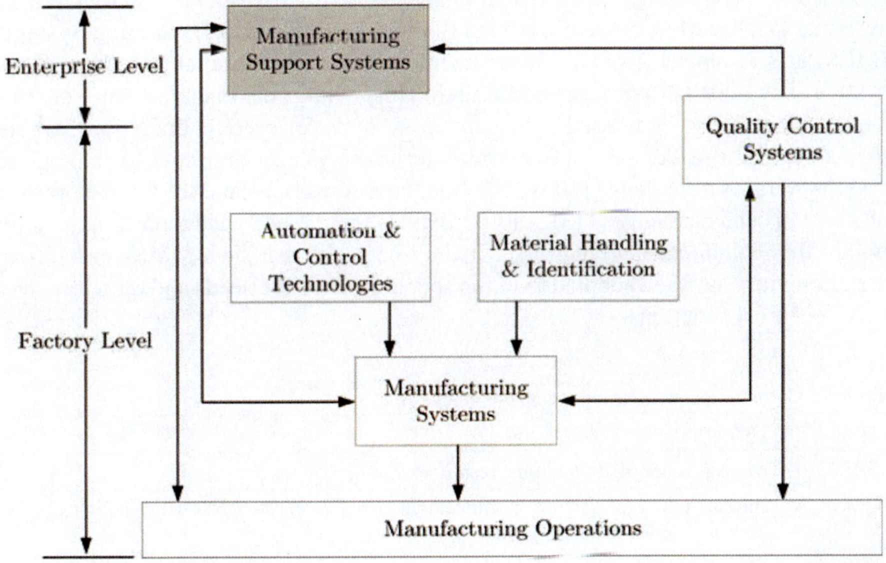

Fig. 1 Technical aspects of a production system [5]

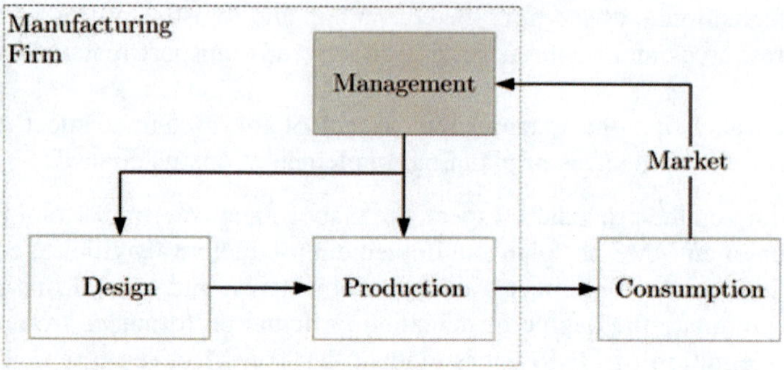

Fig. 2 Functions of a manufacturing firm [7]

and management systems, in the manufacturing firm is to manage the design and production, or manufacture, of products for the market.

2.2 *Manufacturing Management Systems*

Management systems in general attempt to achieve objectives by integrating and coordinating organisational resources [7]. These systems are often realised and implemented as management information systems (MIS). In a manufacturing setting such systems commonly implemented include Enterprise Resource Planning (ERP) systems, Quality Management Systems (QMS), Manufacturing Executing Systems (MES) and Product Lifecycle Management Systems (PLM) amongst others. Such systems have distinct purposes and functionality within the manufacturing organisation. Identifying the needed management systems for specific operations, where they fit into the wider production and enterprise systems and how to implement these systems is a challenge [10–12]. Scheer notes the importance of frameworks for addressing such challenges [13]. Table 1 defines these MMSs and their functionality within the manufacturing enterprise. It shall be noted that such MMSs in industry are often integrated and adapted to fit the specific enterprise needs and therefore may serve different functions.

Table 1 Common manufacturing management systems

MMS	Definition
ERP	Manage enterprise functions, resources and operations [5]
QMS	Interrelated or interacting organisational elements to establish quality policies, objectives and processes [14]
MES	Shop floor management and control including the release of production orders, monitoring work and equipment statuses and inventory control [5]
PLM	Manage products and their data from concept through to their end of life [5]

As can be seen by the MMS functionality descriptions in Table 1, the ERP manages enterprise functions, resources and operations, the QMS manages quality in the organisation, the MES manages and controls the shop floor and the PLM manages the products that flow through the organisation. Each of these systems, along with many others, are seen as subsystems in the wider enterprise system and are imperative in managing and operating modern manufacturing systems.

2.3 Approaches for Architecting and Implementing Manufacturing and Management Systems

The associated challenges with identifying required MMSs, where they integrate into the wider system and how to implement and integrate them into the manufacturing enterprise has been a challenge ever since these systems were first conceptualised. These challenges were formally investigated during the late 1980s and early 1990s under the name of computer integrated manufacturing (CIM) and through the ESPRIT consortium AMICE amongst others. This consortium and their projects aimed to "design, develop and validate an Open System Architecture for CIM (CIMOSA) and to define a set of concepts and rules to facilitate building and operating future CIM systems" [15]. The CIMOSA modelling approach, presented in Fig. 3, is focused on the definition of a reference architecture from which particular instances of this architecture, namely particular architectures, can be built for specific applications and objectives.

Although CIMOSA is somewhat dated nowadays, the approach taken for the development thereof and the architecture itself, along with others from that period

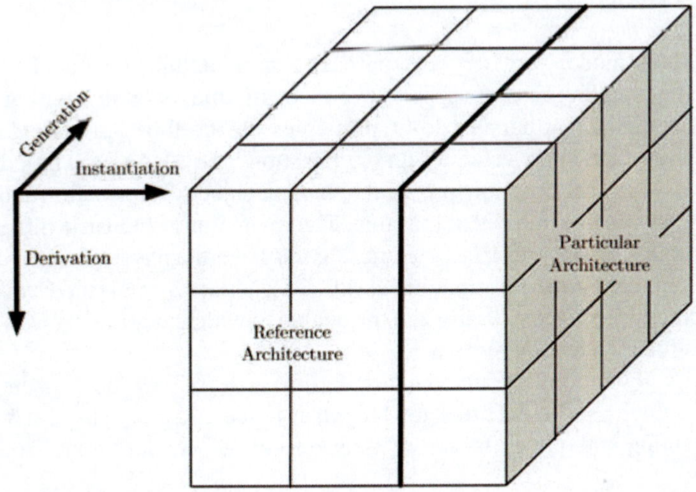

Fig. 3 CIMOSA approach [15]

such as the Reference Model for CIM [16], has been influential in the development of various enterprise and manufacturing architectures, frameworks and models since the 1990s. Some of these that have been influenced by the CIM developments in the 1990s or show similarities to these approaches include ISA-95 standards [17], Reference Architecture Model Industrie 4.0 (RAMI4.0) [18] and the smart manufacturing ecosystem works by the National Institute of Standards and Technology (NIST) [19].

Additional architectures and frameworks of interest in the manufacturing industry include the Architecture of Integrated Information Systems (ARIS) [13], the Generalized Enterprise Reference Architecture and Methodology (GERAM) [20], Industrial Value Chain Reference Architecture (IVRA) [21], Unified Architecture Framework (UAF) [22] and the HORSE framework [23] to name a few.

2.3.1 ISA-95

The IEC 62264 series of international standards, or ISA-95 as they are commonly known, is a series of standards for defining and modelling the interface between enterprise and control activities [17].

ISA-95 defines two hierarchy models that the standards are based on. The functional hierarchy, Fig. 4, and the equipment or organisational physical structure hierarchy, Fig. 5. A host of object and activity models with associated manufacturing terminology are defined for levels 4 and 3. Additional series define the information for lower levels for the specific types of manufacturing control, such as the ISA-88 series for batch control [24].

2.3.2 RAMI4.0

The RAMI4.0 model provides an approach for modelling technical objects, or "assets", and all their relevant aspects over their lifetime, or "vita", within an enterprise to be compliant with industrie 4.0 principles and its ethos. An "asset" is defined as an "object which has a value for an organisation" [18]. RAMI4.0 broadly classifies assets as either information, physical or human objects. Information and physical assets are composable and decomposable. The main aim of Industrie 4.0 as defined according to RAMI4.0 is to take "technical assets from the physical world and virtually represent them in the information world" [18]. Assets are classified by the degree of their communication capability and the degree to which they are "known" by their administrative information system.

The RAMI4.0 model, presented in Fig. 6, aims at describing these assets and their combinations with sufficient precision [18]. This model provides a logical framework for representing enterprise assets to a precise level of granularity and from various viewpoints.

The RAMI4.0 model contains three axes, the architecture layers axis, lifecycle and value stream axis (based on IEC 62890 [25]) and the hierarchy levels axis (based

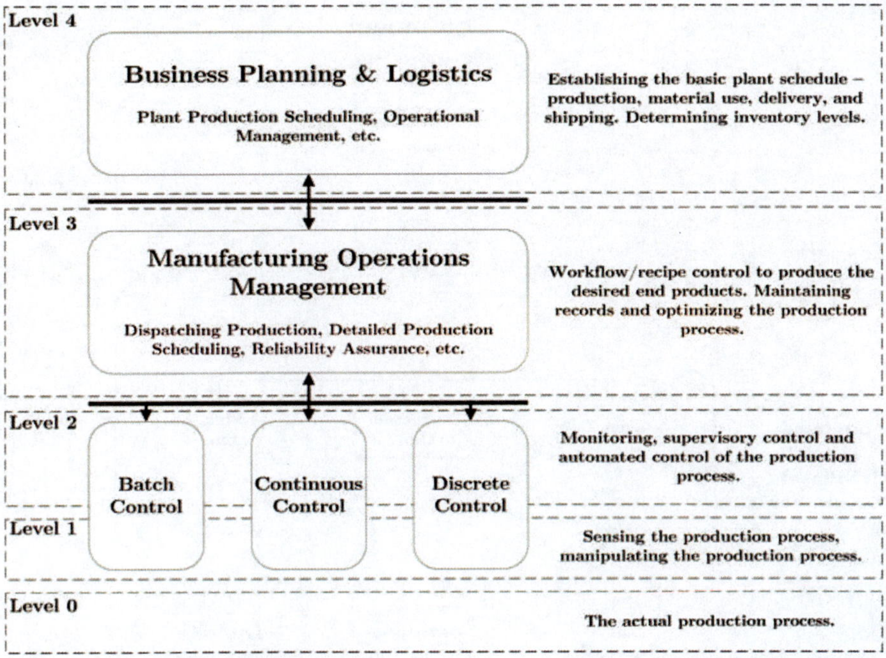

Fig. 4 ISA-95 functional hierarchy [17]

on IEC 62264 [26], previously ISA-95, and IEC 61512 [27], previously ISA-88). Interactions occur at adjacent layers, e.g. between the business and functional layer. Layers can not be skipped but can be passed through [18]. Each layer can contain temporal (lifecycle and value stream) and structural (hierarchy levels) dimensions resulting in a logical representation of enterprise assets for improved administration and interoperability.

2.3.3 NIST Smart Manufacturing Ecosystem

Although not a formal architecture or framework, the NIST smart manufacturing ecosystem and related works provide a holistic view of smart manufacturing, the systems involved and the applicable standards. Smart manufacturing is defined as the "application of networked information-based technologies throughout the manufacturing and supply chain enterprise" [28]. Figure 7 presents the NIST smart manufacturing ecosystem. This diagram provides initial insights into the common MMSs implemented in manufacturing organisations. Three main lifecycles are defined, business, product and production lifecycles. These lifecycles merge during manufacturing, defined as the manufacturing pyramid and analogous to ISA-95. There are

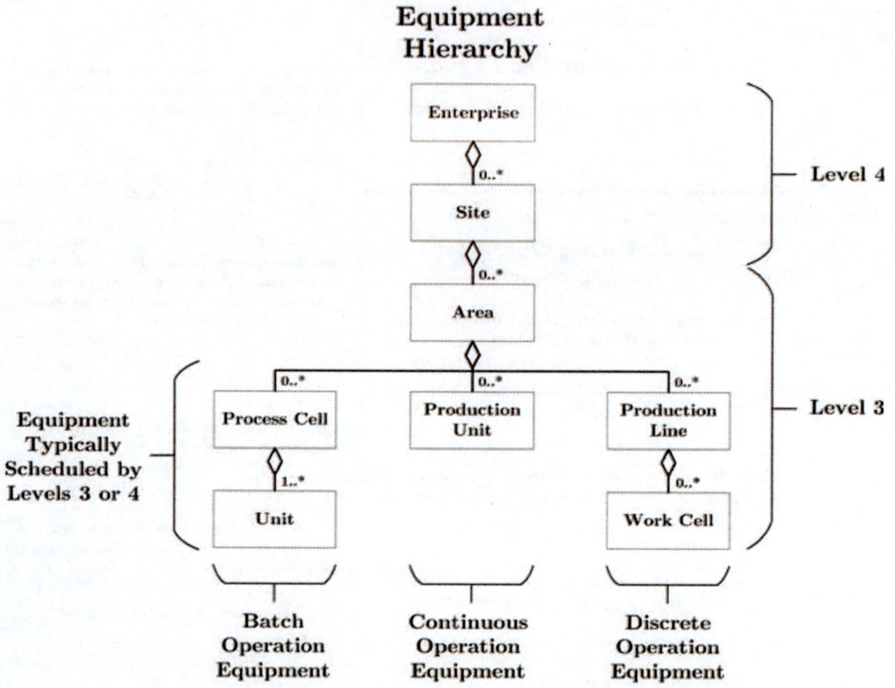

Fig. 5 ISA-95 equipment hierarchy [17]

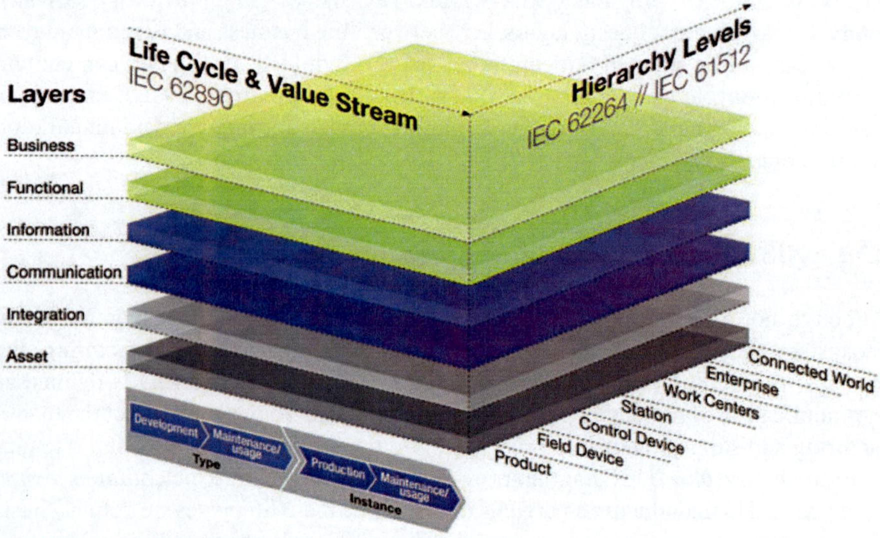

Fig. 6 RAMI4.0 Model [18]

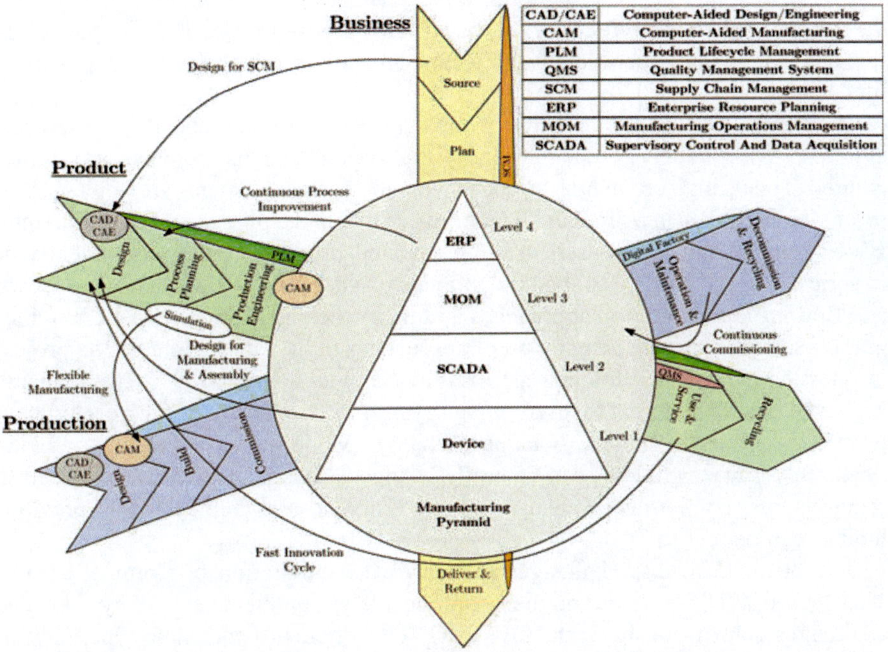

Fig. 7 NIST smart manufacturing ecosystem, adapted from [19]

however many interactions and feedbacks that occur during, before and after manufacturing in these manufacturing enterprise lifecycles. This results in a complex system of interacting and interoperating subsystems and elements.

Such architectures and frameworks that aim to model complex systems often suffer paradoxically from either modelling the system complexity but then being too complex to be understood well by stakeholders or reducing the system complexity in simple to understand diagrams but then losing system information and granularity.

2.3.4 Software and Systems Deployment Strategies

In the domain of software and systems engineering, implementation of the developed system is a critical phase of the engineering process. Implementation and integration are closely linked and go hand in hand, often these terms are used interchangeably. Implementation is the installation and commissioning of system elements into its operational environment. Integration refers to the assembly and operability of these implemented system elements into a whole and functioning system [29]. Implementation and integration occur at the various levels of a system as well as within the wider system of interest following a bottom-up approach. The implementation within the system context is the focus of this paper. This can be viewed as the implementation of a system in its operational architecture and falls within the system deployment lifecycle phase.

Davenport [30] identifies three main approaches to implementing enterprise systems. A big-bang approach whereby the system and its elements are implemented at once, an incremental approach whereby system elements are implemented in small increments in a predefined order over time, and a phased approach whereby system functionality or subsets are implemented in phases either concurrently or in series. The INCOSE Handbook elaborates on these approaches with Top-Down and Bottom-Up integration approaches amongst others [31]. Noting that often integration strategies are a combination of approaches to best fit the integration project in question [31]. An additional approach is the agile approach [31]. As the name suggests, this approach is focused on agility and the capability of the development team and system of interest to adapt to change. Model-based approaches provide improved agility as models can be easily changed, change effects can be traced to requirements, configuration changes can be recorded and elements from previous models can be reused.

Jain et al. [32] note challenges related to the integration of Commerical Off The Shelf (COTS) systems and their components within the enterprise system. The challenges stem from the variability in COTS systems of the same type, lack of standard system architectures, and vendor control lock-out to name a few [32]. Yakimovich et al. [33] note often specialized software, referred to as glueware, is required to effectively integrate COTSs within the enterprise system. This increases cost, time to operation and can reduce overall system quality or lead to unforeseen nonconformances.

2.4 Related Work

In recent years there have been similar investigations into the applicability and use of ArchiMate for manufacturing and management system architecting. Franck et al. [34] investigated applying ArchiMate 3.0 for modelling smart manufacturing operations in accordance with ISA-95. They noted there is a substantial amount of capability in the ArchiMate language to do so, but certain issues do arise when seeking full conformance to ISA-95. Certain deficiencies were identified and solutions were proposed for these. A case study was performed to model a change in the production process of a steel manufacturer [34]. Moones et al. [35] investigated the interoperability of certain MMSs. They utilized ArchiMate to model the integration between manufacturing and PLM and how information is to be exchanged between the ERP and MES systems as defined by ISA-95 [35]. Aldea et al. [36] propose an approach for modelling enterprise architectures in the context of Industry 4.0. Through the application of ArchiMate, it was demonstrated how enterprise data models and systems can be architected. A case study was performed whereby MES metadata was used to generate a model as defined by the architecture model [36].

The literature on implementation and integration approaches for such management systems is somewhat sparse and formal implementation and integration approaches tend to be vendor-specific. Schuh et al. [2] proposed a conceptual process-orientated framework for PLM implementation. Jagoda and Samaranayake [37] proposed a framework for integrating and implementing ERP systems based on Kale's [38] SAP R/3 implementation guide. This framework consists of a three-phased approach, pre-implementation, implementation and post-implementation which they term a stage-gated approach [37]. Le Duigou et al. [39] proposed a framework for integrating PLM systems within small and medium enterprises and applied UML for modelling the PLM system.

3 Methodology

From the review in Sect. 2, it is evident that manufacturing systems are complex and they contain various systems aspects such as technical, managerial, social and financial to name a few. de Weck et al. [9] classify such systems as *engineering systems*. Drawing upon this systems approach we come to the systems engineering concepts of views and viewpoints. These allow us to see the system of interest in different aspects to meet different aims and objectives. ISO/IEC/IEEE 42010 [40] defines an architecture view as the "work product expressing the architecture of a system from the perspective of specific system concerns" and viewpoint as the "work product establishing the conventions for the construction, interpretation and use of architecture views to frame specific system concerns". The ArchiMate 3.1 specification defines these terms simply as "a view is what you see, and a viewpoint is where you are looking from" [41]. The viewpoint frames the concerns ("drivers" in the ArchiMate language) of stakeholders. These system concerns "pertain to any influence on a system in its environment" [40]. A purpose is a type of system concern. System viewpoints that address concerns for MMSs include:

- Business—What capabilities does the MMS provide, does it fulfil a need and which requirements are imposed on the MMS?
- Operational—What functionality does the specific MMS provide, which systems does it cooperate and interoperate with and what information is transferred between these systems?
- Technological—What technological and physical infrastructure is required for the MMS to operate?

An efficient system model contains the required system views to address all stakeholder and system concerns, no more, no less.

The methodology applied for the development of the proposed framework in this research leverages systems engineering principles. As MMSs are, in and of themselves, systems, the application of systems engineering principles is fitting. The systems engineering principles of interfacing, implementation and integration are deemed vital for implementing MMSs effectively.

Applicable modelling languages were reviewed for this application [31, 41–43]. The languages reviewed and their comparison with one another are presented in Table 2. The ease of understanding the modelling language is of high importance as stakeholder review of the models is required and the implementation is performed by humans, many of which will not have a background in modelling and software development.

It is evident from the comparison in Table 2 that ArchiMate is highly applicable. This language is:

- Easy to understand notation, leading to increased stakeholder engagement.
- Capability to model a wide range of aspects such as process, physical, information, applications amongst others.
- Ability to model integrated views and cross-view diagrams through customizable diagram views.

ArchiMate was not intended to model systems in great detail. Languages such as SysML are better suited at this [42]. The ArchiMate language is suited for modelling architectures at a higher level of abstraction, thereafter languages such as the ones reviewed in Table 2 provide greater benefits for detailed and domain-specific design.

Table 2 Comparison of modelling languages

Language	Intended use	Benefits	Drawbacks
UML	Software systems	• Able to model functional and physical aspects • Model-based language	• Requires knowledge of software modelling concepts • Object-orientated focused • Fixed diagrams and views
SysML	Systems	• Applicable to systems in general • Model-based language	• Requires knowledge of the language to understand models • Fixed diagrams and views
IDEF0	Functional modelling	• Very easy to understand • Ability to model high amounts of inputs, controls and outputs • Affinity for decomposition	• No logical constructs, physical infrastructure or requirements • Not model-based
BPMN	Business processes	• Easy to understand • Many software vendors • Model-based language	• Only applicable to processes • Lacks context modelling capabilities
ArchiMate	Enterprise architectures	• Very easy to understand • Model-based language • Integrated with other model-based languages • Customizable views	• Only applicable to high-level modelling

The Archi[1] modelling tool was used to develop the models in this paper in accordance with the ArchiMate 3.1 specification [41], the reader is referred to this specification for the definition of the notation. The ArchiMate modelling language and notation was selected as ArchiMate was developed for modelling enterprise architectures, which incorporate MMS architectures.

3.1 Modelling Approach

The approach defined in Fig. 8 is defined as a heuristic for planning MMS implementations and is focused specifically on the planning phase. This phase is driven by a need to change and improve on the current system architecture and results in

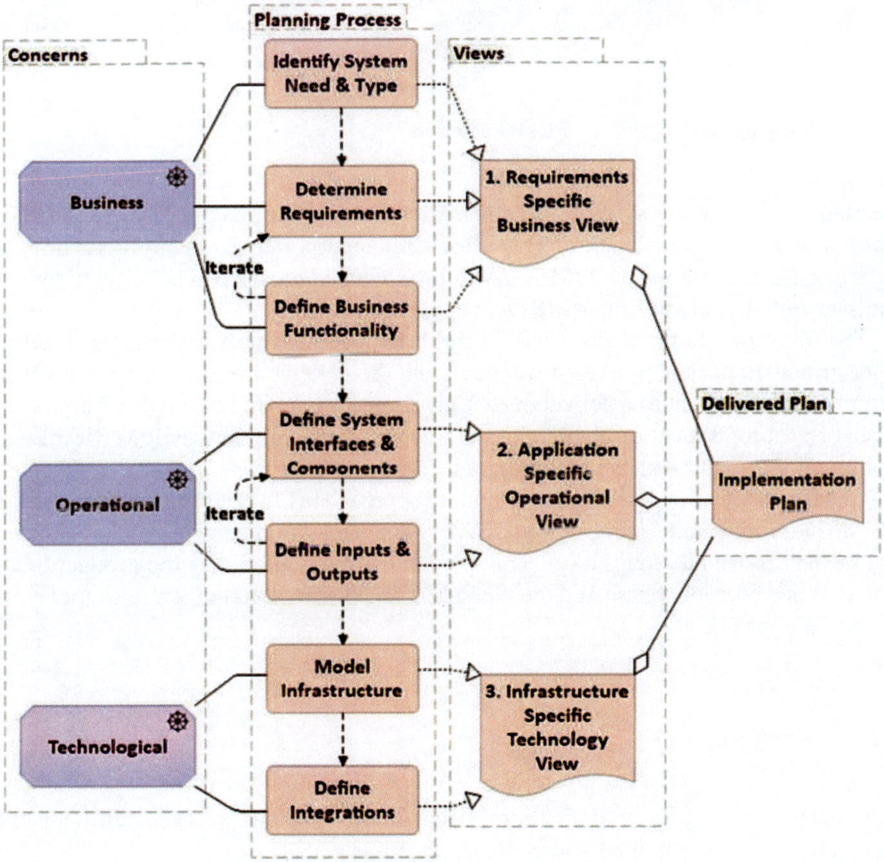

Fig. 8 Manufacturing management system implementation planning approach

[1] https://www.archimatetool.com/.

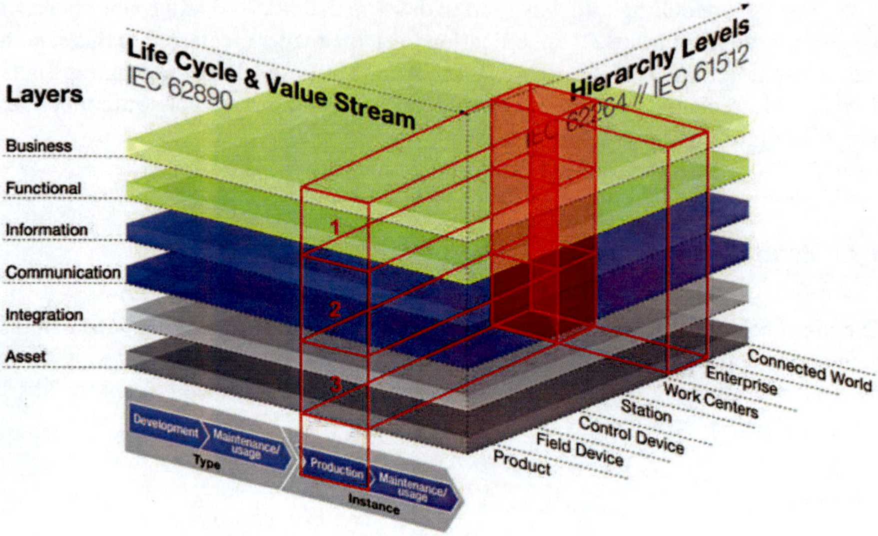

Fig. 9 Mapping views against RAMI4.0 framework

a plan for conducting system implementation. Integral to this approach is the definition of system functionality. This allows for incorporating system modularity, a common aspect of modern MMSs. By doing so, the approach can be adapted for the implementation of specific MMSs and applications.

Reading the diagram from left to right, enterprise concerns drive the need to perform work packages that realize views and these views are aggregated to form an implementation plan as a deliverable. When read from top to bottom this framework begins at a high level of abstraction and decomposes and derives lower level and physical information as the planning process is performed.

The three views output from the planning approach in Fig. 8 are mapped against the RAMI4.0 framework in Fig. 9. The PLM system and its implementation are located at the enterprise hierarchy level. This implementation falls under the production of an instance lifecycle phase and the views 1, 2 and 3 are traced to their relevant layers in the framework.

4 Case Study

Following the approach defined in Sect. 3, an illustrative case study for the implementation of an open-source PLM system is performed.

The business view is presented in Fig. 10. The requirements that drive the need for the installation of a PLM system are defined at the top in purple.

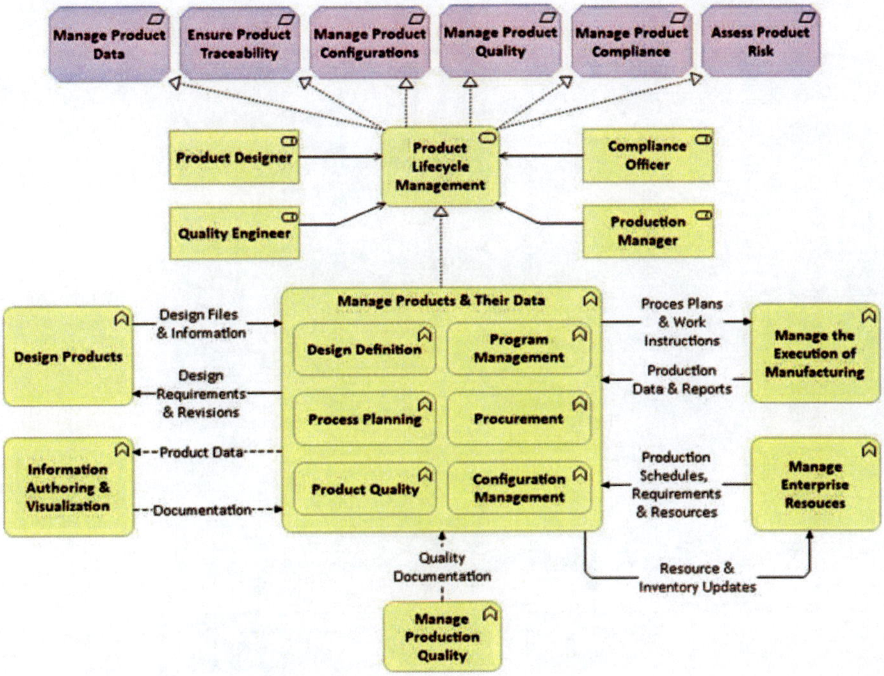

Fig. 10 Requirements specific business view

The business functions in the centre are derived to fulfil these requirements. High-level information flows between the required business function, *Manage Products & Their Data*, and external functions are defined. Figure 10 defines the requirements specific business view and is compliant with the ArchiMate requirements realization viewpoint [41]. Solid full arrows indicate triggering items and dashed full arrows indicate item flows per the ArchiMate 3.1 specification [41]. The reader is referred to this specification for the definition of the other arrows.

The application structure is defined in Fig. 11. This view defines the PLM system and the modules to be installed to fulfil the business functions as defined in the business view in Fig. 9. Interfaces with external enterprise systems are defined. Figures 11 and 12 are extracted from the application-specific operational view and separated into two figures for improved legibility. This view is compliant with the ArchiMate application usage viewpoint [41].

The information inputs and outputs are defined in Fig. 12. This information is transferred across the interfaces defined in Fig. 11. Information from the QMS system is modelled as flows as this information is to be pulled from the QMS system for reference in manufacturing plans and documentation by the PLM system. This is true for the information that flows between the PLM system and the Office client.

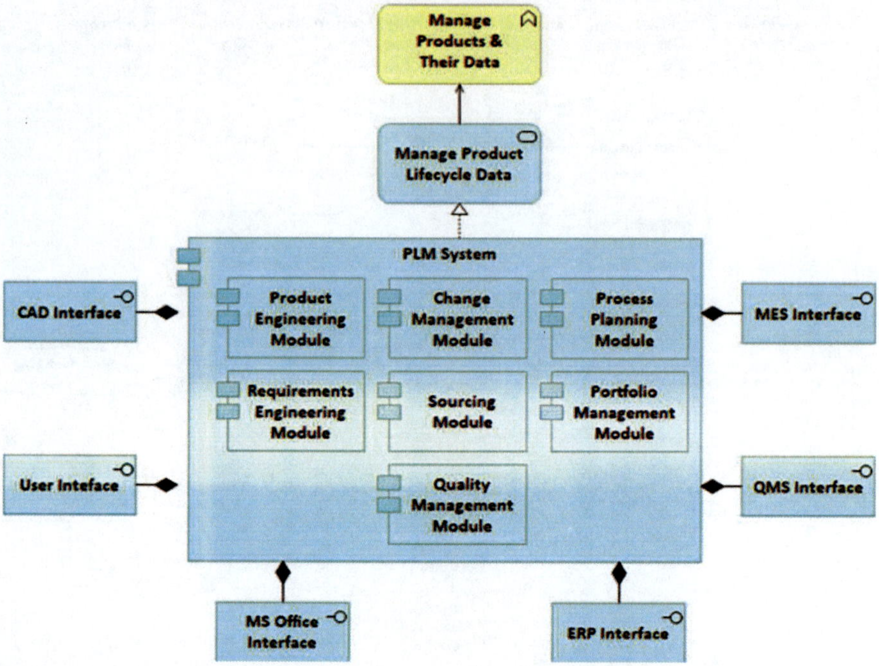

Fig. 11 Application-specific operational view: application components

A further decomposition of these information and data exchanges between these systems and their users should be modelling in UML or SysML sequence diagrams whereby each data item and their sequence can be defined.

The infrastructure specific technology view of the PLM system implementation is defined in Fig. 13. The PLM system interfaces with the other enterprise systems via the network and is to be accessed by users from the web browser of a client computer.

This network shall be secured to reduce the risk of information loss or sabotage. Product data is stored and retrieved from a dedicated database server defined by the *Database Node*. ERP and QMS information is located on the enterprise mainframe and can be accessed by the PLM system over the network. The infrastructure specific technology view in Fig. 13 is compliant with the ArchiMate technology usage viewpoint [41].

Figure 14 depicts the bottom-up mapping to the product quality requirements. The *Quality Management Module* application component is decomposed into the four main functions it performs within the PLM system. These module functionalities are traced to the requirements using the realization relation. This traceability can be exported from the model if required.

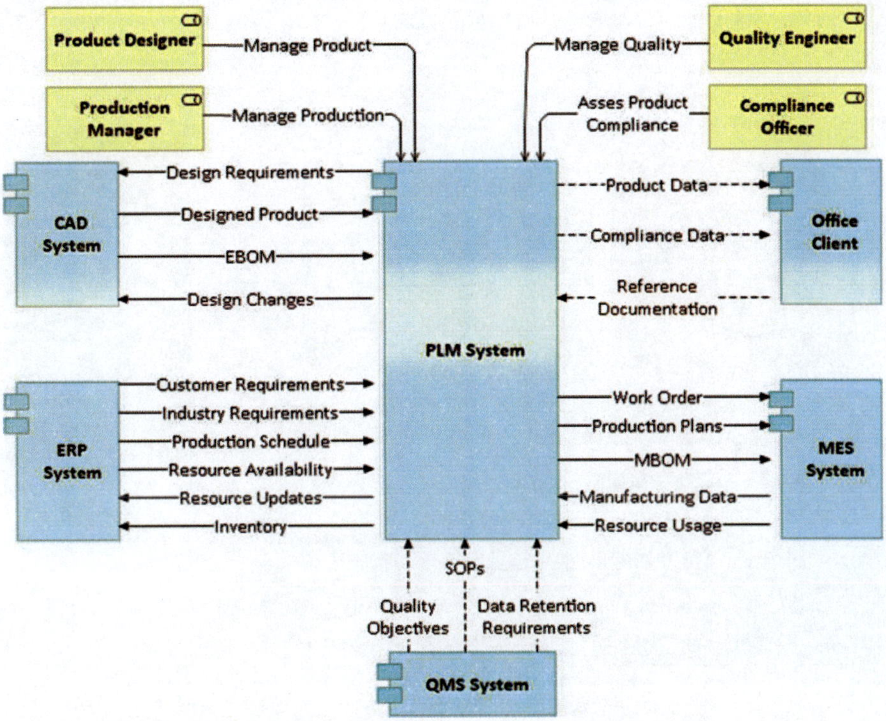

Fig. 12 Application-specific operational view: Information inputs and outputs

5 Discussions

The application of ArchiMate for planning the implementation of MMSs was investigated. This language provides stakeholders with capabilities for modelling enterprise systems and their implementation architectures from multiple different viewpoints. This modelling language is a beneficial tool for small and medium enterprises looking to implement MMSs to improve their competitiveness in local and global markets.

The ArchiMate language is deemed beneficial for use at a management level and engagement with stakeholders, this leads to improved communication and an unambiguous implementation plan. Once stakeholder acceptance has been gained and the project has been approved, the ArchiMate models can be used as a baseline for creating high fidelity and domain-specific system models and project plans to manage the implementation processes. The ArchiMate language has capabilities to model many types of enterprise systems from various viewpoints. Having one model that defines the enterprise systems at a high level is beneficial for integration between these systems. The use of a model-based approach and language as presented in this paper allows for models to be easily changed and updated as systems and their requirements

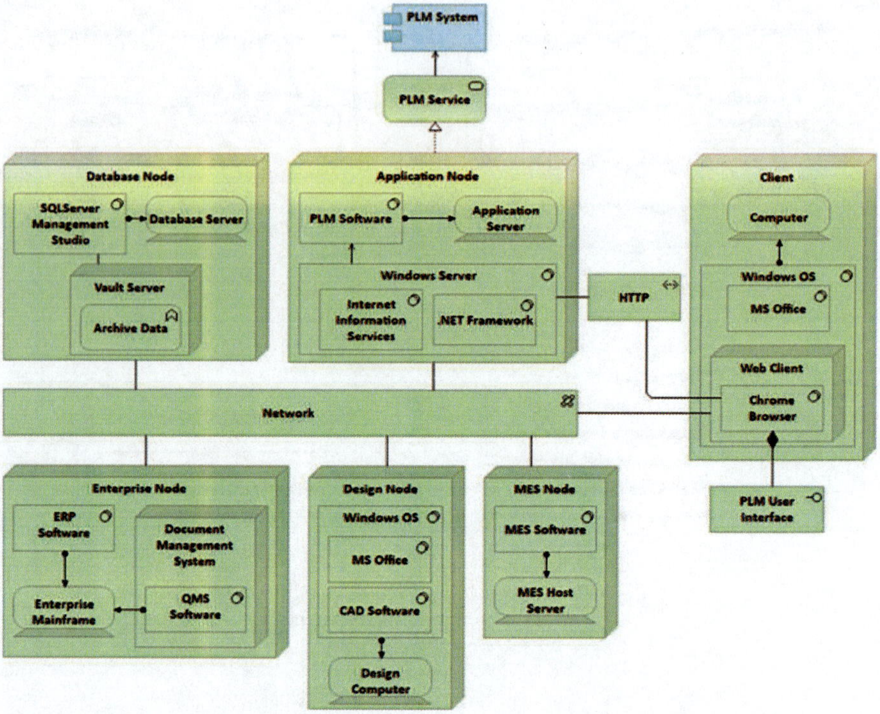

Fig. 13 Infrastructure specific technology view

change. This allows management to rapidly update system configurations and trace to the entities affected by these changes.

Future work should aim at modelling and addressing implementation risks and the development of an integration and verification framework for MMSs. Definition of risk is not formally defined in the ArchiMate language and notation, although the assessment motivation notation can be used. Defining risks during the planning and implementation phases of projects is a vital means for reducing nonconformances and failures in all enterprise and manufacturing systems both during their implementation and operation. MMS integration and verification occur after the system has been implemented. A framework for addressing such integration and verification should be domain-specific and it should integrate with the ArchiMate language and model presented in this paper.

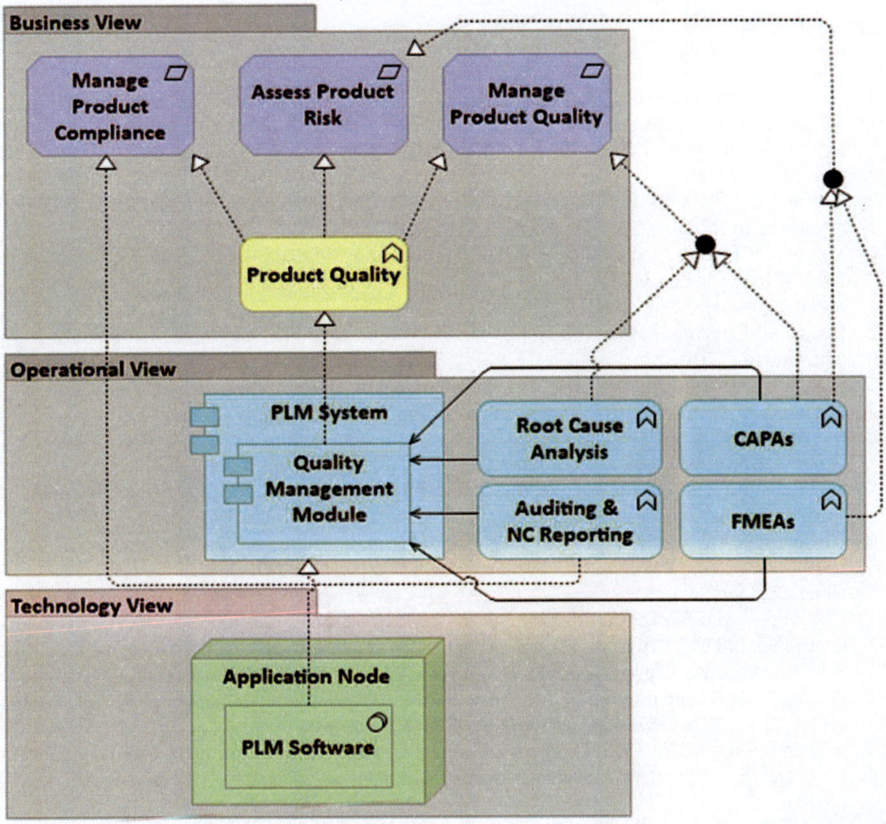

Fig. 14 Exemplar mapping to requirements

6 Conclusions

Implementing modern manufacturing management systems is by no means a trivial task for large enterprises, the challenges associated with such projects are exacerbated for small and medium enterprises. This research investigated the use of model-based techniques for planning the implementation of such systems. The ArchiMate language was deemed relevant to this application and is applied to a case for the planning the implementation of a product lifecycle management system. The use of model-based approaches to not just the design of systems but the implementation thereof provides many benefits and is an area highlighted for further research.

Acknowledgements This work is based on the research funded in part by the National Research Foundation (NRF) of South Africa (Grant number 131356). The authors would like to express their gratitude for this funding.

References

1. Chryssolouris, G.: Manufacturing Systems: Theory and Practice, 2nd edn. Springer (2006)
2. Schuh, G., Rozenfeld, H., Assmus, D., Zancul, E.: Process oriented framework to support PLM implementation. Comput. Ind. **59**(2–3), 210–218 (2008). https://doi.org/10.1016/j.compind.2007.06.015
3. Sumner, M.: Risk factors in enterprise-wide/ERP projects. J. Inf. Technol. **15**(4), 317–327 (2000). https://doi.org/10.1080/02683960010009079
4. Singh, S., Chandra Misra, S., Kumar, S.: Identification and ranking of the risk factors involved in PLM implementation. Int. J. Prod. Econ. **222**(September 2019), 107496 (2020). https://doi.org/10.1016/j.ijpe.2019.09.017
5. Groover, M.P.: Automation, Production Systems, and Computer-Integrated Manufacturing, 4th edn. Pearson (2016)
6. Wu, B.: Manufacturing Systems Design and Analysis, 1st edn. Springer (1992)
7. Hitomi, K.: Manufacturing Systems Engineering, 2nd edn. Taylor & Francis (1996)
8. Suh, N.P., Cochran, D.S., Lima, P.C.: Manufacturing system design. CIRP Ann. Manuf. Technol. **47**(2), 627–639 (1998). https://doi.org/10.1016/S0007-8506(07)63245-4
9. de Weck, O.L., Roos, D., Magee, C.L.: Engineering Systems—Meeting Human Needs in a Complex Technological World, 1st edn. MIT Press (2011)
10. Giachetti, R.E.: A framework to review the information integration of the enterprise. Int. J. Prod. Res. **42**(6), 1147–1166 (2004). https://doi.org/10.1080/00207540310001622430
11. Chen, D., Doumeingts, G., Vernadat, F.: Architectures for enterprise integration and interoperability: past, present and future. Comput. Ind. **59**(7), 647–659 (2008). https://doi.org/10.1016/j.compind.2007.12.016
12. Ben Khedher, A., Henry, S., Bouras, A.: Integration between MES and product lifecycle management. In: International Conference on Emerging Technologies and Factory Automation. ETFA, 2011. https://doi.org/10.1109/ETFA.2011.6058993.
13. Scheer, A.-W.: ARIS — Business Process Frameworks, 3rd edn. Springer (1999)
14. ISO: Quality management systems—Fundamentals and vocabulary (ISO 9000:2015). ISO (2015)
15. ESPRIT Consortium AMICE: CIMOSA: Open System Architecture for CIM, 2nd edn. Springer (1994)
16. Williams, T.J.: Reference model for computer integrated manufacturing from the viewpoint of industrial automation. In: IFAC Proceedings Volumes, vol. 23, pp. 281–291. https://doi.org/10.1016/s1474-6670(17)51748-6
17. ANSI/ISA: Enterprise-Control System Integration Part 1: Models and Terminology (ANSI/ISA-95.00.01), p. 142. ANSI (2000)
18. DIN: Reference Architecture Model Industrie 4.0 (RAMI4.0) (DIN SPEC 91345), p. 40. DIN (2016)
19. Lu, Y., Morris, K., Frechette, S.: Current Standards Landscape for Smart Manufacturing Systems (NISTIR 8107) (2016). https://doi.org/10.6028/NIST.IR.8107
20. IFIP–IFAC Task Force: GERAM: Generalised Enterprise Reference Architecture and Methodology (1999)
21. Industrial Value Chain Initiative: Industrial Value Chain Reference Architecture (IVRA) (2016)
22. Object Management Group: Unified Architecture Framework UAF—Domain MetaModel (DMM)—Version 1.1, p. 235. Available https://www.omg.org/cgi-bin/doc?formal/17-12-02.pdf (2020)
23. HORSE Consortium: HORSE Project—Complete System Design (D2.2a). https://www.horse-project.eu (2020)
24. ANSI/ISA: Batch Control Part 1: Models and Terminology (ANSI/ISA–88.01:1995) (1995)
25. IEC: Industrial-process measurement, control and automation—Life-cycle-management for systems and components (IEC 62890), p. 138. IEC (2020)
26. IEC: Enterprise-control system integration—Part 1: models and terminology (IEC 62264-1), p. 154. IEC (2013)

27. IEC: Batch control—Part 1: Models and terminology (IEC 61512-1), p. 177. IEC (1997)
28. Davis, J., Edgar, T., Porter, J., Bernaden, J., Sarli, M.: Smart manufacturing, manufacturing intelligence and demand-dynamic performance. Comput. Chem. Eng. **47**, 145–156 (2012). https://doi.org/10.1016/j.compchemeng.2012.06.037
29. ISO/IEC/IEEE: Systems and software engineering—System life cycle processes (ISO/IEC/IEEE 15288), p. 105. ISO (2015)
30. Davenport, T.H.: Mission Critical: Realizing the Promise of Enterprise Systems. Harvard Business School Press (2000)
31. INCOSE: Systems Engineering Handbook—A Guide for System Life Cycle Processes and Activities, 4th edn. Wiley (2015)
32. Jain, R., Chandrasekaran, A., Erol, O.: A framework for end-to-end approach to Systems Integration. Int. J. Ind. Syst. Eng. **5**(1), 79–109 (2010)
33. Yakimovich, D., Bieman, J.M., Basili, V.R.: Software architecture classification for estimating the cost of COTS integration. In: Proceedins of International Conference on Software Engineering, pp. 296–302 (1999). https://doi.org/10.1145/302405.302643
34. Franck, T., Iacob, M.E., van Sinderen, M., Wombacher, A.: Towards an integrated architecture model of smart manufacturing enterprises. In: International Symposium on Business Modeling and Software Design, 2018, vol. 309, pp. 112–133, https://doi.org/10.1007/978-3-319-78428-1_6
35. Moones, E., Vosgien, T., Kermad, L., Dafaoui, E.M., El Mhamedi, A., Figay, N.: PLM standards modelling for enterprise interoperability: a manufacturing case study for ERP and MES systems integration based on ISA-95. Lect. Notes Bus. Inf. Process. **213**, 157–170 (2015). https://doi.org/10.1007/978-3-662-47157-9_14
36. Aldea, A., Iacob, M.E., Wombacher, A., Hiralal, M., Franck, T.: Enterprise Architecture 4.0—A vision, an approach and software tool support. In: Proceedings—2018 IEEE 22nd International Enterprise Distributed Object Computing Conference, EDOC 2018, pp. 1–10 (2018). https://doi.org/10.1109/EDOC.2018.00011
37. Jagoda, K., Samaranayake, P.: An integrated framework for ERP system implementation. Int. J. Acc. Inf. Manag. **25**(1), 91–109 (2017). https://doi.org/10.1108/IJAIM-04-2016-0038
38. Kale, V.: Implementing SAP® R/3: The Guide for Business and Technology Managers (2000)
39. Le Duigou, J., Bernard, A., Perry, N.: Framework for product lifecycle management integration in small and medium enterprises networks. Comput. Aided. Des. Appl. **8** (2012). https://doi.org/10.3722/cadaps.2011.xxx-yyy
40. ISO/IEC/IEEE: Systems and software engineering—Architecture description (ISO/IEC/IEEE 42010:2011) (2011). https://doi.org/10.1109/IEEESTD.2012.6170923
41. The Open Group: Archimate 3.1 Specification, p. 206. The Open Group (2019)
42. Lankhorst, M.: Enterprise Architecture at Work, 4th edn. Springer (2017)
43. Object Management Group: OMG Systems Modeling Language (OMG SysML™)—Version 1.6, p. 398. OMG (2019) Available https://www.omg.org/spec/SysML/20181001/sysmldi.xmi (2019)

Duncan W. Gibbons received his Bachelor of Industrial Engineering from Stellenbosch University in 2018. He is currently a Ph.D. candidate at the same department. His research is focused on the qualification and certification of metal additive manufacturing for the aerospace industry through the application of a systems engineering approach.

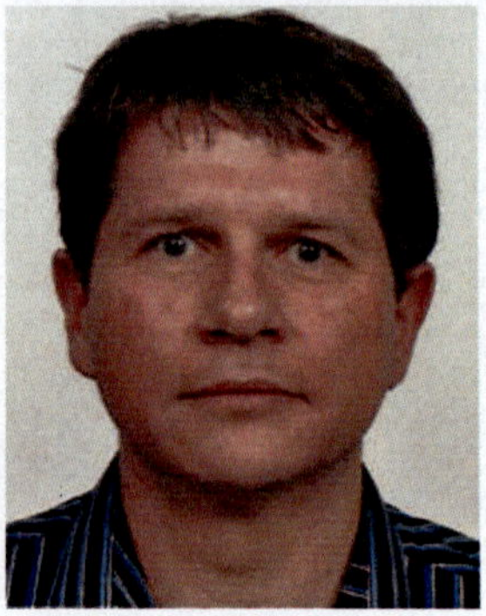

André F. van der Merwe holds a Bachelors of Mechanical Engineering, a Masters of Industrial Engineering and a Ph.D. in Engineering.

Currently a Professor at the Industrial Engineering Department of Stellenbosch University in the field of Resource Efficiency Engineering Management. REEM research group focusses on Commercial Readiness Indicators for Industrie 4.0, which includes Additive Manufacturing, Human-Machine interface and Food Security using Internet of Things.

Overview of Design Dimensions for Ambidexterity in Manufacturing Innovation Management

Q. Gärtner and A. Dorth

Abstract Continuous improvement is a core process within innovation management, as innovation managers are constantly trying to improve manufacturing systems with regards to reliability, variability and productivity. Manufacturing units must be able to deal with complex and volatile environmental circumstances while acting both exploratory and exploitatively to produce incremental and radical innovations. The organizational ability to achieve this is called ambidexterity. To address the growing challenge of ever-increasing manufacturing efficiency while simultaneously introducing radical technological concepts into the manufacturing system, the need for ambidexterity has increased significantly in recent years. Therefore, this paper presents an organizational approach to enable ambidexterity in manufacturing innovation management by identifying the most important organizational dimensions for designing such innovation management and ensuring long-term competitiveness.

Keywords Manufacturing · Innovation management · Ambidexterity

1 Introduction

Manufacturing companies are facing three main challenges in the present day. First, the volatile business environment requires an ever-greater degree of adaptability and flexibility [1], while globalization further intensifies competition for markets and prices [2]. The third major challenge is the accelerating pace of technological change [3].

Especially within manufacturing, these challenges cause great pressure to improve both individual processes and the production system as a whole through the development and integration of innovations [4]. More specifically, two basic types of

Q. Gärtner (✉) · A. Dorth
Institute for Machine Tools and Industrial Management, Technical University of Munich, Munich, Germany
e-mail: quirin.gaertner@iwb.tum.de

A. Dorth
e-mail: alexander.dorth@whu.edu

K. von Leipzig et al. (eds.), *Smart, Sustainable Manufacturing in an Ever-Changing World*, Lecture Notes in Production Engineering,
https://doi.org/10.1007/978-3-031-15602-1_22

innovation are essential to address these challenges and therefore must be distinguished. On the one hand, incremental innovation refers to minor improvements with a high affinity to the existing product or process [5], while radical innovation on the other hand describes a fundamental change [6], resulting in completely new products or processes.

There is a particular complexity in dealing with innovations within manufacturing. The requirements on process stability, product and process quality and the intercompatibility of technologies create an environment, in which incremental innovations represent the natural and thus primarily used lever to increase efficiency [7]. This causes a neglect of radical innovation approaches. Consequently, the organization is confronted with the productivity dilemma described by Adler et al. [8], which ultimately poses a threat to the long-term adaptability and competitiveness [9] of the manufacturing unit.

One important response to this productivity dilemma is referred to as ambidexterity management in organizational science. Regarding manufacturing, ambidexterity describes the capability of dealing with the dilemma between operational efficiency and long-term adaptability while the manufacturing system itself remains operationally stable [10, 11]. Building upon this response, organizational science needs to investigate how manufacturing units can be designed organizationally to implement such ambidextrous manufacturing innovation management (MIM).

2 Objective and Research Methodology

The aim of this paper is to present an overview on relevant ambidexterity design dimensions and subsequently postulate suitable managerial implications to foster radical innovation in manufacturing. The findings of this paper are based on an analysis of the existing scientific literature and aim to answer three main research questions: (*RQ#1*) Which organizational design dimensions have an influence on enhancing ambidexterity, (*RQ#2*) how can these organizational design dimensions be interpreted in a manufacturing context, (*RQ#3*) how can these design dimensions be implemented practically in manufacturing?

To answer these research questions, the method bibliometric literature study has been used according to Ball and Tunger [12]. Initially, a literature screening has been carried out using initial keywords and established literature search engines. The relevance of the resulting publications has then been checked on the basis of the title and the abstract. Further, the keywords have been specified. The bibliometric literature screening has been repeated iteratively until no further information has been found. Finally, relevant publications have been read and analyzed in full.

The remainder of this paper is structured as follows. In Sect. 3, the underlying terms and concepts are defined. Section 4 presents the scientific state of the art including the identification of previously used organizational design dimensions.

Section 5 derives the most relevant design dimensions for organizational ambidexterity in manufacturing based on the findings of Sect. 4. Further, it provides elaboration concerning their interpretation in a manufacturing context. In closing, a critical discussion of the presented concept is conducted in Sect. 6 and a conclusion of the paper and an outlook on further research activities is provided.

3 Definitions

3.1 Manufacturing Innovation Management

According to Schumpeter [13], an innovation in the manufacturing context can be defined as the recombination of forces and objects to either produce differently or produce something different. To define an innovation, two key factors are decisive. First, an invention is required, aiming at improving a product or a process. Secondly, this invention must be followed by a successful implementation into commercial practice [14]. One of the most common distinctions of innovation types is looking at the performance impact of implemented innovations. Incremental innovation refers to improvements with a high affinity to the existing product or process [5], tapping their underlying potential through comparatively minor changes [6]. In contrast, radical innovation describes a leap [4, 5] or fundamental change [6] that results in a breakthrough towards a new product or process (-component) [5].

The identification, development and integration of innovation using management methods enjoys constant attention in the scientific and industrial context. These methods include all organizational and processual aspects to fulfill the aforementioned goals. Following, MIM comprises the continuous contribution of a Manufacturing Unit to a company's innovation management (IM) by designing processes and environments to develop radical and incremental innovation in manufacturing [15]. The entire production network including all direct-value-adding sub-units constitutes the manufacturing unit including planning, research and development and maintenance.

3.2 Ambidexterity

Within the innovation activities of companies, a basic distinction can be made between an exploitative behavior (incremental) and an explorative behavior (radical). While the exploitative strategy aims to increase efficiency in the short term, the explorative strategy intends to secure long-term competitiveness through novel innovation impulses. The organizational ability to simultaneously engage in both innovation activities, is referred to as organizational ambidexterity (see Fig. 1) [4].

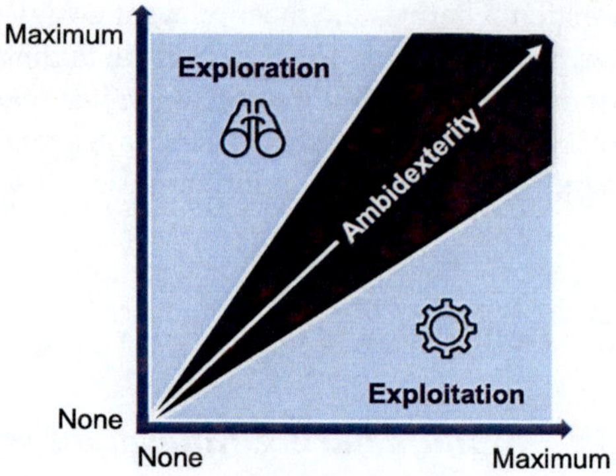

Fig. 1 Illustration of the balance between exploitation and exploration to create ambidexterity

With reference to manufacturing, we define ambidexterity as the ability to deal with the dilemma between operational efficiency and radical or disruptive changes in processes innovation.

While organizational research initially followed the understanding of March [16], solely focusing on the trade-offs between exploration and exploitation, current research recognizes that there are two separate, but reciprocally interacting parts of ambidexterity [4, 17]. For one, the "balance dimension", which encompasses the trade-offs following March's [16] understanding, and on the other, the "combined dimension", describing the synergistic effect of activities in both domains [4, 17].

To achieve both, it is necessary to counteract forces that create imbalance or isolation between both activities at times. For this purpose, a distinction is made between differentiating and integrating managerial actions. Differentiation ensures that the tasks of exploration and exploitation can be carried out without obstructing each other. On the other hand, integration aims at linking both approaches to ensure mutual reinforcement [4]. Consequently, previous scientific works have dealt theoretically and empirically with the identification of design dimensions, to support the establishment of an ambidextrous IM in an organization using this concept.

4 Organizational Design Dimensions

4.1 Design Dimensions in Organizational Management

Within the research field of organizational management, there are numerous approaches for the organizational diagnosis and design of companies to increase their innovative capacity. Table 1 shows an excerpt of the approaches considered and the covered organizational design dimensions. The selected approaches present defined organizational design dimensions and have therefore been selected for further evaluation. Based on the frequency with which the dimensions are highlighted, conclusions can be drawn from a scientific perspective about the relevance.

Table 1 Excerpt of analyzed approaches and the used organizational design dimensions within organizational management

Legend

- Lists same dimension
- Lists similar dimension
- Describes aspects of the dimension

Approaches for Org. Design \ Org. Design Dimensions	Strategy	Structures	Leadership	Culture	Knowledge
Morton (1991)	Lists same dimension	Lists same dimension	Lists similar dimension		
Burke & Litwin (1992)	Lists same dimension	Lists same dimension	Lists same dimension	Lists same dimension	Lists similar dimension
Gavrea et al. (2007)	Lists same dimension	Lists same dimension	Lists same dimension	Lists same dimension	Lists similar dimension
Govender & Parumasur (2018)		Lists same dimension	Lists same dimension	Lists similar dimension	Lists similar dimension
Kumari (2018)	Describes aspects of the dimension	Lists same dimension	Lists same dimension	Describes aspects of the dimension	
Alemu & Shea (2019)		Describes aspects of the dimension	Lists similar dimension	Lists same dimension	
Rüegg-Stürm & Grand (2019)	Lists same dimension	Lists same dimension	Describes aspects of the dimension	Lists same dimension	
Thommen et al. (2020)	Lists same dimension	Lists same dimension	Lists same dimension	Lists same dimension	

The organizational design dimensions of strategy, structure and leadership are mentioned most frequently. Here, the importance of a goal-oriented alignment of the company through an adequate strategy with adaptable structures and purposeful management mechanisms and leaders is emphasized [18–22]. Furthermore, the dimensions of culture [23–25] and knowledge management [19–21] are highlighted. Here, the influence of a vivid corporate culture and the targeted distribution and combination of knowledge to increase the ability to innovate are underlined.

4.2 *Organizational Design Dimensions in Ambidexterity Management*

Furthermore, various organizational design dimensions can be derived from preliminary scientific work in the field of organizational ambidexterity. Although the authors pursue different approaches and goals, commonalities can nevertheless be identified. Table 2 provides an excerpt of approaches considered and the organizational design

Table 2 Excerpt of analyzed scientific approaches and the used organizational design dimensions within organizational ambidexterity

Legend: ● (dark field) Lists same dimension; ● (light field) Lists similar dimension; ◑ Describes aspects of the dimension

Approaches for Ambidexterity Mgmt. / Org. Design Dimensions	Strategy	Structures	Leadership	Culture, Context & Processes	Knowledge	Vision & Values	Awareness & Consensus
O'Reilly & Tushman (2008)	Lists similar dimension	Lists same dimension	Lists same dimension	Describes aspects of the dimension	Describes aspects of the dimension	Lists same dimension	Describes aspects of the dimension
Raisch & Birkinshaw (2008)		Lists same dimension	Lists same dimension	Lists similar dimension			
Simsek (2009)		Lists similar dimension	Lists same dimension	Lists similar dimension			
Andriopoulos & Lewis (2009)	Lists same dimension	Lists similar dimension		Lists similar dimension	Lists similar dimension	Lists similar dimension	
O'Reilly & Tushman (2013)	Lists similar dimension	Lists same dimension	Lists same dimension	Lists similar dimension	Lists similar dimension	Lists same dimension	
Güttel & Konlechner (2014)		Describes aspects of the dimension	Lists same dimension	Lists similar dimension	Lists similar dimension	Lists similar dimension	
Olivan (2019)	Lists same dimension	Lists similar dimension	Lists same dimension	Describes aspects of the dimension	Lists same dimension	Lists same dimension	
Schneeberger & Habegger (2020)	Lists same dimension	Lists same dimension	Lists same dimension	Lists similar dimension			Describes aspects of the dimension

dimensions identified. The presented selection of approaches is based on the point that the authors define specific organizational design dimensions as fields of action to enable ambidexterity in IM.

O'Reilly and Tushman [26] provide a starting point by developing five propositions that support leaders in implementing ambidexterity. Their propositions serve as a basis for selecting the relevant organizational design dimensions and include strategic intent, separate organizational structures, conscious leaders, adequate contexts a continuous communication of knowledge a shared vision and values, as well as clear consensus.

Raisch and Birkinshaw [27] and Simsek [28] extend the design dimensions of structures and leadership by identifying the behavioral context and the organizational environment as an additional important field to foster ambidexterity. Andriopoulos and Lewis [29] identify similar design dimensions as their predecessors with the exception of the dimension leadership. In addition, they highlight the importance of innovation strategy for ambidexterity.

O'Reilly and Tushman [11] complement their previous approach by emphasising the dimensions of culture and context, as well as the transfer of knowledge to create ambidextrous organizations. Güttel and Konlechner [30] further present three dimensions for the targeted re-integration of the separate structures in the sense of O'Reilly and Tushman [26]. These dimensions are top management, project work at staff level and the integrative corporate culture.

Olivan [4] presents an additional approach concerning organizational design for radical technology development. In his balance model, six differentiating or integrating design dimensions are postulated. These include an ambidextrous strategy, separate structures, a separate ecosystem, a unifying vision, unifying knowledge and unifying management. Most recently, Schneeberger and Habegger [31] presented five questions for executives to identify relevant elements within their organization, creating a rather practically focused approach. The dimensions considered here relate to already known dimensions like management, strategy, organizational structure, resource allocation and corporate culture. Complementary, parts of the dimension of consensus and attention to ambidexterity in the organization are highlighted.

5 Organizational Ambidexterity in Manufacturing

To create a structure based on the preliminary work described in Sects. 4.1 and 4.2, the following organizational design dimensions have been derived for ambidextrous IM: Strategy, structure, culture, behavioral context and processes, vision and values, knowledge, and leadership. Furthermore, awareness of the relevance of ambidexterity, and a clear consensus within the management team, is identified as a fundamental prerequisite.

The identified design dimensions are initially independent of manufacturing. A delimitation to manufacturing will only become apparent in the specific content of the respective dimension.

To achieve the organizational target state of ambidexterity, certain intended impacts of the combined measures within a dimension and desired interactions between them need to be ensured. Hence, we propose that the design dimensions as a whole must be utilized differently in terms of their intended organizational impact. In adherence with the most widely accepted view to foster ambidexterity, we propose to divide the design dimensions into differentiating and integrating dimensions.

5.1 Awareness and Consensus

However, for the dimension of *Awareness and Consensus* this classification cannot be clearly made, for which reason it is considered a fundamental prerequisite for the implementation of any kind of ambidextrous IM without a direct direction of effect.

Awareness includes a common understanding of ambidexterity across the company and a common definition of radical and incremental innovation in the manufacturing context.

In addition, there must be a *Consensus* among managers on the importance of achieving an ambidextrous IM, which in turn implies a willingness to change [26].

This can be done within manufacturing by linking ambidexterity with productivity indicators or economic performance indicators, and analyzing Best-practice organizations or competitors. Here, for example, the connection between the organizational design of IM and the number of innovations developed on the process side and their impact on productivity (quality, costs and time) could be clarified using empirical research.

5.2 Design Dimensions Classified as "Differentiation"

5.2.1 Strategy

Within the differentiating design dimension *Strategy*, an organization should aim at establishing a defined, convincing strategic intent that justifies both incremental and radical innovation by addressing their contribution towards the comprehensive strategy and goals of the organization. In the manufacturing context, this may include, for example, a mission statement of the manufacturing unit in which radical innovation approaches are also taken up to complement the existing processes. Further, appropriately formulated strategy guidelines for executives within the manufacturing unit can be mentioned, which are detached from incrementally achievable productivity goals [4]. These guidelines must be addressed and implemented by executives within the operational areas of manufacturing, ensuring the information of the whole manufacturing unit.

5.2.2 Structure

To avoid conflict between incremental and radical activities, *Separate Structures* for exploration and exploitation can be created. Regarding manufacturing this can be done through the organizational structure itself or by executing both activities at different times to achieve a sequential separation [26]. The decision as to where and when incremental or radical innovations are to be created is thereby made in advance, since the choice for implementing separating structure must be made on an overarching organizational level. Structural separation is created by separating the fields into separate functional units, creating spin-offs or temporally separated project units. Meanwhile, sequential separation is possible through a fixed temporal ratio between incremental and radical projects. The best-known example for this is the so-called "Friday for future" [4], which indirect manufacturing areas such as manufacturing planning could dedicate exclusively to exploratory topics.

5.2.3 Culture, Context and Processes

Culture, Context and Processes refer to conditions or procedures that should help teams or individuals to meet the requirements for generating both incremental and radical innovation. An organization can achieve this form of differentiation through the creation of appropriate supporting business-unit contexts, which influence individual-level behavior and enable employees to decide between exploration and exploitation themselves. This constitutes the main distinguishing towards the design dimension *Structure* as the decision is made on an individual instead of an organizational level. A prerequisite for such contextual separation to achieve the intended outcome is a sufficient amount of organizational slack, describing "*the pool of resources in an organization that is in excess of the minimum necessary to produce a given level of organizational output*" [32]. These slack resources include e.g. employees, idle capacity, and capital expenditures.

In manufacturing, approaches and conditions fostering the development of radical innovation initiatives should be emphasized, so that individuals have access to them in addition to existing processes for the integration of incremental innovation. For example, so-called internal "*Innovation Labs*" or external open test environments constitute such business-unit contexts. They are particularly suitable for the manufacturing area, as they provide realistic, process-oriented and changeable environments [33] necessary in manufacturing innovation development. Furthermore, incentive systems that are intended to stimulate the achievement of both sub-goals and the ambidextrous strategy are part of this design dimension.

Culture forms another contributor strengthening the individual's ability to decide on incremental and radical innovation development. Cultural aspects are generally independent of manufacturing but enable ambidextrous IM in general, fostering ambidexterity in manufacturing as well. For example, a cultural basis of diversity can be achieved through the means of knowledge and background by assigning employees to different geographical locations and implementing methods like job-rotation even at management levels. This in turn creates the ability for individuals in the organization to act both explorative and exploitative based on the context. Further, the understanding for both activities is emphasized facilitating integrative measures of both activities.

5.3 Design Dimensions Classified as "Integration"

5.3.1 Vision and Values

A common *Vision* and shared fundamental *Values* should allow employees from incremental and radical oriented divisions or projects to form a common identity uniting both activities in a meaningful way in the long term. In the manufacturing context, this includes, for instance the creation of a vision and the resulting derivation of long-term goals and challenges. To create common values, incremental work philosophy approaches such as the "zero defect strategy" can be established within manufacturing, which in turn can be supplemented with identity-forming aspirations such as "technology leadership" to integrate radical topics.

5.3.2 Knowledge

To integrate exploration and exploitation, interfaces and *Knowledge* transfer between both separated activities need to be implemented. Hence, this design dimension thus includes integrating instruments that connect employees and enable knowledge exchange between incremental and radical activities. In a manufacturing context, this can be implemented on three levels [4, 30]. On the first level, basic information is shared via open channels with an unlimited group of participants. White paper, lectures or theme weeks, for example, are suitable communication instruments for this. On the second level, a direct, bidirectional exchange between individuals in incremental and radical areas can take place. Communities of Practice (CoPs), for example, which informally bundle knowledge from certain sub-areas, are useful instruments for this. On the third level, integration takes place through formalized exchange of knowledge. Here, for example, certain manufacturing areas can be informed at regular intervals by the CoPs about radical or incremental activities.

5.3.3 Leadership

This is to be understood as a joint management that presides over both incremental and radical fields. In the manufacturing context, its task include to balance organizational conditions, to connect departments and managers with each other, to coordinate the allocation of resources, dissolve self-serving behavior and routinize or promote the integration of radical topics. The design dimension leadership can further be considered the most important integrating dimension, as it has a strong influence on other integrating or differentiating dimensions.

6 Conclusions and Outlook

This contribution is an attempt to provide an overview on the various organizational design dimensions and derive the most relevant transferred into a manufacturing context. Furthermore, the presented concept aims at enabling an ambidextrous management approach for manufacturing units to enhance radical innovation capability and to cope with fast changing technological requirements.

In summary, it can be concluded that there are seven dimensions for the organizational design of ambidextrous MIM. The dimensions of strategy, structure, culture, context and processes have a differentiating effect, while the dimensions of vision and values, knowledge and leadership have an integrative effect. The seventh dimension of awareness and consensus is a fundamental prerequisite for achieving ambidexterity within the organization. Focusing on manufacturing, several examples have been highlighted for the practical implementation of each dimension. Therefore, it can be stated that manufacturing units, despite their specific requirements in terms of stability and efficiency, may nevertheless be able to implement an ambidextrous IM.

The presented approach is intended as a starting point to depict how an ideal IM framework for ambidexterity might look like revealing tasks that must be addressed and prompting starting points for further research. Therefore, this paper also aims at constituting an outlook for future research on ambidexterity in MIM. Since the developed concepts of this paper do yet only offer punctual implementation possibilities for the design of an ambidextrous MIM, further research should build upon investigating practical implementation. In addition, the definition and content-related specification of the organizational design dimensions is not yet sufficiently detailed to actively support strategic or operational management of innovation projects. Additionally, research needs to focus on the identification of interlinks and dependencies between managerial implications to successfully implement an ambidextrous MIM.

Acknowledgements The German Research Foundation (DFG) funds this research and development project 431696521 "Bewertung des Wettbewerbspotenzials von Produktionstechnologien". We extent our sincere thanks to the DFG for the generous support.

References

1. Dobbs, R., Manyika, J., Woetzel, J.R.: No ordinary disruption: the four global forces breaking all the trends. Public Affairs, New York (2015)
2. Abele, E., Reinhart, G.: Zukunft der Produktion: Herausforderungen, Forschungsfelder, Chancen. Hanser, München (2011)
3. Iansiti, M., Lakhani, K.R.: Digital ubiquity: how connections, sensors, and data are revolutionizing business. Harv. Bus. Rev. **11**(92), 90–99 (2014)
4. Olivan, P.: Methode zur organisatorischen Gestaltung radikaler Technologieentwicklungen unter Berücksichtigung der Ambidextrie. Dissertation, Stuttgart (2019)
5. Gassmann, O.: Internationales F&E-Management: Potentiale und Gestaltungskonzepte transnationaler F&E-Projekte. Oldenbourg, München (1997)
6. Schallmo, D., Brecht, L.: Prozessinnovation erfolgreich anwenden: Grundlagen und methodisches Vorgehen: ein Management- und Lehrbuch mit Aufgaben und Fragen. Springer, Berlin (2017)
7. Hofer, A., Brandl, F., Bauer, H., Haghi, S., Reinhart, G.: A framework for managing innovation cycles in manufacturing systems. Procedia CIRP **93**(1), 771–776 (2019)
8. Adler, P.S., Benner, M., Brunner, D.J., MacDuffie, J.P., Osono, E., Staats, B.R., Takeuchi, H., Tushman, M.L., Winter, S.G.: Perspectives on the productivity dilemma. J. Oper. Manag. **27**(1), 99–113 (2009)
9. Abernathy, W.J., Utterback, J.M.: Patterns of industrial innovation. Technol. Rev. **80**(7), 40–47 (1978)
10. He, Z.-L., Wong, P.-K.: Exploration versus exploitation: an empirical test of the ambidexterity hypothesis. Organ. Sci. **15**(4) (2004)

11. O'Reilly, C.A., Tushman, M.L.: Organizational ambidexterity: past, present, and future. Acad. Manage. Perspect. **4**(27), 324–338 (2013)
12. Ball, R., Tunger, D.: Bibliometrische Analysen - Daten, Fakten und Methoden: Grundwissen Bibliometrie für Wissenschaftler, Wissenschaftsmanager, Forschungseinrichtungen und Hochschulen. Forschungszentrum Jülich GmbH, Jülich (2005)
13. Metcalfe, S.: J.A. Schumpeter and the Theory of Economic Evolution . Papers on Economics and Evolution, Jena (2012)
14. Schuh, G., Bender, D.: Grundlagen des Innovationsmanagements. In: Schuh, G. (ed.) Innovationsmanagement, 2nd edn., pp. 1–16. Springer, Berlin, Heidelberg (2012)
15. Brandl, F., Kagerer, M., Reinhart, G.: A hybrid innovation management framework for manufacturing: enablers for more agility in plants. Procedia CIRP **72**(1), 1154–1159 (2018)
16. March, J.G.: Exploration and exploitation in organizational learning. Org. Sci. **2**(1), 71–87 (1991)
17. Cao, Q., Gedajlovic, E., Zhang, H.: Unpacking organizational ambidexterity: dimensions, contingencies, and synergistic effects. Org. Sci. **20**(4), 781–796 (2009)
18. Morton, S.: The Corporation of the 1990s: Information Technology And Organizational Transformation. Oxford Univ. Press, New York (1991)
19. Burke, W.W., Litwin, G.H.: A causal model of organizational performance and change. J. Manag. **3**(18), 523–545 (1992)
20. Gavrea, C., Liviu, I., Stegerean, R.: Determinants of organizational performance: the case of Romania. Manage. Mark. **6**(2), 285–300 (2007)
21. Govender, P., Brijball Parumasur, S.: Organizational diagnosis, the stepping stone to organizational effectiveness. Corp. Board Role Duties Compos. **12**(2), 65–76 (2016)
22. Kumari, N.: Organizational diagnosis: a case of Infosys, India. J. Asian **5**(1) (2018)
23. Alemu, D.S., Shea, D.: A path analysis of diagnosis of organizational levels of functionality. Int. J. Educ. Manage. **33**(7) (2019)
24. Rüegg-Stürm, J., Grand, S.: Das St. Gernaller Management-Modell: Management in einer komplexen Welt. Haupt Verlag, Bern (2019)
25. Thommen, J.-P., Achleitner, A.-K., Gilbert, D.U., Hachmeister, D., Jarchow, S., Kaiser, G.: Allgemeine Betriebswirtschaftslehre: Umfassende Einführung aus managementorientierter Sicht. Springer, Wiesbaden (2020)
26. O'Reilly, C.A., Tushman, M.L.: Ambidexterity as a dynamic capability: resolving the innovator's dilemma. Res. Org. Behav. **28**(1), 185–206 (2008)
27. Raisch, B., Birkinshaw, J.: Organizational ambidexterity: antecedents, outcomes, and moderators. J. Manag. **34**(3), 375–409 (2008)
28. Simsek, Z.: Organizational ambidexterity: towards a multilevel understanding. J. Manage. Stud. **46**(4), 597–624 (2009)
29. Andriopoulos, C., Lewis, M.W.: Exploitation-exploration tensions and organizational ambidexterity: managing paradoxes of innovation. Organ. Sci. **20**(4), 696–717 (2009)
30. Güttel, W.H., Konlechner, S.: Ambidextrie als Ansatz zur Balancierung von Effizienz und Innovativität in Organisationen. In: Burr, W. (ed.) Innovation: Theorien, Konzepte und Methoden der Innovationsforschung, pp. 345–371. Kohlhammer, Stuttgart (2014)
31. Schneeberger, S.J., Habegger, A.: Ambidextrie - der organisationale Drahtseilakt. In: Schellinger, J., Tokarski, K.O., Kissling-Näf, I. (eds.) Digitale Transformation und Unternehmensführung, pp. 105–144. Springer, Wiesbaden (2020)
32. Nohria, N., Gulati, R.: What is the optimum amount of organizational slack? A study of the relationship between slack and innovation in multinational firms. Eur. Manage. J. **15**(6), 603–611 (1997)
33. Tisch, M., Ranz, F., Abele, E., Metternich, J., Hummel, V.: Learning factory morphology: study of form and structure of an innovative learning approach in the manufacturing domain. Turkish Online J. Educ. Technol. **10**(2), 356–363 (2015)

Quirin Gärtner holds a M. Sc. in industrial engineering. He is a research assistant in the Department of Mechanical and Process Engineering at Duesseldorf University of Applied Sciences with a focus on innovation, entrepreneurship, and product service systems.

Alexander Dorth holds a B.Sc. in Int. Business Administration from WHU—Otto Beisheim School of Management. He currently pursues a M.Sc. in Management and Technology at the Technical University of Munich, Germany.

Possibilities and Challenges for Human-System Integration in the South African Manufacturing Context

T. W. Defty, K. Kruger, and A. H. Basson

Abstract South Africa, a developing country, over the last decade has seen a decline in production value and employment in the manufacturing sector for many reasons; one being the failure to effectively integrate Industry 4.0 (I4.0) technologies with low-skilled workers to reap the performance benefits from both. In the context of South Africa and other developing economies, creating and retaining jobs for a large, under-skilled workforce is considered a priority. While the I4.0 development focus has predominantly been on enhancing automation, it is evident that human workers still provide dexterity, adaptability and decision-making capabilities that cannot be substituted by current technologies. This paper summarises the possibilities of human-system integration (HSI) for improving worker training and assignment, quality assurance and traceability of manual operations, data acquisition for decision making, and the health and safety of workers. Furthermore, it identifies challenges for effective human-system integration related to information and communication technology infrastructure, technology acceptance, investment strategies, risk and safety policies, and system reconfigurability. Though many challenges exist to realize HSI developments, the possibilities discussed in this paper motivate the future development and evaluation of such HSI applications.

Keywords Industry 4.0 · Manufacturing · Human-system integration

T. W. Defty · K. Kruger (✉) · A. H. Basson
Department of Mechanical and Mechatronic Engineering, Mechatronics, Automation and Design Research Group, University of Stellenbosch, Stellenbosch, South Africa
e-mail: kkruger@sun.ac.za

T. W. Defty
e-mail: 20873077@sun.ac.za

A. H. Basson
e-mail: ahb@sun.ac.za

K. von Leipzig et al. (eds.), *Smart, Sustainable Manufacturing in an Ever-Changing World*, Lecture Notes in Production Engineering,
https://doi.org/10.1007/978-3-031-15602-1_23

1 Introduction

The global manufacturing market has seen an increase in productivity with the adoption of new technologies and developments in the era of the fourth industrial revolution (I4.0) [1]. These developments include advances in intelligent manufacturing, IoT enabled manufacturing, cloud manufacturing enabled by the Internet of Things, cyber-physical systems (CPSs), cloud computing, big data and information and communication technology (ICT) [2]. The CPS concept is one of the key contributing ideas to the integration of digital and physical layers as manufacturing environments move towards intelligent manufacturing strategies. CPSs promote a highly interconnected and integrated manufacturing environment through devices and software. With the adoption of such developments and automation of industrial processes, the manufacturing labour market is experiencing a shift in employment demands from low-skilled workers who perform repetitive activities, to higher-skilled workers who perform flexible activities, to increase labour productivity. Accordingly, Accenture has estimated a 40% increase in labour productivity across 12 developed countries by 2035 due to I4.0 developments [3].

Manufacturing in South Africa (SA), a developing country, in 2018 contributed to 14% of the national gross domestic product (a decrease from 24% in 1980) and one job out of every 10 people [4]. SA has also seen a decline in production volume and demand. With the late introduction of I4.0 developments and innovative industrial strategies in SA, the retrenchment risks and employment difficulties for under-skilled and uneducated workers have increased. The poor adoption of I4.0 developments also threatens the goals of the Integrated Manufacturing Strategy, established in 2002 [5], and the successful fulfilment of Sustainable Development Goal (SDG) 8 (economic growth, and decent work for all) and SDG 9 (sustained industrialization) [6]. To increase labour productivity while compensating for under-skilled workers, the technical and managerial challenges for human-system integration of workers into I4.0 oriented manufacturing systems must be addressed. In doing so, I4.0 developments must augment and enhance workers efforts to create a more competitive and productive workforce [7].

An example of such a development is the BASE administration shell [8] (Fig. 1), which represents the worker in the cyber layer and communicates on behalf of the worker with other systems in the manufacturing environment. The BASE administrative shell presents a potential human-system integration solution for manufacturing.

This paper summarises the challenges and possibilities of integrating workers in manufacturing systems with the support of I4.0 developments. The paper aims to help shape manufacturing developments to achieve sustainable employment goals in South Africa and similar countries.

In Sect. 2, an overview of the South African manufacturing sector and global comparison is presented followed by an overview of the current concepts for workers

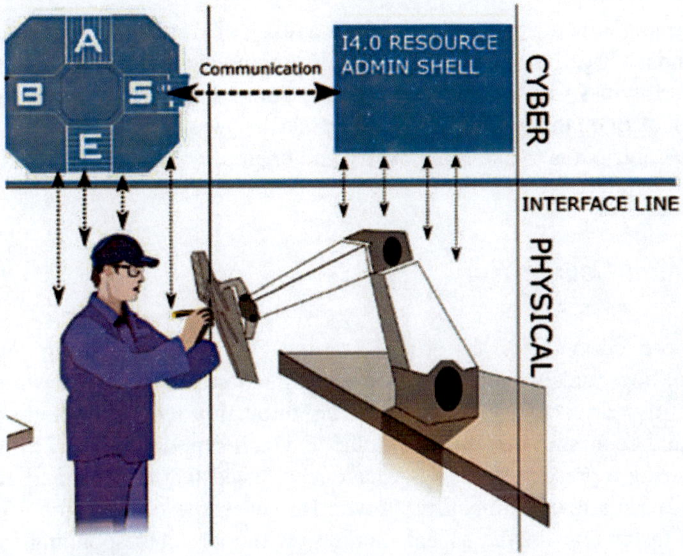

Fig. 1 BASE administrative shell in a CPS [8]

in intelligent manufacturing systems in Sect. 3. In Sect. 4, the possibilities of human-system integration are discussed while the challenges are discussed in Sect. 5. This paper draws conclusions in Sect. 6.

2 Workers in South African Manufacturing

2.1 Industry State and Challenges

The South African manufacturing industry consists of various sectors, including metals, food and beverage, petroleum, clothing and textiles, and automotive. Similarly to the mining and agricultural industry, the manufacturing industry requires substantial manual labour to perform operational tasks. As such, these industries have experienced changes in skills requirements due to automation, requiring more cognitive and complex information processing skills [9]. The overall skill composition of the manufacturing workforce showed a 3.1% increase from 2007 to 2017 for skilled workers (21.8% of the workforce), but a 2.5% decrease in semi-skilled (61.9%) and 0.6% decrease in low-skilled (16.3%) workers [10].

Over the past 20 years, the national education system has performed poorly and left many school-leavers with poor literacy, cognitive and behavioural skills. Although these challenges are addressed by the National Development Plan 2030 (NDP), job retention in manufacturing has declined. As a result of current automation and digitisation trends, a McKinsey & Company study estimated a net loss of 231,000 jobs in manufacturing in SA by 2030, exacerbated by poor education and skill development [11]. The lack of qualified workers for higher skilled jobs in manufacturing therefore, indirectly, contributes to extreme poverty and unemployment.

2.2 *Global Comparison*

The SA production sector has dropped on the global manufacturing competitiveness scale, now ranked 25th [12]. The top-ranked countries have been supported by competitive drivers such as talent-driven innovation (education and skill development) and economic systems and policies which support trade [12]. In contrast, manufacturing representatives in SA have recognised that the cost and availability of labour is SA's main competitive driver. However, the cost of labour in SA has increased faster than in the global markets, without a corresponding increase in labour productivity [12].

3 Workers in Intelligent Manufacturing Systems

Manufacturing companies have typically replicated the standard automation pyramid for the integration of their systems and physical assets (e.g. machinery, conveyor belts) through systems engineering approaches. Within such systems, workers are often assumed to be a 'magic human' who will perform the necessary tasks to ensure that the production operations will continue as required [13]. For any disturbance beyond the ability of the automation system's control, the 'magic human' will be assumed to intervene with the necessary corrective action. This failure to include the human in the control loop of the production system has led to extended operational downtime, quality issues and worker safety concerns. Furthermore, the dexterity, flexibility, awareness and decision-making capabilities of workers are not fully utilised within the system. Therefore, workers often are left performing repetitive and dull tasks.

A key enabler for I4.0 manufacturing goals is human cyber-physical systems (HCPSs), an extension of CPSs. Within HCPSs, human-automation symbiosis requires automation systems to act as an extension of workers, rather than workers as an extension of automation systems. Romero et al. [14] portrayed a conceptual worker for I4.0 operations called Operator 4.0. Operator 4.0 is physically cognitively and socially enhanced to realise the full potential of the worker during manufacturing activities. The Operator 4.0 concept presents a more effectively integrated worker which aims to benefit the productivity of the overall system and the well-being of the worker.

Human-system integration (HSI) can be described as the field of study and design which aims to incorporate all the "human considerations within and across all software and system elements" [15]. HSI can be approached on an organisational level and a technical level, of which this paper addresses the technical level. Effective HSI can be achieved through the development and application of software systems and technological advancements which aid the activities of workers within operational tasks.

Such HSI developments could include the use of augmented, virtual reality devices and other wearables to observe and inform workers of the necessary information to improve their productivity and collaboration with machines. Software developments could further use multi-agent principles to represent humans in the cyber layer of the manufacturing CPS, to improve the delegation and coordination of worker activities.

As a step towards HSI, Sparrow et al. [16] developed an I4.0 digital administration shell for workers in South African manufacturing environments. His BASE administration shell is capable of delegating tasks on behalf of the worker, as well as augmenting the handling of data and processing tasks for the worker. Furthermore, the BASE shell enables interfacing between the software administration shell and the worker. This ensures the worker is included in the control loop. The BASE administration shell outperformed other administrative logistic functions by providing an effective HSI solution for intelligent manufacturing system.

4 Possibilities

Through effective HSI, the following possibilities potentially could be achieved and enhanced within manufacturing operations. These possibilities aim to, on a small scale, benefit manufacturing companies in reaching their goals but on a larger scale, aid the manufacturing sector in realising the SDG and NDP goals.

4.1 Worker Training

The weaknesses in SA's education sector have created a skill deficiency in the manufacturing sector. This deficiency has constrained the rate of adoption of new technologies and processes which aim to enhance productivity. Training periods for workers are typically lengthy and result in further costs to the employer, therefore often discouraging such investment into the development of their workforce.

HSI developments can enhance the training procedures for workers for new production methods, collaboration with automated machines and optimising their activities. As seen in [17] assembly operations were performed by workers using augmented reality headsets. These systems were able to detect the component which had to be assembled and instruct the worker through augmented visualisations to select the correct components in the correct order. Such integration technologies catalysed the learning process for workers unfamiliar with the new products or workstation by acting as an assistant or advisor to the worker.

4.2 Worker Assignment

Resource utilisation is an important metric for manufacturing firms, which includes the assignment, scheduling and delegation of workers on the factory floor. Workers are typically paid fixed rates during operational hours and poor worker utilisation results in a cost to the company with no or little value-adding services. Effective worker assignment ensures that the cost of workers is matched to their value-added activities. Therefore, by increasing worker utilisation, manufacturing firms can reduce costs and increase productivity.

Worker's roles and tasks vary across different manufacturing sectors. The main manufacturing processes can be categorised into continuous, batch and job process methods, or hybrids of these methods. For batch methods, workers are required to perform activities at various workstations/machinery to produce a batch of products. These tasks and roles are repeated for each batch, whereas tasks might differ depending on product variances between batches. For continuous methods, workers are allocated to workstations to perform repetitive tasks for the duration of their shift with slight variance in activities depending on product customisations. Manufacturers who use multiple continuous production lines perform load balancing, by delegating and shifting workers between lines depending on product demand and faults experienced by each line.

The challenge to dynamically allocate workers and enable worker context switching (i.e. moving between workstations for different tasks) to reduce repetitive strain injuries and boredom, is left unresolved. Therefore, constraining workers to repetitive tasks at 'fixed' workstations have remained standard practise since it less complex to manage, but does not utilise the full potential of the workforce. Workers can be left waiting for processes or machine activities to finish before commencing with their allocated task again. Therefore, current production processes are not set up to realise the full benefit of workers and HSI developments to support workers are not fully exploited.

HSI developments that can coordinate and represent the individual schedule and intentions of each worker and integrate such information with the global manufacturing execution system, would enable improved worker management. Workers could be dynamically scheduled and informed in real-time of new tasks and responsibilities at various stages in the production process. Through integrated digital solutions, workers can be dynamically allocated to various workstations depending on production demand, fault handling and bottlenecks as showcased in [18] through a worker holon architecture which resulted in higher productivity, flexibility and of the worker.

4.3 *Quality Assurance*

HSI can effect quality assurance (QA) because QA includes all the systems and procedures which are implemented to ensure that the product requirements are fulfilled during the production stage. Worker's activities and tasks have a direct impact on the quality of the products being manufactured and as such need to fall within the QA procedures. Historically, tracking the activities of workers and their supervisors has been challenging, and humans have been a cause of major quality issues, partly due to a lack of supervision through effective instruction and monitoring. HSI developments could mitigate, or even overcome, some of these challenges through virtual supervision. Such a system would be able to offer real-time instruction depending on the product and production stage. Furthermore, the system would be able to receive and trace feedback from the worker for detected quality concerns, since humans have excellent overall awareness and sensory capabilities.

4.4 *Traceability of Manual Operations*

HSI developments provide a mechanism of maintaining a detailed and (near) real-time digital thread of worker movement and activities.

4.4.1 Costing

Job process methods, seen in small and medium manufacturing enterprises, are typically costed depending on material usage, equipment usage, consumables, overheads and labour to bring the job to completion. Monitoring and tracing manual labour for jobs have typically been performed through legacy systems such as manual timesheets. Integrating the humans into the manufacturing execution system schedule enables accurate job costing for value-added manual labour. A digital thread of manual operations will enable transparency between the manufacturing enterprise and the customer for related costs.

4.4.2 Workflow Optimisation

HSI developments using sensors and wearable devices could provide historic data of worker movements and activities which can be used for strategic decision making and optimisation of workflows and production layouts. For example, wasted time incurred as the worker moves between workstations to perform their various tasks, can be quantified through analysis of the historic data.

4.4.3 Labour Disputes

The South African Labour Department reported 227,040 lost working days for the manufacturing sector in 2018 due to strikes [19], some of which were related to high wage demands and undesirable work conditions. Through improving worker productivity with HSI solutions and effectively monitoring worker performance during operations, employers might reward workers with increased wages.

4.5 Feedback to Control Pyramid

Valckenaers and Van Brussel [20] promoted the design of systems which reflect reality accurately to allow for optimised analysis and decision-making capabilities. As an example, navigation systems have real-time feedback of the current traffic patterns and data which one can use to make optimised routing decisions for travel. Similarly, manufacturing systems require mechanisms for feedback from the physical production system. Such feedback is typically from sensors integrated into the typical automation pyramid, using PLCs and SCADA systems.

Workers have exceptional observational capabilities and situational awareness which are underutilised in current manufacturing systems in South Africa since most have only limited HMI capabilities to receive worker feedback. HSI solutions could enhance the feedback workers can provide to the manufacturing execution systems of the current production state to ensure systems is optimised in decision making and responsiveness to emergent behaviour.

4.6 Worker Health

Wearables and personal health monitoring devices are becoming readily available and cost-effective. These devices can monitor heart rate, blood pressure and movement. Wearables enable HSI solutions that can monitor the health of the worker during operations to detect mental fatigue, over-exertion and risk of injury.

Worker health data can be useful for task delegation. As workers fatigue during operations, the delegation systems can use the fatigue data and optimise decision-making and allow for appropriate breaks or context switching to ensure the worker remains vigilant during operations. Furthermore, higher priority or dangerous tasks can be delegated to workers who are less fatigued to ensure high quality and safety during the operations. Worker health data can also provide a means of ensuring appropriate work environment standards and a healthy workforce.

5 Challenges

This section discusses challenges that the South African manufacturing industry faces to realise the HSI possibilities mentioned.

5.1 ICT Infrastructure

The ICT infrastructure encompasses all software, hardware and communication networks within a manufacturing enterprise required for people and components of the manufacturing systems to interact inside or outside the enterprise. The infrastructure could include sensors on the production floor, human–machine interfaces and the network connection to the manufacturing execution system. This infrastructure clearly plays an important role in the production operations. When considering HSI, ICT infrastructure is also an essential enabler because ICT devices and communication networks must be used to integrate humans into the manufacturing systems. The ICT infrastructure is essential for robust and low-latency communication on the production floor amongst people, devices and software systems.

Many South African manufacturing enterprises have not implemented substantial ICT infrastructure, which will inhibit HSI. The most advanced ICT infrastructure can be seen in the automotive and in the food and beverages manufacturing sectors. Even these ICT developments have largely not integrated workers on the factory floor. Furthermore, many small and medium enterprises (SMEs) do not have the capital to support such ICT. SMEs will therefore first have to invest in ICT infrastructure before they can embark on HSI.

5.2 Technology Acceptance

5.2.1 Organisational Level

On an organisational and strategic level, managers are often hesitant towards the adoption of new technologies and systems due to a lack of cost–benefit evidence. Showcasing the cost–benefit of HSI solutions is challenging since these developments are costly implement for production specific applications. For this reason, and because of the lack of successful industrial "role models", many enterprises are not willing to accept the risk.

Manufacturers in Europe and the Americas have had a greater technology acceptance threshold which has led to greater advancements in manufacturing innovation leading to improved production performance. This acceptance may be partly due to government support in first world countries that allows the development of real-world demonstrations.

5.2.2 Worker Level

To realise the full benefit, workers' cooperation will be required for utilising HSI developments in manufacturing enterprises. However, workers who perform operational activities are often hesitant towards using new HSI technologies due to:

- The unemployment concerns of being replaced by automation systems.
- The steep learning curve to use new technologies and methods.
- The health and safety concerns if required to work in proximity with moving machinery.
- The ethical concerns in having movements, actions and performance monitored and traced.

These concerns vary for workers depending on educational background, demography, job role and labour protection (union membership).

5.3 *Investment Strategies*

Globally, some first world countries like Germany are known for their aggressive investment strategies towards new technologies and manufacturing methods. Consequently, these countries have seen significant growth in production productivity and efficiency.

South African manufacturing enterprises, particularly SMEs, are not aggressive due to capital constraints and not having the capacity to take risks in investing in new technologies. This has been aggravated by the gradual economic decline in manufacturing. Therefore, enterprises strive to keep business at the break-even point, even resorting to retrenching employees to decrease labour costs. These factors have contributed to a stagnant innovation environment in South African manufacturing. Although the South African government is investing in various area of science and innovation, HSI is not a priority in these government programmes.

5.4 *Risk and Safety Policies*

Often when human–machine collaboration is implemented in production applications, these applications require the workers to work near moving machinery (e.g. industrial robots and autonomously guided vehicles). This proximity leads to hazardous environments if not managed correctly with appropriate safety procedures and risk analyses.

To allow for more collaborative activities, HSI aims to create environments where humans and machines can work near one another with fewer safety barriers. For example, an HSI realisation could entail a worker working with a collaborative robot, where the robot can know the intentions and movements of the workers and vice versa to prevent collisions. But such work environments are currently not supported by national risk and safety policies and regulations, which typically still require barriers to mitigate risks. These policies need to be adapted to allow for innovative collaborative workstations and HSI.

5.5 *System Reconfigurability*

ICT systems and HSI must be understandable, user friendly and reconfigurable for manufacturing enterprises to adopt such solutions. HSI software solutions which are complex and designed for only one production application are typically not attractive to managers as they do not have the necessary skilled workforce or capital to change the HSI developments as production demands and workflows change. Therefore, HSI developments and systems must be reconfigurable with short development times and low development costs.

6 Conclusion

This paper suggests that the manufacturing sector in SA could benefit from HSI developments that enhance the operational roles of manual workers. HSI developments have the potential to increase worker wellbeing, employment and productivity. Though many challenges exist to realise these HSI developments, the possibilities summarised in this paper motivate the need for future development and evaluation of such HSI applications.

Acknowledgements The financial assistance of the National Research Foundation (NRF) towards this research is hereby acknowledged. Opinions expressed and conclusions arrived at, are those of the authors and are not necessarily to be attributed to the NRF.

References

1. Pereira, A.C., Romero, F.: A review of the meanings and the implications of the industry 4.0 concept. Procedia Manuf. **13**, 1206–1214 (2017)
2. Zhong, R.Y., Xu, X., Klotz, E., Newman, S.T.: Intelligent manufacturing in the context of industry 4.0: a review. Engineering **3**, 616–630 (2017)
3. Purdy, M., Daugherty, P.: Why artificial intelligence is the future of growth. Accenture (2016)
4. Statistics South Africa: Manufacturing: winners and losers of 2018. http://www.statssa.gov.za/?p=11890. Accessed 27 Aug 2021 (2018)
5. Machaka, J., Roberts, S.: The DTI's new "integrated manufacturing strategy? Comparative industrial performance, linkages and technology. South Afr J Econ **71**, 679–704 (2003)

6. United Nations Department of Economic and Social Affairs: The 17 goals. In: Sustainable development. Department of Economics and Social Affairs and Sustainable Development, https://sdgs.un.org/goals. Accessed 28 Aug 2021 (2015)
7. Maisiri, W., Van Dyk, L.: Industry 4.0 skills: a perspective of the South African manufacturing industry. SA J. Hum. Resour. Manag. **19**, 1–9 (2021)
8. Sparrow, D.E., Kruger, K., Basson, A.H.: The BASE architecture for the integration of human workers into modern manufacturing environments. Laboratory Report, Stellenbosch University, Department of Mechanical and Mechatronic Engineering
9. Bughin, J., Hazan, E., Lund, S., et al.: Automation and the workforce of the future. McKinsey Global Institute. https://www.mckinsey.com/featured-insights/future-of-work/skill-shift-automation-and-the-future-of-the-workforce. Accessed 28 Aug 2021 (2018)
10. Vass, R., Raaidani, P.: Facts and figures on skills in manufacturing, vol. 2, no. 24. Department of Trade Industry (2019)
11. Magwentshu, N., Rajagopaul, A., Chui, M., Singh, A.: The future of work in South Africa: digitisation, productivity and job creation. Mckinsey Global Institute (2019)
12. Deloitte: Enhancing Manufacturing competitiveness in South Africa, Deloitte (2011)
13. Trentesaux, D., Millot, P.: A human-centred design to break the myth of the "magic human" in intelligent manufacturing systems. Stud. Comput. Intell. **640**, 103–113 (2016)
14. Romero, D., Bernus, P., Noran, O., et al.: The operator 4.0: Human cyber-physical systems & adaptive automation towards human-automation symbiosis work systems In: IFIP Advances in Information and Communication Technology, pp. 677–686 (2016)
15. Madni, A.M.: Integrating humans with software and systems: technical challenges and a research agenda. Syst. Eng. **13**, 232–245 (2010)
16. Sparrow, D.E., Kruger, K., Basson, A.H.: An architecture to facilitate the integration of human workers in Industry 4.0 environments. Int. J. Prod. Res. (2021)
17. Paelke, V.: Augmented reality in the smart factory: Supporting workers in an industry 4.0. environment. In: 19th IEEE International Conference on Emerging Technologies and Factory Automation, ETFA 2014, Institute of Electrical and Electronics Engineers Inc. (2014)
18. Leuvennink, J., Kruger, K., Basson, A.: Architectures for human worker integration in holonic manufacturing systems. Stud. Comput. Intell. **803**, 133–144 (2019)
19. Department of Employment and Labour: Strikes in 2018 reaches a high in the past five year. Department of Employment and Labour. http://www.labour.gov.za/strikes-in-2018-reaches-a-high-in-the-past-five-year—department-of-employment-and-labour. Accessed 20 Sep 2021 (2020)
20. Valckenaers, P., Van Brussel, H.: Design for the Unexpected: From Holonic Manufacturing Systems Towards a Humane Mechatronics Society. Butterworth-Heinemann, Oxford, UK (2015)

Travis Defty holds a BEng, is an MEng (Mechatronics) candidate at Stellenbosch University and is a member of the of the Mechatronics, Automation and Design Research Group. He is researching human digital twins for human-system integration in a collabora-tive manufacturing environment.

Karel Kruger obtained his Ph.D. from Stellenbosch University, South Africa. He is a senior lecturer in the Department of Mechanical and Mechatronic Engineering at Stellenbosch University, and is co-leader of the Mechatronics, Automation and Design Research Group.

Anton Basson obtained his Ph.D. in Aerospace Engineering at Penn State University. He was appointed in 1997 as Professor in Mechanical Engineering at Stellen-bosch University and is co-leader of the Mechatronics, Automation and Design Research Group.

Systematization of Technological Capabilities for Connected Adaptive Production

Günther Schuh, Thomas Scheuer, and Jannik Herding

Abstract Increasing networking in production and the complexity of new technologies are presenting companies with ever greater challenges. To survive in this dynamic environment, companies can develop into agile, learning organizations by implementing suitable Industry 4.0 solutions. Many companies have already developed an understanding of the fundamental principles of connected adaptive production and recognized the added value for maintaining long-term competitiveness. However, systematic implementation remains a challenge. Often, there is a lack of deeper understanding of the capabilities to be built to unlock the full potential of Industry 4.0 solutions. In particular, the required technological capabilities regarding resources and information systems in companies reflect a wide range of possible applications. For the targeted selection of measures towards the connected adaptive production and their successful implementation, companies need a structured overview of the required technological capabilities. Therefore, this paper presents a model for systematizing the technological capabilities of manufacturing companies for the realization of connected adaptive production.

Keywords Industry 4.0 · Capability · Digitalization · Adaptivity · Manufacturing

G. Schuh (✉)
Chair of Production Engineering at Laboratory for Machine Tools and Production Engineering (WZL) of RWTH Aachen University, Aachen, Germany
e-mail: g.schuh@wzl.RWTH-Aachen.de

G. Schuh · T. Scheuer · J. Herding
Department of Technology Management at Fraunhofer Institute for Production Technology IPT, Aachen, Germany
e-mail: thomas.scheuer@ipt.fraunhofer.de

J. Herding
e-mail: Jannik.herding@ipt.Fraunhofer.de

K. von Leipzig et al. (eds.), *Smart, Sustainable Manufacturing in an Ever-Changing World*, Lecture Notes in Production Engineering,
https://doi.org/10.1007/978-3-031-15602-1_24

1 Introduction

Especially manufacturing companies are currently facing the challenges of increasing complexity due to a variety of endogenous and exogenous drivers. The exogenous, market-driven complexity consists in particular of growing requirements to handle variant diversity or even individualization of products up to batch size one, increased competition in global markets, and required flexibility due to uncertain forecasts [1–6]. Against this background, endogenous, company-driven complexity inevitably results from historically grown structures, the diversity of IT systems, and the use of the latest technologies [1, 7]. The vision of Industry 4.0 represents a fundamental transformation of industrial value creation and enables the emergence of flexible, highly dynamic, and globally connected value networks [8, 9]. The associated decentralization of autonomous intelligence, decision-making power as well as control of processes offers a solution to balance internal with external complexity [1].

Many companies have already recognized the added value of Industry 4.0 in the context of industrial production. Nevertheless, the comprehensive operational implementation of connected adaptive production (CAP) is often only hesitant, as particularly the scaling of isolated pilot projects to company scale represents a major hurdle [6, 10]. This circumstance is due to a lack of in-depth understanding and actionable solutions as well as no uniform standards [7, 11]. Companies lack knowledge of the required technological capabilities in terms of technical resources and information systems for the systematic design of CAP. However, to cope with the complexity described at the beginning, they must realize production in the sense of the Industry 4.0 vision by implementing suitable infrastructural measures. Thus, there is a need for the structuring of the required technological capabilities that can be applied in practice for the derivation of specific bundles of measures for the implementation of CAP.

Therefore, this paper aims to answer the following research question: "*How to describe and systematize required technological capabilities to realize connected adaptive production?*".

2 Research Methodology

The central research question addresses a problem with high practical relevance and thus requires an interdisciplinary approach. Against this background, the present work is based on the research process of applied sciences according to Ulrich, shown in Fig. 1 [12]. This research process consists of seven steps, of which the first five are examined in this paper. Step A has already been discussed in chapter "Identification of Residual Development Efforts in Agile Ramp-Up Production". Process steps B and C are object to Chapters "Modeling Interactions and Dependencies in Production Planning and Control an Approach for a Systematic Description" and

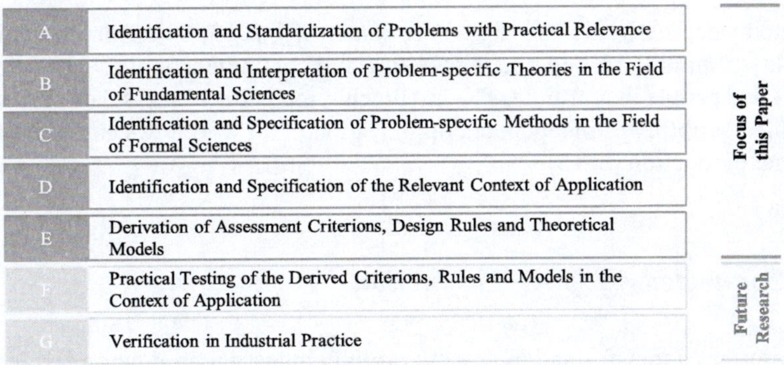

Fig. 1 Methodology of applied science according to Ulrich [12]

"ARTI-Based Holonic Manufacturing Execution System Using the BASE Architecture: A Case Study Implementation", presenting fundamentals and an analysis of relevant literature. Finally, Chapter "Bridging the Gap Between Digital Human Simulation to Standard Performance with Human Work Design" aims to develop the structure for systemizing the required capabilities for a CAP (steps D and E). The paper is concluded by a summarizing discussion of the results in Chapter "Interface Holons in the BASE Architecture for Human-System Integration in Cyber-Physical Systems".

3 Theoretical Background

The goal of this paper is to systematize and describe the technological capabilities required for the realization of CAP. To this end, fundamentals from the overarching topic of Industry 4.0 are presented first, before CAP is elaborated on as a focused area of consideration for the present work. The description of the theoretical background is concluded by an examination of technological capabilities.

3.1 Industry 4.0

The term "Industrie 4.0" has been coined in Germany in 2011 to promote the "informatization" of German industry [6–8]. However, this term is not solely tied to German industry. At the international level, further initiatives are dealing with the same topic of digitally connected industrial production, such as the Industrial Internet Consortium in the USA or the Connected Industries Initiative in Japan. For all of them, Industry 4.0 is understood as the fundamental transformation of industrial value creation through the emergence of flexible, highly dynamic, and globally connected value networks and digital ecosystems [6, 7, 13, 14]. This is enabled by the ability of multimodal ad-hoc networking and communication of cyber-physical systems and people in real-time. The intelligent processing of mass data in real-time offers the possibility of independent optimization of all workflows and processes in industrial production [6, 15].

3.2 Connected Adaptive Production

CAP describes a transformation process result in industrial production and refers to the digitization and networking of production processes, machines, and plants [16]. It is considered as synonymous to the term smart factory [17, 18].

The first aspect of the target state of CAP is interoperability through the networking of elements such as machines, plants, software systems, control systems, and sensors. The objective is to form an end-to-end cyber-physical production system, which is a basic prerequisite for the implementation of Industry 4.0 [19]. Interoperability is made possible by holistic networking and horizontal and vertical integration [6, 19].

Networking in Industry 4.0 breaks down monolithic structures in companies' IT systems and enables the bidirectional exchange of heterogeneous data [6, 16]. This provides the foundation for data-based autonomous adaptation to dynamic changes and self-optimization [6, 19]. These capabilities of production systems in the CAP are reflected in adaptivity as the second aspect of the target state. In the understanding of this paper, adaptivity refers to the holistic data-driven optimization approach of CAP. The term adaptivity subsumes all applications of intelligent behavior for the targeted processing and use of data to generate knowledge, support decisions, and feed optimized parameters back into processes via control or appropriate actuation [3, 8]. The highest level of adaptivity is continuous and autonomous self-optimization [1, 6].

3.3 Technological Capabilities

In human sciences, the concept of capability is closely related to the concept of intelligence. A capability enables a certain form of action or interaction with the environment, whereby actions are guided by the standards and norms specified by the underlying intelligence [20]. Thus, a capability is understood as a possibility, created by certain abilities and properties, to fulfill certain functions and to meet certain requirements, and perform them when needed. The term technology is traditionally understood as the knowledge of the technical and scientific goal-means relationships for solving technical problems [21]. From the link between technology and capabilities, it can be deduced that technological capabilities are those based on mastery of a particular technology [22]. In the context of production, the integration of technologies into resources generates technological capabilities which are thus made utilizable for production. In the CAP environment, data-based technologies are used for a data-driven optimization approach. Therefore, data-based technological capabilities play an important role in the development of the present paper's model.

4 Literature Review

To identify, whether and to which extent the approaches of other authors support the addressed research question, their works are subjected to a structured analysis in four topics. All the approaches are regarding their suitability for the context of this work in general (object area), as well as regarding their suitability to answer the formulated research question. A summary of the discussion will be given in Fig. 2 in the last section.

4.1 Theory of CAP

Within the examined approaches of Baum et al., Pennekamp et al., Schuh et al. (2017), and Schuh et al. (2020), practical use cases and their technological enablers, as well as the measures for implementing a CAP in general, are addressed [16, 17, 23–25]. It becomes clear that far-reaching concepts and technologies for implementing a CAP already exist, but that linking them to form a comprehensive overall solution continues to be a challenge [23]. The target state aspects of interoperability and adaptivity are mostly addressed only indirectly in the approaches examined and, in particular, no concrete technological capabilities for CAP are yet formulated and presented in a structured manner using suitable ordering principles to enable measures to be derived [16, 17, 23–25].

4.2 Approaches for Describing Capabilities for Industry 4.0

The presented work of Geisberger et al. and Bauernhansl et al. only partially fulfill the criteria of the object area, since central technology fields of CAP, such as the digital twin or decision making by assistance systems, are not addressed [1, 18]. Instead, the approaches described deal with the potentials and challenges of cyber-physical systems in manufacturing companies and demonstrate capabilities at a very generic level [1, 18]. Furthermore, a structuring of capabilities in both approaches is based on criteria from the application context and not in the context of the target state aspects defined in the present work [1, 18]. The generic description of capabilities and lack of systematization makes it difficult to derive concrete measures with respect to the infrastructure in the production environment [1, 18]. Furthermore, it is not addressed

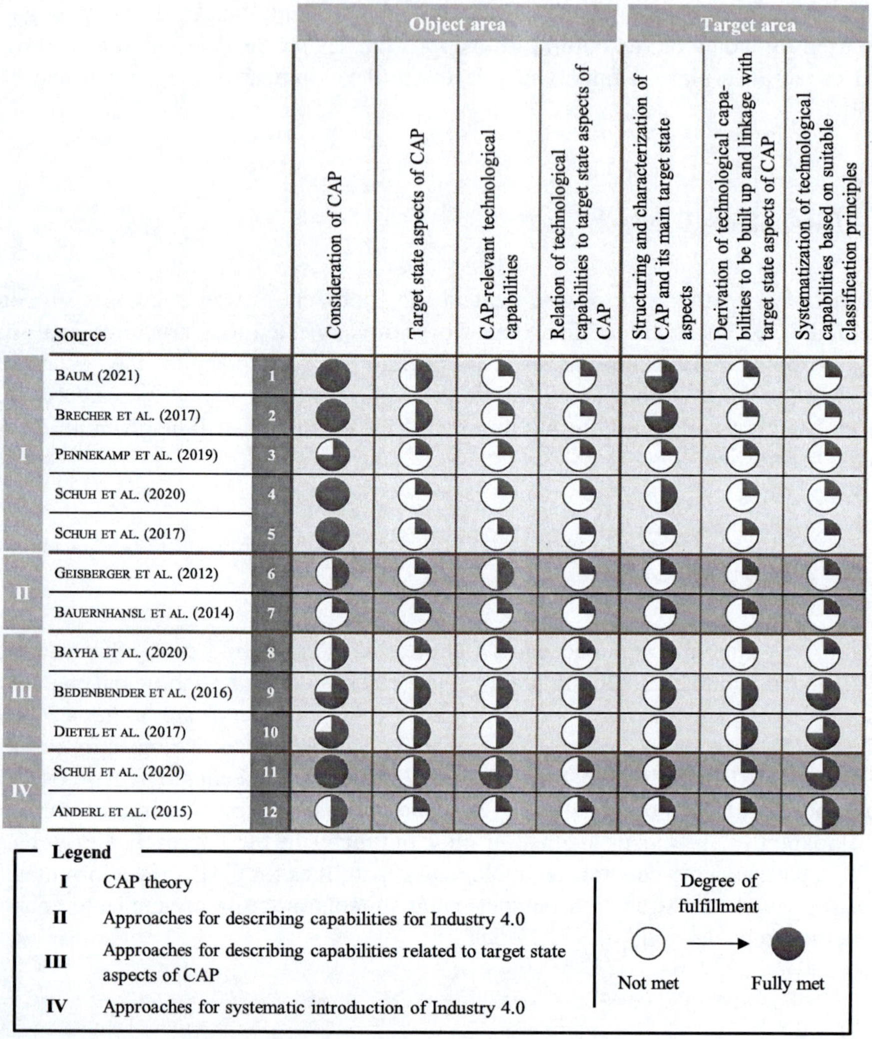

Fig. 2 Results of literature review [1, 6, 11, 16–19, 23–27]

which basic prerequisites must be created for a CAP and which capabilities are required for a holistic optimization approach [1, 18]. The capabilities presented are described at a high level of abstraction and are related in particular to the planning and control of processes and therefore only partly contribute to the derivation of capabilities for the CAP [1, 18].

4.3 Approaches for Describing Capabilities Related to Target State Aspects of Networked Adaptive Production

The publications of Bayha et al., Bedenbender et al., and Dietel et al. have high relevance for the object area, as they all deal with the realization of connected, modular production as a basic prerequisite for Industry 4.0 [19, 26, 27]. It becomes clear that good preliminary work has already been done in the context of standardization approaches for generating interoperability. RAMI 4.0 has a high practical relevance and is used in many approaches as a model for mapping the Industry 4.0 solution space to formulate common standards and requirements. Bedenbender et al. and Dietel et al. use the RAMI 4.0 as a framework for describing requirements regarding interoperability in a networked production [19, 27]. Nevertheless, the object, as well as target domain of the present work, are only partially addressed since the approaches do not contain a structuring of technological capabilities with regard to the target state aspect of adaptivity and a holistic data-driven optimization approach.

4.4 Approaches for Systematic Introduction of Industry 4.0

The publications of Schuh et al. and Anderl et al. deal with the systematic introduction of Industry 4.0 in manufacturing companies [6, 11]. The approach of Schuh et al. is highly consistent with the object area, as the implementation of Industry 4.0 is explained based on a comprehensive description of the topic and the capabilities to be built up. The presented structure enables a stepwise implementation along maturity levels. Anderl et al. develop a step-by-step guide for the introduction of Industry 4.0 with a focus on the methodology rather than the content. The primary goal is the development of new business models [11]. Both approaches support the systematic introduction of Industry 4.0, however, the target state aspects of CAP are only indirectly addressed. Thus, the systematization presented is not suitable for the present work. Furthermore, both works describe capabilities on a generic level and are therefore not suitable for deriving specific measures. Due to their broad scope of consideration in the field of Industry 4.0, the two approaches serve as an orientation for the model development. Nevertheless, capabilities must be formulated in a significantly higher level of detail with regard to the objective of this paper.

4.5 Overview on Examination of Existing Approaches

Existing approaches have been systematically examined regarding the above-defined criteria. Figure 2 summarizes the discussion described in Sects. 1–4.

Based on the examination of approaches shown in section B., it has to be concluded that no suitable approach exists to date for systematizing the technological capabilities required to implement a CAP and thus for answering the overarching research question [1, 6, 11, 16–19, 23–27]. Existing approaches to CAP or related topics focus primarily on single, very specific areas of consideration at a high level of detail or illuminate several topic areas at a more abstract level. As a result, either a complete, broad consideration of the topic or an appropriate level of detail for deriving relevant and non-trivial measures is missed. In addition, there are isolated approaches that make proposals for standardization concerning the target state aspect of interoperability. However, there is a lack of embedding this in the overall context of the CAP as well as of the standardized description of necessary capabilities for a holistic data-driven optimization approach in the sense of adaptivity. In particular, there is no approach, which holistically systematizes technological capabilities with a high level of practical relevance and thus supports the formulation of specific measures for the implementation of a CAP.

5 Results

The central area of consideration of the present work is the CAP. Due to the breadth and technological complexity of the topic, manufacturing companies often lack a deeper understanding and are challenged by the systematic implementation of CAP. The elaborated model is therefore intended to support manufacturing companies in the practical implementation by systematizing the technological capabilities required to realize a CAP. Therefore, those capabilities need to be described in a form that can be operationalized on the one hand, and is generic on the other hand, to ensure that targeted measures can be derived, and the model can still be applied across manufacturing companies.

5.1 Derivation of Model Structure

In the understanding of the present work, systematization includes the structure and classification of the model contents based on an ordering principle [28]. This ordering principle should have a high degree of realism to satisfy the model requirements. Therefore, the model formation takes place in analogy to the VDI guideline 2221 for the solution of complex problems [29]. In this sense, the overall problem is first broken down into subproblems and a solution is then elaborated from the rough to the detailed according to the theory of systems engineering [30].

In Chapter "Modeling Interactions and Dependencies in Production Planning and Control an Approach for a Systematic Description", the target state of CAP has already been presented as decomposed into the aspects of interoperability and adaptivity. The networking of production system participants to generate interoperability

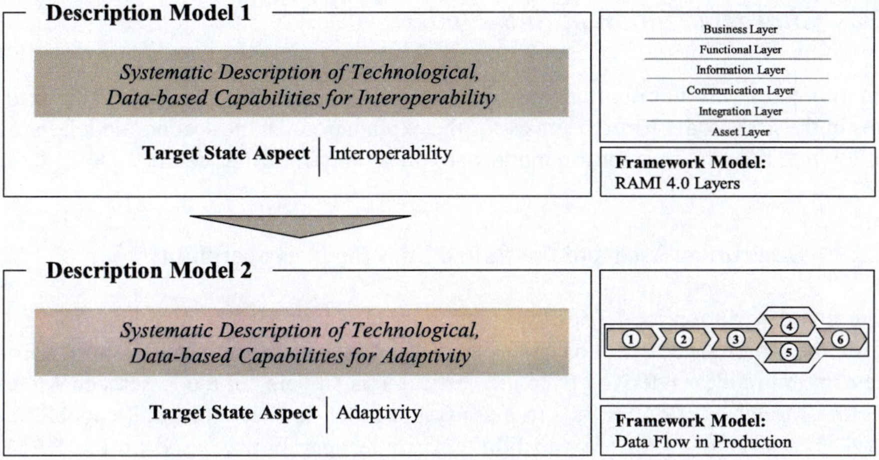

Fig. 3 Two-step model structure for systematization of technological capabilities for the connected adaptive production

is a basic prerequisite for a smooth and automated flow of information [17]. Only through interoperability, a workable production system can emerge in the sense of Industry 4.0 and provide the infrastructure for a holistic data-driven optimization approach, which is reflected by the aspect of adaptivity. Thus, interoperability is a prerequisite for adaptivity. Against this background, two consecutive descriptive models are proposed, which reflect in detail the required data-based technological capabilities for realizing the two target state aspects of interoperability (model 1) and adaptivity (model 2) of the CAP (cf. Fig. 3).

The generation of an interoperable and modular production system for the exchange of data, services, and capabilities in the considered area of production is a necessary basic requirement for the realization of adaptivity. Thus, it must be carried out in the first step. The interoperability sub-model presents the required capabilities. Only if involved products or production units possess those capabilities, the second step of adaptivity can be focused on. Moreover, the interoperability capabilities are required for an understanding of the further ones in the adaptivity model.

Building on the infrastructure created in model 1, capabilities for the practical implementation of applications in the sense of Industry 4.0 can be derived. This happens in the second step with the help of the adaptivity sub-model against the background of the individual objectives.

5.2 *Characterization of Sub-Models*

Interoperability is an important basic prerequisite for Industry 4.0, thus the capabilities in the first model form the basis for the explanations in the second model. In the following, the two constituting models are presented in further detail.

5.2.1 Description Model of the Capabilities for Interoperability

The first description model specifies data-based, technological capabilities for realizing the target state aspect of interoperability. The organizing principle of the first description model is derived from the vertical axis "layers" of the Reference Architecture Model 4.0 (RAMI 4.0) to classify the identified technological capabilities [31]. The structuring of the capabilities for interoperability, based on the RAMI 4.0, enables a high degree of realism. In addition, the recognition of the RAMI 4.0 in practical application contributes significantly to the understanding of the model contents [31]. Thus, this model is considered as a well-suited framework for the interoperability model. Figure 4 gives an overview of the model structure which will suit as a framework for the elaboration of specific capabilities in future work of the authors.

In the understanding of the present work, the CAP represents a network of modular participants in the form of intelligent and interconnected production units and products. The technological capabilities listed in the first description model are inherent capabilities which all participants must possess to participate in the CAP network. The capabilities, therefore, do not refer to a specific use case but reflect in their entirety the fundamental capability of the participants in the production network to establish the state of interoperability. Exemplary, a capability of a production unit on the information layer could be "Industry 4.0 sematic and syntax". Overall,

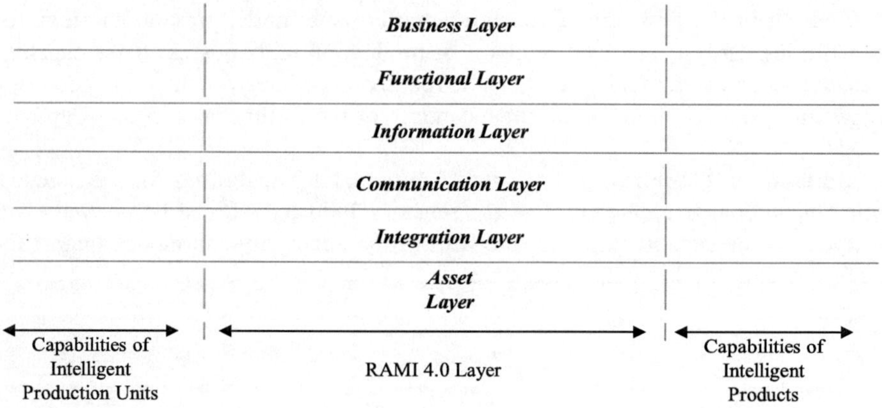

Fig. 4 Framework structure for interoperability capabilities [31]

the systematic description of data-based, technological capabilities provides a clear representation of the required capabilities for interoperability for transfer to corporate practice. The generic model structure and description of the capabilities also allow the model contents to be reused across companies by different users.

By describing data-based, technological capabilities with a high level of detail, the first description model offers a cross-company user group a suitable tool for generating an interoperable infrastructure in the production system under consideration. To this end, users can compare the technical specifications of individual production units or products and the associated capabilities with the required capabilities in the description model and derive corresponding needs for action. Interoperability enables the cross-domain exchange of data and thus provides the basis for the realization of a holistic data-driven optimization approach based on the capabilities in the second sub-model.

5.2.2 Description Model of the Capabilities for Adaptivity

The second description model describes data-based, technological capabilities for realizing the target state aspect of adaptivity. In contrast to interoperability, adaptivity is not assigned a specific condition, but it concerns rather differently complex applications of intelligent behavior in production for the realization of a holistic data-driven optimization approach. The organizing principle of the second description model consists in the derivation of a step-by-step path of data in the CAP from raw data to enriched and usable information as well as the further division of the steps into topic-specific design fields. In the following, different approaches to defining the data flow are introduced.

According to Brecher et al., the path of data begins with the collection of data in the production environment and other data-carrying systems. The data is then made available via suitable interfaces and middleware to further processing systems for data processing, modeling of the digital twin, and finally targeted output by assistance systems [17]. Kiesel et al. specifically define the seven topic areas of "sensor systems & data acquisition", "interfaces & connectivity", "data synchronization & middleware", "data modeling & data analysis", "digital twin in the product lifecycle", "cloud systems & IT architecture", and "digital business models" for the CAP. Topics 1–5 are oriented in chronological order to the path of data in the CAP, while the latter stand for the use of modern information and communication technology solutions as well as the development of data-based business models [32, 33]. Bitsch introduces five levels of digital shopfloor management for the efficient processing, management, and use of large heterogeneous data volumes in the context of Industry 4.0. These include the acquisition, storage (of unstructured data), processing, storage (of structured data), and presentation of data [34]. The approach of Meinel et al. includes a simplified taxonomy of data-based technologies on a so-called smart data platform [34]. This taxonomy presents further potential building blocks for the definition of data flow steps and the subsequent design fields.

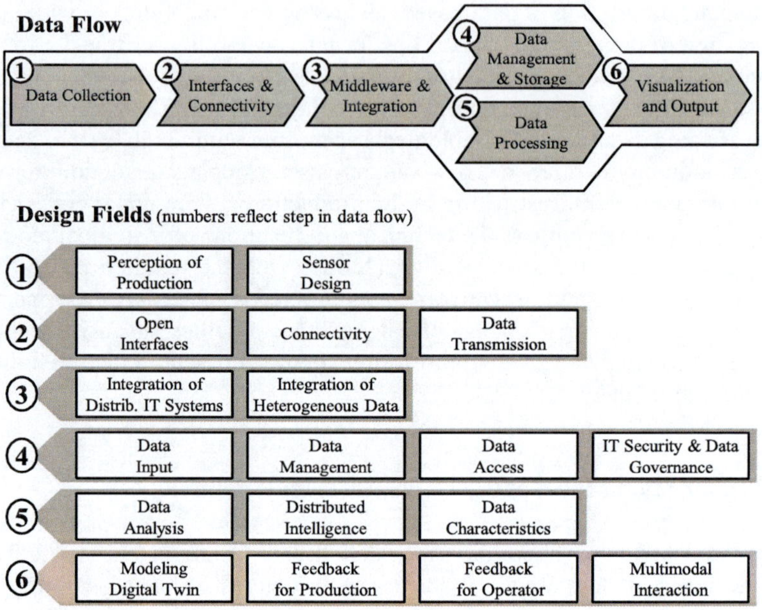

Fig. 5 Framework structure of the adaptivity capabilities [17, 32, 34, 35]

Overall, based on a comprehensive analysis of the previously listed approaches, the definition for the steps of data flow and the underlying design fields is derived [17, 32, 34, 35]. Due to its high practical relevance, the definition of a uniform data flow enables the capabilities to be presented in a comprehensible and structured manner. Figure 5 displays the defined data flow, as well as the design fields which allow a further classification of the adaptivity-related capabilities.

The capabilities listed in the second description model are not those of individual participants in the production system as in the first sub-model but refer to a composite of data-based technologies in information systems and technical resources from the production environment. An exemplary capability in the design field of *Distributed Intelligence* within the *Data Processing* (step 5) could be "Decentralized and continuous pre-processing of data with low latency". The systematic description of the required data-based, technological capabilities for the realization of adaptivity is intended to provide the user group with a comprehensive solution space of capabilities for transfer to business practice. This should simplify the derivation of individual measures for the practical implementation of CAP.

In contrast to interoperability, adaptivity does not describe a specific state, but rather subsumes all applications of intelligent behavior using data-based technologies in order to realize the holistic data-driven optimization approach of the CAP. Against this background, the second description model offers a structured framework for potentially required capabilities to a cross-company user group, which can be used to derive situation-specific bundles of measures. The specific capabilities depend case-specifically on the objective regarding the optimization goal or the application to be implemented in the sense of Industry 4.0.

6 Conclusion

The CAP as the central area of consideration of this paper is understood as a fundamental transformation of industrial production using Industry 4.0 technologies. This enables a holistic data-driven optimization approach and fosters cross-domain collaboration along the order-to-delivery process. Based on this understanding, interoperability as a state of connected and modular production systems and adaptivity as an umbrella term for applications of intelligent behavior in production for the value-creating use of data were defined as the two central target state aspects of the CAP. With this subdivision, a model was designed with the help of systems engineering, which supports manufacturing companies in deriving specific implementation measures by systematizing the capabilities required for a CAP.

The contents presented in the first sub-model are to be understood as the required inherent capabilities of each participant in the network of a CAP and serve to generate a workable production system in the sense of Industry 4.0. By building up the capabilities, distributed intelligent production units and products can communicate with each other, enabling flexible and dynamic planning as well as control of production. The model thus enables a user group to derive action requirements for establishing interoperability by comparing the model contents with the already existing technological capabilities in the specific consideration of individual production units or products. Based on realized interoperability in production, the second sub-model describes the technological capabilities for realizing the target state aspect of adaptivity. The model contents presented are to be understood as the required capabilities of the infrastructure in the production environment to enable the holistic data-driven optimization approach of a CAP. The model provides a structured solution space to a user community as a tool to derive required capabilities when implementing targeted applications in production. The structure of the capabilities presented generates an infrastructure of data-based technologies and thus enables the practical implementation of the vision of Industry 4.0 in the production environment.

Systematization of the capabilities consists first of all in their differentiation regarding the target state aspects of interoperability and adaptivity. The further structuring of the capabilities in the first sub-model based on the vertical levels of RAMI 4.0 enables a clear differentiation as well as an application-oriented description of the individual capabilities at a high level of detail. The structuring of the capabilities in the second sub-model based on the steps in the data flow as well as the analytically derived design fields within the steps enables a clear thematic structuring. This leads to a clear differentiation of the individual capabilities through a high degree of correspondence of the model contents with the observable reality. Altogether, both models show a suitable and application-oriented systematization for capabilities and permit by the generally valid model structures transferable applicability for an enterprise-spreading circle of users. The model structure will serve as a basis for further works of the authors and provides the framework for the allocation of specific capabilities.

With the model developed, the initially formulated research question "*How to describe and systematize required technological capabilities to realize connected adaptive production?*" can finally be answered.

Overall, the developed model helps manufacturing companies to systematically build a CAP, which creates added value by exploiting data from the production environment to support decision-making and the development of data-driven business models. In addition, data from the CAP can be made available quickly across domains via suitable interfaces, and thus contribute to a holistic optimization of processes across different functional areas and an increase in the value-added.

Acknowledgements This research was supported by the European Commission through the H2020 project EPIC (https://www.centre-epic.eu/) under grant No. 739592.

References

1. Bauernhansl, T., ten Hompel, M., Vogel-Heuser, B.: Industrie 4.0 in Produktion, Automatisierung und Logistik: Anwendung, Technologien, Migration, p. 639. Springer, Wiesbaden (2014)
2. Dorst, W., Glohr, C., Hahn, T.: Umsetzungsstrategie Industrie 4.0: Ergebnisbericht der Plattform Industrie 4.0, 100 pp. Berlin (2015)
3. Schuh, G., Reinhart, G., Prote, J.-P., Sauermann, F., Horsthofer, J., Oppolzer, F., Knoll, D.: Data mining definitions and applications for the management of production complexity. Procedia CIRP **81**, 874–879 (2019)
4. Esmaeilian, B., Behdad, S., Wang, B.: The evolution and future of manufacturing: a review. J. Manuf. Syst. **39**, 79–100 (2016)
5. Hu, Y., Li, J., Holloway, L.E.: Towards modeling of resilience dynamics in manufacturing enterprises: literature review and problem formation. In: 4th IEEE Conference on Automation Science and Engineering, pp. 279–284 (2008)
6. Schuh, G., Anderl, R., Dumitrescu, R., Krüger, A., ten Hompel, M.: Industrie 4.0 maturity index: managing the digital transformation of companies. Update 2020. acatech STUDY, Munich. https://www.acatech.de/publikation/industrie-4-0-maturity-index-update-2020/. Accessed Apr 2021 (2020)
7. Roth, A.: Einführung und Umsetzung von Industrie 4.0: Grundlagen, Vorgehensmodell und Use Cases aus der Praxis, p. 278. Springer, Berlin, Heidelberg (2016)
8. Bauernhansl, T., Krüger, J., Reinhart, G., Schuh, G.: WGP-Standpunkt Industrie 4.0, 29 pp (2016)
9. Obermeier, B.: Kompetenzen für Industrie 4.0: Qualifizierungsbedarfe und Lösungsansätze, p. 46. Herbert Utz Verlag GmbH, München (2016)
10. Staufen AG, Staufen Digital Neonex GmbH: Deutscher Industrie 4.0 Index, 48 pp. Eine Studie der Staufen AG und der Staufen Digital Neonex GmbH, Köngen (2019)
11. Anderl, R., Picard, A., Wang, Y.: Leitfaden Industrie 4.0: Orientierungshilfe zur Einführung in den Mittelstand. VDMA-Verl., Frankfurt am Main, CI (2015)
12. Ulrich, H., Dyllick, T., Probst, G.J.B.: Management: Hrsg. von Thomas Dyllick u. Gilbert J. B. Probst. Haupt, Bern & Stuttgart (1984)

13. Plattform Industrie 4.0: Digitale Ökosysteme global gestalten, 8pp. Leitbild 2030 für Industrie 4.0, Berlin (2019)
14. Gausemeier, J., Klocke, F.: Industrie 4.0: Internationaler Benchmark, Zukunftsoptionen und Handlungsempfehlungen für die Produktionsforschung, p. 83. Heinz Nixdorf Institut Universität Paderborn, Paderborn (2016)
15. Kosch, B., Lason, M., Porst, U., Schulz, E.: Industrie 4.0—Status und Perspektiven, 31 pp. Studie, Berlin (2016)
16. Baum, C.: Industrie 4.0—Vernetzte, adaptive Produktion, 32 pp. https://www.ipt.fraunhofer.de/content/dam/ipt/de/documents/Broschueren/Industrie40-VernetzeadaptiveProduktion.pdf
17. Brecher, C., Klocke, F., Schmitt, R., Schuh, G. (eds.) Internet of Production für agile Unternehmen: AWK Aachener Werkzeugmaschinen-Kolloquium 2017, 18. bis 19. Mai, 1. Auflage ed., 460 pp. Apprimus Verlag, Aachen (2017)
18. Geisberger, E., Broy, M.: Agenda CPS: Integrierte Forschungsagenda Cyber-Physical Systems, p. 296. Springer, Berlin (2012)
19. Dietel, M., Heyer, S., Läpke, H.-J.: Industrie 4.0—Die Bedeutung von Interoperabilität im Referenzarchitekturmodell Industrie 4.0 (RAMI 4.0), 55 pp. Leitfaden, Berlin (2017)
20. Worthmann, H. (ed.): Praktische Intelligenz und die Zweiteilung des Wissens. J.B. Metzler, Stuttgart (2019)
21. Pfeiffer, W., Metze, G., Schneider, W., Amler, R.: Technologie-Portfolio zum Management strategischer Zukunftsgeschäftsfelder, 6, durchges, Aufl, p. 145. Vandenhoeck & Ruprecht, Göttingen (1991)
22. Binder, V.A., Kantowsky, J.: Technologiepotentiale: Neuausrichtung der Gestaltungsfelder des strategischen Technologiemanagements, 413 pp. Univ., Diss. V. Binder, J. Kantowsky, Zugl.: St. Gallen. DUV Dt. Univ.-Verl., Wiesbaden (1996)
23. Pennekamp, J., Glebke, R., Henze, M., Meisen, T.: Towards an infrastructure enabling the internet of production. In: 2019 IEEE International Conference on Industrial Cyber Physical Systems (ICPS). 2019 IEEE International Conference on Industrial Cyber Physical Systems (ICPS), Taipei, Taiwan, 06 May to 09 May 2019, pp. 31–37. IEEE (2019)
24. Schuh, G., Prote, J.-P., Dany, S., 2017. Internet of production. In: Schuh, G., Brecher, C., Klocke, F., Schmitt, R. (eds.) Engineering Valley—Internet of Production auf dem RWTH Aachen Campus, 1. Auflage ed. Apprimus Verlag, Aachen (2017)
25. Schuh, G., Prote, J.-P., Molitor, M., Cremer, S.: Internet of production—Steigerung des Wertschöpfungsanteils durch domänenübergreifende Kollaboration. In: Hompel, M., ten, Vogel-Heuser, B., Bauernhansl, T. (eds.), Handbuch Industrie 4.0, pp. 1–24. Springer, Berlin, Heidelberg (2020)
26. Bayha, A., Bock, J., Boss, B., Diedrich, C.: Describing Capabilities of Industrie 4.0 Components, 36 pp. Berlin (2020)
27. Bedenbender, H., Billmann, M., Epple, U.: Welche Kriterien müssen Industrie 4.0-Produkte erfüllen? 20 pp. Leitfaden, Frankfurt am Main (2016)
28. DWDS—Digitales Wörterbuch der deutschen Sprache: Systematisieren. https://www.dwds.de/wb/systematisieren. Accessed 5 April 2021 (2021)
29. Richtlinie 2221 Blatt 1/Part 1—Entwicklung technischer Produkte und Systeme, 56 pp. Modell der Produktentwicklung.
30. Haberfellner, R., de Weck, O., Fricke, E., Vössner, S.: Systems Engineering, p. 470. Springer, Cham (2019)
31. Plattform Industrie 4.0: RAMI 4.0—Ein Orientierungsrahmen für die Digitalisierung, 32 pp. Berlin (2018). https://www.plattform-i40.de/PI40/Redaktion/DE/Downloads/Publikation/rami40-einfuehrung-2018.html
32. Kiesel, R., Pothen, M., Arntz, K., Bergs, T., Schmitt, R.: Interoperabilität als Erfolgsfaktor für die vernetzte adaptive Produktion: Kollaboration von Enablern und Anwendern in sieben interdisziplinären Themenfeldern für die Digitalisierung der Produktion. ZWF Digitalisierung **27**(8–9), 344–347
33. Lossie, K., Pothen, M., Olowinsky, A.: Vernetzte, adaptive Produktion: Internationale Community zur Entwicklung von Anwendungen und Technologien für die Industrie 4.0, 16 pp. Aachen (2021)

34. Bitsch, G.: Digitales shopfloor-management: Ein adaptives informations- und Entscheidungsinstrument im Umfeld von Industrie-4.0-Produktionssystemen. In: Obermaier, R. (ed.) Handbuch Industrie 4.0 und Digitale Transformation, pp. 295–315. Springer, Wiesbaden (2019)
35. Meinel, C., Schneiss, D.: Smart Data—Potenziale und Herausforderungen, 20 pp. Vernetzte Anwendungen und Plattformen für die digitale Gesellschaft (2015)

Günther Schuh is the chair for Production Systematics at RWTH Aachen University. Furthermore, he is the director of FIR e. V. and a member of the board of directors of WZL of RWTH Aachen University and Fraunhofer Institute for Production Technology IPT.

Thomas Scheuer studied Mechanical Engineering and Business Administration at RWTH Aachen University. He works as research associate in Technology Management at Fraunhofer Institute for Production Technology IPT.

Jannik Herding studied Mechanical Engineering and Business Administration at RWTH Aachen University and Tsinghua University Beijing. He worked as research assistant in Technology Management at Fraunhofer Institute for Production Technology IPT.

COVID-19 Manufacturing, Services, Business Models

Did the Covid-19 Pandemic Improve Engineering Education?—A South African-German Perspective

Deborah Blaine, Claudia Fussenecker, Jörg Niemann, and Karin Wolff

Abstract The Covid-19 pandemic has been influencing every aspect of the globalized world since its outbreak. Not only is the economic environment changing rapidly, but also global Higher Education (HE) is facing a disruptive change. The way students are educated and knowledge is transferred and perceived had to be adapted, leading to Emergency Remote Teaching (ERT). Along with the challenges of the ever-changing demands in HE engineering education as a result of digitalisation and Industry 4.0, a new approach to educating future engineers is required. The Covid-19 pandemic has put further pressure on the education system and created new obstacles. However, did something good come out of this pandemic in regard to engineering education? Did students acquire skills which might help them to become better engineers in the future? What happened to the teaching skills of university staff? Did the ability to manage a crisis and adapt to new circumstances improve teaching? What kinds of setbacks were experienced? This paper draws on engineering student and staff survey data from Stellenbosch University, South Africa and Düsseldorf University of Applied Sciences, Germany. The survey takes a holistic approach and considers the incontestable relationship between three key facets of the educator's mandate: to facilitate the development of knowledge, citizenship and skills, by providing cognitive, affective and systemic support aligned to the 'knowing, being, doing' dimensions of the curriculum. The survey then sought to determine how ERT has affected academics in their professional, personal and practical lives. These responses were analysed using the three support dimensions of academic work, *cognitive*, *affective* and *systemic*. The generated data was then analysed to reveal key insights into both shared and differing challenges and successes across the North–South divide, through the lens of Hofstede cultural dimensions.

D. Blaine · K. Wolff
Faculty of Engineering, Mechanical & Mechatronic Engineering, Stellenbosch University, Stellenbosch, South Africa

C. Fussenecker (✉) · J. Niemann
Department of Mechanical & Process Engineering, University of Applied Sciences Düsseldorf, Düsseldorf, Germany
e-mail: Claudia.Fussenecker@hs-duesseldorf.de

K. von Leipzig et al. (eds.), *Smart, Sustainable Manufacturing in an Ever-Changing World*, Lecture Notes in Production Engineering,
https://doi.org/10.1007/978-3-031-15602-1_25

Keywords Engineering education · 4IR skills · Hofstede · Remote teaching · CAS model

1 Introduction

Educating future engineers has not only been a challenge due to Covid-19 since 2020, but also due to the societal, industrial and engineering changes in recent years. There is increasing demand to adequately prepare engineering students for digitalisation and Industry 4.0, where interdisciplinary knowledge is crucial. This increase is evidenced in rising numbers of publications, conferences and funding [1, 2]. The Covid-19 pandemic offers an ideal opportunity to interrogate to what extent staff and students in engineering faculties are equipped to engage in the information communication technology (ICT) space, given the rapid move to online and remote teaching, learning and assessment.

Several themes have emerged within literature about higher education responses to Emergency Remote Teaching (ERT) [3]. These include the adaptation of online learning materials, communication systems and platforms, as well as early lessons such as the importance of clear and transparent communication from the university to students [4]. Broadly speaking, educators are responsible for the holistic support of students in three domains: *cognitive, affective and systemic* [5]. In other words, they are required to support the 'knowing, being and doing' of learning by providing access to a wide range of knowledge, learning experiences, and systems of engagement.

ERT saw faculty members responsible for the quick redesign and adjustment of the curriculum and learning, while having to adapt to new technologies and platforms. The challenge was how to best teach in the new virtual format where students do not react or respond in the same way as before. Here, regular feedback from students has been extremely helpful [6]. From a systemic perspective, literature suggests that the different learning modalities require specific planning, student autonomy and flexibility [4]. A popular strategy has been the 'flipped classroom' mode, with asynchronous and synchronous online sessions. A major systemic constraint in the global context has been inequitable access to devices, data and connectivity [7], despite state or university interventions [8].

Challenges for educators were the development of digital literacy/fluency and netiquette, the management of student expectations as well as resolving technical difficulties on an ongoing basis [4]. Despite state and university crisis management strategies, nobody was completely prepared for the pedagogical challenges that resulted from teaching remotely during a crisis. 18 months down the line, we have evidence of a greater focus on the affective consequences, such as the impact of working from home and increased workload on well-being [9], the need to show compassion, a sense of humour [10] and care for students, and designing responsive teaching strategies.

For engineering education specifically, one "of the main issues raised by industry … (is) that many university graduates lack skills which complement their technical knowledge, such as problem solving, teamwork and the ability to adapt to change" [11]. As one possible solution for this lack of skills, Problem-based Learning (PBL) and Project-based Learning (PjBL) in Science, Technology, Engineering and Mathematics (STEM) has increasingly been adopted. PBL (and PjBL) means that students are given an understanding of the topics by solving "real-world, open-ended problems (or projects), mostly in groups", to give them a feeling for later work experience in industry. These strategies are constructive, self-directed, collaborative and contextual [11], requiring 'inductive' teaching, as opposed to the traditional 'deductive' style where students merely listen and memorize without having the ability to apply it to real world problems [12]. The aim of strategies such as PBL and PjBL is to develop the communication, leadership and teamwork skills, and lifelong learning attributes necessary for a changing industrial environment [13]. However, staff need support and training in implementing inductive teaching methods [14] and practice-oriented learning [15], which have been particularly difficult to implement during ERT.

The Covid-19 pandemic has significant implications for engineering education, given the requirements of industry for our graduates to be flexible problem solvers, technically equipped and capable of dealing with complexity. The current era marks a key moment to reflect on the relationship between engineering education and employability. This paper considers three questions:

- What is the core of good engineering education to support the progression of smart, sustainable and competitive manufacturing?
- How did the Covid-19 pandemic influence engineering education for students and teaching staff in South Africa and Germany in higher education (HE)?
- How did both developments change the design of future engineering education curricula and their focus?

To formulate answers to these research questions, the engineering faculties of the University of Stellenbosch (SUN) as well as the Düsseldorf University of Applied Sciences (HSD) collaborated in collecting and analysing data for a North–South perspective on the future of engineering education. The questionnaires and evaluations come from different channels and could therefore unfortunately not be standardised. However, the results provide a good basis for a first analysis.

2 Theory and Methods

The collaborative project has adopted a mixed-methods, interpretative approach to analysing staff and student responses to questions around the shift to ERT in both the German and South African institutional contexts. Methodologically, the Cognitive-Affective-Systemic (CAS) model has been used as a broader analytical framework to categorise staff and student survey responses in order to compare the types of questions in the two different research sites.

These data are then interrogated using Hofstede's cultural dimensions to illuminate contextual and cultural differences.

2.1 *Hofstede's Cultural Dimensions*

In collaborating on a Global North-South review of the potential for identifying enabling Industry 4.0 development factors that have potential implications for technology-driven work, such as manufacturing, some models such as Hofstede's cultural dimensions may be useful [16, 17]. Although critiqued for not taking specific contextual factors into account, Geert Hofstede's research on organisational culture, economics and management is an often-used source for corporate training programs and research and has been used to evaluate student—of 69 nationalities—academic performance and social integration at a university [18]. His cultural dimensions model is based on data collected in the 1980s from global IBM employees, and has been expanded over the past decades to include 76 countries [16]. The data has been validated through research to include statistics such as the gross national income and its share with poorer nations, saving rates and the sticking to legal obligations and will therefore be used for a first phase analysis. However, a critical view on the limitation of Hofstede's research, such as the reliable scale for the culture dimensions at the level of individuals or organizations should not be excluded [19].

His studies on national and regional differences between societies and organisations generated a model with six cultural dimensions [16, 17]:

- **Power Distance** measures the degree to which a society accepts and expects an unequal (hierarchical) distribution of power. It illustrates the dependence relationship between less powerful members of institutions and organisations and those in positions of authority.
- Under **Individualism versus Collectivism**, the relationship between the individual and the group is analysed as well as the degree of interdependence within the society, and how obligation and loyalty are perceived.
- **Feminism versus Masculinism** relate to how closely society expects traditional gender roles to be followed by its members. In a masculine society, men are expected to be more assertive and competitive, measuring success by wealth and status, while women are expected to be modest, cooperative, and focused on quality of life. Feminine societies do not associate such qualities with particular genders.
- **Uncertainty Avoidance** shows how societies deal with uncertainty and ambiguity and how they try to manage situations. Are they more control-oriented or are they able to just let the future happen?
- **Long Term versus Short Term Orientation** displays how a society is dealing with the past, present and future and if they are stuck with traditions or follow a more pragmatic approach in their decisions.
- **Indulgence versus Restraint** describes the extent a society can control desires and impulses based on their socialization.

The theories will be used to interpret some of the findings of the questionnaires from SUN and HSD as well as for future research on that matter.

3 Germany

3.1 Teaching Staff Survey

The questionnaire which was conducted at HSD in January 2021, encompassed 122 responses from professors, research assistants and lecturers from all faculties. The main idea is to obtain a more realistic view on how the first two semester online were perceived by the ones overseeing the teaching. Questions were focused at (1) Experience in digital teaching, (2) Challenges due to the new teaching environment, and (3) Changes in learning objectives and examination methods.

From a *systemic* perspective, more than 54% had no experience with digital teaching before, but after two semesters 74% felt well trained in digital teaching. When being asked how they coped with shift from onsite to online teaching, 76% of the teaching staff either handled it OK or good. However, a lack of the right technical equipment (70%), the redesign of course material and more time necessary to communicate with colleagues and students were seen as challenges that led to a higher workload (50%).

Teaching was mainly online live (85%), followed by live exercises online (61%) using Microsoft Teams (88%). Further teaching material, such as the script (58%), additional text material (69%), audios (23%), videos (31%), podcasts (9%) and links to the library (70%) and other internet sources (52%) was offered via an online-learning platform Moodle (70%). In general, it was said, when combining live online teaching together with the mentioned asynchronous tools lead to a successful fulfilment (61%) of the targeted teaching goals. However, group work, laboratory experiments, debating and other practical exercises were named as not possible to perform with the available tools.

Further responses demonstrate a more *cognitive* focus. Some of the teaching staff (36%) partly changed all the learning objectives as well as the examination methods (43%) due to the Covid-19 pandemic. The results of the exams have been slightly poorer (18%) but mainly did not change at all.

In the *affective* domain, staff report that social contact was missing and that communication with students was difficult, as they were harder to read, without being in personal contact. However, more than 88% of the responses said that there was a (strong) interaction, which took place in chats, discussions after and during the course, break-out sessions, surveys and direct video calls. The willingness from the students to interact was evaluated as good (26%) or very good (13%) from the teaching staff.

However, 61% of students were partly active or not active at all, also showing the problem of not being able to reach everybody and leading to teaching staff being only partly satisfied with the general interaction. The reasons given were that often no camera or microphone was used to interact. This is due to a possible psychological barrier, especially when classes have high numbers of participants. Low motivation was also named as a possible answer. More than 65% of the teaching staff said, that they often tried to motivate the students to study the provided material in advance and to use the online lecture for questions etc. The result was rather sobering, with only 33% of students recognized to follow that suggestion.

In general, the teaching staff suggested that there should be further trainings on digital competencies in the future including a more enhanced IT support. However, the idea of also offering online exam after the pandemic is interesting.

3.2 Student Survey

After the first online summer (August) semester 2020 a short student survey on the online lecture was conducted in the department of mechanical and process engineering of HSD. 214 students took part in the questionnaire. When being asked if only online teaching would be an option, 39% of the students disagreed. Interestingly, the majority (43%) thought that a mix of online and onsite teaching could be an idea for the future. Only 18% of the participants were satisfied with a fully remote way of teaching. In contrast, 39% said that they would not want to have online lectures only and 43% were between these two viewpoints, saying they could accept a partly remote teaching style, where a few lectures are taught online and the rest on site as usual. The second question allowed free-text answers which then could be divided into Knowledge transfer; Teaching intake; Social aspects; and Other topics.

In the first category, the students had concerns regarding the communication between them and the teaching staff. Asking a question and receiving an answer during lectures seems to be more difficult in an online environment as well. Furthermore, for certain subjects, the practical aspect, such as experiments, is missing. Other problems include lack of access to the university laboratories and more suitable laptops than the ones students have at home. The second category is dealing with the student's intake of the taught material. Here the students mostly reported that their personal discipline and their motivation to study and participate is not as high as during normal times with onsite lectures. The second to last category deals with the social component and the lack of social contact with teaching staff and other students in their field of study. Especially for students starting off their study in these times, meeting new people while just having online classes is very difficult. In addition to that, not knowing the teaching staff in person can also hinder communication. In the last category students mostly criticized the formalities of online teaching, such as not having the right equipment or a good internet connection to participate in online classes. Further, field trips and an insight into different companies are missing. In the end, students also claimed that their orderly daily routine is missing with only online lectures.

However, online lectures can also be beneficial students noted because recorded lectures such as videos allow them to re-wind and fast forward to new points and topics or listen to information again. A disadvantage was the use of different platforms for online teaching; students can easily get confused on how and where to contact the lecturer when there is no consistent approach. Differences in the quality of internet connection and how teaching staff handled the equipment as well as platform lead to problems during the lecture and students missed certain content or had problems understanding.

4 South Africa

4.1 Teaching Staff Survey

The South African Society for Engineering Education (SASEE) conducted an anonymous, qualitative survey across all national members in September 2020. This research formed part of a Stellenbosch University Faculty of Engineering Impact Evaluation project. 60 qualitative responses were received from engineering faculty management, lecturers and postgraduate students in lecturing or tutoring roles, comprising 20,000 words of narrative data. In addition to the survey data, a SWOT analysis was conducted in May 2021 of the faculty engineering academic feedback via forums and semi-structured interviews. The combined data are used to compare results to those in the German study. Both surveys focused on changes in the environment due to ERT, adoption of ICT platforms and strategies, as well as perceived challenges and successes.

The key themes that emerge from the staff survey are initially *systemic*, concerning digital fluency, access to equipment, data and devices, and the workload implied by the changes to learning material. As can be seen in Fig. 1, although relatively equal percentages of positive versus negative perspectives on systemic issues like digital fluency and communication systems are reported, it is clear that practical access to equipment, data and ICT infrastructure represented a major constraint to 72% of all respondents. The national survey indicated the vast range of platforms and communication systems being used across the different institutions, but most lecturers indicated a major concern with student levels of digital fluency [7], poor forum engagement, and lack of equitable access to the internet.

The faculty teaching strategy during the first phase of lockdown was standardised as the asynchronous provision of recorded lectures via the Moodle-based Learner Management System (LMS), with synchronous MS Teams tutorial sessions and peer learning engagement via forums.

From a *cognitive* support perspective, all practical-dependent subject areas required significant redesign, primarily as more theory-focused, conceptual projects. All staff respondents report a significant increase in workload. The relaxation of lockdown regulations saw a shift to an Augmented Remote Teaching, Learning and Assessment (ARTLA) model. From 2021, students were able to attend one contact-based session in smaller groups per subject per week. These sessions were generally a replacement for the synchronous online tutorials or practical, equipment-based, laboratory sessions.

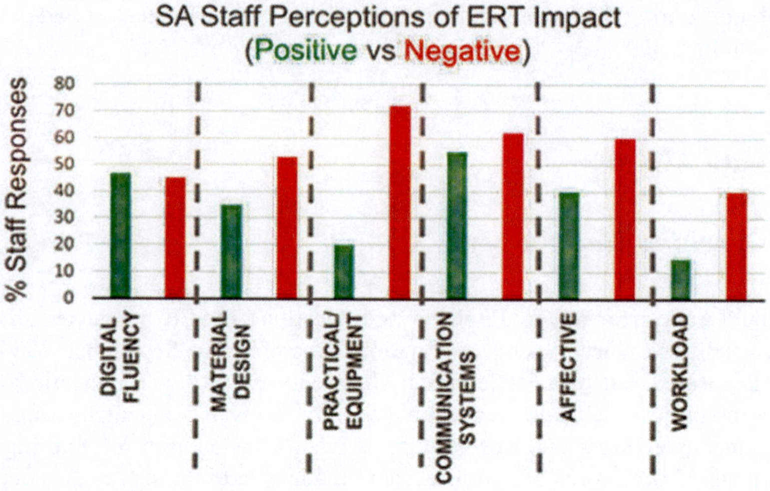

Fig. 1 SA staff perception of ERT impact

By May 2021, staff feedback as part of the faculty SWOT analysis indicates a shift to *affective* concerns around well-being, mental health and burnout, in relation to serious concerns around systemic and affective student engagement.

4.2 Student Survey

After the first full year of ERT and ARTLA, an anonymous faculty-wide survey of all Undergraduate (UG) and Postgraduate (PG) students (n = 4000) was conducted, primarily to determine how students are coping with academic stress. 733 responses were received, and data from questions on 'what do you find most stressful?' and 'recommendations for the faculty' have been used in this paper.

Across the board, the majority of the UG and PG students report *systemic* challenges such as 'information overload' and the navigation of multiple, non-standardised forms of communication: LMS announcements, group emails, subject forums and individual lecturer-student emails. In the South African context, large classes are the norm, so class sizes range from 100 to 1000, with staff generally responsible for up to 350 students. As in the German study, it is evident that students find it difficult to ask questions in the online spaces. For those students with sufficient access, the ability to stop and rewind lectures has been indicated as positive.

One area of particular focus in the faculty case study is the *affective* impact of tertiary education under ERT/ARTLA conditions. Students report feelings of depression, lack of motivation, lack of support and being overwhelmed. What is particularly evident is the second-year students' lack of the traditional network they would have established in their 1st year.

Equally problematic across the national landscape is the level of digital fluency of staff. Student feedback in the faculty case indicates that different staff have been using different platforms or approaches, and that the quality of presentation differs.

5 Discussion

The German and South African results show commonalities with respect to the systemic, cognitive and affective implications for both students and staff in response to ERT.

Systemic Significance of infrastructure in context: Availability of stable and fast network services, as well as equipment that can support online learning is critical in supporting learners and teaching staff. Furthermore, staff would benefit from training and technical support for the online learning environment.

Cognitive The longer-term impact on cognitive growth and learning has yet to be determined. Early indications from the SA PG data do suggest that students with adequate systems and affective support are demonstrating improved independent and peer learning, as well as responsive and innovative approaches to their research.

Affective Lack of social interaction, in and outside of the classroom, led to feelings of isolation and lack of motivation, presenting significant challenges to learning. Innovative approaches to promoting engagement in the online space had some success, but cannot replace the in-person experience.

When considering Hofstede's cultural dimensions for a first analysis, differences between South Africa and Germany can be found under the Long-Term Orientation, Indulgence and Uncertainty Avoidance dimensions, as well as a smaller difference under the Power Distance dimension [17].

Reflecting on these differences, the following points have been noted (Fig. 2):

Power Distance (PD) South Africans are more hierarchically accepting than Germans. This can be explained by considering that Germany has a strong and established middle-class that expects to be consulted and participate in decision-making, while South Africa has one of the highest Gini inequality indexes in the world [20].

Long-Term Orientation (LO) Germans have a more pragmatic orientation, which often proves to facilitate the adaptation to new situations. Whereas South Africa is seen to be more normative, respecting and following tradition.

Indulgence (I) South Africans tend to be more optimistic, have a positive attitude and are more impulsive.

Uncertainty Avoidance (UA) Germans have a higher tendency to avoid uncertainties and they prefer to compensate for higher uncertainty by creating a structured and regulated environment. South Africans are less expecting of a regulated environment and prefer to be left to their own devices.

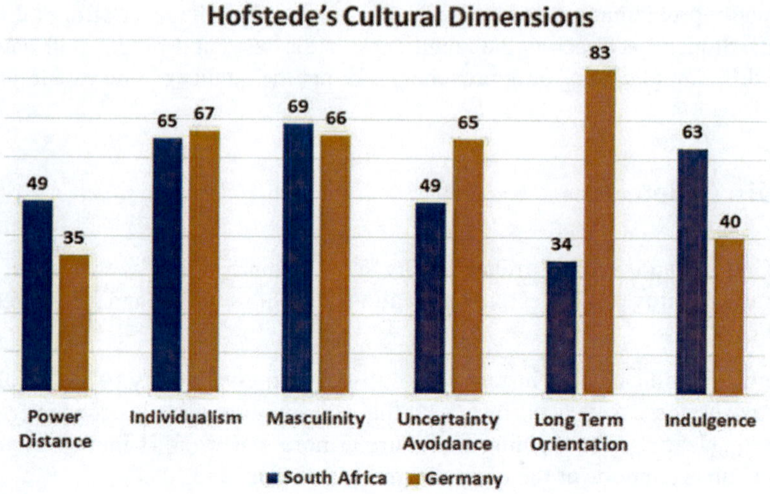

Fig. 2 Hofstede's cultural dimensions on South Africa and Germany [17]

These dimensions have been translated into the potential implications for engineering education, based on the results of this study.

Students' reliance on systemic structure for learning has implications for the long-term development of necessary innovation and adaptability competencies. In Germany, there was a strong drive from the teaching staff for the institution to create a set of rules and regulations (high UA) in order to structure the learning environment. In South Africa, students expected to be instructed by their lecturers on what to do (high PD). There were differences in how UG and PG students responded to the change in learning environment, both in South Africa and Germany. Less experienced UG learners struggled to structure their own time and learning (I) when there was a lack of structure in their environment (low UA), whereas PG students enjoyed the freedom of being able to structure their own time and learning (low PD).

Both in Germany and South Africa, staff reported an increased workload associated with a steep learning curve in becoming familiar with and competent using online learning platforms and approaches. Both environments benefitted as standardised procedures for online engagement and platforms were established institutionally (low UA, high PD). South Africa and Germany both have slightly higher individualism indices and societies that align more with masculine characteristics. There was a balanced approach to meeting the needs of the group (students and staff) with a drive to provide the tools required for success.

6 Concluding Comments

The research team is fully aware of the fact that the design of experiments used for this paper does not lead to a fully representative study due to the already mentioned points of criticism. However, the team from South Africa and Germany plans to set up a common research project in the future. This shall include some questionnaires, more students and more researchers involved. With regard to the teaching staff as well as the student skills and whether the challenges posed by the Covid-19 pandemic improved the skills on both sites, a long-term analysis including the time after the pandemic is needed. Further, the assumption made by Hofstede on cultural dimensions and the evaluation of South Africa and Germany shall be extended to further studies (Schwartz, Trompenaars, Globe, Lewis, etc.) and compared to the contemporary experience and further evaluated. This will provide structure for further research in this regard.

When considering the conference topic "Competitive Manufacturing" it is essential to note that successful engineering education is the foundation to meeting the needs of our future. The ongoing changes due to Industry 4.0 as well as the rapid adaptations made because of the Covid-19 pandemic showed that engineers need more than only technical expertise. Technological knowledge is critical, but also the 'broader' competencies, the so-called "Soft-Skills" are essential for educating prosperous engineers for the twenty-first century and handling challenges.

References

1. Cruz, M.L., Saunders-Smits, G.N., Groen, P.: Evaluation of competency methods in engineering education: a systematic review. Eur. J. Eng. Educ. **45**(5), 729–757 (2020)
2. Niemann, J., et al.: ELIC—Teachers as a Medium to Build a New Generation of Skilled Engineers, (COMA) Conference 2019, Stellenbosch, South Africa (2019)
3. Hodges, C., Moore, S., Lockee, B., Trust, T., Bond, A.: The difference between emergency remote teaching and online learning. EDUCAUSE Rev. 3. https://er.educause.edu/articles/2020/3/the-difference-between-emergency-remoteteaching-and-online-learning (2020)
4. Gelles, L.A., Lord, S.M., Hoople, G.D., Chen, D.A., Mejia, J.A.: Compassionate flexibility and self-discipline: student adaptation to emergency remote teaching in an integrated engineering energy course during COVID-19. Educ. Sci. **10**, 304 (2020)
5. Tait, A.: Planning student support for open & distance learning. Open Learn.: J. Open Distance e-Learn. **15**(3), 287–299 (2000)
6. Swartz, B.C., Gachago, D., Belford, C.: To care or not to care-reflections on the ethics of blended learning in times of disruption: the ethics of care & academic development. S. Afr. J. High. Educ. **32**, 49–64 (2018)
7. Czerniewicz, L., et al.: A wake-up call: equity, inequality and Covid-19 emergency remote teaching and learning. Postdigital Sci. Educ. **2**(3), 946–967 (2020)
8. Bundesministerium für Bildung und Forschung (BMBF): Überbrückungshilfen für Studierende. https://www.bmbf.de.de/wissenswertes-zur-ue-berbrueckungshilfe-fuer-studierende11509.html. Last accessed 23 Nov 20.
9. Lewis, C., Wolff, K., Bekker, B.: Supporting project-based education through a community of practice: a case of postgraduate renewable energy students. World Trans. Eng. Technol. Educ. **19**, 35–40 (2021)

10. Yen, T.F.: The performance of online teaching for flipped classroom based on COVID-19 aspect. Asian J. Educ. Soc. Stud. **8**(3), 57–64 (2020)
11. von Solms, S., Nel, H.: STEM project based learning: towards improving secondary school performance in mathematics and science. 2017 IEEE AFRICON (2017)
12. Felder, R.M., Silverman, L.K.: Learning and Teaching Styles in Engineering Education (1988)
13. Mahanija, Md K.: Evaluation of an engineering program: a survey to assess fulfillment of industry requirements (2009)
14. Abrandt Dahlgren, M.:Higher Education (1998)
15. Capraro, R.M., Slough, S.W.: Why PBL? Why STEM? Why now? An introduction to STEM project-based learning: an integrated science, technology, engineering, and mathematics (STEM) approach (2009)
16. Hofstede, G. et al.: Cultures and Organizations: Software of the Mind, 3rd edn. McGraw-Hill Education, ProQuest Ebook Central. https://book-central-proquest-com.ezp.hs-duesseldorf.de/lib/bibl/detail.action?docID=6262353 (2010)
17. Hofstede-Insights, Country Comparison Germany–South Africa. https://www.Hofstede-in-sights.com/country-comparison/germany,south-africa/,last. Accessed 27 Aug 21
18. Rienties, B., Tempelaar, D.: The role of cultural dimensions of international and Dutch students on academic and social integration and academic performance in the Netherlands. Int. J. Intercult. Relat. **37**, 188–201 (2013). https://doi.org/10.1016/j.ijintrel.2012.11.004
19. Venaik, S., Brewer, P.: Critical issues in the Hofstede and GLOBE national culture models. Int Mark Rev. 30 (2013). https://doi.org/10.1108/IMR-03-2013-0058
20. Worldbank: Data on South Africa, https://www.worldbank.org/en/country/southafrica/overview. Last accessed 17 Sept 21

Deborah Blaine holds a Ph.D. (EngSci) from the Pennsylvania State University, USA and a BEng (Mech) from Stellenbosch University, South Africa. She is an associate professor at Stellenbosch University, Mechanical & Mechatronic Engineering.

Claudia Fussenecker holds a M.A. in European Management from the FH Bund and a Diploma in Industrial Management from EUFH, both Brühl. She is a scientific assistant focusing on Engineering Education and Services and Life Cycle Management.

Karin Wolff holds a Ph.D. in Education from the University of Cape Town, with a focus on Engineering Curriculum and Practice. She is the Engineering Teaching and Learning Advisor at Stellenbosch University, South Africa.

Typology and Implications of Equipment-as-a-Service Business Models in the Manufacturing Industry

L. Holst and V. Stich

Abstract In the manufacturing industry, technological developments around cyber-physical systems create completely new possibilities for generating value for customers. An essential part of these developments are Equipment-as-a-Service (EaaS) business models, which promise growth with existing customers even in saturated markets and in which the interests of manufacturers and providers are aligned. However, different types of EaaS business models must be differentiated in manufacturing industry, especially with regard to the risk transfer from the customer to the provider. In this paper, a typology for different EaaS types is developed based on grounded research. Furthermore, practical implications are derived that should help with the necessary EaaS transformation, that has been previously outlined by the EaaS typology.

Keywords Equipment-as-a-Service · Industrial subscription · Manufacturing industry · Typology

1 Introduction

The Covid 19 crisis once again showed the world its natural limits. For individuals and companies alike, the ability to adapt to radically changing environmental conditions became a decisive survival criterion [1]. Covid-19 thus accelerated two global industrial trends to an unexpected extent: on the one hand, the need for consistent digitization of existing value creation activities [2], and on the other hand, the paradigm shift in industry caused by the increased scarcity of resources, from the "throwaway economy" to a circular economy geared to the conservation of resources.

L. Holst (✉) · V. Stich
Institute for Industrial Management (FIR) at RWTH Aachen University, Aachen, Germany
e-mail: lennard.holst@fir.rwth-aachen.de

V. Stich
e-mail: volker.stich@fir.rwth-aachen.de

K. von Leipzig et al. (eds.), *Smart, Sustainable Manufacturing in an Ever-Changing World*, Lecture Notes in Production Engineering,
https://doi.org/10.1007/978-3-031-15602-1_26

The manufacturing industry has also been dealing with the digitalization of existing value creation activities in the course of Industry 4.0 for several years [3].

However, despite the in some cases considerable investments in the development of digital infrastructures and forecast annual revenue increases of 2.5 percent on average as a result of digitization or Industry 4.0 in 2014 in Germany [4], no breakthrough in business with digital solutions is yet discernible across the board in 2020 [3]. According to Schuh et al. the lack of monetarization of digitization in mechanical and plant engineering is primarily due to the focus on product innovations while at the same time adhering to the existing, transactional business model based on the principle of "money for products" or "money for service hours" [3]. However, differentiation in the competitive environment of the manufacturing industry is increasingly no longer taking place solely through innovations in products and their functionalities [5] but rather through the ability to monetize the usage phase of smart, connected products via Equipment-as-a-Service (EaaS) business models or synonymous Industrial Subscription business models [6, 7].

EaaS business models consist of four main fundamentals [7]:

1. Recurring revenue mechanisms such as pay-per-use or pay-per-outcome.
2. A value proposition geared to individual customer success (e.g. increasing productivity).
3. Digital performance systems networked via the Internet of Things, consisting of integrated bundles of products, services and software.
4. Long-term customer relationship based on a trustworthy partnership.

In an EaaS business model a customer pays constantly for the solution provided, not for the machine itself. Thus, the provider has to bear the financial risks by interim financing high-valued physical assets. Through a contractual link and recurring payment flows (e.g., pay-per-use, pay-per-outcome, etc.), the providers' interest in enabling the customer to use the resources more efficiently and effectively increases due to the assumption of various risks.

The technical requirements for EaaS are digital, continuously improving product-service-systems consisting of products, financial and industrial services, software and, where applicable, consumables, which are provided to ensure and enhance customer performance [7–9]. EaaS are beneficial in certain financial, performance and risk criteria for the provider as well as for the customer but come along with certain disadvantages.

Table 1 shows important advantages and disadvantages of EaaS business models for manufacturing companies from a provider and a customer perspective [6, 7].

However, there are various types (implying various combinations of advantages and disadvantages) of EaaS in the industry that have not yet been fully differentiated [3]. A typology is supposed to help better understand the complexity of EaaS in the manufacturing industry to shorten the design of a company specific EaaS business models. Therefore, the first research objective is to characterize the most important types of EaaS. Furthermore, EaaS completely change the way companies must operate to generate revenues. Therefore, the second research objective is to derive overarching implications for purposeful EaaS transformation.

Table 1 Advantages and disadvantages of EaaS business models in the manufacturing industry based on [6, 7]

	+	–
Provider	• Growth with existing customers • Differentation, no price competition	• Risk transfer from customer • Transformation necessity in almost all areas
Customer	• Opex instead of Capex • Risk transfer to provider • Performance-increase over time	• Lock-in-effect • Know-how loss/data transfer

2 Methodology

Since the EaaS business model is a new business concept in the manufacturing industry and since the results of this research are intended to provide implications for practitioners, an applied research approach is chosen for data collection and analysis. Therefore, we use the grounded theory according to Glaser a. Strauss to collect data as a theoretical sample, identify recurring structures and derive practical implications [10]. In this approach, different research methods for collecting and analyzing primary data are combined: expert interviews, observations, and focus groups.

Table 2 provides an overview of the applied research that took place between the years 2019 and 2021.

Table 2 Applied research methodology

Company	Interview partner role	Applied research methods
Machine tool manufacturer	Head of subscription Head of process management	3 interviews, 1 company visit, 2 joint workshops between 2019 and 2020
Printing press manufacturer	Head of customer success management Operative CS manager	4 interviews, 1 observation of a customer success meeting with customers between 2019 and 2021
Machine tool manufacturer	Operative CS manager	1 interview in 2020
Software-as-a-service	Director marketing campaigns and channels	3 interviews, 1 joint workshop in 2020
Software-as-a-service	Success management lead	2 interviews in 2019
Various manufacturing companies	Various roles	Focus group in the context of a joint remote workshop in 2020

3 Typology of Eaas

Based on the applied research in the manufacturing industry four types of EaaS business models could be derived. The various types differ primarily in terms of the assumption of specific customer risks, pricing mechanisms, and the scope of the service offering [11, 12] (Fig. 1).

3.1 Type 1: Availability-Based EaaS

In the availability-based EaaS business model the guaranteed availability of the equipment is sold to the customer usually for a fixed monthly price. The provider ensures uptime and guarantees a specific service-level. The value proposition consists of the sharing or assumption of the "investment risk" and the "performance risk" by the provider [7, 11].

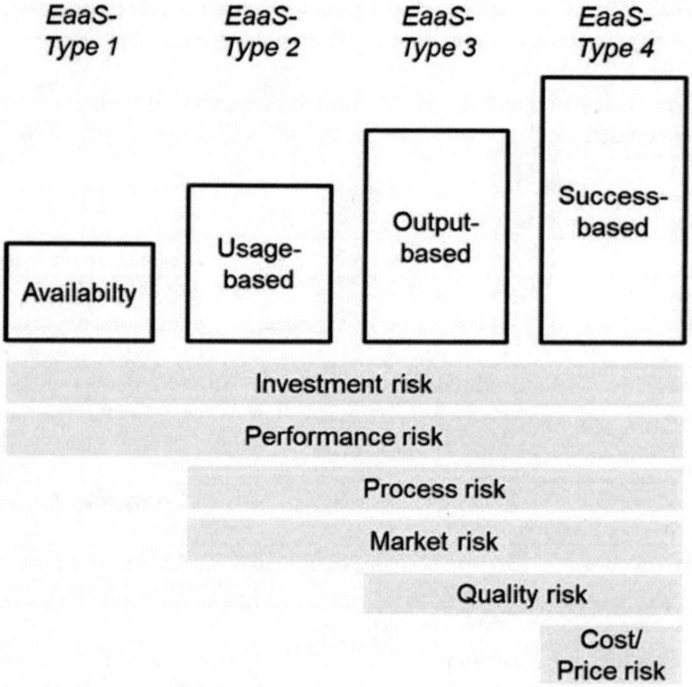

Fig. 1 Types of EaaS business models within the manufacturing industry based on [12]

3.2 Type 2: Usage-Based EaaS

In the usage-based EaaS business model, e.g., implemented via the pay-per-use revenue mechanisms, the customer pays a usage fee for a defined unit of measurement, such as the active time of spindle operation of a milling machine, depending on the individual intensity of usage. In practice, the revenue mechanism of usage is often combined with the revenue mechanism of availability, i.e., a monthly fixed base price. In addition to the base charge, the variable usage charge is levied when the machine or system is in operation. The provider assumes part of the customer's "process risks", e.g., in order to achieve a high usage intensity through efficient work processes at the customer's site. In addition, the "market risk" is shared, i.e., if the customer does not achieve his sales targets, machines and equipment are used less intensively and thus do not generate any income for the supplier [11, 12].

3.3 Type 3: Output-Based EaaS

In the output-based EaaS business model, e.g., realized via the pay-per-outcome mechanism, the customer pays depending on the actual production result, such as a printed, defect-free sheet of paper or a cut dimensionally accurate sheet. In this way, the supplier also assumes the "quality risk" and tries to minimize the customer's scrap. This type of yield mechanism is therefore often accompanied by far-reaching integration of the supplier into the customer's value creation, e.g., by providing consumables or logistics solutions for material supply or removal. Whereas in the usage-oriented EaaS business model, the aim is to achieve a high intensity of usage, the goal is now to increase output while minimizing input resources, so that the supplier assumes part of the "productivity risk" [7, 11].

3.4 Type 4: Success-Based EaaS

In the success-based EaaS business model, e.g., realized via the revenue mechanism pay-per-cost-savings, the customer pays depending on the realized economic success, such as e.g., saved costs of the customer. Thus, the provider takes over parts of the "cost or price risk" of the customer. For this yield mechanism, there must be potential for economic success both on the customer side, e.g., by reducing operating costs in the areas of energy, materials or personnel, and on the supplier side, e.g., through product innovations for lower energy consumption of the machines operated. The saved costs can serve as a basis for this yield mechanism. Thus, in practice, this characteristic is only suitable for special performance systems with high saving or profit potential. To ensure economic success, the supplier integrates itself intensively into the customer's value creation processes, sometimes far beyond its own original performance system [11].

4 Implications of EaaS

As the complexity of the EaaS types increases from type 1–4, due to the more extensive risk and performance transfer from the customer to the supplier, so does the need for digital and organizational transformation in companies in the manufacturing industry [5, 7]. For this reason, the most important findings on overarching implications for a successful EaaS transformation are discussed below [13, 14] (Table 3).

A: Service Offering
In EaaS business models, the provider's generation of value for the customer changes. Whereas previously in a transactional business model only a value proposition could be formulated, in EaaS the monetary- or performance-based linkage occurs with the customer's successful usage phase. Due to the often shorter-cycle contract structures, the provider must deliver value on an ongoing basis, as the customer could withdraw from the contract if performance is poor.

B: Engineering
In the transactional business, planned obsolescence (e.g. rapid wear) often dominates in order to generate money in the after-sales business. The EaaS business model therefore places new demands on the design and engineering for the development of EaaS specific products that allow easy maintenance and further improvements during the usage phase. Furthermore, engineering or design in the transactional model are often dominated by the provision of many functions for the possibility of usage. In EaaS business models, where there is a continuous, data-based connection to the customer, further customer-specific differentiation can take place primarily via software releases or sometimes even with hardware upgrades.

Table 3 Implications of EaaS

	Transactional business model	EaaS business model
A: Service offering	Value proposition (a priori)	Value delivery (a posteriori)
B: Engineering	(Planned) Obsolescence	Design for EaaS
C: Pricing	Cost-, competitive-based	Value-based
D: Industrial services	Profit center	Cost center
E: Sales	Provider success	Customer success

C: Pricing

Whereas in the transaction business, services are primarily priced based on cost-based or competition-based procedures, in EaaS, pricing must be more value-based. For this purpose, providers must transfer the specific risks that are assumed in addition to the reference case of the transaction business into a value-based pricing system that is understood by the customer. In addition, new pricing models that are linked to usage are necessary to create continuous incentives for performance increases as well as opportunities for monetary growth. In most cases, base fees are charged to hedge the increasing risks that are transferred to the suppliers [14].

D: Industrial Services

In today's transactional business, industrial service business is often organized as a profit centre. Spare parts and maintenance services often generate a significant share of revenue and profit. In EaaS business models, the logic of service provision is reversed, so that providers themselves become customers of their own services. In EaaS business models, industrial services are no longer directly remunerated, so the provider strives for efficient service provision and the targeted use of digital technologies, such as predictive maintenance, to increase efficiency [3].

E: Sales

In the transactional business, sales departments are primarily incentivized to increase the sales of physical goods or individual services or service contracts. The success of the provider is at the centre of the sales approach. In EaaS business models, on the other hand, a complete reorientation of the classic sales approach must take place, since the success of the provider is directly dependent on the success of the customer, as the four EaaS types showed. Thus, there is a need for the development of a so-called Customer Success Management in addition to the classic sales approach in order to create a unit that is completely incentivized on the realization, monitoring and continuous increase of the individual customer success [15–17].

5 Summary and Limitations

EaaS models offer a wide range of potentials for customers and providers in the manufacturing industry. However, the various EaaS types make a systematic transformation of manufacturing companies difficult, as they differ specifically in terms of revenue mechanics and risk transfer. With the EaaS typology derived in this work, it was possible to describe a first practicable model for EaaS types based on previous research results [12] and put them into the current context of challenges in the manufacturing industry. Furthermore, five specific implications for a EaaS transformation could be described. These become more pronounced as the complexity of the EaaS types increases. The aim is to provide an aid for theory and practice to enable EaaS to be implemented more quickly and systematically. This paper naturally faces certain limitations. For example, the research method is based on a small number of practical use cases, which would have to be expanded in the future to include an empirical study. Furthermore, the findings must be continuously supplemented or adapted to current developments.

Acknowledgements The research and development project Future Data Assets that forms the basis for this report is funded within the scope of the 'Smart Data Economy' technology program run by the German Federal Ministry for Economic Affairs and Climate (BMWK) and is managed by the DLR project management agency. The authors are responsible for the content of this publication.

References

1. Leach, M., Macgregor, H., Scoones, I., Wilkinson, A.: Post-pandemic transformations: how and why COVID-19 requires us to rethink development. In: World Development, 138 (2021)
2. Soto-Acosta, P.: COVID-19 Pandemic: shifting digital transformation to a high-speed gear. Inf. Syst. Manag. **37**(4), 260–266 (2020)
3. Schuh, G., Frank, J., Jussen, P., Rix, C., Harland, T.: Monetizing industry 4.0: design principles for subscription business in the manufacturing industry. In: Proceedings, 2019 IEEE International Conference on Engineering, Technology and Innovation (ICE/ITMC). Sophia Antipolis Innovation Park, France, 17–19 June 2019, Valbonne Sophia-Antipolis, France, pp. 1–9. IEEE, Piscataway (2019)
4. Koch, V., Kuge, S., Geissbauer, R., Schrauf, S.: Industrie 4.0. Chancen und Herausforderungen der vierten industriellen Revolution. Hrsg.: Strategy &. New York (2014)
5. Tzuo, T., Weisert, G.: Subscribed: why the subscription model will be your company's future—and what to do about it (2018)
6. Stojkovski, I., Achleitner, A.-K., Lange, T.: Equipment as a service: the transition towards usage-based business models (2021)
7. Holst, L., Schuh, G., Stich, V., Frank, J.: Customer success management in the subscription business of manufacturing companies: towards a task-oriented reference model. In: Herberger, D., Hübner, M. (eds.) Proceedings of the Conference on Production Systems and Logistics (2021)
8. Schuh, G., Wenger, L., Stich, V., Hicking, J., Gailus, J.: Outcome economy: subscription business models in machinery and plant engineering. Procedia CIRP **93**(1), 599–604 (2020)
9. Claasen, M., Blum, C., Osterrieder, P., Friedli, T.: Everything as a service? In: Meierhofer, J., West, S. (Eds.) Introducing the St. Gallen IGaaS Management Model, 2nd Smart Services Summit, Thun, pp. 61–65 (2019)
10. Glaser, B.G., Strauss, A.L.: The discovery of grounded theory: strategies for qualitative research, 4. paperback printing ed. Aldine, New Brunswick, 271 pp (1967)
11. Porter, M.E., Heppelmann, J.E.: How smart, connected products are transforming companies. Harvard Bus. Rev. **93**, 1–19 (2015). https://hbr.org/2015/10/how-smart-connected-products-are-transforming-companies.
12. Stoppel, E., Roth, S.: The conceptualization of pricing schemes: from product-centric to customer-centric value approaches. J. Revenue Pricing Manag. **16**(1), 76–90 (2017)
13. Mansard, M., Cagin, J.-M.: Reaping the recurring benefits of industry 4.0: a manufacturing executive playbook for business model transformation towards new revenue streams. Zuora, Roland Berger (2020)
14. Lah, T., Wood, J.B.: Technology-as-a-Service Playbook: How to Grow a Profitable Subscription Business. Point B Inc., [United States] (2016)
15. Eggert, A., Ulaga, W., Gehring, A.: Managing customer success in business markets: conceptual foundation and practical application. SMR **4**(2–3), 121–132 (2020)

16. Hilton, B., Hajihashemi, B., Henderson, C.M., Palmatier, R.W.: Customer success management: the next evolution in customer management practice? Ind. Mark. Manag. **90**, 360–369 (2020)
17. Hochstein, B., Rangarajan, D., Mehta, N., Kocher, D.: An industry/academic perspective on customer success management. J. Serv. Res. **23**(1), 3–7 (2020)

Lennard Holst, M.Sc. is head of the department of Service Management at the Institute for Industrial Management (FIR) at the RWTH Aachen University since 2021. Lennard Holst leads various research and industrial projects in the fields of subscription business management, digital business model transformation and industrial service management

Prof. Dr.-Ing. Volker Stich is head of the Institute for Industrial Management (FIR) at the RWTH Aachen University since 1997. Prof. Dr.-Ing. Volker Stich worked for the St. Gobain Automotive Group for 10 years and led the management of European plant logistics. In addition, he was responsible for the worldwide coordination of vehicle development projects

Investigating the Shattering Index of Coal Fines Briquettes Produced with Natural Binders

R. Nemukula and D. M. Madyira

Abstract Coal is the primary source of energy in South Africa today. In the extraction and processing of coal, significant amounts of coal fines waste are generated. Currently, waste coal fines have no economic value. As a result, they are disposed of in landfills or underground. Furthermore, if coal fines are not properly managed, they have a negative environmental impact including spontaneous combustion, groundwater contamination, acid mine drainage, and wind-driven air pollution. When inhaled, coal fines, can cause lung disease (black lung disease) and other health problems. However, coal fines have the potential to be used as a source of energy. The aim of this work was to produce coal fines briquettes using varying ratios of natural binders such as cactus. The integrity of the briquettes was assessed using shattering index. The briquettes produced using 10% cactus paste by mass exhibited the best shattering index of 86%, which is sufficient for good storage and handling of the briquettes. Further work will be required to assess the combustion behaviour of the briquettes for domestic and industrial energy applications.

Keywords Shattering index · Compressive strength · Moisture content · Coal fines · Compaction pressure · Binder · Cactus · Briquette

1 Introduction

Coal is a highly abundant natural resource with a wide range of applications such as electricity generation, steel production, cement production, liquid fuel production etc. Approximately 6.9 billion tons of coal is extracted worldwide annually [1]. The geology of the coal deposit and its distance from the surface determine the mining method. In South Africa alone, during the extraction process of coal, over 10 million tons of waste in the form of coal fines is generated per annum [2]. These coal fines are

R. Nemukula (✉) · D. M. Madyira
Department of Mechanical Engineering Science, University of Johannesburg, Johannesburg, South Africa
e-mail: 216031063@student.uj.ac.za

K. von Leipzig et al. (eds.), *Smart, Sustainable Manufacturing in an Ever-Changing World*, Lecture Notes in Production Engineering,
https://doi.org/10.1007/978-3-031-15602-1_27

a low-to no-value by-product of coal mining operations that is unavoidable. Furthermore, coal fines can cause significant environmental pollution and health risks if not properly managed [3]. To control coal fines, dust suppression systems are commonly employed in coal-fired thermal power plants [4]. Moreover, despite the fact that coal fines have no monetary value, they must be monitored to avoid spontaneous combustion. These coal residues are usually dumped in slurry impoundments, which has a negative impact on the land and the ground water [5].

South Africa is currently experiencing electricity shortages characterised by frequent load shedding and reduction. The South African power utility, is therefore failing to meet demand leading to a negative impact on the economy. This is partly driven by population increase and industrialization growth in different sectors. The growing demand for energy has necessitated the need for more efficient use of both renewable and non-renewable resources. Non-renewable resources, such as coal can potentially deplete at a faster rate if not properly managed. Strategies are therefore needed to more efficiently utilize these available natural resources.

Additionally, coal fines can have serious environmental consequences if they are not properly managed. If inhaled, coal fines have the potential to cause lung diseases (black lung disease) and other health issues [6]. Furthermore, coal fines contain sulphur, which can contaminate groundwater as it can be washed to water streams [7]. Wind-driven dust pollution and spontaneous combustion explosions may affect residents living near coal mines [8]. The utilization of coal fines for energy can decrease the amount of coal mining waste produced by 10% [9]. The goal is to reduce the risks of coal fines disposal to the environment and human health.

One way to extract value from the coal fines is to convert them into briquettes that can be used for domestic cooking and other industrial applications. Due to the nature of coal particles, which cannot easily agglomerate, stand-alone briquettes made from coal fines alone have been reported to have low shattering index [10]. The addition of biomass and binders to coal fines during briquetting process has resulted in briquettes with improved shattering index and compressive strength [11]. This improves briquette handling and storage. Previous research has shown that waste biomass and binders can be included in the briquette composition to improve combustion behaviour [12]. Binders such as starch, cassava pulp, sawdust, soybean residue, and coal tar have been used to improve the briquettes properties [10]. Manyuchi et al. [1] established the co-briquetting of coal fines of particle size of less than or equal to 2 mm with sawdust and molasses as binder. They found that increasing the sawdust and molasses concentrations increased compressive strength by 8% while the shattering index decreased by 146% [1]. Taulbee et al. [13] investigated the co-briquetting of coal fines with biomass and reported that a shattering index less than 0.80 is considered poor, 0.80–0.85 marginally acceptable, 0.85–0.90 acceptable, 0.90–0.95 good, and greater than 0.95 excellent. Therefore, coal fines briquettes with a minimum shattering index of at least 85% to endure everyday handling and storage. It is also important to use a binder with fairly low impact on environmental pollution. Shuma et al. [14] report that cactus bonded biomass briquettes have low emissions of CO_2, CO, NO_x and SO_2. However, the briquettes were found to have low shattering index. However, the methodology demonstrated the possibility to use natural binders (cactus) in the production of briquettes using loose biomass.

This study therefore proposed to use cactus as a binder in the production of briquettes with coal fines targeted at domestic cooking and other industrial applications. The success of this effort will be determined by measuring the shattering index and compressive strength of the produced briquettes. The use of coal fines and cactus binder will also aid in the alleviation of energy issues in underprivileged communities by providing them with briquettes for domestic cooking. Furthermore, significant reduction of deforestation in rural areas where people rely on wood for energy is anticipated.

2 Research Methodology

2.1 Aim of the Investigation

The aim of this work was to investigate the effectiveness of using cactus as a binder in the production of coal fines briquettes with shattering index above 85% to enable handling and storage.

2.2 Material Description

Coal fines with particle sizes less than 2.36 mm derived from anthracite coal with carbon content greater than 87% were used. The coal fines were collected from a coal mine located in the Mpumalanga Province of South Africa. The cactus used as the binder was collected from an open garden in a suburb of Johannesburg, South Africa.

2.3 Sample Preparation

For briquetting, the coal fines were pulverized and sieved to particle sizes ranging from 0 to 2.36 mm. Thereafter, they were sun-dried for 48 h to ensure they are completely dry. The cactus leaves were then de-thorned and cut down to 4 mm sized pieces before blending. Dried coal fines were then mixed with cactus at varying concentrations. Binder concentrations of 10, 15, 20, 25, 30, 35, and 40% weight fractions were applied. The coal/binder mixture was then briquetted using a 55 ton hydraulic piston press into donut type briquettes with an outside diameter of 95 mm and an inside diameter of 50 mm. The briquettes were allowed to homogenise at room temperature for 48 h. The briquettes were then dried in a convection oven at

100 °C for 3 h using a Barnstead Thermolyne 6000 furnace prior to shattering index drop test and compression testing.

2.4 Experiment Procedure

2.4.1 Briquetting Procedure

A sample size of 0.350 kg (350 g) was used for each briquette to ensure that all the briquettes have the same mass. The briquettes were formed at compaction pressure ranging from 1.6 to 8 MPa.

2.4.2 Determination of the Shattering Index

A briquette sample with a known mass was dropped three times onto a concrete surface from a height of 2 m. The mass of the briquette before and after the drop test was used to determine the shattering index. The shattering index was determined according to the methodology used by Richards [15]. Figure 1 shows the test rig for shattering index measurement.

Fig. 1 Shattering index test rig

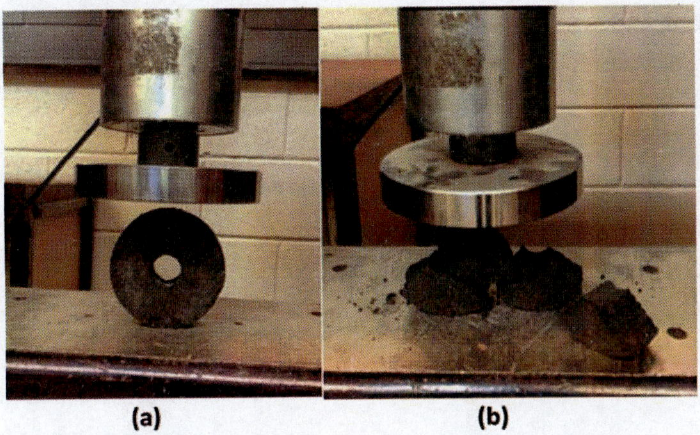

Fig. 2 Crush testing of prepared briquettes

2.4.3 Determination of the Compressive Strength

The briquette samples had an average thickness of 35 mm, and a diameter of 95 mm. Compression testing was conducted on an Instron testing machine with a load cell of 100 kN capacity. Figure 2 shows the configuration of the applied test. The compression strength for such a test is defined as a load per unit length (briquette thickness) of the load line. The crosshead speed (strain rate) in this test was set to 3 mm/min.

3 Results

Table 1 shows the variation of moisture content with varying concentration of cactus binder in briquette at a constant compaction pressure of 8 MPa. As shown in Table 1, an increase in binder fraction led to an corresponding increase in moisture content. This indicates the importance of drying the briquettes after manufacture to achieve good handling and combustion behaviour. It is important to note the low moisture content of 2.5% in the coal fines only sample.

These results are presented in graphical form in Fig. 3 depicting the relationship between binder concentration and moisture content. It is worth noting that when no binder is used, the briquette sample has the lowest moisture content of 2.5%. Furthermore, the lowest cactus percentage of 10% used in this investigation corresponds to the lowest moisture content of 4% by weight while the maximum moisture content of 8% was recorded for 40% cactus fraction. These results also confirm the need for drying to achieve good combustion behaviour.

The durability rating of the briquettes was expressed using the percentage of the oven dried mass and of the material remaining after the drop test. The percentage weight loss is calculated using the Eq. 1.

Table 1 Moisture content for varying cactus ratio

Briquette type	Mass (After compaction) (kg)	Mass (Oven dried) (kg)	Moisture content (%MC)
40% cactus	0.295	0.220	7.50
35% cactus	0.275	0.195	8.00
30% cactus	0.320	0.265	5.50
25% cactus	0.315	0.260	5.50
20% cactus	0.315	0.265	5.00
15% cactus	0.310	0.265	4.50
10% cactus	0.290	0.250	4.00
0% cactus	0.190	0.165	2.50

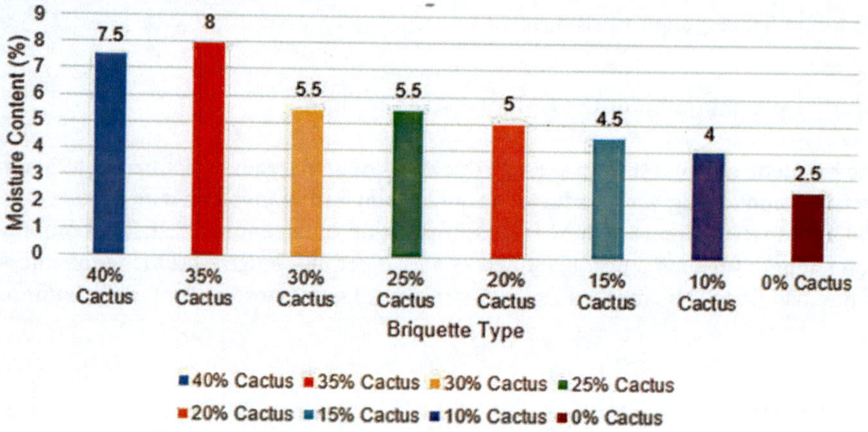

Fig. 3 Moisture content as a function of cactus ratio

$$\%WL = \frac{W_1 - W_2}{W_1} \times 100 \tag{1}$$

where $\%WL$ is the percentage weight loss, W_1 is the mass of the oven dried sample and W_2 is the mass of the sample after the drop test.

The shattering index was then determined using Eq. 2.

$$Shattering\ Index = 100 - \%WL \tag{2}$$

The shattering index constantly increased as the weight percent of the binder concentration was decreased, and the weight percent of the coal fines was increased as can be seen in Table 2.

Figure 4 shows the variation of shattering index of coal fines briquettes with cactus as a binder at varying binder concentration and constant compaction pressure. The

Table 2 Shattering index for varying cactus ratio

Briquette type	Mass (Oven dried) (kg)	Mass (After drop test) (kg)	Percentage weight loss %	Shattering index (%)
40% cactus	0.220	0.095	56.818	43.18
35% cactus	0.195	0.1175	39.744	60.26
30% cactus	0.265	0.1275	33.772	66.23
25% cactus	0.260	0.1625	17.756	82.24
20% cactus	0.265	0.1725	18.798	81.20
15% cactus	0.265	0.160	19.071	80.93
10% cactus	0.250	0.160	13.5135	86.48
0% cactus	0.165	0.105	40.000	60.00

results show a decrease of shattering index with increasing binder concentration. The highest shattering index of 86.48% was recorded for cactus ratio of 10%. However, this result seems to suggest an optimum cactus fraction between 0 and 10%. The 0% cactus fraction (100% coal fines) briquetted recorded a shattering index of 60%. According to literature [13], for briquettes to endure everyday handling a minimum durability of 85% is required. Therefore, it is important to keep the binder concentration at relatively low percentage of 10% to ensure that the coal fines briquettes have a high durability so that they can be easily handled and transported.

The compressive strength was measured for constant compaction pressure and varying binder weight percent. For each briquette sample, the maximum load at

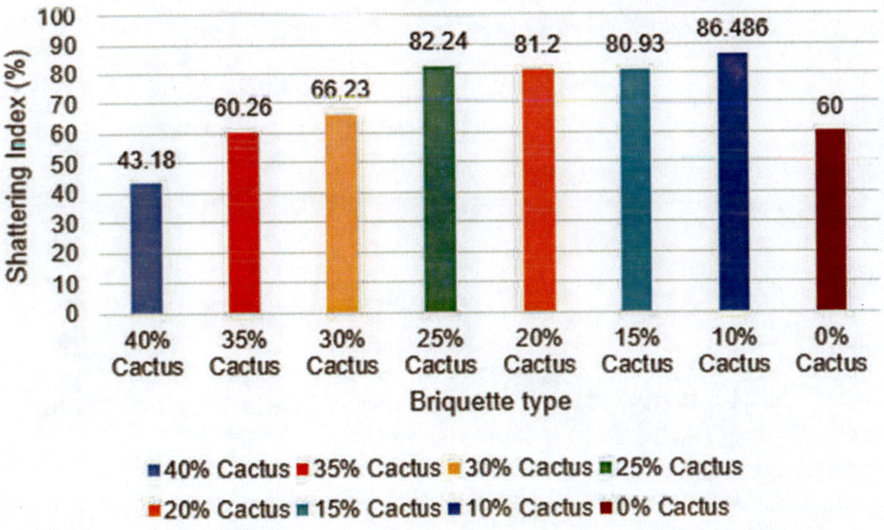

Fig. 4 Shattering index versus cactus fraction

which the briquette breaks was recorded and used to compute the strength of the briquette as shown in Table 3.

Figure 5 shows the variation of the compressive strength of the briquettes with cactus ratio. The briquette with 0% cactus recorded a strength of 4.6 N/mm. An increase in cactus fraction led to an increase of strength to 5.53 N/mm at 10% cactus fraction. This represents 16.8% improvement in strength compared with coals fines only sample. Further increase of cactus ratio resulted in a reduction of compressive strength to even below the binderless briquette. At 40% cactus fraction, the briquette strength was 2.29 N/mm. In terms of strength, the 10% cactus fraction also appears to be the optimum. This is in agreement with the results of the shattering index tests as well.

Table 3 Compressive strength as a function of cactus ratio

Briquette type	Width of Briquette (mm)	Extension (mm)	Load (N)	Compressive strength (N/mm)
40% cactus	33	31.754	75.629	2.292
35% cactus	33	31.554	88.880	2.693
30% cactus	33	31.254	100.109	3.033
25% cactus	33	31.154	135.332	4.101
20% cactus	33	31.053	137.957	4.181
15% cactus	33	31.004	164.255	4.977
10% cactus	33	30.903	182.630	5.534
0% cactus	33	30.754	155.085	4.610

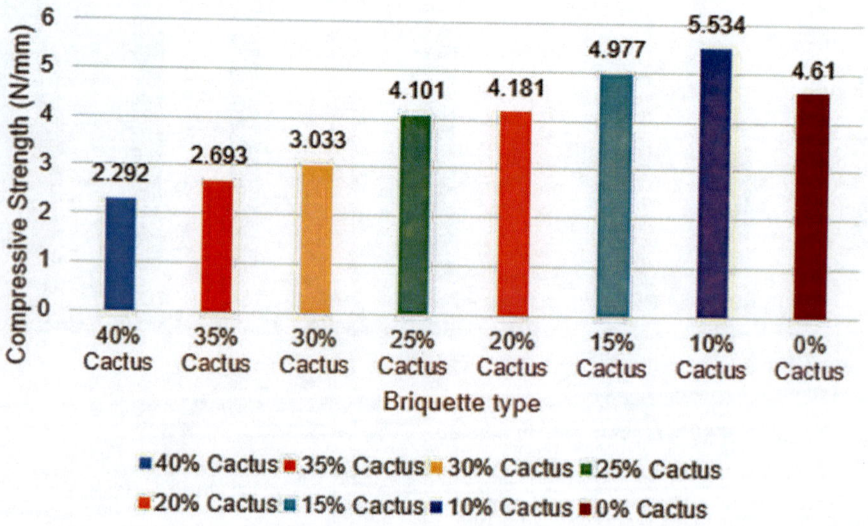

Fig. 5 Compressive strength as a function of cactus ratio

The compaction pressure was also varied to study its effect on briquette properties. Table 4 presents the results obtained for a cactus fraction of 10%.

The results are also presented graphically in Fig. 6. The shattering index is seen to significantly depend on the compaction pressure used to produce the briquettes. As the compaction pressure increases, the shattering index also increases. The maximum shattering index of 85.86% was reported for a compaction pressure of 8 MPa. The maximum pressure in this investigation was limited by the capacity of the hydraulic press used and the dimensions of the briquette selected for the test. Further tests would be required to determine whether there is an existing limiting pressure on the compaction of coals fines using cactus as a binder.

Furthermore, the compressive strength of the briquettes with the lowest porosity were tested at varying compaction pressure. These were deemed to be suitable for transportation and they meet the required durability for daily handling and transportation. Table 5 tabulate the results obtained.

Table 4 Shattering index at varying compaction pressure

Compaction pressure (MPa)	Mass (Oven dried) (kg)	Mass (After drop test) (kg)	Percentage weight loss %	Shattering index %
1.6	0.235	0.070	70.212	29.790
3.2	0.220	0.055	75.000	25.000
4.8	0.225	0.150	33.333	66.670
6.4	0.225	0.165	26.667	73.330
8	0.198	0.170	14.141	85.859

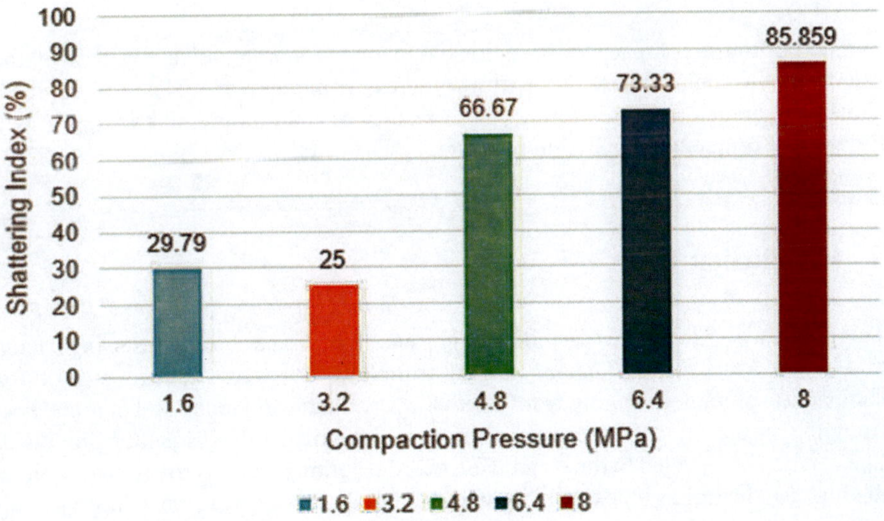

Fig. 6 Shattering index (%) versus compaction pressure

Table 5 Compressive strength of briquettes at varying compaction pressure

Compaction pressure (MPa)	Extension (mm)	Load (N)	Compressive strength (N/mm)
1.6	30.617	69.492	2.106
3.2	28.956	85.633	2.595
4.8	30.971	97.608	2.958
6.4	30.442	147.926	4.859
8	30.261	152.640	5.044

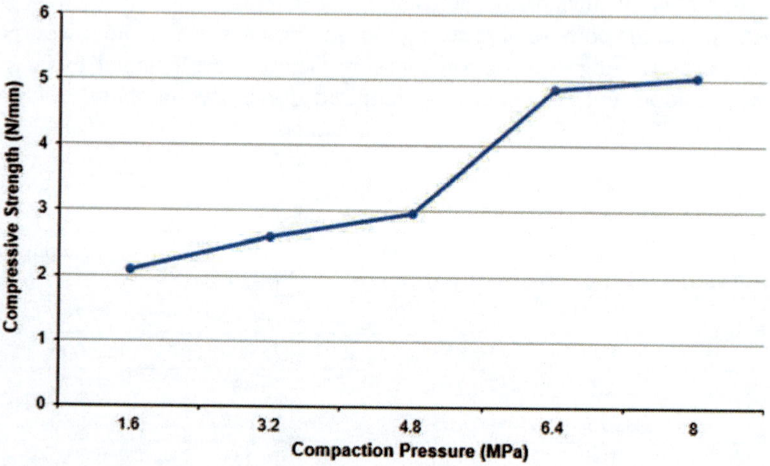

Fig. 7 Compressive strength as a function compaction pressure

The optimum compaction pressure based on the compressive strength of the briquettes was found to be 8 MPa with a compressive strength of 5.04 N/mm. Figure 7 shows a graphical representation of the results of the compressive strength as a function of briquette compaction pressure.

4 Conclusion

This work investigated the feasibility of producing coal fines briquettes using cactus as a binder. Mass ratios of cactus ranging from 0 to 40% were experimented on for compaction pressures ranging from 1.6 to 8 MPa. Shattering index and compressive strength was used to characterize the structural integrity of the briquettes produced. The results obtained led to the conclusion that the optimum cactus fraction to achieve the highest shattering index is 10% which led to a 17% improvement in shattering index compared with the coal fines only briquettes. The maximum achievable shattering index was 86.48% corresponding to a cactus fraction of 10%. The recommended compaction was found to be 8 MPa. However, there is room to improve on this using higher capacity compaction machines. There is further need to investigate the possibility of optimum ratios between 0 and 10% cactus fraction. Additionally, the sensitivity of the compaction to moisture content need should be investigated. It is also important to conduct combustion and emissions tests to burning behaviour and check for any toxic elements.

References

1. Muzenda, E.: Potential uses of South African coal fines. In: International Conference on Mechanical, Electronics and Mechatronics Engineering, Abu Dhabi (2014)
2. Borowski, G., Hycnar, J.J.: Utilization of fine coal waste as a fuel briquettes. Int. J. Coal Prep. Utlization **33**, 194–204 (2013)
3. Ramudzwagi, M., Tshiongo-Makgwe, N., Nheta, W.: Recent developments in benefication of fine and ultra fine coal-review paper. J. Clean. Prod. **276**, 122693 (2020)
4. Marshak, S.: Earth Portrait of a Planet. Toroto: W. W. Norton (2019)
5. Shuma, R., Madyira, D.M.: Emissions comparison of loose biomass briquettes with cow dung and cactus binders. Procedia Manuf. **35**, 130–136 (2017)
6. Rauner, S., Bauer, N., Dirnaichnerm, A., Mutel, C., Luderer, G.: Coal-exit health and environmental damage reductions outweigh economic impacts. Nat. Clim. Change **10**(4), 149–168 (2019)
7. Finkelman, R.B., Wolfe, A., Hendryx, M.S.: The future environmental and health impacts of coal. Energy Geosci., pp. 99–112 (2021)
8. Jha, E., Dutta, S.K.: Optimization of binder for improving strength and shatter index of briquettes for BOF dust using design experiments. Int. J. Eng. Adv. Technol. **9**(1), 6282–6286 (2019)
9. Sen, R., Wiwatpanyaporn, S., Annachhatre, A.P.: Influence of binders on physical properties of fuel briquettes produced from cassava rhizome. Int. J. Environ. Waste Manag. **17**(2), 158–175 (2016)
10. Tosun, Y.I.: Clean fuel-magnesia bonded coal briquetting. Fuel Process. Technol. **88**(10), 977–981 (2007)
11. Zakari, I.Y., Ismaila, A., Sadiq, U., Nasiru, R.: Investigation on the effects of addition of binder and particle size on the high calorific value of solid biofuel. J. Nat. Sci. Res. **3**(12), 30–34 (2013)
12. Manyuchi, M.M., Mbohwa, C., Muzenda, E.: Value addition of coal fines and sawdust to briquettes using molasses as a binder. S. Afr. J. Chem. Eng. **4**, 2–5 (2018)
13. Taulbee, D., Hodgen, R., Aden, N.: Co-briquetting of coal and biomass. In: International Pittsburgh Coal Conference Organized by the University of Pittsburg, Pittsburgh (2012)
14. Shuma, R., Madyira, D.M.: Emissions comparison of loose biomass briquettes with cow dung and cactus binders. Procedia Manuf. **35**, 130–136 (2019)
15. Richards, S.R.: Physical testing of fuel briquettes. Fuel Process. Technol. **25**(2), 89–100 (1990)

Mr. Rofhiwa Nemukula is a final year student in Mechanical Engineering at the University of Johannesburg. He also serves as a teaching assistant for Advanced Manufacturing Systems 4A at the Department of Mechanical Engineering Science, Faculty of Engineering and the Built Environment at the University of Johannesburg.

Prof. Daniel M. Madyira is an Associate Professor in the Department of Mechanical Engineering Science, Faculty of Engineering and the Built Environment at the University of Johannesburg. He is a registered professional mechanical engineer with Engineering Council of South Africa (ECSA). His wide research interests include biomass value chain development for renewable energy applications.

Kanban in Software Development—The Role of Leadership and Metrics

C. Fagarasan, C. Cristea, C. Mihele, O. Popa, D. Ciceo, and A. Pisla

Abstract Traditional software development organizations have rigid processes that can be difficult to change. One of the most critical challenges is the management processes that can ensure the customer's desired level of quality and optimize the organization's resources. Kanban represents one of the most well-known approaches that bring Lean principles, philosophies, and tools into software organizations. This study analyses the leadership behind the Lean methodology, based on which the Kanban software implementation method has been created, by doing a literature review focusing on lean leadership principles that can optimize the software delivery and implementation. Lean practices can be applied in almost any domain like accounting, services industry, product services, and even software development. Two basic concepts stay at the core of the Lean model that can be employed at any organization level. The first focuses on eliminating waste by optimizing processes and systems, and the second focuses on maximized customer value. Consequently, it can be affirmed that Lean is focused on maximizing customer value with a minimum amount of waste. Furthermore, a case study is presented that shows the Kanban implementation at a software development company. The practical case of the Kanban adoption is analyzed, highlighting the link between the leadership principles and the process development flow. Finally, key performance indicators are identified and instituted to guarantee the continuous improvement of the software delivery

C. Fagarasan (✉) · C. Mihele · O. Popa · A. Pisla
Design Engineering and Robotics Department, Industrial Engineering, Robotics and Production Management Faculty, Technical University of Cluj-Napoca, 103-105 Muncii Avenue, Cluj-Napoca, Romania
e-mail: office.cristif@gmail.com

A. Pisla
e-mail: adrian.pisla@muri.utcluj.ro

C. Cristea
Electrical Machines and Drives Department, Faculty of Electrical Engineering, Technical University of Cluj-Napoca, 26-28 G. Baritiu Street, Cluj-Napoca, Romania
e-mail: ciprian.cristea@emd.utcluj.ro

D. Ciceo
International Airport "Avram Iancu" Cluj, 149-151 Traian Vuia Street, Cluj-Napoca, Romania

K. von Leipzig et al. (eds.), *Smart, Sustainable Manufacturing in an Ever-Changing World*, Lecture Notes in Production Engineering,
https://doi.org/10.1007/978-3-031-15602-1_28

model. The software development projects driven by the Kanban project management approach reported essential improvements such as waste reduction, team communication, and collaboration enhancement, work in progress diminishment, productivity-boosting, better delivery predictability, and project visibility. Therefore, the type of leadership influences the software development projects' performance. The findings suggest that Kanban effectively enhances the software development delivery model and team performance.

Keywords Leadership · Kanban · Lean practices · Software development · Software implementation · Agile transformation

1 Introduction

It is widely accepted the adoption of a new way of working, regardless of the domain, can take years, and there is a possibility that the management commitment will be lost in time [1]. The leadership principles in the Lean methodology are designed to sustain and improve employee performance by eliminating any activities from the production process that do not add value for the customer. Even if this sounds simple, it is proven that identifying wasteful activities can be challenging [2]. During the last decade, Lean has become more popular, being tied with Agile practices, and introduced in many industries. Lean has its origins in manufacturing, and the concept was introduced around the 1980s when the Toyota Production System was rebranded [3]. The main challenges that Toyota faced were overproduction, a large inventory of car parts that were not used, and the lack of visualization of the manufacturing process. The change was possible because one of Toyota's manufacturing principles was the continuous quest for excellence. The heart of the lean manufacturing system is eliminating waste, or Muda, as it is known in Japan [4]. However, even if the Lean methodology is widely known and understood, experts and various practitioners still analyze the factors contributing to a successful Lean transformation. Salonitis and Tsinopoulos [5] mention in their findings that "organizational culture and ownership" and "management commitment and capability" play a significant role in Lean adoption. In a similar study, it was determined that "Leadership and Management" is one of the essential parts of the Lean transformation [6].

A survey made by Alefari and Salonitis [1] presents the critical factors of Lean adoption, and in Fig. 1, the study's overall results can be observed. In their studies, Jones and Womack [7] presented all the lean foundations by building the House of Lean, highlighting that the lean transformation is about the organization, not only specific departments. The senior management is responsible for coordinating and optimize all individual activities and operations present in Fig. 2, being highly involved in the organizational change. Software development organizations are achieving success by implementing processes that leverage principles of the Lean methodology.

Fig. 1 Lean adoption—critical factors [1]

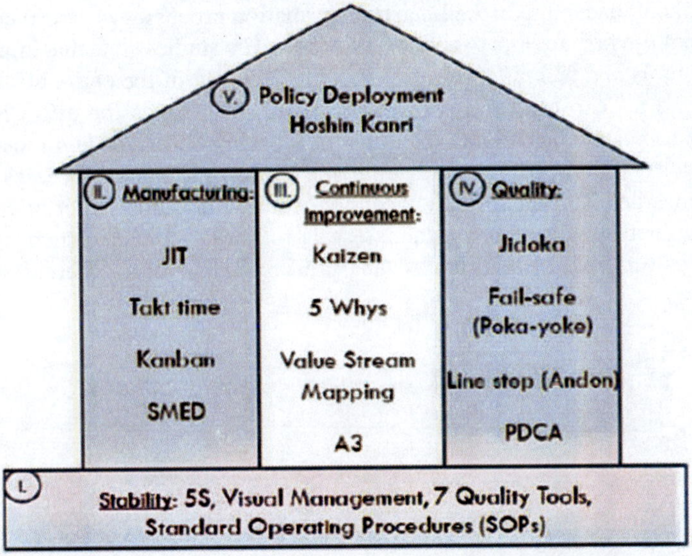

Fig. 2 House of lean [7]

In 2004, David J. Anderson introduced the Kanban method to software development by applying its principles within Microsoft's Information Technology (IT) team. His objectives were to improve the team's workflow, create explicit policies, enhance collaboration, and introduce a continuous improvement mindset by implementing feedback loops [8]. This paper analyses the main leadership principles that contributed to the Lean methodology popularity by conducting a literature review and presenting the key findings. In the second part of the paper, the presented case study shows how the Lean methodology is now applied in the software development industry by leveraging the Kanban implementation method.

2 The Role of Leadership in the Lean Methodology

Lean leadership can be considered a more servant type of leadership. The emphasis is not on a strong personality or aggressive decision-making skills, as it may be seen by traditional management. The mission of the Lean leader is to guide the people to find the most efficient way of doing the work by creating conditions that promote teamwork and collaboration. Leaders must communicate the organization's vision so that teams understand the high-level objectives of the company and offer support when required to ensure continuous and efficient transposal of the vision into reality. Dombrowski and Mielke analyzed different Lean leadership approaches [9] and concluded that there are five fundamental principles of Lean leadership, presented in Fig. 3.

The role of leadership in the Lean transformation process was described by Mann [10], one of the first authors to analyze this topic. His studies state that implementing the Lean tools and techniques represents only 20% out of the entire effort required to complete a Lean transformation. The rest of the 80% is the effort invested in changing mindsets, behaviors, and ways of working. Lean leaders must establish the long-term vision of the organization and value-supporting processes to deliver that vision. The lean transformation cannot occur without the senior management's direct involvement through regular intervals to check on commitments. The link between culture and change is also analyzed by Mann in a different study [11],

Fig. 3 Five principles of lean leadership [9]

and the results show that the culture change results from the management system change. The leadership efforts should be focused on changing how the organization operates by focusing on visual work processes, routines, and specific expectations. The culture shift is critical, but it is the last step in the transformation process. The usage of the Kanban method empowers the team to identify what can be improved, recognize the efficiency, and implement the continuous improvement mindset. This is a prerequisite in a Lean transformation process, and this is the core principle that drives organizational change. It is leaders' responsibility to instill this mindset in all employees. A strategy to implementing the continuous improvement process (CIP) is empowering employees to suggest improvements daily and put them into practice. In time, daily improvement will become a natural behavior of all employees, which should be driven by personal wishes for personal development [12]. In one of the early studies about the Toyota Production System, Imai [13] describes the difference between innovation and continuous improvement process (CIP) by presenting twelve attributes that show the changes necessary for continuous improvement. These can be seen in Fig. 4. Employees need to be given more space to experiment with their ideas about process improvement, and leaders need to support their implementation in the long term. In conclusion, the continuous improvement process affects the organizations' daily work and enhances the communication between leadership and employees.

The role of leadership within the lean methodology is to ensure continuity of the organization's continuous improvement culture and behavior. Leaders must start with improving themselves personally and professionally and promote the same trend for all the organization members. Leaders have to serve as examples and pass the master on the company's problem-solving activities to their employees through coaching. An

Innovation	Characteristic	CIP
Short-term, dramatic	**1. Effect**	Long-term, undramatic
Big steps	**2. Pace**	Small steps
Intermittent	**3. Timeframe**	Continuous incremental
Abrupt and volatile	**4. Change**	Gradual and constant
Select few champions	**5. Involvement**	Everybody
Individualism	**6. Approach**	Collectivism
Scrap & rebuild	**7. Mode**	Maintain & improve
New inventions	**8. Spark**	State of the art
Large investment but little effort to maintain	**9. Requirements**	Little investment but big effort to maintain
Technology	**10. Effort orientation**	People
Results, profit	**11. Evaluation**	Process, improvements
Better for fast growth economy	**12. Advantage**	Better for slow growth economy

Fig. 4 Attributes of continuous improvement process [13]

example is an improvement through failure, assessing mistakes, eliminating the root cause, and impacting final products delivered to customers. Lean leadership always must prioritize long-term objectives versus short-term goals when in conflict. Cote [14] stated that it is preferred to use short-term goals to achieve ultimate objectives. Moreover, to ensure the Kanban implementation method's success, leaders need to offer continuous training to all employees, transparency, and persevere in using all communication channels to update the organization about the Kanban principles, constantly refreshing the perception about the change objective.

3 Lean and Kanban—Applying Lean Principles in the Development of Software Products

The development teams which practice Kanban can visualize the workflow, set limits for the work in progress, and measure their performance to shorten the cycle time. Principles of the Kanban software development method are similar to the Lean ones.

As part of Fig. 5, a visual board can be observed where a simplified software development process is presented. In the "To do" status, all the activities that need to be addressed by the team are listed. Usually, these are ordered from top to bottom by priority. Each team member can assign themselves items from the list and start working on them. According to the "limit work in progress" principle, every column from the board should have a limited number of items that can be dragged into the queue at a particular time, proportional to the team capacity. This enables the continuous delivery of customer value as the team is focusing on a limited number of activities at a time [15]. There can be identified three main parts of the Kanban implementation that need to be considered:

- The software development team who is doing the project work.
- The process used to deliver the project work.
- The product that is being delivered through the project.

Within these significant parts, the principles of Kanban must be incorporated through introducing routines, practices, and team events that nurture a performance-based team structure.

4 Case Study: A Practical Implementation of the Kanban Framework

There are several motivating factors that can lead to the decision of adopting the Kanban implementation method and these can be to improve team communication, to reduce the cycle time and time-to-market, to reduce costs by identifying and removing waste, to create transparency on the work in progress, and finally, to introduce a

Fig. 5 Kanban principles and framework [15]

customer focused mindset. However, this is not an easy task, and a change leader is needed to support the Kanban implementation. There are several basic steps that can be used as a guideline by software development teams to adopt the Kanban method:

- Provide training for teams who are about to implement the Kanban implementation, to promote a change in mindset, by clearly exposing the benefits of the change. Additionally, basic training should be provided to the executive management team.
- Map the steps of the current development process.
- Allow the teams to experiment the new implementation method, by promoting a learn by doing way of working.
- As aground rule, set continuous improvement as a team goal.
- Respect the current team organization.
- Ensure that senior leadership is aware of the change, and they are committed to support the Kanban adoption.

For this case study, a software development team of nine members was chosen with the following configuration: four software engineers, two quality assurance engineers, one business analyst, one designer, and one project manager. The team is cross-functional; it has all the required skills to deliver working software to the customer. The steps taken for the Kanban adoption are presented in the sub-chapters below. This team should be working as one unit, by setting a common vision and mission, and having clear goals for all their activities.

4.1 Provide Training and Clearly Explain the Benefits of Adopting Kanban

One of the biggest challenges of the teams' adopting Kanban is the lack of experience with the new implementation method [16]. Therefore, it is important to prepare the change of mindset by organizing training sessions for team members and senior leadership, to avoid falling back to the old ways of working. If the leadership team supports and promotes the Kanban adoption, then the resistance to change will be significantly smaller.

4.2 Identify the Current Development Process

For this activity, the project manager should take the initiative and facilitate a brainstorming session where the entire team participates in mapping out all the stages of the current development process. The session's objective is to iron out the current process and not to come with ideas on how to improve it. An example of a development workflow can have the following stages: Backlog, Analyze, Design, Review, Build, Integrate and Test, Done. This workflow can be implemented in one of the issue management systems available on the market and in this way, every requirement will go through these stages and the time spent in each phase can be measured. When passing from a phase to another, usually there are some informal handoffs happening between team members. This can be avoided by involving the whole team early in the development process and will ensure awareness with all the team members. Introducing a regular release cycle and regular timeboxed ceremonies like release refinement, planning, review and inspect and adapt sessions to support the identified workflow is key to ensure predictability and sustainability. The end client should be invited to these sessions to enhance the customer-team collaboration which is an essential factor in delivering software products in short cycles.

4.3 Allow the Teams to Experiment with the New Implementation Method

The first cycles should be experimental ones, so team have no pressure and the new way of working can be learned by doing. After the first release, the team can regroup in an inspect and adapt session where it is discussed what can be improved in the next one. Shorter cycle times are preferred.

4.4 Visualize the Workflow

When all the process stages are listed, the next step is to make them visible on a board. The workflow can be displayed on a physical or a virtual board. Usually, software development teams use tools like Jira, an Issue Management System produced by Atlassian, where virtual boards can be built. All the work items that are currently in progress should be added to the new board to be visible to all team members.

4.5 Set Ground Rules

A kick-off meeting should be organized where all the team members participate. The way of working is discussed, so all sub-processes are explicit. There might be sign-offs needed for work items to pass from a stage to another. All these details need to be very clear and defined in a place accessible by everyone, with clear roles and responsibilities. During this phase, team ceremonies need to be planned in terms of cadence and recurrency. The type and cadence of meetings need to be scoped, depending on the project needs.

4.6 Ensure There is a Professional Kanban Leader Within the Team

If there is a lack of guidance, the purpose and theory behind the Kanban method can be misunderstood. Communication between team members, with the end client or within the organization can be an issue at the beginning of the adoption so the coach is there to promote over communication instead of under communication. The adoption of Lean can be seen as a significant change within the organization. The lessons learned within the change management literature can be applied when implementing lean. As part of his studies, Kotter [17] identified the leading causes of change failures that future leaders can use as a checklist to ensure the success of any change process. Some of the leadership errors that can lead to failures can be:

- Lack of a sufficient sense of urgency or creating too much complacency.
- Lack of a powerful guiding coalition.
- Lack of a vision or understanding the power of a vision.
- Grossly under-communicating the vision.
- Failure to remove obstacles to the new vision.
- Failure to create short-term wins.
- Declaring victory/success too soon.
- Neglecting to anchor changes firmly in the corporate culture.

For the adoption to be successful, the team should set continuous improvement and delivery as their goals, and these principles should be applied to daily activities. This is a mindset change, which is one significant factor of the Kanban adoption. The training provided may not enough, so the leader should work individually with the team members to motivate and change people mindsets, which is not an easy endeavor, and it usually is time consuming. This happens thorough education and by clearly communicating the benefits of adopting Kanban. On the other hand, the transformation needs to be understood by the leadership team and the Kanban transformation leader needs to have their full support.

4.7 *Manage the Flow of Work Items*

For each stage of the workflow, limits should be set according to the team size. All the team members should discuss and decide the maximum amount of work items that can be in parallel in the same stage. This supports the continuous delivery of business value to the customer. Figure 6 show an example of a Kanban board with work in progress limits.

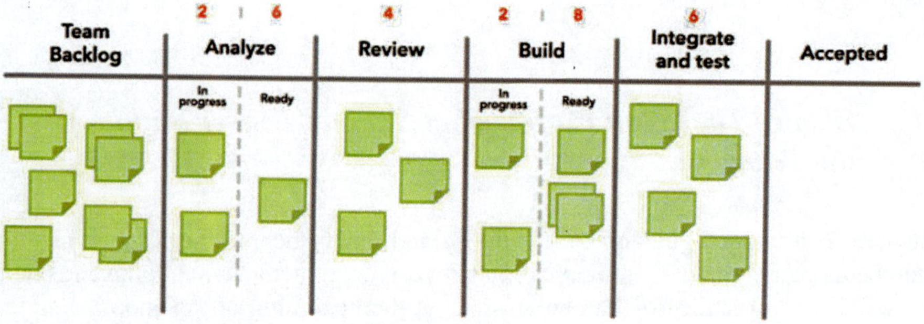

Fig. 6 A team Kanban board [19]

4.8 Implement Key Performance Indicators

Metrics can offer support in tracking the team's progress and productivity and setting KPIs can help the team to determine if their performance is increasing or decreasing. The implementation of the metrics is mandatory as it serves as a personal assistant for the transformation leader and the team that helps is identifying bottlenecks and pain points in the development process. The main conventional metrics used in Kanban are described in the next chapter of the current paper. In a study presented by Kirovska and Koceski [18], there are presented some unconventional metrics that show, among others, a 23% decrease of the difference between the estimated and real time of working tasks. This can suggest that by implementing Kanban, the efficiency of the team can be improved together with their predictability and reliability.

5 Key Performance Indicators Used in Kanban for Measuring the Delivery of Software Products

Traditional software development methods rely on precise requirements and defined scope. That leads to a baselined scope, cost, and time, which makes monitoring and controlling a project straightforward. On the other hand, when developing software in an Agile way, the project's scope is continually evolving, making progress tracking difficult. Agile teams focus on delivering business value through working software, which is the primary measure of success [20]. However, in Kanban, several metrics can help with reporting on project progress. These are usually tracked automatically as part of the project management system used for issue management.

5.1 Cycle Time

In a workflow, there are multiple stages through which a work item will pass, and these are usually captured as statuses in a Kanban board. The cycle time is represented by the number of days a work item spends in a workflow stage. This can help a software development team to identify the bottlenecks and find a solution to optimize the development workflow. Queues can appear in the workflow, for example tasks waiting in the Review column. This can impact the overall team performance.

5.2 Lead Time

In connection to the cycle time, the lead time is represented by the number of days it takes for a work item to reach from start to finish. The metric can be calculated with an emphasis on the customer and can be defined as the interval between the time the customer requested a feature and the time it was delivered. The first two presented metrics are the most important ones because they reveal the speed through which features are delivered to the end customers.

5.3 Throughput

The throughput measures the team efficiency and productivity, and it is defined as the number of work items delivered during a time unit. This metric offers a better understanding of a software development team capacity and its velocity.

5.4 Work in Progress (WIP)

The WIP metric is defined by the total number of work items that are currently in progress in all workflow stages. In combination with other metrics, WIP can provide valuable insights. For example, the Little's Law can be applied to calculate the Average Cycle Time by using the following formula:

$$Average\, Cycle\, Time = WIP / Average\, Throughput \tag{1}$$

5.5 Tact Time

The tact time metric is a key performance indicator that measures the frequency of work items delivered. The tact time can be calculated by using the following formula:

$$Tact\, time = Average\, Cycle\, Time / WIP \tag{2}$$

This metric can reflect the team's velocity and can assist the development team in their future planning efforts.

5.6 Monte Carlo Simulations

Work estimation is a challenge that every software development team experiences. Monte Carlo simulations are valuable for obtaining probabilistic forecasts regarding KPIs, such as throughput or cycle time, by using historical data of the workflow.

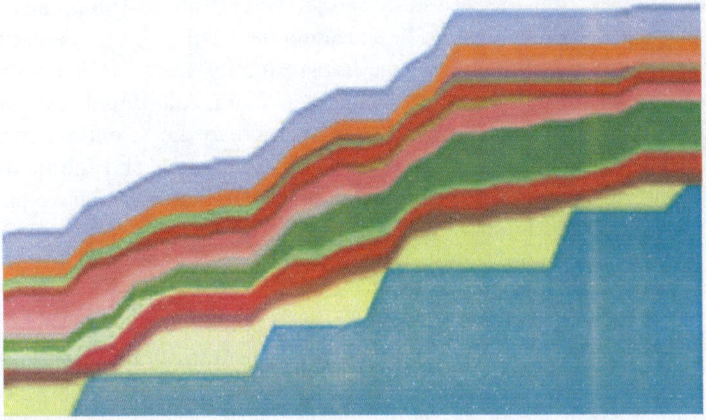

Fig. 7 Cumulative flow diagram [20]

5.7 *Cumulative Flow Diagram (CFD)*

Concerning WIP, the diagram presented in Fig. 7 can help visualize the work items in each workflow status. Each color from the graph below represents a status from the software development process.

These are the most important Kanban metrics that can be used to increase the productivity of a software team and the efficiency of the development workflow.

6 Conclusions

A significant factor in the Kanban adoption is played by the senior leadership team of the organization. The Kanban transformation leader needs to lay down all benefits and disadvantages of the Kanban adoption to the senior management, and their buy-in needs to be assured. Moreover, the leader needs to inspire the team members to adopt the new continuous improvement mindset, leading by example. The responsibility of the team maturity falls on the Kanban transformation leader as well. The needs of training and coaching need to be continually assessed and acted upon, as this is one of the main challenges that the teams face. The primary motivating factors for implementing Kanban discussed throughout the paper are:

- Team communication and collaboration improvements.
- An increase in team efficiency and productivity, by improving the software development workflow.
- Reducing development cycle times and time-to-market by implementing the Kanban metrics.
- Transparency withing the organization.

The steps presented in the case study can serve as a practical guideline for any software development team that would like to adopt the Kanban implementation method while underlying the importance of the transformation leader role and the need for continuous support from the senior management team. Additionally, the importance of the Kanban metrics was underlined in relation to the relentless improvement mindset that Kanban promotes. The transformation leader and the metrics implemented as part of the Kanban transformation are key factors and have proven to be a must in any Kanban adoption.

References

1. Alefari, M., Salonitis, K., Xu, Y.: The role of leadership in implementing lean manufacturing. Procedia CIRP **63**, 756–761 (2017)
2. Ikonen, M.: Leadership in Kanban software development projects: a quasi-controlled experiment. In: LESS 2010, vol. 65, pp. 85–89 (2010)
3. Womack, J., Jones, D., Roos, D.: The Machine that Changed the World. Scribner Book Company, New York (1990)
4. Liker, J.-K.: The Toyota Way 2nd edition: 14 Management Principles from the World's Greatest Manufacturer. McGraw-Hill Education, New York (2020)
5. Salonitis, K., Tsinopoulos, C.: Drivers and barriers of Lean implementation in the Greek manufacturing sector. Procedia CIRP **57**, 189–194 (2016)
6. Zargun, S., Al-Ashhab, A.: Critical success factors for lean manufacturing: a systematic literature review—an international comparison between developing and developed countries. Adv. Mater. Res., 668–681 (2014)
7. Jones, D.T., Womack, J.: Lean Thinking: Banish Waste and Create Wealth in Your Corporation. Free Press, New York (1996)
8. McKinney, J.: Optimal Software Development: Waterfall. Capstone Seminar, University of Denver, Scrum & Kanban (2016)
9. Dombrowski, U., Mielke, T.: Lean leadership—fundamental principles and their application. Procedia CIRP **7**, 569–574 (2013)
10. Mann, D.: The missing link: lean leadership. Front. Health Serv. Manage. **26**, 15–26 (2009)
11. Mann, D.: Creating a Lean Culture, 3rd edn. Productivity Press, New York (2015)
12. Dombrowski, U., Mielke, T.: Lean leadership—15 rules for a sustainable lean implementation. Procedia CIRP **17**, 565–570 (2014)
13. Iman, M.: Kaizen: The Key to Japan's Competitive Success. McGraw-Hill Education, New York (1986)
14. Cote, D.: Winning Now Winning Later. HarperCollins Leadership, Nashville (2020)
15. Ahmad, M., O., Oivo, M., Markkula, J.: Kanban in Software Development: A Systematic Literature Review, Software Engineering and Advanced Applications (SEAA), pp. 9–16 (2013)
16. Ahmad, M.O., Markkula, J., Oivo, M., Kuvaja, P.: Usage of Kanban in software companies: an empirical study on motivation, benefits and challenges. In: Proceedings of the 9th International Conference on Software Engineering Advances (2014)
17. Kotter, J.: Why Transformation Efforts Fail. Harvard Business Review (January edition) (2007)
18. Kirovska, N., Koceski, S.: Usage of Kanban methodology at software development teams. J. Appl. Econ. Bus., 25–34 (2015)
19. Knaster, R., Leffingwell, D.: SAFe Distilled. Pearson Addison-Wesley, New Jersey (2020)
20. Karunanithi, K.: Metrics in Agile and Kanban, Software Measurement Techniques, Software Measurement Project (2016)

Cristian FĂGĂRĂŞAN is an Engineering and Management Ph.D. student at the Technical University of Cluj-Napoca, Romania, focusing on IT project management and Agile methodologies.

Ciprian Cristea holds a Ph.D. in Industrial Engineering, and he is an Assistant Professor at the Technical University of Cluj-Napoca, Romania, also focused on project management emerging practices and techniques.

Cristina Mihele is Engineering and Management Ph.D. student, at the Technical University of Cluj-Napoca, Romania with a focus on differences between management and leadership, with an emphasis on the present leadership principles.

Ovidiu Popa is Engineering and Management Ph.D. student, at the Technical University of Cluj-Napoca, Romania with a focus on introducing the adaptive software project management methods into the automotive industry.

David Ciceo is the Director of the International Airport "Avram Iancu" Cluj-Napoca and President of the Romanian Airports Association. He holds a Ph.D. in Airport Management attained at Technical University of Cluj-Napoca.

Adrian PÎSLĂ is the Director of Siemens PLM Training Centre at Industrial Engineering, Robotics and Production Management Faculty, Technical University of Cluj-Napoca, Romania.

Extended Reality

Application of Augmented Reality for the Training in the Field of Refrigeration and Air-Conditioning

F. Bellalouna and R. Langebach

Abstract Augmented Reality (AR) technology is one of the key technologies in the digital transformation area. AR is the technology that enables the overlaying of physical objects with computer-generated virtual perceptible data in real time to provide an interactive user experience in the real environment. This technology is currently used in many business fields such as engineering, education, medicine, logistics and transport either for experimentation or productivity. Due to the rapid development of display hardware and mobile devices in terms of performance, as well as the rapid development of the range of functions, the implementation of powerful augmented reality applications for industrial applications has become easier. This paper presents a use case of an AR training application for an ammonia screw compressor asset using the Video-See-Through AR method. The intention of this use case is to show the benefits and potential of AR technology for teaching and training complex technical systems in terms of increasing learning efficiency. The paper discusses the gathered experiences in the course of the use case and during the use of the AR training application in the university and in the industry.

Keywords Enter extended reality (XR) · Augmented reality (AR) · Video see-through AR · Screw compressor

1 Introduction

Progress in information and communication technologies (ICT) largely eliminates economic restrictions and creates completely new possibilities for action. This leads to innovations that were previously simply unimaginable [1]. This process is referred to as "digital transformation" and is defined by the MIT Center of Digital Business as "using new digital technologies (e.g. social media, mobile communication devices, integrated sensors, cloud computing) to significantly improve and increase the performance of business processes (e.g. improving the customer experience,

F. Bellalouna (✉) · R. Langebach
University of Applied Sciences, Karlsruhe, Germany
e-mail: fahmi.bellalouna@h-ka.de

K. von Leipzig et al. (eds.), *Smart, Sustainable Manufacturing in an Ever-Changing World*, Lecture Notes in Production Engineering,
https://doi.org/10.1007/978-3-031-15602-1_29

rationalizing operations or creating new business models) [2]. Augmented Reality (AR) is one of the key technologies of digital transformation in the industrial and non-industrial environment. AR is the technology that extends the physical world through computer-generated, perceptible digital objects to enable an interactive user experience in the real world. Due to the rapid development of display hardware, new interaction devices and tracking systems, virtual and augmented reality applications are being developed now, which could only have been created in large research laboratories a few years ago [3]. AR has meanwhile found its way into many industries and business processes such as trade, sales, e-commerce, tourism, production, logistics, and training. The main advantages of AR applications are that users can be fully integrated in the processes and systems through high-performance display, user tracking and UI functions and that users feel part of an experience. The use of AR helps to improve the spatial perception of complex technical and scientific systems in comparison to a simple 2D desktop presentation and thus, helps to depict complex behaviours in a more tangible and understandable manner. Several scientific studies have shown that AR enhances human sensory skills and reduces the cognitive effort required to imagine complex systems and products [4]. Scientists from the István Széchenyi University in Hungary have examined the increase in learning efficiency through AR in higher education and have come to the following results: 30% more student activities and teamwork in the classroom and 50% better understanding of the lesson content [5]. The potential of AR technology to increase the efficiency of business processes and to generate new business models has already been recognized by companies from both the IT and industrial sectors. The great interest of companies in AR is reflected in the market figures. The market research experts all agree on the enormous growth potential for the AR market. Forecasts speak of a global AR market turnover of 70–75 billion dollars in 2023. Up to 2.5 billion AR devices are to be sold worldwide by 2023, which creates enormous opportunities for the development of innovative AR applications [6]. Despite this trend, AR applications are currently often in the experimental stage in the industrial environment and in many cases, they have not yet been transferred to actual business processes. According to a study by the Capgemini Research Institute from 2018, 55% of the approximately 600 companies surveyed worldwide that deal with AR technology described themselves as experimenters and 45% as implementers, in Germany the figure is 62% and 38%, respectively [7].

This article describes an application for the implementation of an AR application to visualise the structure and function of an ammonia screw compressor for air-conditioning application in the field of education and training. The application was implemented as a cooperation between the University of Applied Sciences Karlsruhe and the company BITZER Kühlmaschinenbau GmbH. During the implementation of the use case, the following questions were in the foreground:

- Required implementation effort of the AR application.
- Performance and usability of the implemented AR application.
- Added value and benefits for the own business processes and the future viability of AR technology.

This presented work has an experimental character and rather than a fundamental research. The developed AR applications were implemented as a prototype by the students of the University of Applied Silences Karlsruhe during the AR laboratory. The gained knowledge in this work can be considered as suggestions for further research works.

2 Extended Reality—XR

The rapid technological development in the field of computer graphics has led to the emergence of a number of applications in recent years that combine reality with virtuality through simulation techniques. The result is the creation of an artificial environment, in which the distinction between reality and virtuality is not possible through deliberate illusion. Depending on the reality and virtuality content, these applications are classified into Virtual Reality (VR), Augmented Virtuality (AV) or Augmented Reality (AR). The umbrella term eXtended Reality (XR) was introduced to denote the various application forms. The "X" is also used in the literature as a placeholder for "A" (augmented) or for "V" (virtual). The transition between reality and virtuality was mapped by Paul Milgrim's reality-virtuality continuum [8] (Fig. 1). At the left end of the continuum, the environment to be viewed is only displayed in real life, while one at the right end is only displayed virtually. Applied to the example with the screw compressor (Fig. 1), the real exhibited compressor is located on the far left in the "Real Environment" area (Fig. 1a) and the virtual 3D graphic model on the far right in the "Virtual Environment" (VR)" (Fig. 1d) of the continuum. Mixed Reality (MR) is a continuum that stretches between reality and virtuality. The proportion of reality decreases continuously, while that of virtuality increases accordingly or vice versa [3]. As far as the real environment is in the focus of the consideration and is expanded by virtual content, this is an augmented reality environment (AR). The virtual visualization of the refrigerant flow through the real screw compressor is an AR application in the attached example (Fig. 1b). If, on the other hand, the focus is on the virtual environment and is enhanced by real content, the term augmented virtuality is used (AV). In relation to the illustrated example (Fig. 1c), the consideration and positioning of the graphic model for the screw compressor and the flanged electric motor in the real space on the frame is an augmented virtuality application (AV).

Figure 2 shows the basic structure of an eXtended Reality system [9]. The focus of an XR system is the XR hardware, which has XR-capable graphics and computing processors. The XR application and rendering are carried out on the XR hardware. The XR application is the core application for performing a certain process task with the help of XR (VR, AR, AV). Rendering realistically reproduces the XR environment depending on the viewing position and the user interaction data. For this purpose, viewing angles, 3D sound and object movements, for example, are rendered in real time. For user interaction with the XR application, input components are required that record the interaction data. The interaction data can be visual, auditory, tactile or haptic. The data acquisition can take place by means of sensors such as cameras,

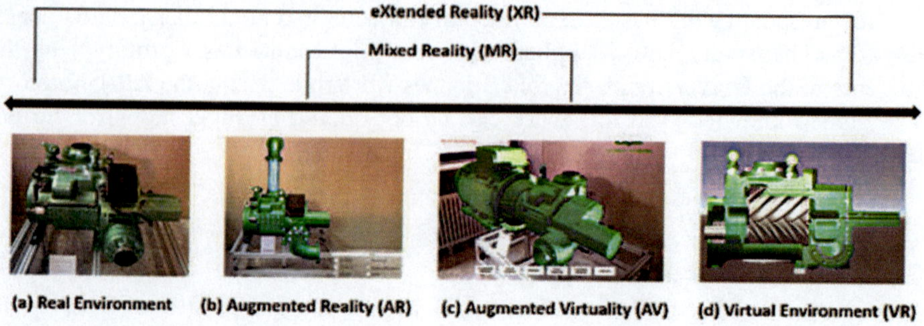

Fig. 1 Reality-virtuality continuum

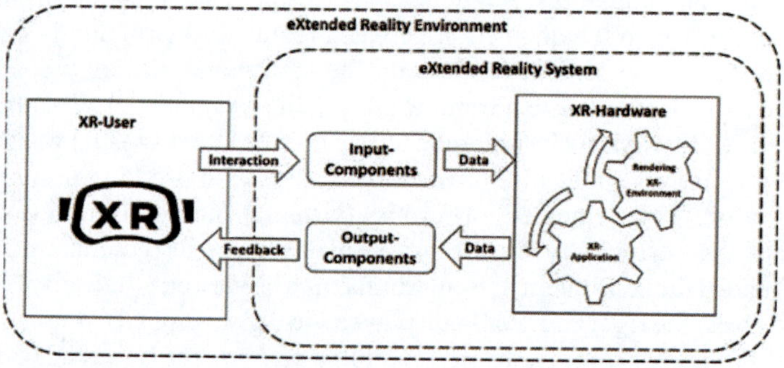

Fig. 2 Structure of the eXtended reality sytem

microphones, as well as acceleration and position sensors. The XR environment is rendered with the help of output components. These are output devices for audio or visual information such as glasses, screens, screens or loudspeakers. There can also be haptic feedback by means of force feedback or proprioception via a motion platform.

3 Augmented Reality (AR)

According to Dörner et al. [3], AR is a direct, interactive and real-time capable extension of the perception of the real environment by virtual content for any senses which is based on reality as much as possible in its expression and appearance, so that, in extreme cases, a distinction between real and virtual (sensory) impressions are no longer possible. AR expands the perception of the real environment with virtual elements by combining spatially defined digital content with the real world in real time. This leads to a merging of reality with virtuality [3]. A common example of an application of AR is the insertion of additional information during the live broadcast of a soccer game. Offside lines or goal distances during a free kick are displayed digitally on television [10]. In theory, however, AR can be combined with every sense that humans can perceive, e.g. visual AR for the perception of virtual objects, auditory AR for the perception of noises, tactile AR for the haptic perception of virtual objects and vestibular AR for the perception of movements and accelerations [3]. The following common methods are currently used in practice to implement AR applications:

- Video-See-Through AR: with Video-See-Through AR, the virtual content is superimposed on a video recording of the real environment in real time.
- Optical-See-Through AR: In the context of Optical-See-Through AR, the virtual content is displayed on a transparent projection surface.
- Projection-based AR: For Projection-based AR, the surfaces of real objects are used as a projection surface for displaying virtual content with different colour lighting.

4 Review of Related Works

In this section an overview about the relevant related research works focusing on the use of the AR methods in the engineering field will be given without claim of completeness. Danielsson et al. [11] give in their paper an overview of the current knowledge and future challenges of augmented reality smart glasses (ARSG) for use by industrial operators especially during the assembly processes. This paper has investigated ARSG from an operator perspective and has come up to the following conclusions: lack of standards for the use of ARSG for the assembly instruction; ARSG can cause efficiency losses; limited sensors and visual recognition of the ARSG. In the paper by Lotsaris et al. [12] an augmented-reality-based framework for supporting human workers in human–robot collaboration manufacturing environment is presented. The developed AR application provides the user with handy tools to interact with the mobile robot platform, gives direct instructions to it and receives information about the robot through an OHMD AR device. According to the author, the testing of the AR tools within an automotive industry use case has shown a significant reduction of the process lead time. Mourtzis et al. present in their paper [13] an AR-based proposal, to handle and display production scheduling and machining data. The AR application can be fed with real production data (e.g. machines data) and digital production data and visualize these data to the production managers in interactive and intuitive way using OHMD AR devices. In the paper by Vorraber et al. [14] a study to evaluate AR-based maintenance processes supported by OHMD AR device is discussed. A total of 12 test runs conducted with 6 maintenance engineers from the automotive industry were analyzed based on video recordings and interviews. According to the author, the engineers could complete their tasks approx. 20% faster than without the AR application.

5 Significance of AR Technology for Scientific Partner Bitzer Kühlmaschinenbau GMBH

The refrigeration and air conditioning technology, for which BITZER develops, produces and sells essential components such as refrigerant compressors, heat exchangers and pressure vessels, is unknown to many people due to their work in the background. In 2009, BITZER began communicating about its product and application topics using AR to highlight its products that are not immediately visible to the public. BITZER is considered to be the forerunner of the current AR projects in refrigeration and air conditioning technology. In 2014, BITZER gave the first AR product presentations on its stand at the world's leading trade fair for refrigeration technology, Chillventa. Since 2019, the new BITZER headquarters in Sindelfingen have a showroom, in which, for the most part, traditional, physical cut-away models of products are deliberately avoided and focuses much more on digital solutions. Some AR projects are currently running in cooperation with the authors with the aim of expanding the SCHAUFLER Academy's range of training courses to digital teaching content. The SCHAUFLER Academy is the company's own international training centre at their location in Rottenburg-Ergenzingen. Future AR projects are planned to focus on training, service and commissioning of these complex refrigeration and air conditioning systems, offering an increasing and supportive role for BITZER on trade fairs and other events, whether virtual or in person.

6 Use Case: AR App as a Tool for Visualization of the Structure and Functions of a Screw Compressor in Training

6.1 Motivation

The components used in refrigeration and air conditioning systems have inner workings that do not necessarily reveal themselves. Small subtleties and optimizations of the machines and corresponding refrigeration cycles make the difference between achieving climate policy goals such as increasing energy efficiency and reducing the carbon footprint or not. In order to explain these quite complex machines, elaborately built sectional models allow a look into the inner structure. The SCHAUFLER Academy use those so-called cut-away models for training purposes. They are very popular and always generate a lot of interest from visitors—for BITZER products, in particular, and for the refrigeration and air conditioning industry, in general. The big advantage of real cut-away models is that they make technology physically tangible. There are disadvantages, such as the costs of creating those models, as well as the time to be invested to keep adapting them to the current state of development. In addition, there is a trend towards higher power, for example for district cooling and

Fig. 3 Ammonia open screw compressor (*Courtesy of* BITZER)

heating, and thus, larger machines. Resulting in models that are becoming increasingly unwieldy to handle. Furthermore, the BITZER compressors used on trade fairs in some cases do not have the actual inner workings and just embody the outer shell, due to weight. A now obsolete exhibition model of an ammonia open screw compressor (Fig. 3) was donated to the University of Applied Sciences Karlsruhe in order to carry out an AR project in cooperation with BITZER. The aim is to digitally recreate the entire inner workings and structure of the compressor using AR.

6.2 *AR Application for the Visualization of the Structure and Functions of a Screw Compressor in Education and Training*

As part of the cooperation between the University of Applied Sciences Karlsruhe and the SCHAUFLER Academy, an AR application was developed which demonstrates the benefits and potential of AR technology for teaching and knowledge transfer of complex technical systems in the field of refrigeration and air-conditioning. The ammonia screw compressor offers a suitable application example for this, as 2D documents, such as books, training documents or videos, cannot make its complex functions visible in a simple and accessible way. Screw compressors are typically used in refrigeration technology for large displacement volumes at low to medium pressure ratios. The name is derived from its working principle that employs typically two rotors—male and female part—that rotate against each other similar to a screw (Fig. 4). The helically designed rotors form the working chambers within the tooth flanks, which reduce in size by continuous rotation towards the pressure side. Due to their design, these screw compressors have a built-in volume ratio that is influenced by the geometric design of the outlet on the pressure side. Another speciality of screw compressor is the so-called bleeding hole which occurs in commonly used screw profiles at the tooth base while rotating against the second rotor. This undesirable geometric opening allows the compressed refrigerant to flow back to the suction side and built leakage. Hence, screw compressors are operated with a high quantity of oil in the refrigerant flow, in order to minimize the associated loss of efficiency due to internal leakage. The continuously forming working chambers as well as the location and nature of the bleeding hole are very difficult to explain, even with the help of a non-moving cut-away model. An AR application, which superimposes and visualizes the computer-generated assembly structure and functionality of the screw compressor via a 2D image (e.g. product catalogue, product flyer) or a physical closed model (e.g. the real screw compressor 3D print model) should significantly reduce this effort. Furthermore, the application should offer an AR experience that improves learning efficiency in terms of increasing of the motivation through greater involvement of the training participants and thus, activation of long-term memory of the audience. The AR application was developed as a Video-See-Through AR application and can be used via a mobile iOS and Android device. In order to be able to explore the functionalities of the screw compressor interactively and intuitively, the AR application offer six main functions:

Function for **Show/Hide components** of virtual data of the screw compressor's selected components. In addition, the housing can be displayed transparent, in order to make the internal components of the screw compressor visible.

Function for **sectional views** for visualization of movable cutting planes in order to virtually display the screw compressor's internal assembly structure (Fig. 4). The sections can either be selected statically in commonly used half-section view or dynamically and thus, individually movable towards all coordinate axes.

Animation function for displaying, in a tangible and understandable way, the complex main functions of the screw compressor such as refrigerant flow, rotation and interaction of the rotors, compression process, and capacity control (Fig. 5). Some of the animations can be interactively controlled.

The **Views** function allows special views, such as the exploded view (Fig. 6) or the view of the compression workspace, to be selected and partially influenced interactively. The function for the **context-dependent presentation of information**

Fig. 4 AR function for the section view

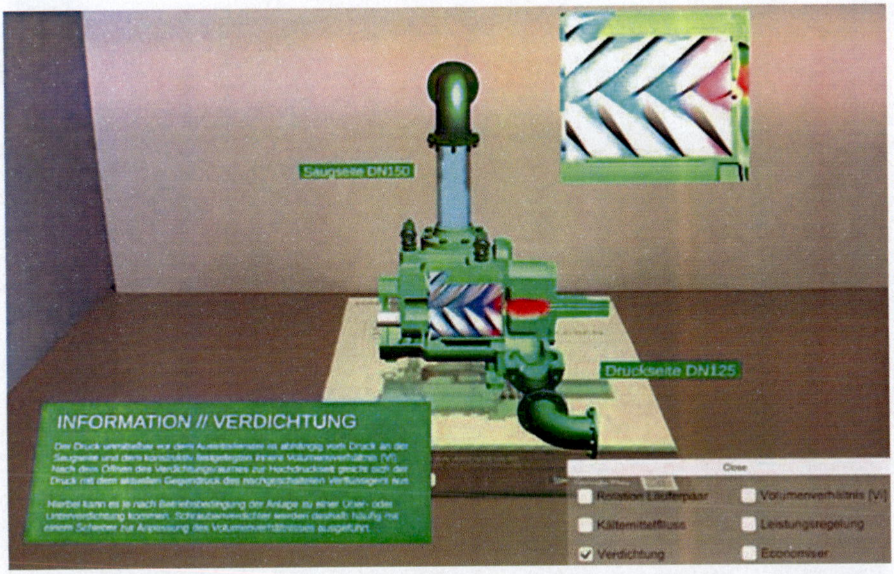

Fig. 5 AR function for the animation of the refrigerant flow (AR app based on the flyer of the compressor)

extends the displayed virtual 3D models of the selected components with additional information from literature or user manual such as text descriptions or operational diagrams, some of which are animated (Fig. 7). The context-dependent information is presented depending on the view selected in the application depending on the context. The purpose of this function is to offer a better understanding of the content displayed in the AR application.

With the help of the **knowledge** function, the teaching progress can be reviewed via an interactive multiple-choice test. Knowledge gaps can be identified, it is possible to repeat the course as often as necessary.

6.3 Evaluation of the AR Application in Education and Training at BITZER

The generated app was provided for interested users via QR code on flyers and displays at the SCHAUFLER Academy. This enables the user during the testing period to discover the compressor and to explore its components as described before. With different levels of details, depending on interest and previous knowledge different target groups can be addressed. The short knowledge questionnaire motivates users to improve individual learning progress regarding the individual components, the fluid (refrigerant) flow and machine-related diagrams. Overall, the digital cutaway model offers much more information and explanations than real

Fig. 6 AR function for the exploded view

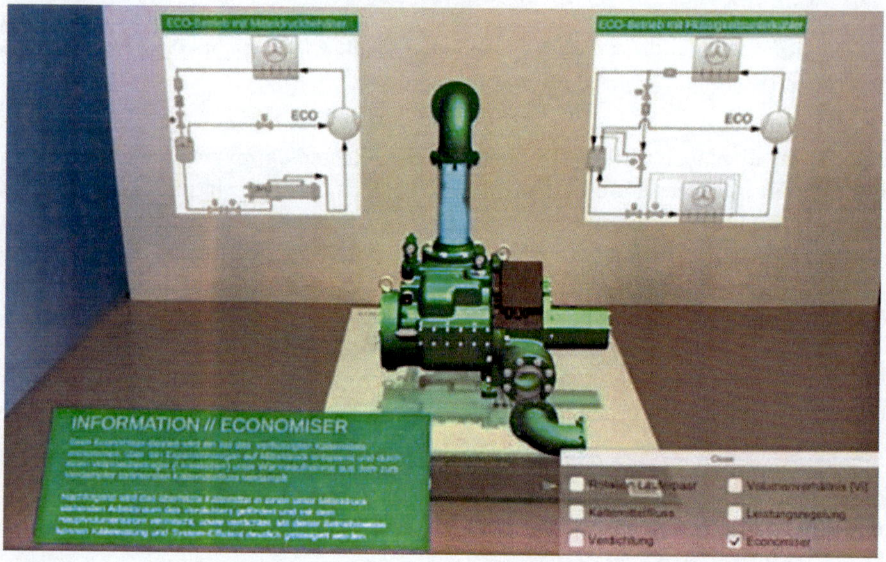

Fig. 7 AR function for animation of the refrigerant flow and for context-dependent information display

cutaway models can cover. A great advantage is that the model is available virtually everywhere and can be integrated into online seminars, presentations, customer and supplier meetings at any time. There is also a decisive advantage with regard to the weight of real models. The AR app also works in targeting with smaller, much more manageable plastic models—for example from the 3-D printer, opening up new ways, especially in marketing.

The models and contents are customizable, so that the platform may also offer future developments and customization options. A future replacement of today's operating instructions is conceivable. The flexibility and the application possibilities are diverse and can provide interested schools, universities, institutions and parties with models without great efforts. Knowledge tests and gamification are possible, as well as simple explanations. Overall, AR cutaway models can be seen as a success and as a worthy replacement of classic cutaway models. As shown here, AR applications offer great added value and substantially improve motivation and further interest in refrigeration and air conditioning technology of all target groups—especially young people.

7 Experience Report

7.1 Experience Report at BITZER Company

The AR project carried out so far, positively and it outreaches all expectations in terms of its outcomes will certainly offer undreamt-of potential in the future, especially if the processes are further standardized and automated. Ongoing AR projects are well received in BITZER branches around the world thanks to German and English language capability. The content can be expanded and optimized at any time through the use of standardized software. Digital cutaway models can definitely be created much faster and more cost-effectively. AR offers enormous potential to convey technically complex or potentially dangerous and invisible teaching content for the user in a playful but realistic way. Similar to the additional training of pilots in aviation, simulation and AR offer great opportunities in technical and industrial environments. In the future, these digital applications will certainly become even more important in education, training and other forms of knowledge transfer.

7.2 Experience Report Within Refrigeration and Compressor Lectures at the University of Applied Sciences Karlsruhe

The lectures on the basics of refrigeration and compressor technology at the University of Applied Sciences Karlsruhe use the AR app for visualizing the structure and function of a screw compressor as well as the videos that can be easily extracted from the app. The feedback from students on these new formats is extremely positive and underpins the approach of modern, digitally supported teaching. As a pleasant side effect, the image of outdated (unfortunately often dusty) cut-away models is dramatically improved and acceptance from the students' point of view is significantly increased. Difficult-to-understand content can be explained in a more accessible way to the students. Interest and attention are increased perceptible just by offering the use of personal electronic devices during the class—for example their own mobile phone.

Acknowledgements The discussed work and their out-comes were achieved during the cooperation project between the University of Applied Sciences Karlsruhe and the Schaufler Academy of the company BITZER Kühlmaschinenbau GmbH.

References

1. Schellinger, J., Tokarski, K.O., Kissling-Näf, I.: Digital Transformation und Unternehmensführung: Trend und Perspektiven für die Praxis. Springer Gabler (2019)
2. Fitzgerald, M., Kruschwitz, M., Bonnet, D., Welch, M.: Embracing digital technology: a new strtegic imperative. MIT Sloan Manag. Rev. **55**(2) (2014)
3. Dörner, R., Broll, W., Grimm, P., Jung, B.: Virtual and Augmented Reality (VR/AR): Grundlagen und Methoden der Virtuellen und Augmentierten Realität. Springer Vieweg (2013)
4. Liu, Y., Lather, J., Messner, J.: Virtual reality to support the integrated design process: a retrofit case study. In: International Conference on Computing in Civil and Building Engineering, Orlando, pp. 801–808 (2014)
5. Horvat, N., Škec, S., Martinec, T., Lukačević, F., Perišić, M.: Comparing virtual reality and desktop interface for reviewing 3D CAD models. In: International Conference on Engineering Design, Delft, 1923–1932 (2019)
6. DigiCapital.: For AR/VR 2.0 to live, AR/VR 1.0 must die (2019). 15 Jan. 2019, Available at https://www.digi-capital.com/news/2019/01/for-ar-vr-2-0-to-live-ar-vr-1-0-must-die/.. Access on 12th Jun 2020
7. Augmented and Virtual Reality in Operations: A guide for investment, Capgemini, 2021, https://www.capgemini.com/research-old/augmented-and-virtual-reality-in-operations/. Access on 25th May 2021
8. Milgram, P., Takemura, H., Utsumi, A., Kishino, F.: Augmented reality: a class of displays on the reality-virtuality continuum. In: Telemanipulator and Telepresence Technologies, SPIE vol. 2351, pp. 282–292 (1994)
9. What is XR? (2021) https://xr4all.eu/xr/. Access on 28th Aug. 2021
10. Augmented Reality or Mixed Reality? 2021, https://www.augmented-minds.com/en/augmented-reality/what-is-augmented-reality/. Access on 28th Sept. 2021
11. Danielsson, O., Holm, M., Syberfeldt, A.: Augmented reality smart glasses for operators in production: survey of relevant categories for supporting operators. In: Proceedings of 53rd CIRP Conference on Manufacturing Systems. Procedia CIRP, vol. 93, pp. 1298–1303 (2020)
12. Lotsaris, K., Fousekis, N., Kouks, S., Aivaliotis, S., Kausi, N., Michalos, G., Makris, S.: Augmented Reality (AR) based framework for supporting human workers in flexible manufacturing. In: Proceedings of 8th CIRP Global Web Conference—Flexible Mass Customisation. Procedia CIRP, vol. 96, pp. 301–306 (2021)
13. Mourtzis, D., Siatras, V., Zogopoulos, V.: Augmented reality visualization of production scheduling and monitoring. In: Proceedings of 13th CIRP Conference on Intelligent Computation in Manufacturing Engineering, CIRP ICME '19. Procedia CIRP, vol. 88, pp. 151–156 (2020)
14. Vorraber, W., Gasser, J., Webb, H., Neubacher, D., Url, P.: Assessing augmented reality in production: remote-assisted maintenance with HoloLens. In: Proceedings of 13th CIRP Conference on Intelligent Computation in Manufacturing Engineering, CIRP ICME '19. Procedia CIRP, vol. 88, pp. 139–144 (2020)

Fahmi Bellalouna obtained his Ph.D. degree in computer integrated Mechanical Engineering from the Ruhr University Bochum. In 2015 he was appointed Professor in Mechanical Engineering and Mechatronics at Karlsruhe University of Applied Sciences

Robin Langebach obtained his Ph.D. degree in refrigeration and cryogenics from Technical University of Dresden. Since 2019 he has been taken over the newly established professorship for compressor technology at Karlsruhe University of Applied Sciences endowed by THE SCHAUFLER FOUNDATION

Augmented Reality for Operators in Smart Manufacturing Environments: A Case Study Implementation

T. Gramberg, K. Kruger, and J. Niemann

Abstract The fourth industrial revolution, or Industry 4.0, is gaining momentum globally and in various sectors. Promising smart and highly-connected digital and physical systems, the vision of Industry 4.0 relies on the effective integration of some key enabling technologies—amongst which is Augmented Reality (AR). AR is a digital medium that involves virtual objects and computer-generates elements that appear in one or more layers over the real-world environment. The integration of production data with customized AR experiences promises great benefits for operators in the manufacturing sector and beyond. This paper describes the case study implementation of an integrated AR experience for operators in a smart manufacturing environment. The case study uses a Microsoft Hololens II AR device, with Vuforia Studio software, to integrate an operator with the operation and control of a Fischertechnik Learning Factory 4.0 miniature manufacturing system. Within the case study, the implementation demonstrates an integrated AR experience capable of identifying components of the real manufacturing system using marker recognition and shape screening as well as visualizing multi-media data objects. Furthermore, an outlook of visualizing real-time IoT production data and sending operator commands via the AR system to affect the manufacturing system operation is given. In light of the case study results, the paper also discusses the possibilities and challenges for the deployment of AR in real manufacturing environments.

Keywords Augmented reality · Smart manufacturing · Human-systems integration

T. Gramberg · J. Niemann
Department of Mechanical and Process Engineering, Duesseldorf University of Applied Sciences, 40476 Duesseldorf, Germany

K. Kruger (✉)
Department of Mechanical and Mechatronic Engineering, Stellenbosch University, Stellenbosch 7600, South Africa
e-mail: kkruger@sun.ac.za

K. von Leipzig et al. (eds.), *Smart, Sustainable Manufacturing in an Ever-Changing World*, Lecture Notes in Production Engineering,
https://doi.org/10.1007/978-3-031-15602-1_30

1 Introduction

In the past decade, the fourth industrial revolution and the accompanying digital transformation have been strongly focused on in research and by companies across industries. With the presentation of the results of Kagermann's working group in [1] in 2013, the term "Industry 4.0" has found wide acceptance in this context for Europe, and beyond. From the northern American area rather the term "Smart Manufacturing" was coined as the fourth industrial revolution [2, 3], therefore both terms are used synonymously in the context of this work.

The focus of Industry 4.0 is the networking and communication capability of individual entities into a common cyber-physical system. In this context, it is also referred to as Internet-of-things (IoT) capable objects that are interconnected, exchange data and can be controlled in a decentralized manner. The overall goal is to enable an automated, intelligent manufacturing concept that should lead to a paradigm shift with new technology and opportunities to increase efficiency in the production environment [3].

One of these key technologies is Augmented Reality (AR), which is a digital medium that involves virtual objects and computer-generated elements that appear in one or more layers over the real-world environment [4]. Compared to virtual reality (VR), AR allows the user to perceive the environment and merely provides additional superimposed information in the form of digital objects, whereas the use of a VR environment is fully immersive. Milgram's Reality-Virtuality Continuum [5], which is shown in Fig. 1, serves to classify these terms. Mixed reality serves as an umbrella term that covers the spectrum within real and virtual reality.

By overlapping virtual and real reality, users are provided with new opportunities to increase efficiency and transfer knowledge in different environments. In [6], the expansion of human–machine learning opportunities through the combination of AR and learning factories is described, based on the increased flexibility and rapid adaptability of the digital content. In [7], a process model is supplementary presented on how knowledge transfer in the context of AR can act in the form of digital assistance and thus lead to fewer errors in maintenance cases.

In the industrial environment, AR offers various application scenarios that can have a positive impact on all stages of the value chain. For example, product visualizations in marketing and sales achieve competitive advantages [8]; faster development times and planning phases can be realized by evaluating 3D prototypes in a real

Fig. 1 Reality-virtuality continuum (based on [5])

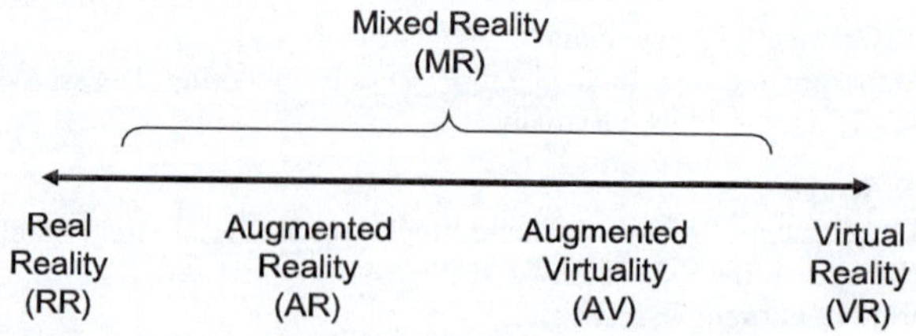

environment [9]; process flows in logistics and manufacturing can be made less error-prone through visual instructions [10]; on-boarding processes can be accelerated by learning from digital objects [11]; and maintenance services can be implemented more efficiently not only with preprogramed digital assistance but also through the possibility of remote assistance and the display of real-time data [12]. In a study by Microsoft and Hypothesis [4] on versatile application examples for AR in the manufacturing sector, especially with the head-mounted device (HMD) of Microsoft's HoloLens, impressive process optimizations and efficiency increases were found. Hincapié et al. [13] also describes that AR is gaining momentum due to advancing hardware and can better reconcile the real and virtual worlds as well as better handle complex environments.

It remains partly unclear, however, how an overall approach and a feasible implementation recommendation for AR applications might look like. This paper describes a practical case study in a smart manufacturing environment. Several functional possibilities are tested across two research sites and then evaluated for a recommended course of action.

The paper concludes with a SWOT analysis summarizing the insights gained from the case study, combined with general strengths and challenges, as well as the current market situation for AR. This is intended to serve as a basis for decision-making and orientation for operators for the implementation of AR applications in real smart manufacturing environments.

2 Case Study Description

For the description of the basic technical function of the case study, the diagram presented by Porter and Heppelmann [14] in Fig. 2 is illustrated. It represents how to merge the real and the digital world using AR.

In this diagram the AR experience is described in six essential steps. First, the object, in this example an industrial robot, is recognized by the AR system and then connects to an external program that simulates a digital replica of the object, called digital twin (DT). In parallel, the physical robot sends sensor data to this digital replica. Once the industrial robot is detected, digital content can be provided to the display device for visualization or instruction and superimposed on the real world.

If the operator interacts with the AR content, this information can be given to the decentralized replica and then on to the physical robot, enabling control [14].

Derived from that overview and various application examples from industry on AR, three focal points were defined for the paper, which were to be practically implemented in the form of a case study:

- Identifying components of the real manufacturing system
- Visualizing multi-media objects
- Integrating and interacting with real-time data from IoT capable devices.

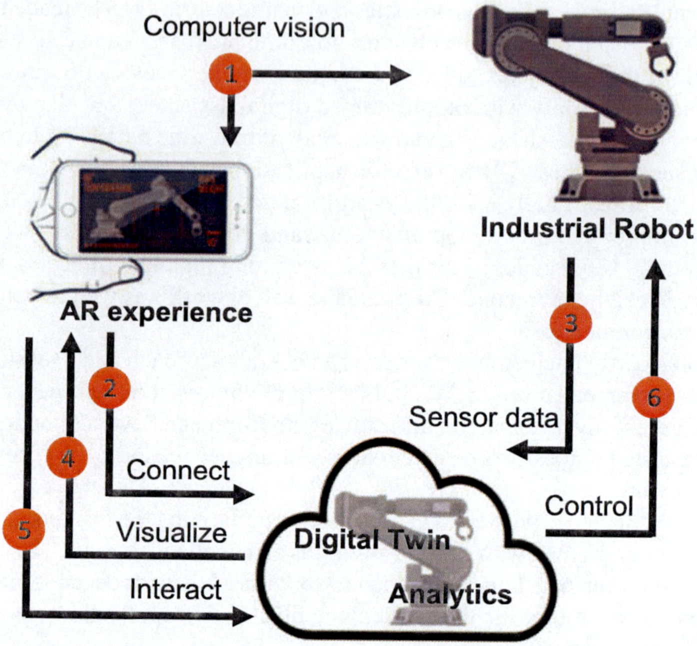

Fig. 2 Merging real and digital worlds (based on [14])

For the implementation, the application is to be considered cross-locational for the two participating research institutes in Stellenbosch (South Africa) and Düsseldorf (Germany). The hardware available to both locations is the HoloLens II as an AR device and the Fischertechnik learning factory for simulating a smart manufacturing environment.

3 Developing An Augmented Reality Experience

The third chapter describes the applied resources for creating the AR content and illustrates the three central focal points that were implemented as part of the case study. In the process, essential functional possibilities for operators in a smart manufacturing environment were tested.

3.1 Applied Resources for Creating the AR Content

To create the AR content within the case study a test version of the PTC's Vuforia Studio software is mainly used, as it is considered a low-code development platform, promises fast implementation results, and is designed for different operating systems and thus also for the HoloLens. To display the created AR content, the corresponding Vuforia View program must be installed on the AR system. Both programs access a common cloud-based experience server [15].

With reference to Fig. 2, the connectivity, visualization and instruction, as well as the interactions are enabled by the two Vuforia programs in the case study. For the collection of sensor data, the IoT platform Thingworx also from the company PTC is needed. To represent the integrated real-time data as AR content, it can usually be dragged and dropped from Thingworx to Vuforia Studio [15]. Since the factory model communicates by MQTT protocol, an MQTT extension kit is also required as a bridge to Thingworx, which can be implemented by the program Faircom Edge [16].

The integration of the different components within the AR environment in this case study is illustrated in Fig. 3.

It involves applying the basic technical function of an AR environment from Fig. 2 to the actual situation of the case study with the resources deployed and demonstrates how the six essential steps are implemented. The integration and control of the real-time data from the sensors of the factory model at steps (3) and (6) are implemented by the IoT platform Thingworx and Faircom Edge as MQTT bridge. The remaining four steps are enabled by the interaction of the two programs Vuforia View and Studio. Through this interaction, some functions can be tested and reproduced digitally in separate instances. However, a full digital replica and simulation of the physical object cannot be represented by a concrete digital copy. For this reason, the case study does not refer to a DT, but only to digital analytics. According to Kruger et al. [17] a DT "ideally creates a highly accurate digital model of the physical system in cyberspace [… and] can accurately replicate and simulate the behavior of the physical system".

3.2 *Recognizing the Factory Model With the AR System*

One of the key requirements in the execution of AR is the recognition of the physical object by the AR device. To enable this, markers or the surface can be scanned, among other things, which then initiate the AR experience [18].

In the context of the case study, scanning the markers proved to be an effective way to initiate the AR application. Figure 4 shows the physical factory model with a unique marker in (a) which can be detected from the AR system and initiate the pre-developed AR content in (b). Further methods were also tested, such as image screening of the Fischertechnik logo and screening of surface contours, either of the entire factory model or of individual components of the factory. However, these approaches were less effective and are therefore not explained further.

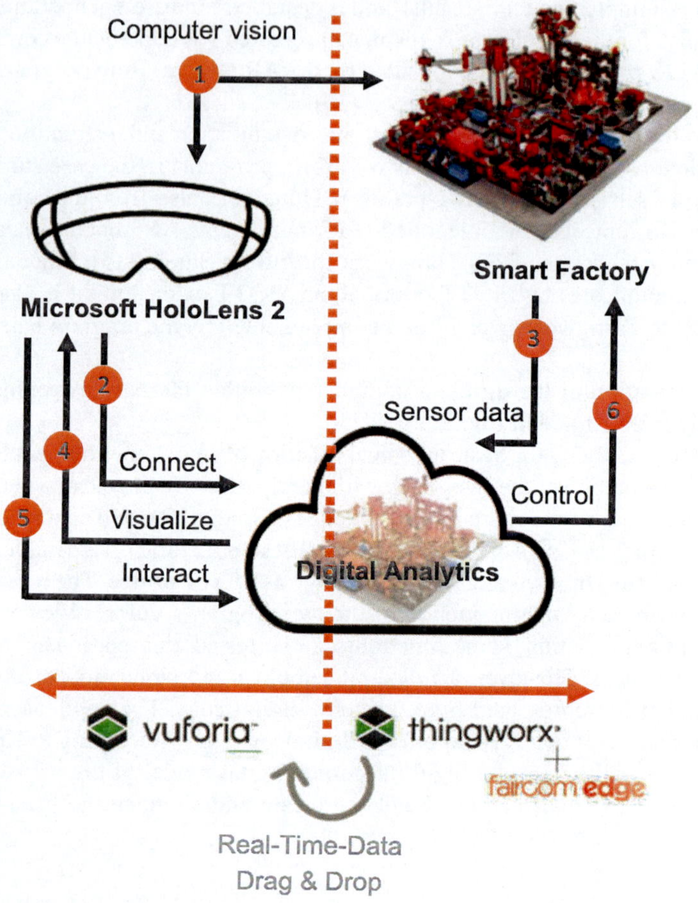

Fig. 3 Architecture of the AR environment from the case study (based on [14])

3.3 Visualizing Digital Multi-media Objects in the Real World

With the visualization of additional information in the field of view, the actual added value is created by simplifying the user's processing. With AR, a mental bridge between the digital and the physical world is built. Information no longer must be read from a separate document or screen and then applied to the situational environment, but is inserted directly as digital objects. This reduces the cognitive distance and cognitive load of the operator described in [19].

As partially shown in Fig. 5, various multi-media digital objects were superimposed on the physical factory model in the case study. In addition to a digital control dashboard to show and hide individual content, a scalable factory component and an assembly video were also inserted. Furthermore, the entire factory model can be

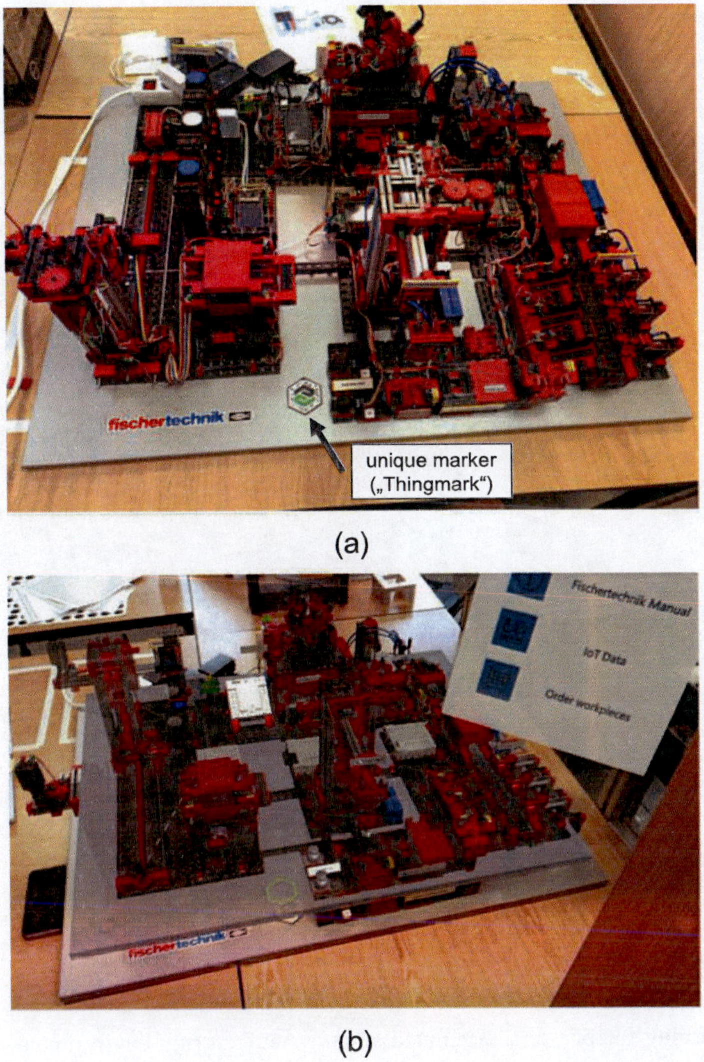

Fig. 4 **a** Marker scanning for initiating the AR experience in (**b**)

over-laid onto the real factory, as already illustrated in the previous Fig. 4b. An operating manual has also been added, which can be superimposed using the dashboard as well as accessed by voice command [20].

The digital objects are intended to simulate assistance for an operator of the factory, who can exemplarily view additional information of the learning factory and thus ensure an easier handling. To reduce the cognitive load, printed documents or a separate screen are no longer required here. Ideally, the disassembly of the component would also be illustrated directly on the 3D model instead of in a separate

Fig. 5 Superimposed digital objects

video. However, due to compatibility issues this is only possible with the PTC's Creo Illustrate software [21]. An additional advancement for the control of multi-media objects, besides gesture and voice control, would be the integration of eye-tracking. In the industrial environment, this offers further additional advantages for the operator, such as simplified use and optimized data representation [22].

3.4 *Real-Time Data Exchange and Interaction*

In the learning factory, as a simulated smart manufacturing environment, the individual controllers communicate via MQTT protocol. This also includes the measurement data collected by the integrated environmental sensor. This data can be addressed individually via defined *topics*, and both receive information and send commands in the form of *payloads* [23]. One goal for the case study was defined as the visualization of the environmental sensor data and the initiation of an ordering process in the factory model. For the implementation, the MQTT data was transferred by the Faircom Edge program [16] as an MQTT extension kit to the IoT platform Thingworx. Here, the data can be stored as so-called *Things* and then transferred to the sister program Vuforia Studio. In Vuforia Studio, this data is then added, for example, as *properties* of a 3D gauge widget, enabling the display of real-time data such as temperature. In addition, the data from Thingworx can also be integrated as *events* that allow interaction with the physical model, sending *payloads* to a subscriber [24]. However, due to the use of the test version of the software, the transfer of the custom created *Things* is not possible, thus only simulated data can be used [25].

An alternative possibility for the integration of MQTT data was also explored—by directly connecting the Fischertechnik factory via the JavaScript interface at Vuforia Studio. For this purpose, a Web socket server was configured, since Vuforia Studio is web browser-based. The desired control could indeed be applied within the preview mode at Vuforia Studio, but not in the operational mode with the HoloLens—this option will thus not be discussed further in this paper.

4 Evaluation of the Augmented Reality Application

Through the processing of the case study, the general research work on the theoretical foundations and the market situation of industrial AR (IAR), it is possible to identify possibilities, but also challenges. In order to evaluate these, the characteristics are collected in the four-field matrix of a SWOT analysis (Fig. 6).

The foundation of the SWOT analysis is based on linking the internal perspective as strengths and weaknesses with the influences resulting from environmental developments in order to serve as a basis for decision-making [26]. By superimposing digital content, IAR enables the rapidly increasing amount of IoT data to be visualized in an operational manner [27]. One of the main strengths of IAR is that it reduces the cognitive load by eliminating the mental translation of information into the environment [19]. Based on the application examples from industry, it is demonstrated that competitive advantages can be achieved with IAR, e.g. in sales [8] or through shorter development times [9]. This is contrasted by a potentially high effort and competence required to create the IAR application [28]. With regard to surveys [28, 29] and during the creation of the case study, immature functional features and dependencies on other programs have also become clear in some cases, so that the technology is considered not ready for wide adoption.

According to market statistics on the predicted market volume [30] and cross-industry investment levels [8], IAR is considered to have high potential. Some application examples already show impressive efficiency increases [4]. However, the industry-wide adoption has not yet occurred and, contrary to expectations, is a long time coming, as described in [31]. In addition, there are concerns about data privacy in IAR applications tracking the employees [32], as well as a potential shortage of skilled workers with the know-how to successfully implement IAR [28].

5 Conclusion

Due to the dynamic and complex nature of smart manufacturing, it must be individually examined which type of visualization can be useful. There is no blanket

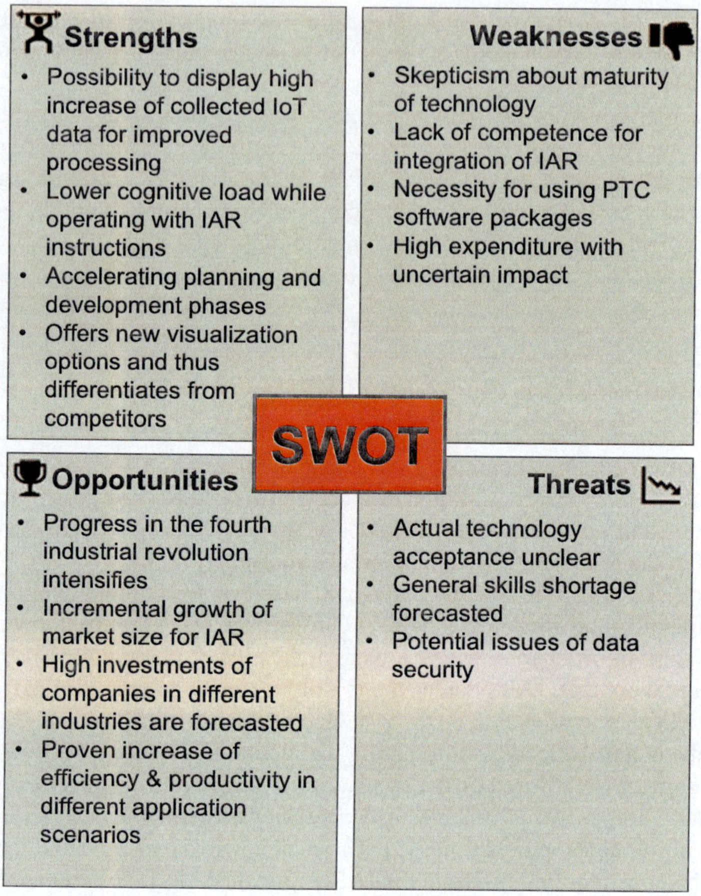

Fig. 6 SWOT analysis for evaluation the AR experience for operators

or "one-system-fits-all-solution" for the integration of IAR [3]. Nevertheless, the findings from the case study and the SWOT analysis can serve as orientation in the strategy process for the use of IAR. For the challenges identified in the context of this analysis, there are also approaches, e.g., from the Boston Consulting Group [27], for managing the risks in implementing IAR-IoT applications.

The case study was limited to using a trial version, so that especially the integration of real-time data and the bidirectional interaction with the physical factory could not be fully tested. However, with the planned use of a full version, these functions should be enabled so that further research can build on and extend the findings of this paper.

In summary, the realized functional possibilities with a reliable connection to the real-time data of the factory model show that AR can play a significant role in Industry 4.0. Even if integration cannot be recommended across the board, enterprises should at least investigate the possibilities of IAR more intensively, as also specified in detail in [19] and [27].

Acknowledgements The financial assistance of the National Research Foundation (NRF) towards this research is hereby acknowledged. Opinions expressed and conclusions arrived at, are those of the authors and are not necessarily to be attributed to the NRF.

References

1. Kagermann, H., Wahlster, W., Helbig, J.: Implementation recommendations for the future project Industry 4.0: securing Germany's future as a production location (own translation). Technical report, acatech—Deutsche Akademie der Technikwissenschaften e.V., Munich (2013)
2. Smart Manufacturing Leadership Coalition, Implementing 21st Century Smart Manufacturing: Workshop Summary Report, Washington D.C. (2011)
3. Thoben, K.-D., Wiesner, S., Wuest, T.: "Industrie 4.0" and smart manufacturing—a review of research issues and application examples. Int. J. Autom. Technol. **11**, 4–19 (2017). https://doi.org/10.20965/ijat.2017.p0004
4. Microsoft & Hypothesis Group, Mixed Reality Intelligence, Manufacturing Edition, https://info.microsoft.com/rs/157-GQE-382/images/Mixed_Reality_Intelligence_Manufacturing_Edition.pdf. Accessed 31.08.2021
5. Milgram, P., et al.: Augmented reality: a class of displays on the reality-virtuality continuum. Telemanipulator Telepresence Technol. **2351**, 282–292 (1994)
6. Juraschek, M., Büth, L., Posselt, G., Herrmann, C.: Mixed reality in learning factories. Procedia Manuf. **23**, 153–158 (2018). https://doi.org/10.1016/j.promfg.2018.04.009
7. Kovacs, K., Ansari, F., Geisert, C., Uhlmann, E., Glawar, R., Sihn, W.: A Process Model for Enhancing Digital Assistance in Knowledge-Based Maintenance (2019). https://doi.org/10.1007/978-3-662-58485-9_10
8. Harvard Business Review Staff.: Augmented reality in the real world. In: Harvard Business Review, vol. 95(6), p. 59 (2017)
9. BMW AG, https://www.press.bmwgroup.com/deutschland/article/detail/T0317125DE/im-muenchner-pilotwerk:-bmw-group-setzt-augmented-reality-bei-prototypen-ein?language=de. Accessed 24.08.2021
10. Glockner, H., et al.: https://www.dhl.com/content/dam/downloads/g0/about_us/logistics_insights/csi_augmented_reality_report_290414.pdf. Accessed 04.08.2021
11. Airbus S.A.S, https://www.airbus.com/newsroom/press-releases/en/2017/11/airbus-develops-worlds-first-mixed-reality-trainer-for-a350-xwb.html. Accessed 04.08.2021
12. TK Elevator GmbH, https://www.tkelevator.com/global-en/newsroom/thyssenkrupp-unveils-latest-technology-to-transform-the-global-elevator-service-industry-microsoft-hololens-for-enhancing-interventions-20928.html. Accessed 31.08.2021.
13. Hincapié, M., Mauricio & Caponio, A., Rios, H., Mendívil, E.: An Introduction to Augmented Reality with Applications in Aeronautical Maintenance (2011)
14. Porter, M. E., Heppelmann, J.E.: How does augmented reality work. In: Harvard Business Review, vol. 95(6), p. 58 (2017)
15. PTC Inc., https://www.ptc.com/de/products/vuforia/vuforia-studio. Accessed 17.08.2021
16. Faircom, https://docs.faircom.com/doc/c-treeEDGE_DevelopmentGuide/ThingWorxAlwaysOnPlug-in.htm. Accessed 21.08.2021
17. Kruger, K., Redelinghuys, A.J.H., Basson, A.H., Cardin, O.: Past and future perspectives on digital twin research at SOHOMA. In: Borangui, T., Trentesaux, D., Leitao, P., Cardin, O., Lamouri, S. (eds.) Service Oriented, Holonic and Multi-Agent Manufacturing Systems for

Industry of the Future: Proceedings of SOHOMA 2020. Studies in Computational Intelligence, vol. 952, pp. 81–98 (2021)
18. Pentenrieder, K.: Augmented Reality based Factory Planning. Dissertation at Technical University Munich, https://mediatum.ub.tum.de/doc/652443/652443.pdf. Accessed 06.08.2021
19. Porter, M.E., Heppelmann, J.E.: Why every organization needs an augmented reality strategy. Harv. Bus. Rev. **95**(6), 46–57 (2017)
20. PTC Inc., http://support.ptc.com/help/vuforia/studio/en/#page/Studio_Help_Center%2FEyewearExperiences.html. Accessed 01.09.2021
21. PTC Inc., https://www.ptc.com/de/products/creo/illustrate. Accessed 01.09.2021
22. Schramm, A., Kruger, K., Niemann, J., Grote, W.: Possibilities, limitations and considerations for eye tracking in industrial environments: experience from a case study. In: 7th International Conference on Competitive Manufacturing, pp. 390–396 (2019)
23. Github.com (eds.) https://github.com/fischertechnik/txt_training_factory/blob/master/TxtSmartFactoryLib/doc/MqttInterface.md. Accessed 21.09.2021
24. PTC Inc., http://support.ptc.com/help/vuforia/studio/en/#page/Studio_Help_Center%2FAppAndDevicePropPanel.html%23. Last accessed 17.08.2021
25. PTC Inc., https://community.ptc.com/t5/TingWorx-Developers/How-to-connect-Thingworx-IOT-platform-Hosted-Trial-30day-and/td-p/688998. Accessed 18.08.2021
26. Bea, F.X., Haas, J.: Strategic Management (Own Translation), 5th edn. UVK Publisher, Stuttgart (2009)
27. Boston Consulting Group, www.bcg.com/de-de/publictions/2020/unleashing-the-power-of-data-with-iot-and-augmented-reality. Accessed 06.08.2021
28. Boston Consulting Group, www.bcg.com/publications/2018/augmented-reality-is-camera-next-big-thing-advertising. Accessed 21.08.2021
29. PwC, https://www.pwc.com/us/en/industrial-products/publications/assets/augmented-virtual-reality-next-manufacturing-pwc.pdf. Accessed 04.05.2021
30. Statista, https://www.statista.com/statistics/897587/world-augmented-reality-market-value/. Accessed 06.08.2021
31. Deloitte, https://www.deloitte-mail.de/custloads/141631293/md_1778806.pdf?sc_src=email_4507339&sc_lid=189197778&sc_uid=fBzbJJ7iQY&sc_llid=279. Accessed 05.08.2021
32. Bosch Rexroth AG, https://www.boschrexroth.com/en/xc/trends-and-topics/directions/interacting-with-contents. Accessed 10.08.2021

Till Gramberg is completing his master's degree in international industrial engineering at Düsseldorf University of Applied Sciences. During a semester abroad, he is writing his master's thesis at Stellenbosch University

Karel Kruger obtained his Ph.D. from Stellenbosch University, where he is now a lecturer at the Department of Mechanical and Mechatronic Engineering and co-leader of the Mechatronics, Automation & Design research group

Jörg Niemann obtained his Ph.D. in manufacturing engineering from the University of Stuttgart. He is currently a Professor at the University of Applied Sciences, Düsseldorf, and head of the FLiX Centre for Life Cycle Excellence

Augmented Reality Combined with Machine Learning to Increase Productivity in Fruit Packing

M. van der Westhuizen, K. H. von Leipzig, and V. Hummel

Abstract This paper studies the benefits of using Augmented Reality and Machine Learning in the agricultural industry for the purpose of fruit classification. During fruit classification colour plays a vital role in determining fruit quality and attractiveness. It is for this reason that technology in agriculture is being adopted that can visually grade produce. Currently, the study of Augmented Reality and Machine learning technologies in the Agricultural sector is limited, specifically cornering productivity improvement resulting from the implementation of these technologies. Therefore, technology which offers collaboration between employees and visual technology, in the form of Augmented Reality using the HoloLens 1, was studied. Augmented Reality requires strong data analytical support because the effectiveness of Augmented Reality is directly proportional to the quality of the information utilised. To ensure accurate data analytics Machine Learning was used. To analyse the use of Augmented Reality and Machine Learning in agriculture these two technologies were used to classify avocados in terms of both fruit grade and size. Machine Learning was implemented using Microsoft Azure which was used to grade the fruit. This was done by providing 1053 photos of avocados to Microsoft Azure from which the Machine Learning algorithm could learn how the fruit was to be graded. To determine the size of the avocado the number of pixels and the distance of the avocado from the HoloLens was used. An Augmented Reality and Machine Learning prototype was implemented, and the time taken to pack an avocado box was taken. It was found that there was a packing speed increase of 29.87% and a decrease in the variation of this speed by 96.2% when the prototype was implemented. Doing a t-test it was quantified that the increase in packing speed was statistically significant. Therefore, it can be concluded that the use of Augmented Reality and Machine Learning can be used to aid employees to improve tasks in the agricultural industry.

M. van der Westhuizen · K. H. von Leipzig (✉)
Department of Industrial Engineering, University of Stellenbosch, Stellenbosch, South Africa
e-mail: kvl@sun.ac.za

V. Hummel
Logistics Management, ESB Business School, Reutlingen University, Reutlingen, Germany
e-mail: vera.hummel@reutlingen-university.de

K. von Leipzig et al. (eds.), *Smart, Sustainable Manufacturing in an Ever-Changing World*, Lecture Notes in Production Engineering,
https://doi.org/10.1007/978-3-031-15602-1_31

Keywords Augmented reality · Machine learning · Avocado fruit · Productivity · Hololens · Microsoft azure

1 Introduction

In agricultural industries, especially those focused on the cultivation of fresh fruit, produce is judged based upon visual parameters [1, 2]. This is because the colour of fruit is an important factor in determining fruit quality [1, 2]. Also, the visual attractiveness of fruit significantly impacts sales [1]. To classify fruit, employees' judge the fruit based on various parameters such as size, colour, blemishes on the fruit skin, etc. [3]. Currently, fruit classification is done manually resulting in there being a significant human component in this part of the industry [3]. This is however changing with the development and adoption of expensive automated classification equipment [4]. This new classification technology is becoming increasingly adopted for use in this industry as there are significant productivity and quality benefits for those with capital to adopt new visual equipment [5]. The downside, however, is that 10% of the employees working in the agricultural sector in South Africa could be lost [6]. It is not only South Africa that will be affected, as it is estimated that 50% of the global workforce is involved in the agricultural sector [7, 8]. These employees are especially vulnerable because they perform routine manual work which could be automated with smart machines with advanced visual technological capabilities [9].

Advances in automation are not the only new technological developments, fortunately. Technologies like Augmented Reality (AR) can assist employees by improving their judgment, decision-making, efficiency, and effectiveness [9]. In doing so the employees are more valuable, and thus there is a greater incentive for organizations to retain labour. The use of AR to aid employees can add a significant amount of value to an industry like the avocado industry for fruit classification. This is because the avocado industry has potential for significant quality improvements with 50% of all avocados harvested being discarded, and of those sold at outlets 40% of avocados have a significant portion of flesh bruised or damaged [10]. These quality issues lead to both significant levels of waste and a reduction in consumer confidence. Also, the avocado industry is still a predominantly manual industry with only a moderate amount of automation being used in practice [11]. However, the industry is evolving and therefore it can be tested whether manual processes can be improved with AR to provide an alternative to automation globally.

2 Literature

Industry 4.0 technologies are advancing quickly, and in doing so provide an opportunity for constructive change in almost all sectors and industries globally [12]. Before implementation however, it is imperative that one understands not only these technologies, but also the implications of using them. Two technologies of particular concern for this project are AR and Machine Learning. Research studying the effects of implanting AR and ML in agriculture are limited [13]. Also the research that has been done on AR specifically in the agricultural sector did not focus on the productivity improvement resulting from the implementation of AR technologies [7, 13–15]. Therefore, this paper explores the use of AR and ML in the agricultural sector with a focus on the productivity improvement that these technologies can provide.

2.1 Augmented Reality

Augmented Reality is a technology used to alter the user's perspective of the real world by layering atop it images, sounds, or vibrations [6]. To ensure that Augmented Reality is not limited to specific technologies, it should conform to the following criteria: combining real and virtual experiences, no delay in interaction, and must be able to process and project 3D imagery [16]. The information presented to the user enables more informed decision making due to the value offering of AR [5].

The three main value offerings, as can be seen in Fig. 1, are real time monitoring and feedback (RTMF), the contextualisation of relevant information, and being a visual aid during task execution [6, 14, 17–20]. AR enables RTMF by taking data from the environment and giving relevant information back to the user in real time [14]. Contextualisation is enabled by providing the user with the right information at the right time and in the right context. In doing so information is framed in a way that is logical, easy to understand, and relevant to the user in their current environment [20]. AR is a visual aid by using real time contextualised information and presenting it to the user to guide or provide aid during task execution [19].

The value offered by AR enables proper and timely decision making using real time data. This results in an increase in the quality and productivity of task execution and improved resource management [18, 20, 21]. During the avocado packing process AR empowers employees who pack avocados by guiding them during the packing process. Avocados are packed based on fruit size and quality as outlined by various regulatory bodies [22, 23]. The benefits of AR will enable avocado packing quality

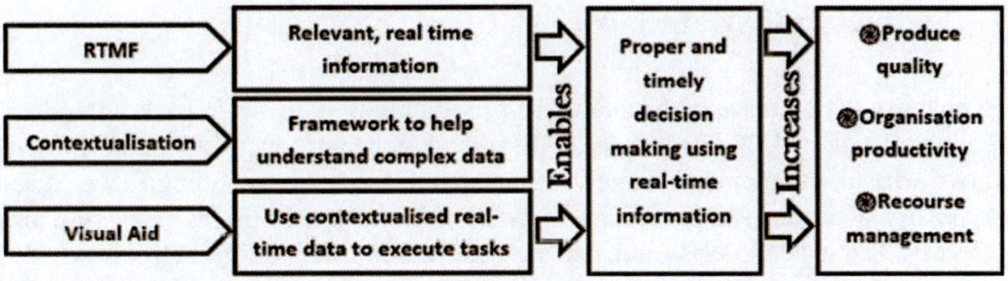

Fig. 1 Augmented reality value offering [6, 14, 17–21]

and productivity to be improved as packers will be aided to make more accurate and faster decisions.

The AR device captures the image where after the image needs to be processed and analysed. The analysis of the image data is important because the value of the insights and conclusions drawn, are directly correlated to the degree that the image data is analysed, and key insights are extracted [24]. Machine Learning is valuable in an AR system because it aims to find patterns or key insights from the data captured [24]. It is for the purpose of analysis, that Machine Learning forms an integral part of vision systems that are being deployed in the real world [25].

2.2 *Machine Learning*

Machine Learning (ML) is focused on creating computer systems that can improve on their own using historical data without changing core processes, as well as the focus on the body of knowledge surrounding the laws that govern learning systems [25, 26]. A critical aspect of ML and why it is so valuable, is the ability of ML algorithms to uncover hidden or unconventional relationships [27]. ML can do this because the algorithms will use all the variables available to determine relationships without being told what those relationships are. It is also this ability of ML to find relationships by itself that it has become easier to implement ML than to manually code algorithms [25]. It is now easier to develop a system where algorithms develop based on training, from inputs received, then to manually develop software, based off anticipated expectations [25]. Thus, these algorithms can identify patterns without preconceived biases as to what patterns exist and how these patterns relate to each other. If correctly trained, pattern recognition is valuable in a fruit classification system because the fruit can be classified objectively. ML will recognise patterns of correctly classified fruit based off calculated patterns allowing for consistent classification of fruit. This, connected with AR, will create a system where ML analytical skills can be conveyed to the user of AR to enable increased efficiency and effectiveness in the packing of fruit.

3 Methods and Materials

To quantify the increase in the quality and productivity of avocado packing a prototype was developed. The prototype, as discussed in the introductory section, consists of two primary components, namely AR and ML. These two technologies were used to classify avocado fruit according to grade, referred to simply as class, and size category. The avocado class and size are determined based off guidelines provided by the South African government [22]. Quality factors determine if the fruit is graded as a class 1, class 2, or unclassified (usually referred to as a class 3). Avocado weights determine the size category which are even numbers from 4 to 32. These numbers represent the number of avocados packed into a standard avocado box. To classify avocados into size and class categories a prototype was developed in three phases. The first phase was the experimental set-up, the second phase was the AR development, and the third phase was the integration of the ML component.

3.1 *Augmented Reality Device*

To select an appropriate pilot device the three most popular and sophisticated AR devices available were compared [28]. The three devices, namely HoloLens 1, HoloLens 2, and Magic Leap 1 were compared using different factors which were used as selection criteria. The factors were chosen based on the capabilities of each device as well as the support provided. The following factors were selected: development support, cost, ergonomics, display area, and resolution.

The three were compared using the analytic hierarchy process (AHP) method, which assigns relative weights to the different factors used as selection criteria [29]. The product of the AHP weights, the scores given, and the reasoning behind the scores can be seen in Table 1. It was determined that the HoloLens 1 would the best device for the prototype. It should however be noted that were cost not a factor or the HoloLens 1 not already available, the HoloLens 2 would have been considered the best device. Therefore, for the purpose of this paper which is to test the benefit of AR with ML the HoloLens 1 was selected.

3.1.1 Marker

On the sorting table a marker was also present. The reason for the marker was that the distance measured by the HoloLens 1 (HoloLens) was not always accurate [30]. This means that there may be noise present, with 1 m being read as 1.2 or 0.8 m. This variability was tested by taking 30 sample measurements, and the results were as follows: 10 measurements had more than a 5% deviation; 4 had more than 10%; and 1 had over 20%. These results indicated that there is significant noise in the prototype system.

By introducing another object of known size, the size of the avocado could be determined relative to this object. The size calculation now being a relative measure, and not an absolute measure, made the calculation more accurate. This was confirmed when a linear regression was used to test the accuracy of predicting the weight of avocados with and without a marker present. With a marker the R-squared (R^2) value improved from 55.22% to 65.27%. The R^2 was determined by comparing the actual weight of the avocados to the estimated weights. The estimated weights were derived from using the avocado's distance and the number pixels that make up the avocado as well as the marker's distance and pixel number if it was present. More information regarding regression and R-squared can be found in the next sub section.

Table 1 A table showing the AHP weighting per factor, score per factor for each device, and the total scores [28, 29, 31–33]

Factor	AHP weight	HoloLens 1		HoloLens 2		HoloLens 3	
		Reason	Score	Reason	Score	Reason	Score
Development support	0.37	Microsoft released well documented tutorials; community documentation is lack luster	0.70	Microsoft released well documented tutorials; fair amount of community documentation	0.80	Tutorials available but not as thorough as those by Microsoft; community documentation is lack luster	0.50
Cost	0.27	HoloLens 1 is available so protoy0e development cost would be R0	1.00	HoloLens 2 currently costs ~ R56 000	0.30	Magic Leap One currently costs ~ R37 000	0.54
Ergonomics	0.15	Heavy device with fragile frame that can cause discomfort when worn for extended periods of time	0.50	Weighty device with knobs to increase comfort that can cause some discomfort when worn for extended periods of time	0.70	Well balance with pads to increase comfort that can cause some discomfort when worn for extended periods of time; has a cable which may be in users way	0.80
Display area (degrees)	0.12	34° display vision versus 60° human vision	0.57	52° display vision versus 60° human vision	0.87	50° display vision versus 60° human vision	0.83
Resolution	0.09	1280 × 720	0.42	2048 × 1080	1	1300 × 1300	0.76
total	1.00	Most appropriate	0.71	Second best	0.68	Third	0.62

3.1.2 System Logic

The HoloLens was programmed in such a way that it would identify all the avocados in its field of vision that were of a predetermined size and class. By packing for a set size and class boxes of avocados, with uniform characteristics, could be packed. To achieve this, the operational facilities of the HoloLens needed to be understood. The way the HoloLens perceives a point in its environment can be seen in Fig. 2. This

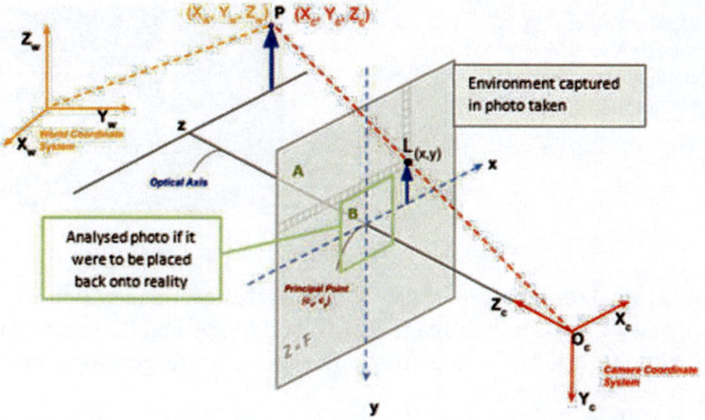

Fig. 2 Figure showing how x, y, and z co-ordinates are utilised using the HoloLens 1 [34]

figure only explains how the HoloLens perceives the surrounding environment. In the figure a point can be represented by point P, as seen from the HoloLens, which is located at point Oc, which has an x, y, and z co-ordinate, representing height, width, and depth parameter as seen from the HoloLens.

The HoloLens simply sees a rough 3D canvas with protrusion, such as point P in Fig. 2, at various depths. The HoloLens cannot know what those protrusions are as it can only know 3D holograms that have been placed by the HoloLens.

Some parts of the rough canvas can be labelled however, so that when a human operator views their environment through the HoloLens, they will have more detail about their surroundings. This will occur when an image of the environment is taken by the HoloLens, analysed and information is projected onto the environment. That image will be analysed using ML software. The analysis identifies the avocados, classes them, and provides x and y co-ordinates of the avocados in the image. The x and y co-ordinates of the avocado in the image were used to get the x and y co-ordinates of the placement of the label of the avocado.

To translate the x and y co-ordinates of the avocado in the image to x and y co-ordinates from the HoloLens perspective, it is first necessary to "remember" which area of the environment was photographed. To "remember" a curser was always present when viewing the environment through the HoloLens. This curser was the middle point of that which was currently being viewed through the HoloLens. The curser was important because its co-ordinates were stored when an image was taken, as can be seen by the critical point in Fig. 2. The image that was taken is represented by the plane A in Fig. 2 and the image A in Fig. 3. After the image is taken, it is analysed using ML, and in doing so became an information rich image.

The information rich image contains the placement of the relevant objects in that image and information about each object. When the information rich image was projected back onto the original image it was scaled differently. Therefore, the information rich image is represented by the plane B in Fig. 2 and the image B in

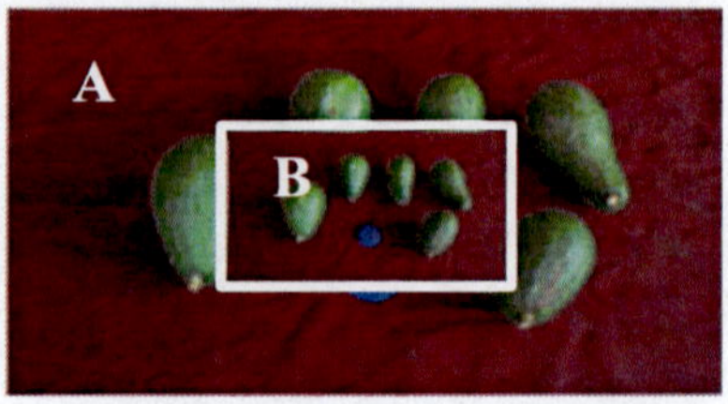

Fig. 3 Figure showing of the original image A and data rich image B, captured using the HoloLens, are scaled

Fig. 3. It was found through experimentation that the information rich image tended to be smaller than the original image. So, although the middle point of the image matches the middle point of the environment, where the image was taken, the rest of the image is scaled differently.

To counter this, the information rich image needed to be scaled so that it overlays the original image perfectly. This was another reason why the marker of a known size was introduced. If the markers size is known, the information rich image of the environment could be scaled so that the image of the marker is the same size in the actual marker. Once this was done both images observed through the HoloLens should also be the same size. In practice the whole image was not scaled but the co-ordinates of labels that needed to be placed on real world objects were to get the correct x and y co-ordinates. Once this was done the labels were then pushed back untill they collided with the canvas seen by the HoloLens. This canvas had protrusions which indicated objects, such as avocados, seen by the HoloLens. The location of the z-coordinate of the label, after being pushed back untill it collided with the avocado, was used to get the z- value of the avocado.

The information rich image also contains the parameters of a bounding box around the avocados that have been identified. Using only the pixels in the bounding box the green pixels were grouped together to determine the number of pixels in the avocado. The way a computer interprets colour is in terms of a RGB scale in the form of a 3-dimensional vector (r, g, b). Each of the variables can have a variable value from 0 to 255 as can be seen in Fig. 4. Each unique configuration of the RGB vector indicates a new colour is represented, with white being (255, 255, 255) and black being (0, 0, 0). It was calculated that in the RGB vector that, if the g component is bigger than the r component, then it would be a pixel in the avocado. 4 samples were taken, as seen in Fig. 4, to show that g needed to be greater than r to be considered a pixel on the avocado. The first sample is off the avocado, so the r component was larger than the g component. The other three samples were taken from pixels on the avocado and each time the g component was found to be larger than the r component. Therefore, all the pixels with a g variable larger than the r variable was counted in the bounding box to get the number of pixels in the avocado. The same was done with the marker except the b variable had to be larger than the r variable.

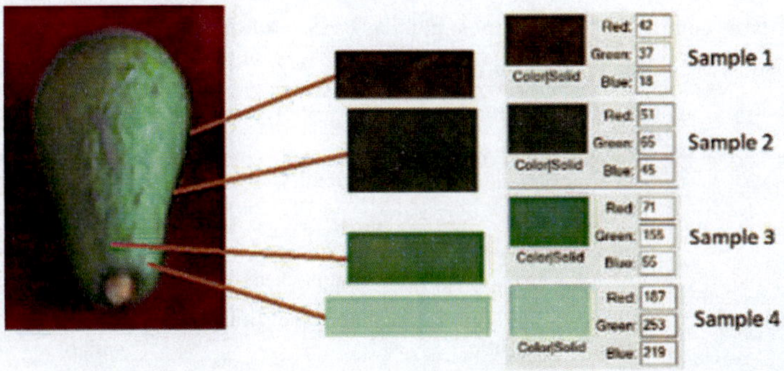

Fig. 4 Figure showing 4-pixel samples taken and the RGB value of each pixel

3.2 *Machine Learning Component*

Computer vision and ML algorithms are already complex, and this complexity is expected to increase in the future [35]. Software platforms provide computer vision and ML software solutions to developers who do not have the knowledge, skills, or time to program the required algorithms [36]. The platforms provided are more than just software as a service (SaaS) because hardware is also used to run and manage some ML computations [36]. Therefore, the services provided are referred to as platform as a service (PaaS).

PaaS can be defined as the provision of technology, hardware and software, as a service for the purpose of enabling developers and independent software vendors to host software or software as a service (SaaS) solution [36]. So, to develop a prototype to test AR with ML without the risk of not having the skills nor time to finish the project PaaS was used. The purpose of this project is to determine the feasibility of using visual technology for fruit. Once this has been established future work may be what machine algorithms will be bested suited for fruit classification.

The three PaaS industry leaders are Amazon, Google, and Microsoft who provide Amazon Web Service Recognition, Google Cloud Vision, and MS Azure Computer Vision (MS Azure) [37]. Of these three services the only one that does not have significant drawbacks is MS Azure, as can be seen in Table 2. Azure also has a significant number of resources available online which ensures that the knowledge that is required to finish the prototype on time is available. The response time and the accuracy of classification of avocados are important both of which will be best accomplished by MS Azure. It is for these reasons that MS Azure was selected platform as the platform of choice.

The computer vision platform offered by Azure is called Custom Vision (CV). The CV platform offered by Azure was used which utilizes ML for object recognition. The disadvantage of using a PaaS is that the ML algorithm that is utilised is unknown so the accuracy of the prototype cannot be pinned to a specific ML algorithm [37]. Avocados were pre-sorted carefully and then images of the pre-sorted avocados were

Table 2 A table comparing the three most popular PaaS solutions [35–39]

Characteristics	PaaS contenders		
	Amazon web service recognition	Google cloud vision	Microsoft azure
User interface	User friendly	User friendly	Complex (resources available)
Cost	Cost effective	Pay-as-you-go (only pay for what is used)	Vague with a free trial and $100 available
Online resources	Little (existing are complex with little useful information)	Available (but limited)	Significant amounts available
HoloLens 1 integration	Complex (almost no resources to assist development)	Possible	Facilitated (Azure + HoloLens Microsoft products so built-in compatibility)
Response Time	Medium	Medium	Fast (comparatively)
Other significant disadvantages	Limited image size (may result in poor computer vision capabilities)	Low "noise" tolerance (may result in poor computer vision capabilities)	None

taken with the HoloLens and sent to Azure's Custom Vision platform. These images were used to train the system so that it could, using ML, classify the avocados into class 1, 2, and 3. 1053 images of avocados were used to train the system.

The prototype developed parameters that needed to be evaluated to determine if the prototype build is sufficient for drawing conclusions. The accuracy of class classification is determined from precision and recall. Precision is the percentage of predicted positives which are true positive [40]. Recall is the percentage of true positives that are predicted as positive [40]. Using 1053 images of avocados the precision and recall scores were determined. The precision score was 85% and the recall score was 81%. The f1 score, the accuracy using both precision and recall, seen in Eq. 1 is calculated to be 83%. The time taken from when an image was taken untill an information rich image was sent back to the HoloLens was approximately 13 s.

$$
\begin{aligned}
f_1 &= 2 \times \frac{Precision \times Recall}{Precision + Recall} \\
&= 2 \times \frac{85\% \times 82\%}{85\% + 82\%} = 83\%
\end{aligned}
\tag{1}
$$

3.3 Prototype System

The 13 s delay by the MS Azure platform as a service means the prototype does not classify avocados on a continual basis, but that there is a discontinuity during each classification event. This is different to how the avocados are currently being packed, as shown by Fig. 5. Currently the employees pack an avocado box untill it is fully packed with an even number of the same class. However, the current HoloLens prototype takes 13 s to classify the avocados and only the avocados in the field of vision when the photo is taken are classified. Therefore, fewer avocados will be classified then necessary to fill a box, and there will be a 13 s delay between iterations. To solve this problem and test the functionality of the prototype the times when an avocado box was being packed were isolated. This was done so that only packing times were examined. In this way the current system and the HoloLens system behave similarly, and a comparison can be made. By stitching the times together to pack different iterations till a box is packed, the time taken using the HoloLens could be determined.

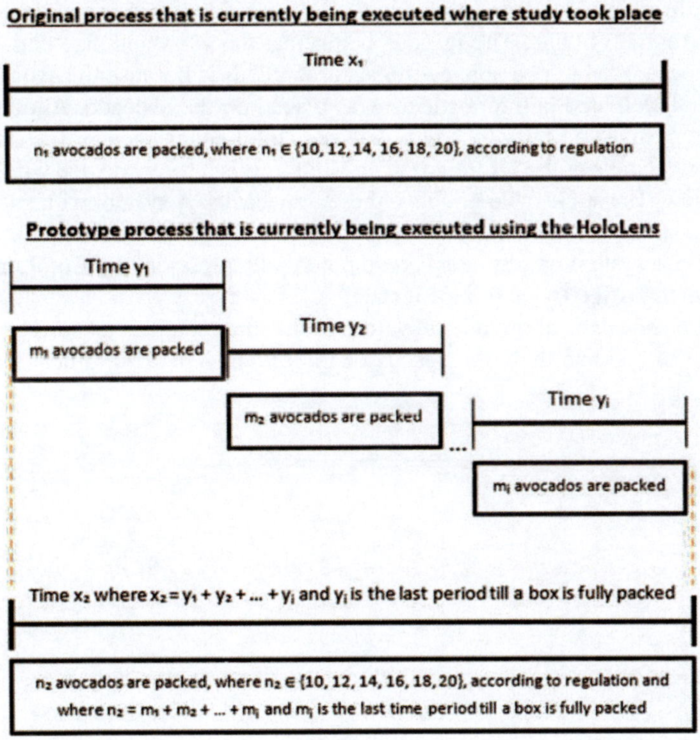

Fig. 5 Figure showing how the avocados packing times were measured both with and without the HoloLens

4 Results and Discussion

The current packing system which does not utilise the HoloLens classifies 73% of avocados according to class, as explained in sub Section "Machine Learning Component" and 59% according to size, as explained in sub Section "System Logic". The size and class was determined using guidelines provided by the South African government as explained in Section "Introduction". With the utilization of the HoloLens these figures improve to 83% and 73% respectively. Showing that, with a more data centric approach, the quality of the of task of packing avocados can be improved. Given that the quality of avocado packing could be improved using the HoloLens it was then necessary to test the productivity using the HoloLens. Using the method discussed in the previous section, the HoloLens was used to pack 30 boxes of avocados. Two other samples were also taken. The first was simply the time taken to pack 30 different boxes without the use of the HoloLens, or other external aids. The second sample was the time taken to pack 30 boxes of avocados that had already been sized and classed correctly, and the avocados that needed to be packed had a sticker placed on the avocado. The results generated can be seen in Fig. 6.

The utilisation of stickers represented a saturation that was more ideal than that of the current HoloLens. The reasons being that the class and size classifications can be expected to both be above 90% accuracy. This is because the avocados were studied and weighted before a sticker was placed on the avocado. Another reason why stickers may currently be better than the HoloLens is because the stickers are not as cumbersome as the HoloLens. The HoloLens is a heavy device with a limited field of view. The stickers do not have these drawbacks. A system with stickers also does not have the problem of having to stitch different pack times together till a box is filled. So, a system with stickers could potentially represent the HoloLens system after it has improved to a more ideal state.

Figure 6 shows the 30 individual values of the three samples taken. The averages were used to compare how the speeds of the different packing times of the three

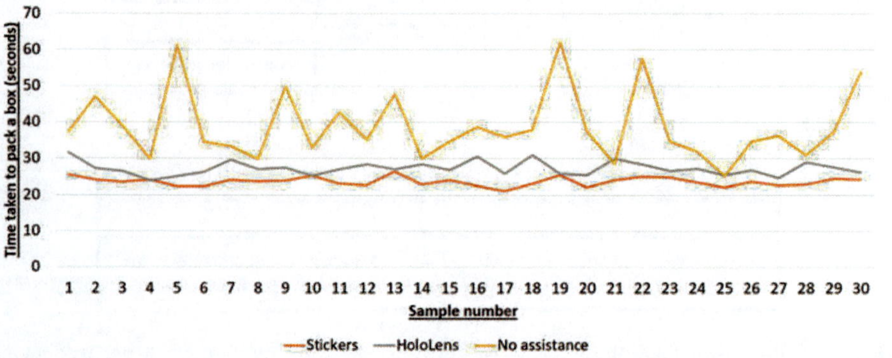

Fig. 6 A graph depicting the time taken by a trained packer to pack 30 boxes of avocados using: no assistance, a HoloLens with the help of Microsoft Azure, or stickers

different samples differ. The average packing times are 23.69 s, 27.31 s, and 38.94 s when stickers, HoloLens, and no assistance are used respectively. Using these averages we can determine that when stickers are used the packing times are 13.26% and 39.17% faster compared to when the HoloLens and no assistance is used respectively. The HoloLens was 29.87% faster than when no assistance is used.

When the avocados were packed with no assistance the times taken to pack a box fluctuated significantly. This can be seen by the fact that 2 data points captured took longer than 60 s while 4 others were below 30 s. The variance was also calculated and found to be 93.97 s. The reason for the large deviation was observed to be due to uncertainties while packing. It was observed that while packing occurred these long packing times were caused when a box needed to be repacked or avocados had to be searched for to be packed. The repacking of boxes occurred when a full box was packed but there were either significant gaps between some of the avocados or a box contained an odd and not an even number of avocados. During repacking some avocados needed to be taken out of the packed box and replaced till both the gaps had been reduced significantly and the box contained an even number of avocados. Long packing times are also caused when the right avocados need to be searched for. The problem, as mentioned in the introduction, is that avocados appear very similar so it is often difficult to judge which avocados should be packed. Thus, when most of the avocados of a similar size have been packed that were near the packer, other avocados further away had to be looked for. But, with avocados looking so similar it is often difficult to find the avocado of the right size.

The two above mentioned difficulties were not present when the HoloLens or stickers were used. This is because the packer is directed towards avocados that are of a similar size. This leads to the elimination of repacking and a significantly reduced search time. The variation was calculated to be 1.41 s^2 and 3.57 s^2 when packing occurred with the aid of stickers and the HoloLens respectively. The variation is significantly less than 93.97 sec^2 which is the variance when avocados were packed with no assistance. With the stickers not having a reduced field of vision, the identification of avocados are more seamless which could explain why the packing time with stickers is faster and has less variation compared to when packing occurred with the HoloLens.

A t-test was constructed to test if there is a statistically significant difference (SSD) between the packing times when the HoloLens and stickers were used to pack avocados. The t-test showed that there is a SSD between the two samples and thus the packing times of when stickers is used compared to when the HoloLens is used is meaningfully faster. However, the packing time difference is only 13.26% which is a small difference when comparing the packing time to when no assistance was used. This means that although the utilisation of stickers is meaningfully faster, the HoloLens is only 13.26% away from an ideal state.

5 Conclusion and Recommendations

In conclusion, the use of AR and ML to classify and pack avocados lead to a significant increase in both packing quality and packing speed, and a decrease in packing speed variation. This means that if the technology is applied it could benefit the user by having more uniformly packed boxes packed at a faster rate. This should increase the sales price of boxes of avocados sold and reduce the labour cost per box. The reduced variation in packing speed could potentially facilitate increased levels of monitoring and control of the supply chain. This could reduce supply chain waste. Ultimately, this prototype thus shows that by applying AR and ML to fruit classification and packing would not only affect the user but potentially the whole industry in a positive way.

References

1. Bandoim, L.: How Self-Driving Tractors And AI Are Changing Agriculture (2019). https://www.forbes.com/sites/lanabandoim/2019/04/27/how-self-driving-tractors-and-ai-are-changing-agriculture/?sh=2eadf8657fa1 (Accessed 06 Dec 2020)
2. Wegren, S.K.: The 'left behind': Smallholders in contemporary Russian agriculture. J. Agrar. Chang., (2018). https://doi.org/10.1111/joac.12279
3. Dautovic, G.: Automation and Job Loss Statistics - 2020 Overview | Fortunly," *Fortunly*, Jun. 30, 2020. https://fortunly.com/statistics/automation-job-loss-statistics#gref (Accessed Dec. 06, 2020)
4. South African National Planning Commission, "our future-make it work executive summary executive summary" (2012)
5. Anthony, R., de Belen, J., Nguyen, H., Filonik, D., Del Favero, D., Bednarz, T.: A systematic review of the current state of collaborative mixed reality technologies: 2013–2018. AIMS Electron. Electr. Eng. **3**(2), 181–223 (2019). https://doi.org/10.3934/electreng.2019.2.181
6. Schueffel, P.: *The Concise FINTECH COMPENDIUM*. Fribourg, School of Management (2017)
7. Abdullah, F.A., Samah, B.A.: Factors impinging farmers' use of agriculture technology. Asian Soc. Sci. **9**(3), 120–124 (2013). https://doi.org/10.5539/ass.v9n3p120
8. Arslan, A.: How old is the average farmer in today's developing world?. IFAD (2019). https://www.ifad.org/en/web/latest/blog/asset/41207683 (Accessed 28 Jun 2020)
9. M. C.-B. F. H. Laurence Morvan., Ovanessoff, A.: A Responsible Future for Immersive Technologies. (2019). Accessed 29 Jul 2021. [Online]. Available: https://www.accenture.com/us-en/insights/technology/responsible-immersive-technologies
10. Schaffer, B., Wolstenholme, B.N., Whiley, A.: The Avocado: Botany, Production and Uses (2013)
11. Fourie, I.: Current Farming Practices in The Avocado Farming Industry (2020)
12. Tantawi, K.H., Sokolov, A., Tantawi, O.: Advances in Industrial Robotics: From Industry 3.0 Automation to Industry 4.0 Collaboration (2019). https://doi.org/10.1109/TIMES-iCON47539.2019.9024658
13. Caria, M., Sara, G., Todde, G., Polese, M., Pazzona, A.: Exploring smart glasses for augmented reality: a valuable and integrative tool in precision livestock farming. Animals **9**(11), 1–16 (2019). https://doi.org/10.3390/ani9110903
14. Phupattanasilp, P., Tong, S.R.: Augmented reality in the integrative internet of things (AR-IoT): application for precision farming. Sustain **11**(9) (2019). https://doi.org/10.3390/su11092658
15. Jhuria, M., Kum, A.: Image processingg for smart farming: detection of disease and fruit grading. 2013 IEEE 2nd Int. Conf. Image Inf. Process. IEEE ICIIP **2013**, 521–526 (2013)

16. Azuma, R.T.: A survey of augmented reality. Presence Teleoperators Virtual Environ. **6**(4), 355–385 (1997). https://doi.org/10.1561/1100000049
17. Pilati, F., Faccio, M., Gamberi, M., Regattieri, A.: Learning manual assembly through real-time motion capture for operator training with augmented reality. Procedia Manuf. **45**, 189–195 (2020). https://doi.org/10.1016/j.promfg.2020.04.093
18. Pivoto, D., Waquil, P.D., Talamini, E., Finocchio, C.P.S., Dalla Corte, V.F., de Vargas Mores, G.: Scientific development of smart farming technologies and their application in Brazil. Inf. Process. Agric. **5**(1), 21–32 (2018). https://doi.org/10.1016/j.inpa.2017.12.002
19. Palmarini, R., Erkoyuncu, J.A., Roy, R., Torabmostaedi, H.: A systematic review of augmented reality applications in maintenance. Robot. Comput. Integr. Manuf. **49**(June 2017), 215–228 (2018). https://doi.org/10.1016/j.rcim.2017.06.002
20. White, G., Cabrera, C., Palade, A., Clarke, S.: Augmented reality in IoT. *Lect. Notes Comput. Sci. (including Subser. Lect. Notes Artif. Intell. Lect. Notes Bioinformatics)*, **11434**(LNCS), 149–160 (2019). https://doi.org/10.1007/978-3-030-17642-6_13
21. Gupta, M., Abdelsalam, M., Khorsandroo, S., Mittal, S.: Security and privacy in smart farming: challenges and opportunities. IEEE Access **8**(March), 34564–34584 (2020). https://doi.org/10.1109/ACCESS.2020.2975142
22. Bickerton, G.B.: Department of agriculture. Gov. Gaz. **1990**(119), 201–202 (1990). https://doi.org/10.1126/science.os-1.8.78
23. Melban, K.: Food Safety Certification Key to Future Marketability of California Avocados., **i**, 2017–2018 (2017)
24. Brown, M.: Smart Farming—Automated and Connected Agriculture (2018). https://www.engineering.com/DesignerEdge/DesignerEdgeArticles/ArticleID/16653/Smart-FarmingAutomated-and-Connected-Agriculture.aspx (Accessed 06 Dec 2020)
25. Jordan, M.I., Mitchell, T.M.: Machine learning: trends, perspectives, and prospects. Science (2015). https://doi.org/10.1126/science.aaa8415
26. Alloghani, M., Al-Jumeily, D., Mustafina, J., Hussain, A., Aljaaf, A.J.: A Systematic Review on Supervised and Unsupervised Machine Learning Algorithms for Data Science (2020)
27. Mullainathan, S., Spiess, J., Lipkovich, I., Dmitrienko, A., D'Agostino, R.B.: Machine learning : an applied econometric approach sendhil mullainathan and jann spiess. Stat. Med., (2017)
28. Software Testing Help (2020) 10 BEST Augmented Reality Glasses (Smart Glasses) In 2021. https://www.softwaretestinghelp.com/best-augmented-reality-glasses/ (Accessed 18 Jan 2021)
29. Viana Vargas, R.: Using the Anaytic Hierarchy Process (AHP) To Select and Prioritize Projects in a Portfolio (2010)
30. Liu, Y., Dong, H., Zhang, L., El Saddik, A.: Technical evaluation of HoloLens for multimedia: a first look. IEEE Multimed., (2018). https://doi.org/10.1109/MMUL.2018.2873473
31. TWNKLS: AR glasses comparison matrix (2018). https://twnkls.com/en/blogs/whitepapers/ar-glasses-comparison-matrix/ (Accessed 18 Jan 2021)
32. Bhise, V.D.: *Ergonomics in the Automotive Design Process* (2016)
33. Statista: Household disposable income in South Africa from 1990 to 2019. *Statista* (2019). https://www.statista.com/statistics/874035/household-disposable-income-in-south-africa/ (Accessed 16 Aug 2021)
34. wuff: python - How to get Z-distance (not depth) from camera to object for every pixel in the frame?—blender stack exchange. *StackExchange* (2020). https://blender.stackexchange.com/questions/180391/how-to-get-z-distance-not-depth-from-camera-to-object-for-every-pixel-in-the-f (Accessed 10 Aug 2021)
35. Jed Ng, "Top 10 Computer Vision APIs: AWS, Microsoft, Google and more | by Jed Ng | Rakuten RapidAPI | Medium," *Medium*, Jun. 14, 2019. https://medium.com/rakuten-rapidapi/top-10-computer-vision-apis-aws-microsoft-google-and-more-fe6fe9a9bc8c (Accessed Apr. 15, 2021)
36. Beimborn, D., Miletzki, T., Wenzel, S.: Platform as a Service (PaaS). WIRTSCHAFTSINFORMATIK (2011). https://doi.org/10.1007/s11576-011-0294-y
37. Cummaudo, A., Vasa, R., Grundy, J., Abdelrazek, M., Cain, A.: Losing Confidence in Quality: Unspoken Evolution of Computer Vision Services (2019). https://doi.org/10.1109/ICSME.2019.00051

38. Microsoft: Cloud Computing Services | Microsoft Azure. https://azure.microsoft.com/en-us/ (Accessed 28 Apr 2021)
39. Foley, M.J.: Microsoft's HoloLens 2: Why it's really all about the cloud | ZDNet. (2019). https://www.zdnet.com/article/microsofts-hololens-2-why-its-really-all-about-the-cloud/ (Accessed 16 Apr 2021)
40. Nicholson, C.: Evaluation metrics for machine learning—accuracy, precision, recall, and F1 defined | pathmind. Path Mind (2020). https://wiki.pathmind.com/accuracy-precision-recall-f1 (Accessed 06 May 2021)

Markus van der Westhuizen holds a B.Eng degree in Industrial Engineering from Stellenbosch University. He is currently pursuing a M.Eng and M.Sc from the Universities of Stellenbosch and Reutlingen respectively.

Konrad von Leipzig obtained both his B.Eng and M.Eng degrees in Industrial Engineering from Stellenbosch University. He later also completed a B.Com degree. He has been a lecturer at the Department of Industrial Engineering at Stellenbosch University since 1987. He forms part of the team undertaking the initialisation and realisation of the Learning Factory in Stellenbosch.

Vera Hummel Prof. Dr.-Ing., Dipl.-Ing., has been a professor at the ESB Business School, Reutlingen University since 2010. Previously she held leading positions at the Fraunhofer IPA in Stuttgart, the working area Industrial Engineering, and the Graduate School of Advanced Manufacturing Engineering at the University of Stuttgart. She is currently leading the expert group in logistics at Reutlingen University.

Enterprise Design

Measuring the Impact of Sustainability of Product-Service-Systems

D. Kretschmar, J. Niemann, C. Deckert, and A. Pisla

Abstract Sustainability will be one of the most important influences on corporate activities. As a result, and due to the technological innovations product-service-systems can be seen as a sustainable opportunity. They enable the effective and efficient use of products and thus the saving of resources as well as their decoupling of value generation. Regarding this, the research question which should be answered in this essay is "How can an impact model be designed that shows the sustainable impact of product-service-systems?" For this purpose, the most important sustainability indicators are identified in the automotive sector, and a business model approach is developed that brings together sustainability aspects and product-service-system requirements. Subsequently, a first attempt is made to transfer sustainability indicators to the business model approach via an impact model.

Keywords Product-service-system · Business model · Corporate sustainability · Sustainable indicators · Impact model

1 Introduction

Sustainable thinking in a business context will play an increasingly important role and will be one of the most important trends in the upcoming years. Besides social

---

D. Kretschmar · J. Niemann (✉) · C. Deckert
Düsseldorf University of Applied Sciences, Düsseldorf, Germany
e-mail: joerg.niemann@hs-duesseldorf.de

D. Kretschmar
e-mail: dominik.kretschmar@hs-duesseldorf.de

C. Deckert
e-mail: carsten.deckert@hs-duesseldorf.de

A. Pisla
Technical University of Cluj-Napoca, Cluj-Napoca, Romania
e-mail: Adrian.Pisla@muri.utcluj.ro

K. von Leipzig et al. (eds.), *Smart, Sustainable Manufacturing in an Ever-Changing World*, Lecture Notes in Production Engineering,
https://doi.org/10.1007/978-3-031-15602-1_32

and political impacts, market logics and customer needs and thus entrepreneurial thinking will be influenced [1, 2].

The need for a sustainable approach to our economy is highlighted in the excessive use of resources. One indicator is the "Earth Overshoot Day". It is computed by dividing the planet's biocapacity by humanity's ecological footprint. The overshoot day 2021 of Germany was the 5th of May and South Africa reached it at the 4th of July [3]. Taking a closer look into the sustainable development goals, both still have problems in responsible consumption and production [4].

The solution to this problems can be found not only in technological innovation, but also in innovative business models (BM) [5, 6]. The trend towards servitization can be seen as an indicator here. In this context, previously purely physical products have become product service bundles [7]. The added service varies purposefully over the different phases of the product life cycle [8]. These are also called product-service-systems (PSS), an aggregation of products and services to deliver value propositions to customers [9]. This PSSs are enabler to reach new markets and customers. This is also shown by the fact that service has played an increasingly important role in the value creation of companies, for years [10]. Not at least due to digitization and the related technologies, the scope of services offered by companies were pushed strongly forward [11, 12].

Regarding this, the design of PSSs have already been extensively studied [13, 14], as have their BM [15]. The same applies to sustainable BM [16, 17], also with regard to Industry 4.0 [18], of which PSSs can be seen as a sub-area. The sustainable impact of PSSs is mentioned many times, but only in few cases all sustainable dimensions are considered integratively and dynamically [19, 20], and both the PSS reference or the BM approach is missing.

This leads to the research question: "How can an impact model be designed in such a way that shows the sustainable impact of product service systems?".

2 Research Approach

With regard to the research question, the research approach is divided into three main blocks:

1. Identification of a suitable framework for corporate sustainability and of the most important indicators.
2. Design of a suitable BM approach with focus on the inclusion of the sustainability indicators and the PSS-requirements
3. Combining them in a first theoretic try via an impact model.

Finally, the results are critically reviewed, and further research directions are presented.

For a common understanding, the most important definitions will be given first, and then a rough overview of network/system theory and system modelling is given.

2.1 *Basic Definition and Holistic View*

Sustainability is based on the concept of three pillars: ecology, economy and social [21]. Many definitions are derived from this. For example, Ladrum says in [22] that in the business environment, all current and future stakeholders act in such a way, that the long-term existence of the company and the associated social, economic, and ecological systems are ensured. So do Sander and Woods who talk about a triple bottom line of people, planet, and profit [23]. Brandstotter even includes the sustainability in his definition of PSS, with the addition of "PSS ties to reach the goals of a sustainable development, which means improved economic, environmental and social aspects" [24]. Thus, PSS can serve as an approach to decouple value creation from increased resource consumption [25, 26], to combining economic growth and sustainability [15, 27].

Business models have a large variety of characteristics [28–30] and thus their visualisation. For the purpose of this paper, the definition, of an abstracted view of all aspects of a company, which are necessary to do business, is sufficient (cf. [29, 31–33]). The most widely used BM visualisation tool is the "Business Model Canvas" [31]. If BMs are considered as a strategic tool with a future perspective, scenarios can also play an important role [34].

Network and system concepts have been discussed many times in relation to BM. For example as a value network and active system [35] or as an value constellation [28]. Sustainability should also be seen as a system, at least due to the three dimensions [21] as well as PSS is considered to be one of them.

2.2 *Methods for System Modeling*

System theory allows us to focus on the dynamics of business organizational systems and enable us to quickly adapt the organization to changing environmental conditions [36]. System Dynamics is a simulation method for analysing complex, nonlinear systems in a business context [37]. The benefits of the System Dynamics approach, to methodologically support the modelling and analysis of business systems, have been widely demonstrated [38–40]. Especially in systems, characterized by dynamic complexity and unpredictability, they are used to model and test strategies for sustainable development and change. For this purpose, dependencies are determined in feedback loops and are visualized by objects. This causal loop diagrams are the most important step in building a system dynamics model [41, 42]. This usually results in complex systems for which software tools are used.

There are many other mathematical methods [43] such as the cross-impact analysis [44], however, the explanation and consideration would go beyond the scope of this work.

3 Corporate Sustainability Indicators

In many countries there is a corporate social responsibility (CSR) directive, for example the European CSR Reporting Directive 2014/95/EU, but there are no regulations or standards on how companies must report about their CSR. Nevertheless, there is an increasing number of frameworks to standardize and make it comparable, due to the increasing importance of corporate sustainability.

3.1 Corporate Sustainability Frameworks

A comparison among the most common frameworks shows that the Global Reporting Initiative (GRI) framework is the most comprehensive one to measure the corporate sustainability [45]. 92% of the world's 250 largest companies report on their sustainability performance and 74% of these use the GRI standards to do so [46]. In addition, 35 countries already apply the GRI standards in their own sustainability policies. Thus, the GRI standard and the indicators they contain are seen as most appropriate.

3.2 Identification of the Most Important Sustainability Indicators

The most important sustainability indicators are determined on the basis of the topic-specific standards GRI 200—economic [48], GRI 300—ecological [49] and GRI 400—social [50] (see lowest level of Fig. 1). The lower-level topics are clustered, and each identified by the last digit of the hundreds (e.g., 201, 202 …). The indicator can be clearly identified by the individual digits after the crossbar (e.g., 201–1, 201–2 …). To ensure a homogeneous group, the selection of companies was based on a single industry. Due to the importance for the German market [51, 52], the 12 largest OEMs and 11 tier one suppliers [53] of the automotive industry are analyzed [54]. To maintain a global perspective, attention has been paid to include the most important global companies, too.

The indicators were determined by a document analysis [55, 56] of the respective CSR report with the following attributes related to the GRI requirements: If an indicator is given or is referred to in another report, this is displayed with "1" (available). A "0.5" is displayed, if the report is only partially complete, not correctly structured or the requirement is not fully met but a valid conclusion is possible. If this is not possible or if no information is provided, it is displayed with "0" (missing) (cf. Also [47]). The average of all reports indicates its relevance. Based on the Net Promoter Score, all indicators reported on more than 70% are considered important (see Table 1).

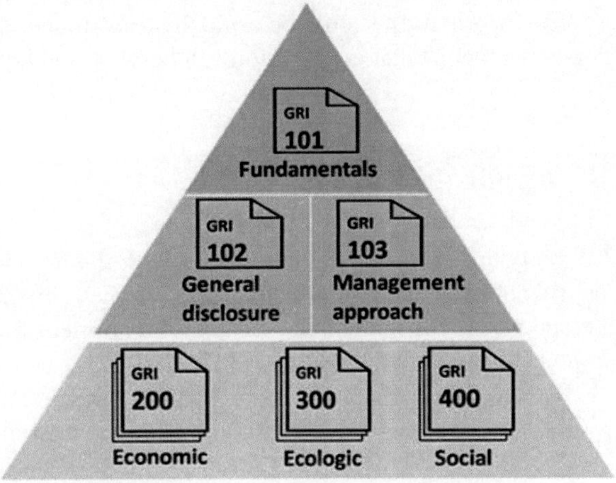

Fig. 1 Structural design of the GRI standard [47]

Table 1 The most reported sustainability indicators in the automotive industry

GRI-NR	GRI Standard	Overall ranking (%)
201–1	Direct economic value generated and distributed	96
201–3	Defined benefit plan obligations and other retirement plans	74
205–2	Communication and training about anti-corruption policies and procedures	73
201–2	Financial implications and other risks and opportunities due to climate change	73
203–1	Infrastructure investments and services supported	72
302–1	Energy consumption within the organization	94
305–1	Direct (Scope 1) GHG emissions	89
305–2	Energy indirect (Scope 2) GHG emissions	89
305–7	Nitrogen oxides (NOX), sulfur oxides (SOX), and other significant air emissions	78
302–4	Reduction of energy consumption	76
306–2	Waste by type and disposal method	74
308–1	New suppliers that were screened using environmental criteria	74
403–1	Occupational health and safety management system	93
416–1	Assessment of the health and safety impacts of product and service categories	86
404–2	Programs for upgrading employee skills and transition assistance programs	83
403–4	Worker participation, consultation, and communication on occupational health and safety	76
403–6	Promotion of worker health	76
403–5	Worker training on occupational health and safety	72

The information should be critically questioned, as the lack of directives in CSR reporting means that information can be provided or omitted for marketing reasons.

4 Applicability to PSS-BM

To apply the identified indicators to a PSS, the Business Model Canvas is chosen as approach for visualization [31]. Based on the criticism in which only the economic perspective is considered, the sustainability dimension was supplemented in the triple layered Business Model Canvas [57].

Next to this, various efforts have been made to include the sustainability dimensions in a BM [16, 17] also with a dynamic approach [19]. On the other hand, there is a PSS-specific [14] BM approach [58] but without sustainability dimensions.

This led to difficulties in using existing BM approaches and made a new design necessary.

4.1 Sustainable Business Model for PSS

The Business Model Canvas [31], the Value Proposition Canvas [59] and the Smart Service Canvas [58] were combined and supplemented by the three sustainability dimensions and a multistakeholder perspective (see Fig. 2).

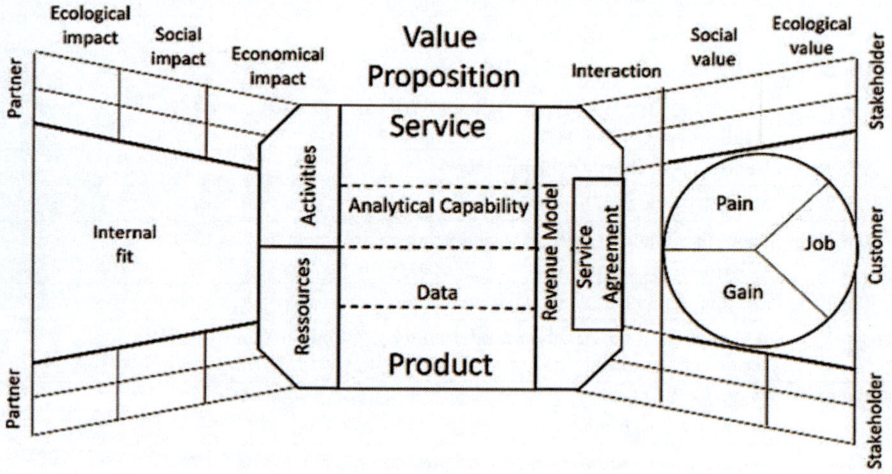

Fig. 2 Sustainable business model for PSS

In the adapted sustainable BM for PSS, equal weight is given to the service and the product as a 'value proposition' and linked through the data and its analysis capabilities.

On the right side of Fig. 2, the stakeholders are integrated through the revenue model and the service agreements. The customer is emphasized here as the most important stakeholder. This fulfilled multistakeholder perspective and sustainability dimensions are also integrated. The economic dimension is covered by the customer perspective. On the left side, the external partners and internal partners are integrated via the activities and resources. The three sustainability dimensions are covered.

4.2 *Demonstration of the Impact Model*

To demonstrate that an impact model can be set up with the identified indicators and the adapted BM.

A generic causal loop diagram for PSS was adopted [60], supplemented with the identified indicators and mapped on the sustainable business model for PSS, with the focus on the value proposition. To maintain clarity, the indicators are limited on those which are reported on more than 90%. The diagram does not claim to be complete and serves only as a first demonstration that the indicators can be integrated into a PSS-BM via an impact model.

The sustainable indicators are highlighted in green. Positive impacts are indicated by green arrows and negative impacts by red arrows (Fig. 3).

The economic indicator "direct economic value generated and distributed" were placed in the revenue model and is positively influenced by the "sales of PSS", "sales of service in PSS", "sales of product in PSS" and "sales of cxisting product".

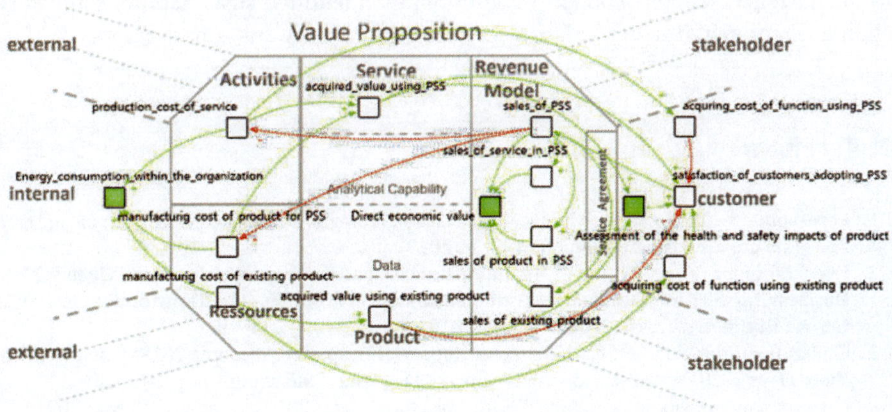

Fig. 3 Impact model (based on [60]) for a sustainable business model for PSS

The environmental indicator "energy consumption within the organization" is positively influenced by the "production cost of service", "manufacturing cost of product for PSS" and the "manufacturing cost of existing product".

The social indicator "assessment of the health and safety impacts of product and service categories" is positively influenced by "acquired value using existing product" and "acquired value using PSS". It has a positive impact on """satisfaction of customer adopting PSS".

5 Conclusions

Due to the increasing importance of sustainability for companies, the measurement of the impact on sustainability is becoming more and more important. Therefore, a system method could be seen as holistic approach. For its development and due to the CSR reporting directive, the most important indicators were identified based on the GRI framework and by reviewing the CSR reports from the automotive industry.

Next to this, a business model could be designed that matches the aspects of sustainability by integrating the ecological, economic, and social dimension as well as the multistakeholder perspective, with the specific requirements of a PSS, through the integration of product, data, analytical capabilities, and service. This has not been practically validated at this point.

The challenge was to combine the sustainability indicators and the sustainable business model for PSS via an impact model and keep it manageable and clear. The impact model shows that this is generally possible, without claiming to be complete and without practical validation at this stage.

Thus, it can be stated that a theoretical answer to the research question is given, but a practical validation of this still must be given by further research. The same applies to the designed business model. In addition, the identified sustainability indicators should be compared with monitoring systems in industry and other sectors.

References

1. Papasabbas, L., Pfuderer, N., Muntschick, V.: Neo-Ökologie. Der wichtigste Megatrend unserer Zeit, 1st edn. Trendstudie Zukunftsinstitut (2019)
2. Lüdeke-Freund, Froese, T. (eds.): 4th International Conference on New Business Models. New Business Models for Sustainable Entrepreneurship, Innovation, and Transformation. ESCP Europe Berlin, Berlin, 01.—03.07.2019 (2019)
3. Global Footprint Network National Footprint and Biocapacity Accounts 2019: Country overshoot Days 2020. When would Erth Overshoot Day land if the world's population lived like… https://www.overshootday.org/about-earth-overshoot-day/ (2019). Accessed 22 Mar 2021
4. Sachs, J., Schmidt-Traub, G., Kroll, C., Lafortune, G., Fuller, G.: Sustainable Development Report 2019. Bertelsmann Stiftung and Sustainable Development Solutions Network, New York (2019)

5. Brooks, S., Wang, X., Sarker, S.: Unpacking green IS: a review of the existing literature and directions for the future. In: vom Brocke, J., Seidel, S., Recker, J. (eds.) Green Business Process Management, pp. 15–37. Springer, Berlin Heidelberg (2012)
6. Boudreau, M.C., Chen, A., Huber, M.: Green IS: Building sustainable business practices. Inf. Syst.: A Global Text **2008**, 1–17 (2008)
7. acatech: Smart Service Welt-Reports 2018, München. https://www.acatech.de/publikation/smart-service-welt-2018-wo-stehen-wir-wohin-gehen-wir/download-pdf?lang=de (2018). Accessed 11 Jan 2021
8. Schweitzer, E.: Lebenszyklusmanagement investiver Produkt-Service Systeme. In: Aurich, J.C. (ed.) Produkt-Service Systeme. Gestaltung und Realisierung, pp. 7–13. Springer, Berlin (2010)
9. Goedkoop, M.J., van Halen, C.J.G., te Riele, H. R. M. und Rommens, P. J. M.: Product Service systems, Ecological and Economic Basics (1999)
10. Dachs, B., Biege, S., Borowiecki, M., Lay, G., Jäger, A., Schartinger, D.: The servitization of European Manufacturing Industries. https://mpra.ub.uni-muenchen.de/38995/2/MPRA_paper_38995.pdf (2012). Accessed 8 Jan 2021
11. Bruhn, M., Hadwich, K.: Service Business Development. Springer Fachmedien Wiesbaden, Wiesbaden (2018)
12. KVD e. V., FIR e. V.: KVD-Service-Studie 2020 (2020)
13. Salwin, M., Kraslawski, A., Lipiak, J.: State-of-the-Art in Product-Service System Design. In: Panuwatwanich, K., Ko, C.-H. (eds.) The 10th International Conference on Engineering, Project, and Production Management. Lecture Notes in Mechanical Engineering, pp. 645–658. Springer Singapore, Singapore (2020)
14. Richter, A., Glaser, P., Kölmel, B., Waidelich, L., Bulander, R.: A Review of Product-service System Design Methodologies. In: Proceedings of the 16th International Joint Conference on e-Business and Telecommunications. 16th International Conference on e-Business, Prague, Czech Republic, 26.07.2019–28.07.2019, pp. 115–126. SCITEPRESS—Science and Technology Publications (2019–2019). https://doi.org/10.5220/0007917201150126
15. Reim, W., Parida, V., Örtqvist, D.: Product–Service Systems (PSS) business models and tactics—a systematic literature review. Journal of Cleaner Production (2015). https://doi.org/10.1016/j.jclepro.2014.07.003
16. Geissdoerfer, M., Vladimirova, D., Evans, S.: Sustainable business model innovation: a review. J. Cleaner Prod., (2018). https://doi.org/10.1016/j.jclepro.2018.06.240
17. Lüdeke Freund, F., Carroux, S., Joyce, A., Massa, L., Breuer, H.: The sustainable business model pattern taxonomy—45 patterns to support sustainability-oriented business model innovation. Sustainable Production and Consumption (2018). https://doi.org/10.1016/j.spc.2018.06.004
18. Khan, I.S., Ahmad, M.O., Majava, J.: Industry 4.0 and sustainable development: a systematic mapping of triple bottom line, circular economy and sustainable business models perspectives. J. Cleaner Prod., (2021). https://doi.org/10.1016/j.jclepro.2021.126655
19. Cosenz, F., Rodrigues, V.P., Rosati, F.: Dynamic business modeling for sustainability: Exploring a system dynamics perspective to develop sustainable business models. Bus. Strat. Env. (2020). https://doi.org/10.1002/bse.2395
20. Lee, S., Geum, Y., Lee, H., Park, Y.: Dynamic and multidimensional measurement of product-service system (PSS) sustainability: a triple bottom line (TBL)-based system dynamics approach. J. Cleaner Prod., (2012). https://doi.org/10.1016/j.jclepro.2012.03.032
21. Systematisierung der Nachhaltigkeitsdimensionen. In: Hauff, M. von (ed.) Nachhaltige Entwicklung, pp. 159–180. De Gruyter Oldenbourg (2014)
22. Landrum, N.E., Edwards, S.: Sustainable business. An executive's primer, 1st edn. Strategic management collection. Business Expert Press, New York, N.Y. (222 East 46th Street, New York, NY 10017) (2009)
23. Sanders, N.R., Wood, J.D.: Foundations of sustainable business. Theory, function, and strategy. Wiley, Hoboken, NJ (2020)
24. Brandstotter, Haberl, Knoth, Kopacek: IT on demand—towards an environmental conscious service system for Vienna (AT). In: 2003 EcoDesign 3rd International Symposium on Environmentally Conscious Design and Inverse Manufacturing. 2003. 3rd International Symposium on

Environmentally Conscious Design and Inverse Manufacturing—EcoDesign'03, Tokyo, Japan, 08.12.2003–11.12.2003, pp. 799–802. IEEE (2003–2003). https://doi.org/10.1109/ECODIM.2003.1322776
25. Vezzoli, C., Kohtala, C., Srinivasan, A., Diehl, J.C.: Product-service system design for sustainability. Green leaf, Sheffield (2014)
26. Kjaer, L.L., Pigosso, D.C.A., Niero, M., Bech, N.M., McAloone, T.C.: Product/service-systems for a circular economy: the route to decoupling economic growth from resource consumption?. J. Indus. Ecol., (2019). https://doi.org/10.1111/jiec.12747
27. Tukker, A.: Eight types of product–service system: eight ways to sustainability? Experiences from SusProNet. Bus. Strat. Env. (2004). https://doi.org/10.1002/bse.414
28. Wirtz, B.W.: Business Model Management. Design—Instrumente—Erfolgsfaktoren von Geschäftsmodellen, 4th edn. Springer Gabler, Wiesbaden (2018)
29. Schallmo, D.: Bestehende Ansätze zu Business Model Innovationen. Springer Fachmedien Wiesbaden, Wiesbaden (2015)
30. Gassmann, O., Frankenberger, K., Csik, M.: Der St. Galler Business Model Navigator. 55 Karten zur Entwicklung von Geschäftsmodellen. Hanser, München (2018)
31. Osterwalder, A., Pigneur, Y.: Business model generation. A handbook for visionaries, game changers, and challengers. Wiley, New York (2013)
32. Gassmann, O., Frankenberger, K., Csik, M.: Geschäftsmodelle entwickeln. 55 innovative Konzepte mit dem St. Galler Business Model Navigator, 2nd edn. Hanser, München (2017)
33. Amit, R., Zott, C.: Value creation in E-business. Strategic Manag. J. (2001). https://doi.org/10.1002/smj.187
34. Fink, A., Siebe, A.: Szenario-Management. Von strategischem Vorausdenken zu zukunftsrobusten Entscheidungen. Campus Verlag, Frankfurt, New York (2016)
35. Allee, V.: Value Networks and Evolving Business Models for the Knowledge Economy. In: Holsapple, C.W. (ed.) Handbook on Knowledge Management, pp. 605–621. Springer, Berlin Heidelberg, Berlin, Heidelberg (2003)
36. Gillenkirch, R.: Systemtheorie. https://wirtschaftslexikon.gabler.de/definition/systemtheorie-49100/version-272341 (2018). Accessed 21 Feb 2021
37. Kortzfleisch, G.V., Krallmann, H.: Industrial Dynamics. Kern, W. (Hrsg.): Handwörterbuch der Produktionswirtschaft, pp. 725–733 (1979)
38. Bianchi, C.: Dynamic Performance Management, vol. 1. Springer International Publishing, Cham (2016)
39. Torres, J.P., Kunc, M., O'Brien, F.: Supporting strategy using system dynamics. Euro. J. Oper. Res., (2017). https://doi.org/10.1016/j.ejor.2017.01.018
40. Moellers, T., Burg, L. von der, Bansemir, B., Pretzl, M., Gassmann, O.: System dynamics for corporate business model innovation. Electron Markets (2019). https://doi.org/10.1007/s12525-019-00329-y
41. Morecroft, J.: Rationality in the analysis of behavioural simulation models. Management Science, 900–916 (1985)
42. Morecroft, J.: A critical review of diagramming tools for conceptualizing feedback system models. Dynamica **8**, 20–29 (1982)
43. Höck, M., Voigt, K.-I.: Operations Management in Theorie und Praxis. Gabler, Wiesbaden (2008)
44. Gordon, T.J., Hayward, H.: Initial experiments with the cross impact matrix method of forecasting. Futures (1968). https://doi.org/10.1016/S0016-3287(68)80003-5
45. Saeed, M.A., Kersten, Logist, W.: Supply chain sustainability performance indicators—a content analysis based on published standards and guidelines. Logistics Res., (2017). https://doi.org/10.23773/2017_12
46. Blasco, J.L., King, A.: KPMG Survey of Corporate Responsibility Reporting 2017. https://home.kpmg/content/dam/kpmg/xx/pdf/2017/10/kpmg-survey-of-corporate-responsibility-reporting-2017.pdf (2017). Accessed 21 Sept 2021
47. Carsten Deckert (ed.): CSR und Logistik. Spannungsfelder Green Logistics und City-Logistik, 2nd edn. Gabler Verlag (2021)

48. GRI 201- 207: Economic. https://www.globalreporting.org/how-to-use-the-gri-standards/gri-standards-english-language/. Accessed 27 Sept 2021
49. GRI 301–308: Ecological Standards. https://www.globalreporting.org/how-to-use-the-gri-standards/gri-standards-english-language/ (2016). Accessed 27 Sept 2021
50. GRI 401—419: Social Standard. https://www.globalreporting.org/how-to-use-the-gri-standards/gri-standards-english-language/ (2016). Accessed 27 Sept 2021
51. Statistisches Bundesamt: Umsatz der Automobilindustrie in Deutschland von 2010 bis 2020 (in Millionen Euro) [Graph]. In Statista. https://www.bmwi.de/Redaktion/DE/Textsammlungen/Branchenfokus/Industrie/branchenfokus-automobilindustrie.html (2021). Accessed 19 Jul 2021
52. VCI: Umsätze der wichtigsten Industriebranchen in Deutschland in den Jahren von 2018 bis 2020 (in Milliarden Euro) [Graph]. In Statista. https://de.statista.com/statistik/daten/studie/241480/umfrage/umsaetze-der-wichtigsten-industriebranchen-in-deutschland/ (2021). Accessed 13 Oct 2021
53. Top-100 Automobilzulieferer 2019. https://www.automobil-industrie.vogel.de/top-100-automobilzulieferer-2019-d-43147/ (2020). Accessed 25 Oct 2021
54. Forbes: Größte Automobilhersteller weltweit nach Gewinn im Jahr 2020 (in Millionen US-Dollar) [Graph]. In Statista. https://de.statista.com/statistik/daten/studie/256312/umfrage/gewinn-der-weltweit-fuehrenden-automobilhersteller/ (2021). Accessed 25 Oct 2021
55. Handbuch für Organisationsuntersuchungen und Personalbedarfsermittlung, Berlin. https://www.orghandbuch.de/OHB/DE/ohb_pdf.pdf?__blob=publicationFile&v=28 (2018). Accessed 18 Jun 2021
56. Scott, J.: A matter of record. Documentary sources in social research. Polity Press, Cambridge (1990)
57. Joyce, A., Paquin, R.L.: The triple layered business model canvas: a tool to design more sustainable business models. J. Cleaner Prod. (2016). https://doi.org/10.1016/j.jclepro.2016.06.067
58. Poeppelbuss, J., Durst, C.: Smart service canvas—a tool for analyzing and designing smart product-service systems. Procedia CIRP (2019). https://doi.org/10.1016/j.procir.2019.04.077
59. Osterwalder, A., Pigneur, Y., Bernarda, G., Smith, A.: Value Proposition Design. Entwickeln Sie Produkte und Services, die Ihre Kunden [wirklich] wollen; beginnen Sie mit…; [die Fortsetzung des Bestsellers Business Model Generation!, 1st edn. Campus-Verlag, Frankfurt am Main (2015)
60. Coreynen, W., van Witteloostuijn, A., Vanderstraeten, J.: Toward servitized research: an integrated approach for sustainable product-service innovation. Sustainability (2021). https://doi.org/10.3390/su13158422

Dominik Kretschmar M.Sc., is a research assistant at the Department of Mechanical and Process Engineering at Düsseldorf University of Applied Sciences with a focus on sustainable business models and product service system

Prof. Dr.-Ing. Dipl. Wirt. Ing. Jörg Niemann is head of the Life Cycle Excellence (FLiX) research center and the FMDauto—Institute for Product Development and Innovation at Düsseldorf University of Applied Sciences.

Prof. Dr.-Ing. Carsten Deckert is Professor of Innovation and Production at Düsseldorf University of Applied Sciences.

Prof. Dr.-Ing. Adrian Pisla is professor at the Design Engineering and Robotics Department at the Technical University of Cluj-Napoca.

People, Process, Master Data, Technology: Data-Centric Engineering of Manufacturing Management Systems

Thomas Gittler, Lasse Plümke, Francesco Silani, Pietro Moro, Lukas Weiss, and Konrad Wegener

Abstract Production enterprises increasingly revert to digitization and data to gain insights and transparency in manufacturing environments. With the growing adoption of the Industrie 4.0 paradigm, the digital transformation of factories is in full swing. IIoT platforms, production management and manufacturing execution systems are installed in more and more production plants, which requires costly acquisition and customization. A significant number of implementation projects fail in intermediate or late stages due to disregard of master data and movement data design. Master data design implies intricate shop floor process reengineering, whereas movement data design relates to a data-centric approach between systems and process engineering. This study proposes a framework for the digital transformation of manufacturing systems from a data-centric engineering perspective. It specifically considers data-related pitfalls and their avoidance for digitization initiatives, and proposes a multi-step approach to align digital transformation projects with their expected benefits during design, development and implementation. It is intended to support, challenge and guide practitioners for the initiation and execution of digital manufacturing excellence.

Keywords Digital transformation · Manufacturing · Data-centric engineering · Digitization · Master data

L. Plümke · F. Silani · P. Moro · K. Wegener
Institute of Machine Tools and Manufacturing (IWF), ETH Zurich, Zurich, Switzerland
e-mail: francesco.silani@ch.abb.com

P. Moro
e-mail: pmoro@student.ethz.ch

K. Wegener
e-mail: wegener@iwf.mavt.ethz.ch

T. Gittler (✉) · L. Weiss
Inspire AG, Zurich, Switzerland
e-mail: gittlert@ethz.ch

L. Weiss
e-mail: weiss@inspire.ethz.ch

K. von Leipzig et al. (eds.), *Smart, Sustainable Manufacturing in an Ever-Changing World*, Lecture Notes in Production Engineering,
https://doi.org/10.1007/978-3-031-15602-1_33

1 Introduction

Rising requirements for quality and growing cost pressure in increasingly complex value-creation networks pose a challenge to production enterprises. As a remedy, small and large companies increasingly adopt digitization principles and technologies, as revealed in a study by McKinsey in 2018 [1]. According to their survey, 9 out of 10 have named digitization as a key topic on their strategy roadmaps. Additionally, the Gartner institute has very recently published a survey conducted among board members and CEOs, of which 80% see data analytics as a top priority for 2021 [2]. In manufacturing settings, data representations of shop floor parts, assets, processes and transactions are expected to deliver information and insights. These digital representations, commonly referred to as digital shadows or digital twins, provide two significant benefits: The digital shadows deliver a detailed description of actual shop floor transactions, allowing for drill-down, aggregation and inferences. They also imply the idea of decoupling investigation from analysis, enabling continuous monitoring and removing ambiguities from measurements. These two aspects, in return, drive efficiency, value-add or process stability, according to Kagermann et al. [3]

To achieve this broad range of targets and enablers, digitization systems need to be implemented. This not only refers to the installation of a system, but to the fundamental redesign of the setup of shop floor transactions. Though digitized processes may not fundamentally change the way value is created in the factory in view of manufacturing processes, they offer inherently different advantages over what is possible with paper-based transactioning and execution of shop floor orders, as well as the associated information collection and analysis. Hence, assets, parts, processes and shop floor transactions all require scrutiny and a detailed concept for their transition to the digital factory.

Independent of their size or sector, manufacturing companies experience difficulties with the setup and implementation of connected data streams in their value chain operations [4, 5]. As a result, a large majority tends to pursue digitization ambitions with pilot or lighthouse project. These project types, aimed at gaining experience or proving a concept, often do not achieve implementation maturity. McKinsey has conducted a survey in 2018, showing that the vast majority of organizations conduct pilot projects, whereas the rollout and integration of fully-fledged solutions lags far behind [1]. Pilot projects have a crucial drawback: They tend to result in isolated applications or satellite systems, given their nature of pursuing low-hanging fruit and requiring little depth of integration. This makes their working mechanism fundamentally opposing to that of the digital transformation, resulting in only small or non-existing benefits with regard to quality and delivery [6]. Overall, pilot projects and their slim success probability significantly reduce the likelihood of a vast digital transformation in industry. Additionally, large transformation projects come with large Capital Expenditure (CAPEX), block a significant amount of internal resources, demand skills and know-how often foreign to firms, and require many departments to involve and fundamentally change some of their day-to-day working mechanisms.

The interdisciplinarity and the concatenation of processes across multiple departments create a strong failure risk potential. Therefore, the trade-off between pilot projects and large transformation initiatives is mostly decided in favour of pilot projects, despite their low benefits and impact. Consequently, there is a need for operational frameworks, combining experience with formalization to support the large-scale digital transformation of value chains in manufacturing enterprises.

This manuscript describes a data-centric engineering approach to the implementation and integration of manufacturing management systems. It focuses specifically on transactional data generated on the shop floor allowing to realize transparency and data analytics benefits, as well as master data used to describe, detail and manage shop floor operations. It emphasizes the interplay between stakeholders, the actual processes, their meta-descriptions and finally the technology employed to unite these factors. This study represents the essence of experiences and outcomes of multiple digital transformation projects in Switzerland-based global manufacturing enterprises.

2 Theoretical Background

To build up the required background knowledge the following chapter gives an overview of the digital transformation of manufacturing companies and the data related to shop floor processes.

2.1 *Manufacturing Management Systems*

Manufacturing management support through digital systems has known its advent in the 80s of the previous century, where the idea and concept of Computer-integrated Manufacturing (CIM) emerged. Whereas CIM was mostly based on the idea of dedicated and isolated solutions to fulfil a sole, precisely-defined task, Manufacturing Management Systems (MMS) follow the principle of integration and consolidation of functions. This concept is deep-seated in the directive 5600 of the Association of German Engineers (VDI) [7]. It describes a specific type of MMS, the so-called Manufacturing Execution System (MES). The definition and creation of MES had begun in the context of CIM, and was predominantly coined by Kletti [8]. Today, MMS come in many different shapes and sizes. In general, they fulfil a set of shop floor functions or transactions, including data acquisition, information management, material management, quality management and performance analysis. Concurrent expressions and definitions are Manufacturing Operations Management (MOM), Manufacturing Intelligence Platform (MIP), dedicated Industrial Internet of Things (IIoT) platforms, and cloud-based Manufacturing Integration and Intelligence (MII) platforms or applications.

The incentives for the digital transformation of factories lie in the increased transparency of operations, allowing to precisely pinpoint deficiencies and efficiently manage shop floor processes. MMS promise to deliver accordingly. However, it would be a severe underestimation to limit their introduction to merely technological aspects. Besides their technological complexity, interfaces with adjacent systems and entanglements with shop floor processes call for an intricate reengineering of organizational processes to comply with digitization boundary conditions.

2.2 Data and Data Types

Data is a crucial factor for digitalization and digital transformation activities. The increasingly harnessed potentials of statistical modelling fundamentally change the types and structures of recorded data in manufacturing systems. But not only data records of events and transactions, but also process descriptions and definitions rely on data objects, containers and relations. These form the basics and backbone of digital transformation projects, resulting in a significant and direct impact on cost and benefits. Changes in data format, type, frequency and volume specifications are comparable to hardware changes in product development: Change efforts and cost increase exponentially with development progress.

Following the definition by Legner and Otto [9], one may distinguish between two main types of data in manufacturing companies: master data and movement data. Master data is the most important basic data in manufacturing companies on different business levels of a company. They are used to describe assets, entities, objects, process plans, instructions and recipes. These mainly static data sets remain unchanged by operative processes. While the digitization of a company is strongly connected with the build-up of appropriate master data, every digitalized manufacturing process creates traces in the form of movement data. In some literature, this is also referred to as transaction data. In contrast to master data, movement data is created and changed dynamically during business processes.

2.3 Shop Floor Process Reengineering

In the context of enterprise transformations, the term business process reengineering is an important factor. The digital transformation of manufacturing environments makes no exception, in that shop floor processes are subject to changes or improvements [10]. In the context of this manuscript, one should distinguish the notions of *manufacturing processes* and *shop floor processes* (Fig. 1). Manufacturing processes describe the actual mechanical machining or assembly processes that cause a value-add to a specific part, e.g. milling, grinding or gluing. Shop floor processes are the transactions to execute a fabrication order, such as factory logistics, information

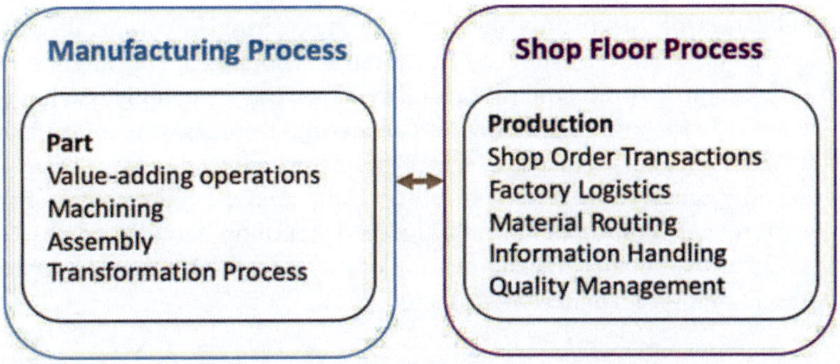

Fig. 1 Distinction of manufacturing process and shop floor process scopes

handling, material routing, or quality management. In simplified terms, the distinction between manufacturing and shop floor processes is the comparison of parts vs. production.

There are different levels of process reengineering, depending on the depth of changes or transformations. The very basic level comprises a thorough process definition, allowing to pinpoint where deficiencies and waste prohibit productivity and efficiency. It in return allows to take counteractive measures, which are commonly referred to as *value capture* [11]. The next higher level describes any change, add-on or improvement within the process boundaries, and is termed as *value extension.* They are at the intersection of function-specific and core-business projects, and often leverage best-practices from similar or adjacent industries. On the next step-up, refinements and enhancements go beyond the process scope, and deal with *value-chain redesign* [12]. They usually pursue the expansion of existing systems or programs to effectively optimize the primary value chain across multiple processes and departments. The highest form of shop floor process reengineering is the *digital value-chain transformation.* It may not only optimize the value chain, but redefine value and its creation.

2.4 Data Centric Engineering

Mark Girolami of the Alan Turing Institute recently described data-centric engineering as the transformation of [...] manufacturing [...] disciplines by the growing intersection of engineering science and data science [13]. Despite data-centric engineering not being a fundamentally new concept, it nonetheless merits to obtain the appropriate, scientific research frame to foster emergences in new mathematics, computational methods and models for engineering companies across all sectors.

Data-centric engineering describes the convergence and combination of fundamental physical laws with data-derived, empirical evidence. While the combination of physical laws and data from experiments is as established as science itself, the formalization of specifically data-centric engineering is comparably new. Its importance stems from the tendency of systems becoming increasingly complex. A figurative example is the process capability index C_{pk} commonly used in quality management. Measuring, calculating and accessing a process capability is a rear-view mirror on past occurrences, in which trends or tendencies are interpolated to derive potential scenarios for the near future. Data-centric engineering in manufacturing environments considers not only the feedback loop between process and its data representation, but also the engineering of a system that delivers an appropriate data representation of its underlying process.

3 Methodology

The methodology presented hereafter can be considered a sequel to the data-driven requirements collection process introduced by Lorenz et al. in 2018 [14] and subsequently extended by Gittler et al. [15]. It highlights the pitfalls and high-effort areas which are all too often overlooked and puts them into the overall context of digital manufacturing excellence.

3.1 Implementation Framework

In MMS integration projects, a major issue is that the acquired data does not contain the information required to deliver the promised benefits. To consider the aspect of data accordingly, there are two central factors: the *movement data* harvested from processes, assets, parts and people, and the *master data* which describes how processes, assets, parts and people should interact in order to deliver the desired material and data output. Due to the fact that the master data are closely related to the shop floor process, the following framework suggests extending the paradigm to *people, process, master data, technology.* This means a specific set of human resources is required on different enterprise levels, a well-defined and controlled process, and master data reflecting a precise blueprint of an ideal process. Only if all those preconditions are fulfilled the last step, the technological implementation can be performed. Taking a critical look at shop floor organization, it is straightforward to detect a coexistence and concurrence between (Fig. 2):

(1) What is defined as or would be the optimal organization ("As-Should factory"),
(2) what is described in master data and process organizations ("As-Defined factory"), and
(3) what actually happens on the shop floor ("As-Is factory").

The divergence between these three hidden factories is compensated by the human factor in the loop. In the As-Is factory, human shop floor workers are able to cope

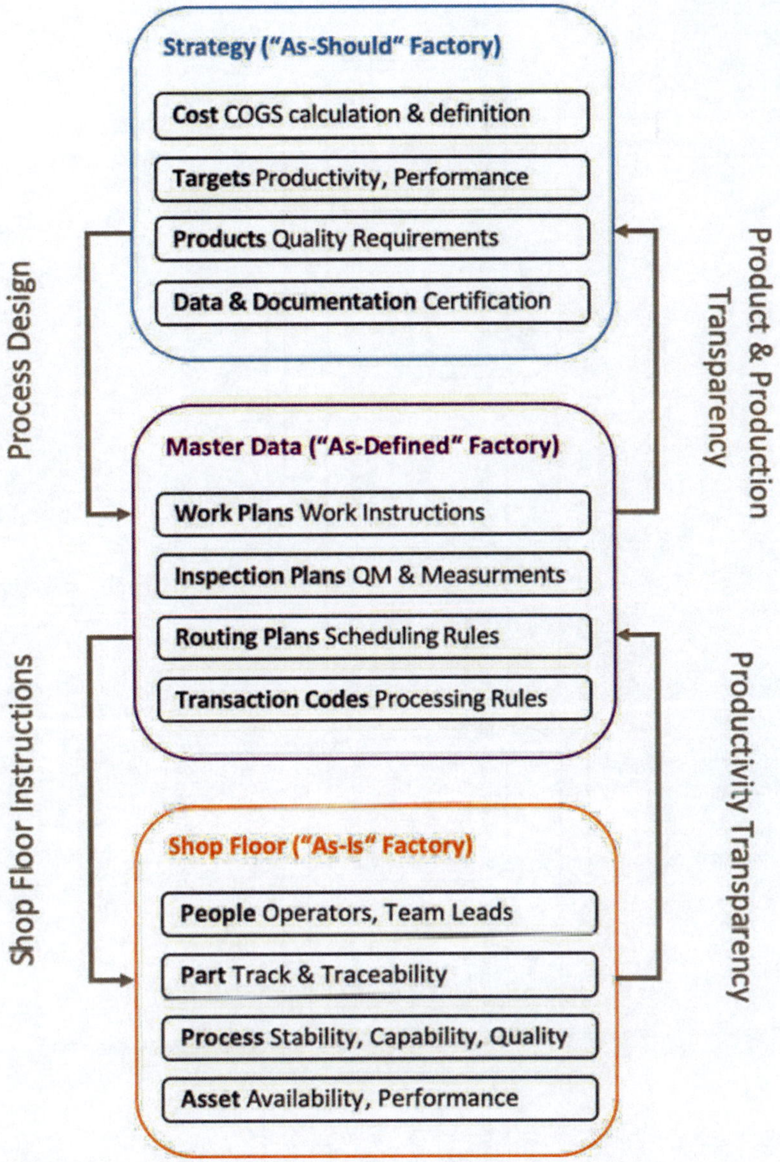

Fig. 2 A major challenge in the digital transformation: divergence of the three "hidden" factories: as-should, as-defined and as-is

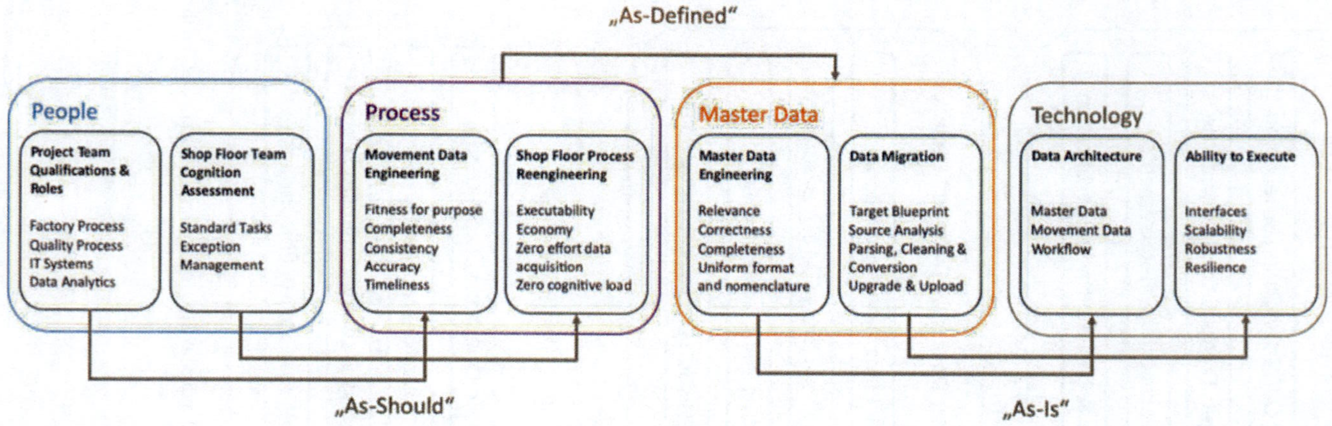

Fig. 3 Data-centric end-to-end engineering framework for the digital transformation in manufacturing

even with loosely or undefined problems or issues. The description of process, routing and quality plans not always includes all necessary instructions, and may not even follow an ideal process. Technological restraints, boundary conditions imposed by systems, or generally accepted fuzziness in master and meta data create a divergence between the optimal and the defined process. In a conventional factory organization, these divergences explain an important part of the difference between productivity, quality, time and cost targets and the corresponding, actual factory output.

Digitally transformed value chains merge these three factories, as process definitions need to be followed closely. With a focus shift to precise workflow definitions, the departments creating master data need clear instructions or a blueprint on the ideal process—and in return, need to provide unambiguous process descriptions for their shop floor colleagues. These preconceptions pave the way for a digital representation of all transactions and processes of the shop floor, directly translating into digital productivity, production and product transparency. Based on this fundamental prework, the appropriate technological platform, solution or system can be selected, adapted and implemented.

3.1.1 People

People are the starting point, not only from a project settings point of view. The digital transformation of traditional manufacturing companies leads to the challenge that projects require qualified personnel not only in manufacturing-related, but also IT and data topics. Besides the traditional areas of factory processes and quality assurance, specific understanding of manufacturing IT systems, data processing and advanced analytics is necessary. Starting with the project team, it needs to be assured that the right skills, capabilities and enterprise-specific knowledge are present. This also includes external expertise that may provide support in any project stage. While a project team can include or even consist of external partners with deep knowledge of the general principles of manufacturing and digitalization, it is beneficial to have the advisory and support of experienced shop floor specialists within the company. This is fundamental to ensure that the hidden "As-Is" factory is effectively and efficiently discovered and assessed. With a project team covering the appropriate know-how, set up and prepared for the endeavor, the focus should be directed towards the people in the factory. An assessment of the required cognition of shop floor workers for standard and exception processes needs to be conducted. The shop floor colleagues' use of their cognition allows for interpretation of fuzzy, imprecise or missing instructions, and perform operations that are in the best interest of the organization—even if they do not comply with working plans. Erroneous or missing instructions are countered by communication between workers, team leaders and experts. For the digital transformation, this is a challenge of large magnitude, and needs to be assessed and tackled at first hand. It is important to distinguish between a standard process, free of deviations or ambiguity, in which the process and role of the worker is clearly defined and followed—and the exception management, in which unforeseen circumstances (e.g. asset or resource breakdown, quality issues) require the intervention of the operator. The standard process needs to be free of cognitive loads, unless specifically formalized for value-adding tasks.

3.1.2 Process

With the "As-Should" factory defined, two parallel tasks in view of processes ensue. The *movement data engineering* relates to the task definition of which data is needed to digitally represent the relevant aspects of today's and future shop floor transactions, whereas the *shop floor process reengineering* is a direct consequence of the cognition assessment, allowing to reduce cognitive loads in standard processes. Their tasks and duties need to be integrated into process reengineering such that the required data is created without putting any creating effort or additional load for the operators.

The motivation for movement data engineering stems from the issue that data created and acquired in shop floor processes is either cumbersome to aggregate and analyze, and also often not fully understood. Lagging and lacking synchronization due to incoherent or incompatible data formats, descriptions and types are the natural enemy of straightforward analytics, and in return hinder the thorough extraction of informational content. Potential decision taking processes need to be supported by the available data and the corresponding information flow. Often, as much data as possible is collected without any regard for defined data formats, descriptions and types. This leads to unsynchronized data streams, containing ambiguous labels and meta data, leading to misinterpretation and fuzziness, and additionally creating considerable efforts for the processing and usage of the data. To prevent this from happen the project team engineers the movement data. This means implementing tailor made data streams that fit to the desired purpose with regard to their amount, consistency, accuracy and time-granularity, ultimately delivering the desired benefits for managing, analyzing and optimizing the factory. Only after a successful movement data engineering and process reengineering the full potential of data-driven manufacturing processes can be realized.

The justification and importance of data-centric engineering lie in the increasing complexity of processes and systems. Their design, engineering and operations requires knowledge beyond the scope of current theory. Moreover, process variances are introduced by intrinsic factors (process flaws, people and erroneous data), as well as extrinsic factors (stochastic environmental influences, variance and fuzziness of constants and parameters). Data-centric engineering delivers the means to understand where fundamental laws of physics and factory organization remain valid—and where data analytics advance the understanding of manufacturing management.

3.1.3 Master Data

The reengineering of shop floor processes as well as the definition of type, structure and quality of movement data as a process output imply an important follow-up process: The structuring of type, layout and quality of master data as the process input, in order to obtain the appropriate product and data output. This is an inverse approach to the actual shop floor transaction, where master data is the actual process input. Based on the designed and desired movement data, master data must be thoroughly conceived to produce data that is useful for further analyses and adoptions to optimize the manufacturing process. High quality master data, the "As-Defined" factory, represents the optimal process, the "As-Should" Factory, as closely as possible and serves as the foundation for the shop floor operations, the "As-Is" factory.

This reconciliation and matching of the three hidden factories can be achieved using the following master data management framework: As a first step the initial state including different data sources and systems is analyzed and then extended with expert knowledge about the desired process to assess the general integration of the "As-Should" factory in master data. This may refer to the business targets prioritization (e.g. quality, cost, on-time delivery), the level of detail, the automation rate targets, or the product and process variance, especially with configurable and customizable material and assemblies. Here, manufacturing companies face the challenge that master data, as the input of the processes, must fulfill not only the previously outlined quality criteria for master data. Data must be present in finer granularity: "Tighten housing bolts" becomes "tighten all 16 casing M8×22 bolts of strength class 10.8 with 32 Nm in crosswise order, starting at 12 o'clock". This ensures that the precise task definition is not left to the responsibility and effort of the operator, who has to engage in information procurement, interpretation and validation.

As a next step, the data structures for the "As-Defined" factory are constructed to represent and respect the shop floor reality, the "As-Is" factory, and at the same time conventions and standards for the data that fulfil all required criteria are set. Lastly, all existing data are transferred into the new data structure and missing information due to the required higher level of detail is added. This level of detail, correctness and accuracy enables the straightforward integration of tools and systems that support the operator to make the standard process effortless. Besides this digital representation of operations and workflows on the instruction side, the processing and analysis is both a challenge and a value driver. The emergence of advanced analytics and modeling solutions and tools provide support and ease-of-use for the recognition of reoccurring patterns and activities on the shop floor. These patterns can be as simple as the comparison of planned times and actual times for operations and manipulations. However, they can also go as far as parameter inference of high-variance products and processes, and their impact on quality or productivity. This of course requires the acquisition of process data in conjunction with the appropriate meta data. However, it is the very foundation for feedback loops that continuously compare planned and performed processes, and suggest adjustments for optimization purposes. These statistical tools, paired with advanced analytics and artificial intelligence, have started and will continue to provide cognitive abilities of manufacturing systems, further minimizing variance and divergence of processes. The hereby growing requirements for master data and their administration, also globally referred to as Master Data Management (MDM), especially in the context of optimizing and future self-optimizing systems need to be integrated before moving any further.

3.1.4 Technology

Despite digital technology being often considered to be at the center of the digital transformation, it is the enabling layer that should be targeted as the potentially last step of digital transformation projects. Whereas the internal resources necessary to perform the analysis of people, process and master data are often undervalued by organizations, the necessary CAPEX for digital technologies is often times overvalued. Expenditures related to the acquisition of software, systems and customization are an important factor in the return and amortization assessment. However, their magnitude in comparison with the required internal resources should be put into perspective: As a generalized rule of thumb, software and systems, consulting and training, as well as customization each represent a third of the overall CAPEX allocation for a profound digital transformation of a manufacturing system. The often overlooked, but nonetheless important factor, is the necessary internal resources, which amount roughly to the equivalent of the allocated CAPEX. To ensure an effective project with an efficient resource usage, it is paramount to start internally with people, process and master data, before effectively engaging in cash-out for systems and solution procurement.

For the functional assessment of the appropriate technology or system, besides the industry-specific boundary conditions, two main factors follow the proposed framework: the accommodation of the defined *data architecture* stemming from the master data step, and the system's *ability to execute* on the defined processes and the resulting data structure. The data architecture needs to accommodate or integrate relevant data transfer from Enterprise Resource Planning (ERP), asset and resource management, quality management and other satellite systems. The notion of interface not only refers to the purely technological data exchange, but also concerns logical structures and working mechanisms, allowing systems to effectively collaborate. This logical reconciliation especially includes the ingestion, storage and accessibility of inserted master data and recorded movement data. The replication of logic in multiple systems needs to be avoided imperatively, as it drives complexity, cost and inefficiency—and it may ultimately lead to ineffectiveness. The ability to execute is the shop floor-facing factors that allow for the interfacing and integration of assets, and the intuitive, efficient information handling to and from workplaces in the factory. Moreover, solutions need to be able to grow in multiple dimensions: Scale-up of systems, future integration of emerging technologies, and robustness. Considering technology as the very last step is paramount. Many requirements heavily impacting the appropriate technology choice arise from the process and master data step. For instance, if detail levels of working or routing plans need a finer granularity, work order steps to be planned, transferred and processed can rapidly multiply. Movement data design can also imply the connectivity of assets which currently operate standalone or in digitally isolated systems. The variance of interfaces, data types and format, transmission protocols and functionalities or features can only be constituted conclusively with the closing of the shop floor process, movement data and master data engineering, as well as the definition of the data migration target format.

4 Discussion

The proposed framework was developed for and applied in a digital transformation project of manufacturing activities of a Swiss machinery enterprise with multiple plants in global locations. The project, initially planned and designed to be a pure systems project, gradually reached a setup in which the adjacent notions of people, process and master data became pivotal. In the absence of an established best-practice, and considering the substantial shortcomings of the *people, process, technology* paradigm, an appropriate structure and framework was designed. The framework was applied in projects of different length and magnitude with more than five Swiss manufacturing companies. The people have proven to be the initial and crucial point. Equal focus and effort on digitalization and shop floor processes in view of team composition and skills represented is critical. With fluctuation and demand-dependent resource allocation, it has become clear that understaffing the digital staff members results in shortcomings of the data-centric engineering aspects, whereas expert shortage for shop floor processes causes a lack of problem–solution-fit. With flaws or gaps in the people or process parts of the framework, the following tasks suffer from disorientation, ambiguity and ultimately purposelessness. In view of the master data aspect, the result from people and process heavily impacts the resulting quality, and especially the necessary rework. Comparable to production engineering, rework effort and cost increase tenfold with every major progress step. Especially in the high-variance environment with configurable material, the migration logic of legacy satellite systems to uniform, standardized process plans in the ERP system have highlighted this consideration. On the technology side, a lack of structure to accommodate all details of the engineered master data, as well as a failure to deliver the desired movement data can challenge the feasibility of an entire project, even when close to completion. For instance, a missing synchronization between work order steps and an automated production cell recipe demands complex logic to be integrated between MMS and the production cell control system. Hence, it is of utmost importance to have clear, concise and correct results of process and master data before specifying solution and system factors.

5 Conclusions

The framework proposed in this study is intended to serve as a contribution to practitioners in the field, supporting the formalization and synthetization of digital transformation endeavors in manufacturing enterprises. It has its roots and its target audience in high variance, high value-add production systems, given its development and application context with Swiss manufacturers as a high-cost production ground. With high-variance manufacturers, the volume and divergence of master data is naturally higher than for manufacturers of uniform and standardized goods.

In the future, the separate steps presented will be detailed and tailored to specific industries, in order to provide further and thorough guidance, especially to those inexperienced with either the process or IT side of manufacturing systems. A strong influencing factor also stems from the degree of maturity exhibited by the respective firm, especially in view of master data quality and the current systems' ability to execute. For the technology aspect, it is therefore a crucial follow-up question how to decide on systems integration versus systems replacement in the presence of legacy solutions and platforms.

References

1. McKinsey.: Digital Manufacturing—Escaping Pilot Purgatory. Digital McKinsey (2018)
2. Gartner Institute.: Top Priorities for IT: Leadership Vision for 2021 (2021)
3. Kagermann, Wahlster, W., Helbig, J.: Recommendations for Implementing the Strategic Initiative INDUSTRIE 4.0. Final report of the Industrie 4.0 WG/April (2013)
4. Bracht, U., Masurat, T.: The digital factory between vision and reality. Comput. Ind. **56**(4), 325–333 (2005)
5. Radziwon, A., Bilberg, A., Bogers, M., Madsen, E.S.: The smart factory: Exploring adaptive and flexible manufacturing solutions. Procedia Eng. **69** (2014)
6. Wegener, K., Kunz, A., Bochmann, L., Bänziger, T.: Industrie 4.0 für den Maschinen- und Anlagenbau, pp. 121–145 (2016)
7. Verein Deutscher Ingenieure.: VDI Guideline Manufacturing Execution Systems (MES) 5600 (2007)
8. Kletti, J.: Manufacturing Execution System—MES. Springer Science & Business Media (2005)
9. Legner, C., Otto, B.: Stammdatenmanagement. In: WISU—Das Wirtschaftsstudium, 236th edn., pp. 562–568. Lange Verlag, Düsseldorf (2007)
10. Doumeingts, M.: Business Process and shop floor Re-Engineering. In: Camarinha-Matos, L. M. (ed.) Re-engineering for Sustainable Industrial Production. Springer, Dordrecht (1997)
11. Lepak, D.P., Smith, K.G., Taylor, M.S., Lepak, D.P., Smith, Keng, et al.: Value creation and value capture: a multilevel perspective, **32/1**, 180–194 (2017)
12. Arora, S., Kumar, S.: Reengineering: a focus on enterprise integration. Interfaces **30**(5), 54–71 (2000)
13. Girolami, M.: Introducing data-centric engineering. Data-Centric Eng. **1** (2020)
14. Lorenz, R., Lorentzen, K., Stricker, N., Lanza, G.: Applying user stories for a customer-driven industry 4.0 transformation. In 16th IFAC Symposium on Information Control Problems in Manufacturing INCOM (2018)
15. Gittler, T., Lorenz, R., Weiss, L., Wegener, K.: Analytical user story clustering supporting requirements and synthesizing efforts for the digital transformation. In: International Conference on Competitive Manufacturing (COMA), vol. 1, pp. 39–46 (2019)

Thomas Gittler works on manufacturing-related Industry 4.0 topics in his doctoral studies at the Institute for Machine Tools and Manufacturing (IWF) of the Swiss Federal Institute of Technology Zurich (ETHZ). He also accompanies the digital transformation of manufacturers.

Lasse Plümke is Mechanical Engineering graduate from ETHZ and an MBA candidate at Collège des Ingénieurs. In his studies he focused on manufacturing-related data in the digitalization process. Currently, he works in the digital transformation of production.

Francesco Silani is a project manager in the area of digitization at ABB Turbocharging in the OPEX & Investments Team, specialized in MES solutions. He holds a Master of Science degree in Mechanical Engineering from the Swiss Federal Institute of Technology Zurich (ETHZ).

Pietro Moro is a M.Sc. candidate in Mechanical Engineering at the Swiss Federal Institute of Technology Zurich (ETHZ). His research is focus on machine tools and manufacturing processes, and the digitalization of production with MES.

Lukas Weiss is a senior researcher with an industrial background leading a group focused on conceptual issues, innovative materials, energy efficiency and digitalization of machine tools at a research institute related to ETH Zürich.

Konrad Wegener is Head of the Institute for Machine Tools and Manufacturing (IWF) at the Swiss Federal Institute of Technology Zurich (ETHZ). His research is focused on machine tools, manufacturing processes and process chains, as well as the development, evaluation and optimization of production machinery.

Derivation of Requirements for the Formation of Collective Target Systems for Technology-Based Cooperation Between Manufacturing Corporates and Startups

G. Schuh and B. Studerus

Abstract Disruptive innovations are putting incumbent manufacturing companies under increasing pressure to defend their competitive position in globalized markets. To withstand this pressure, they can form cooperation agreements with startups aiming for the creation of technical innovations and, thus, ensuring access to technologies and growth. Due to organizational differences and an insufficient explication of cooperation objectives, these cooperation pose a major challenge for both partners. In this paper, the authors discuss the status-quo in the formation of entrepreneurial target systems and, thereby, systematically derive corporate as well as startup-specific cooperation deficits. Based on the analysis of creating individual target systems, a first attempt is taken to elaborate requirements for the development of a model to form collective target system for the cooperation between corporates and startups. Subsequently, model characteristics for the derivation of joint targets and requirements are discussed to enable a comparison between corporates and startups. The development of a concept for a requirements comparison based on a collective cooperation target system supports corporates and startups to ensure the fulfilling of the competitive advantage.

Keywords Cooperation · Collaboration · Requirements · Target · Corporate · Startup · Target system

1 Introduction

Emerging competitors, changing customer needs and disruptive technological developments have radically changed the market environment of established companies

G. Schuh (✉)
Laboratory for Machine Tools and Production Engineering (WZL), RWTH Aachen University, Aachen, Germany
e-mail: g.schuh@wzl.rwth-aachen.de

G. Schuh · B. Studerus
Fraunhofer Institute for Production Technology IPT, Aachen, Germany
e-mail: bastian.studerus@ipt.fraunhofer.de

K. von Leipzig et al. (eds.), *Smart, Sustainable Manufacturing in an Ever-Changing World*, Lecture Notes in Production Engineering,
https://doi.org/10.1007/978-3-031-15602-1_34

in recent years [1–3]. In order to survive in highly dynamic markets and intensified competitive environments in the long term, established manufacturing companies are therefore faced with the challenge of further expanding their existing business while at the same time opening up new areas of business [4–6]. Cost reductions and efficiency improvements alone are no longer enough to ensure the company's market competitiveness [2, 3, 7]. Instead, novel strategies and approaches for adapting to market changes need to be developed to ensure the company's ability to innovate, compete and thrive in the future [8–10]. However, rigid organizational structures [11, 12], long decision-making paths, little freedom to make a decision and a poor culture of error impede the renewal of the company [13, 14]. Thus, external stimuli are required, which is why manufacturing corporates are increasingly often entering into cooperative ventures with young companies, so-called startups, in order to cope with the speed and radical nature of innovations and to drive organizational change [2, 6, 15–18].

Established companies expect to gain access to innovations, new business models and ways of working through cooperation and thereby improve their competitiveness [1, 16, 19–21]. Startups with product innovations in the manufacturing environment, on the other hand, are faced with the challenge of surviving in technologically demanding environments due to a lack of available resources [2, 22, 23]. In addition to the general scarcity of resources, lack of market access, industry experience and industry knowledge are often causes of failure for young ventures [20, 24]. In order to ensure access to valuable tangible and intangible resources and, thus, to meet the growth ambitions, startups are increasingly often entering into cooperation with corporates [6, 12]. Established corporates not only provide startups with a sufficient capital base and access to resources, but also with valuable sector and industry knowledge [2, 25].

In practice, however, cooperation represents a major challenge for both companies due to the differences between the partners [25–27]. As a result, a majority of all cooperation between corporates and startups fail [28, 29]. Reasons for this are the unclear, yet insufficient target definition of the individual cooperation partners, the derivation of corresponding requirements as well as the significant efforts to coordinate the individual targets within the organizations [6, 10, 21, 29, 30]. Often incompatibilities or contradictions between the individual targets only become apparent in the course of the cooperation and then lead to its failure [31, 32]. Transparent requirements and target-oriented cooperation are seen as a necessary prerequisite for the success of cooperation [10].

Therefore, this paper aims to investigate the previously motivated issue of deriving requirements to form target systems for technology-based cooperation of corporates and startups, subsequently addressing the following research question:

What are the requirements for the design of cooperation between corporates and startups based on the formation of a collective target system?

Within the scope of the present paper, the requirements for the formation of collective target systems and the present required sub-models shall be derived. The starting point of this work is the presentation of the practical background for the contextual classification in Section "Introduction". Section "Research Methodology" presents the chosen research methodology, which was used to derive the identified deficits and requirements. Section "Theoretical Background" focuses on the theoretical background aiming to outline the theoretical basics and contextual understanding. This is followed by a literature review in Section "Literature Review", which gives a brief examination of existing approaches for startup cooperation as well as for the development of target systems. The reviewed approaches are investigated to identify deficits in the creation of collective target systems and to derive textual requirements for the intended model. Based on this, the requirements are derived, subsequently discussed, and presented in Section "Results". The final section presents an overall conclusion and gives an outlook on future research.

2 Research Methodology

As the paper at hand focuses on an issue with practical significance, the research methodology follows the process of applied science by Ulrich [33]. Figure 1 shows that the structural approach aims for the development of models, shaping the future through description, explanation as well as configuration of selected areas of application [33]. Hence, the methodology by Ulrich comprises seven successive steps, five of which are covered in this paper, as depicted in see Fig. 1 [33]. Testing and verification, steps 6 and 7, in the industrial practice are by means of research focus not part of this paper.

By framing the practical problem from past and recent industrial practice, the first chapter of this paper covers step A. Step B is covered through Sections "Research Methodology" and "Theoretical Background", wherein fundamental theories and

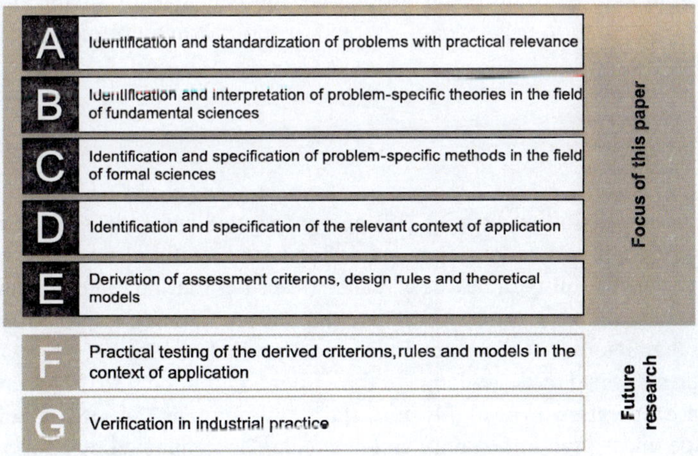

Fig. 1 Process of applied science [33]

hypotheses are presented to form the theoretical background. Building on this, Section "Literature Review" investigates and reviews existing approaches for the creation of target systems and an according derivation of deficits for the creation of target systems—process step C. Subsequently, requirements based on the deficits are being identified and discussed. The discussion of the requirements in Section "Results" enables a derivation of recommended courses of action for the development of a practical applicable solution, covering the process steps D and E.

This paper concludes with a brief summary and the discussion of directions for future research as an outline for steps F and G in Section "Conclusion and Future Research".

3 Theoretical Background

Based on the clarified problem and objective, this section shall give an overview of the theoretical principles and relevant definitions. First of all, the fundamentals of the organization theory of startups and corporates are presented, and their context-specific characteristics are described (see Section "Fundamentals of Organization Theory"). Following, the fundamentals of entrepreneurial target systems will be described in Section "Target System". Finally, the fundamentals of targets as well as the description of a strategic cooperation target space concludes the theoretical basis in Section "Characteristics of Entrepreneurial Targets".

3.1 Fundamentals of Organization Theory

The cooperation of two enterprises follows the fundamentals of organization theory. Thus, both cooperation partners are characterized to form a uniform understanding within this work:

3.1.1 Startup

Startups are per definition [10, 34, 35] newly founded enterprises. With regards to their brevity of existence, startups are defined by specific characteristics. Firstly, startups rely on a liability of newness, which describes the short age of existence and the pronounced degree of novelty [36]. Second, startups are characterized by an owner centricity that determines the potential of success to a high degree, whilst relying on the specific skills and the knowledge of the owner and founder [36, 37]. Furthermore, startups are operating in a highly uncertain environment [24, 38]. Decisions are being made under high uncertainty and hence, are accompanied by a high risk [39]. Because of its small size, a startup is forced to operate with limited financial and human resources [36, 37]. The liability of size makes it unlikely for a startup to

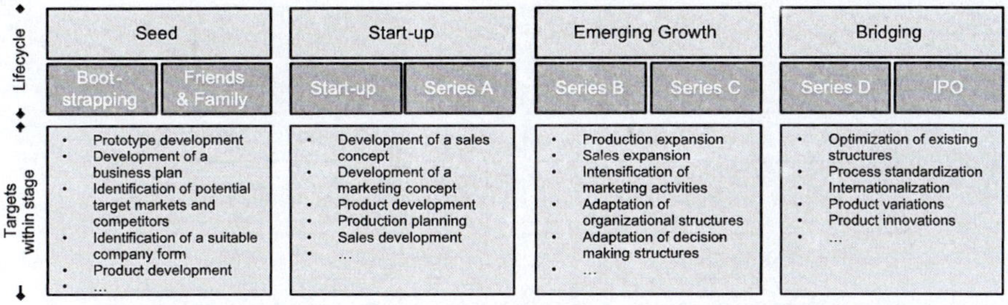

Fig. 2 Cooperation targets of startups in respective life cycle phases [35, 41–45]

survive a longer period of economically less successful business, but the small size also allows short communication channels and fast response times [40]. Yet another characteristic of a startup is the liability of adolescence, which describes outstanding scaling as well as growth ambitions [36, 46]. Due to the strong growth in sales and employees, structures and processes must be continuously adapted to the state of development and therefore pose a great challenge [46].

Startups are described in their development using life cycle models. In addition to differentiating between established corporations and startups, these life cycle models serve to identify strategic goals, tasks and risks that ensure the survival and growth of the company under consideration in various development phases [36]. In general, a distinction is made between four development phases, each of which is followed by financing phases. For the purposes of this paper, the development phases of a startup are divided into Seed, Start-up, Emerging Growth and Bridging [35, 41–43, 45]. The phases and the associated targets for cooperation are displayed in Fig. 2. Within these phases, startups pursue distinct targets to meet the overarching ambition of establishing a corporation. The Seed phase is characterized by technical and economic feasibility targets [41], whereas a market development strategy is elaborated in the Start-up phase [45]. During the Emerging Growth phase, the startup pursues an aggressive expansion strategy with the aim of establishing itself in the market and expanding its market position [45, 47]. Lastly, the Bridging phase is characterized by sustainable growth [43] and is intended to prepare the transition from startup to corporate. Derived from the main targets, tangible sub-targets are pursued in each phase, which are also pursued by startups within cooperation (see Fig. 2).

3.1.2 Corporate

The term "corporate" initially relates to large companies [48, 49]. In the literature, there are several definitions for corporates described as public companies [50–52]. Besides the description based on the economic key figures, corporates in the context of this paper are defined as enterprises operating on renowned markets with established products and well-known business models [17, 53, 54]. Thereby, corporates are

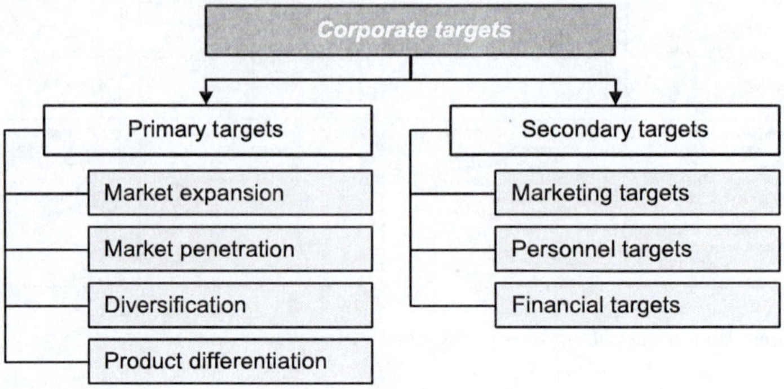

Fig. 3 Cooperation targets of corporates [57–59]

well-established market players, focusing on the increase of efficiencies, the improvement of quality and the reduction of costs for existing processes [54]. Pursuing the goal of long-term business success, corporates strive to combine established business areas with emerging ones and, thereby, foster dynamic technology developments [55].

Whereas the targets of startups are highly dependent on the corresponding life cycle phase, corporates enter cooperation pursuing strategic corporate targets, which can be distinguished in primary and secondary targets. As Fig. 3 shows, the primary targets are priority cooperation targets and can be assigned to market expansion, diversification, market penetration and product differentiation following Ansoff [56–58]. Secondary targets are accompanying targets such as marketing targets, personnel targets or financial targets and are pursued in parallel with the primary targets [57]. Furthermore, they have less strategic relevance than primary targets [57] and, thus, are not sufficient as the sole component of a cooperation.

3.2 *Target System*

The development of targets and corresponding target systems must comply with specific requirements in order to ensure that the targets are achievable [60]. Hence, the formulation of targets needs to be specific, measurable, timed, realistic, but yet ambitious [60, 61]. Moreover, target systems of enterprises need to meet the requirements of integrity [62], freedom of redundancy [63], independence [62] as well as simplicity [62]. Primary target of every enterprise is the preservation and successful development of the company [64]. This defines the overarching target and thus implies that all further business targets must focus on the development of comparative competitive advantages [57]. In their relationship to each other, individual targets form a target system [65]. A target system summarizes the total sum of all targets, which are relevant in a certain decision situation [63]. According to Friese, the individual cooperation targets of a company, corporate as well as startup, can be structured and sorted by main target, strategic sub-targets and operational sub-targets [66]. Further, a collective target system describes the aggregation of multiple individual target systems in order to form a unique and cross-partner target system for cooperation [67, 68]. Target relations and the order of individual targets in a collective target system are influenced by the power structures, the economic environment, the legal framework and social as well as cultural norms [68]. Based on this, Eisenführ and Nitzsch derive five requirements of a target system [62, 63]: Integrity, freedom of redundancy, measurability, independence, and simplicity. The relation of single targets in a target system is widely discussed in the literature.

3.3 Characteristics of Entrepreneurial Targets

In general, the relation of individual targets can be classified in three segments [57, 69, 70]: Indifferent, complementary, or competing. In their nature as individual or collective targets, indifferent targets do not affect each other's degree of fulfilment [57]. Two targets are complementary if the fulfilment of one target simultaneously increases the fulfilment of another target [68]. Whereas competing targets describe targets where the fulfilment of one target leads to a deterioration of the degree of fulfilment of the other target [68].

3.4 Strategic Cooperation Target Space

Aside from a target system, corporates and startups in collaborations usually also have a strategic focus [71, 72]. This focus can be interpreted as a target space and enables the comparability of target systems of different organizations. The previous explanations have shown that both startups and corporates develop target systems for cooperation, which need to be compared and then also elaborated in the context of a cooperation. For the purpose of achieving holistic comparability, the target space for cooperation comprises five different fields for organizations. These fields can be understood as strategic dimensions and include targets from product development [70], technology development [64], business model development [73, 74], market development [75] as well as organizational development [76, 77].

3.5 Technology-Based Cooperation

Cooperation is a possibility of obtaining access to new technologies [64]. Therefore, technology-based cooperation is characterized by an overarching ambition of the corporate to get access to a technology outside of the mastery scope and an overarching ambition of the startup to develop and scale the technology nucleus. Brodbeck classifies strategic targets as a effectiveness-oriented selection of technology in cooperation and sees especially technologies with a high importance for the achievement of competitive advantage as attractive [78]. Strategically, cooperation can be seen as a tool for core competence-oriented corporate management with the target to build up (technology) competencies [79]. Corporates profit from technology-based cooperation through a rapid reduction of a technological backlog [80]. In parallels, startups can realize technological economies of scale [72, 81].

4 Literature Review

Based on the research methodology by Ulrich, the following chapter analyses and critically reflects relevant approaches for the design of target systems and the identification of requirements [33]. The goal of this comprehensive literature analysis is to create the foundation for the discussion of the formation of target systems and corresponding requirements for cooperation between corporates and startups. Within this scope of consideration, 13 relevant approaches are initially identified. Following, these identified approaches are analyzed regarding their objective handling of the research question within the object area as well as target area. Concluding, the major deficits of the analysis are elaborated. The overview represents a comprehensive literature review and systematic derivation of deficits for the further discussion.

Following, a short summary of the literature review is given, whereas Fig. 4 provides an overview for the evaluation of the identified approaches. Generally, the identified approaches can be divided into two categories: Approaches for the definition of cooperation requirements and approaches for the modelling of entrepreneurial target systems. Within these two categories, the relevant literature is analyzed regarding seven focus evaluation areas. The first area covers cooperation in innovation and corresponding target systems for successful innovation. The second and third focus area take the consideration of corporates and startups, respectively, regarding the formulation of cooperation targets into account. Following, the fourth target area analyses the individual targets and requirements of both cooperation partner, corporates and startups equally. Moreover, the focus of the analysis in fourth place lies on an equivalent consideration of both cooperation partners. This is especially relevant since startups are smaller, less powerful, and often not established in the markets like corporates are. Within focus area 5, the analysis evaluates the recording and structuring of requirements and targets within a cooperation between corporates and startups. Focus area six primarily examines the creation of collective target systems in cooperation, while focus area seven and eight analyze the evaluation and comparison of requirements for structured cooperation.

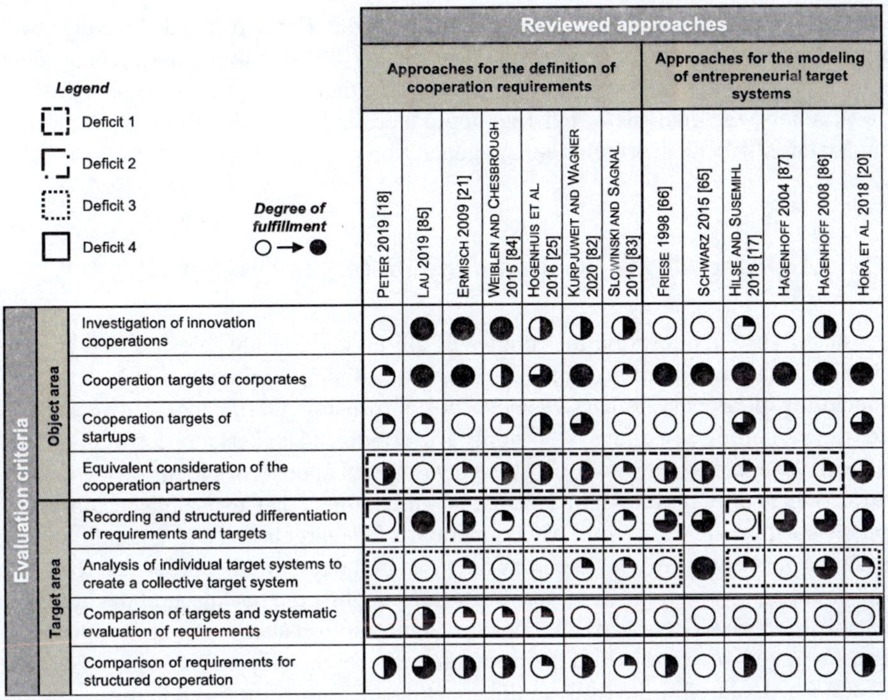

Fig. 4 Literature review of relevant approaches

4.1 Approaches for the Definition of Requirements

There are several approaches for the definition of requirements in cooperation existing. Reviewing existing approaches, it can be found that most approaches for the definition of requirements in cooperation are taking innovation or cooperation for the purpose of innovation into account [21, 25, 82–85]. It is remarkable that within the second focus area, all reviewed approaches are considering the requirements for corporates, whereas only Kurpjuweit and Wagner are also focusing on cooperation requirements of startups [82]. The approach of Kurpjuweit and Wagner describes individual milestones of the startup as a prerequisite for a cooperation project, but neglects a comprehensive investigation of the link between life cycle phases and the requirements for the cooperation design [82].

As a result of the one-sided contemplation, all reviewed approaches only insufficiently represent an equivalent consideration of the two cooperation partners. Hence, the creation of collective target systems and derivation of collective requirements is not adequately elaborated. Regardless of the preceding formation of a collective system of requirements, some of the approaches considered compare the requirements for establishing structured cooperation. Here, especially Lau formulates requirements based on different dimensions of cooperation and assigns these to specific cooperation targets [85]. Hogenhuis, van den Henden et al. [25], Kurpjuweit and Wagner [82] as well as Slowinski and Sagal [83] define requirements as decision rules for entering cooperation and formulate them using the stage-gate process. In summary, the analysis of existing approaches indicates that there is no uniform understanding of the formulation and generation of requirements.

4.2 Approaches for the Modelling of Target Systems

As target systems describe the totality of all targets that are relevant in a specific decision-making situation [60], the review of existing approaches shows that the modelling of target systems is a major focus of research. Of the reviewed approaches for the modelling of target systems only Hagenhoff [86] takes cooperation for innovation into account. However, all of the investigated approaches focus on the cooperation targets of corporates and describe tangible targets that an established corporate pursues within the framework of a cooperation. The targets and corresponding target systems of startups are considered by only two of the reviewed approaches. Hilse and Susemihl are evaluating the targets of startups against the background of successful establishment of cooperative relationships, while Hora et al. investigate specific cooperation targets of startups and their potential effects on cooperation design [16, 20]. Moreover, Hora et al. investigate the specifics of both startups and corporates and thereby consider both cooperation partners equivalently. All other approaches for the modelling of target systems do not consider two asymmetric cooperation partners equivalently. However, several approaches are enabling a structured description of cooperation targets [65, 66, 85, 87], whereas others are defining requirements as deciding factors for cooperation [25, 82, 83]. There was no approach identified deriving requirements for cooperation from cooperation targets. The formation of collective target systems for the uniform design of collaborations is supported by two of the reviewed approaches: While Schwarz differentiates target constellations between enterprise and cooperation level, Hagenhoff considers individual and collective targets separately [65, 86]. Concluding, none of the analyzed approaches to modelling entrepreneurial target systems considers the derivation or preparation of requirements for corporates and startups.

The evaluation of the reviewed approaches supports the identified problem of this paper, which serves as a starting point for the development of requirements for the cooperation design between corporates and startups. It has to be noted that there is no systematic identification of requirements for the design of cooperation taking company-specific target systems into account. Previous research clearly shows the importance of startup's development phases for cooperation design. Furthermore, collaborations are usually viewed from a corporate perspective and evaluated in terms of how established companies ensure their ability to innovate. The evaluation of the compatibility of individual targets and the transformation into collaborative targets is only partially done. In the literature reviewed, the influence of company-specific targets on cooperation design is predominantly investigated and described empirically. Lastly, none of the identified approaches presents a systematic procedure for deriving requirements from collective target systems and evaluates the significance of the requirements for cooperation design.

4.3 Derivation of Theoretical Deficits

In the previous section, existing approaches for the description of relevant requirements for the design of cooperation between corporates and startups were analyzed and evaluated. Based on the motivation for this research (see Section "Introduction"), the review indicates that there is no universal solution to the research question in current research. However, theoretical deficits emerge, which will be discussed in more detail below.

The review shows that cooperation between corporates and startups are predominantly observed and evaluated from the corporate perspective. There is **no equivalent consideration of corporate targets and startup targets described**. However, to ensure a fair exchange relationship and to avoid opportunistic behavior, it is imperative that targets and requirements for the design of cooperation be formulated from a mutual perspective. (Deficit 1).

As previously discussed, targets and corresponding requirements form the basis for cooperation between corporates and startups. However, among the reviewed approaches there is **no systematic differentiation of targets and requirements**. Thus, no requirements for the cooperation can be delineated from the collective targets and both partners, corporates as well as startups, tend to behave opportunistic. (Deficit 2).

Emerging from the perspective of observation, it can be noted, that there is **no analysis of partner-individual target systems in order to form a collective cooperation target system** conducted. This shows that the assessment of the compatibility of individual targets and the transformation into collaborative targets is only performed partially. Nevertheless, the definition of a common vision is mentioned in almost all described approaches as a prerequisite for successful cooperation. (Deficit 3).

Finally, **none of the approaches considered enables a comparison of the targets and a systematic evaluation of derived requirements** for the structured design of a collective cooperation target system. (Deficit 4).

4.4 Interim Summary

In the previous chapter, existing approaches for the development of collective target and requirement systems for technology-based cooperation between corporates and startups were analyzed and evaluated. As a result, it can be concluded that there is no approach for systematic derivation of requirements for the formation of collective cooperation target systems taking corporates and startups equally into consideration. The reviewed literature can be categorized in approaches that focus on the definition of cooperation requirements and approaches that focus on the modeling of entrepreneurial target systems. However, these approaches revealed theoretical deficits which address major challenges in the formation of collective target systems to enable the cooperation of corporates and startups. It is particularly important to note that neither the targets of startups are considered equivalently in their life cycle phases, nor is there a systematic procedure for developing a collective target system.

5 Results

Building upon the literature review and the derivation of theoretical deficits in Sections "Literature Review", "Results" aims to derive the requirements. Thereby, the authors differentiate between formal requirements for the operationalization and textual requirements for the formation of collective target systems. Formal requirements enable the derivation of requirements for the formation of collective target systems. These findings enable a further handling of the requirements and set the foundation for a consecutive conception of a solution model for the design of collective target systems in cooperation between corporates and startups. The results are further explained and represent step E in the process of applied science by Ulrich [33].

The requirements derived in this paper fall into two categories. On the one hand, formal requirements and, on the other, textual requirements. According to the research methodology of applied science, there is a need for standardization of the identified problem. This can be achieved with the help of model theory, serving to simplify complex correlations in realty [88]. The **formal requirements** follow the model theory and determine the focus of the elaboration. Following an adaptation of Patzak, the formal requirements for an applicable theoretical model are empirical and formal accurate, usable as well as the suitable for a specific purpose [88].

Textual requirements for the formation of collective target systems emerge from the practical motivation of this research (Section "Introduction"), the object and target area (Sections "Approaches for the Definition of Requirements" and "Approaches for the Modelling of Target Systems") as well as from the identified theoretical deficits (Section "Derivation of Theoretical Deficits"). Against this background, the derivation of textual requirements presents a prerequisite for a further discussion and marks the result of this paper. As the previous analysis shows, the derived textual requirements outline the linkage of the practical problem and the theoretical deficits, simultaneously spanning the object and target space for developing a solution model. Following, the textual requirements for the formation of a collective target systems as a prerequisite for the successful establishment of cooperation between corporates and startups are introduced.

5.1 Systematic Description and Structured Delineation of Cooperation Requirements and Targets

Within the framework of a cooperation, every company can pursue several targets at the same time, between which various relations exist. The knowledge of these relations is the prerequisite for a meaningful order of the target contents and the structure of an entrepreneurial target system. Due to the inconsistent understanding of target systems [65, 86] of corporates and startups in the literature (see Fig. 4), there is a need for a uniform recording and structured delineation of requirements and targets for a cooperation. In addition, there is a need for a generally valid description of partner-specific cooperation targets based on the derived target system to provide the user with a manageable orientation for target formulation.

5.2 Systematic Analysis of Individual Target Systems to Create a Collective Target System

As described in detail, there is a need for comprehensive comparison of individual target systems of the cooperation partners both in business practice and in the scientific literature. Cooperation between corporates and startups can only be successful in the long run if individual target systems are highly compatible [65, 66] and a collective target system [10] has been formulated for the cooperation project. Therefore, there is a requirement to transparently compare individual target systems and to identify target conflicts in order to define a collective target system.

5.3 Formulation of the Startup Target System Depending on the Life Cycle Phases

The environment in which startups operate is characterized by a high degree of uncertainty. The targets and needs of a startup change continuously depending on the respective development and financing phase [35, 36, 41, 43, 45]. In order to take this dynamic into account when designing the cooperation, the influence of the life cycle phases on the respective targets and requirements of a startup must be recorded. The startup target system depending on the respective life cycle phase should serve the user as an orientation aid for the formulation of the collective target system for cooperation. Hereby, it can be ensured that the collective target system focuses on the essential targets of the partner with the weaker negotiating position.

5.4 Determination of the Effective Relationships Between Requirements and Targets in the Cooperation

In order to determine relevant requirements for the design of collective target systems for cooperation design, it is necessary to analyze the effective relationships between requirements and targets [85]. In general, there is a lack of a systematic concept in the literature so far (see Fig. 4), which examines the importance of the requirement for the target achievement and evaluates the relevance of the requirements. For this reason, solutions need to determine the cause-effect relationships between requirements and targets and finally determine the relevance of individual requirements for the collective target system.

5.5 Evaluation of the Requirements Compatibility and Importance for the Cooperation Design

The practical problem outlined in Section "Introduction" shows that different expectations of the cooperation partners are often the reason for the failure of a cooperation. The requirements of the individual cooperation partners are not communicated transparently and, thus, lead to conflicts and disputes within the framework of the cooperation [10]. The solution to be developed shall therefore not only examine the relevance of the requirements, but also evaluate the compatibility of the requirements in a target system among each other. This will enable the user to identify competing requirements and to derive recommendations for action for the design of cooperation based on collective targets. For this purpose, the requirement for the formation of collective target systems is to identify competing requirements and to derive recommendations for action for the design of cooperation.

5.6 Equivalent Consideration of Cooperation Targets from Both the Corporate and Startup Perspectives

Unequal dependency structures during the initiation and implementation of cooperation as well as information asymmetries are the cause of opportunistic behavior. As a result, mostly established companies exploit their dominant position [89] and common targets are not achieved. In order to ensure a fair exchange relationship as well as the equality of the cooperation partners in the design of collective target systems and consequently to increase the chances of success of the cooperation, the equivalent consideration of the cooperation partners is a necessary requirement. Requirements and goals for the cooperation project should be formulated from both the corporate and the startup perspective and considered equally in the analyses.

This ensures a fair exchange relationship as well as the equality of both cooperation partners and increases the prospects for success by reducing the information asymmetries.

6 Conclusion and Future Research

The preceding discussion shows that the formation of collective target systems for technology-based cooperation between corporates and startups can only be done by considering explicit requirements. To enable the further development of a model for the formation of collective target systems, the conclusion and the need for future research summarize the result of the derivation and give an outlook.

6.1 Conclusion

Cooperation between startups and corporates offer numerous advantages for both partners. Startups can meet their scaling aspirations through access to valuable knowledge and resources, while corporates can ensure their innovative capabilities through access to new products and markets. Section "Section "Introduction" shows that cooperation is entered to achieve these targets. However, there is a lack of a collective target system, which is why individual and opportune target pursuit impedes cooperation success. To elaborate a solution, it is initially necessary to develop an understanding of the problem. In accordance with the research method by Ulrich, the theoretical foundations are therefore first outlined in Section "Theoretical Background". Thus, these basics allow for a comprehensive literature review in Section "Literature Review". The review concludes that current approaches inadequately address the problem raised and do not provide a solution for the design of a cooperation target system from a mutual perspective. This highlights the need to derive requirements for a design of a collective target system for cooperation between corporates and startups based on the deficits. The derivation of textual requirements enables the identification of recommendations for action that outline the potential and focus of the solution development. This paper points out that the requirements for the design of a collective target system for technology-based cooperation do not only address an analysis of individual targets, but above all open up the identification of compatible targets and requirements derived from them as a solution space.

6.2 Future Research

The paper at hand provides a profound basis on which further research can be built. According the process of applied science by Ulrich, the developed model needs to be further improved and tested within a practical context. The elaboration of the requirements has shown that the formulation of targets and requirements for a cooperation in particular is highly dependent on the cooperation partners and their respective capabilities. Therefore, a process for defining cooperation targets and requirements should be developed in further research work to facilitate the formulation of context-specific targets and requirements. The topic presented in this paper is currently in the focus of a doctoral thesis at Fraunhofer Institute for Production Technology IPT. In the context of this thesis, the requirements set the foundation for the development of a solution model, which is applied in bilateral projects to ensure the feasibility in industrial practice.

References

1. Becker, W., Ulrich, P., Botzkowski, T., Fibitz, A., Stradtmann, M.: Kooperationen zwischen Mittelstand und Start-up-Unternehmen, p. 280. Springer Fachmedien Wiesbaden, Wiesbaden (2018)
2. Steiber, A., Alänge, S.: Corporate-startup collaboration: effects on large firms' business transformation. EJIM ahead-of-print, **1** (2020)
3. Goeke, C.: Unternehmenskooperation und Branchentransformation, p. 261. Springer Fachmedien, Wiesbaden (2009)
4. Ohlhausen, P., Warschat, J.: Kooperation—Zusammenarbeit zwischen Unternehmen. In: Bullinger, H.-J., Warschat, J. (eds.) Forschungs- und Entwicklungsmanagement, pp. 29–46. Vieweg+ Teubner Verlag, Wiesbaden (1997)
5. Benner, M.J., Tushman, M.L.: Exploitation, exploration, and process management: the productivity dilemma revisited. AMR **28**(2), 238–256 (2003)
6. Löher, J., Paschke, M., Schröder, C.: Kooperationen zwischen etabliertem Mittelstand und Start-ups (2017)
7. Hajizadeh-Alamdary, D., Kuckertz, A.: Corporate Entrepreneurship als neues Unternehmertum? Warum große Unternehmen externe Innovationsimpulse suchen und sich mit kleinen Start-ups vernetzen. In: Keuper, F., Schomann, M. (Eds.), Entrepreneurship heute. Unternehmerisches Denken Angesichts Der Herausforderungen Einer Vernetzten Wirtschaft, pp. 4–25. Logos Verlag, Berlin (2015)
8. Lee, S.M., Olson, D.L., Trimi, S.: Co-innovation: convergenomics, collaboration, and co-creation for organizational values. Manag. Decis. **50**(5), 817–831 (2012)
9. Steiber, A.: Technology management: corporate-startup co-location and how to measure the effects. J. Technol. Manag. Innov. **15**(2), 11–22 (2020)
10. Wrobel, M., Preiß, K., Schildhauer, T.: Kooperationen zwischen Startups und Mittelstand. Alexander von Humboldt Institute for Internet and Society, Berlin, Online-Ressource, Learn. Match. Partner (2017)
11. Freeman, J., Engel, J.S.: Models of Innovation: startups and Mature Corporations. Calif. Manage. Rev. **50**(1), 94–119 (2007)
12. Katila, R., Rosenberger, J.D., Eisenhardt, K.M.: Swimming with sharks: technology ventures, defense mechanisms and corporate relationships. Adm. Sci. Q. **53**(2), 295–332 (2008)
13. Wessel, M.: Why big companies can't innovate. Harvard Bus Rev **09** (2012)
14. O'Reilly, C.A., Tushman, M.L.: Organizational ambidexterity: past, present, and future. AMP **27**(4), 324–338 (2013)
15. de Faria, P., Lima, F., Santos, R.: Cooperation in innovation activities: the importance of partners. Res. Policy **39**(8), 1082–1092 (2010)

16. von der Oelsnitz, D.: Kooperation: Entwicklung und Verknüpfung von Kern-kompetenzen. In: Zentes, J., Swoboda, B., Morschett, D. (eds.) Kooperationen, Allianzen und Netzwerke, pp. 183–210. Gabler Verlag, Wiesbaden (2003)
17. Hilse, H., Susemihl, I.: Erfolgreiche kooperationen von corporates und start-ups: wie david und goliath gemeinsam die geschäfte von morgen entwickeln. Organisations-entwicklung: Zeitschrift für Unternehmens-entwicklung und Change Management **37** (1), 18–24 (2018)
18. Peter, L.: Gestaltungsbereiche für Grossunternehmen zur Kollaboration mit Startups: das startup- collaboration-model. Die Unternehmung—Swiss J. Bus. Res. Practice **73**(3), 193–212 (2019)
19. Minshall, T., Mortara, L., Valli, R., Probert, D.: Making "Asymmetric" partnerships work. Res. Technol. Manag. **53**(3), 53–63 (2010)
20. Hora, W., Gast, J., Kailer, N., Rey-Marti, A., Mas-Tur, A.: David and Goliath: causes and effects of coopetition between start-ups and corporates. Rev. Manag. Sci. **12**(2), 411–439 (2018)
21. Ermisch, R.: Management Strategischer Kooperationen im Bereich Forschung und Entwicklung, p. 293. Springer Fachmedien, Wiesbaden (2009)
22. Dowling, M., Helm, R.: Product development success through cooperation: a study of entrepreneurial firms. Technovation **26**(4), 483–488 (2006)
23. Brettel, M., Engelen, A., Heinemann, F., Kessell, A.: Marktorientierte Unternehmenskultur als Erfolgsfaktor in jungen Wachstums-unternehmen. Z. Betriebswirth **78**(11), 1197–1220 (2008)
24. Giardino, C., Unterkalmsteiner, M., Paternoster, N., Gorschek, T., Abrahamsson, P.: What Do We Know about Software Development in Startups? IEEE Softw. **31**(5), 28–32 (2014)
25. Hogenhuis, B.N., van den Hende, E.A., Hultink, E.J.: When should large firms collaborate with young ventures? Res. Technol. Manag. **59**(1), 39–47 (2016)
26. Ojaghi, H., Mohammadi, M., Yazdani, H.R.: A synthesized framework for the formation of startups' innovation ecosystem. JSTPM **10**(5), 1063–1097 (2019)
27. Spender, J.-C., Corvello, V., Grimaldi, M., Rippa, P.: Startups and open innovation: a review of the literature. EJIM **20**(1), 4–30 (2017)
28. Liesebach, J.N.: Innovationsmanagement in Unternehmenskooperationen Erfolgsfaktoren für ressourcenintensive Start-Ups in Ko-operationen mit Großunternehmen. Dissertation, Wuppertal (2017)
29. Bode, C., Bogaschewsky, R., Eßig, M., Lasch, R., Stölzle, W.: Supply Management Research. Springer Fachmedien Wiesbaden, Wiesbaden (2019)
30. Friedli, T., Schuh, G.: Die operative Allianz. In: Zentes, J., Swoboda, B., Morschett, D. (eds.) Kooperationen, Allianzen und Netzwerke, pp. 487–514. Gabler Verlag, Wiesbaden (2003)
31. Scherle, N., Boven, C., Stangel-Meseke, M.: Scheitern in internationalen Unternehmenskooperationen. In: Kunert, S. (ed.) Failure Management. Springer, vol. 10, pp. 249–270. Berlin Heidelberg, Berlin, Heidelberg (2016)
32. Bronner, R., Mellewigt, T.: Entstehen und Scheitern Strategischer Allianzen in der Telekommunikationsbranche. Schmalenbachs Z betriebswirtsch Forsch **53**(7), 728–751 (2001)
33. Ulrich, H., Dyllick, T., Probst, G.J.B.: Management: Hrsg. von T. Dyllick u. G. J. B. Probst, p. 364. Haupt, Bern and Stuttgart (1984)
34. Achleitner, A.-K., Braun, R.: Entrepreneurial Finance. In: Faltin, G. (ed.) Handbuch Entrepreneurship, vol. 57, pp. 1–20. Springer Fachmedien Wiesbaden, Wiesbaden (2014)
35. Ripsas, S., Tröger, S., Nöll, F.: Deutscher Startup Monitor. KPMG, Berlin (2015)
36. Engelen, A.: Marktorientierung junger Unternehmen. Gabler, Wiesbaden (2008)
37. Brettel, M., Heinemann, F., Sander, T., Spieker, M., Strigel, M., Weiß, K.: Erfolgreiche Unternehmerteams: Teamstruktur—Zusammen-arbeit—Praxisbeispiele, 1, Aufl Gabler Verlag / GWV Fachverlage GmbH Wiesbaden, Wiesbaden (2009)
38. Ries, E.: Lean Startup: Schnell, risikolos und erfolgreich Unternehmen gründen, 5, Auflage, p. 256. Redline Verlag, München (2017)
39. Bruderl, J., Schussler, R.: Organizational mortality: the liabilities of newness and adolescence. Adm. Sci. Q. **35**(3), 530 (1990)
40. Blank, S.C.: Why the lean start-up changes everything. Harvard Bus. Rev.: HBR **91**(5), 64–72 (2013)

41. Hahn, C.: Finanzierung und Besteuerung von Start-up-Unternehmen: Praxisbuch für erfolgreiche Gründer, p. 274. Springer Gabler, Wiesbaden (2014)
42. Zinke, G., Ferdinand, J.-P., Groß, W., Möring, J., Nögel, L., Petzolt, S., Richter, S., Robeck, M., Wessels, J.: Trends in der Unterstützungslandschaft von Start-ups—Inkubatoren. Akzeleratoren und andere, Berlin (2018)
43. Reichle, H.: Finanzierungsentscheidung bei Existenzgründung unter Berücksichtigung der Besteuerung: Eine betriebswirtschaftliche Vorteilhaftigkeitsanalyse. Zugl.: Hagen, FernUniv., Diss., 2010. Gabler Verlag/Springer Fachmedien Wiesbaden GmbH Wiesbaden, Wiesbaden (2010)
44. Bogott, N., Rippler, S., Woischwill, B. (eds.): Im Startup die Welt gestalten. Springer Fachmedien Wiesbaden, Wiesbaden (2017)
45. Bogott, N., Rippler, S., Woischwill, B.: Phasen von Startups. In: Bogott, N., Rippler, S., Woischwill, B. (eds.) Im Startup die Welt gestalten, pp. 111–119. Springer Fachmedien Wiesbaden, Wiesbaden (2017)
46. Olson, P.D.: Entrepreneurship and management. J. Small Bus. Manage. **25**(3), 7 (1987)
47. Jäger, C.C., Heupel, T.: Management Basics: Grundlagen der Betriebswirtschaftslehre—dargestellt im Unternehmenslebenszyklus, p. 403. Springer Gabler, Wiesbaden (2020)
48. Hirst, S.: The Case for Investor Ordering. The Harvard Law School Program on Corporate Governance Discussion Paper 2017–13 (227) (2018)
49. Güttler, K.: Formale Organisations-strukturen in wachstumsorientierten kleinen und mittleren Unternehmen. Zugl.: Aachen, Univ., Diss., 2008, 1. Aufl. ed. Gabler Verlag/GWV Fachverlage GmbH Wiesbaden, Wiesbaden (2009)
50. Bundesministerium der Justiz und für Verbraucherschutz, 2015. §267 Umschreibung der Größenklassen
51. Europäische Kommission: Definition der Kleinstunternehmen sowie der kleinen und mittleren Unternehmen (2003)
52. Günterberg, B.: Unternehmens-größenstatistik: Unternehmen, Umsatz und sozialversicherungspflichtig Beschäftigte 2004 bis 2009 in Deutschland, Ergebnisse des Unternehmensregisters (URS 95) (2012)
53. Blank, S.G., Dorf, B.: The startup owner's manual: The step-by-step guide for building a great company. K & S Ranch, Pescadero, Calif (2012)
54. Mercandetti, F., Larbig, C., Tuozzo, V., Steiner, T.: Innovation by collaboration between startups and SMEs in Switzerland. TIM Rev. **7**(12), 23–31 (2017)
55. Bleicher, K.: Das Konzept Integriertes Management: Visionen—Missionen—Programme, 9. aktualisierte und erweiterte, Auflage, p. 714. Campus Verlag, Frankfurt, New York (2017)
56. Meldrum, M., McDonald, M.: The Ansoff Matrix. In: Meldrum, M., McDonald, M. (eds.) Key Marketing Concepts, pp. 121–126. Macmillan Education UK, London (1995)
57. Heinen, E.: Das Zielsystem der Unternehmung: Grundlagen betriebs-wirtschaftlicher Entscheidungen, p. 281. Gabler Verlag, Wiesbaden, s.l. (1966)
58. Eckert, R.: Business Model Prototyping: Geschäftsmodellentwicklung im Hyperwettb-ewerb. Strategische Überlegenheit als Ziel. Springer Science and Business Media.
59. Thommen, J.-P., Achleitner, A.-K., Gilbert, D.U., Hachmeister, D., Kaiser, G.: Allgemeine Betriebswirtschaftslehre, 8, vollständig überarb, Auflage, p. 586. Springer Gabler, Wiesbaden (2017)
60. Edvardsson, K., Hansson, S.O.: When is a goal rational? Soc Choice Welfare **24**(2), 343–361 (2005)
61. Schuh, G., Stich, V.: Konzeptentwicklung in der Produktionsplanung und -steuerung. In: Schuh, G., Stich, V. (eds.) Produktionsplanung und -steuerung 2. Springer, vol. 3, pp. 149–415. Berlin Heidelberg, Berlin, Heidelberg (2012)
62. von Nitzsch, R.: Entscheidung bei Zielkonflikten. Gabler Verlag, Wiesbaden (1992)
63. Eisenführ, F., Weber, M.: Rationales Entscheiden, Vierte, neu bearbeitete, Auflage, p. 417. Springer, Berlin Heidelberg, Berlin, Heidelberg, s.l. (2003)
64. Schuh, G., Klappert, S.: Technologie-management. Springer, Berlin Heidelberg, Berlin, Heidelberg (2011)

65. Schwarz, D.: Zielsysteme und Erfolgs-faktoren von Kooperationen im Destinationsmanagement. Ein Beitrag zum effektiven Kooperationsmanagement von Tourismusorganisationen durch theorie- und empirie-geleitete Exploration. Universität Trier, Trier, Online-Ressource (2015)
66. Friese, M.: Kooperation als Wettbewerbs-strategie für Dienstleistungsunternehmen, Gabler Edition Wissenschaft ed. Deutscher Universitätsverlag, Wiesbaden, s.l (1998)
67. Marz, O., Baum, M., Schimitzek, P., Kramer, E.: Organisationspsychologie—die Ressource Mensch im Fokus. In: Marz, O., Baum, M., Schimitzek, P., Kramer, E. (eds.) IT-Investitionen verstehen und bewerten. Springer, vol. 84, pp. 63–80. Berlin Heidelberg, Berlin, Heidelberg (2019)
68. Laux, H., Gillenkirch, R.M., Schenk-Mathes, H.Y.: Entscheidungstheorie, 9., vollst. überarb. Aufl. (ed.), p. 594. Springer Gabler, Berlin (2014)
69. Heinen, H.: Ziele multinationaler Unter-nehmen: Der Zwang zu Investitionen im Ausland. Gabler Verlag, Wiesbaden (1982)
70. Pahl, G., Beitz, W., Feldhusen, J., Grote, K.-H.: Konstruktionslehre: Grundlagen erfolg-reicher Produktentwicklung; Methoden und Anwendung, 7, Aufl Springer, Berlin, Heidelberg (2007)
71. Dyer, J.H., Singh, H.: The relational view: cooperative strategy and sources of interorganizational competitive advantage. AMR **23**(4), 660–679 (1998)
72. Albers, S., Gassmann, O.: Handbuch Technologie- und Innovationsmanagement. Gabler Verlag, Wiesbaden (2005)
73. Osterwalder, A., Pigneur, Y.: Business Model Generation: Ein Handbuch für Visionäre, Spielveränderer und Herausforderer, 1, Auflage, p. 286. Campus Verlag, Frankfurt, New York (2011)
74. Wirtz, B.W.: Business Model Management: Design—Process—Instruments, 2nd ed. 2020 ed., p. 317. Springer International Publishing; Imprint: Springer, Cham (2020)
75. Porter, M.E.: Wettbewerbsvorteile: Spitzenleistungen erreichen und behaupten = (Competitive Advantage), 8, durchges, Aufl, p. 688. Campus-Verlag, Frankfurt am Main (2014)
76. Despres, C.: Clients, consultants and the social cognition of organizational change. Strat. Change **3**(1), 29–44 (1994)
77. Bowman, C., Asch, D.: Strategic Management. Macmillan Education UK, London (1987)
78. Brodbeck, H.: Strategische Entschei-dungen im Technologie-Management: Relevanz und Ausgestaltung in der unternehmerischen Praxis. Verl. Industrielle Organisation, Zürich, V, 167 S (1999)
79. Strautmann, K.-P.: Ein Ansatz zur strategischen Kooperationsplanung. Zugl.: München, Univ., Diss., 1993, München (1993)
80. Wolfrum, B.: Strategisches Technologie-management, p. 450. Gabler Verlag, Wiesbaden (1991)
81. Abele, T.: Verfahren für das Technologie-Roadmapping zur Unterstützung des strategischen Technologiemanagements. Stuttgart, Univ., Diss, Heimsheim, Zugl. (2006)
82. Kurpjuweit, S., Wagner, S.M.: Startup Supplier programs: a new model for managing corporate-startup partnerships. Calif. Manage. Rev. **62**(3), 64–85 (2020)
83. Slowinski, G., Sagal, M.W.: Good practices in open innovation. Res. Technol. Manag. **53**(5), 38–45 (2010)
84. Weiblen, T., Chesbrough, H.W.: Engaging with startups to enhance corporate innovation. Calif. Manage. Rev. **57**(2), 66–90 (2015)
85. Lau, F.: Modell zur typenbasierten Gestaltung von firmeninternen Inkubatoren. Dissertation, 1. Auflage ed., Aachen (2019)
86. Hagenhoff, S.: Innovationsmanagement für Kooperationen. Göttingen University Press, Göttingen (2008)
87. Hagenhoff, S.: Kooperationen: Grundtypen und spezielle Ausprägungsformen (2004)
88. Patzak, G.: Systemtechnik—Planung komplexer innovativer Systeme. Springer, Berlin Heidelberg (1982)
89. Dyerson, R., Pilkington, A.: Gales of creative destruction and the opportunistic incumbent: the case of electric vehicles in California. Technol. Anal. Strategic Manag. **17**(4), 391–408 (2005)

Prof. Dr.-Ing. Dipl.-Wirt. Ing. Günther Schuh
(*1958) holds the chair for Production Systematics at RWTH Aachen University. Furthermore, he is director of the FIR e. V., member of the board of directors of the WZL of the RWTH Aachen University and the Fraunhofer Institute for Production Technology (IPT).

Bastian Studerus
(*1991) holds a M.Sc. degree in Business Administration and Mechanical Engineering from the RWTH Aachen University and Linköpings Universitet. Since 2018, he is a research fellow (Dr.-Ing.) in the department Strategic Technology Management at the Fraunhofer Institute for Production Technology (IPT).

Resources Collaboration and Optimization in Industry 4.0 Environments

Elif Ocakci, Anca Draghici, and Jörg Niemann

Abstract In order to act competitively, efficiently and economically on the market, it is of great importance for companies to continuously optimize their capacity planning. In this context, developments in the field of Industry 4.0 have steadily gained in importance in recent years. Many companies are increasingly investing their human and machine resources towards Industry 4.0 readiness to increase the performance of their entire value creation process. Production systems can, for example, re-order materials themselves or request repairs. The operational planning processes are therefore subject towards these technological changes. The paper examines these changes with focus on capacity planning processes under Industry 4.0 and how this affects the ability to forecast future production resources requirements. The paper will evaluate the economic added value resulting from investments in Industry 4.0-based capacity planning and what the technological developments mean for the capacity optimization of human resources and machines in such production environments.

Keywords Industry 4.0 · Capacity planning · Production resources

1 Introduction and Motivation I4.0

Industrial companies are currently facing challenging times due to internationalization, growing competitiveness, changes in supplier and customer markets and innovation in technology [1]. The so-called digital transformation is progressing

---

E. Ocakci
Department of Strategy and Business Development, Continental Teves AG, Guerickestraße 7, 60488 Frankfurt am Main, Germany

A. Draghici
Faculty of Management in Production and Transportation, Politehnica University of Timisoara, 14 Remus str., 300191 Timisoara, Romania

J. Niemann (✉)
Faculty of Mechanical and Process Engineering, FLiX Research Center for Life Cycle Excellence, Duesseldorf University of Applied Sciences, Muensterstrasse 176, 40476 Duesseldorf, Germany
e-mail: joerg.niemann@hs-duesseldorf.de

K. von Leipzig et al. (eds.), *Smart, Sustainable Manufacturing in an Ever-Changing World*, Lecture Notes in Production Engineering,
https://doi.org/10.1007/978-3-031-15602-1_35

rapidly and uninterruptedly. The attack on what already exists has begun [2]. Since Konrad Zuse in 1941 took significant steps into the computer age and in 1950 around 8000 worldwide Computer systems existed, today all areas of life are permeated with information and communication technology. The development of the mass market for broadband internet and the introduction of smartphones, tablets and wearables caused the number of networked objects to explode. Digitization not only enables new intelligent products and new forms of internet-based services (so-called smart services), it also changes industrial production. The term Industry 4.0 (i4.0) stands for a new level of organization and Control of complex value creation structures. Obviously, the industry is undergoing a profound change known as the fourth industrial revolution. Intelligent technology forms the basis for Industry 4.0 systems that are networked with one another via the "Internet of Things". Of importance will be the intelligence of the systems, the adaptive, robust, predictive and will enable particularly user-friendly systems. Fascinating, unmanageable possibilities open for designing new business models and innovative value creation structures as part of future-oriented corporate management in the age of digitization. In sum, this shows that there is a great need for productivity-enhancing, but also flexibility-promoting measures and approaches in the industry. Industry 4.0 is seen as one of the answers to these megatrends and the associated challenges. Therefore, it is not surprising, that according to a study the following key statements are made [3]:

- Industry 4.0 leads to a higher production and resource efficiency of -18%.
- Industry 4.0 enables new, often disruptive digital business models.
- Digitized production and services generate an additional €30 billion per year for the German economy.
- Digitization of the product and service portfolio is the key to sustainable Corporate success.
- Horizontal cooperation enables better fulfilment of customer requirements.
- Industry 4.0 transforms the entire company.
- The integrated analysis and use of data is the core capability within Industry 4.0 [3, 4].

2 Production Capacity and Operational Optimization

2.1 Production Systems

Production systems focus on the management of production environments in order to achieve the business objectives. Therefore, the entire value chain will be considered, from supplier to the customer. The input (e.g. knowledge, materials, financial resources and more) will be transformed during the value creation process (i.e. manufacturing) to an intermediate or end product. Associated processes such as transport or quality measures assist and sustain the manufacturing process. The propulsions

Labor	Multifactor productivity
Increased supply and productivity • Increased labor-force participation • Better and faster matching of workers with employers • Increased productivity of workers in the labor force	R&D and product development • Better use of data leads to new inventions • Faster product development cycles enabled by better testing and quality control Operations and supply-chain optimization • Real-time monitoring and control of production lines • Better logistics routing through path optimization and prioritization Resource management • Improved energy efficiency through intelligent building systems • Increased fuel efficiency • Decreased waste of raw materials
Capital	
Improved asset efficiency • Preventive maintenance decreases downtime and reduces expenditure on maintenance • Increased use of assets	

Fig. 1 Improvement opportunities by categories (modified according to 17])

for that are humans and technical resources like machines or IT-systems. The relation between the several elements of the production system is determined by the use of adequate methods and therefore impinge on the process and structural organization [5]. The production system defines what guidelines to follow when developing, implementing and maintaining processes in the coherence of production. As a holistic system it will unfold its full effected when all elements and processes correlate. Future production systems will focus more on performance measurement by means of precise data and their connection. Sensors will observe the entire production process, from incoming goods through manufacturing to shipping. The sensor data will be connected and stored with additional relevant data from other sources in a single data warehouse. Thanks to that connectivity, fact-based, target-oriented decisions will be enabled and guesswork will be eliminated. Through the access to the comprehensive and topical information, reporting can be automatized and employees as well as managers will be enabled to identify optimisation potentials and/or to take precise actions [6, 7]. Figure 1 shows different categories and leverages in production systems induced by the implementation of Industry 4.0.

Smart factories are going to be capable of profitably producing custom specific items in an agile way, and smart assistance systems will help workers to focus on really value-added activities rather than routine tasks.

2.2 *Flexibility and Transformability*

Above all, companies in general are challenged by the need to steadily adjust the many elements of manufacturing environments and production systems [8]. Due to the influence of dynamic and volatile markets in the automotive industry, this can only be achieved by the development and utilization of specific methods, technologies, resources and organizational structures. The production environment must be able to react immediately, economically and situationally on influencing factors. The key factor to this is the enabling of the production system and thus the production environment to more changeability. So far there is no consistent use of terms in the literature. Elkins, Huang and Alden, for instance, use the term "agile", but refer also to the overall issue of changeability [9]. For convenience, this paper will follow the methodology of Nyhuis et al. [10]. In the context of quality, changeability can refer to the ability of a production environment to adapt to new quality influencing circumstances such as an increase in production volume. Since Nyhuis focused on the production level, he distinguished additionally five different degrees of changeability. Two of them are flexibility and transformability. Because the distinction of the three other terms has only a minor impact on this work, solely flexibility and transformability will be discussed subsequently. Transformability, however, should not be confused with flexibility. Flexibility enables a production environment to react on changes by using predefined measures which have predefined capacities and limits [10]. For instance the capability to react on a supplier failure with predefined measures. Whereas, transformability enables a production environment to counter divergencies with measures beyond the predefined set of measurements [11]. The divergencies can be bridged without defining the solution in advance [12]. Hence, transformability leads to the condition that solutions arise from the circumstances and the borders of capacities become flexible. This leads to an efficient utilization of existing resources.

Flexibility and transformability are independent from each other. If a production environment defined to bridge divergencies flexibly and to a large extent, then—despite its inefficiency—the attribute of transformability becomes dispensable [5]. Production environments nowadays contain the component of flexibility but show a lack of transformability. Considering the automotive industry, suppliers are often contractually forced to have a leeway to increase their output, in case another supplier of the same part/component failures to deliver timely. The use of new technologies, such as additive manufacturing or business analytics, would establish a transformable production environment which enable the anticipation of supplier failures and other divergencies. This is just one example of many. The literature offers a wide range of examples for projects in the automotive industry [1, 4, 13].

2.3 Transformation Enablers

In order to encounter the challenges in automotive industry, it becomes necessary to enable optimal response capacity of the production environment. In the literature there exists a larger number of transformation enablers [13–15]. Focusing on the production environment level, five main transformation enablers can be listed [9]. The production environment is characterized by the methods and organizational structure, which are arranged by the production system. Input factors like manpower or machines will be transformed to output factors, such as products and services. That transformation process is impacted by internal and external factors. In order to encounter and react to these challenges, the transformation enablers are needed [16]. Universality describes the design and dimensioning of objects. The object thereby, fulfils from the beginning various requirements and can undertake different tasks without hinging on anything [9]. A common example is a machine which can process several different product variants or even whole units without being set up. Mobility is important to reconfigure (add/remove) objects. An object with this attribute has no spatial constraints. It can be moved without great effort, such as a machine with wheels [5]. Scalability is the ability to scale something up or down. This can be related to technologies, humans, space and many other things [9]. A non-physical example is the extending of the working time in order to use open capacities. An object or system which consist of several elements or functions which are independent from each other fulfils the requirement of modularity [5]. For instance, a removable robotic arm which is replaced by another one that fulfils the same technical standards. Compatibility enables objects internally and externally to network with other objects regarding materials, information, energy and many other things [9]. A software with many standardized interfaces fulfils exactly that attribute. The difference between modularity and compatibility is that the former focuses on the exchange of modules while the latter enables the networking between two objects. Additionally, these transformation enablers have different degrees. A machine that can process a high number of different elements has a higher extent of universality than a machine with a low number of processable elements. Also, not every enabler can be used in every case. In the context of real estate, mobility is hardly not realizable. Furthermore, the enablers correlate. Often modular objects need also to fulfil the requirement of compatibility, so that the replacing module fits into the interface. Nowadays, many objects in the production environment fulfil these attributes partially. With the activities in the framework of Industry 4.0, available prerequisites will be used to develop the production environments with a holistic approach. This will contribute to hedging and enhancing the competitiveness of producing companies [8].

3 Industrial Changes in I4.0

3.1 Humans and Machines

Humans will continue to be an essential component of the Industry 4.0. In the earlier beginnings of industrialization mainly muscle power that was required to operate machines. In contrast, today it is the "mental power" [17, 18]. This can be justified by the fact that humans are now relieved of heavy activities by the machines and robots. The production workers in smart factories are thus increasingly taking over tasks of monitoring, maintaining and controlling the machines used (so called "assisting operators."). As a result of the increasing interactions between operators and machines the time spent on secondary tasks can be significantly reduced. The smart data is no longer transmitted manually, but directly via the communication channels. In addition, errors such as manual typing errors will be avoided. As a result a significantly accelerated production process is being established. Additional advantages are the associated reduction in errors when transferring data as well as the simpler control of the processing status. Additionally, smart factories also offer support for employees through so-called assistance systems. A wide range of variants are currently available. Such systems include "augmented reality" glasses, smartwatches or tablets. "Augmented reality" refers to computer-assisted perception or representation that expands the real world with virtual objects. With the help of this, employees can obtain information, employees can intervene much faster and better in the production process. However, these systems can be used not only in the event of machine malfunctions, for example but also as a support for an employee in order to visualize the next step with the help of video. Thus, the productivity of the employees can be increased, which leads to a significantly better productivity rate. Furthermore, assistance systems can also improve the health of employees. For example, workbenches that can be automatically adjusted in height can help to ensure the back does not have to be subjected to additional strain due to the correct working position. The consequence of this is indirect cost savings, as the sickness rate of the employees is thus much lower [18]. The assistance systems even offer another advantage, which relates to the demographic change and the need/demand to work up to a later retirement age. Workplaces can be adapted more individually and more flexibly in line with the age of the workforce. This turns out to be an advantage because the average age in companies is also rising. One hurdle, however, is mastering these technical tools and assistance systems.

3.2 *Means of Production*

With the help of the networked machines, efficiency is increased on various levels and costs are reduced. Thanks to constant communication between machines and tools, relevant data can be called up and checked at any time without having to be directly present at the machine. This also enables remote diagnostics by specialist personnel, who can check the machine from any location. In case that assistance systems are included information can be instantly sent to other machines or processes to already start set-up processes and preparation processes. This accelerates the production process by shortening the throughput time by reducing waiting time. Further on, machines might be even able to take over these processes independently to a certain extent as a "self-configuration, self-calibration, and self-monitoring" function [18]. The efficiency of a machine is also increased by the tools used, since the current tool wear is always monitored. This helps to ensure that the tools can be used precisely and optimally until the end of their useful life. In addition, the tools are set up more accurately as the machine continuously checks the settings and readjusts the machine coordinates according to the tool data.

3.3 Capacity Planning

As the previous sections have already shown, Industry 4.0 can significantly increase the efficiency and productivity of companies. An important factor is capacity planning. The application of extensive analyses by software programs enables a more accurate planning of capacities in terms of times, volume and parts needed. On the one hand, this is due to the large amount of data collected throughout the production process. This data reflect the real capacity utilization of a machine in a given period. Production planning systems (PPS) usually use a top-down approach. This means that in the case of a change in the purchase order at short notice, a complete reschedule of the material requirements planning is necessary. Production scheduling and capacity planning must also start again from the beginning. Industry 4.0 technologies enable a quick and autonomous system adaption by processing the data directly and adapting all system parameters and resources instantly. If complications arise in the new capacity planning, various measures must be checked. In the background a direct check is possible due to the communication between the machines. By a glance of the responsible employee on a monitor, for example, on which all machines are displayed with their current availability, production status and maintenance, a decision can be made directly. For planning, however, the capacities of linked AGVs (autonomous guided vehicles) must also be taken into account.

4 Financial Models and Savings Potentials

An important aspect for companies tackling the topic of Industry 4.0 is the financing. For entrepreneurs, the question arises as to how the technology can be financed and whether this is economically viable at all. The challenge is that new programs or systems have such fast innovation cycles that a long-term commitment to a particular technology is usually not profitable. This is due to the continuous progress of technology. For this reason, smart financing solutions are also required. One of the most important solutions is leasing. The contracts can vary and range from pure hardware and/or software leasing up to full service concepts. Modern business models allow to pay for machine utilization on "pay per use" basis. This business models include the advantage for the user to limit expenditures in case of lower product demand from the market side. Another advantage is that the capital commitment costs are kept low. If, for example, ten lathes are leased, they are directly available, but do not have to be paid directly according to their value. Figure 2 summarizes the range of different financial options and business models available in many branches of mechanical engineering.

Industry 4.0 offers therefore large potentials for additional services. These so calls smart services base on data retrieved from processes and various other production resources. Added value for additional services is created by the merge of these different data sources. The combination of different (electronic) data sources (from

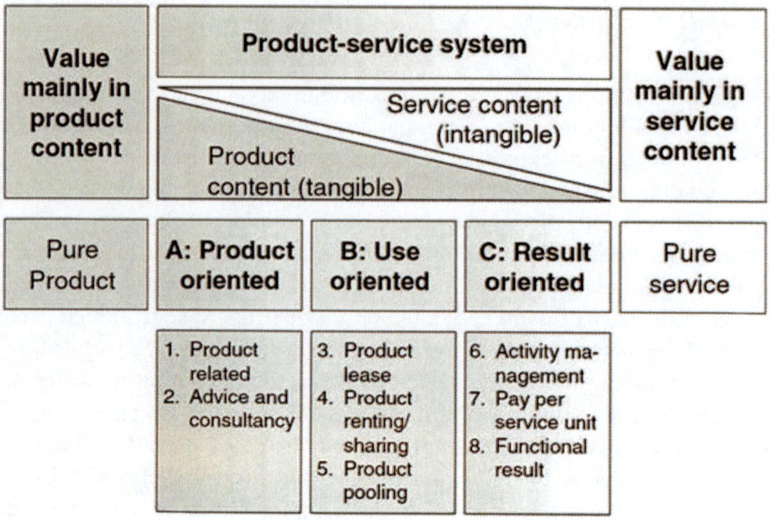

Fig. 2 Business models enabled by Industry 4.0 (modified according to 17)

different production resources) paves the way for the increase in productivity. Figure 3 shows three basic different forms how the value-added is created.

Industry 4.0 technologies enable services with a clear technical focus. Such services are characterized by ensuring a (more) reliable machine operation by expanded functionalities concerning their surveillance. A second group of services addresses a savings potential that lies in the processes itself. The enhanced technologies and data sources allow deeper process insights and by this open opportunities for process optimisations. The third approach is to analyse the user behaviour and the

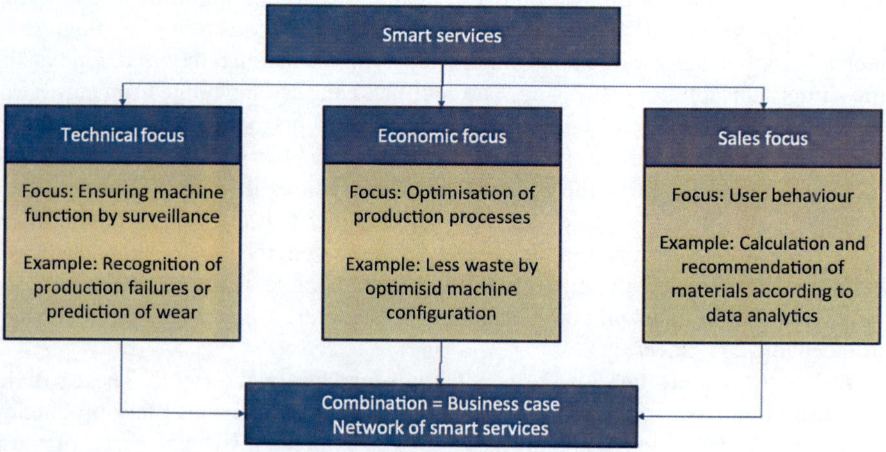

Fig. 3 Saving potentials and business models

Table 1 Estimated benefits of industry 4.0 (enhanced according to Edh Mirzaei et al. [1])

Cost type	Effects	Potentials (%)
Inventory	• Reduction of safety stocks • Avoidance of bullwhip and burbridge effects	−30 to −40
Production	• Increase in overall equipment effectiveness (OEE) • Process control loops • Improvement in vertical and horizontal labor flexibility	−10 to −20
Logistics	• Higher degree of automation (milk run, picking,…)	−10 to 20
Complexity	• Enhanced span of control • Reduced trouble shooting	−60 to −70
Quality	• Real-time quality control loops	−10 to −20
Maintenance	• Optimization of spare part inventories • Condition-based maintenance (process data, measurement data) • Dynamic prioritization	−20 to −30

modes of machine utilization. By this the entire supply chain around the production resources can be addressed and optimised (e.g. material supply). The implementation of Industry 4.0 technology needs high investments by companies, but will also enable savings in various areas of production. Table 1 shows the potential savings according to Edh Mirzaei et al. [1].

Inventory costs can therefore be reduced by up to 40%, as the permanent mapping of inventory, stock movements and safety stock levels can be reduced as well as optimize the ordering cycles. Furthermore, 10–20% of costs can be saved within production. These result from the flexibility within the production process. The key factors here are the employees and machines, which can now be deployed more flexibly according to their workload. Through the use of automation within the logistics process, for example by means of AGVs, for example, can save around 10–20% here. The greatest savings potential is offered first and foremost by complexity costs. However, these costs are not directly tangible in practice, however, but only arise indirectly. Among these indirect areas include, for example, data processing, which has been improved by Industry 4.0, or packaging and customer service. Through Industry 4.0 these processes are automated. For customer service, for example, there are already systems that can answer customers' most frequently asked questions in an automated manner. Within quality, costs can be reduced in particular through end-to-end testing of the products can be reduced. Real-time information on the quality status of workpieces and of the workpieces and the tools, it is possible to intervene in the process much faster and more and the proportion of defective parts can be reduced. Last but not least, the table shows a cost savings potential of 20–30% in maintenance. This value can be achieved through improved stockkeeping of spare parts and tools. This makes it easier to prioritize machine maintenance and more targeted. At best, the machines to be serviced report in advance so that they can be serviced at the right time and do not hold up the production process. It should be noted that many of the above cost factors are correlated with each other. For example, manufacturing costs cannot be reduced if the machines do not communicate with each other and with the systems in real time. This communication serves as the basis for servicing the machines in good time or for providing machine tools or making them available right in time for replacement. If this is not possible, maintenance costs cannot be reduced.

5 Conclusion and Outlook

In addition to the fundamentals of Industry 4.0, this paper focuses on cost optimizations that are possible under Industry 4.0. In particular flexibility and capacity planning. Based on the examples, it can be determined that an investment of the companies in digitization is certainly economically worthwhile for both large and small companies. As a result, the organization is generally no longer controlled centrally, but decentrally via cloud-based systems. These allow real-time access to the production from anywhere. Lengthy processes in capacity planning as well as in lead times can be significantly shortened by making them more flexible. The onset of digitization will also enable companies to respond more agilely to current market events. Especially with the background of the corona pandemic, the market's ever-increasing desire for individuality, this agility must be ensured in order to keep up with the competition. As a rule, the cost savings cannot be attributed precisely to a single factor. This is because changes affect several factors at the same time. However, it has been shown that even changes in sub-processes can reduce costs can be reduced across the entire company. Ultimately, intelligent systems and the masses of data can be used to improve forecasting capabilities in both the short and long term. The direction of Industry 4.0 can therefore be described as positive at this point in time, provided that the people continue to play an important role in this process. Meanwhile, the future of further industrial revolutions is still uncertain. Since a large of companies have neither started nor fully completed the implementation of Industry 4.0. A look in the direction of Industry 5.0 should only be taken in the very distant future.

References

1. Edh Mirzaei, N., Hilletofth, P., Pal, R.: Challenges to competitive manufacturing in high-cost environments: checklist and insights from Swedish manufacturing firms. Oper. Manag. Rev. Res. **14**, 272–292 (2021)
2. Durakbasa, N., Gençyılmaz, G. (Eds.).: Digitizing Production Systems. Springer International Publishing (2021)
3. PWC-Study, Industrie 4.0—Chancen und Herausforderungen der vierten industriellen Revolution, 10/2014
4. Pistorius, J.: 'Industrie 4.0—Schlüsseltechnologien für die Produktion: Grundlagen Potenziale Anwendungen. Springer-Vieweg Verlag GmbH, Deutschland, Berlin (2020)

5. Steegmüller, D., Zürn, M.: Wandlungsfähige Produktionssysteme für den Automobilbau der Zukunft. In: Bauernhansl, T., ten Hompel, M., Vogel-Heuser, B. (eds.) Industrie 4.0 in Produktion, Automatisierung und Logistik. Springer Vieweg, Wiesbaden (2014). https://doi.org/10.1007/978-3-658-04682-8_5
6. Gupta, V., Ulrich, R.: How the Internet of Things Will Reshape Future Production Systems (2017). Available at: https://www.mckinsey.com/business-functions/operations/ourinsights/how-the-internet-of-things-will-reshape-future-production-systems
7. Irniger, A.: Difference Between Digitization, Digitalization and Digital Transformation (2017). Available at: https://www.coresystems.net/blog/difference-betweendigitization-digitalization-and-digital-transformation Accessed 29nd Sept. 2021
8. Benanav, A.: The Future of Automation and Work. Verso, London, New York (2020)
9. Elkins, D.A., Huang, N., Alden, J.M.: Agile manufacturing systems in the automotive industry. Int. J. Prod. Econ. (2004). https://doi.org/10.2516/ogst/2013131
10. Nyhuis, P., Heinen, T., Brieke, M.: Adequate and economic factory transformability and the effects on logistical performance. Int. J. Flex. Manuf. Syst. (2007). https://doi.org/10.1007/s10696-007-9027-3
11. Rogalski, S.: (2011) 'Flexibility Measurement in Production Systems: Handling Uncertainties in Industrial Production. Springer-Verlag GmbH, Deutschland, Berlin (2018)
12. Zäh, M.F., Sudhoff, W., Rosenberger, H.: Bewertung mobiler Produktionsszenarien mit Hilfe des Realoptionsansatzes. ZWF Z. für wirtschaftlichen Fabrikbetrieb **98**(12), 646–651 (2003). https://doi.org/10.3139/104.100716
13. Burggräf, P., Schuh, G.: Fabrikplanung: Handbuch Produktion und Management 4. Springer-Verlag GmbH, Deutschland, Berlin (2021)
14. Koren, Y.: General RMS characteristics. Comparison with dedicated and flexible systems. In: Dashchenko, A.I. (eds.) Reconfigurable Manufacturing Systems and Transformable Factories. Springer, Berlin, Heidelberg (2006). https://doi.org/10.1007/3-540-29397-3_3
15. Wiendahl, H.P., et al.: Changeable Manufacturing—Classification, Design and Operation. CIRP Ann. Manuf. Technol. (2007). https://doi.org/10.1016/j.cirp.2007.10.003
16. Hinrichsen, S., et al.: Versatile assembly systems—requirements, design principles Bibliography 120 and examples. In: Hochschule Ostwestfalen-Lippe: 4th International Conference Production Engineering and Management 2014. Proceedings: September 25 and 26, 2014, Lemgo, Germany (2014)
17. Niemann, J., Pisla, A.: Life Cycle Management of Machines and Mechanisms. Springer Nature, Switzerland (2021)
18. Niemann, J., Fussenecker, C., Schlösser, M., Ocakci, E.: The workers perspective -eye tracking in production environments. In: Acta Technica Napocensis, Series Applied Mathematics, Mechanics and Engineering, vol. 64(1 -S1) (2021)

Elif Ocakci holds a Masters in Industrial Engineering and a diploma in business administration. She works as head of strategy and communication at Continental in Frankfurt, Germany.

Anca Draghici holds a Ph.D. in Mechanical Engineering and Management. She is full professor at Politehnica University of Timisoara, Romania.

Jörg Niemann obtained his Ph.D. degree in Manufacturing Engineering. He is professor in Business Management and Mechanical Engineering at the Düsseldorf University of Applied Sciences, Germany.

Production Systems and Maintenance

Investigation of Predictive Maintenance Algorithms for Rotating Shafts Under Various Bending Loads

C. I. Basson, G. Bright, J. Padayachee, and S. Adali

Abstract Condition monitoring plays an important role with regard to forecasting structural failure of shafts in Advanced Manufacturing Systems (AMS). Repairs affect the downtime of machines considerably due to scheduled maintenance. The waste product of scheduled maintenance are parts that are treated as exhausted components for disposal. The disposed parts contain Residual Useful Life (RUL). Operational costs due to scheduled maintenance can be reduced through Condition Monitoring (CM) parameters that are utilized in Predictive Maintenance (PdM). The study utilizes logistic regression Machine Learning (ML) algorithm to predicting specific classification markers as a relevant step to monitor the health of a bright-steel shaft under various bending loading. Various loading conditions were monitored and compared, employing Principal Component Analysis (PCA). Predictions were tested by utilizing K-Means and DBSCAN clustering techniques. The Logistic Regression (LR) machine learning algorithm was employed to determine the prediction accuracy under various loads. A shaft was rotated under various bending loads which followed the experimental methodology of the R.R. Moore fatigue test. The goal of the experiment was to determine the prediction accuracy under various loads. Prediction scores for K-means clustering showed a overall decrease in accuracy in the increase of cluster numbers and the prediction accuracy showed increases and decreases for DBSCAN clustering with the increase of loading for various cluster selection.

Keywords Predictive maintenance · Condition monitoring · Rotating machines

C. I. Basson (✉) · G. Bright · J. Padayàchee · S. Adali
Department of Mechanical Engineering, University of KwaZulu-Natal, Berea, South Africa
e-mail: bassonc@ukzn.ac.za

G. Bright
e-mail: brightg@ukzn.ac.za

J. Padayachee
e-mail: padayacheej@ukzn.ac.za

S. Adali
e-mail: adali@ukzn.ac.za

K. von Leipzig et al. (eds.), *Smart, Sustainable Manufacturing in an Ever-Changing World*, Lecture Notes in Production Engineering,
https://doi.org/10.1007/978-3-031-15602-1_36

1 Introduction

1.1 Technical Monitoring for Rotating Machines in AMS

Rotating machines experience vibrations and stress loading during continuous operation. Shafts are susceptible to material failure with regard to resonance frequency, fatigue loading, centrifugal forces, residual stresses and material defect distribution. Faults in machinery have a significant effect on component health. Machine faults include cracked shafts, unbalance rotating eccentricities, bends and misalignments [1]. Applied forces and bending moments affect the overall dynamic efficiency of a rotating machine. The increase of bending loads influence an ascending trend in terms of higher vibrational response [2].

Current PdM systems typically focus on fixed structures and do not provide diagnostic tools for interchangeability and reconfigurability in AMS. Dynamic and real-time prognostic forecasts of component health according to reconfigurable structures and components are of key interest to allow for rapidly updating failure predictions [3]. Data driven technical monitoring provides the most accurate information to derive statistical and physical properties for failure conditions when compared to classical failure prediction methods [4].

Symptoms before component failure for rotating machines are identified in terms of high amplitude vibration, excessive noise, the rise of operational temperature and the presence of smoke, in order of occurrence. Acoustic emissions and the presence of fluctuating vibrations are predominantly the first indicators of health deterioration of machinery.

Condition monitoring is primarily focused on vibration and acoustic health diagnostics [5].

1.2 Research Contribution

The paper focused on the PCA characteristics of the vibration response of a stainless steel shaft under various bending load conditions to develop a b. K-Means and DBSCAN clustering was utilized to determine the data cluster characteristics as basis for PdM. The shaft was stressed by means of a pure bending load employing R.R. Moore fatigue test conditions. The cluster characteristics were compared to illustrate vibration variation under loading conditions. The main contribution of the research is summarized as follows:

- Investigating the validity of vibration response employing R.R Moore fatigue test loading conditions to determine health prognosis characteristics.
- Analyzing PCA data clustering characteristics utilizing K-Means and DBSCAN clustering for R.R Moore fatigue test conditions.

- Experimental feasibility of proposed investigational method for PdM strategies for R.R Moore fatigue testing conditions.

2 Literature Review

2.1 *PdM Strategies*

PdM possesses some advantages when compared to other maintenance strategies specifically corrective and preventive maintenance. The advantages of PdM are: replacement of equipment close to imminent failure; minimization of component waste; optimization of cumulative operation time; reduction in maintenance cost and time due to catastrophic failure. Although PdM possesses capabilities to optimize production, drawbacks must be taken into consideration when designing for a PdM system. The following drawbacks of PdM were identified: the prediction of failure and RUL possesses statistical inaccuracies and may result in unnecessary repairs and maintenance; PdM algorithms demand data access, quality data and data integration between different operational components, and the requirement for processing infrastructure to utilize PdM is directly proportional to the quantity of big data especially in large production environments [6].

PdM strategies consists of two categories namely: statistical based and condition based maintenance [7]. Condition based maintenance is defined as the process of monitoring a system in terms of the health of the current machine or its components inherent failure mechanisms to optimize and determine the needs of maintenance and operating costs [8]. Statistical based maintenance takes into consideration the historical operational data and does not require a strong knowledge about the characteristic failure mechanisms [9].

2.2 *Current CM and PdM Systems that Have the Potential for Implementation*

Rotating shafts are susceptible to fatigue failure. Fatigue is caused by continuous cyclic application of variable loads. Typical fatigue failure is caused by micro-crack propagation that leads to structural failure. The R.R Moore fatigue test provides a testing procedure to determine the fatigue limit of rotating shafts under pure bending loads [10]. The disadvantage of standard dead-load rotating-bending fatigue machines is that these test benches can only deal with a constant amplitude fatigue tests. A test bench was proposed by [11] that utilizes a servo-control system in conjunction with a load cell to manipulate the loading cycle.

ML is employed to solve association, clustering and classifications for determining fault characteristics in rotating machinery. ML is a vital tool utilized in PdM to compute big data and improve on the accuracy of failure predictions. Artificial Neural Network (ANN) is a sub-division of ML and employs three basic layers: an input layer, a hidden layer and an output layer. The neural network layers provide rapid associations and pattern recognition [12]. A Genetic Algorithm (GA) in conjunction with ANNs was employed by [13], using the sensory data from accelerometers and acoustic sensors in the diagnostic of mechanical bearings. An ANN was utilized by [14] to predict and compensate for the backlash error in machine centers. While there are many different ML methods to approach PdM problems, LR was used in the study incorporating K-means and DBSCAN clustering.

Machine faults through vibration signals are also detected by means of PCA. PCA possesses the capability of reducing multiple Degrees of Freedom (DOF). The PCA method is advantageous when dealing with multiple sensory systems with various data that requires simultaneous processing. The reduction of DOF into fewer components allows for rapid data integration into other processing algorithms for e.g. ML and Fourier Transforms for signal recognition and classification [15]. K-means clustering is frequently used as a data clustering algorithm in executing unsupervised ML tasks [16].

3 Experimental Layout

3.1 Problem Definition/Formulation

The PCA vibration characteristics were unknown for a shaft with various bending loads under R.R Moore test conditions. Vibration data was required for multiple experiment repeatable sequences. Different loading conditions were to be defined and monitored. Data characteristics were generated through PCA incorporating K-Means and DBSCAN clustering algorithms. Bending loads were identified through clustering attributes. Correlation between logistic regression prediction results and clustering attributes were unknown.

The test bench was designed to replicate the R.R Moore fatigue test whereby the shaft is loaded under various pure bending conditions as shown in Fig. 1. The design incorporated 3-axis accelerometers directly attached to the shaft bearings and the structural layout was adapted from a design describe by [17]. A fault detection and prognosis system incorporating a vibrational sensory system was developed by [18] and adopted in the study. An experimental setup was built as shown in Fig. 1.

3.2 Method of Analysis

The experimental procedure utilized acceleration data from accelerometers to detect changes under bending loading conditions of a rotating shaft as shown in Fig. 2.

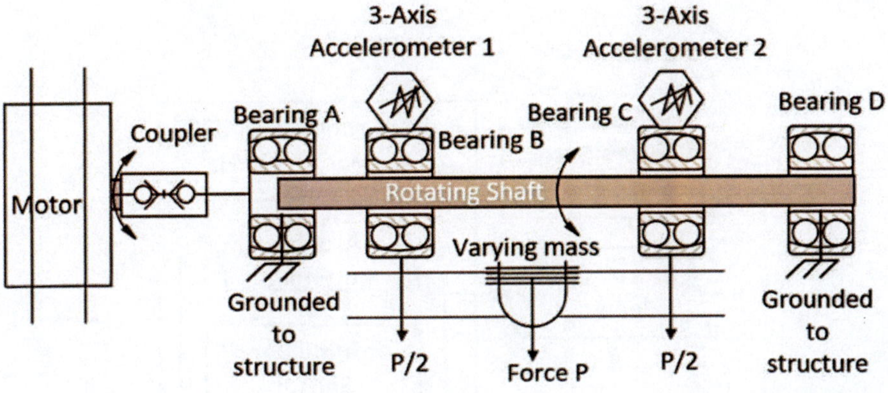

Fig. 1 Experimental setup

The process was initialized through a data retrieval algorithm. A signal was sent from a Python code to Arduino to read data at 100 Hz from the accelerometers. The value was sent to an Arduino Mega 2560 microcontroller, and was stored in a Pandas database. The retrieval algorithm was repeated until 300 readings were recorded for further processing. The data file was processed through a 2 dimensional PCA using Python to reduce the DOF. K-Means and DBSCAN clustering were employed to determine component characteristics of the reduced data [19, 20]. Cluster categories were allocated to the raw data and processed through a LR algorithm. A confusion matrix was generated as an output, which represents the predictions made. Seven variations of loading were applied using a bending loading range of 0–590 g as shown in Table 1. The experiment was run for 30 s at 100 Hz.

4 Results and Discussion

4.1 Vibration Response

Raw data was received from accelerometers for variations 1–7. Figure 3 illustrates the noisy sample data from variation 1 with zero added weights for two accelerometers in the x, y and z direction for accelerometer 1 only. The data preparation used a standard scalar algorithm to smooth out the noisy signal as shown Eq. (1). The computation of applicable statistics of the sample set is generated through Python. The scaling and centering of data occurs independently. The smoothing of data was only employed to determine the principal components. A 2-dimensional PCA was performed on the smoothed data as shown in Fig. 4. The noisy data were used as testing and training samples.

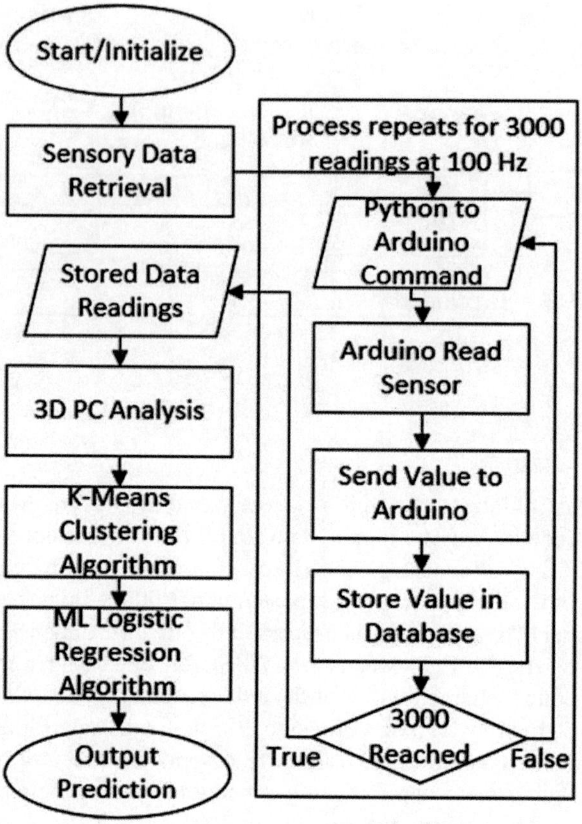

Fig. 2 Pseudocode for monitoring system

Table 1 Experimental operation parameters

	Var 1	Var 2	Var 3	Var 4	Var 5	Var 6	Var 7
Mass	0 g	120 g	200 g	275 g	320 g	475 g	590 g
PC1 (%)	33	36	29	28	35	25	32
PC2 (%)	24	21	27	24	20	22	25

$$z = \frac{x - u}{s} \tag{1}$$

The above equation has the following variables with z being the standard score, x is the sample, u is the mean and s is the standard deviation.

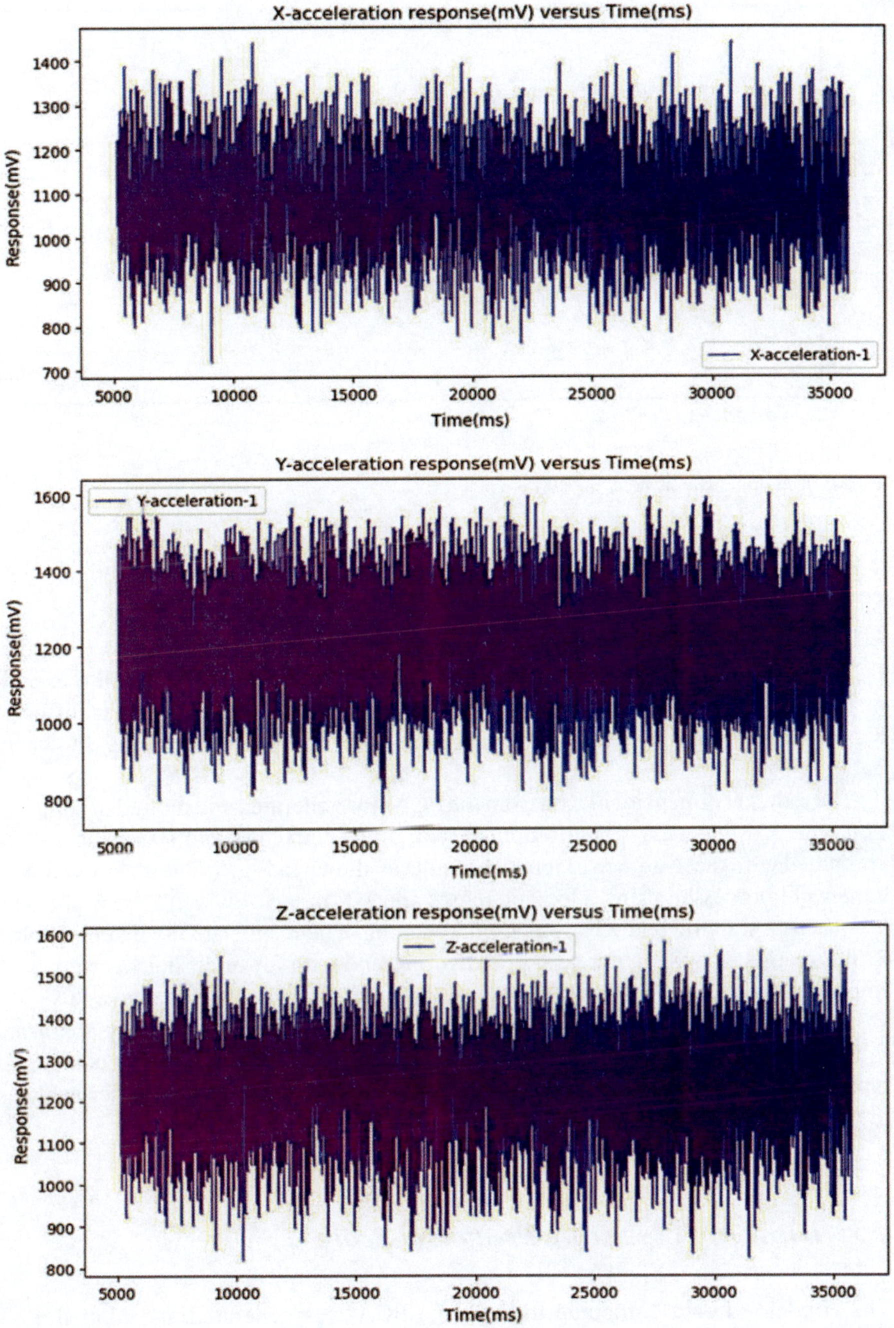

Fig. 3 Vibration response for no added weights for accelerometer 1 for X, Y and Z axis

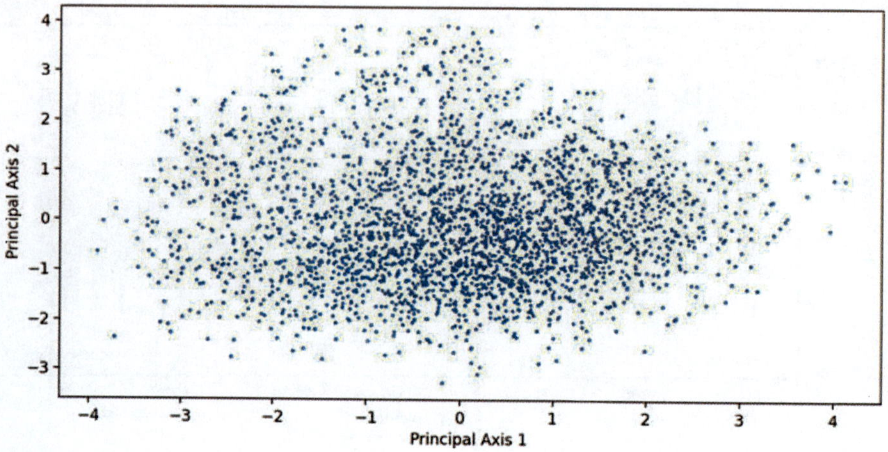

Fig. 4 Two dimensional PCA for 590 g weights

4.2 K-Means Clustering Prediction Score

PCA reduced the dimensionality of the data to 2 Principal Components (PC1 and PC2). Proportion of variance for PC1 and PC2 was determined across all 6 experimental variants. The variance describes the shape and the shape of data is different for all of the variants. The variation in varience for PC1 and PC2 is shown in Table 1.

The reduced dimensional data from the PCA was generated and plotted in conjunction with K-Means clustering outputs. Two, three, four, five and six clusters were computed with the 2-dimensional PCA results as shown in Fig. 5. The clusters parameters were processed using a logistic regression ML mode of which 75% of the data was used as training and 25% was used as testing. The results are tabulated in Table 2. The results posses an increasing prediction trend with fewer clusters as well as an improved prediction with higher loads. Good predictions scores are above 80%.

Further testing with respect to fault conditions is required to identify abnormal vibration symptoms that is caused by faults scenarios to rotating components. Fault clusters can be identified and cluster centroids can be compared for further analysis to predict life expectancy.

4.3 DBSCAN Clustering Prediction Score

The condensed data computed through the PCA was generated and plotted with DBSCAN clustering outputs. DBSCAN employs epsilon and sample thresholds for computing clusters. Epsilon characterizes neighbour data points with a minimum distance and minimum samples indicate the amount of data points in the epsilon

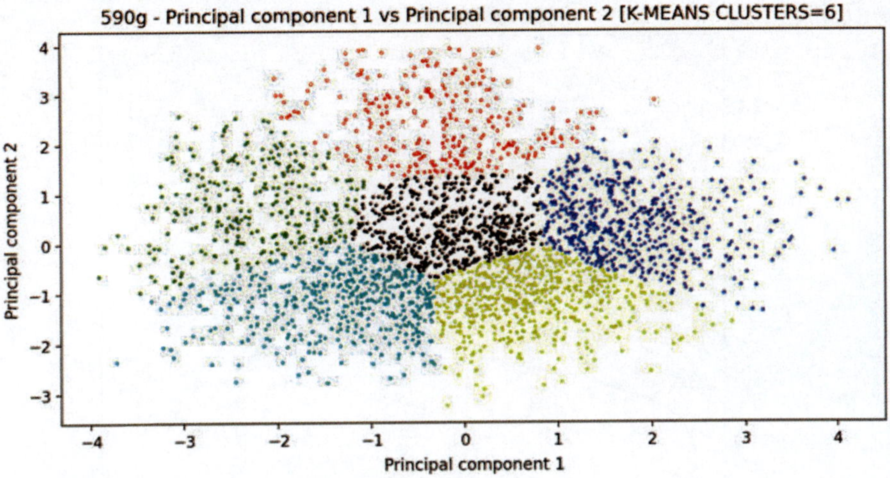

Fig. 5 K-Means clustering (6 clusters) for 590 g weights

Table 2 Prediction scores for all 7 experiments and K-Means clusters (2–6 clusters)

Prediction score					
Variant	2 Clusters (%)	3 Clusters (%)	4 Clusters (%)	5 Clusters (%)	6 Clusters (%)
1	99	96.87	96	94.6	95.2
2	98.5	95.5	91.2	91	91
3	99.4	91.8	90.8	84.7	83.7
4	98.4	90.1	86.1	86.7	85.3
5	99	91	89.3	82.1	78.7
6	98.9	90.1	86.1	83.2	78
7	98.8	95.3	94.3	92.9	91.1

length radius required to be classified as a core point. Data sets can be uniquely identified through an Epsilon threshold characteristic graph as seen in Fig. 6.

Epsilon and minimum sample values were chosen as seen in Table 3. Two, three, four, five and six clusters were computed with the 2-dimensional PCA results as shown in Fig. 7. The same clusters parameters were processed as with the K-means clustering algorithm using a logistic regression ML mode. The results for DBSCAN clustering are shown in Table 4.

Primarily DBSCAN generates two distinct visual "data zones". Data zone 1 as indicated by the colour green, represents the vibrational data close to normal operation. Data zone 2 as indicated by the colour blue represents the vibrational data outside of normal operation. Further experimentation with respect to failure conditions is required to identify abnormal vibration symptoms that causes damage to

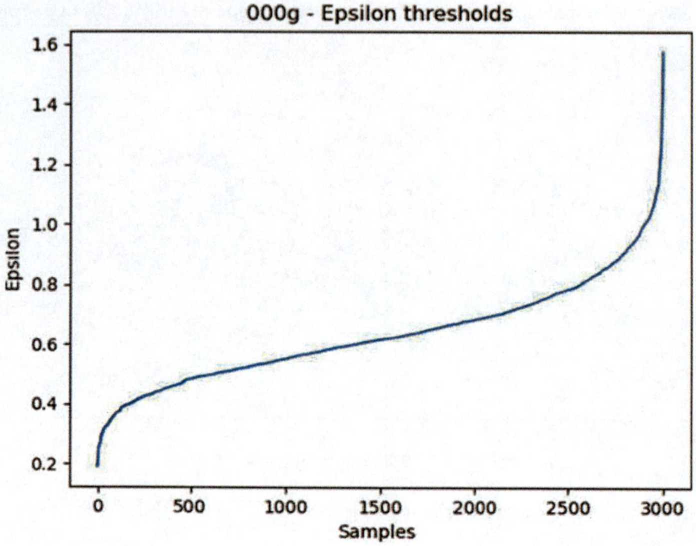

Fig. 6 Epsilon thresholds for DBSCAN

Table 3 DBSCAN parameters

Prediction score for 000 g					
Variant	2 Clusters	3 Clusters	4 Clusters	5 Clusters	6 Clusters
Epsilon	0.28	0.26	0.255	0.25	0.215
Min samples	20	20	20	20	20

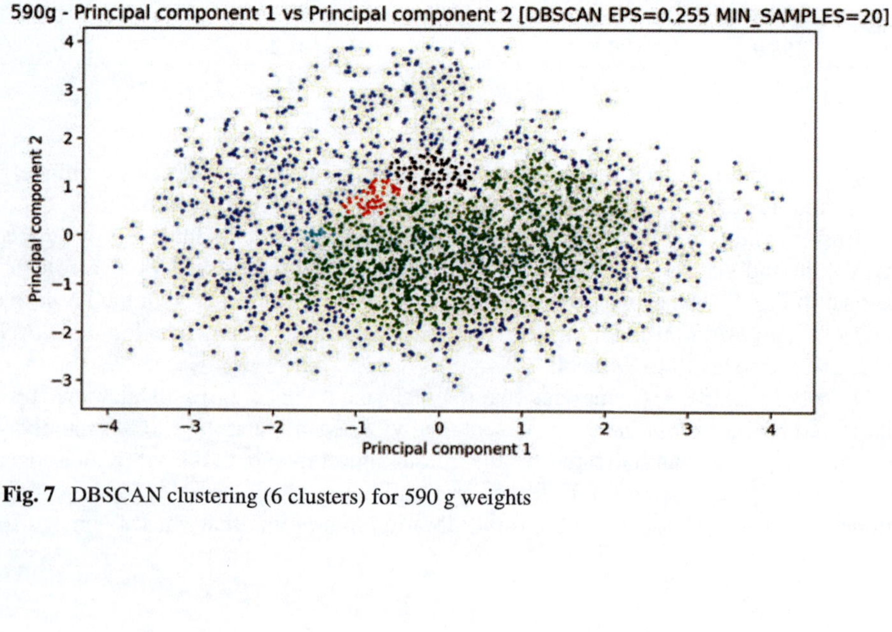

Fig. 7 DBSCAN clustering (6 clusters) for 590 g weights

Table 4 Prediction scores for all 7 experiments and DBSCAN clusters (2–6 clusters)

Prediction score					
Variant	2 Clusters (%)	3 Clusters (%)	4 Clusters (%)	5 Clusters (%)	6 Clusters (%)
1	89.4	86	83.4	83.7	77.5
2	87.7	83.6	83.1	82.5	68.8
3	93.3	90.7	90.4	89.5	71.66
4	89.6	87	86.2	84.3	79
5	91.9	86.1	86.2	81	85
6	90.4	87.3	83.7	78.7	76.3
7	91.5	86.8	83.1	76.5	74.4

rotating components. When failure conditions are known a failure distribution curve can be generated and component life expectancy can be examined further.

5 Conclusions and Recommendations

The study described a conceptual development of a condition monitoring system to observe the vibration response from R.R Moore loading conditions of a stainless steel shaft. Results illustrated that predicable features were achieved by increasing the bending load. A decrease in accuracy for K-means clusters was determined. The DBSCAN clustering experiements showed an overall increase of accuracy for 2 clusters and 5 clusters and an overall decrease of accuracy for 6 clusters. There was no overall accuracy change for 3 clusters and 4 clusters with respect to DBSCAN. Identifiable parameters were accurately classified through clustering and accurate classification comparisons were verified by means of a logistic regression ML algorithm. The variation in accuracy of the number of clusters for DBSCAN is the result of the variation of the shape of data during montioring for different variants.

Further studies can include higher frequency ranges for the classification of predominant identifiable parameters. The study indicated that higher loadings could improve on prediction outputs, therefore higher loading forces could be used to improve identifiable parameters. Further study is required for multiple attempts until failure to be conclusive. Other materials and shaft geometry are also potential variables that can be tested for further investigation.

References

1. Lees, A.W., Friswell, M.I.: Where next for condition monitoring of rotating machinery? Adv. Vibration Eng. **5**(4), 263–277 (2006)
2. Guo, D., Shi, X., Wang, Y., Sun, G.: Effect of shaft manufacturing bending deviation on dynamic response of geared rotor system. Adv. Mech. Eng. **8**(10), 1–13 (2016)

3. Dong, Y., Xia, T., Xiao, L., Xi, L.: Real-time prognostic and dynamic maintenance window scheme for reconfigurable manufacturing systems. In: *ASME 14th International Manufacturing Science and Engineering Conference*, Fairfield (2019)
4. Ye, X.W., Su, Y.H., Han, J.P.: A state-of-the-art review on fatigue life assessment of steel bridges. Math. Problems Eng., **2014**(Special Issue), 1–13
5. Lin, T.R., Tan, A.C.C., Howard, I., Pan, J., Crosby, P., Mathew, J.: Development of a diagnostic tool for condition monitoring of rotating machinery. In: *Proceedings of ICOMS Asset Management Conference*, Gold Coast (2011)
6. Li, Z.: Industry 4.0 - Potentials for Predictive Maintenance. In: *6th International Workshop of Advanced Manufacturing and Automation*, Manchester (2016)
7. Moya, M.C.C.: The control of the setting up of a predictive maintenance programme using a system of indicators. Omega **32**(1), 57–75 (2003)
8. Koons-Stapf, A.: Condition Based Maintenance: Theory, Methodology, & Application, *Reliability and Maintainability Symposium*, Tarpon Springs (2015)
9. Leone, G.: An algorithm for data-driven prognostics based on statistical analysis of condition monitoring data on a fleet level. In: *IEEE International Instrumentation and Measurement Technology Conference*, Pisa (2015)
10. Sepahpour, B.: A Practical Educational Fatigue Testing Machine. In: *121st ASEE Annual Conference and Exposition*, Indianapolis (2014)
11. Nogueira, R.M., Meggiolaro, M.A., Castro, J.T.P.: A Fast-Rotating Bending Fatigue Test Machine. In: *24th ABCM International Congress of Mechanical Engineering*, Curitiba (2017)
12. Jiří, K., Kuca, K., Blazek P., Krejcar O.: Application of Artificial Neural Networks in Condition Based Predictive Maintenance, Recent. In: *Developments in Intelligent Information and Database Systems*, Cham, Springer International Publishing, pp. 75–86 (2016)
13. Saxena, A., Saad, A.: Evolving an artificial neural network classifier for condition monitoring of rotating mechanical systems. Appl. Soft Comput. **7**(1), 441–454 (2007)
14. Wang, K., Li, Z., Braaten, J., Yu, Q.: Interpretation and compensation of backlash error data in machine centers for intelligent predictive maintenance using ANNs. Adv. Manuf. **3**(2), 97–104 (2015)
15. Plante, T., Stanley, L., Nejapak, A., Yang, C.X.: Rotating machine fault detection using principal component analysis of vibration signal. Anaheim, IEEE AUTOTESTCON (2016)
16. Ding, C., He, X.: K-means clustering via principal component analysis. In: *ICML '04: Proceedings of the twenty-first international conference on Machine learning*, New York (2004)
17. Arebi, L., Gu, F., Ball, A.: A comparative study of misalignment detection using a novel Wireless Sensor with conventional Wired Sensors, *25th International Congress on Condition Monitoring and Diagnostic Engineering* (*COMADEM 2012*), Huddersfield (2012)
18. Saxena, M., Bannett, O.O.: Bearing fault evaluation for structural health monitoring, fault detection, failure prevention and prognosis. In: *12th International Conference on Vibration Problems, ICOVP*, Guwahati (2015)
19. Chris D., He, X.: Cluster structure of K-means Clustering via Principal Component Analysis: *Advances in Knowledge Discovery and Data Mining*, pp. 414–418. Berlin, Springer (2004)
20. Hsu, J., Wang, Y., Lin, K., Chen, M., Hsu, J.H.: Wind turbine fault diagnosis and predictive maintenance through statistical process control and machine learning. IEEE Access **8**, 23427–23439 (2020)

Christian Basson holds an M.Sc. Eng degree in Mechanical Engineering from the University of KwaZulu-Natal. He is currently doing his Ph.D. in the Department of Mechanical Engineering. His specific focus and area of research are robotics and reconfigurable manufacturing technologies.

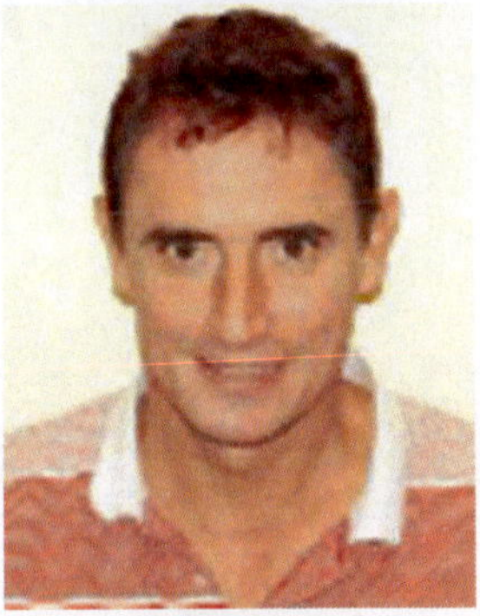

Glen Bright graduated with a B.Sc. (Mechanical Engineering), M.Sc. (Engineering) and Ph.D. (Engineering) degree at the Ex-University of Natal in Mechatronics, Robotics and Advanced Manufacturing Systems. Professor of Mechatronics, Robotics and Advanced Manufacturing Systems since 2002.

Jared Padayachee holds B.Sc. Eng, M.Sc.Eng. and Ph.D. (Engineering) degrees from the University of KwaZulu-Natal. His research focus is on the mechanics of robotic mechanisms.

Sarp Adali holds B.Sc. degree from Middle East Technical University (Turkey), Ph.D. from Cornell University (USA) and is a Professor at UKZN. His research areas include design optimization, composite materials and structures, nanomechanics and variational methods.

Analysis and Modelling of the Track Quality Index of Railways

Stefan Laubscher and Johannes L. Jooste

Abstract Public transportation is vital to the sustainability of urban areas. The benefits associated with rail transportation make it an attractive public transport option for commuters. Efficient maintenance and asset management is required to ensure the condition of railway infrastructure remains at a sufficient level for railway operations to continue without delays or disruptions. A known cause of deteriorating railway tracks is ineffective or absence of maintenance practices. Technological advancements cause an increase in available data in various industries. The railway industry is no exception to this, with geometry measurements such as track quality index (TQI) of permanent way being an example of high volumes of generated data in the railway industry. TQI is an indicator of track condition based on five rail geometric irregularities. Changes in track geometry cause track irregularities, which reduce the quality of railway tracks. TQI data pose challenges during the analysis thereof due to the large number of data entries. This paper investigates the track geometry parameters that are measured as part of the TQI. A quantitative approach is followed to propose a method for analysing TQI data. A case study is conducted on segments of the railway network in Cape Town, South Africa to evaluate the application of the proposed analysis method. The proposed method makes use of data mining techniques along with ArcGIS software and network spatial analysis that takes the network characteristics of a railway network into account. The proposed method of analysis aims to assist railway technicians with their maintenance decision-making to ultimately improve the predictive maintenance and asset management strategies currently in place.

Keywords Railway maintenance · Track quality index · Network spatial analysis

S. Laubscher · J. L. Jooste (✉)
Department of Industrial Engineering, Stellenbosch University, Stellenbosch, South Africa
e-mail: wyhan@sun.ac.za

S. Laubscher
e-mail: 19851472@sun.ac.za

K. von Leipzig et al. (eds.), *Smart, Sustainable Manufacturing in an Ever-Changing World*, Lecture Notes in Production Engineering,
https://doi.org/10.1007/978-3-031-15602-1_37

1 Introduction

The importance of reliable public transport in urban areas are well-known. An effective public transport system provides the population with access to most parts of a city [1]. Rail transportation is seen as a dependable method of commuting since it is not affected by weather factors or traffic [2]. The incorporation of efficient railway transportation systems into populated areas increases the mobility of commuters and lowers individual traffic in urban and suburban areas [3]. Rail transportation also contributes to a country's economy, as economic opportunities are closely associated with the ability of people to move around [4].

Maintenance of the railway is required to ensure that the track condition does not deteriorate to a level where derailment or other problems become likely [5]. Incorrect or delayed maintenance activities might cause railway tracks to degrade to the extent where preventive maintenance cannot be applied and repairs would be required [6]. Railway maintenance activities include rail grinding, rail replacement, track stabilisation, tamping, ballast injection, and sleeper replacement to correct effects such as deformation, geometry changes, and abrasion [7].

A variety of track quality evaluation indicators are currently used globally. None of these indicators is universally standardised upon [8, 9].

Furthermore, various approaches exist for analysing the indicator data. Studies by [10] and [11] are examples of initial mechanistic analysis models where railway track degradation predictions are made by means of observation and historical data. Scholars have also applied a variety of statistical models for analysing track degradation. Examples of deterministic linear-, exponential-, multistage- and multivariate regression models are provided by [12–15], respectively. Stochastic modelling techniques that have also been explored include a bivariate Gamma process [16], autoregressive moving average modelling [17] and Petrie Nets [18]. Artificial intelligence (AI) techniques, such as artificial neural networks [19], adaptive neuro-fuzzy inference systems [20] have also been used, while [21] explored the application of principle component analysis (PCA) and [22] used survival modelling and analysis. Disadvantages of these techniques include; the inability to cater for uncertainty (e.g., mechanistic models), the combination of different evaluation indicators into one (e.g., statistical models), loss of information (e.g., PCA), identical track assumptions (e.g., survival analysis) and high cost of computing power (e.g., AI).

Track quality index (TQI) is one of the indicators which is widely used as a measure of the condition of railway infrastructure, including in South Africa. TQI is a statistical summary of five track geometry parameters that indicate the condition of a railway track [23]. TQI parameters are measured by a track geometry inspection vehicle that can provide real-time measurement data on-board and post-processed reports afterwards. TQI is used as a health index that is used to evaluate the quality of railway tracks. This it typically done by comparing the TQI value to standards and limits to determine the quality of the track and identify areas which should be prioritised.

South Africa has around 22,000 km of railway track in total [24]. Taking into consideration that TQI data measurements are taken over sections of 50, 100, and 200 m, that results in 110,000–440,000 measurements, each providing information in the form of several parameters. This amounts to countless data points that pose challenges when analysis and interpretation is performed. Current methods of analysing and interpreting TQI data are time-consuming and unstructured, which creates an opportunity to improve the methods and utilise TQI data to its full potential.

In this paper, a framework for analysing TQI data is proposed that would assist railway technicians in effectively using track geometry parameter measurements. This framework incorporates network spatial analysis and includes visual representation for improved interpretation. The objective of this framework is to assist railway technicians in improving maintenance decision-making. A further objective is to facilitate a shift from corrective and reactive maintenance to more preventive maintenance practices that would ultimately improve the reliability and sustainability of railway tracks.

2 Background

In this section an overview is provided about the parameters that form part of the TQI and about network spatial analysis. This serves as basis for the reader to comprehend the novel analysis approach to TQI data.

2.1 *Track Quality Index*

The track quality index, which is seen as a health index to determine the condition of railway tracks, consist of five geometry parameters. Equation 1 provides the function for calculating the overall TQI of a railway:

$$TQI = \sum_{i=1}^{5} \sigma_i \tag{1}$$

The five underlying track geometry parameters which TQI consists of are: vertical alignment, horizontal alignment, gauge, superelevation, and twist [23]. Geometry parameter exceedances are measured and recorded for each track segment and are categorised as priority 1, priority 2, or priority 3 exceedances according to their urgency. Priority 1 exceedances are urgent and require immediate attention, while priority 2 and priority 3 exceedances are less urgent.

2.1.1 Vertical Alignment

Vertical alignment, also known as longitudinal level, measures the vertical relationship between three points on one rail. Any irregularities are determined by measuring the distance between the running surface and a reference line that remains constant. Vertical alignment issues generally occur in areas where the surface below the rails cannot support the load distributed by the rails or sleepers, which means one spot on the rail carries a larger load than the rest [25].

2.1.2 Horizontal Alignment

Horizontal alignment is a measure of the degree to which the rails are aligned in the intended direction. Deviations are measured from a reference line to a point on the inside of the left- and right-hand railway tracks. The measured difference between the actual alignment and what it is supposed to be, constitutes a horizontal alignment deviation and is caused by faulty sleepers, base plates, fastenings, or excessive rail stress [25, 26].

2.1.3 Gauge

Gauge is defined as the distance between two rails measured at a right angle at a location 14 mm below the running surface of the rail [25, 27]. Locations with excessive side wear, incorrect or missing fastenings, or wide gauges on sharp turns are prone to gauge issues, which lead to increased dynamic loads [25, 28].

2.1.4 Superelevation

Superelevation describes the distance in vertical elevation between a point on one rail and a point on the opposing rail, with the elevation difference being measured along a line that runs perpendicular to the centre line of the track. Superelevation is important in curves, as it is required that the inside rail of the track has a lower elevation than the outside rail to compensate for centrifugal forces [29].

2.1.5 Twist

Twist constitutes the algebraic difference between two superelevation measurements taken at a fixed distance from one another and is usually expressed in mm/m as a ratio between the two measurement points [27, 30]. A vehicle with a rigid body that runs on a track with one wheel not making contact with the track is usually indicative of a twist issue [25].

2.2 *Network Spatial Analysis*

Spatial analysis is used to identify and describe the tendencies of a data set with relation to geometry and is performed with mathematical expressions and relations [31]. While conventional spatial analysis assumes a Euclidian distance and continuous plane between points, network spatial analysis assumes space that is constrained to a network. Therefore, network spatial analysis provides more accurate results than conventional spatial analysis when analysing network phenomena [32]. Overestimation of clusters or densities is usually a consequence of applying conventional spatial analysis to network phenomena [33]. In the remainder of this paper network spatial analysis is applied to TQI data which is shown to provide a more accurate representation of exceedance clusters to assist with the implementation of preventive maintenance practices.

3 Methodology

The aim of applying network spatial analysis is to analyse and visualise TQI exceedances to identify problem areas, or so-called *hotspots*, along the railway network. Spatial analysis along networks (SANET) is used in combination with the ArcGIS software to perform the analysis and visualisation.

3.1 *Spatial Analysis Along Networks (SANET)*

SANET is defined as a toolbox containing various methods that can be used to analyse events along networks. Events which occur directly on a network, known as on-network events, and events which occur next to a network, known as alongside-network events, are grouped together under along-network events. Network spatial analysis can be applied to any network system, which includes streets and pedestrian ways, rivers and canals, gas and oil pipelines, and electrical and telephone wires to name a few. It also has the ability to analyse locational events that are influenced by shortest-path distances. A third feature of this method is the ease with which it can be applied to a three-dimensional network, since a three-dimensional space has the same topological data structure as a two-dimensional space. Another advantage of network spatial analysis is the ease with which analytical derivation can be controlled when implemented on a network compared to when it is implemented on a planar space [34].

The SANET toolbox contains various network spatial analysis methods, of which the network kernel density estimation (KDE) method is selected for this analysis. Previous research was performed in which the application of conventional and network spatial analysis to traffic accidents was compared to determine which method

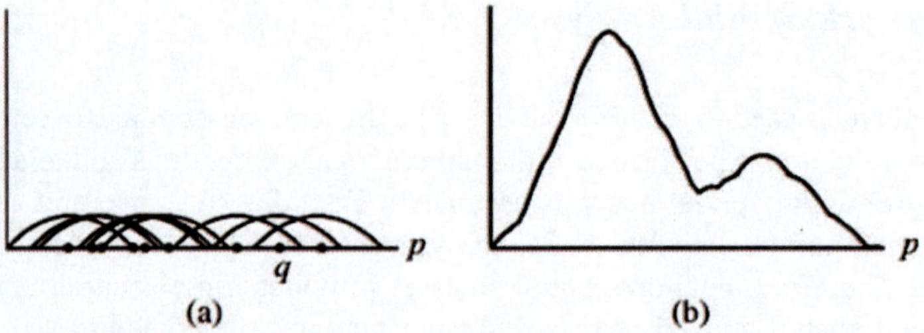

Fig. 1 Kernel density estimation: **a** small hills with points as their centres, **b** density function produced by heaped hills

is better suited to a network application. The authors concluded that network KDE is more reliable due to lower levels of bias [35]. Planar KDE does not consider the density of the road network into account and tends to overdetect clusters on a network [35, 36].

Network KDE is used to identify clusters with a high density of points along a network. It produces a density function by placing a point at the location of a certain event, and a curved line (like a small hill) is produced with the black point as its centre. Figure 1 is an illustration of how the network KDE function is created.

3.2 ArcGIS

Geographic information systems (GIS) are able to capture, store, inspect and visualise data entries that are referenced to a specific geographic position. It is a system operated on a computer that can establish and investigate spatial relationships, trends and patterns between data points [37]. ArcGIS is a geospatial processing program used to create maps with geographic data and analyse any data with a locational component. Two components of ArcGIS, ArcMap and ArcScene, are used in this study. ArcMap is the main component of ArcGIS used for displaying and editing maps and allows the user to manipulate and analyse spatial data. ArcScene enables the user to visualise data and analysis outputs in three dimensions by extracting height information from feature geometry, attributes or layer properties, and also includes analysis tools. ArcMap is used in this study to create, edit and prepare the locational data prior to the analysis process, while ArcScene is used to visualise the outputs generated by the SANET analysis.

3.3 Track Quality Index Data Analysis Framework

The Track Quality Index Data Analysis Framework (TQI-DAF) is developed to guide technicians when analysing track geometry information. The framework is a formal

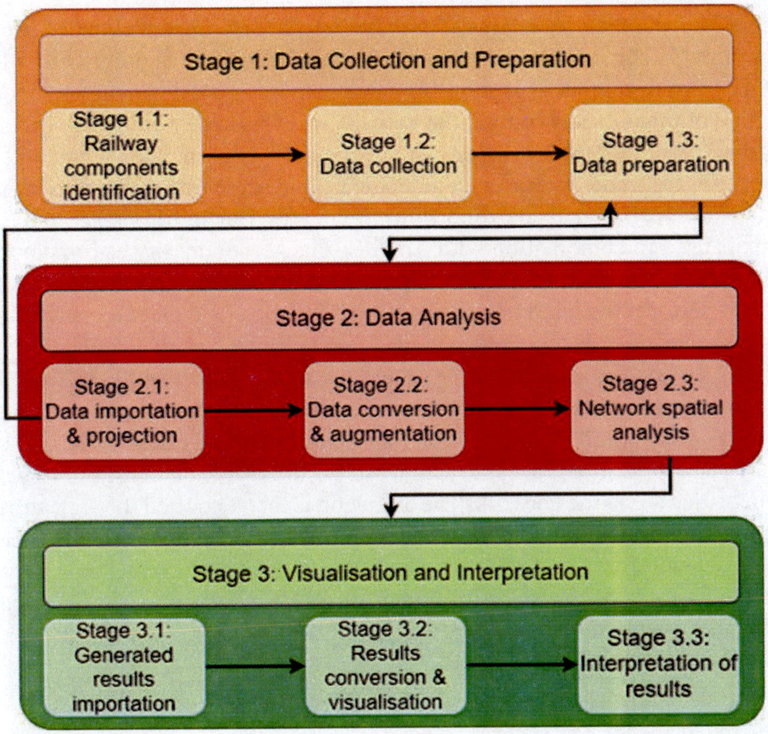

Fig. 2 TQI-DAF

representation of the proposed methodology that provides the steps that must be followed when analysing track geometry information. The TQI-DAF consists of three stages, each stage containing three substages. Figure 2 is an illustration of the TQI-DAF that shows the sequence of the different stages.

Stage 1 covers the collection and preparation of data. Stage 1.1. covers the identification of railway components for analysis. Railway track infrastructure consists of several components, which includes ballast, clearance, geometry, overhead track equipment, and wear information. Measurements and reports are generated for each of these elements of railway infrastructure. The objective of the analysis determines what data is needed and the railway components should be selected in accordance with the objective. In Stage 1.2 the collection, sorting and storing of the data for the identified components are done. A data collection strategy is developed detailing the process of collection sorting and storing. Key elements of the strategy should include the source of the data, mode of collection, duration and resources required and ethical and confidentiality considerations concerning the data. Stage 1.3 follows with the process of data wrangling to ensure the data is in a reliable format for analysis. This process includes steps to combine data sources, removing empty fields or duplicates and publishing the data in a suitable format for analysis.

Stage 2 covers the data analysis. In Stage 2.1 the data importation and projection are performed. This is done by using the ArcGIS software as described in Section "ArcGIS". An important aspect of this stage is to ensure the format and coordinate projection of input data is compatible with the analysis to be performed. In Stage 2.2 data conversion and augmentation is performed. This allows for alterations of the data set after the initial importation in Stage 2.1 and screening of the data in ArcGIS. The network spatial analysis follows in Stage 2.3. During this stage the SANET toolbox is used in combination with ArcGIS. The kernel density estimation method of SANET, as described in Section "Spatial Analysis Along Networks (SANET)", is used to perform the analysis. The result of this stage is the actual network kernel density estimation.

Stage 3 covers the visualisation and interpretation of the results. In Stage 3.1 the density estimation data files are imported into the ArcMAP software, as described in Section "ArcGIS", and configured for visualisation. In Stage 3.2 the density estimations are visualised. The process to achieve this is to extrude the results to visualise the point densities along the network. The extrusion is then exported through an attribute selection process and imported in the ArcScene software (Section "ArcGIS") to illustrate the densities on a map. The result of this stage is three visual representations: the three-dimensional extrusion, the densities on a basemap and a hotspot map. The final Stage 3.3 consists of the interpretation of the results.

4 Case Study

The TQI-DAF is applied in a case study to the southern line of the City of Cape Town's railway network, which runs from Cape Town Station to the end of the line at Simon's Town. For the case study the TQI exceedance data of several segments with different surroundings are investigated as proof of the applicability of the TQI-DAF. The TQI-DAF Stages 1.1–3.1 were performed as described in Section "Spatial Analysis Along Networks (SANET)". The results and interpretaion—Stages 3.3 and 3.4—are presented in the next sections to demonstrate the outputs of the TQI-DAF and its value towards decision making. Outputs are interpreted individually or combined with each other to gain insights on the track condition. Combining the outputs of the proposed framework with existing reports and practices would provide further valuable knowledge about the condition of the railway track.

4.1 Three-Dimensional Extrusion

The first output of the framework is a three-dimensional extrusion, which shows the exceedance density profile of each parameter on a segment. This visualisation is created from the network KDE analysis results in ArcScene. The density clusters are extruded to clearly indicate areas with more detected exceedances. This visual

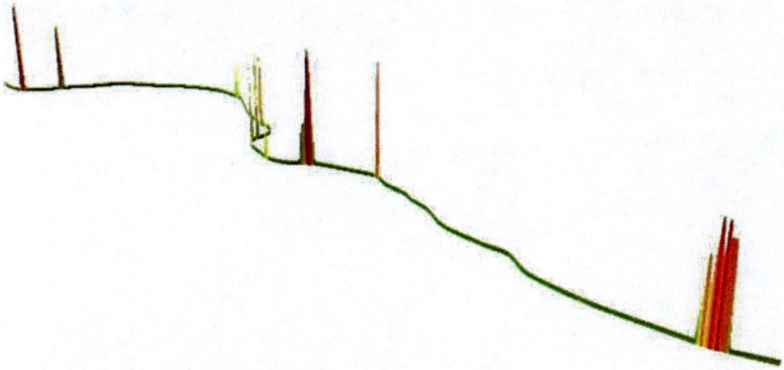

Fig. 3 Three-dimensional extrusion

representation enables the user to identify high-density areas with ease. Figure 3 is an example of a three-dimensional extrusion of horizontal alignment exceedances on a track segment.

From Fig. 3, it is evident that there are areas along the segment where exceedances occur in clusters.

4.2 Exceedance Density Points on Basemap

Another output of the methodology is a visualisation of the density points combined with a basemap. This visualisation is created from the three-dimensional extrusion visualisations by selecting the attribute of interest and exporting it to ArcMap. In this case, the selected attribute is the average value of each kernel density point, which represents the height of the kernel density function. In ArcMap, these average values are visualised according to size, with green indicating a low-density value and red indicating a high density value. Combining the density points with a basemap of the area enables the user to pinpoint the exact location of the exceedance clusters, while identifying possible causes of these exceedances in the surroundings of the clusters. Figure 4 illustrates the combination of density points with a basemap, which shows the exact location of exceedance clusters. The most prominent exceedance clusters are circled.

With the use of ArcMap, further investigation might reveal external factors responsible for these exceedance clusters.

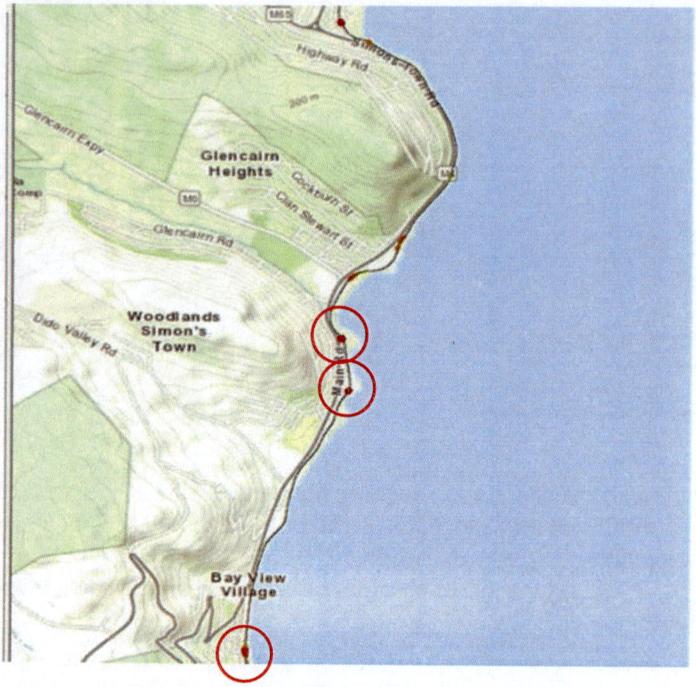

Fig. 4 Exceedance densities on a basemap

4.3 *Interpretation of Methodology Outputs*

In this case study, each track geometry parameter is analysed individually for each segment. This is done so one can determine which, if any, geometry parameter(s) is the main cause(s) of detected exceedances. It provides a detailed view of the track condition and prevents the user from drawing any misleading conclusions from combined data outputs. During the interpretation of the outputs, it is often seen that exceedance clusters of different geometry parameters occur in the same locations. The outputs of different geometry parameters are then combined to confirm this observation, which leads to the identification of truly problematic areas.

Exceedances are also analysed separately according to urgency, i.e. priority 1, priority 2 and priority 3 exceedances are initially analysed individually. The priority exceedances are also combined on occasion should the user require to do so. Priority 1 exceedance outputs are used to identify areas that require urgent corrective work. Ideally, priority 2 and priority 3 exceedance outputs should be used for planning the preventive maintenance strategy since they do not require urgent attention.

5 Conclusions

An investigation into the current use of TQI data revealed that it is not used to its full potential. TQI data contain valuable information about the condition of a railway track and, if used effectively, could assist railway authorities in their efforts to ensure that railway tracks operate at improved levels of reliability.

In this study, a methodology to analyse TQI data is proposed. The methodology is developed based on the nature of network data to ensure it produces accurate and reliable outputs. Network spatial analysis is applied to railway geometry data since it assumes space that is constrained to a network, along with geospatial processing programs that are used to visualise the analysis outputs.

The proposed methodology is applied to real data in a case study to determine its feasibility. The generated outputs enable the user to identify areas along the rail network that are problematic, and plan accordingly. Maintenance technicians are assisted in their decision-making, as the maintenance strategy and schedule can be produced based on the methodology outputs. Better maintenance planning, with the use of the proposed methodology, will improve the overall condition of a railway network. This would result in a shift from corrective maintenance work to a maintenance strategy that is more focused on predictive work.

Although the framework provides new insights into the condition of railway tracks, only one network spatial analysis method is applied. Future research should focus on expanding the framework with additional network spatial analysis methods, and the SANET toolbox in particular. Further work should focus on applying the TQI-DAF for more rail track segments to gain more insight from its practical use into the results and the interpretation thereof. This would allow for identifying opportunities to transition to more predictive maintenance. Analysis of priority 2 and priority 3 (less severe) exceedances with the TQI-DAF should also be explored to provide improved decision support for maintenance planning and scheduling.

References

1. Schaffner, C.: The Importance of Public Transportation for Sustainability in Arctic Cities, Arctic PIRE (2017)
2. The Economic Times: Definition of 'Rail Transport', The Economic Times (2020)
3. Polom, M., Tarkowski, M., Puzdrakiewicz, K.: Urban transformation in the context of rail transport development: the case of a newly built railway line in Gdansk. J. Adv. Transp. (2018)
4. Rodrigue, J.P.: Chapter 3—Transportation, Economy and Society, The Geography of Transport Systems (2020)
5. Railway Technical: Train Maintenance, The Railway Technical Website, Available from http://www.railway-technical.com/trains/train-maintenance/ [22 Oct 2021] (2019)
6. Hamid, A., Gross, A.: Track quality indices and track degradation models for maintenance-of-way planning. Transp. Res. Record (1981)
7. AGICO Group: Railway Track Maintenance, The AGICO Group (2020). Available from http://www.railroadfasteners.com/railway-track-maintenance.html [21 Oct 2021]

8. Sadeghi, J.M., Askarinejad, H.: Development of track condition assessment model based on visual inspection. Struct. Infrastruct. Eng. **7**(12), 895–905 (2011)
9. Fazio, A.E., Corbin, J.L.: Track quality index for high speed track. J. Transp. Eng. **112**(1), 46–61 (1986)
10. Shenton, M.J.: Ballast deformation and track deterioration. Track Technology, pp. 253–265 (1985)
11. Sato, Y.: Japanese studies on deterioration of ballasted track. Veh. Syst. Dyn. **24**, 197–208 (1995)
12. Westgeest, F., Dekker, R., Fischer, R.: Predicting rail geometry deterioration by regression models. Adv. Safety Reliab. Risk Manag., pp. 926–933 (2012)
13. Sadeghi, J., Askarinejad, H.: Development of improved railway track degradation models. Struct. Infrastruct. Eng. **6**(6), 675–688 (2010)
14. Ahac, M., Lakusic, S.: Tram track maintenance-planning by gauge degradation modelling. Transport 30 **4**, 430–436 (2015)
15. Guler, H., Jovanovic, S., Evren, G.: Modelling railway track geometry deterioration. vol. 2 of Transport 164. Institution of Civil Engineers, Institution of Civil Engineers—Transport 164 (2011)
16. Mercier, S., Meier-Hirmer, C., Roussignol, M.: Bivariate gamma wear processes for track geometry modeling, with application to intervention scheduling. Struct. Infrastruct. Eng. **8**(4), 357–366 (2012)
17. Quiroga, L., Schneider, E.: Modelling of high speed railroad geometry ageing as a discrete-continuous process. Stochastic Modelling Techniques and Data Analysis International Conference (2010)
18. Andrews, J., Prescott, D., De Rozieres, F.: A stochastic model for railway track asset management. Reliab. Eng. Syst. Saf. **130**, 76–84 (2014)
19. Sadeghi, J., Askarinejad, H.: Application of neural networks in evaluation of railway track quality condition. J. Mech. Sci. Technol. **26**(1), 113–122 (2012)
20. Dell'Orco, M., Ottomanelli, M., Caggiani, L., Sassanelli, D.: New decision support system for optimization of rail track maintenance planning based on adaptive neurofuzzy inference system. Transp. Res. Record: Journal of the Transportation Research Board **2043**, 50–53 (2008)
21. Lasisi, A. and Attoh-Okine, N.: Principal component analysis and track quality index: a machine learning approach. Transp. Res., (2018)
22. Khouy, I.A., Larsson-Kraik, P., Nissen, A., Juntti, U., Schunnesson, H.: Optimisation of track geometry inspection interval. J. Rail Rapid Transit **228**(5), 546–556 (2014)
23. Liu, R., Xu, P., Sun, Z., Zou, C., Sun, Q.: Establishment of track quality index standard recommendations for Beijing Metro. Discrete Dyn. Nature Soc., **15** (2014)
24. Department of Transport: Welcome to Rail Branch (2020), Available from https://www.transport.gov.za/rail [21 Oct 2021]
25. Zaayman, L.: Track condition monitoring and analysis for effective maintenance planning. Magazine South African Inst. Civil Eng., (2011)
26. Sadeghi, J.: Development of railway track geometry indexes based on statistical distribution of geometry data. J. Transp. Eng., 693–700 (2010)
27. Soleimanmeigouni, I., Ahmadi, A., Kumar, U.: Track geometry degradation and maintenance modelling: a review. J. Rail Rapid Transit **232**, 73–102 (2018)
28. Zarembski, A.M., Attoh-Okine, N., Einbinder, D., Thompson, H., Sussman, T.: How track geometry defects affect the development of rail defects. American Railway Engineering and Maintenance of Way Association, 1081–1096 (2016)
29. Powell, A.F., Grabe, P.J.: Exploring the relationship between vertical and lateral forces, speed and superelevation in railway curves. J. South African Inst. Civil Eng. **59**, 25–35 (2017)
30. Muinde, M.S.: Railway Track Geometry Inspection Optimisation, Maintenance Engineering Master's Thesis, Lulea University of Technology (2018)
31. Mayhew, S.: A Dictionary of Geography. Oxford University Press (2009)
32. Ying, V.: Methods for Spatial Analysis on a Network, Master of Science in Statistics, University of California (2013)

33. Yamada, I., Thill, J.F.: Comparison of planar and network k-functions in traffic accident analysis. J. Transp. Geogr. **12**, 149–158 (2004)
34. Okabe, A.: Spatial analysis along networks. Encyclopaedia of GIS (2017)
35. Shafabakhsh, G.A., Famili, A., Akbari, M.: Spatial analysis of data frequency and severity of rural accidents. Int. J. Transp. Res., (2016)
36. Steenberghen, T.: Spatial clustering of events on a network. J. Transp. Geogr. **18**, 411–418 (2010)
37. Micalizio, C.: GIS (Geographic Information System), National Geographic (2017)

Stefan Laubscher obtained a B.Eng in Mechanical Engineering from Stellenbosch University. He is currently in the process of completing his MEng in Engineering Management, also at Stellenbosch University.

Johannes Jooste is a Professional Engineer and holds a PhD in Industrial Engineering from Stellenbosch University. He is currently a senior lecturer in the Department of Industrial Engineering and focusses on asset- and asset management related research.

A Project Management Framework for the Modernisation of Passenger Railway Depots in Developing Countries

Adquate L. Masikati, Johannes L. Jooste, and Cornelius Fourie

Abstract Passenger railway organisations in developing countries have been operating old rolling stock for decades. With the advancement and complexity of modern technology, these organisations are required to renew their fleet. However, there are no guidelines that can help these organisations to implement a disruptive technological transition from old, dilapidated depots to the desired state that can support new technology in new rolling stock fleets. In this paper a project management framework is developed for modernising passenger railway depots to support the new rolling stock fleet and realise efficiency gains. To realise efficiency gains, the railway depots are viewed as large technical systems. Data was collected by conducting a systematic literature review and interviews with subject matter experts from industry. The project environment for railway depots is multidisciplinary hence the framework is developed to be robust for accommodating different project landscapes. Due to the complexity of railway systems, the framework offers the necessary flexibility to accommodate different project dynamics. The feedback mechanism in the framework ensures project flexibility without compromising on control. The goal of modernisation is to achieve sustainability through several objectives, namely supportability, interoperability, systems performance improvement, maintenance optimisation and productivity improvement. These objectives are integrated in a customised work breakdown structure which forms part of the framework and which facilitates concurrent execution of projects by different teams in a closely coordinated decentralised framework. With the developed project management framework, the process of modernisation becomes integrated, the delivery performance is improved and lean methodologies are easily implemented thereby improving the competitiveness of the modernisation plan.

Keywords Modernisation · Project management · Railway

A. L. Masikati · J. L. Jooste (✉) · C. Fourie
Department of Industrial Engineering, Stellenbosch University, Stellenbosch, South Africa
e-mail: wyhan@sun.ac.za

C. Fourie
e-mail: cjf@sun.ac.za

K. von Leipzig et al. (eds.), *Smart, Sustainable Manufacturing in an Ever-Changing World*, Lecture Notes in Production Engineering,
https://doi.org/10.1007/978-3-031-15602-1_38

1 Introduction

The establishment of passenger railway depots was determined by economic conditions of individual countries. As time progressed, new technologies were invented rendering existing technologies obsolete, operational costs increased, technical standards and asset management principles and practices were introduced resulting in a need to modernise these depots [1]. Modernisation was done by additions and changes on an ad hoc basis without stopping depot operations. The modernisation of passenger railway depots did not change the kind of work, but did change the resources and methods of performing the work. The advent of disruptive technologies in the transport sector such as automobiles in the 1920s marked the end of railway monopoly in passenger transport and was followed by the great depression in the 1930s [2]. However there has been a renaissance of railway transport that is motivated by the need to address climate change, population growth and the need to exploit inherent competitive strengths in railway genetic technologies.

The first railway in Africa was established in 1853 and in South Africa in 1860 [3]. Passenger railway business in Africa was marred by unsubsidised operating costs for passenger transport, high maintenance costs, high costs of renewal, lack of skills, lack of strategic planning and political instability. The idea of concessions tried to resuscitate the railway business in some countries but it was not an outright success because of poor commercial structures, unsupportive institutional frameworks, poor decision-making and assessments at the stages of project identification and preparation [3].

Africa is still trying to establish itself in modern rail transport and there is a gap between the current status and the regional vision. The railway industry is being transformed by high-speed rail resulting in inertia within developmental activities in conventional rail [4].

After years of poor maintenance and low investment, many railways in developing countries are underutilised and require renewal. The Passenger Railway Agency of South Africa (PRASA) among other organisations are mandated with providing passenger railway transport but have been challenged over the years with outdated rolling stock technology in their fleet. With the advancement and complexity of modern technology, these organisations are required to renew their fleets for passenger rail but there are no guidelines that can assist them to transform from the old railway depots to modernised ones that offer efficient technical support to accommodate new rolling stock technologies.

Railway depots in these countries are often characterised by dilapidated infrastructure, technological incompatibilities and incapacitation, hence they require modernisation.

Modernisation is driven by both the internal and external environment which influence how organisations respond to change. Since the way an organisation performs is influenced by its technical resourcefulness, organisations modernise to align capabilities with their vision for sustainability [5]. In order to meet the required levels of productivity and operational performance in terms of reliability, availability, maintainability and safety (RAMS), organisations change by adopting new technology to bridge technological gaps in a disruptive transition.

Railway organisations are not only consumers but also producers of some of their replacement parts. Hence, they must choose a way of manufacturing them. Tailored replacement parts are manufactured to meet the demand for which they are produced [6]. With the advancement of manufacturing methods, information and communication technologies, there is a need to shift from traditional processes to more advanced technology-based processes. Rolling stock of railway organisations consist of different generations of technology and designs. In order to mass-produce customised replacement parts with different specifications, the production system must respond in real-time by adopting Smart Manufacturing Systems principles. Implementation of Smart Manufacturing Systems require interoperability between systems for communication and information sharing. Consequently, modernisation of machine shops in railway depots must be adaptable and responsive which are important metrics of Smart Manufacturing. The goal of Smart Manufacturing is to increase the productivity of a production firm using a technology-driven approach to optimise production processes [7].

Railway depots are further asset intensive environments which require an asset-centric approach in the identification, design, construction, operation and maintenance of assets for organisations to remain competitive. Asset management is important to organisations with assets such as rolling stock, maintenance and production equipment. Modernisation is a strategic asset management activity that consists of different life cycle stages of an asset to ensure that the asset meets its service requirements according to modern ways of doing work [8]. Asset management plays a systematic and coordinated role in ensuring organisations remain competitive in their operations. According to [9], modernisation is a branch of asset management that seek to achieve a targeted level of sustainability. However, there is a lack of guidelines that can be used by passenger railway organisations to modernise their depots taking advantage of technological advancement to bridge technological gaps in a disruptive transition for sustainability.

To address a lack of guidelines for technological transformation of passenger railway depots in developing countries, this paper presents results of research conducted to conceptualise and develop a railway depots' modernisation framework. The conceptualisation is based on a systematic literature review of 42 papers and data collection through subject matter expert interviews. The development of a framework followed, where an iterative process was used to confirm that the developed framework had all the required specifications to meet the goals of modernisation and fill the gap in literature.

The developed framework was presented to four industry practitioners to confirm its validity and to gain feedback regarding its usefulness and support for decision-making.

2 Development of a Modernisation Framework for Passenger Railway Depots

The modernisation of passenger railway depots is carried out within the framework of large technical systems in a projectised environment. The contexulisation phase of the research produced a modernisation transitional framework which provides a holistic overview of modernisation. This framework serves as guide for modernisation concerning passenger railway depots. This framework is based on insights and themes collected during the systematic literature review which was confirmed and expanded on with semi-structured interviews with subject matter experts. These included a senior project manager, an asset care manager, a rolling stock manager and depot supervisors from PRASA. The modernisation transitional framework is shown in Fig. 1 and further expanded on in the next sections.

2.1 *Modernisation Transitional Framework*

The modernisation transitional framework systematically maps all dimensions of passenger railway depots modernisation into a coherent structure as shown in Fig. 1. The migration from the old system to the desired system require the analysis of the two systems' relationships. This migration can be carried out in cycles, iterations, stages or linearly depending on relationships, complexity, fragility and number of system interfaces. The modernisation transitional framework is a roadmap composed

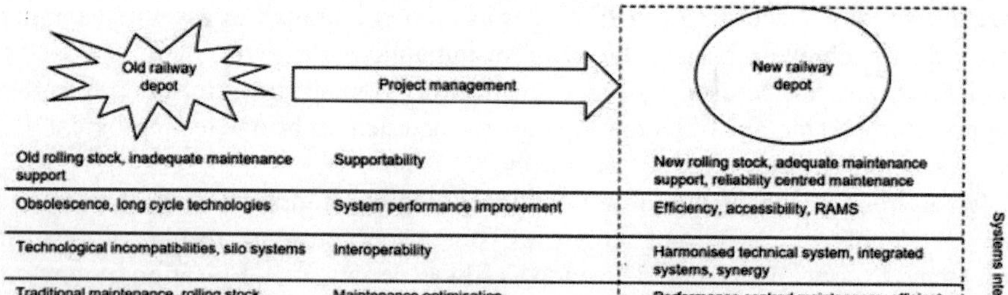

Fig. 1 Modernisation transitional framework

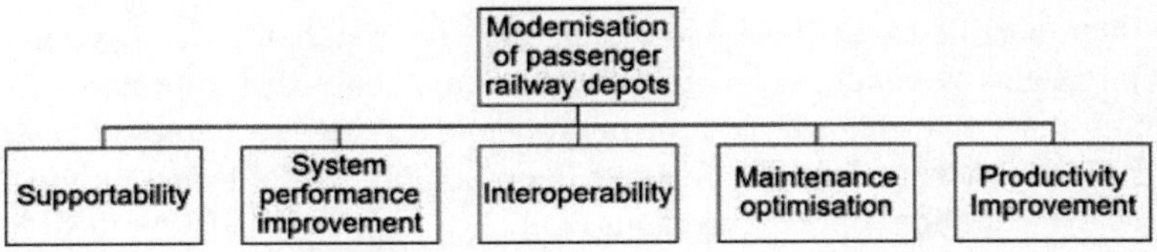

Fig. 2 Top-level requirements WBS

of push factors from the old state of depots, goals for modernisation, systems migration and integration to facilitate modernisation of railway depots in a project environment through a number of objectives, namely; supportability, system performance improvement, interoperability, maintenance optimisation and productivity improvement to achieve sustainability.

The objectives are the top-level requirements of modernisation which constitute the top static level of an integrated work breakdown structure (WBS) as shown in Fig. 2. The WBS facilitate a holistic approach to modernisation in a multidisciplinary project environment so that the project deliverables contribute to a common goal. The integrated WBS allow different project teams to operate independently but in a closely coordinated framework.

2.1.1 Supportability of New Rolling Stock

Railway organisations rely on their depots to sustainably maintain their rolling stock. Investment in maintenance ensures smooth operation, improved quality of work and enhanced performance and safety of rolling stock. The technical resourcefulness of a firm determines the quality of maintenance work. To ensure that there is effective and efficient maintenance, the support resources are designed according to the design attributes of the maintained systems. The demand for support resources is determined by the failure patterns of the maintained systems.

Maintenance demands can only be met when there are adequate support resources in terms of tooling and equipment to carry out the work. Businesses such as the transport industry thrives on reliability hence maintenance must evolve around reliability.

To achieve the required levels of reliability, a reliability centred maintenance strategy is adopted to exploit different maintenance strategies and customise them for different components depending on criticality and reliability thresholds. The support equipment to carry out different maintenance tasks must then be provided.

2.1.2 Interoperability

Interoperability facilitates the integration of systems to enhance overall performance. Operational and strategic objectives such as smart manufacturing require systems to be interoperable to facilitate communication. Interoperability is a systems property and means by which autonomous systems are composed to provide a service. Autonomy is a key characteristic of interoperable systems whereby a system can function independently but can cooperate interoperably with other systems to provide complex services. Interoperable systems facilitate integration and offer flexibility that make businesses agile in a large technical system environment. The addition of new systems or software to existing systems or components must ensure that these systems can communicate during run time so that operations are not obstructed by technical barriers. Interoperability exploits opportunities for synergies between systems by abandoning the traditional silo or monolithic systems which operate in isolation. Service-oriented systems can be loosely coupled to achieve interoperability from a holistic and systematic perspective at the business and application levels.

The internet provides an efficient interface for net-centric systems but rate of evolution of these systems poses obsolescence challenges. Technical standards facilitate technical harmonisation to ensure interoperability.

2.1.3 Maintenance Optimisation

Maintenance optimisation is carried out to improve the availability of a system. Maintenance is important for organisations that operate complex assets whose failure can be devastating.

Organisations are incorporating a performance-centric approach which adopts the principles of a risk-based maintenance philosophy. The performance-centric approach uses condition monitoring to report on the condition and performance of an asset according to predefined business protocols. This approach to maintenance help to improve the productivity of an asset by reducing the time it takes to restore its functionality. Decisions regarding repairing or replacing faulty components are made based on a number of factors such as obsolescence, availability of spare parts, cost and repair effectiveness. Trade-off decisions are also made depending on the criticality of the equipment or component whether to maintain preventively or correctively for availability maximisation or cost minimisation.

2.1.4 System Performance Improvement

Performance assessment of an asset against benchmarks is key to performance improvement decision-making. The perfomance of an asset can be measured by incoporating technological intelligence in physical assets for the measurement and analysis of asset data. Smart manufacturing systems help organisations attain high levels of perfomance with respect to quality, agility and productivity. Production equipment have mandatory performance targets. However, perfomance improvement can be implemented at the design and construction stage according to standards such as EN50126 which faciltate implementation of RAMS for railway systems.

Perfomance is measured from a machine level and aggregates to the overall performance of a whole production system. Performance of large systems such as railway depots are determined by technical aspects such as interoperability, interconnections and standardisation which facilitate systems integration for communication and data exchange in real time.

2.1.5 Productivity Improvement

Productivity improvement seeks to achieve optimum output from a machine by eliminating waste. Waste (such as breakdowns, bottlenecks, idleness, defects, accidents and motion) lowers the capacity utilisation of machinery resulting in reduced productivity. Modernisation of the machine shops brings the latest technology for sustainable operation. Controlled processes such as machining have feedback loops to facilitate data driven decision-making which is an objective of smart manufacturing technology.

Machine productivity (local optimisation) can be enhanced through automation whereas section level productivity can be optimised by reducing work in progress and lead times through optimising machine layout. The increase in productivity borrows Lean methodologies in the identifcation and elimination of waste.

2.1.6 Systems Integration

The autonomy of individual systems and their interdependence forms a coherent system that identifies the organisation beyond individual system identities. To realise efficient gains in the operation and use of a product or system, there is a need to incorporate many functions into a product or system to meet operational requirements. Integration of subsystems seeks to achieve a holistic and coherent system that delivers high levels of performance in terms of capacity, safety, timeliness and resource usage optimisation.

Integration takes place at different levels; physical system integration (for communication) and application integration (for cooperation). The integration of technologies in a manufacturing environment increases performance with the objective of achieving Smart Manufacturing. Systems that are developed in an agile environment are integrated and tested at the sprint or product increment level during development. Aspects like functional integration are drawn from the WBS. Functional integration takes note of functional definitions hence there is a need to observe standards. An integrated WBS facilitate successful and timely integration of a complex system.

2.1.7 Systems Migration

Different systems migration strategies are used in a project environment using different project management life cycle models to deliver the intended product.

The most used approaches are cycles, iterations, stages and linear. Different project management lifecycle models are applied for the required migration approach. The most used project management life cycle models are the traditional project management life cycle model, the agile project management life cycle model and the progressive elaboration life cycle model together with their variants.

The integrated WBS which is formed by project objectives is deliverable-oriented. From the interviews conducted, it was evident that the project planning and ideation phase is the most important phase to project success. Since the WBS is the backbone for project planning and control, its importance is emphasised. The framework was consequently developed to meet the requirements of the WBS to facilitate depot transformation by implementing appropriate migration strategies.

2.2 *Developed Multidisciplinary Project Management Framework (MPMF)*

The developed MPMF framework follows a logical sequence of process steps that are classified into phases to facilitate an evolutionary transformation of railway depots by reshaping the operational system and activities that support goal realisation.

The feedback loop is for planning, improvement and aligning the project management life cycle model with the project scope and landscape to enhance the quality of the deliverable. Due to the uniqueness of different projects, the framework facilitate tailoring to suit different project requirements. The framework has a value-centric approach that delivers according to project objectives and goals according to quality requirements. Finally, the MPMF framework exhibits robustness because the multidisciplinary nature of the project environment does not require a single approach to managing the project but different approaches can be applied depending on various projects' level of complexity and tailored to suit project characteristics. The developed framework is shown in Fig. 3.

The project is formulated in the project ideation and definition phase. Ideas to solve a problem or exploit an opportunity are evaluated. Decisions made in this phase concerns the inclusion or exclusion criteria for different options. The business case is produced to justify the project idea and that idea is broken down into a hierarchical framework—the WBS. The project is then described in non-technical terms in a project overview statement and a project proposal is written. Once the proposal is approved, the project manager starts organising and preparing for the project in the subsequent phase.

The scope of the project is defined to have a systematic overview of the work to be done. The level of complexity of the work to be done determines the project landscape. A suitable project management life cycle model is selected based on the landscape. The project characteristics are assessed to tailor the selected life cycle model to suit the unique requirements of the project. Once the tailoring aligns the life cycle model with the project requirements, execution work commences.

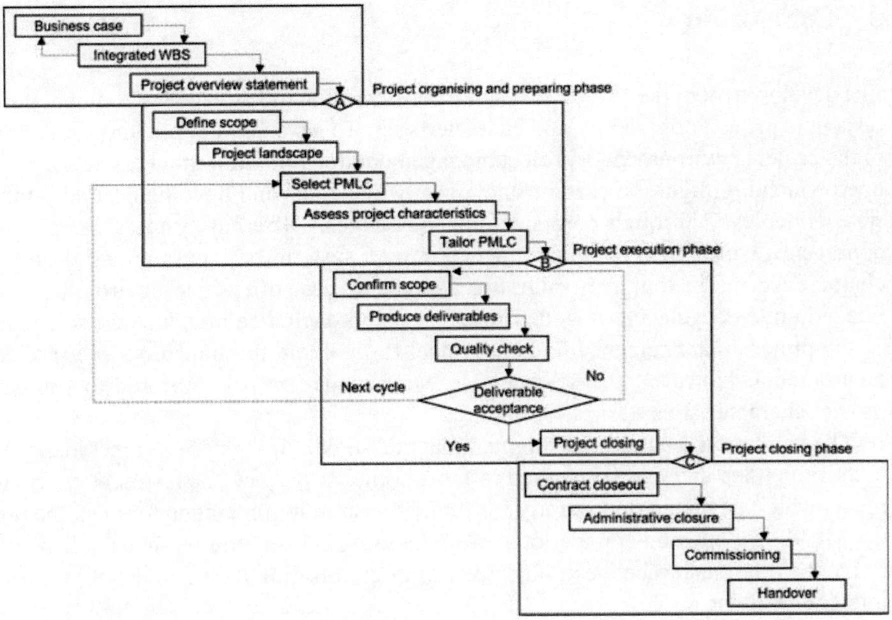

Fig. 3 MPMF framework

Since the project is deliverable-oriented, the scope is confirmed before producing the deliverables. The deliverables are examined for quality to check if they meet the need for which the project was undertaken. If the quality is below the acceptable level, the deliverable is reproduced as indicated in the execution phase in Fig. 3. If the quality is acceptable, the project is closed for projects with one delivery cycle. For projects with more than one delivery cycle, the project manager selects the life cycle model for the subsequent cycle and repeat the process until the project is complete. Once the project is complete, the contracts are closed. The performance measurement result from the project phases is used to audit the project and close administrative work. The deliverable, depending on its nature, is then commissioned and handed over to operations.

The framework was validated by means of face validation with four subject matter experts from industry, among them was the permanent way specialist, a senior project and contracts manager, a senior depot manager and an engineering manager for a rolling stock department. Feedback from the validation interviews provided evidence in support of the framework and its usefulness in decision-making concerning modernisation of passenger railway depots.

3 Conclusion

The development of the MPMF takes into consideration the project environment with respect to project complexity and characteristics. To have an in-depth understanding of the project environment, a project management transitional framework was developed which highlight the current state of railway depots and how the desired future state is achieved through corresponding objectives. Different systems' migration approaches would then be applied depending on system (both current and desired) characteristics. The four main migration approaches used in a project environment are linear, iterative, cycle and staged. The selected approach then influence the selection of the project management life cycle model to facilitate the transition in a project environment. However, the selected life cycle model must be tailored to suit the project characteristics.

The validation feedback highlighted that the MPMF has generic project management principles and can be applied to most complex project management environments which create an opportunity for future research into methods for customising the MPMF for specific applications. More research is also required to establish and refine the relationship between objectives of modernisation based on the objectives' interconnectedness.

References

1. Blumenfeld, M., Wemakor, W., Azzouz, L., Clive, R.: Developing a new technical strategy for rail infrastructure in low-income countries in Sub-Saharan Africa and South Asia. Sustainability **11**(4319), 1–23 (2019)
2. Beaumont, T.: Communist Trade Unionism and Industrial Relations on the French Railways, 1914–1939, Fellow Travellers, 1st edn. Liverpool University Press, Liverpool (2019)
3. Bullock, R.: Off Track Sub-Saharan African Railways, Washington, DC (2009). Available at: www.worldbank.org, Accessed: 16 Nov 2020
4. Schuitmaker, R.: Energy consumption and carbon dioxide emissions: focus on passenger rail services, 6th edn, Paris, The International Energy Agency (IEA) and the International Union of Railways (UIC) (2017)
5. Soto, J.L.-L., Ganges, L.S.-Y.: Exploring the modernisation processes of railway workshops, IRHA International Conference, 1–19 (2006)
6. Lu, Y., Riddick, F., Ivezic, N.: The paradigm shift in smart manufacturing system architecture, IFIP Advances in Information and Communication Technology, pp. 767–776. Springer, (2016)
7. Helu, M., Libes, D., Lubell, J., Lyons, K., Morris, K.-C.: Enabling smart manufacturing technologies for decision-making support. Proceedings of the ASME Design Engineering Technical Conference, 1–10 (2016)
8. El-Akruti, K., Dwight, R.: A framework for the engineering asset management system. J. Qual. Maint. Eng. **19**(4), 398–412 (2013)
9. Dwight, R., El-Akruti, K.-O.: The role of asset management in enterprise strategy success, ICOMS Asset Management Conference, Engineering Commons, 68–76 (2009)

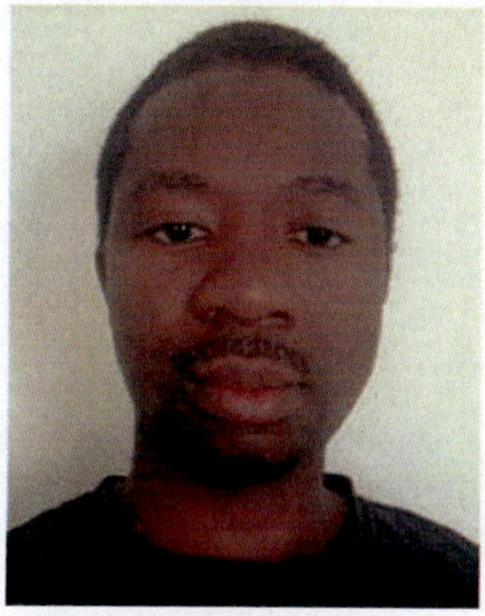

Adquate L. Masikati holds a B.Eng in Industrial Engineering and is currently a second year M.Eng (Industrial Engineering) student at the University of Stellenbosch.

Wyhan Jooste is a Professional Engineer and holds a Ph.D. in Industrial Engineering from Stellenbosch University. He is currently a senior lecturer in the Department of Industrial Engineering and focusses on asset- and asset management related research.

Cornelius (Neels) Fourie is a Professional Engineer and holds a Ph.D. in Industrial Engineering from Stellenbosch University. His academic experience covers a span of thirty-two years with research focused on project management and railway engineering, among others.

A Gamified Learning Approach Using Systems Modelling for Understanding the Effects of Asset Management Decision-Making

Ilicia M. Van Breda, Johannes L. Jooste, and Vera Hummel

Abstract Engineering asset management is a broad and multidisciplinary field with a variety of interconnected facets. Asset management (AM) encompasses a larger spectrum of risk, finance, and maintenance factors that make managing assets a complex and integrated multidisciplinary process. An industrial and educational challenge is to provide holistic training and insight in a comprehensible way for developing learner understanding and knowledge about the larger system's effect and practical implications of decision-making across the multidisciplinary facets of AM. In this paper, it is argued that gamification may present a suitable alternative for addressing this challenge. Arguments are presented in support of gamification, based on the success of existing games within industrial environments and the respective state of the art. The paper presents the design and development of a simulation-based game of AM using an architecture for bidirectional learning for serious game development. The game model is developed using system dynamics (SD). The paper further explores the applicability of SD methods to model an AM game. Two software modelling tools (*Vensim* and *AnyLogic*) are used to explore the benefits and limitations of SD methods to simulate the dynamics being considered. The strength of the SD method resides in its ability to enhance and simplify learning in complex systems. The paper concludes with an outlook of progress with the development of an AM simulation-based game.

Keywords Asset management · System dynamics · Gamification · Serious gaming · System modelling

I. M. Van Breda · J. L. Jooste (✉) · V. Hummel
Department of Industrial Engineering, University of Stellenbosch, Stellenbosch, South Africa
e-mail: wyhan@sun.ac.za

I. M. Van Breda
e-mail: 19801858@sun.ac.za

V. Hummel
e-mail: vera.hummel@reutlingen-university.de

I. M. Van Breda · V. Hummel
ESB Business School, Reutlingen University, Reutlingen, Germany

K. von Leipzig et al. (eds.), *Smart, Sustainable Manufacturing in an Ever-Changing World*, Lecture Notes in Production Engineering,
https://doi.org/10.1007/978-3-031-15602-1_39

1 Introduction

'Literature related to AM clarifies that modern forms of AM go beyond the management of physical assets and consider any item, thing, or entity that has potential or actual value to an organisation as an asset to the organisation [1]. AM is not a new idea. For a long time, people from all industries have been managing assets and advancing the discipline. At the centre of AM is decision-making. While some managers are willing to concede that their organisations are complex networks of interrelated systems, their thought patterns and behaviour usually give little weight to this notion [2]. Instead of understanding the larger system of operations and incidents that may have contributed to the original problem, some managers frequently fall into the single problem mentality by splitting a problem into smaller parts [3]. The reality might be that the root cause and collective solution may altogether be in another part of the system. This also highlights one of the main challenges in AM, which is to determine the long-term decision effects taken at a prior point in time in the asset lifecycle on other functions in the organisation [4].

Invariably, simulation studies demonstrate how powerful systems are and how important it is for managers to understand the complexity and interdependent nature of these systems. Two practical examples within industrial environments that demonstrate the complexity and interdependent nature of systems, be it supply chain management or AM, are discussed later in the paper. These examples also highlight the benefits of combining simulation with serious gaming to create an interactive learning experience. This paper firstly explores the theoretical background surrounding AM, simulation, and gamification, and secondly reports on the development of a serious game for AM using both simulation and system's modelling.

2 Theoretical Literature Overview

The following section provides an overview of the literature about AM, simulation, and gamification in education.

2.1 Asset Management

AM is defined as the coordinated activity of an organisation to realise value from its assets [1]. AM, a broad and multidisciplinary field, has seen a significant rise in popularity among researchers and organisations aiming to improve their asset performance and operations [5, 6]. One of the most significant developments in the field of AM was the paradigm change that AM is more than an extension of maintenance [7] but rather an entirely separate activity [8] requiring active lifecycle management [4].

Today, AM is accepted as a holistic and collaborative asset lifecycle approach that encompasses various interconnected and multidisciplinary facets [7]. However, most organisations still find themselves making asset-related decisions reactively and in isolation. These decisions are taken based on the limited information that is available locally. Each function in an organisation has its own staff, speciality, leadership, and way of operating [7], and this has caused silo perspectives to develop wherein no one seems to be able to translate local decisions effects to the larger system of operations. The reality in many organisations, is that managers who make strategic decisions are far removed from the operating assets and the reality of everyday operations in each individual function.

Literature highlights the integrative nature of AM and draws on the knowledge of multiple disciplines from risk analysts to safety personnel. Due to the integrative, multidisciplinary nature of AM, training employees in a manner that is both comprehensible and promotes their understanding and knowledge of the practical implications of their AM decisions on the larger system of operations remain a challenge.

Although the concepts of systems thinking are intuitively appealing in the context of AM, they are often difficult to illustrate or teach [3]. With traditional teaching and training methods showing limited skills improvement [9] and the industry now demanding interdisciplinary training [10], the need for new innovative teaching and training approaches that easily translate knowledge into skills within realistic work environments are important. It is essential that AM planning and training extend beyond theoretical knowledge to real-world contexts and skills development [11].

One of the main challenges in the AM domain is determining the implications of AM decisions prior to their implementation on other interconnected functions in the organisation. For example, a cost-reduction effort may cause more damage if the decision-maker cannot balance the associated cost, risks, and performance outcomes. AM decision-making plays a significant role in improving asset performance, and decreasing risks and expenditures such as maintenance costs.

Determining the prior implications of decisions on production, long-term planning, and the associated risks, for example, might help avoid decisions being taken based only on limited, locally available information. Therefore, knowing and having the ability to identify the causal relationships between functions could help improve AM decision-making, promote interdisciplinary collaboration, and boost employees' problem-solving and system thinking capabilities.

2.2 *Gamification in Education and Training*

Educational systems of today face significant challenges around student motivation and engagement [12, 13]. From tertiary level education to industry training workshops, modern educational methods and requirements have evolved to suit the needs of today's technologically inclined generation also referred to by Marc [13] as digital natives. It is via new technological developments enabling more digitised learning environments that gamification has become one of the most notable technological developments to create immersive and engaging learning experiences for human engagement [14].

Gamification, is defined as the use of game design elements in non-gaming contexts [15]. Gamification and its potential applications are gaining significant traction within industrial settings and has become the interest of various researchers [15, 16]. Gamification originated because researchers saw an opportunity to utilise game design elements for purposes beyond that of entertainment. Innovative digital forms of simulation such as serious games have proven to be a valuable education and training method (e.g., aviation training simulations).

Ritterfeld [17] defines serious games as any form of interactive computer-based game software for one or multiple players to be used on any platform, which has been developed with the intention to be more than entertainment. Gamification and serious games are often used interchangeably but differ based on the intention for which they are built. Both, however, are similarly based using gaming strategies and game elements in non-game contexts [15].

2.3 Simulation in Education and Training

According to [18], simulation is the imitation of the operation of a real-world process or system over time. Simulation makes it possible to study, experiment, analyse and manipulate different system scenarios that would otherwise be impossible to carry out practically in the real world [18]. Simulation-based training is gaining momentum in other fields (supply chain management, operations management, AM, etc.) thanks to advances in technology and increased computing power [19].

Simulation-based training would enable employees to visualise the overall impact a decision made at a prior point in the AM lifecycle, and the effect it would have in the future. In doing so, strategic planning between the respective departments and disciplines impacted through the decision can be done before the decision is actualised. This, in turn, would encourage future interdisciplinary communication and teamwork. Despite the benefits of game-based training simulations, didactic teaching methods are still the dominating training and teaching approach used in industries, although these teaching methods are proven to be inefficient and viewed by many as a way of merely transmitting factual information to a large audience with no guarantee that effective learning will result [20].

Game-based training simulations are primarily used for educational purposes [21]. They are designed to help end-users gain a better understanding and insight into the system under review because of its ability to simulate and visually illustrate quantitative effects of user-defined input parameters throughout the simulated system in real-time. Simulation, like computer games, use graphics such as 2D and 3D animation to simplify the model and captivate the audience. Simulation analysts require a fundamental understanding and knowledge of the system to effectively simulate it [22], further emphasising the educational value and improved training alternative that game-based training simulations offer.

Different modelling alternatives exist that can be used to develop a graphical interactive game-based simulation of AM. These alternatives are Discrete Event Simulation (DES), System Dynamics (SD) and Agent-Based Simulation (ABS) [23]. DES is used when simulating a real-world system where changes occur at discrete points in time [24], whereas ABS models simulate systems as individual interacting agents. SD, on the other hand, is an effective simulation-based modelling technique that was initially meant to invoke systems thinking by endogenising relevant variables and mathematically connecting causally linked variables [25].

There is precedence in applying SD to AM, as systems thinking and causal loop diagramming allows researchers to move from conceptual understanding of unidimensional problems to a completed systems model containing equations, each with their appropriate numerical inputs. Once computerised, these models offer ways of systematically testing policies. Stock and flow diagrams are adopted to help develop an AM simulation-based game later in the paper. Firstly, the following section applies theory to practice, by using the concepts discussed in this section to formulate the conceptual model of an AM simulation-based game.

3 Conceptual Model

The following section provides an overview of the conceptual game design and the processes used during the game design. Figure 1 illustrates the SD simulation process followed in this paper.

Firstly, the main strategic variables that form the foundation of AM were determined. There exists an abundance of fragmented AM information, data and research

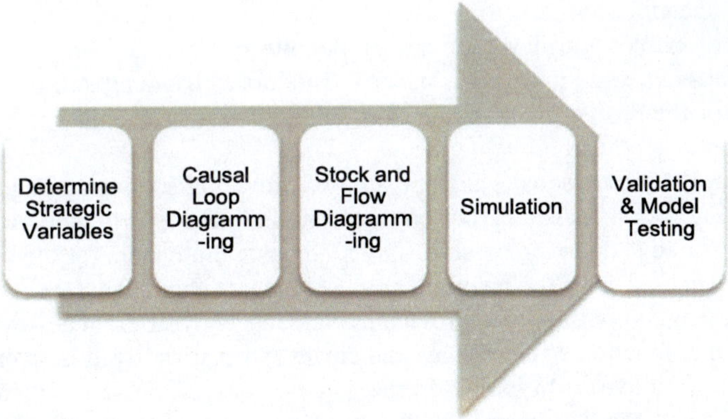

Fig. 1 System dynamics simulation modelling process, adapted from [17]

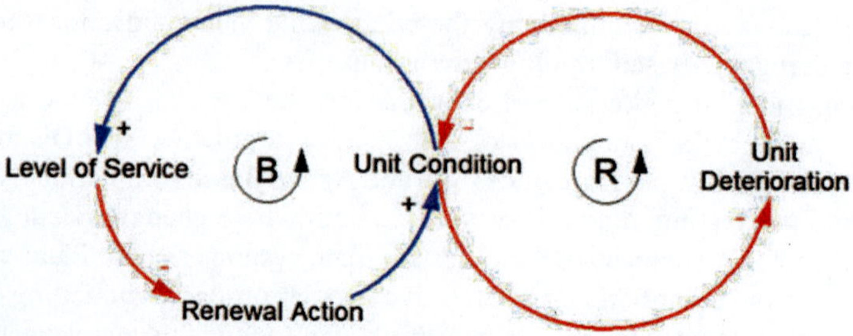

Fig. 2 CLD of strategic variables, adapted from [17]

in the AM scholarship. Various AM sources were identified and consolidated to formulate a collective overview of the interdependencies found in AM.

Four key strategic variables found in the AM scholarship include unit condition, unit deterioration, renewal action and level of service. These four key strategic variables form the basis of the CLD of the primary model, which is developed as part of the conceptual game design. Secondly, the identified strategic variables are translated, expanded, and formulated into the main CLD. An example of a CLD adapted from [17] is shown in Fig. 2.

A unit condition is positively influenced by an increased level of service and vice versa. Renewal action increases as the level of service declines, and the unit condition improves as a result of the renewal action. For example, the quicker maintenance staff responds to a unit breakdown, the faster the condition of the unit improves. The more units are utilised and overused, on the other hand, the faster their state deteriorates, and the unit condition degrades with time. The main CLD was then translated into stock and flow diagrams using simulation modelling software.

The conceptual game design was also guided in partial by an architecture developed by [26] for bidirectional learning games in laboratory performance improvement. The architecture aims to support game designers in the creation of serious games for specific scenarios in a faster, more effective, and efficient manner. The architecture enables the development of flexible games that can be configured, adjusted, tailored, and improved as needed. Therefore, the architecture was adapted and used for the design of the game to provide a more structured approach to the development.

Designing an educational game capable of providing motivated engagement is a challenging task because learner engagement and active participation are key to the success of any educational game. Targetted game motivators for the simulation-based game chosen from the taxonomy by [13] include challenge, feedback, control and roleplaying. It is essential to provide participants with different achievable game objectives that function as challenges and create extrinsic motivation in participants and encourage active participation in the game. Feedback provides extrinsic motivation in the game and indicates the distance to the desired state or goal. Control is incorporated through the player's granted ability to make decisions in the game; and

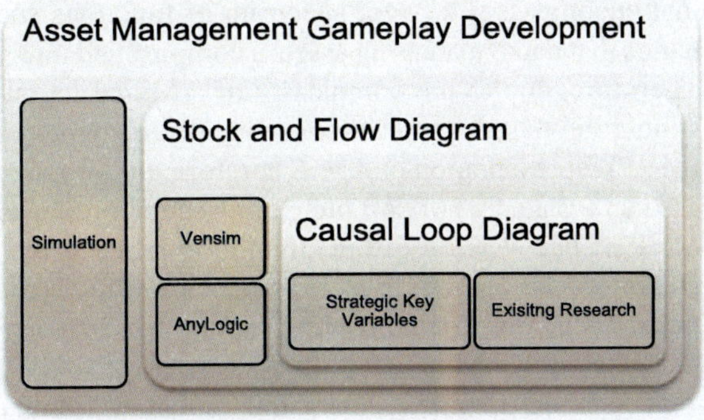

Fig. 3 AM gameplay development

the input mechanisms through which the player interacts with the game. A combination of an intrinsic or extrinsic design is followed in the design to ensure a holistic game experience whilst the primary focus is placed on the learning content and knowledge to skills transfer.

Two software modelling tools (*Vensim* and *AnyLogic*) are used to explore the benefits and limitations of SD methods to simulate the dynamics being considered. *Vensim* is an industrial-strength simulation software for improving the performance of real systems. *Vensim* is used for developing, analysing, and packaging dynamic feedback models. *Vensim* is used to create the numerous CLD's used in the game development.

AnyLogic, a simulation software package, was chosen as a suitable platform for designing and developing a simulation-based game because of its numerous functional capabilities. Several *AnyLogic* components are used in the model to create and illustrate the relationships and interdependencies found in AM through stock and flow diagrams. Figure 3 illustrates a development framework showing the various components and software used in the game development.

4 Model Construction

This section covers the development and construction of the simulation-based game and game environment.

4.1 Simulation-Based Game Model

To improve the holistic understanding of AM and assist decision-makers in understanding the effect (or implications) of AM decision-making across the entire AM lifecycle, a higher abstraction level design is appropriate. AM is a system's process that encompasses a larger spectrum of functions such as risk, finance, and maintenance that make managing assets a complex and integrated multidisciplinary process. Therefore, due to the complexity of AM and its multiple integrated functions, the simulation-based game is consequently constructed as a minimal viable product (MVP) representation of AM. Therefore, the simulation-based game is never considered an exhaustive or complete and accurate representation of AM, but rather as an MVP or first version of the simulation-based game that can be published, altered, and improved upon. While the game is simplified compared to real-world AM operations, it still aims to realistically capture the time delays, costs, and other parameters characterising AM.

The model requires a basic theoretical understanding of AM and its operations. Although the model does not provide the theoretical basis for understanding AM, it builds upon existing knowledge and is more suitable for participants who have basic pre-existing knowledge in the field of AM. Therefore, the model is developed on the assumption that the players or future players have had some exposure or introduction to AM. Further, a common theme from the AM scholarship is that the primary focus in AM is on constantly maintaining a balance between the conflicting drivers: cost, risk and performance. The simulation-based game, therefore, incorporates elements representing these drivers.

Based on the design principles and motivators discussed in the previous section, a design is followed in which the game consists of multiple roleplayers, requiring teamwork and collaboration between the players to reach the overall game objective. Each player is given a role that has its own unique function and sub-objectives. The game concept is based on each player taking on a managerial role of an AM-related organisational function. The four roles are operations manager, maintenance manager, risk manager and finance manager. The four players, each in one of these roles, address key areas found in AM. However, the collaboration between roleplayers is essential towards achieving the primary game objective. The challenge design motivator is incorporated through this collaboration aspect.

Subsequently, players of the game become the roleplayers and need to make decisions relating to their function whilst considering the potential implications of their decisions on the larger system of operations. The interdependent and interdisciplinary nature of the model is what makes it challenging. Challenges and difficulty are incorporated into the model not as game levels that increase systematically as the player progresses but through the interdependent nature of the role players decision inputs. Each roleplayer's decision carries different consequences for the others, and the main goal is to maintain a balance between cost, risk and performance. Participants are provided with the simulation game along with roleplayer briefing documents describing the four roleplayer functions. Each role player has required decision inputs that are automatically set to a default value that the participants can alter.

Key performance indicators (KPI) are defined for each role player to determine the AM performance in each function. These KPIs revolve around cost, risk and performance, which are at the centre of AM. KPIs for the role of the maintenance manager, for example, include the basic failure, replacement and repair rate.

4.2 Game Environment

The game environment consists of three interactive pages. One includes a tutorial page providing players with an introduction to the game, its purpose, the game setup, general playing procedure, settings and the goal.

The purpose of the simulation-based game is to experience systematic decision-effects and to introduce the need for interdisciplinary collaboration across various AM functions. The game setup consists of the four role player functions who need to make decision inputs at the start of the game. The goal is to minimise the overall cost and to maintain a target level of working units at a time.

The game behaviour is determined by the underlying stock and flow diagrams, parameters, variables, collections, functions, events and options lists of the underlying simulation-based model.

The second page illustrates and animates the constructed stock and flow diagram of the game. The game time unit is in days. Figures 4, 5, 6 and 7 illustrate the results, wherein players are prompted to insert their decision inputs.

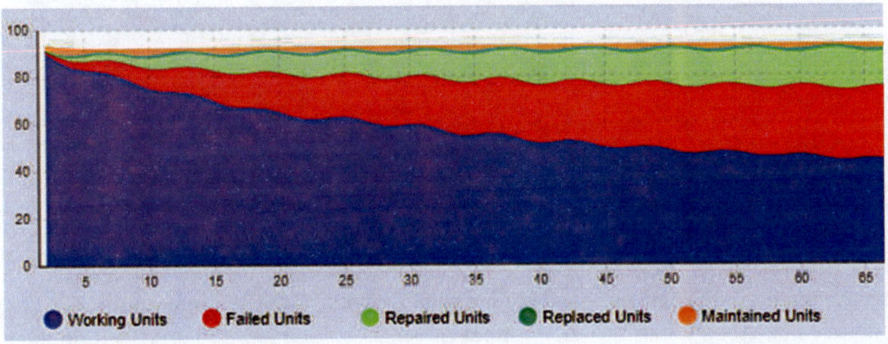

Fig. 4 Partial screenshot of maintenance role player output

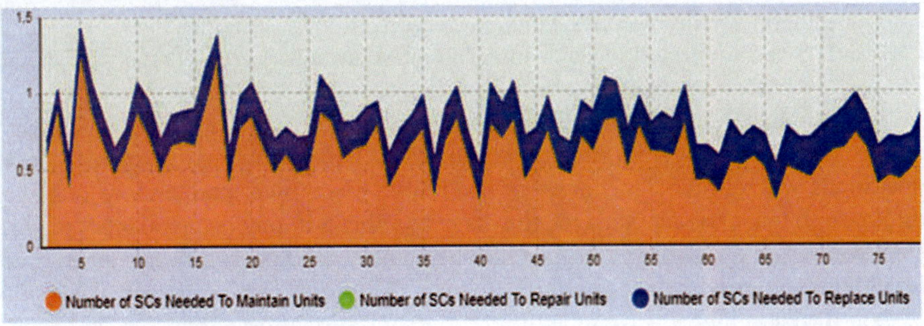

Fig. 5 Partial screenshot of operations role player output

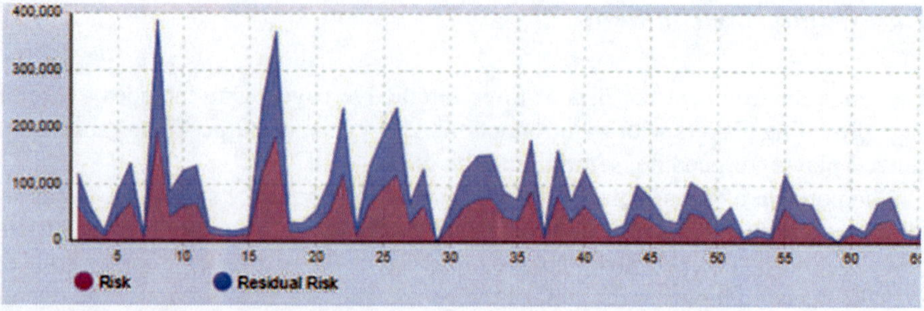

Fig. 6 Partial screenshot of risk role player output

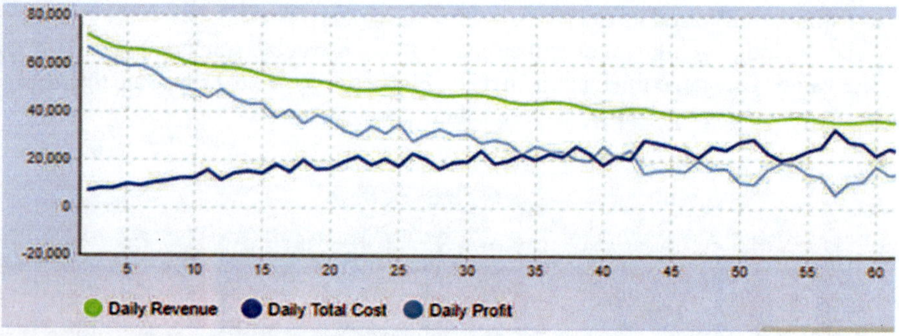

Fig. 7 Partial screenshot of finance role player output

Decision inputs for the role of the maintenance manager include the basic failure, replacement and repair rate. Decision inputs for the role of the operations manager include decisions regarding the number of units maintained, repaired and replaced service crew operators. Decision inputs regarding the risk manager's role include decisions regarding the level of risk to the operations, for example, the probability of system failures. Lastly, the finance manager's decision inputs include the daily revenue generated from working units. The results of their decisions inputs are simulated over a certain period of days and illustrated through a graph (Figs. 4, 5, 6 and 7).

The visualisation aspect of the simulation provides participants with a visual graph representation of the system and progression of each KPI. Similarly to the real-world operations wherein the decisions that need to be made by a maintenance manager differs from that of a risk manager, the game requires different decision input parameters from each of the four roleplayers. The game can be paused at any given time, providing players with the opportunity to visually observe and investigate the results of their decision inputs across the different organisational functions. On the second page of the game interface, players can observe and explore the interdependencies of variables and discuss the results after each game round is completed.

The interaction between players and the consequences of their decisions combined with the dynamic effects of the game environment help facilitate the improved understanding of the consequences of AM decision-making. Input decisions are dynamic and the output results continiously change over time. The change of each decision input changes the results of each metric. This illustrates the interdependant nature of decisions in AM. The output of each graph is the result not only of the individual role player decision inputs but of the collective decisions made by all four role players. The following section provides future work recommendations.

5 Future Work

A simulation-based game was developed using system dynamics and simulation to address the AM training need against the background of other successful gamification-based approaches. The development followed the four-stage gamification process introduced by [27]. Verification and validation of the developed game still need to be conducted using focus group discussions consisting of subject matter experts in the field of AM and gamification.

Finally, the resulting effectiveness of the application will be determined and implemented into a real-world setting for validation. A future work recommendation that falls outside the current scope of the model development is to incorporate bidirectional learning functionalities within the game. The game was developed as an MVP representation of AM. Subsequently the mathematical formulations and equations are simplified, but could be improved upon.

6 Conclusions

A rich AM scholarship and practitioner best practices are available in support of a gamification approach whereby AM principles and decision-making can be explained and illustrated through simulation-based game. A simulation-based game was developed using system dynamics and simulation to address the AM training need against the background of other successful gamification-based approaches. Systems thinking methods such as causal loop and stock-and-flow diagrams were used to construct a game of causal relationships governing AM activities and build a holistic and general representation of AM and its various interdependent functions. These diagrams were used to facilitate the development of simulation-based game for AM. The paper presented the game outline, the game development's current state, and future work recommendations.

References

1. British Standards Institution: Asset Management: BS ISO 55000:2014. BSI Standards Ltd., London (2014)
2. Stone, F.: Deconstructing silos and supporting collaboration. Employ. Relat. Today **31**(1), 11 (2004)
3. Goodwin, J.S., Franklin, S.G.: The beer distribution game: using simulation to teach systems thinking. J. Manag. Develop. Optimising Equipment Lifecycle Decisions. CRC Press. **1**(16), 18–29 (1994)
4. Campbell, J.D., Jardine, A.K., McGlynn, J.: Asset management excellence (2016)
5. Too, E., Betts, M., Kumar, A.: A strategic approach to infrastructure asset management. In: *Faculty of Built Environment and Engineering Infrastructure Theme Conference 2006*, pp. 163–171. Queensland University of Technology, Faculty of Built Environment and Engineering (20006)
6. Pedram S., Elsawah, S.: A literature Review of System Dynamics Modelling for Asset Management (2019)
7. IAM: Asset management: An anatomy. 1, 2, 3, 4, 15, 16, 1, 18, 19, 20, 21, 22, 23, 24, 26, 59 (2015)
8. Hastings, N.A.J.: Physical asset management: With an introduction to ISO55000. Springer (2015)
9. Cachay, J., Wennemer, J., Abele, E., Tenberg, R.: Study on action-oriented learning with a learning factory approach. Procedia-Soc. Behav. Sci. **55**, 1144–1153 (2012)
10. Abele, E., Metternich, J., Tisch, M., Chryssolouris, G., Sihn, W., ElMaraghy, H., Hummel, V., Ranz, F.: Learning factories for research, education, and training. Procedia CiRp **32**, 1–6 (2015)
11. Jooste, J.L., Louw, L., von Leipzig, K., Conradie, P.D., Asekun, O.O., Lucke, D., Hagedorn-Hansen, D.: Teaching maintenance plan development in a learning factory environment. Procedia manuf. **45**, 379–385 (2020)
12. Kiryakova, G., Angelova, N., Yordanova, L.: Gamification in education. In: Proceedings of 9th International Balkan Education and Science Conference (2014)
13. Marc, P.: Digital natives, digital immigrants. On the Horizon **9**(5), 1–6 (2001)
14. Majuri, J., Koivisto, J., Hamari, J.: Gamification of education and learning: A review of empirical literature. In: Proceedings of the 2nd international GamiFIN conference, GamiFIN 2018. CEUR-WS (2018)
15. Deterding, S., Dixon, D., Khaled, R., Nacke, L.: From game design elements to gamefulness: defining "gamification". In: Proceedings of the 15th international academic MindTrek conference: Envisioning future media environments, 9–15. 30, 31–33; R. Y. Zhong, Q. Dai, T. Qu, G. Hu, G. Q. Huang, RFID-enabled real-time manufacturing execution system for mass-customisation (2011)
16. McGonigal, J.: Reality is broken: why games make us better and how they can change the world. Penguin. **5**(30), 31 (2011)
17. Rashedi, R., Saad, D., Hegazy, T.: Infrastructure rehabilitation planning: combined system dynamics and optimisation methods (2021). [online] Open.library.ubc.ca. Available at: https://open.library.ubc.ca/cIRcle/collections/52660/items/1.0076400 [Accessed 5 September 2020]
18. Banks, J.: December. Introduction to simulation. In Proceedings of the 31st conference on Winter simulation: Simulation—a bridge to the future-Volume 1, pp. 7–13 (1999)
19. Kincaid, J., Westerlund, K.: Simulation in education and training. Proceedings—Winter Simulation Conference, pp. 273–280 (2010). https://doi.org/10.1109/WSC.2009.5429337
20. Walklin, L.: Teaching and learning in further and adult education. Nelson Thornes **4** (1990)
21. Chwif, L., Barretto, M.R.P.: Perspectives on simulation in education and training: simulation models as an aid for the teaching and learning process in operations management. In: Proceedings of the 35th conference on Winter simulation: driving innovation, pp. 1994–2000(2003)

22. Banks, J. (ed.): Handbook of simulation: principles, methodology, advances, applications, and practice. Wiley (1998)
23. Maidstone, R.: Discrete event simulation, system dynamics and agent-based simulation: discussion and comparison. System **1**(6), 1–6 (2012)
24. Farrington, P.A., Nembhard, H.B., Sturrock, D.T., Evans, G.W.: Introduction to simulation.
25. Richardson, G.P.: Reflections on the foundations of system dynamics. Syst. Dyn. Rev. **27**(3), 219–243 (2011)
26. Von Leipzight T.: An architecture for bidirectional learning (2021)
27. Rojas, D., Kapralos, B., Dubrowski, A.: The missing piece in the gamification puzzle. In: Proceedings of the First International Conference on Gameful Design, Research, and Applications, vol.135–138, pp. 36, 50, 51, 75, 82, 135. (2013)

Ilicia van Breda is an Industrial Engineering graduate at the University of Stellenbosch. She is currently pursuing a dual master's degree at the University of Stellenbosch and ESB Reutlingen.

Wyhan Jooste is a Professional Engineer and holds a Ph.D. in Industrial Engineering from Stellenbosch University. He is currently a senior lecturer in the Department of Industrial Engineering and focusses on asset- and asset management related research.

Vera Humme Prof. Dr.-Ing., Dipl.-Ing., has been a professor at the ESB Business School, Reutlingen University since 2010. Previously she held leading positions at the Fraunhofer IPA in Stuttgart, the working area Industrial Engineering, and the Graduate School of Advanced Manufacturing Engineering at the University of Stuttgart. She is currently leading the expert group in logistics at Reutlingen University.

Supply Chain

Opportunities for Visualising Complex Data by Integrating Virtual Reality and Digital Twins

G. S. da Silva, K. Kruger, and A. H. Basson

Abstract A digital twin is a digital representation of a physical system and reflects selected aspects of the reality of the physical system in a digital environment. Digital twins are typically used to support data led decision-making. The digital twins of complex systems collect a large amount and variety of data from the physical system, which presents the challenge of integrating and visualising the data to understand the physical twin's own actions and its interactions with its environment. Virtual reality (VR) is visualisation technology that has become more easily available over the past few years. VR places the user in a 3-D visual environment which, using the proper equipment, they can interact with and manipulate items. This paper shows that the integration of digital twins with VR offers opportunities such as improved data navigation, data interpretation, collaboration during data interpretation and, thereby, decision making. The paper also explores the challenges that integration brings, including avoiding potentially overwhelming the user with data, navigating through the data, lack of industry experience and potential negative effects on users' well-being. The paper concludes with suggestions for further research into the integration of virtual reality and digital twins.

Keywords Digital twin · Virtual reality · Complex systems

G. S. da Silva · K. Kruger · A. H. Basson (✉)
Department of Mechanical and Mechatronic Engineering, Stellenbosch University, Stellenbosch, South Africa
e-mail: ahb@sun.ac.za

G. S. da Silva
e-mail: 20935927@sun.ac.za

K. Kruger
e-mail: kkruger@sun.ac.za

K. von Leipzig et al. (eds.), *Smart, Sustainable Manufacturing in an Ever-Changing World*, Lecture Notes in Production Engineering,
https://doi.org/10.1007/978-3-031-15602-1_40

1 Introduction

The fourth industrial revolution (Industry 4.0) has seen the introduction of new concepts such as cyber physical systems (CPSs) and the Internet of Things (IoTs) [1]. The concept of a digital twin (DT) is used in this context and, in essence, a DT is a digital representation of a physical system that reflects selected aspects of the reality of the physical system in a digital environment. The DT is an entity on its own, but is linked to the physical system [2]. A DT is expected to support predictions about and operation of the physical model throughout its lifecycle [3]. DTs are being implemented in various contexts, including manufacturing and production.

The implementation of DTs typically results in the recording of a vast amount of data from the physical system, both historical and real time. A key ability of DTs is to provide data in a consistent format [4]. The data is collected with the intention of converting the data into information that can be used to make decisions related to the physical system. Conveying such information to a user is the focus in this paper.

Conventional data visualisation methods traditionally use computer monitors to view 2-D and 3-D graphics. The latter requires that users visualise 3-D graphics from a 2-D medium, which makes it more challenging to visualise aspects such as scale and depth. The user's ability to interact with the data is also typically based on 2-D devices like a computer mouse.

Recent developments have increased the accessibility of visualisation tools like virtual reality (VR) [5]. VR is a digital artificial environment that creates an experience that human-senses will perceive to be real [3]. VR is also able to provide an interactive simulation that understands a user's prompts and responds accordingly to the behaviour of the user [6]. With developments in VR, users are now able to enter immersive environments and interact with these virtual environments.

Although VR has typically been associated with the entertainment and gaming industry, it is beginning to grow in use within other industries [7]. The information obtained through integrated communication channels can be visualised using VR [8]. Previous studies recorded that 93% of decision-makers indicated that the use of an immersive scenario using VR is more effective than using conventional 2-D visualisation methods [9].

This paper considers the opportunities and challenges associated with the integration of VR and DTs to visualise complex DT data. The combination of VR and DTs is, at this stage, relatively unexplored.

The paper is structured as follows: Section "Related Work" summarises related work that considered DTs, using VR to visualise data, and integrating VR and DTs. Section "Opportunities in Using Virtual Reality for Digital Twins" presents opportunities for using VR and DT technologies together and Section "Challenges of Virtual Reality for Digital Twins" highlights some of the challenges associated with the integration of DTs and VR. Section "Conclusion and Future Work" briefly considers the effectiveness of VR with DTs and provides conclusions and plans for future work.

2 Related Work

The DT concept has seen much development over the past few years. There are several studies that explore the definition of the DT concept, while others apply the concept, for example to create an implementation of DTs in a real-world context for data collection, monitoring and system design. A few typical implementations are given below, but a more complete review is not possible here.

An architecture was developed for DTs that represent building facilities at the University of Cambridge. The facilities were used as a case study to evaluate this architecture [10, 11]. The purpose of this DT was to obtain information of the physical facilities on the campus, using virtual models of the facilities. The DT interfaced with a service layer, where various functions were provided, such as pump anomaly detection, ambient environment monitoring and maintenance optimisation.

Tronrud Engineering, with the help of Siemens, developed a DT of a packaging machine. This DT allowed designers, engineers and programmers to collaborate on the same project simultaneously. The machine was designed virtually, and the use of a DT of the machine allowed the developers to run simulations of the machine and its behaviour. The use of the DT dramatically reduced the development and assembly time of the machine [12].

In another study, a DT was developed for energy optimisation of an assembly line. The assembly line was fitted with many sensors that were used for measuring machine activity and energy consumption. The energy consumption was 2.7 times less after the DT implementation without significantly affecting the throughput [13].

The concept of using VR for data visualisation is a new concept, still in its infancy, but has been developing recently. However, some studies have investigated the use of VR to visualise some types of data, for example factory processes or 3D models.

An experiment was conducted to investigate how users reacted to using VR compared to a conventional PC display [14]. The study considered three scenarios where participants were asked to navigate data and complete tasks. The first scenario was using a desktop display with a gaming console controller, the second was using a VR headset with a gaming console controller, and the third was using the VR headset with the gesture interface, Leap Motion. The study showed that participants ranked the use of VR with gestures and a controller higher than the conventional desktop setup. The study concluded that the use of a VR environment could aid with the visualising of large datasets [14].

There are several studies that consider the use of VR in other industries. One study reviews the implementation of VR with a digital factory [6]. Another mentions previous studies with VR use in application areas such as medicine, learning, training, assembly, maintenance and safety [15].

Another study considered how VR can be used to aid in faster and better designs of human–robot collaboration (HRC) workspaces [3]. The study indicated that the use of VR is beneficial in the analysis of a complex system, including that 3D interactive and immersive VR aids with identifying useful information more effectively than using a two-dimensional computer screen. The use of a DT and VR was studied by

[5]. Much like the previously mentioned study, this study focused on a co-simulation and communication architecture that was implemented in a HRC workspace. The engineers made use of VR to visualise the robot's DT. The robot's DT in VR behaved realistically which aided the engineers during the design phase.

3 Opportunities in Using Virtual Reality for Digital Twins

This section first considers the opportunities offered by VR for data visualisation and decision making. Thereafter, the integration of DTs and VR is considered as a means to enhance the DT data visualisation and the associated decision-making process.

3.1 Data Visualisation and Decision Making

The use of three dimensions for visualising data is more beneficial than only using two dimensions [16]. These benefits are enhanced when using VR.

Humans have an effective natural pattern recognition process [17]. The use of VR allows users to move around more freely within the data. They are, thus, able to navigate the data more intuitively which aids this pattern recognition process [18, 19]. Better pattern recognition allows a user to gain more useful insight when analysing data and thereby make better decisions compared to using conventional visualisation methods.

The main purpose of VR is to provide the user with an experience that is close to reality. VR's benefits are most evident when considering dynamic and immersive visualisation [6]. With the current developments in VR technology, a user is able to visualise data in a high level of detail. Users are able to get close to the data and also view the data in its entirety if desired. This allows users to view the "bigger picture" of the data, and also view each data point individually. This rage of perspectives greatly contribute to the data driven decision making process.

In a VR environment, data is represented more accurately and realistically with regards to the distances and relations between points, which brings the perception of users closer to reality [16] than with the use of conventional methods.

VR introduces the opportunity for collaborative data visualisation, where multiple users interact with and visualise data together. Users are able to communicate more effectively with one another and information can be presented to the relevant stakeholders in a manner where different stakeholders can visualise the same data and create more effective solutions together [9].

From being in an immersive VR environment, users are able to complete data visualisation tasks, and draw conclusions quicker than using conventional methods [20]. VR has been shown to reduce the time taken for verification and validation processes [9]. While completing these tasks, there is also less error by the user [21].

Fewer errors in the data visualisation process should result in fewer errors during the decision-making process.

The above show that VR holds significant opportunities to aid in the data driven decision making process, by enabling faster and better decisions.

3.2 Digital Twin Integration

As mentioned previously, a central role of a DT is to record data, convert this data to useful information and make the information available to a user. The integration of VR and DTs holds promise in better converting data to information by providing more context to data. VR also holds promise in making the information "more available" to users, as indicated by the previous section. Some of the advantages of VR can be achieved with DT data in off-line approaches, where the data is transferred in an "as and when needed" method. This section, however, considers going beyond that and establishing a near real time interface between DTs and a VR environment.

DTs can update the virtual models in near real time and, therefore, integration with VR will allow users to monitor changes with the enhanced perception provided by VR. This ability will be useful, inter alia, during operation. Multidimensional views can be annotated and changed in near real time, as the sensed data changes. Integration adds in these views a real time temporal dimension to the spatial, colour, texture and opacity dimensions employed in VR.

DTs often include a virtual spatial model of the physical system. The immersive environment of VR will provide users with a more accurate and complete spatial representation of the physical system. VR will allow users to view and move around in these virtual models to better sense the spatial mapping of the physical system. Integrating VR and DTs will allow for near real time updating of the virtual model as the physical model changes. The visualisation of physical changes can also be used during the design phase if a DT is developed in parallel (or even before) the physical system. Integrating VR with the DT offers the opportunity to aid the design process by rapidly visualising the intended physical system, reducing design time and costs [5, 6, 8].

A particular opportunity offered by integrating DTs and VR is the enhanced ability to contextualise non-spatial information by overlaying information (e.g. text or graphs) over the physical representation. This will allow the user to view non-spatial information (e.g. the temperature of a motor driving a conveyer belt) in the context where it was recorded. For example, the colour of a motor that is exceeding a predetermined temperature threshold, can even by changed to draw the user's attention to it.

The use of VR also facilitates a virtual hands-on interaction with the DT. Users can interact more intuitively with DT data in 3-D through VR, as outlined in the previous section.

Standardising the integration of DTs and VR will allow for synergy between the developers of DTs and VR developers. The DT developers can specialise in collecting and organising the data, while the VR developers can focus on communicating the information to users. The two teams will have to collaborate in the (dynamic) transfer of data from the DT to the VR environment (and possibly the reverse).

4 Challenges of Virtual Reality for Digital Twins

The previous section showed that integrating DTs with VR provides significant opportunities to enhance the data driven decision making process. This section considers some of the challenges that have to be overcome to realise the above opportunities. These challenges include the data visualising process itself, as well as some challenges that result from the integration of VR and DTs.

4.1 Data Visualisation and Decision Making

A concern with data visualisation in general, and more specifically with VR, is the possibility of information overloading [18]. Previous studies have indicated that a large amount of information being presented at once to a user could have a negative impact on the user's ability to visualise the data and make decisions [6]. With VR, a user will be immersed into the data and can easily become overwhelmed by the amount of data that is presented to them.

Navigating through the presented data in a VR virtual space is another challenge [16, 20]. This difficulty in navigating the data could be due to the learning curve associated with VR. It is a new technology and not many people have been exposed to it, resulting in not many knowing how to use it effectively. Another reason for these navigation difficulties could be that there is no consensus on how VR data visualisation scenes should be navigated, in contrast to navigating 2D monitor representations [19].

VR is a new technology that is still developing and there is limited experience in the developer community. Therefore, the development of VR environments could be more time consuming than using conventional methods [9]. This challenge, however, might be resolved with further development of VR technology, simplifying the interfaces and users becoming more familiar with the technology.

Lack of experience with VR also leads to the challenge of deciding how to represent the different types of information so that users are best able to visualise and interpret it effectively. Therefore, a thorough design and testing process must be followed to develop the VR environment and also how the information is to be displayed in the environment.

Perspective distortion is another challenge associated with using VR for data visualisation [16, 20]. When users visualise data in VR, it is possible that their perception is distorted in some areas of their field of view. A reason for this in a typical VR headset could be due to the fisheye lenses used in the headsets. Such lenses distort the pixels that are on the outer edge of the lens, compared to the pixels closer to the centre.

Occlusion can also be seen as a challenge [16, 20]. Occlusion occurs when one object in a 3-D environment obscures another object behind it. This could affect the data visualisation process, as the user might not be able to see all the desired data points in an environment. This challenge can be partially negated by allowing easy navigation of VR scenes.

Other challenges include the safety and comfort of a user that is using VR equipment [22]. Users might experience a multitude of symptoms when using VR equipment for extended periods, such as motion sickness, disorientation, nausea, sweating and headaches. When using VR, a person's brain is receiving signals from their eyes saying that they are moving, but their inner ear is not detecting any movement. This discrepancy in movement detection can result in the abovementioned symptoms [15].

The cost of a VR system can also be a challenge associated with the technology. VR is only recently becoming more easily available, with the result that VR systems (including headsets and computers with high-power graphics capabilities) are relatively expensive. However, it is reasonable to expect the more widespread use of VR to lead to reduced equipment costs.

4.2 Digital Twin Integration

As DTs record a vast amount of data, it is challenging to decide what DT information is to be displayed in the VR environment. This challenge arises from the variety of data, the risk of overloading the user and the lack of industry experience of using VR in this role. This leads to the challenge of allowing the user to decide what information they would want to visualise and how to achieve the functionality of allowing the user to decide.

DTs and VR are fairly new technologies that are both constantly developing. These "moving targets" exacerbate the challenge of developing a stable interface between the two technologies. To realise the opportunity for bidirectional communication by the integration of the two technologies, effective and logical process that facilitate timely information transfer between VR and a DT must be found.

Another challenge is with regards to the computational power requirements for VR, which has to render two different, stereoscopic scenes for the eyes in a VR headset [19]. In addition, data is to constantly be imported into the scene. Substantial computational capacity is therefore required, especially for visualising complex, voluminous and real time DT data.

5 Conclusion and Future Work

DTs are emerging in Industry 4.0 where large quantities of complex data are being recorded. Current data visualisation techniques are limited in providing the means to interpret this data effectively and make subsequent decisions. It is evident that the use of VR to visualise complex DT data has opportunities and challenges. These opportunities and challenges affect various aspects related to DTs.

With the use of VR, users are able to visualise and navigate represented DT data in more accurate and intuitive ways. They are also able to draw conclusions and complete data visualisation tasks quicker and with fewer errors. VR also allows for multiple users to collaborate when visualising data, allowing better levels of communication to be achieved. The opportunities for the integration of DTs and VR mentioned will have benefits for both technologies as they can now be used to complement one another. These are a few of the opportunities associated with the use of VR.

The challenges with using VR include the amount of computational power required for VR systems, the novelty of the technology and the lack of industry experience. The user's wellbeing when using VR is also a challenge, and measures are required to aid in reducing any possibility of users experiencing unpleasant symptoms when using VR for extended periods. The information transfer between VR and a DT does pose a significant challenge for effectively using the two technologies together.

Although there are challenges associated with VR, the opportunities that the technology presents, especially in the process of data visualisation and decision making for complex DT data, would be valued. VR has the potential to advance the data visualisation process to ensure that data is correctly analysed, and the correct decisions are made for DTs and their physical systems.

Further research is required to evaluate these opportunities and determine their effect on the data visualisation process. The planned future work on this topic is to implement and integrate a DT with a VR data visualisation system. The study will use as context a digital twin of university facilities, to support decisions for facilities management.

Alternative process routes [4].

Acknowledgements The financial assistance of the National Research Foundation (NRF) towards this research is hereby acknowledged. Opinions expressed and conclusions arrived at, are those of the authors and are not necessarily to be attributed to the NRF.

References

1. Redelinghuys, A.J.H., Kruger, K., Basson, A.: A six-layer architecture for digital twins with aggregation. Stud. Comput. Intell. **853**(August 2019), 171–182 (2020)

2. Grieves, M., Vickers, J.: Digital Twin: mitigating unpredictable, undesirable emergent behavior in complex systems. Transdisciplinary Perspect. Complex Syst., pp. 85–113 (2017). https://doi.org/10.1007/978-3-319-38756-7_4
3. Malik, A.A., Masood, T., Bilberg, A.: Virtual reality in manufacturing: immersive and collaborative artificial-reality in design of human-robot workspace. Int. J. Comput. Integr. Manuf. **33**(1), 22–37 (2020)
4. Kritzinger, W., Karner, M., Traar, G., Henjes, J., Sihn, W.: Digital Twin in manufacturing: a categorical literature review and classification. IFAC-PapersOnLine **51**(11), 1016–1022 (2018)
5. Havard, V., Jeanne, B., Lacomblez, M., Baudry, D.: Digital twin and virtual reality: a co-simulation environment for design and assessment of industrial workstations. Prod. Manuf. Res. **7**(1), 472–489 (2019)
6. Chandra Sekaran, S., Yap, H.J., Musa, S.N., Liew, K.E., Tan, C.H., Aman, A.: The implementation of virtual reality in digital factory—a comprehensive review. Int. J. Adv. Manuf. Technol. **115**(5–6), 1349–1366 (2021)
7. Hu, M., Luo, X., Chen, J., Lee, Y. C., Zhou, Y., Wu, D.: Virtual reality: a survey of enabling technologies and its applications in IoT. J. Netw. Comput. Appl., **178**(November 2020) (2021)
8. Kovar, J., Mouralova, K., Ksica, F., Kroupa, J., Andrs, O., Hadas, Z., 2017, Virtual reality in context of Industry 4.0, International Conference on Mechatronics - Mechatronika, ME 2016, Prague, Czech Republic.
9. Akpan, I.J., Shanker, M.: A comparative evaluation of the effectiveness of virtual reality, 3D visualization and 2D visual interactive simulation: an exploratory meta-analysis. Simulation **95**(2), 145–170 (2019)
10. Lu, Q., Parlikad, A.K., Woodall, P., Ranasinghe, G.D., Xie, X., Liang, Z., Konstantinou, E., Heaton, J., Schooling, J.: Developing a digital twin at building and city levels: case study of West Cambridge Campus. J. Manag. Eng. **36**(3), 05020004 (2020)
11. Vivi, Q.L., Parlikad, A.K., Woodall, P., Ranasinghe, G.D., Heaton, J.: Developing a dynamic digital twin at a building level: using Cambridge campus as case study. Int. Conf. Smart Infrastruct. Constr. 2019 (ICSIC): Driving Data-Informed Decision-Making, **2019**, 67–75 (2019). https://doi.org/10.1680/icsic.64669.067
12. Siemens, Virtual commissioning with the digital twin | Industry | Siemens Global. [Online] Available at: https://new.siemens.com/global/en/company/stories/industry/tronrud-engineering-digital-twin.html, Accessed 11 Sep 2021
13. Karanjkar, N., Joglekar, A., Mohanty, S., Prabhu, V., Raghunath, D., Sundaresan, R.: Digital twin for energy optimization in an SMT-PCB assembly line. In: Proceedings - 2018 IEEE International Conference on Internet of Things and Intelligence System, IOTAIS 2018, pp. 85–89 (2019)
14. Andersen, B. J. H., Davis, A. T. A., Weber, G., Wunsche, B. C., 2019, Immersion or diversion: Does virtual reality make data visualisation more effective?, ICEIC 2019 - International Conference on Electronics, Information, and Communication, Auckland, New Zealand.
15. Liagkou, V., Salmas, D., Stylios, C.: Realizing virtual reality learning environment for industry 4.0. Procedia CIRP **79**, 712–717 (2019)
16. Gracia, A., González, S., Robles, V., Menasalvas, E., Von Landesberger, T.: New insights into the suitability of the third dimension for visualizing multivariate/multidimensional data: a study based on loss of quality quantification. Inf. Vis. **15**(1), 3–30 (2016)
17. Donalek, C., Djprgovski, S.G., Cioc, A., Wang, A., Zhang, J., Lawler, E., Yeh, S., Mahabal, A., Graham, M., Drake, A., Davidoff, S., Norris, J.S., Longo, G.: Immersive and collaborative data visualization using virtual reality platforms. In: Proceedings—2014 IEEE International Conference on Big Data, IEEE Big Data 2014, pp. 609–614. Washington, United States of America (2015)
18. Erra, U., Malandrino, D., Pepe, L.: Virtual reality interfaces for interacting with three-dimensional graphs. Int. J. Human-Comput. Interaction **35**(1), 75–88 (2019)
19. El Beheiry, M., Doutreligne, S., Caporal, C., Ostertag, C., Dahan, M., Masson, J.B.: Virtual reality: beyond visualization. J. Mol. Biol. **431**(7), 1315–1321 (2019)

20. Filho, J. A. W., Rey, M. F., Freitas, C. M. D. S., Nedel, L.: Immersive visualization of abstract information: an evaluation on dimensionally-reduced data scatterplots. In: 25th IEEE Conference on Virtual Reality and 3D User Interfaces, VR 2018—Proceedings, pp. 483–490. Reutlingen, Germany (2018)
21. Raja, D., Bowman, D. A., Lucas, J., North, C.: Exploring the Benefits of Immersion in Abstract Information Visualization. In: Proceedings—8th Immersive Projection Technology Workshop, pp. 61–69. Ames, Iowa (2004)
22. Drouhard, M., Steed, C. A., Hahn, S., Proffen, T., Daniel, J., Matheson, M.: Immersive visualization for materials science data analysis using the Oculus Rift. In: Proceedings—2015 IEEE International Conference on Big Data, IEEE Big Data 2015, pp. 2453–2461. Santa Clara (2015)

Gabriel Santos da Silva holds a B.Eng, is an MEng candidate at Stellenbosch University and is a member of the of the Mechatronics, Automation and Design Research Group. He is researching virtual reality to aid in data visualisation for digital twins.

Karel Kruger obtained his Ph.D. from Stellenbosch University, South Africa. He is a senior lecturer in the Department of Mechanical and Mechatronic Engineering and is co-leader of the Mechatronics, Automation and Design Research Group at Stellenbosch University, South Africa.

Anton Basson obtained his Ph.D. in Aerospace Engineering at Penn State University. In 1997, he was appointed as Professor in Mechanical Engineering at Stellenbosch University and is co-leader of the Mechatronics, Automation and Design Research Group.

Prototypical Blockchain Application for Mapping Complex Products in Dynamic Supply Chains

F. Dietrich, L. Louw, and D. Palm

Abstract The blockchain technology represents a decentralised database that stores information securely in immutable data blocks. Regarding supply chain management, these characteristics offer potentials in increasing supply chain transparency, visibility, automation, and efficiency. In this context, first token-based mapping approaches exist to transfer certain manufacturing processes to the blockchain, such as the creation or assembly of parts as well as their transfer of ownership. This paper proposes a prototypical blockchain application that adopts an authority concept and a concept of smart non-fungible tokens. The application enables the mapping of complex products in dynamic supply chains that require the auditability of changeable assembling processes on the blockchain. Finally, the paper demonstrates the practical feasibility of the proposed application based on a prototypical implementation created on the Ethereum blockchain.

Keywords Blockchain · Supply chain management · Smart contract · Ethereum

1 Introduction

The blockchain technology can be defined as a technology to process and verify data transactions based on a distributed peer-to-peer network. It uses cryptographic

---

F. Dietrich (✉) · L. Louw
Department of Industrial Engineering, Stellenbosch University, Stellenbosch, South Africa
e-mail: fabian.dietrich@reutlingen-university.de

L. Louw
e-mail: louisl@sun.ac.za

F. Dietrich · D. Palm
ESB Business School, Reutlingen University, Reutlingen, Germany
e-mail: Daniel.Palm@Reutlingen-University.DE

D. Palm
Fraunhofer Institute for Manufacturing, Engineering and Automation, Stuttgart, Germany

K. von Leipzig et al. (eds.), *Smart, Sustainable Manufacturing in an Ever-Changing World*, Lecture Notes in Production Engineering,
https://doi.org/10.1007/978-3-031-15602-1_41

procedures, consensus algorithms, and back-linked blocks to make transactions practically unchangeable [1]. The technology was introduced for the first time by the pseudonym Satoshi Nakamoto in 2008, who published the famous Bitcoin white paper and thus introduced the blockchain technology with the aim of changing the traditional financial sector and making trusted third parties superfluous [2]. Through this original intention, blockchain technology was initially strongly linked to new potentials regarding financial applications [3–5]. In 2013, Vitalik Buterin published the Ethereum white paper and therefore extended the idea behind Bitcoin. Compared to Bitcoin, the Ethereum protocol moves far beyond using blockchain just as a currency. Ethereum is a blockchain with an embedded fully fledged Turing-complete programming language [6]. Turing-completeness describes a mathematical concept and is a measure of the computability of a programming language. A Turing-complete language design includes complex constructs such as loops and conditions, which enable the creation of general purpose programs [7]. Thus, Buterin [6] coined the term 'smart contract' with blockchain-based decentralised applications. Decentralised applications form the basis for blockchain-based use cases outside the financial sector. In this context, first approaches came up, adopting the immutable, decentralised, and secure characteristics of blockchain technology to enable or increase transparency, automation, visibility, and disintermediation in supply chains [8].

1.1 Related Works and Rational of the Paper

In a supply chain context, it requires to connect physical assets to unique identifiers or 'digital profiles' on the blockchain [9], which is also known as 'tokenising of assets' [10]. Thereby on-blockchain tokens can represent all kind of assets such as currencies, stocks, properties, and coupons [6]. Fundamentally, one can distinguish between two types of tokens, fungible token and non-fungible token:

- Fungible Token (FT): Different units of a FT are interchangeable and have no unique properties.
- Non-fungible Token (NFT): Each unit of a NFT is unique from another, allowing the tracking of their ownership.

Based on these technical properties, FTs are particularly suitable to represent cryptocurrencies while NFTs, as the term 'non-fungibility' suggests, are intended to clearly identify unique digital or non-digital assets [11]. This makes the adoption of NFTs a suitable concept for mapping assets throughout supply chains on the blockchain [12].

Ethereum established token standards for both FTs and NFTs to increase the acceptance of different tokens in the Ethereum blockchain. These token standards include minimum specifications of required functions to allow an implementation. Furthermore, it is possible to add functions that are not part of the standard in order to specify tokens for their respective application [13]. Particularly NFTs that are used for complex applications require to extend the minimum specifications defined in token standards with further attributes and functions. Arcenegui et al. [14] refer to such tokens as 'smart NFTs'.

Smart NFTS with their possibility of linking tokens to requirements represent the technological foundation for mapping complex products consisting of several components on the blockchain. However, as the investigation of current blockchain projects in supply chain management shows, there only exist approaches that enable an effective mapping of simple products without compositional changes [8]. Nevertheless, particularly for dynamic supply chains with complex products that include compositional changes the avoidance and detection of overproduced, cloned, and tampered counterfeit types represents a present research gap. Specifically, the problem of incorporating tampered counterfeit parts in assemblies introduces a vulnerability that must be prevented by means of holistic mapping approaches [15].

A first architecture proposed by Westerkamp et al. [12] adapts the NFT standard ERC-721 for enabling the mapping of assembly processes by adding 'creation requirements' to the NFTs. These requirements ensure that the NFT of an assembly can only be created if the creator owns the NFTs of the required parts for this assembly. This NFT architecture represents an important foundation to enable a transparent and secure mapping of assembly processes on the blockchain. However, the immutability of BCT induces that only a static mapping of product relationships can be generated through this approach. Changes in the 'creation requirements' cannot be added subsequently without re-deploying all affected smart contracts. This makes it considerably more difficult to maintain such NFT construct when mapping dynamic supply chains with complex products [16]. Additionally, Watanabe et al. [17] point out that token-based traceability systems in a supply chain context require the possibility for each network participant to easily confirm the history of the circulation of tokens related to each product. However, the existing token standards merely focus on a secure input and interface design, but do not consider an efficient way of conducting history searches [17]. This paper introduces a prototypical blockchain application that adopts smart NFTs to enable a blockchain-based ecosystem allowing to holistically map complex products in dynamic supply chains. In this context, holistical mapping refers to the mapping of the core supply chain events, object creation/removal, object transformation, object aggregation/disaggregation and object transactions [18]. The application uses a private Ganache Ethereum network and a ReactJS user interface.

2 Prototypical Application

Decentralised applications are subject to a special logic. Changes to the state of the smart contract are transmitted in form of transactions, which must be confirmed by the blockchain network. Figure 1 illustrates the deployment and working of Ethereum smart decentralised applications. To simplify the procedure, the figure does not include the illustration of the mining process. First, Client 1 creates a smart contract in a high-level language. The smart contract is compiled into machine-level byte code where each byte represents an operation, and is then uploaded to the

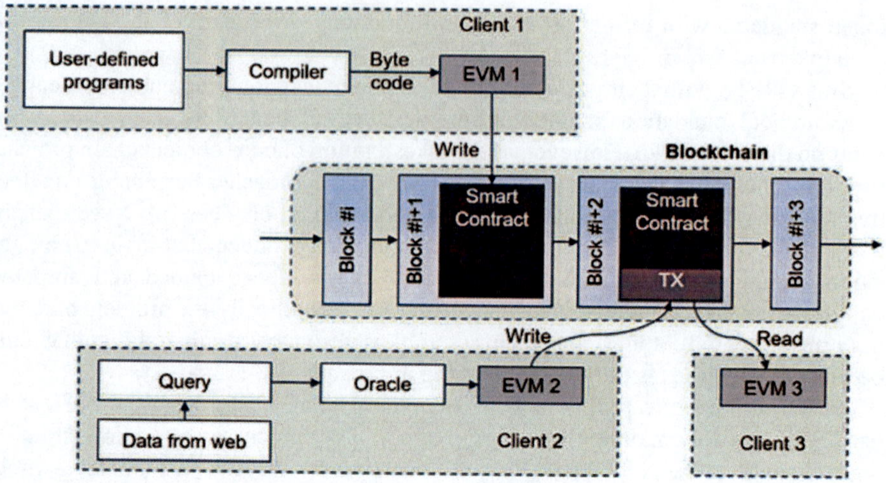

Fig. 1 Smat contract transaction logic (based on Hu et al. [19])

blockchain in the form of a transaction by the Ethereum Virtual Machine 1 (EVM 1). A miner picks this transaction up and confirms it in Block #i + 1. Once Client 2 has interacted via a web interface with the smart contract, the EVM 2 queries the data from the web and embeds it into the transaction TX and deploys it to the blockchain. After the confirmation of the transaction TX, the new state of the contract is updated in Block #i + 2. Client 3 has to synchronise at least to Block #i + 2, to see the changes caused by transaction TX.

In addition to taking this transaction-based logic into account, the prototypical application combines two concepts within one smart contract. First, a concept for an dynamic authority management, which allows supply chain participants to be dynamically involved or excluded in the application. Second, a smart NFT concept that establishes 'token blueprints' in order to link the creation of smart NFTs to requirements.

2.1 Authority Concept

In the physical world, only certain entities of a supply chain are able to create certain parts. This means, that only the entity 'owning' the process of creating a certain part in the physical world is able to have access to the function of the smart contract to create the virtual identity of that part. To guarantee this, the smart contract functions must be linked with the respective authority before creating a token on the blockchain network [16].

In priciple, a smart contract allwos any type of authority concepts. Decentralised and open supply chains with an even distribution of power can be virtually represented, as well as regulated supply chains structures with only one central authority. The protoyical application is based on a central authority principle. The deployer of the smart contract automatically becomes the application's administrator with the ability to add addresses (public keys) of supply chain partners to be involved in the application. Added partners are then part of the supply chain ecosystem and able to receive, create and send tokens. Listing 1 shows the extract from the corresponding source code allowing the deployer of the smart contract to add suppliers to the application.

Function 1: Add Supplier

1. **function addSupplier**(string memory

   _supplierName, string memory
   _supplierContact, address
   _supplierAddress) public onlyAdmin {
2. Supplier memory supplier = Supplier(_supplierAddress, _supplierName, _supplierContact, true);
3. supplierDetails[_supplierAddress] = supplier;
4. allSuppliers.push(supplier);
5. emit SupplierAdded(_supplierAddress);
6. }

Listing 1. Source code for adding suppliers

2.2 *Smart NFT Concept*

When mapping products with changeable configurations on the blockchain, it is necessary to still ensure the immutability of the NFT itself in order to guarantee the integrity of the blockchain-based system. Strictly speaking, when aggregating or transforming an object, the object itself never changes, instead it can be considered as an input object that is consumed in order to produce a new output object [18]. Adapting this approach to NFTs, the change in a token underlies the *process* that causes the change and not in the final token itself. Therefore, prototypical application includes the establishment of 'token blueprints' that represent the manufacturing process. A token blueprint can be seen as a function within a smart contract that defines a token's structure and its requirements to be met when minting it. Each blueprint consists of a unique identifier (token blueprint ID) and a token structure definition. To ensure the uniqueness of each token blueprint, the token blueprint ID is generated by hashing the content of the token structure definition. Therefore, the token blueprint ID is a logical result of its content. Additionally, each blueprint is connected to an owner, which refers to the blockchain account address that has

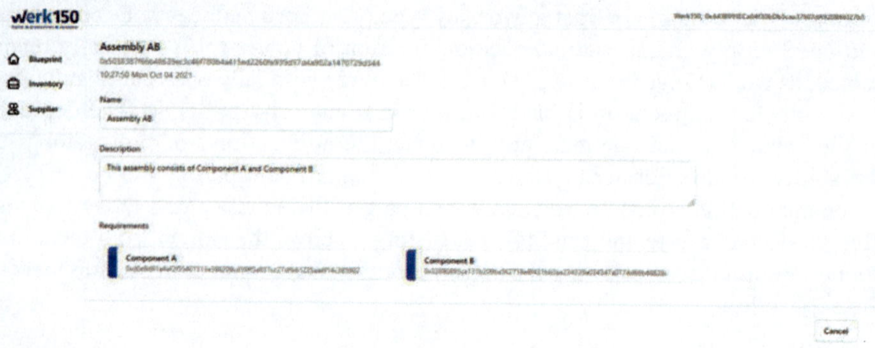

Fig. 2 Blueprint of an assembly with consisting of two components

the permission to mint tokens with the blueprint. After the successful creation of a token blueprint, the blueprint function allows all authorised suppliers to add tokens minted by existing blueprints to the requirements of subsequent token blueprints. Figure 2 illustrates this situation by means of the user interface. For exemplary purposes, the application represents a token blueprint for the 'Assembly AB', which consists of 'Component A' and 'Component B'. The creation requirements ensure that the creator of 'Assembly AB' owns both a token minted with a token blueprint for 'Component A' and 'Component B'. As long as these conditions can be met, the owner of the blueprint for the 'Assembly AB' can create any number of tokens of the same product type with identical technical properties, but which are clearly identifiable by their unique IDs. Similar to other blockchain-based approaches in supply chain management, it is possible to attach these IDs respective products in the physical world, for example via QR codes [9, 12].

The function for sending tokens enables token owners to send them to any address on the Ethereum network. Since the prototypical application is not an established global token standard on the Ethereum blockchain, all addresses can receive tokens, however, these are only visible to authorised participants in the smart contract (see Sect. 2.1) and only they can perform further actions such as sending, aggregating or transforming. Previous approaches such as the approach by Westerkamp et al. [12] use the ERC-721 NFT standard on the Ethereum blockchain, which only defines tokens to include a unique identifier (*tokenId*) and the blockchain account address owing the token (*owner*) [14]. Since the NFTs in the prototypical application predominantly aim to increase the transparency of supply chains, this application extends the '*owner*' attribute by '*history*', an attribute reflecting not only the current owner of the token but also its whole ownership and event history. Like this, each token 'carries' an own tracking record allowing to verify the tokens' history without the need of accessing the blockchain's metadata. Figure 3 shows an exemplary history using the example of 'Component A'.

A logical coupling of aggregated tokens ensures that all tokens are being constrained to the same place at the same time when merged together. If the owner of

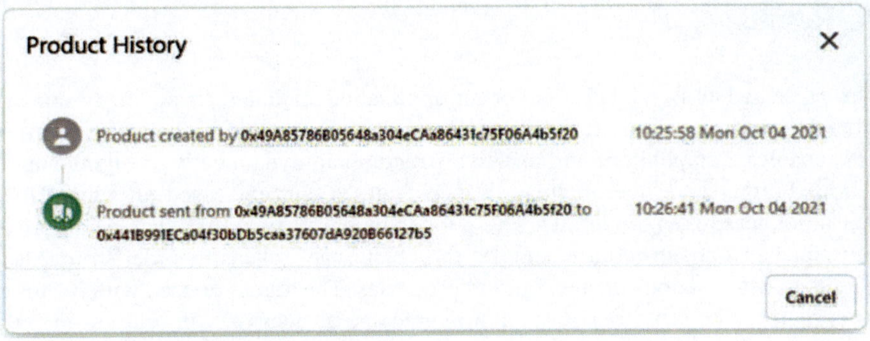

Fig. 3 Integrated token history

'Component AB' sends the token to a new owner, this owner strictly speaking also owns 'Component A' and 'Component B' that are logically coupled to 'Component AB'. In addition to the event history of each token, the smart contracts of the prototypical application also maps each token's composition. Figure 4 shows the composition of 'Assembly AB'. Therefore, each token's composition can be displayed directly within the smart contract reducing the computing-intensive effort that is required in ERC-721-based NFT approaches in order to generate an auditability of linked tokens [17].

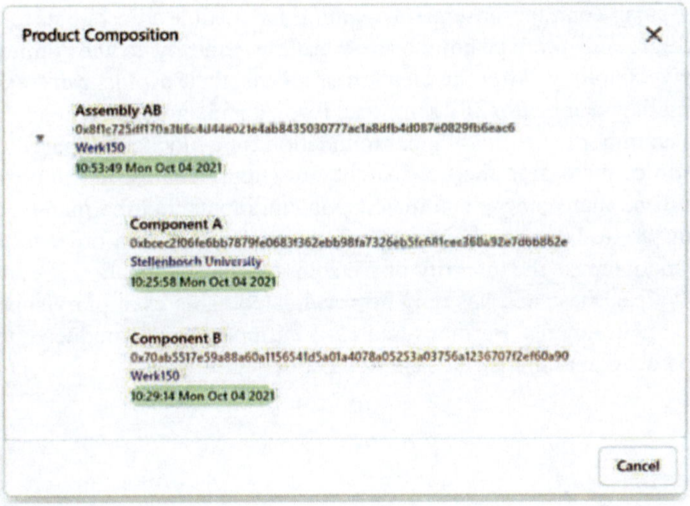

Fig. 4 Integrated token composition

3 Results

The presented prototypical blockchain application includes an authority concept whereby supply chain participants can be added or removed in the smart contract. This enables a mapping of authoritative structures in dynamically changing supply chains. Furthermore, the application incororates a concept based on smart NFTs that introduces token blueprints. These token blueprint functions enable authorised participants to attach requirements to the creation of tokens in order to map any kind of transformation and aggregation processes. The tokens created with the token blueprints can be sent, transformed or aggregated as necessary in order to map the core supply chain events of their physical counterparts. In addition, the application comprises mapping functions in order to map each token's history and composition within the smart contract. This creates token transparency and the associated supply chain transparency without the need for additional evaluation of the blockchain's metadata.

4 Conclusion

The presented blockchain application represents the first apporach that enables a holistical mapping of complex products in dynamic supply chains. The embeddment of authority and blueprint functions in the smart contract proofs that after the deployment of the smart contract subsequent dynamic and flexible adjustments to the supply chain structure and product composition are not contrary to the immutability of blockchain technology. After the creation of tokens, their owners can freely dispose of them and they can be flexibly sent, transformed and aggregated at any time. This represents an important technological foundation for a blockchain-based ecosystem with the aim of increasing the transparency of supply chains. In the next step, the solution will be increasingly automated. Manual inputs and the manual triggering of functions are to be carried out preferably by IoT devices in order to reduce the error rate and increase the integrity of the solution. The application is currently still in a prototypical phase and has only been validated using exemplary supply chains in research environments. Further research is currently being conducted to evaluate the solution using real industrial case studies.

References

1. Gentemann, L.: Blockchain in Deutschland – Einsatz, Potenziale, Herausforderungen (2019). www.bitkom.org. Accessed 10 Oct 2019
2. Nakamoto, S.: Bitcoin: A Peer-to-Peer Electronic Cash System (2008). https://bitcoin.org/bitcoin.pdf. Accessed 20 Oct 2019

3. Grinberg, R.: Bitcoin: an innovative alternative digital currency. Hastings Sci. Technol. Law J. 159–208 (2011)
4. Kaplanov, N.M.: Nerdy money: bitcoin, the private digital currency, and the case against its regulation. Consum. Law Rev. 111–174 (2012)
5. Sorge, C., Krohn-Grimberghe, A.: Bitcoin: Eine erste Einordnung, DuD - Datenschutz und Datensicherheit, pp. 479–484 (2012). https://doi.org/10.1007/s11623-012-0164-9
6. Buterin, V.: Ethereum White Paper: A Next Generation Smart Contract & Decentralized Application Platform (2013). https://blockchainlab.com/pdf/Ethereum_white_paper-a_next_generation_smart_contract_and_decentralized_application_platform-vitalik-buterin.pdf. Accessed 20 Oct 2019
7. Lee, D.K.C., Deng, R.H. (eds.): ChinaTech, mobile security, and distributed ledger. Academic Press, London (2018)
8. Dietrich, F., Ge, Y., Turgut, A., Louw, L., Palm, D.: Review and analysis of blockchain projects in supply chain management. Procedia Comput. Sci. **180**, 724–733 (2021). https://doi.org/10.1016/j.procs.2021.01.295
9. Abeyratne, S.A., Monfared, R.P.: Blockchain ready manufacturing supply chain using distributed ledger. Int. J. Res. Eng. Technol. **5**, 1–10 (2016). https://doi.org/10.15623/ijret.2016.0509001
10. de Filippi, P., Schuppli, B., Choi, C., Reyes, C., Tapscott, D., et al.: Regulatory Framework for Token Sales: An Overview of Relevant Laws and Regulations in Different Jurisdictions Coalition of Automated Legal Applications. Blockchain Research Institute (2019)
11. Dowling, M.: Is non-fungible token pricing driven by cryptocurrencies? Finance Res. Lett. (2021). https://doi.org/10.1016/j.frl.2021.102097
12. Westerkamp, M., Victor, F., Küpper, A.: Tracing manufacturing processes using blockchain-based token compositions. Digital Commun. Netw. (2019). https://doi.org/10.1016/j.dcan.2019.01.007
13. Antonopoulos, A.M., Wood, G.A.: Mastering Ethereum—Building Smart Contracts and DApps. O'Reilly Media, Tokyo (2019)
14. Arcenegui, J., Arjona, R., Román, R., Baturone, I.: Secure combination of IoT and blockchain by physically binding IoT devices to smart non-fungible tokens using PUFs. Sensors (Basel, Switzerland) **21** (2021). https://doi.org/10.3390/s21093119
15. Collier, Z.A., Hassler, M.L., Lambert, J.H., DiMase, D., Linkov, I.: Supply chains. In: Kott, A., Linkov, I. (eds.) Cyber Resilience of Systems and Networks, pp. 447–462. Springer International Publishing, Cham (2019)
16. Dietrich, F., Turgut, A., Palm, D., Louw, L.: Token-based blockchain solutions for supply chain strategies. In: Production at the Leading Edge of Technology, vol. 10, pp. 689–698 (2020). https://doi.org/10.1007/978-3-662-62138-7_69
17. Watanabe, H., Ishida, T., Ohashi, S., Fujimura, S., Nakadaira, A., Hidaka, K., Kishigami, J.: Enhancing blockchain traceability with dag-based tokens. In: International Conference on Blockchain (Blockchain), pp. 220–227 (2019). https://doi.org/10.1109/Blockchain.2019.00036
18. Kennedy, A., Southall, M., Morgan, G., Traub, K.: EPC Information Services (EPCIS) Standard, GS1 AISBL (2016)
19. Hu, Y., Liyanage, M., Mansoor, A., Thilakarathna, K., Jourjon, G., Seneviratne, A.: Blockchain-Based Smart Contracts—Applications and Challenges (2018)

Fabian Dietrich is Ph.D. Candidate (Industrial Engineering) at Stellenbosch University, South Africa and works as Research Associate at ESB Business School, Germany. He holds a M.Eng. (Industrial Engineering) from Stellenbosch University and a M.Sc. (Digital Industrial Management and Engineering) from the ESB Business School of Reutlingen University.

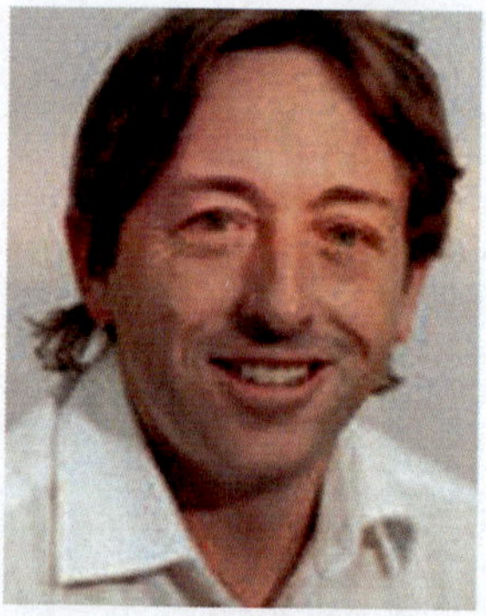

Louis Louw has been an associate professor at Stellenbosch University since 2018. Previously Prof. Dr. Ing. Louis Louw spent twelve years in industry as a business consultant and senior manager, before joining the university in 2014. His research interests are in the field of Digital Operations & Supply Chain Management, and he is a member of the International Association of Learning Factories.

Daniel Palm is Professor for Supply Chain Management at the ESB Business School. Univ.Lektor Prof. Dr. Techn. Dipl. Ing. Daniel Palm is head of the teaching and research center for Value-Added and Logistics Systems at Reutlingen University and the Reutlingen Center Industry 4.0, RZI 4.0—a cooperation of the Fraunhofer Institutes IPA and IAO as well as Reutlingen University.

Development of a Procedure Model to Compare the Picking Performance of Different Layouts in a Distribution Center

Dorit Schumann, Cihan Cevirgen, Julian Becker, Omar Arian, and Peter Nyhuis

Abstract An efficient order picking process is required to realize short-term and on-time deliveries in distribution logistics. The selection of a suitable warehouse layout, as well as an efficient picking strategy needs to be dependent on an evaluation of the picking performance. The paper presents a procedure to compare variable layouts of distribution centers regarding their picking performance by using a simple data set. In particular, the implementation of pick face zones is investigated. By picking out of a pick face zone, replenishment for this zone is necessary as an additional process step. This picking strategy is used in practice, but there is no quantitative way to assess the impact on the picking performance depending on the layout of the distribution center. A calculation method to determine the picking performance theoretically is developed by using process times based on movement data and average distances to the layout zones. Furthermore, the comparison of the layouts with pick face zone and without pick face zone is focused, and a description of a procedure to determine the optimal size of the pick face zone is given. A further aspect of developing the procedure model is selecting the articles to be stocked in this zone. For these articles, picking time savings are expected due to shorter distances to the pick face zone. However, the additional replenishment processes must be considered by comparing the picking performance. The developed calculation method compares the picking times between layouts with pick face zone and without pick face zones. A case

D. Schumann (✉) · C. Cevirgen · P. Nyhuis
Institute of Production Systems and Logistics, Leibniz University Hannover, Garbsen, Germany
e-mail: Schumann@ifa.uni-hannover.de

C. Cevirgen
e-mail: Cevigren@ifa.uni-hannover.de

P. Nyhuis
e-mail: nyhuis@ifa.uni-hannover.de

J. Becker · O. Arian
Continental AG, Hannover, Germany
e-mail: julian.becker@conti.de

O. Arian
e-mail: omar.arian@conti.de

K. von Leipzig et al. (eds.), *Smart, Sustainable Manufacturing in an Ever-Changing World*, Lecture Notes in Production Engineering,
https://doi.org/10.1007/978-3-031-15602-1_42

study in the automotive industry is conducted to apply the generated procedure. The example shows improvements in picking performance through pick face zones. Finally, the presented process model enables the comparison of different layouts regarding their picking performance and provides an approach to determine pick face zones.

Keywords Picking performance · Layouts · Procedure model · Warehouse management · Distribution · Logistic processes

1 Introduction

Companies operate in increasingly dynamic market environments with global supply chains, which increases the complexity of influencing parameters of logistic systems. Equally, the demands on logistic performance and logistic costs are increasing, too [1, 2]. The highest costs within logistics are personnel and systems costs due to distribution operations: transportation and warehousing [3]. Consequently, suitable warehouse layouts are required for efficient logistic processes. The most labor-intensive process in warehouses with manual systems is order picking. The organization of order picking processes impacts the logistic performance of distribution centers. Especially the customers expect of short-term, and on-time deliveries allow only a short time window for order picking [4]. Mainly time is required for transporting the products from the storage location to the loading dock. Thereby, routing methods, warehouse zoning, and picking strategies have a significant influence on picking performance. There are many academic approaches to find the optimal combination of layout, routing, order execution, and take-out strategies. However, for these decision problems, there is still a lack of simple quantitative evaluation methods for companies. One practical approach is the implementation of a pick face zone in the warehouse layout. This zone can be reached more quickly and stores articles with a high access frequency. That is a way to reduce the picking time. Nevertheless, the additional necessary replenishment process for this zone must be considered by evaluating the picking performance. An applicable method to calculate the picking time to compare different warehouse layouts is described in this paper. Besides, it investigates if pick face zones enable an improvement of picking performance and presents a procedure model to determine the size of this zone and the right articles to store in this zone.

2 Picking Methods and Performance

Order picking is an essential task of intralogistics and means the collection and consolidation of required quantities of articles in the right quantity, at the right time, and the right place [5]. The picking process is divided into the following process steps [6]: provision of goods in storage units, movement of the picker, removal of the required quantity of goods, consolidation of the removed goods according to the picking order, transport of the picked order to the loading dock and, if necessary, return the opened storage unit.

The efficiency of a picking system depends mainly on the organization and the used operating strategies. For instance, the layout design of a warehouse has an influence regarding the structure and arrangement of storage areas. Also, handling methods have a significant impact on the picking process. During the planning of a new picking system, the selection of suitable strategies can minimize the necessary investments and future operating costs [7]. The picking strategies can be divided into the categories of placing, execution, take-out, and path [5]. Placing strategies determine on which storage locations and in which zones which articles are stored and provided. The integration of a pick face zone is categorized as part of this strategy. The execution strategy regulates the order processing, and intends a balanced utilization in the picking zones. A parallel order execution can achieve consolidation effects compared to single order execution. The path and sequence of the picking locations is determined by path strategies. Finally, take-out strategies regulate the removal of goods according to certain principles (e.g., FIFO).

The picking performance shows the efficiency of a picking system and is expressed in grab units per hour or positions per hour. The performance results from the reciprocal value of the picking time for one unit: The picking time is the sum of base time t_{base}, path time t_{path}, setup time t_{setup}, and grip time t_{grip} (Eq. 1) [8]. The base time means the time spending of the picker in the base station (loading area) at the beginning and the end of the process and takes up approx. 5–10% of the total picking time. The path time results from the covered distance of the picking process and is significantly influences by the layout. The time for taking out the articles from the storage places and deposit those into the collection units is summed up as the grip time (approx. 5–10% of the total time). A substantial timeshare of 10–35% is caused by the setup time. Activities for information, searching, waiting, or additional handling activities occupied the setup time. For exact calculation methods of the separate timeshares, we refer to foundational books [5, 8, 9]. The total picking time results from the sum of the timeshares:

$$t_P = t_{base} + t_{grip} + t_{path} + t_{setup} \left[\frac{s}{position} \right] \tag{1}$$

The literature provides comprehensive calculations and rules to determine the picking performance depending on the respective picking strategy, and to achieve an improvement of it. Regarding the placing strategy, Richards [10] recommends a classification according to the parameters volume and frequency of the articles in an ABC-XYZ analysis. If the search for the shortest way is initiated in the path strategy, the Travelling Salesman Problem is consulted as classical optimization problem [11]. In addition, analytical models are available for a one-dimensional or two-dimensional movement of in-storing and picking processes under the consideration of specific assumptions [4, 12]. With the focus on the economic and ergonomic performance, the authors of [13] evaluate different rack layouts. In [4, 14] a literature review for different design options of warehouse layouts were discussed. In regard to improve the picking performance, there are several investigations on certain cases, like in [15] for the dimensioning of the conveyor pick face in an inverse picking system or in [16] a Zone-Picking-Network. None of them are feasible for the topic of this paper and they focused parts-to-picker-systems. Therefore, there is a research gap regarding a quantitative assessment of picking performance to compare different layouts, especially with integrated pick face zones in a picker-to-parts picking system. In the following, a concept is developed that allows an approximate assessment of improvements in picking performances through pick face zones by using a simple data set from a sufficiently large period.

3 Development of a Procedure Model

This section describes the development of the procedure model. The model relates to a conventional picking system, where the picker comes with the order collection units to the stationary access units in the storage. First, the layout options of a warehouse are identified, then an evaluation method of picking performance is presented. In the last section of the chapter, the procedure model for determining the pick faces is explained.

3.1 Options of Warehouse Layouts

Since layout affects picking performance, different layout options should be compared based on their picking time. The zoning of the warehouse into ABCD-sections can be done as horizontal or vertical divisions: Vertical according to levels of a rack (A at the bottom, C/D at the top)—horizontal as individual zones in the area. Furthermore, a pick face zone can be additionally integrated into the layout. A pick face zone provides faster accessibility of articles. However, there is a limited number of storage locations in the pick face zone, so replenishment processes are required for the articles stored there. The pick face zone can be set up either as a pick face area or as a ground-level pick face. The ground-level pick face means that only the lowest level of the rack is designated as the pick face. With the pick face area, all storage locations within the zone are designated for the pick face. The following Figs. 1 and 2 show the layout options schematically. The sizes of zones A, B, C, and D depend on the distribution of the article spectrum concerning quantity share and picking frequency. Vertical zoning can only take place in the rack zone. Therefore, a distinction between block and rack storage is required. For horizontal zoning, the storage type is not relevant. In both layout options, a horizontal D-zone is set up for rarely picked special parts. In the vertical layout, the A-articles are stored in the lower levels of the rack, the B-articles in the middle levels, and the C-articles in the

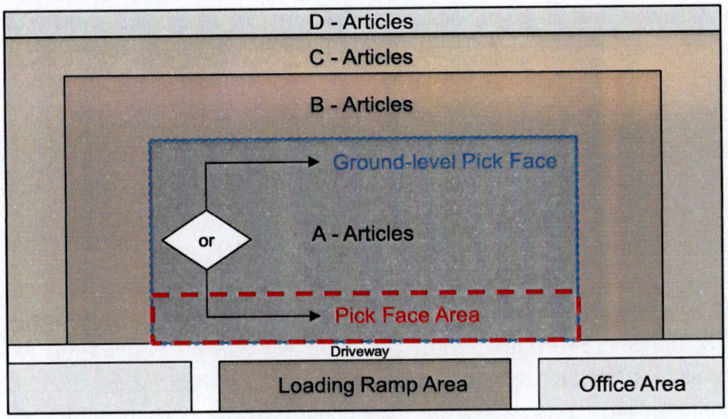

Fig. 1 Horizontal layout with pick face options

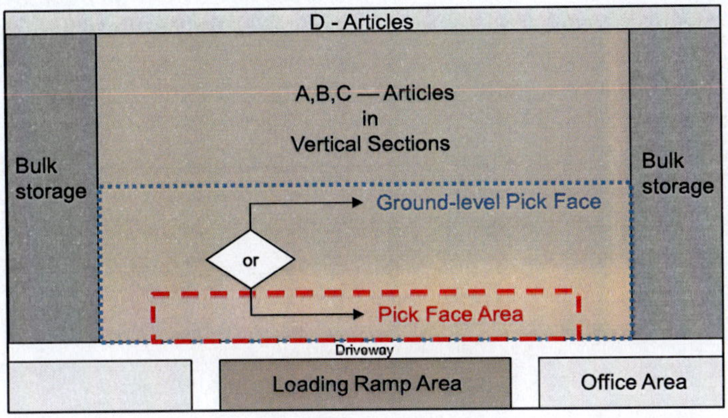

Fig. 2 Vertical layout with pick face options

top levels. The pick face zones are shown as dashed lines. The size of these zones is determined with the developed procedure. Whether and in which design these zones lead to an improvement of the picking performance is checked in the calculation model. The calculation method and the procedure model are explained below.

3.2 *Calculation of Picking Performance*

As already mentioned in Sect. 2, the picking performance results from the picking time consisting of the different time elements (see Eq. 1). Because the number of picked positions does not change for the comparison, the absolute time can be calculated. Several article positions are to be picked in different quantities at their respective storage locations in one picking process. Thus, different pick positions are approached per picking process. Depending on the equipment and the location of the storage location, further process steps are required at this pick position (e.g., raising the forklift in case of a storage location on higher shelf levels). When the storage unit is reachable, the required quantity must be taken out. The grip time per location is calculated with the quantity of picked items multiplied by the timeshare for picking one item.

The highest timeshare is caused by the path time. Therefore, the path distance is required to determine the path time. Each storage location has a different distance from the ramp. However, if the access frequency within a storage zone is equally distributed, the average path length for the zone can be considered [5, 9]. For a symmetrically designed storage layout, the start- and endpoints are defined at the base of the symmetry axis. Accordingly, the path distances to the right and left of the symmetry axis are identical. Finally, the determination of only one value is required for the average path distances of a zone (see Fig. 3). For the rectangular zones, the center point can be easily determined. The path results out of the horizontal and vertical coordinate distance plus the path component for driveway and ramp. Zones B and C result in an L-shape due to the division at the symmetry axis. Therefore, the centroid of the area must be formed by dividing it into partial areas. A detailed description of the calculation method can be found in the literature [17]. The total path results from the roundtrip to the first pick position in one zone, plus the path between the pick positions. This results in one-third of the relevant average path length [5, 9]. Equation 2 shows the path length for one picking process starting in Zone A:

$$s_{n_A} = d_{A1} = d_{A2} = 2 \cdot \left(\frac{b_A}{4} + l_{Driveway} + \frac{l_{Ramp}}{2} + \frac{l_A}{2} \right) + (n-1) \cdot \frac{1}{3} \cdot \left(\frac{b_A}{4} + \frac{l_A}{2} \right) \tag{2}$$

where $s_{n\_A}/d_{A1}/d_{A2}$ = path length A-zone [m]; b_A = width A-zone [m]; $l_{Driveway}$ = length driveway [m]; l_{Ramp} = length ramp [m]; l_A = length A-zone [m]; n = number of picking positions.

To compare the picking performance of the different layout options, the total picking time for all parts in the observation period is considered. In addition to the size, layout, and ABC-zones, the data for the picked articles per transport order (usually one truck) in the observation period is required for the calculation. Consequently, the frequency of picking and the average picking quantity of one article can be derived. Furthermore, the ABC-classification of the article must be known for information in which storage zone this article is stored. As described, several pick positions are approached per picking process. Thus, the picking time of one article is divided by the number of picking positions approached. The calculation is simplified by calculating the number of picking processes required to pick the total quantity of one article, taking into account the average number of pick positions. This simplification is possible because the total number of required processes for picking the total quantities of all articles in the observation period is not dependent on the

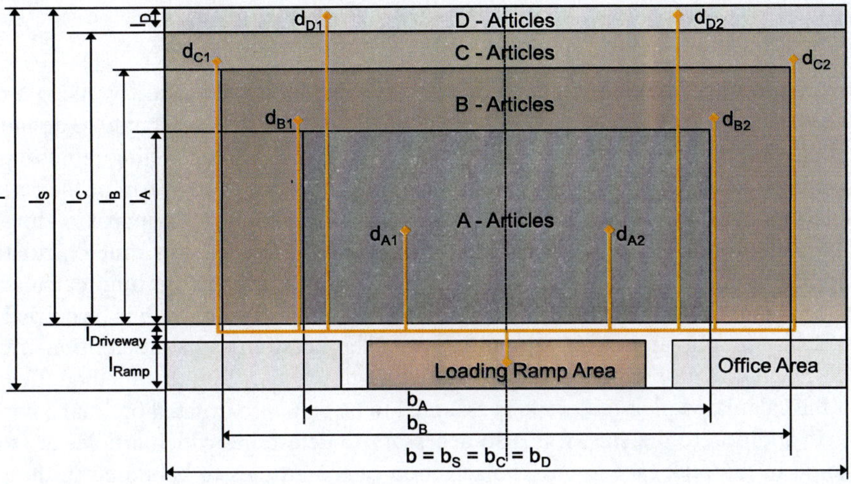

Fig. 3 Path distance calculation of zones

combination of articles in one process. The calculation of the number of necessary picking processes per article number is obtained by dividing the total quantity of the article by the product of average pick quantity and the average number of picking positions. This results in the calculation of the picking time of an article as follows:

$$t_{p\_article} = \left(n_{PP} \cdot \left(t_{base,total} + t_{path,n}\right) + n \cdot \left(t_P + m_{pos} \cdot t_A\right)\right) \cdot (1 + \eta_{dead}) \quad (3)$$

$t_{p\_article}$ = picking time of an article [s]; n_{PP} = number of picking processes; $t_{base,total}$ = total base time per picking process [s/PP]; $t_{path,n}$ = path time per picking process [s/PP]; n = number of picking positions; t_P = timeshare per pick position [s/position]; m_{pos} = pick quantity per pick position [AU/position]; t_A = grip time per article unit [s/AU]; η_{dead}= dead time factor

The total picking time for each layout option is considered for the comparison. This is the sum of picking times of all articles. Consideration at the article level is necessary to determine the pick face zone. If the article is stored in the pick face zone, the required time for replenishment processes must also be taken into account. The number of required replenishment processes per article is calculated by dividing the total demand for the article in the observation period by the capacity of storage units per one article in the pick face zone. The total replenishment time for an article is calculated as follows:

$$t_{P\_article\_PF,total} = t_{P\_article\_PF} + t_R \cdot n_R \quad [s] \quad (4)$$

$T_{P\_article\_PF,total}$ = picking time of the article when stored in the pick face [s]; $t_{P\_article\_PF}$ = picking time of the article [s]; t_R = replenishment time per process [s]; n_R = number of replenishment processes

3.3 Procedure Model to Determine Pick Face Zones

In order to determine the improvement of the picking performance by using a pick face zone, it is important to examine the dimensions. It is also essential to define the articles to be stored in the provided pick face zone. The dimensioning represents an optimization problem. However, this procedure model does not determine the optimal solution, but it presents a practical determination possibility for approximating the optimal solution. In order to estimate the required distances for time calculation, the size of the pick face zone must be fixed in advance of the time calculation. Several iteration steps with differently assumed dimensions must thus be carried out to determine the approximate optimal size. The iteration steps are continued until the calculated time saving does not change significantly. With this method, the size of the calculated pick face zone is assumed to be an approximated optimal solution.

In addition to the size, it is also necessary to determine which articles are to be stored in the pick face zone. The selection of the articles to be stored in the pick face zone is done by sorting the articles in order of the highest savings of picking times. Therefore, the storage capacity of this zone must be determined based on the initially selected dimensions. The calculation model calculates the total picking time for each article for both options—stored or not stored in the pick face zone. Then all articles with savings in picking time are filtered and sorted in descending order of the saving size. For each article, only a defined number of storage units can be provided in the pick face zone. Now the capacity of the pick face zone (number of storage units) is crucial. The described procedure must be performed for the ground-level pick face as well as for the pick face area. With the determined dimensions in each case, the total picking times can be calculated and compared. Several calculation runs must be performed to compare horizontal and vertical layouts, as well as for the dimensioning and design of the pick face zone. In the following case study, the iteration steps for determining the size of the pick face zone become clear.

4 Case Study in the Automotive Industry

The procedure for comparing layout options regarding their picking performance was tested as part of a case study in the automotive industry. Within the case study, different warehouses were considered. This paper focusses only one warehouse in this case study. The necessary input data for the developed model are listed below:

- Layout with zones
- Size of warehouse for distance calculation
- Process times
- ABC-XYZ- classification of the articles
- Report with article numbers, transport orders, pick quantities, capacity of storage units.

For this purpose, different layout configurations were compared for several warehouse locations. Furthermore, the calculation method was validated by comparing the calculated picking times with the real picking times in the current layout configuration. The transport orders for picking of an observation period of 12 months in the past served as the data base. The case study compared several distribution warehouses using different picking strategies and layouts. The process model was transferred into a tool. As an example for one warehouse location, the results for the different layout options are shown in this paper. The presented warehouse has a storage area of about 20.000 m^2 and has an outbound of around 3 Million parts per year. The pick face zone is dimensioned as a percentage of the total warehouse stock. This is shown on the X-axis of the diagram (Fig. 4). The Y-axis shows the absolute picking time to pick all parts. For this warehouse, a vertical warehouse layout results in a lower picking time than a horizontal layout while no pick face zone is installed. Still, it is obvious that the picking time changes depending on the size of the pick face zone. After performing several iteration steps, an approximately optimal solution for the size dimensioning can be identified. In the example, a vertical layout results in only minor time savings due to a pick face zone, whereas a horizontal layout results in significant savings. If a horizontal layout is integrated, the highest possible time savings are achieved with a ground-level pick face covering about 7.5% of the total warehouse stock. If the ground-level pick face is too large, time savings are no longer achieved since the horizontal distance takes up a more significant proportion of the time than the savings due to the elimination of the lift height. Due to the more significant influence of the horizontal path distance, the picking time also increases more slowly after the optimum of the pick face area. With approx. 11%, the pick face area leads to significant time savings. Converted into personnel time, this means the saving of one employee.

In the model, it should be noted that the picking time is subject to several influencing factors. Thus the existing ABC article classification with the selected allocation strategy, as well as the structure of the transport orders has a substantial influence on whether the mechanism of a pick face zone leads to a performance improvement. The other use cases in the case study generated similar results. This demonstrated the feasibility of the model in a collaboration with the company.

5 Conclusions

This procedure model describes how different warehouse layouts can be evaluated in terms of picking performance. A calculation methodology was developed that verifies time savings depending on the dimensioning of a pick face zone. The case study shows that an improvement of the picking performance is possible by setting up a pick face strategy. Further research is needed on the interactions between a pick face zone and other strategies. The developed procedure model is feasible for a simple picking process in a manual warehouse. Furthermore, the concept has to be extended for more complex article and other technological warehouse structures. It

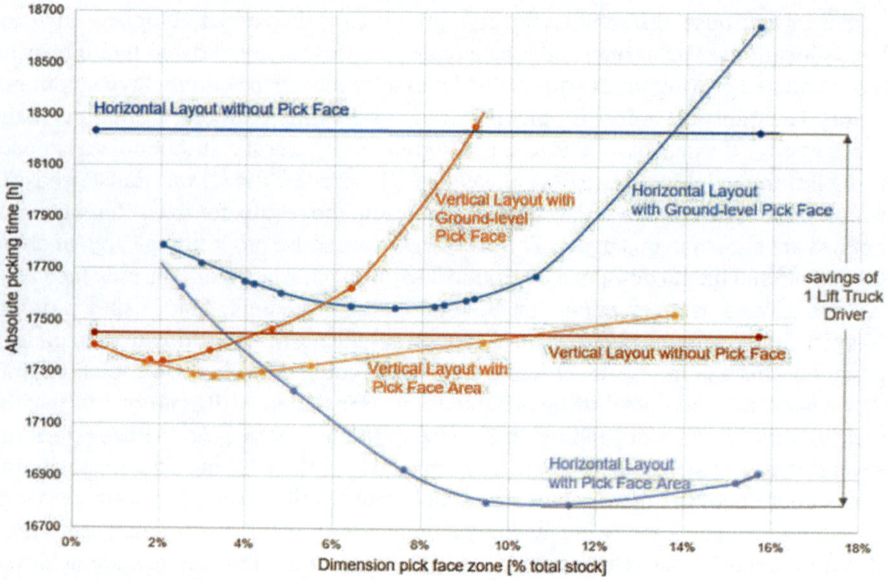

Fig. 4 Comparing picking time of different layouts exemplary for one warehouse

may be possible to derive general rules regarding appropriate occupancy and pick face strategies considering picking performance with the help of new research and further case studies. However, this calculation method allows for simple conditions an evaluation of different layout options of a warehouse in terms of picking performance with relatively little data. The case study shows that the developed model is a practical solution.

References

1. Abele, E., Reinhart, G.: Zukunft der Produktion: Herausforderungen, Forschungsfelder, Chancen, p. 244. Carl Hanser Fachbuchverlag, München (2011)
2. Windt, K., Hülsmann, M.: Changing Paradigms in Logistics, p. 16. Springer, Berlin Heidelberg (2007)
3. Enslow, B.: By the Numbers: Logistics: Benchmark Report, p. 93. Aberdeen Group, Boston (2005)
4. De Koster, R., Le-Duc, T., Roodbergen, K.J.: Design and control of warehouse order picking: a literature review. Eur. J. Oper. Res. **182**(2), 481–501 (2007)
5. Gudehus, T., Kotzab, H.: Comprehensive Logistics, 2, Softcover reprint of the original 2nd ed. 2012 ed., p. 912. Springer Berlin, Berlin (2016)
6. Schönsleben, P.: Integrales Logistikmanagement: Operations und Supply Chain Management, p. 834. Springer Vieweg, Berlin (2020)
7. Schmidt, T.: Innerbetriebliche Logistik, p. 180. Springer Berlin Heidelberg, Berlin (2019)
8. Martin, H.: Warehousing and Transportation Logistics: Systems, Planning, Application and Cost Effectiveness, p. 681. Kogan Page Limited, London

9. ten Hompel, M., Sadowsky, V., Beck, M.: Kommissionierung: Materialflusssysteme 2 - Planung und Berechnung der Kommissionierung in der Logistik, p. 295. Springer, Berlin (2011)
10. Richards, G.: Warehouse Management: A Complete Guide to Improving Efficiency and Minimizing Costs in the Modern Warehouse, p. 513. Kogan Page, London (2018)
11. Churchman, C.W., Ackhoff, L.A., Arnoff E.L.: Operations Research, p. 590. Oldenbourg, Wien-München (1961)
12. Le-Duc, T.: Design and Control of Efficient Order Picking Processes, p. 166. Erasmus Research Institute of Management, Rotterdamm (2005)
13. Calzavara, M., Glock, C.H., Grosse, E.H., Persona, A., Sgarbossa, F.: Analysis of economic and ergonomic performance measures of different rack layouts in an order picking warehouse. Comput. Ind. Eng. **111**(7–8), 527–536 (2017)
14. Wisser, J.: Der Prozess Lagern und Kommissionieren im Rahmen des Distribution Center Reference Model (DCRM), p. 194. KIT Scientific Publishing (2009)
15. Füßler, D., Boysen, N.: Efficient order processing in an inverse order picking system. Comput. Oper. Res. **88**(2), 150–160 (2017)
16. Ho, Y.-C., Lin, J.-W.: Improving order-picking performance by converting a sequential zone-picking line into a zone-picking network. Comput. Ind. Eng. **113**(3), 241–255 (2017)
17. Hagedorn, P., Wallaschek, J.: Technische Mechanik: Band 1: Statik, p. 337. Verlag Europa-Lehrmittel, Haan-Gruiten

Dorit Schumann M.Sc. (*1994) studied production engineering and logistics at Leibniz University Hannover. Since 2021, she works as a research associate in the field of production management at the Institute of Production Systems and Logistics (IFA) at Leibniz University Hannover.

Cihan Cevirgen M.Sc. (*1990) studies industrial engineering at Leibniz University Hannover. Since 2018, he works as a research associate in the field of factory planning at the Institute of Production Systems and Logistics (IFA) at Leibniz University Hannover.

Dr.-Ing. Julian Becker (*1985) studied industrial engineering at Leibniz University Hannover, subsequently worked as a research associate at the Institute of Production Systems and Logistics (IFA) and completed his doctorate in engineering. Since 2017, he has been working in the area of supply chain management and logistics at Continental AG.

Dipl.-Ing. oec. Omar Arian (*1983) studied industrial engineering and management at the Technical University of Hamburg-Harburg. Since 2016, he has been working as a project manager for logistics optimization projects at Continental AG.

Prof. Dr.-Ing. habil. Peter Nyhuis (1957) worked as a research associate at the Institute of Production Systems and Logistics (IFA). After completing his doctorate in engineering, he received his habilitation before working as a manager in the field of supply management. He is heading the IFA since 2003.

Feasibility Assessment of 5G Use Cases in Intralogistics

F. Dietrich, M. Angos Mediavilla, A. Turgut, T. Lackner, W. Jooste, and D. Palm

Abstract The fifth mobile communications generation (5G) can lead to a substantial change in companies enabling the full capability of wireless industrial communication. 5G with its key features of providing Enhanced Mobile Broadband, Ultra-Reliable and Low-Latency Communication, and Massive Machine Type Communication will support the implementation of Industry 4.0 applications. In particular, the possibility to set-up Non-Public Networks provides the opportunity of 5G communication in factories and ensures sole access to the 5G infrastructure offering new opportunities for companies to implement innovative mobile applications. Currently there exist various concepts, ideas, and projects for 5G applications in an industrial environment. However, the global rollout of 5G systems is a continuous process based on various stages defined by the global initiative 3rd Generation Partnership Project that develops and specifies the 5G telecommunication standard. Accordingly, some services are currently still far from their final performance capability or not yet implemented. Additionally, research lacks in clarifying the general suitability of 5G

F. Dietrich (✉) · T. Lackner · W. Jooste
Department of Industrial Engineering, Stellenbosch University, Stellenbosch, South Africa
e-mail: fabian.dietrich@reutlingen-university.de

T. Lackner
e-mail: Thorge.Lackner@Student.Reutlingen-University.DE

W. Jooste
e-mail: wyhan@sun.ac.za

F. Dietrich · M. Angos Mediavilla · T. Lackner · D. Palm
ESB Business School, Reutlingen University, Reutlingen, Germany
e-mail: mario.angos_mediavilla@reutlingen-university.de

D. Palm
e-mail: Daniel.Palm@Reutlingen-University.DE

A. Turgut
Steinbeis Innovation gGmbH, Stuttgart, Germany
e-mail: Ali.Turgut@Reutlingen-University.DE

D. Palm
Fraunhofer Institute for Manufacturing, Engineering and Automation, Stuttgart, Germany

K. von Leipzig et al. (eds.), *Smart, Sustainable Manufacturing in an Ever-Changing World*, Lecture Notes in Production Engineering,
https://doi.org/10.1007/978-3-031-15602-1_43

regarding frequently mentioned 5G use cases. This paper aims to identify relevant 5G use cases for intralogistics and evaluates their technical requirements regarding their practical feasibility throughout the upcoming 5G specifications.

Keywords 5G · Use cases · Feasibility assessment · Intralogistics · Non-public networks

1 Introduction

The fifth mobile communication generation (5G) represents the further development of the fourth mobile communications standard (4G) commonly known as LTE (Long Term Evolution) [1]. While 4G aims to meet the need of people for internet access, 5G aims to specifically meet the increasing needs of the industry for internet access [2]. Therefore, 5G technology promises to improve the Internet of Things (IoT) communication in order to increase the efficiency of industrial processes [2]. The International Telecommunication Union (ITU) and its particular Radiocommunication Sector (ITU-R) defines the following key capabilities of 5G technology [2]:

- Enhanced Mobile Broadband (eMBB): Provides data rates of up to 20 Gbits/s for down and 10 Gbits/s for upload.
- Massive Machine Type Communication (mMTC): Large-scale possibility to connect IoT devices with a density of up to 1 million connections/m^2 and traffic density of 10 Mbit/s per m^2.
- Ultra-Reliable Low-Latency Communication (uRLLC): Enables latencies of less than 1 ms air-interface round-trip time.

In addition, 5G provides Quality of Service (QoS) for critical industrial applications such as the mobility management service in industrial environments and the high precision positioning service [3]. To ensure the development of new telecommunication standards, the 3rd Generation Partnership Project (3GPP) provides so called *releases* forming a foundation for developers and engineers regarding the implementation of features as well as the addition of new functionalities for subsequent *releases* [4]. 3GPP established the transition from 4G to 5G in Release 15 with Non-Standalone (NSA) networks. NSA 5G networks leverage existing LTE core networks and only carry out the communication between the terminal and the antenna based on 5G protocols [5]. In June 2020, 3GPP initiated Release 16 (R16) to enable first standalone (SA) 5G network solutions. SA networks are completely based on a 5G core without the need of leveraging 4G network elements [5]. Release 17 (R17) is planned for June 2022 [4] with further enhancements relating to the capability of 5G and its services. Currently, only R16 and R17 are specified by 3GPP. Release 18 and further *releases* are in the preparation phase and so far only defined regarding their timeline, but not with respect to its 5G specifications [6]. Experiences with

R16 have shown that it takes around 18 months until the specifications are commercially available in 5G infrastructure. Another difference to 4G is the establishment of Non-Public Networks (NPNs). A NPN is a local area network based on 5G New Radio technology for dedicated wireless connectivity in a specific region. This offers companies to create and manage their own local 5G networks [3]. One advantage of a NPN is the exclusive coverage in industrial facilities, making them independent of public networks and offer its owner the possibility of total control (intrinsic control) to guarantee user transparency and security [3]. In particular with respect to 5G NPNs, the industry has high expectations to enable use cases in the context of Industry 4.0 [7]. However, the establishment of 5G is a complex continuous process that is characterised by the already mentioned *releases* [4]. The high expectations of the 5G technology are oriented towards the technological target state of the 5G technology. This creates a distorted picture of the actual current performance of the technology leading to difficulties when assessing the feasibility of industrial use cases [8]. In this context, the early development of prototypes in test environments is necessary to gather knowledge about the feasibility of intralogistics use cases before the 5G technology reaches its technological target state. In particular, development and transfer centres such as academic research institutions with learning factories play an important role in testing and investigating 5G use cases in its early stages [9]. The literature review methodology of this paper is structured into four phases. In the first phase, the databases Scopus, Google Scholar, IEEE Xplore, and the libraries of Stellenbosch and Reutlingen Universities are searched with the keywords '5G applications', '5G use cases', 'intralogistics'. In the second phase, the found intralogistics use cases are investigated to define their technical requirements. Thirdly, the key characteristics of the current 5G *releases* R16 and R17 with regards to intralogistics use cases are identified. Finally, the use case requirements are assessed for their feasibility with the 5G technology with respect to R16 and R17. The paper consequently aims to identify relevant and feasible intralogistics use cases and their possible applications for initial research activities.

2 Literature Review

Several publications exist defining and summarising important use cases in an industrial environment that are frequently connected to 5G technology. For example [10] consider 5G as important enabler technology for use cases in the context of Industry 4.0. This publication subdivides the suitability of 5G technology for three general use case types: Time critical and reliable processes, non time critical communication, and remote control. The authors do not consider the holistic requirements of the respective 5G use cases and only mention individual requirements such as latency, broadband, or reliability. Moreover, the authors only mention the necessary target state of 5G's capabilities when describing the technical requirements of the use cases. A further publication [11], summarises the use cases provided by 3GPP and derive their key characteristics in terms of NPN attributes such as QoS customisation, autonomy, isolation, and security. However, this publication does not consider an assessment of the use cases with respect to the 5G key capabilities of 5G technology described in the introduction. A survey by [12] provides an extensive collection of 5G use cases with the aim of covering all possible areas of wireless communication. It includes usage scenarios outside industrial environments, such as automated traffic control and open air festivals. Compared to previous publications, this survey only covers a partial collection of industrial use cases and focuses primarily on the capabilities latency, data rate and device density. However, these properties are not consistently specified for all usage scenarios and the respective use cases are not evaluated according to their technical feasibility. According to the aforementioned publications, the use and suitability of wireless communication technologies in an industrial environment depends on the technical requirements of the respective application. These requirements can differ between the application cases resulting in different requirements, e.g. in terms of broadband, latency, or machine communication [10–12]. To generate a first decision-making foundation regarding the suitability of wireless communication technologies, technical requirements of use case categories can be classified based on their core requirements. [1] provides an extensive classification of possible industrial use cases in the fields of discrete manufacturing, logistics, process automation, diagnostics and maintenance, augmented reality (AR), and functional safety. Subsequently, [1] evaluates the feasibility of the identified use cases based on reliability, latency, and data rate. As a result, the publication points out, that many use cases cannot be implemented with the state-of-the-art wireless communication standards and additionally addresses the overstressing of the 2.4 GHz band as reason to further research wireless alternatives. However, the publication does not include 5G technology in the evaluation procedure. A recent study conducted by [13] specifies the technological characteristics of 5G and summarises the standardisation and release process by 3GPP while pointing out the release specifications with industrial relevance and compares them to wireless communication technologies such as 4G, Wi-Fi, and Bluetooth. In this study, the technical properties only refer to the 5G services itself and are only briefly translated to the requirements of concrete industrial use cases. In the remainder of this paper a holistic overview of important 5G related intralogistics use cases are provided and these are evaluated for its feasibility and its implementation potential with respect to 5G and its *releases* defined by 3GPP. Relevant use cases are also identified together with its technical requirements. Lastly, the use cases are assessed for its feasibility with respect to 5G and its technical specifications TS 22.261 [14, 15] of R16 and R17.

3 Classification of 5G Use Cases

In this section the relevant intralogistics 5G use cases are summarised. Then, the identified use cases are specified with requirement parameters as foundation for the evaluation process. The 5G performance parameters of the current R16 and the upcoming R17 are listed and finally, the use case requirements are compared to the performance parameters of the 5G R16 and R17. Table 1 presents a list with relevant intralogistics 5G use cases and their corresponing applications based on [1], the 3GPP Service Level [13–15] and [16–18]. These use cases are supplemented with applications identified within industrial cooperation with Small and Medium-sized Enterprises (SMEs) and workshop activities in the course of the German project 5G4KMU. The use cases presented in Table 1 place different technical requirements on the 5G network. Table 2 lists and defines key requirement categories for use case feasibility identified by 3GPP TS 22.104 [19–21]. Table 3 shows a summary of the use cases and their technical requirements. The use cases presented in Table 1 are specified with parameters representing specific requirements to make the use cases feasible. The parameters represent a summary based on the specifications of [1, 13, 17, 18, 20, 22–24].

The global rollout of 5G systems is a continuous process based on various release stages defined by 3GPP. Some services are currently not at their final performance capability or not yet implemented [4]. The feasibility evaluation of the use cases presented in Table 1 requires an investigation of the 5G performance parameters in the course of the upcoming release notes R16 and R17, and are presented in Table 4.

4 Results

The assessment of the feasibility of intralogistics use cases with respect to the technical requirements of R16 and R17 is presented in Table 5. All use cases state high requirements in terms of the availability of the network. Due to the specified availability of 99% in R16, none of the use cases are feasible. Nevertheless, availability is not a critical parameter for all use cases. Time-uncritical use cases such as Factory Monitoring/Flexible Factory Layout are feasible and can already be implemented for test purposes in learning factories under R16. However, in use cases such as AGVs, UAVs and Crane Systems with a need for mobility and safety, the availability and accuracy parameters are considered necessary. In the upcoming R17, all requirements up to availability class 7 can be met. Especially for mobile applications, all use cases require an accuracy of at least 0.5 cm, which is higher than the 1000 cm specified in R16. For the use cases AGVs, UAVs and Crane Systems, the required parameters even cannot be met with R17, which would only allow a localisation accuracy of a maximum of 20 cm. For the required latency, only the use case, Robot Motion Control places strict requirements in the sub-ms range. These latencies of 1 ms or less will only be achieved in R17. The use cases, Factory Monitoring/Flexible Factory Layout and AR/VR in Smart Factories can require data rates up to the Gbit/s-range, depending on the application. Therefore, these use cases are only partially feasible in R16. The device density in highly IoT-oriented use cases such as Indoor Item Tracking and Condition Monitoring can also pose challenges within learning factories. Densities of 20 devices/m^2 or more may require infrastructure upgrades to ensure stable connectivity of all devices. As reference, the density in

Table 1 5G use cases and identified applications [1, 13–18]

Use case	Description	Identified applications
Augmented and virtual reality (AR/VR) in Intralogistics	3D visualisation of resources for warehouse planning or pick-by-vision applications. AR applications require large data rates transmitted with low-latencies to synchronise the VR with the physical reality without any time delay	• Warehouse Planning (3D visualisation of resources and assets) • Pick-by-vision (targeted item picking, virtual warehouse navigation by visual routing system)
Automated guided vehicles (AGVs)	AGVs that transport products within the production/warehouse. The services of 5G enable a reduction of on-board computation power (optimisation of battery lifetime) and can support wireless communication of AGVs to enable a centralised fleet management	• AGV (Reduction of on-board processing power, localisation and navigation, autonomous tugger train platooning, remote control, and fleet management in the edge-cloud)
Unmanned aerial vehicles (UAVs)	Visual inventory control and management, camera inspection and monitoring, and indoor item delivery. UAV require ultra-high reliable communication. Furthermore, the 5G services enable a localisation and the outsourcing of on-board hardware to reduce the UAV's weight	• Inventory Management (item search and buffer stock maintenance, cycle counting, stock taking) • Inspection and Monitoring (inspection of racks in high-bay warehouses, pallet placements, facility inspections, monitoring of hazardous or non-accessible areas)
Condition monitoring	Real-time condition monitoring of AGVs, UAVs, and technical devices/wearables for maintenance and diagnostic reasons	• Condition Monitoring (real-time monitoring of robot conditions, e.g. battery lifetime of AGV and UAV fleets)
Crane systems	Wireless control and remote access to crane systems for precise loading and unloading of goods with high requirements regarding accuracy to reduce cable wear in rotating movements	• Crane Control (wireless control of mobile crane systems and static ceiling cranes for precise loading and unloading of heavy items)

(continued)

Table 1 (continued)

Use case	Description	Identified applications
Factory monitoring and flexible factory layout	Holistic process and asset monitoring through connected sensors and visual systems to control and manage processes and to monitor product states (e.g. temperature) and modular production systems	• Digital Shadow of Factory (process monitoring, process quality control and cycle counting) • Localisation determination of modular production and warehouse infrastructure
Indoor item tracking	Tracking and Tracing of items in warehouses to optimise processes and enable a just-in-time/just-in-sequence delivery. Item tracking requires a high device density and an accurate localisation service	• Tracking and Tracing (position determination of items, just-in-time and just-in-sequence delivery of items to assembly stations)
Robot motion control (Tactile Internet)	High precision robot-human-interaction with real-time tactile feedback in closed-loop control applications. Control of acceleration, pose, speed, grippers and tools or a combination of these actuators. Motion control in closed-loops place high requirements in terms of ultra-reliable low-latencies	• Motion Control (closed-loop and high precision control of actuators such as grippers and remote real-time human-machine interaction for high precision assembly processes) • Precise unloading/placing of items in non-accessible or hazardous areas (chemistry or laboratory, clinically hygienic environment)

NPN provided by Nokia depends on the AirScale System Module. In test networks, this AirScale System Module can typically manage up to 8000 devices. The learning factory 'Werk150' at Reutlingen University is approx. 800 m^2 large, which results in 10 devices/m^2. In case of increased connectivity needs of 10 devices/m^2 or more, currently the Nokia's AirScale system modules can be upgraded up to 60,000 users.

In closing the presented results are transferred to an implementation framework shown in Fig. 1. The proposed framework consists of four application implementation phases including Concept Development, Research Environment Prototyping, Industrial Implementation, and Industrial Optimisation. It is evident from Fig. 1 that there currently exists a gap of approximately 18 months between the specified 5G *releases* and their service availability in 5G infrastructure.

Table 2 Requirement definition

Requirement [20]	Definition/description
Accuracy	Closeness of the measured position of the User Equipment (UE) to its true position value. The accuracy can describe the accuracy either of an absolute position or of a relative position [19]
Availability	Percentage of time when a system is able to provide the required data within the performance targets or requirements [19]. Divided into availability classes from 1 to 7 [22]: 90% = 1; 99% = 2; 99.9% = 3; 99.99% = 4; 99.999% = 5; 99.9999% = 6; 99.99999% = 7
End-to-end latency	Duration between the transmission of a data packet from the source node and the successful reception at destination node [21]. At initialisation of positioning systems, the latency is also defined as the Time to First Fix [19]
UE speed	Closeness of the measured magnitude of the User UE's velocity to the true magnitude of the UE's velocity [19]
Data rate	Describes the data rate per second per user. A distinction can be made between peak data rate and user experience data rate (minimum achievable data rate for a user in real network environment) [21]
Density	Amount of devices for which the system can determine related data per time unit, and/or a specific update rate [19]

Table 3 The 5G use cases with respective technical requirement parameters based on [1, 13, 17, 18, 20, 22–25]

Requirements Use cases	Accuracy (cm)	Availability class	Latency (ms)	UE speed (m/s)	Data rate (up to)	Density (UE/m^2)
AR/VR in logistics	n/a	≥ 3	10	3	Gbit/s	0.02–0.03
AGVs	< 5	≥ 5	15–20	< 10	Mbit/s	0.1
UAVs	< 10	≥ 6	10–40	< 14	Mbit/s	0.1
Condition monitoring	< 50	4	100	< 10	kbit/s	10–20
Crane system	< 10	≥ 6	15–20	< 5	Mbit/s	0.1
Factory monitoring and flexible factory layout	50	4	> 100	< 10	Gbit/s	n/a
Indoor item tracking	< 50	2	< 200	< 10	Mbit/s	10–20
Robot motion control (Tactile Internet)	n/a	≥ 5	0.25–1	< 10	Mbit/s	5

Table 4 The 5G key characteristics and targeted performance for intralogistics scenarios based on [13, 15, 26, 27]

Requirement	Standalone non-public network			
	Release 16		Release 17	
Accuracy (cm)	1000		300–20	
Availability (%)	99–99.9999		99–99.999999	
End-to-end latency (ms)	4 > x > 1		< 1	
UE speed (m/s)	< 10		< 10	
Peak data rate (Gbit/s)	Downlink: 20	Uplink: 10	Downlink: 20	Uplink: 10
User experience data rate (Gbit/s)	Downlink: 1	Uplink: 0.5	n/a	n/a
Density (devices/m^2)	1		1	

Table 5 The 5G intralogitics use cases and identified applications (+ Use case is feasible; o Feasibility depends on the prioritisation and accuracy of one important parameter; – Use case is not feasible)

Use case	Feasibility release 16	Feasibility release 17
AR/VR in smart logistics	o	+
AGVs	–	o
UAV	–	o
Condition monitoring	–	+
Crane systems	–	o
Factory monitoring and flexible factory layout	–	+
Indoor item tracking	–	+
Robot motion control (Tactile Internet)	–	+

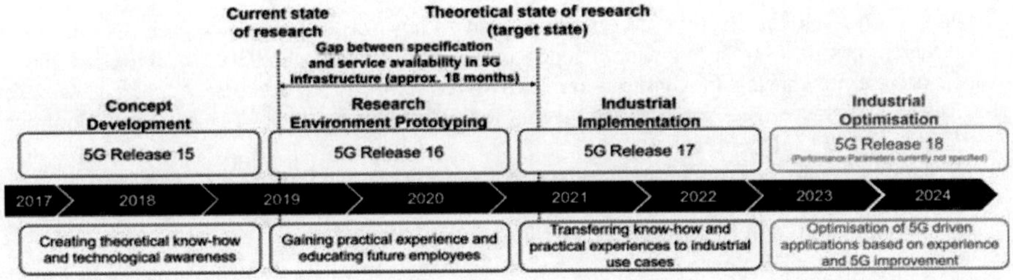

Fig. 1 Implementation framework for 5G use cases by means of learning factories

5 Opportunities for Further Research

The 5G technology is frequently mentioned as a suitable solution to meet the technical requirements of several intralogistical use cases. As this paper shows, none of the specified use cases are fully feasible in R16. The availability of 99% in R16 represents an obstacle regarding the feasibility of the identified use cases, in particular with higher safety and mobility requirements. In terms of localisation accuracy, the requirements are higher than what R16 will provide. Even with R17, the requirements of AGVs, UAVs and Crane Systems cannot be achieved. However, use cases with high localisation requirements could still be successfully implemented if alternative technologies are used and only the data transmission takes place via 5G itself (e.g. Factory Monitoring). As the results indicate, a large part of the identified use cases can be implemented with R17. However, there currently exists a significant gap between the specified 5G *releases* and their service availability in 5G infrastructure. In this context, research environments play a crucial role to test the use cases on a small scale and thus gain initial experience before they are ready for an industrial implementation in the course of the upcoming 5G *releases*.

References

1. VDE. Funktechnologien für Industrie 4.0. ITG AG Funktechnologie 4.0. https://www.vde.com/resource/blob/1635512/acf5521beb328d25fffda9fc6a723501/positionspapier-funktechnologien-data.pdf. Accessed Mar 2021
2. Li, Z., Wang, X., Zhang, T.: 5G+ How 5G Change the Society. Springer, Singapore (2021)
3. Aijaz, A.: Private 5G: the future of industrial wireless. IEEE Ind. Electron. Mag. **14**, 136–145 (2020)
4. 3GPP. Releases. https://www.3gpp.org/specifications/67-releases. Accessed Mar 2021
5. Shetty, R.S.: 5G Mobile Core Network. Apress, Bangalore (2021)
6. 3GPP. Release 18. https://www.3gpp.org/release18. Accessed Sept. 2021
7. Panetta, K.: 5 Trends Drive the Gartner Hype Cycle for Emerging Technologies. https://www.gartner.com/smarterwithgartner/5-trends-drive-the-gartner-hype-cycle-for-emerging-technologies-2020/. Accessed Mar 2021
8. Jabagi, N., Park, A., Kietzmann, J.: The 5G revolution: expectations versus reality. IT Professional **22**, 8–15 (2020)
9. Nokia Oyj: Nokia deploys 5G private networking for 5G4KMU across five German research centers. https://www.nokia.com/about-us/news/releases/2020/12/02/nokia-deploys-5g-private-networking-for-5g4kmu-across-five-german-research-centers/. Accessed Mar 2021
10. Rao, S.K., Prasad, R.: Impact of 5G technologies on industry 4.0. Wireless Pers. Commun. **100**, 145–159 (2018)
11. Ordonez-Lucena, J., Folgueira Chavarria, J., Contreras, L.M., Pastor, A.: The use of 5G non-public networks to support industry 4.0 scenarios. In: IEEE Conference on Standards for Communications and Networking, pp. 1–7 (2019)
12. Navarro-Ortiz, J., Romero-Diaz, P., Sendra, S., Ameigeiras, P., Ramos-Munoz, J.J., Lopez-Soler, J.M.: A survey on 5G usage scenarios and traffic models. IEEE Commun Surv Tutorials 905–929 (2020)
13. VDMA: 5G im Maschinen- und Anlagenbau. Leitfaden für die Integration von 5G Produkten in Produkt und Produktion. https://ea.vdma.org/leitfaden5g. Accessed May 2021
14. 3GPP. TS 22.261 V17.5.0—Service requirements for the 5G system—stage 1 (Release17). https://www.3gpp.org/ftp//Specs/archive/22_series/22.261/22261-h50.zip. Accessed 06 Mar 2021
15. 3GPP. TS 22.261 V16.13.0—Service Requirements for the 5G system—Stage 1 (Release 16). https://portal.3gpp.org/desktopmodules/Specifications/SpecificationDetails.aspx?specificationId=3107. Accessed May 2021

16. Wawrla, L., Maghazei, O., Netland, T.: Whitepaper—Applications of Drones in Warehouse Operations. https://ethz.ch/content/dam/ethz/special-interest/mtec/pom-dam/documents/Drones%20in%20warehouse%20opeations_POM%20whitepaper%202019_Final.pdf. Accessed Sept 2021
17. Fellan, A., Schellenberger, C., Zimmermann, M., Schotten, H.D.: Enabling Communication Technologies for Automated Unmanned Vehicles in Industry 4.0 (2018)
18. Purucker, P., Schmid, J., Hob, A., Schuller, B.W.: System requirements specification for unmanned aerial vehicle (UAV) to server communication. In: 2021 International Conference on Unmanned Aircraft Systems (ICUAS), pp. 1499–1508. IEEE (2021).
19. 3GPP. TR 22.872—Technical Specification Group Services and System Aspects; Study on positioning use cases; Stage 1 (Release 16). https://portal.3gpp.org/desktopmodules/Specifications/SpecificationDetails.aspx?specificationId=3280. Accessed May 2021
20. 3GPP. TS 22.104 V16.5.0—5G; Service requirements for cyber-physical control applications in vertical domains. https://www.etsi.org/deliver/etsi_ts/122100_122199/122104/16.05.00_60/ts_122104v160500p.pdf. Accessed May 2021
21. Jiang, D., Liu, G.: An overview of 5G requirements. 5G Mobile Communications, pp. 3–26 (2017)
22. Gray, J., Siewiorek, D.P.: High-availability computer systems. IEEE Computer, pp. 39–48 (1991)
23. Lu, Y., Richter, P., Lohan, E.S.: Opportunities and challenges in the industrial internet of things based on 5G positioning. In: International Conference on Localization and GNSS, pp. 1–6 (2018)
24. Sick and Fraunhofer IIS. Mobile Bedienterminals mit Not-Halt-Einrichtung. 5G im Maschinen- und Anlagenbau. https://ea.vdma.org/viewer/-/v2article/render/47636048. Accessed May 2021
25. Yang, G., Lin, X., Li, Y., Cui, H., Xu, M., Wu, D., Rydén, H., Bin Redhwan, S.: A Telecom Perspective on the Internet of Drones: From LTE-Advanced to 5G, pp. 1–8 (2018)
26. IEEE Spectrum. 3GPP Release 15 Overview. https://spectrum.ieee.org/telecom/wireless/3gpp-release-15-overview. Accessed Mar 2021
27. Mohyeldin, E.: Minimum Technical Performance Requirements for IMT-2020 radio-interface(s). https://www.itu.int/en/ITU-R/study-groups/rsg5/rwp5d/imt-2020/Documents/S01-1_Requirements%20for%20IMT-2020_Rev.pdf. Accessed Mar 2021

Fabian Dietrich is Ph.D. Candidate (Industrial Engineering) at Stellenbosch University, South Africa and works as Research Associate at ESB Business School, Germany. He holds a M.Eng. from Stellenbosch University and a M.Sc. from the ESB Business School of Reutlingen University.

Mario Angos Mediavilla works as Research Associate at Reutlingen University, Germany. He holds a M.Sc. in Mechanical Engineering with a focus on product development and construction technology from the University of Stuttgart, Germany.

Ali Turgut studied Industrial Engineering (BEng) at the Esslingen University, Germany and Kettering University, USA. He works at Steinbeis Innovation gGmbH and is completing a part-time master's in digital business engineering (M.Sc.) at Reutlingen University, Germany.

Thorge Lackner is currently pursuing a M.Eng. and M.Sc. double degree at Stellenbosch University, South Africa and Reutlingen University, Germany. His area of research interest is 5G in the industry. In 2020, he received his B.Eng. degree from Esslingen University, Germany.

Wyhan Jooste is a Professional Engineer and holds a Ph.D. in Industrial Engineering from Stellenbosch University. He is currently a senior lecturer in the Department of Industrial Engineering and focusses on asset- and asset management related research.

Daniel Palm is Professor for Supply Chain Management at the ESB Business School. Univ.Lektor Prof. Dr. techn. Dipl.-Ing. Daniel Palm is head of the teaching and research center for Value-Added and Logistics Systems at Reutlingen University and the Reutlingen Center Industry 4.0, RZI 4.0 - a cooperation of the Fraunhofer Institutes IPA and IAO as well as Reutlingen University.

Manufacturing Systems

Carbon Capture and Utilization in Cement Industry—Aspects of the Production of E-Fuels by Upcycling Carbon Dioxide

Anika Wacht, Stefan Kaluza, and Philipp Fleiger

Abstract This work deals with the process chain from captured carbon dioxide (CO_2) from a cement plant towards the production of e-fuels, including arising challenges and opportunities with the objective to reduce CO_2 emissions. First, it is demonstrated to what extend a certain Carbon Capture and Utilization (CCU) process chain can contribute to the reduction of CO_2 in an industrial sector like the cement industry. In the cement industry in particular, there is a high demand in reducing emissions due to the irreplaceable raw material limestone. Then, an overview of applicable technologies for the required processes within the process chain from CO_2 capture to e-fuel is given. Thereafter, mass and energy balances for the process chain are calculated based on the emissions released from a Best Available Techniques (BAT) cement plant: The resulting amounts of educts, products and electricity are calculated. Furthermore, an evaluation of the selected process chain and the calculations is carried out with the focus on technical aspects, showing that the production of e-fuels as a CCU measure in the cement industry represents a promising option to reduce CO_2 emissions.

Keywords Carbon capture and utilization (CCU) · CO_2 upcycling · Cement industry · Process chain · E-fuel

A. Wacht (✉) · S. Kaluza · P. Fleiger
Department of Mechanical and Process Engineering, University of Applied Sciences of Duesseldorf, Duesseldorf, Germany
e-mail: anika.wacht@study.hs-duesseldorf.de

S. Kaluza
e-mail: stefan.kaluza@hs-duesseldorf.de

P. Fleiger
e-mail: philipp.fleiger@hs-duesseldorf.de

K. von Leipzig et al. (eds.), *Smart, Sustainable Manufacturing in an Ever-Changing World*, Lecture Notes in Production Engineering,
https://doi.org/10.1007/978-3-031-15602-1_44

1 Introduction and Motivation

The burden on our environment from greenhouse gas emissions caused by the production and consumption of products culminates in different phenomena of climate change that can be observed across the planet. This is, for instance, reflected in the rise of global temperature, resulting in extreme weather events, forest fires, heat waves and floods. In order to halt this rapid increase, it is necessary to find technical solutions to help reducing emissions and to achieve the political objectives of the Paris Climate Convention [1]. Among other components, CO_2, which is released from fossil fuel combustion, accounts for most of the greenhouse gas emissions. Reducing CO_2 emissions by shifting from fossil fuels to renewable energy sources is an essential aspect of achieving the climate change objectives. However, there will remain processes in the future in which unavoidable amounts of CO_2 are produced due to the lack of alternative process options, e.g., in the cement industry as a result of the irreplaceable raw material limestone. Therefore, this work focuses on the reuse of the CO_2 emissions from the cement industry as a feedstock for the production of e-fuels. The process chain of the production of e-fuels is analysed by giving an overview of available technologies. Key indicators are calculated and possible challenges as well as supporting factors are identified. One key contribution is the evaluation of a specific CCU process chain in terms of technical and economic aspects.

CO_2 Abatement due to the Utilization of CCU With regard to the cement industry in particular, the unavoidable CO_2 emissions released within the process of cement production can be reused by the utilization of green hydrogen to produce e-fuels. With this, the amount of CO_2 released is basically halved, which is shown in Fig. 1. This means that with the application of CCU emissions are released only once instead of twice without CCU.

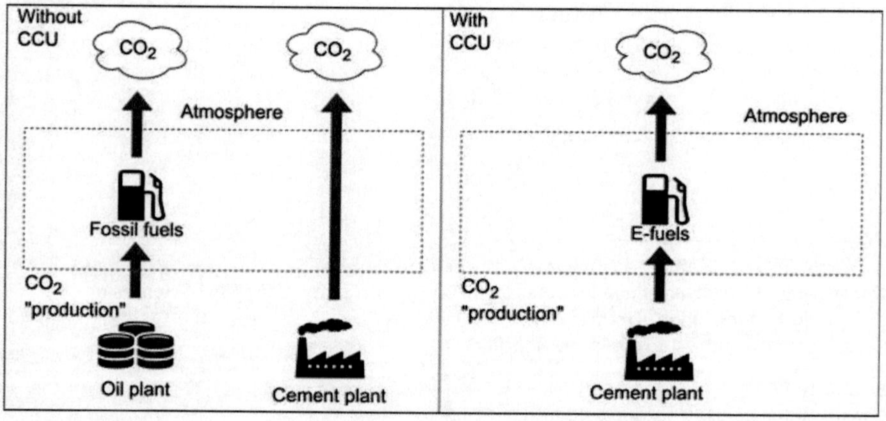

Fig. 1 CO_2 emissions with and without CCU

Table 1 CO_2 amounts resulting from a cement plant, that produces 3000 t of clinker per day

	Without CCU	With CCU
CO_2 from cement plant [t/d]	2600	260
CO_2 from fuel combustion [t/d]	2420	2136
Total [t/d]	5020	2396
CO_2 Abatement [t/d]	2624	

Table 1 exemplary shows the CO_2 savings via CCU. The amount of CO_2 is based on a Best Available Techniques cement kiln that produces 3000 t of clinker per day. In both cases with and without CCU, an amount of 764 t of fuel is considered to be burned (this is the amount of e-fuel, that could be produced with the CO_2 captured from a BAT cement plant). In the case without CCU, CO_2 is released from the cement plant as well as from the combustion of fossil fuel. Thus, 764 t of fuel are responsible for 2420 t of CO_2, considering an emissions factor of 3.168 t of CO_2 per tonne of fuel [2]. This quantity is emitted in addition to the quantity from the cement plant. In the case with CCU, the amount of CO_2 released by combustion of 764 t of petrol is also emitted, but this has been produced from the CO_2 previously captured from the cement plant. Thus, the maximum of 10% of the original amount of CO_2 is released from the cement plant due to CO_2 capture and the subsequent conversion process. The saved amount of CO_2 through CCU results in 2122 t per day, which is more than half of the amount of CO_2 released without CCU.

2 Process Chain and Respective Technologies

The process chain, that is needed within the concept of CCU and in particular for the production of e-fuels, comprises the capture of CO_2 from the flue gas of a cement kiln, its further processing into syngas (mixture of carbon monoxide and hydrogen) by means of a reverse water–gas shift reaction (RWGS), and the subsequent synthesis of e-fuels by utilizing the Fischer–Tropsch-Synthesis (FTS). Figure 2 shows a flow diagram of the necessary process steps.

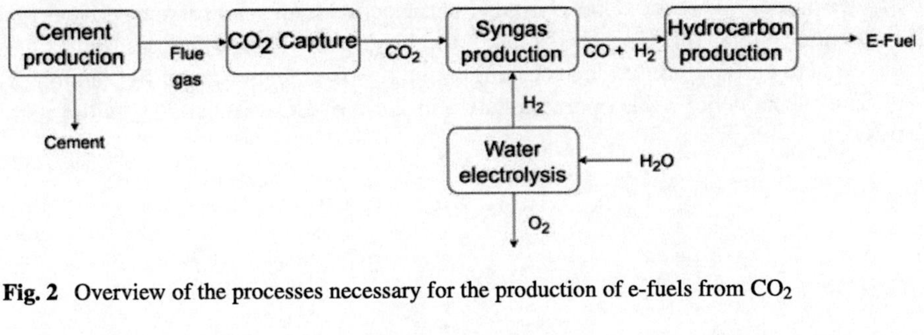

Fig. 2 Overview of the processes necessary for the production of e-fuels from CO_2

In the following paragraphs, the required processes are described in more detail. Furthermore, a specific technology for the implementation of the respective process step is outlined.

Cement Production and CO_2 Source Within the process of cement production, the raw material limestone $CaCO_3$ is calcined to CaO, which is part of the cement clinker, and CO_2. Clinker represents one main component of cement. The process of calcination is carried according to the following Eq. (1):

$$CaCO_3 \rightarrow CaO + CO_2 \quad (1)$$

The reaction makes clear, that CO_2 is not only released due to the burning process, but also due to the irreplaceable raw material limestone. The cement industry already invests in research and development as well as the application of different measures to reduce CO_2 emissions, e.g., through the reduction of the amount of cement clinker, the use of alternative fuels or the concept of CO_2 capture and its storage (Carbon Capture and Storage (CCS)) [3–5]. Nevertheless, there is still the need for the cement industry to further reduce its CO_2 emissions by applying innovative concepts, such as the concept of CCU [4, 6].

CO_2 Capture The produced CO_2 is captured for its further utilization as feedstock in the overall CCU process. CO_2 capture is realised via chemical absorption using aqueous amine solutions (MEA CO_2 Capture) as a solvent, which is currently considered as the most mature option for CO_2 capture [8]. The principle of the process comprises the absorption of CO_2 in an amine solution by formation of carbamates and the subsequent desorption of the loaded washing solvent to release the CO_2.

Hydrogen Production For the further conversion respective upcycling of CO_2, hydrogen is required. Hydrogen is produced via Alkaline Electrolysis using aqueous potassium or sodium hydroxide solution as electrolyte, which is the most established technology to decompose water [7, 8]. The process of electrolysis follows according to the overall Eq. (2).

$$2\,H_2O \rightarrow 2\,H_2 + O_2 \quad (2)$$

Application of CO_2 and H_2 as Syngas For the actual hydrocarbon respective e-fuel production, syngas serves as feedstock. For this purpose, part of the captured CO_2 is converted to carbon monoxide according to the reversed water–gas shift reaction (3) [9]. The subsequent FTS is operated with a mixture of CO_2, CO and H_2 with in-situ RWGS.

$$CO_2 + H_2 \rightarrow CO + H_2O \quad (3)$$

Hydrocarbon Production via Fischer–Tropsch-Synthesis (FTS) The actual process of hydrocarbon production is realised via the FTS, patented by Franz Fischer and Hans Tropsch in 1929 [10].

The complex reaction network in the FTS can be simplified to the following CO-based reaction (4) [8]:

$$CO + 2\,H_2 \rightarrow (-CH-) + H_2O \tag{4}$$

The product of the process is a mixture of hydrocarbons with various hydrocarbon chain lengths, together with various by-products such as olefins or alcohols. The mixture needs to be refined via distillation and absorption columns to produce the desired product fraction [11].

3 Results and Discussion

The required process steps and the respective technology for an implementation are figured out and shown in the flow diagram in Fig. 3.

The individual processes are each commercially available and industrially applicable, but the processes are not yet seamlessly coordinated.

Scalability respectively multiplication, in every process is given, but the linkage in between has to be established on an industrial scale. Therefore, the operation parameters are investigated as shown in Table 2. Synoptically, the operation parameters do not represent a challenge. The required pressure for the FTS ranges from 20 to 25 bar, which can be met both by the CO_2 as well as by the hydrogen.

Then, basic mass and energy balances are calculated for the overall coupled process. Figure 4 shows a fundamental mass flow diagram and visualizes the results of the stoichiometric calculation. The mass flows of the respective substance are shown,

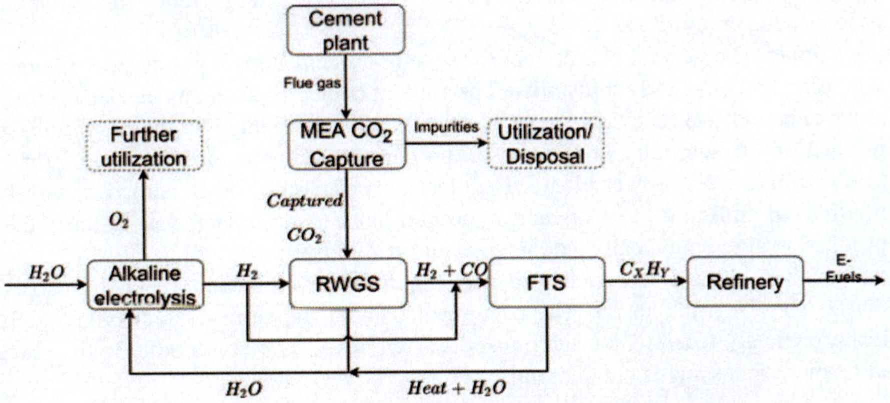

Fig. 3 Process chain with the respective technologies selected for each process step

Table 2 Process parameters of the respective process step in the CCU chain

Parameter	MEA Capture	AEL	RWGS	FTS
Press. [bar]	~1	60	1–25	20–25
Temp. [°C]	40–60 124–140	60–90	800	220–350

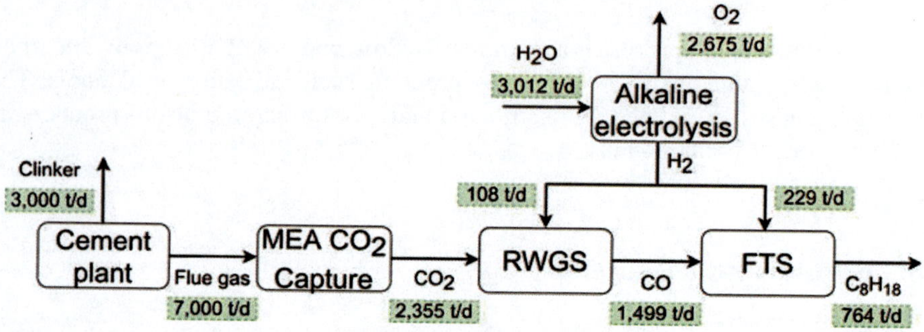

Fig. 4 Fundamental mass flow diagram including water electrolysis, CO_2 capture and CO_2 conversion for the production of e-fuels with the respective amount

considering a stoichiometric production, where losses due to decreased efficiency or chemical equilibrium limitations, for instance, are neglected.

The overall amount of electrical energy, that is required for the complete process, ends up in about 12 GWh to produce 764 t of octane (due to the wide range of hydrocarbon chain length, octane was chosen as benchmark hydrocarbon for the calculations).

The objective of this work is the reduction of CO_2 emissions. Therefore, it is obligatory to produce this large amount of electricity necessary in a renewable, climate-friendly way, where the net CO_2 emissions are zero. With the required amount of electricity and the produced amount of fuel, the theoretical efficiency of the overall process adds up to ~70%.

Although the general coupling of the several process steps is possible, challenges regarding transport and supply arise. The amount of electrical energy needed is obligatory to be met via renewable generation. This is accompanied by the decentralised production of electricity in contrast to the current, still predominantly centralised production in a few power plants. The electricity grid must be expanded massively for this. In addition, the CO_2 and hydrogen have to be transported to the e-fuel production site, which can be done via a gas grid in the best case.

However, there is no such grid yet, which is why transport has been carried out by ship or truck so far. The capture of CO_2 at the cement plant is generally technologically feasible, but the transport of the respective commodity to the place of further processing is challenging.

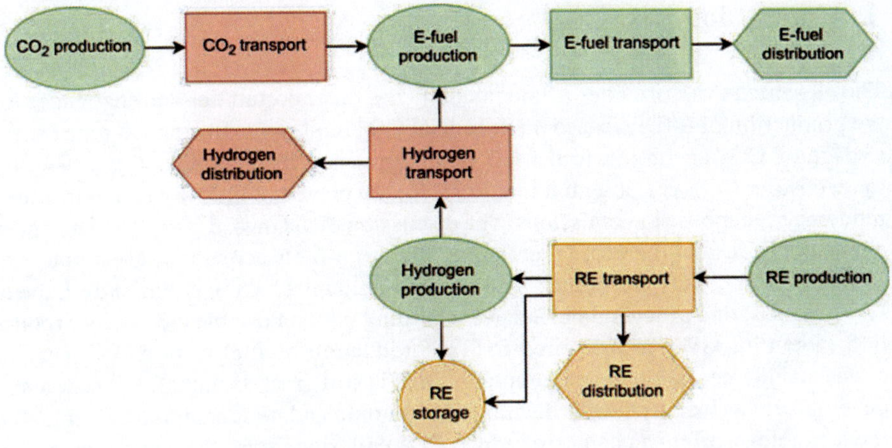

Fig. 5 Qualitative mapping of the complexity of sector coupling; green: already developed, yellow: already developed but in need of expansion, red: development required

This can be summarised in the need for infrastructure expansion. Figure 5 exemplary shows the complex structure behind dependencies among the different sectors such as production and transport.

For example, the production of renewable energy (RE) generally does not represent a challenge, but the transport does. The same applies to hydrogen production and transport. In general can be said, that large-scale implementation of CCU will require both the expansion of the electricity as well as the construction of a CO_2 gas grid.

Besides technical aspects, economic and political aspects need to be considered as well. With the application of the concept of CCU, e.g. high investment costs or operating costs will arise, which leads to initial higher costs for the e-fuel compared to fossil fuels. This could discourage both the industry from investing in the processes and the consumer, who would have to buy the synthetic fuel at a high price. Therefore, a political framework to ensure security both for the industry as well as the consumer would be required to establish CCU. The framework has to be established for phrasing out the use of fossil fuels and for substituting them with alternatives [12, 13].

Nevertheless, based on the previous aspects, it can be said that the concept of CCU is a promising option to reduce CO_2 emissions at the cement plant. Synthesising e-fuels from CO_2 using renewable electricity is a perfect solution to replace or supplement fossil fuels as drop-in fuel without major changes regarding distribution technology.

4 Conclusion

The presented work provides a fundamental view on opportunities and challenges for the production of e-fuels based on industrial CO_2 emissions. In contrast to the mere storage of CO_2 according to the CCS concept, upcycling provides the possibility to reconsider CO_2 as a potential feedstock for the chemical industry simultaneously reducing greenhouse gas emissions. The discussion of the investigations and balances show that CCU and the subsequent production of e-fuels constitutes great potential in reducing the CO_2 emissions: The emitted amount of CO_2 is basically halved. Nevertheless, due to several challenges regarding e.g., renewable electricity production (about 12 GWh are required for the production of fuel from a BAT cement plant) and the necessary sector coupling, there is still great demand for research and development as well as the clarification of economic and political issues. Eventually, the technology offers a chance to reduce CO_2 emissions from the cement plant and further supports the transition from fossil to renewable fuels.

Acknowledgements We acknowledge the support of VDZ Technology gGmbH and the funding given by Gerd-Wischers-Foundation.

References

1. United Nations: The Paris Agreement (2021). https://unfccc.int/process-and-meetings/the-paris-agreement/the-paris-agreement. Last accessed 26 May 2021
2. Umweltbundesamt.: Emissionsfaktoren für fossile Brennstoffe (2021). https://www.umweltbundesamt.de/publikationen/. Last accessed 26 May 2021
3. Verein Deutscher Zementwerke, V.H.: Dekabonisierung von Zement und Beton - Minderungspfade und Handlungsstrategien. Düsseldorf (2020). https://www.vdz-online.de/dekarbonisierung. Last accessed 29 Sept 2021
4. IEA: Cement (2021). https://www.iea.org/fuels-and-technologies/cement. Last accessed 29 Sept 2021
5. Equinor: Nothern Lights CCS (2021). https://www.equinor.com/en/what-we-do/northern-lights.html. Last accessed 29 Sept 2021
6. Cembureau, T.E.: Cementing the European Grean Deal—Reaching Climate Neutrality Along the Cement and Concrete Value Chain by 2050 (2020). https://lowcarboneconomy.cembureau.eu/carbon-neutrality/. Last accessed 29 Sept 2021
7. Voldsund, M., et al.: CEMCAP Framework for Comparative Techno-Economic Analysis of CO_2 Capture from Cement Plants—D3.2. CEMCAP Project (2018). https://www.sintef.no/projectweb/cemcap/results/. Last accessed 26 May 2021
8. Dietrich, V.: Power-to-liquid via synthesis of methanol, DME or Fischer–Tropsch-fuels: a review. Energy Environ. Sci. **13**, 3207–3252 (2020)
9. Daza, Y.A.: CO_2 conversion by reverse water gas shift catalysis: comparison of catalysts, mechanisms and their consequences for CO_2 conversion to liquid fuels. RSC Adv. **6**, 49675–49691 (2016)
10. Fischer, F.T.: Verfahren zur Gewinnung mehrgliedriger Paraffinkohlewasserstoffe aus Kohlenoxyden und Wasserstoff auf katalytischem Wege (Nr. 484337), Deutsches Reich Reichspatentamt (1929). http://www.fischer-tropsch.org/primary/_documents/patents/DE/de484337.pdf

11. James, O.O.: Reflections on chemistry of the Fischer-Tropsch synthesis. RSC Adv. **2**, 7346–7366 (2012)
12. Bergk, F., Fehrenbach, H., Lambrecht, U., Räder, D., Fucks, A.-L.S., Schwarz, S., Wolf, P.: Beitrag strombasierter Kraftstoffe zum Erreichen ambitionierter verkehrlicher Klimaschutzziele in Baden-Württemberg—Kurzgutachten im Auftrag des Ministeriums für Verkehr BW (2018). https://vm.baden-wuerttemberg.de/fileadmin/redaktion/m-mvi/intern/Dateien/PDF/181126_Klimaschutz_Kurzgutachten_Strombasierte_Kraftstoffe_ifeu_ZSW.pdf. Last accessed 28 Sept 2021
13. IOGP (Coordinator): The potential for CCS and CCU in Europe—Report to the Thirty Second Meeting of the European Gas Regulatory Forum 5–6 June 2019. https://ec.europa.eu/info/sites/default/files/iogp_-_report_-_ccs_ccu.pdf. Last accessed 28 Sept 2021

Anika Wacht holds a BEng in the field of Energy and Environmental Engineering from the University of Applied Sciences in Duesseldorf (HSD) and is currently completing her MSc in Simulation and Experimental Technology. Besides, she works as a research assistant at Fraunhofer UMSICHT.

Stefan Kaluza received his Ph.D. from the Ruhr-University Bochum in 2009. After heading a research group at Fraunhofer UMSICHT, he became Professor of Industrial Chemistry and Catalysis at the University of Applied Sciences in Duesseldorf (HSD).

Philipp Fleiger is part-time professor for Cement Process Technology at the University of Applied Sciences in Duesseldorf (HSD) and head of the 'Digital Process Support'-group at VDZ. His work covers various aspects of digitalization and decarbonation along the value chain of cement and concrete.

A Procedure to Achieve Single Minute Exchange of Die for Cold Roll Forming

T. Pillay, Glen Bright, Christian Basson, and Avern Athol-Webb

Abstract Intricate and complex production systems are incorporated across vast platforms, ranging from hand-held computer devices to the much larger scaled aerospace industry. Given the contrast of manufacturing environments, the cost implications to business organisations are significant due to machine downtime. Non-Value-Added activities such as tool change-over and tool setup need to be streamlined, contributing to the efficiency of production in an Advanced Manufacturing System (AMS). This paper considers the cold roll forming (CRF) process within the automotive industry and aims to examine the implementation of a revised forming procedure to achieve tool change-over in a singular minute. A South African automotive accessories manufacturing company was considered where tool change-over of the tube mill is achieved in 500 min. The application of a unique die design is aimed at replacing the multiple rollers with a singular tool to address the non-value-added activity of tool change over and addressing the disadvantages of the conventional CRF process. The Cage forming operation was selected as a possible approach in achieving a low-cost flexible and AMS to produce ERW tubes.

Keywords Tool change-over · Single minute exchange of die (SMED) · Flexible manufacturing systems

1 Introduction

Product variety has been identified as one of the foremost competitive edges for manufacturing companies to meet customer's diverse demands [1]. Consequent to

T. Pillay · G. Bright (✉) · C. Basson · A. Athol-Webb
University of Kwazulu Natal, Kwazulu Natal, South Africa
e-mail: brightg@ukzn.ac.za

C. Basson
e-mail: bassonc@ukzn.ac.za

A. Athol-Webb
e-mail: atholwebba@ukzn.ac.za

K. von Leipzig et al. (eds.), *Smart, Sustainable Manufacturing in an Ever-Changing World*, Lecture Notes in Production Engineering,
https://doi.org/10.1007/978-3-031-15602-1_45

the economic volatility more companies have adopted “Lean principles” to optimise production rates and maximise profitability. According to Toyota Production System (TPS), waste is regarded as a task or elements that add no value from the client’s perspective. Seven types of waste generally identified are [2]: (1) Over-production, (2) Inventory, (3) Material transportation, (4) People’s movement, (5) Unnecessary operations (over-processing), (6) Waiting and (7) Product defects.

Cold Roll Forming (CRF) is an immensely productive process that is adopted in a wide range of manufacturing industries which is spread throughout the world. The forming operation is a high-volume manufacturing process that plastically deforms metals by using a series of dies. The absence of heat results in an obligation for numerous stacks of rollers to add a minuscule amount of forming to the metal strip, so that the desired cross-section is achieved at the end of the lengthy process [3].

Cage forming is a system that is a continuous forming process [4]. Groups of singular simple rollers, supporting inner and outer rollers of the material are used to achieve the continuity of the forming process, as seen in Fig. 1. The objective of the setup is to save on high tooling costs, setup and change-over-time. The saving is achieved by not having to change all the rollers to produce varying geometric products.

Pending the precise and intricate process of cold roll forming, the challenge lies in achieving a tool change-over and setup without compromising on product quality and standards. The proposed solution was to adopt a constant forming ideal by replacing the multiple rollers required to achieve the desired geometry. Where tool change-over and setup is concerned, a singular die is interchanged, providing a significant advantage in achieving tool change-over in a singular minute.

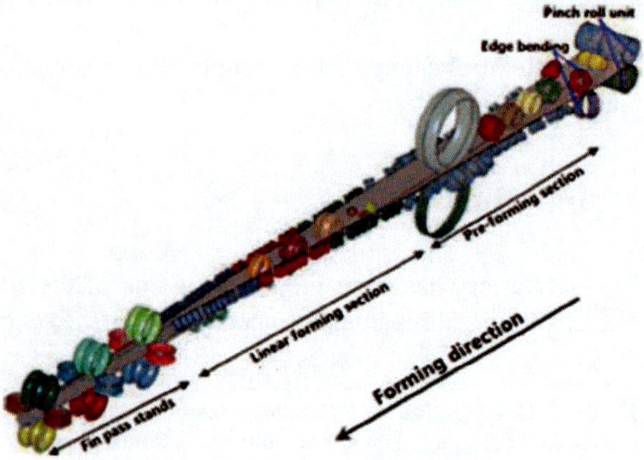

Fig. 1 Roller arrangement of the Cage Forming process [5]

2 Literature Review

2.1 Tool Change-Over

A machine change-over is the time required to arrange a machine or a process from producing the last good product from the previous production run to the first good product of the next production run [6]. Preparing for the change-over involves single or multiple changes where parameters, inputs, components and aspects are modified to support the production of goods for the new production demand [7]. Tool change-overs can potentially be very time consuming, different manufacturing organisations must adopt a variety of lean procedures to reduce time lost during tool change-over. Overall Equipment Effectiveness (OEE) is a quantitative measure absorbed into manufacturing processes for observing and controlling the productivity of production equipment [8] (Fig. 2).

The three main factors of OEE are:

- Availability—considers downtime losses.
- Performance—considers speed losses.
- Quality—considers losses due to quality.

OEE is the ratio of fully productive time to planned production time, and is calculated from the below equation [9]:

$$\text{OEE} = \frac{B}{A} \times \frac{D}{C} \times \frac{F}{E}$$

where;

$$\text{Availability} = \frac{B}{A} = \frac{Operating\,time}{Planned\ production\,time}$$

$$\text{Performance} = \frac{D}{C} = \frac{Ideal\,cycle\,time}{Operating\,time/Total\ pieces}$$

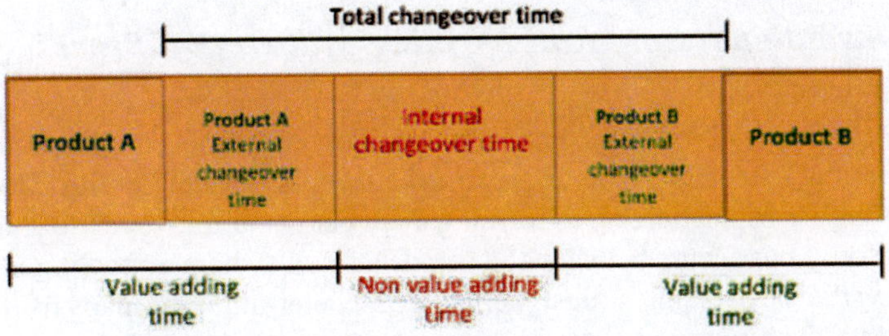

Fig. 2 An overview of the definition of change-over time and the two different components

Fig. 3 Three elements of SMED

$$\text{Quality} = \frac{F}{E} = \frac{Good\ pieces}{Total\ pieces}$$

2.2 SMED

In the 1950's Shigeo Shingo developed SMED as part of the Toyota Production System. The idea of SMED was essentially to reduce the setup time on a machine. The term "Single minute" does not suggest that all start-ups and change-overs should take only one minute, but that they can be performed in less than 10 min (singular minute) [10] (Fig. 3).

Shingo divided the setup into two parts [11]: (1) *Internal setup;* the setup operation that can be performed only when the piece of machinery is shut down (attaching or removing of dies) (2) *External setup;* the setup operation that can be performed while the piece of machinery is still running.

2.3 Conventional Cold Roll Forming Mill Design Process

CRF is a continuous bending exercise, where the bending occurs incrementally in multiple forming steps, from an undeformed strip to a finished profile. The forming process is highly sophisticated, as forming does not occur due to the forming stands (die/tools) only but occurs between the stands as well (where no tooling is present). Concerning the tool design, the designer must first determine how many forming steps are necessary. The required steps are dependent on the shape of the cross-section, thickness of material, material properties and tolerance [12].

Halmos [3] details the step by step procedure to designing a CRF mill. The following highlights the basics of a CRF design process:

- **Orientation of Forming**: Prior to calculating the number of passes required for forming, the orientation of the finished section must be determined. Given the simplicity of the tube geometry and manufacturing nature of ERW tubes, welding generally occurs at the top of the formed section, thus forming should ideally occur symmetrically. The simpler the geometry of product, the fewer the required number of passes.
- **Number of Passes**: Numerous attempts have been made to introduce science into roll forming. Halmos, has developed an empirical formula to serve as a guideline to determine the number of passes required, this serves as a mere guideline, simulation and software packages tend to be more accurate and aid in the design process
- **Strip Width**: To accurately determine the strip width of a coil the final cross-section must be divided into straight and curved elements. During forming, it is assumed that the theoretical neutral axis of the bent element moves from half of the thickness location towards the inside radius. Consider Fig. 4,

$$L = \frac{\pi}{180}(R_i + \text{kt})\alpha$$

where; *L*, length of the curved element. *t*, thickness of the material. R_N, neutral axis radius. R_i, inside radius. *k*, "*k*" factor (Determined from tables and theoretical equations).

For a circular tube section, the cross-section will be divided into four equal parts.

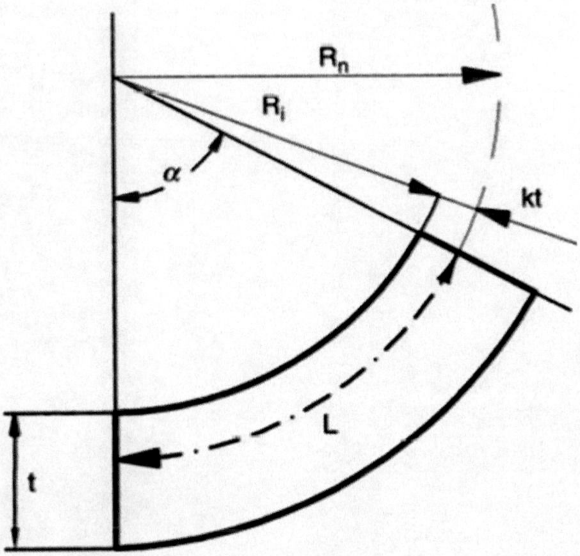

Fig. 4 Length of curved element

3 Methodology

3.1 Change-Over and Setup Process

The different sized tubes manufactured at the organisation under investigation are:

- 76.2 mm, round and oval.
- 63.5 mm, round and oval.
- 50.8 mm, round and oval.
- 38.1 mm, round and oval.

All tubes are produced in stainless steel and mild steel. The variety of tubes and the production process sets the stage for the implementation of SMED techniques. A total of 22 rollers are required to complete a successfull tool change-over. Each changeable roller is required to be aligned and adjusted manually by an operator to the millimeter. This ensures a consistent and accurate weld gap for prior to the welding process, as seen in Fig. 5.

Depending on the diameter of the tube to be manufactured, specific components and dies are selected, transported and loaded by forklifts, then assembled and fastened by experienced individuals. A case study regarding tool change-over, conducted over one month for TM2 is highlighted in Table 1. The average change-over for the month is 519 min and average setup scrap is 12.1 m.

The largest contributor to the change-over downtime is concerning tooling (rollers, spacers and fasteners), contributing 24% of the total change-over. The nature of the

Fig. 5 Formed tube prior to the welding process

Table 1 Details of change-over conducted in a month

Change-over	Time required (min)	Diameter change	Scrap (m)
1	507	38.1–50.8	4.4
2	576	50.8–76.2	12.8
3	403	63.5–50.8	14.7
4	588	50.8–76.2	19.5

change-over can be conducted as a project. J.M Juran applied the 80:20 rule to quality control and realised that 80 percent of problems stem from 20% of the possible causes [13]. Juran stated that the main point is to focus on the vital few problems instead of the trivial many, to make the most significant improvements to quality.

Theoretically, if the specific components to (rollers, spacers and fasteners) manufacture a tube can be changed in a singular minute, the overall change-over has the potential to be completed in 331 min. Though 331 min is a considerable time to lose to production, achieving this target will be a huge achievement in cost saving. This justifies the research into the single forming die design.

3.2 *Forming Die*

The cage forming operation improves the flexibility of electric resistance welding pipe manufacturing by dopting sevral groups of smaller cage rolls [5]. After consideration to the pareto chart it is obvious that emphasis should be placed on the number of forming tools required for the forming process. Figures 6 and 7 of a computer aided drawing (CAD) shows the prototype concept. Details of the individual components contributing to the assembly are:

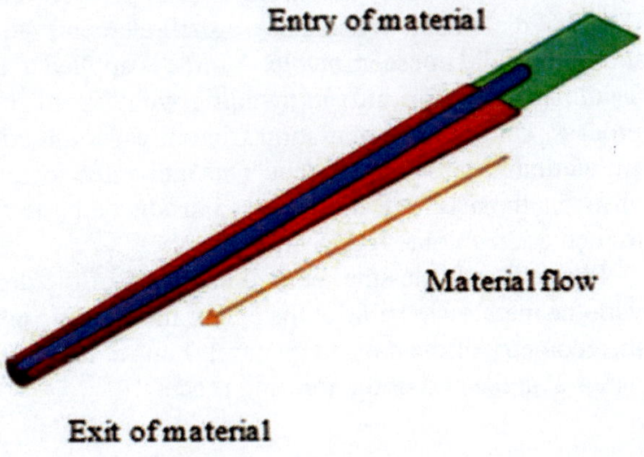

Fig. 6 Isometric view of the continuous forming assembly

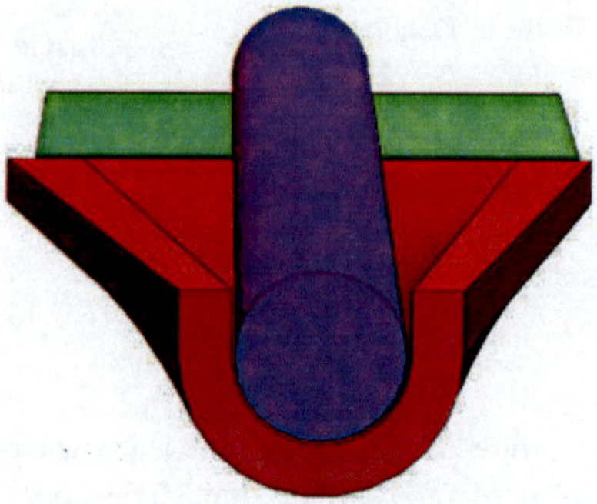

Fig. 7 Front view of continous forming die assembly

- Forming die—Red component.
- Mandrel—Blue component.
- Metal strip to be formed—Green component.

The colours in the assembly are for illustration purposes only to distinguish the different parts that make up the assembly.

Figure 7 illustrates the front view of the proposed continuous forming process. The die (represented in red) replaces the multiple forming rollers required for deformation. As the metal strip (represented in green) moves forward, the material is formed into the geometry governed by the top surface of the die and the bottom surface of the mandrel.

3.3 *Simulation of the Forming Process*

The continuous forming die assembly was modelled in Solidworks and imported into Siemens NX to create the simulation. To generate the simulation and results, symmetrical parts were modelled, as seen in Fig. 8. The strip is modelled as a 0.5 mm thick solid element. When using a shell element, the model did run faster, however the element did not seem stable. A force is applied to the front of the strip to simulate a pulling force, this aids in avoiding winkling of the material during the forming process. During this simulation friction was omitted to introduce simplicity to the simulation for proof of concept. This makes the feeding mechanism extremely unrealistic as the nature of the process introduced by the forming die concept will tend to increase frictional forces dramatically.

Figure 8 indicates the meshed elements of the die, mandrel and the metal strip, with the metal strip being at the end of the forming process. The strip is governed by the geometry of the die and mandrel. Figure 9 indicates the geometry of the formed material at the end of the forming process.

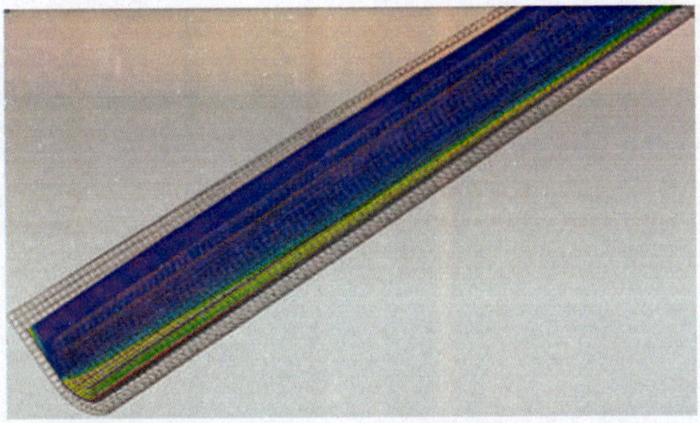

Fig. 8 Meshed elements of the simulated model

Fig. 9 Front view of simulation

4 Prototype Trial Run's and Results

Prototype Trial Run 1

The arrangement of the roller being utilised to force the material through the die was a failure. Though the contact between the material and roller was sufficient the material was not able to overcome the fictional force generated by the contact of the material and the sleeve. The theory of using a single roller to ***force*** the material through will not be considered as a solution. Figure 10 highlights the trial run conducted. As indicated by the image, the material was not able to move past the roller (Table 2).

Prototype Trial Run 2

The roller was removed. The idea of having a pulling force was considered as opposed to a pushing force. A winch was used to pull the material through the die. The cable of the winch was attached to the material in order to allow for the cable and material to pass in between the mandrel and sleeve, once sufficient material passed the exit point of the machine the winch hook was attached to the material to continue with the rest of the process, as seen in Fig. 11. Specifications of the winch are: 230 V @ 50 Hz, 2.2A, 500 W and 4/8 m/min.

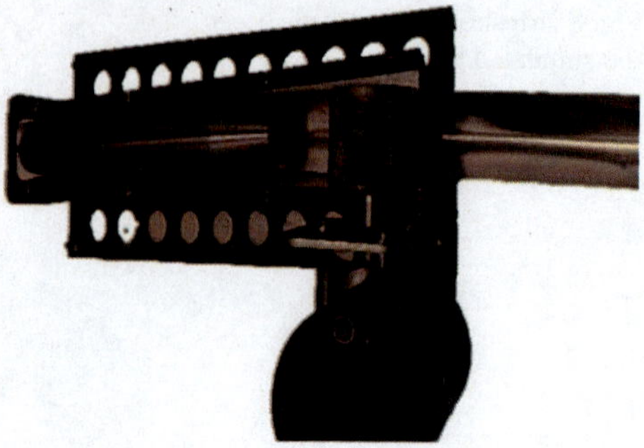

Fig. 10 Mandrel and machine arrangement of trial run1

Table 2 Details and functionality of the prototype components

Component	Details	Function
Die/ sleeve	400 mm x 3 mm (thickness) 304SS. Sleeve is flat faced at entry of the material and gradually transforms along its length into a semicircle of 30 mm exit of the material	Used to house and form the material
Mandrel	Material, EN8 mild steel. At entry the part is tapered, and changes to the geometry of the die	Used to keep the material in contact with the sleeve
Motor	DC motor. 70 rpm at 12 V	Used to drive the roller
Roller	The roller is not geometrically specific with a knurled finish	Used to move the material through the sleeve
Supports	Die; 5 mm thick mild steel powder coated Mandrel; 5 mm thick 409ss	Keeps the parts in their desired positions
Material	Shimstock brass, 0.25 mm thick. Material was cut at 78 mm width	Used to confirm the machine design

Fig. 11 Winch attached to the material

Fig. 12 Material after forming

5 Discussion

Trial Run 1: The design consideration of having a single roller to replace the multiple rollers on the current CRF process cannot be considered. Though the current process uses a pushing force, for the purpose of the unconventional forming machine design a pulling force will need to be used.

Trial Run 2: The material was pulled through the sleeve. The winch was able to overcome the fictional forces created by the material and sleeve contact. From Fig. 12, it is evident that the material was not being pulled through the sleeve in the desired orientation. The mandrel was not able to keep the material perfectly straight to allow for a perfectly symmetrical shape. A fully circular tube was not achieved.

6 Conclusion

A new technique was developed to manufacture cold roll formed ERW tubes. This was introduced to address the loss of production due to tool change-over. The notion of achieving a tool change-over in a singular minute, as stipulated by SMED was considered, as this is a significant reduction to the existing tool change-over process.

Based on the prototype trials conducted, the continuous forming die concept is yet to be proved concerning the impact on tool change-over and the ability to produce defect free useable tubes. Research still needs to go into the concept, the idea needs to be deemed successful through intense scrutiny of the material achieved from the prototype trials without any external assistance from the machine.

7 Recommendations

A thorough FEA and simulation needs to be conducted to determine if the proposed die concept satisfies the requirements of a conventional cage forming process and produces parts that are free of defects like a flexible roll forming process.

The sleeve/die, does not keep the material in the desired position for a perfectly round tube. As with the existing cold roll forming process, a set of guide rollers will have to be adopted into the machine design. This will not affect the change-over process, the guide rollers will accommodate all diameters.

References

1. Nazarian, E., Ko, J., Wang, H.: Design of multi-product manufacturing lines with the consideration of product change dependent inter-task times, reduced change-over and machine flexibility. J. Manuf. Syst. **1**(29), 35–46 (2010)
2. Assaf, R., Haddad, T.: An application of single minute exchange of die approach in an aluminium profiles extrusion production system. Int. J. Sci. Res. Innov. Technol. **4**(7), 14–22
3. Halmos, G.T.: Roll Forming Handbook (Manufacturing Engineering and Materials Processing). CRC, Taylor & Francis Group (2006). ISBN 0-8247-9563-6
4. Data M—Sheet Metal Solutions. [Online]. Available: https://www.datam.de/en/products/software-division/roll-forming/h9-copra-rf-cageforming. Acedido 9 Sept 2021
5. Chen, W., Jiang, J., Li, D., Zou, T., Peng, Y.: Flower pattern and roll positioning design for the cage roll forming process. J. Mater. Process. Tech. **264**, 295–312 (2019)
6. Madhav, R., Marnewick, A., Nel, H., Pretorius, J.: Managing Change-Over Waste in Manufacturing Plants When Using Single Minute Exchange of Dies. IEOM Society International, Rabat, Morocco, 2017.
7. Vermaak, T.: Critical Success Factors for the Implementation of Lean Thinking in South African Manufacturing Organisations. Thesis (2008)
8. Clements, E.J., Sinwaney, V., Singh, R.K.: Measurement of overall equipment effectiveness to improve operational efficiency. Int. J. Process Manage. Benchmarking **8**(2), 246–261 (2018)
9. Singh, R.K., Clements, J.E., Sonwaney, V.: Measurement of overall equipment effectiveness to improve operational efficiency. Int. J. Process Manage. Benchmarking **8**(2), 246–259 (2018)
10. Ulutas, B.: An application of SMED Methodology. Int. J. Mech. Aerosp. Ind. Mechatron. Manuf. Eng. **5**(7), 1194–1197 (2011)
11. Dave, Y., Sohani, N.: Single minute exchange of dies: literature review. Int. J. Lean Think. **3**(2), 27–37 (2012)
12. Lindgren, M.: Cold roll forming of a U-channel made of high strength steel. J. Mater. Process. Technol. **186**, 77–81 (2007)
13. Juran, J.: Juran's Quality Handbook. The McGraw-Hill Companies, New York City (1998)
14. Soyaslan, M.: The effects of roll forming pass design on edge stresses. Sigma J. Eng. Nat. Sci. **36**(3), 677–691 (2018)

T. Pillay is currently completing a M.Sc. degree in Mechanical Engineering at the University of Kwazulu Natal in South Africa. He works as a Development Engineer within the automotive industry.

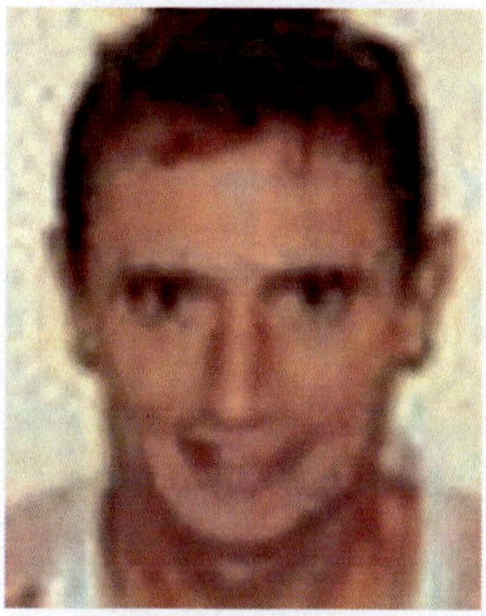

Prof. Glen Bright is a Professor of Mechatronics, Robotics and Advanced Manufacturing Systems since 2002 and a holder of the James Fulton Chair in Mechanical Engineering at the University of Kwazulu Natal.

Christian Basson is currently doing his Ph.D. in Integrated Technical Monitoring for Composite Based Robot Centric Reconfigurable Manufacturing Systems (RMS).

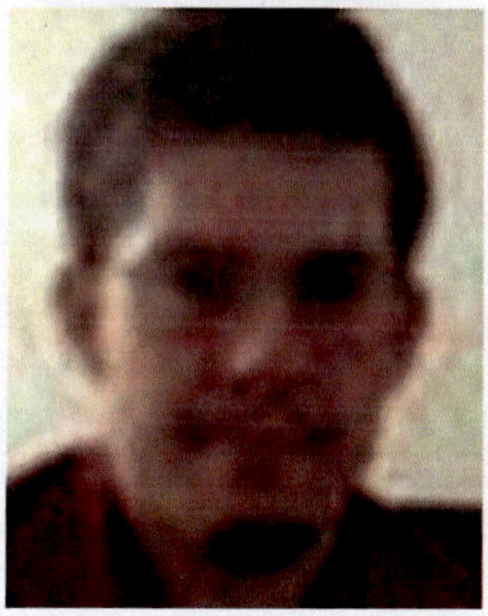

Avern Athol-Webb comes from industry with experience in production engineering. Previous research background was in vibration topics.

Machine Learning for Soft Sensors and an Application in Cement Production

Marcel Stöhr and Thomas Zielke

Abstract Soft Sensors are predictors for measurements that are difficult or impossible to obtain by means of a physical sensor. A soft sensor delivers virtual measurements based on several or many physical measurements and a mathematical or numerical model that incorporates the physical knowledge about the interdependencies. In industrial processes, control, optimization, monitoring, and maintenance can benefit from the application of soft sensors. In recent years, Machine Learning (ML) has proven to be very effective for building soft sensors for industrial applications. We work on modelling dynamical processes for which the measurements of the physical variables are available as time series over long periods. In this article we present work on the comparison of different ML methods for modelling dynamic processes with the objective of predicting certain output variables. The main application focus of our work is on the optimization of the cement production process. We compare approaches using Recurrent Neural Networks (RNN) and Convolutional Neural Networks (CNN) using real process data from a cement production. The cement sector is the third-largest industrial energy consumer. There is a high potential for energy savings and for the reduction of CO_2 emissions by optimizations of the production process. We present first results of our research that primarily is aimed to improve the robustness of a soft sensor for grain size in the cement production process.

Keywords Soft sensors · Machine learning · Dynamic processes · Cement production · Robustness · Artificial intelligence

---

M. Stöhr (✉) · T. Zielke
Düsseldorf University of Applied Sciences, Düsseldorf, Germany
e-mail: marcel.stoehr@hs-duesseldorf.de

T. Zielke
e-mail: thomas.zielke@hs-duesseldorf.de

K. von Leipzig et al. (eds.), *Smart, Sustainable Manufacturing in an Ever-Changing World*, Lecture Notes in Production Engineering,
https://doi.org/10.1007/978-3-031-15602-1_46

1 Introduction

The objective of a soft sensor is the estimation of a quantity, which cannot or not easily be measured directly. This description may also match a simple resistance thermometer. However, the term soft sensor is only used for inferential measurements that are based on several or many physical measurements and a mathematical or numerical model that incorporates the physical knowledge about the interdependencies. In the literature, the first industrial applications of soft sensors are described in the context of chemical process operations. Rao et al. [1] state two objectives for developing soft sensors: (1) *providing near optimal values for important non-measurable control variables associated with product quality to improve real-time control systems; (2) providing the interpretation of the important process variables for operators to enhance the interaction between chemical processes and human operators* (...).

The objectives of our work are very similar. The processes we deal with are dynamical and the physical measurements are represented by time series of continuous values. We use methods of machine learning to achieve the objectives.

Data-driven approaches for the development of soft sensors have been used for more than twenty years [2]. Early applications of neural networks for this purpose are described in [3–6]. Deep Neural Networks (DNN) allow more complex models that potentially can lead to an improvement of the prediction accuracy [7, 8]. Recurrent Neural Networks (RNN) and Convolutional Neural Networks (CNN) are specific types of DNN. They are particularly apt to capture the information on the dynamics [9, 10] of the available measurements. A recent review on soft sensors in industrial processes [11] provides an excellent overview of the field, far beyond the scope of the brief introduction given here. In modern factories all operational events and all measurements are digitally recorded. This is also the case for the cement mill that we aim to optimize. Figure 1 is an illustration of the cement production process. The mill is filled with fresh and coarse material. After grinding, the material is split by an air separator into new coarse material and the finished product. Since direct measurements inside a cement ball mill are hardly possible, the grain size of the material, a crucial parameter, has to be determined offline. This is a manual operation with a relatively low sampling rate typically about two hours [12]. Based on the measured distribution of the current grain size, a machine operator can adjust the air separator. Ball mills have a high specific grinding energy demand [13]. The reduction of grinding iterations by means of a real-time estimation of the grain size can significantly lower the power consumption.

Previous research has shown that the grain size can be estimated using soft sensors developed with a data-driven approach [12, 14]. A related example is the prediction of the content of free lime (f-CaO) in cement clinker [15]. In practice, a sustainable deployment of this technology has not been achieved yet. The main problems are long-term drifts of process parameters, insufficient robustness with respect to situations not covered by the training data, and lack of transparency of the model behaviour

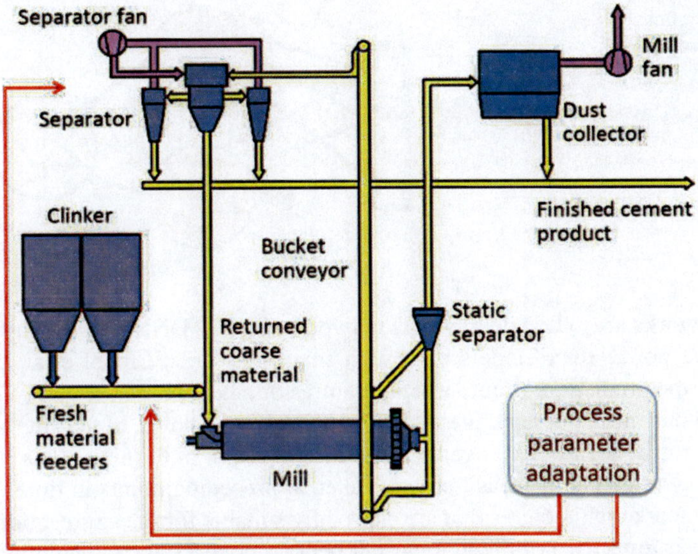

Fig. 1 Cement production process flowchart. (Adapted from en.wikipedia.org/wiki/Cement_mill)

for the responsible process operators. These problems are typical for many applications of artificial intelligence in industry [16]. In Sect. 2 we briefly introduce the RNN and the CNN architectures. Our comparative experiments, described in detail in Sect. 4, use current and historic data from a cement mill and from the operation of a gas fired absorption heat pump [7].

2 Neural Networks for Time Series Prediction

2.1 *Deep Neural Networks (DNN)*

Artificial neural networks (ANN/NN) are an established methodology for modeling complex input–output relationships. Figure 2 shows a feedforward neural network, a type of network architecture that is widely used for regression or classification problems. The example network has a three-dimensional input vector **x**, one output value y, and one hidden layer with two neurons. Using a machine learning algorithm, e.g., backpropagation, the network can be trained to produce certain values for y, depending on the given input vectors **x**. The training process can be described as a data-driven optimization of the free network parameters, typically the values for the weights and biases. The network architecture and its trained parameters of the network, data-driven optimized values for the weights and biases, represent a numerical model of the relationship between the inputs and the outputs. A feedforward neural network may have many layers and a large number of free parameters.

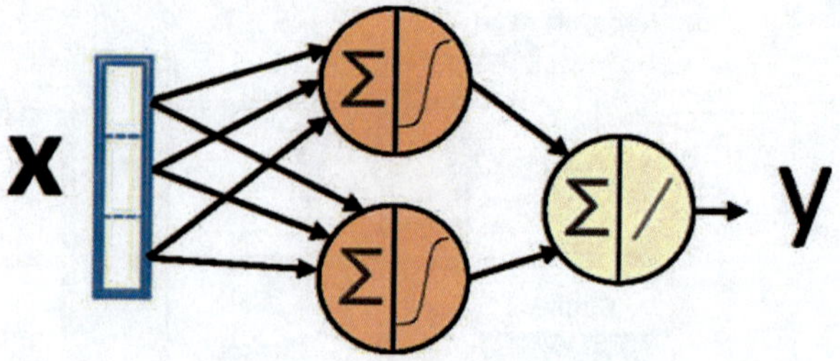

Fig. 2 Example of a feedforward network

Such networks are called deep neural networks (DNN). DNN models have a higher expressive power than models built with small networks. Lippel et al. [7] used a DNN for the prediction of output temperatures of a heat pump.

The input measurements are time series data. As shown in [7], the prediction accuracy can be greatly improved when the input vector of the network is augmented by some aggregation of measurements taken at preceding points in time. There are neural network architectures that are especially suitable for the aggregation of information over time. Two of them, recurrent neural networks and convolutional neural networks are briefly described in the following paragraphs.

2.2 Recurrent Neural Networks (RNN)

The typical characteristics of the RNN architecture are feedback loops, at least one. This gives the RNN the capacity to update the current state based on past states and current input data [17]. That is, unlike a feedforward NN the output of a RNN depends on the current input and an internal state, which can also be called a memory. Practical implementations of the feedback loop are based on 'unfolding'. As is illustrated in Fig. 3 only a certain number of past points in time are used by the 'memory'. In our example, only the directly preceding measurement (*t-1*) is used together with the current measurement at time *t*. For our experiments we use a special type of RNN, a so-called 'long short-term memory' network (LSTM). LSTM networks have been widely and successfully used in various applications. An explanation of LSTM is beyond the scope of this article. We refer the interested reader to a review paper by Yu et al. [17].

2.3 Convolutional Neural Networks (CNN)

A convolutional neural network is a special type of feedforward network. CNNs have proven extremely successful for image analysis. The CNN architecture is biologically inspired, since it is known that the visual system is based on 'receptive fields' that play a similar role as filter banks in classical computer vision systems [18]. Recently

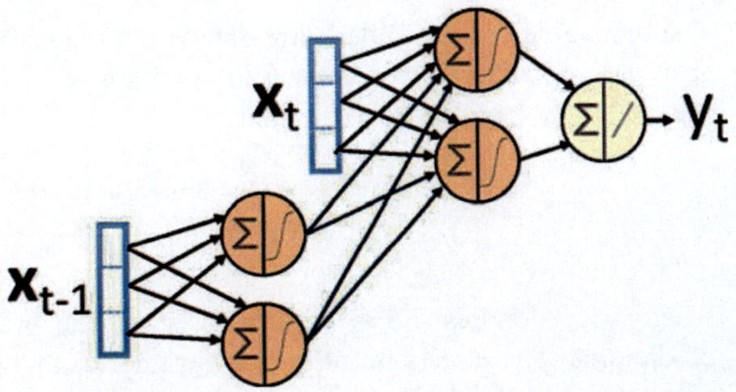

Fig. 3 A simple unfolded RNN with one hidden layer. x is a three-dimensional input vector and y is the output value at time t

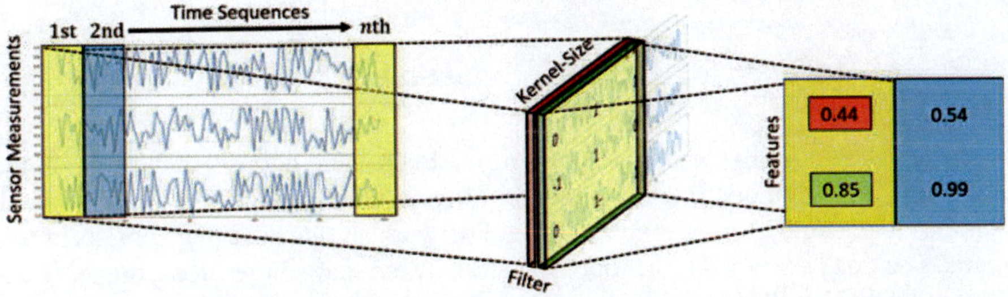

Fig. 4 First convolutional layer of a trained CNN for three 1D time sequences. The kernel slides over all time sequences. For every filter the dot product of the input and the kernel weights is calculated. After the first calculation (yellow) the kernel slides a fixed step size forward and calculates the next value (blue)

this network architecture has been widely used to process 1D signals from production processes [12, 15]. A CNN typically consists of four types of layers: convolutional, pooling, flatten and fully connected layer. The convolutional layer calculates for each filter a dot product of the input vector and kernel weights. This is followed by a so-called pooling layer that reduces and aggregates the raw data. In case of an ‘average pooling layer’ the amount of data is reduced by averaging the respective inputs. A ‘max pooling layer’ reduces the amount of data by selecting the largest value. After each calculation this filter window slides forward across the layer inputs. The fully connected layer at the end of the CNN, will predict the target value (Fig. 4).

3 Auxiliary Methods

3.1 Metrics

The most common used error metric is the mean squared error (MSE), which weights outlier stronger. The mean absolute percentage error (MAPE) is also a good metric to compare models on different datasets. Using MAPE has the advantage that the absolute values of the underlying data is not made explicit. In this article the mean absolute percentage error is used to compare results between models and datasets [19]:

$$MAPE = \frac{\sum \frac{|y_{true} - y_{pred}|}{y_{true}}}{n} \cdot 100 \qquad (1)$$

where y_{true} denotes the ground truth, y_{pred} the prediction and n the number of predictions. There are a number of other metrics that could also be considered as an alternative to MSE [20].

3.2 *Autoencoder for Anomaly Detection*

An often-overlooked problem are possible anomalies and outliers in the data. In the following two methods for the detection of anomalies and outliers are described. The so called Autoencoder is special neural network architecture that consists of two parts. The first part encodes the input sequence by reducing dimensions, the second is a decoder that ideally reproduces the input data. Using the training data as reference, outliers or anomalies are detected by a reconstruction of the test data and calculating the deviations between input and output data of the Autoencoder [21].

4 Experiments

The comparison of the different neural network architectures is made with real process data from a cement production plant and from the operation of a gas-fired absorption heat pump [7]. Hyperparameter optimization for the number of filters (CNN) and cells (LSTM) is performed for both models and datasets. For the heat pump the objective is to predict the outlet temperatures for heating T_h and cold-water T_c circuit based on five input variables such as the volume flow rate of used gas, inlet temperature and volume flow rate of the heating and cold-water circuit. In the cement production process the task is to predict two parameters of the Rosin–Rammler-Sperling-Bennett RRSB distribution [22, 23]. In the following sections the results for the hyperparameter search and the reaction of the models on outliers are shown. For this article a CNN like LeNet-5 is used which consists of two convolutional layers followed by a pooling layer and three fully connected layer with a size of 120, 64 and 2 [24]. In both experiments the prediction models are the same and are built according to Table 1. The datasets differ in the number of training data, validation data, and number of features, while the number of targets is two for both models. The heat pump dataset consists of 1.2 million training data points and 950 thousand validation data points each with five input parameters. In contrast, the

Table 1 Concrete model structures for CNN and LSTM used in the experiments. The fully connected layers form the end for both models. F_{convA} and F_{convB} denote the number of filters and $Cell_A$ and $Cell_B$ describe the number of LSTM cells

CNN		*LSTM*	
Layer	Output	Layer	Output
Convolutional	(1200, F_{convA})	LSTM	(1200, , $Cell_A$)
Avg. pooling	(600, F_{convA})	LSTM	($Cell_B$)
Convolutional	(600, F_{convB})	–	–
Avg. pooling	(300, F_{convB})	–	–
Flatten layer	(300 $\cdot$ F_{convB})	–	–
Both			
Layer		Output	
Fully connected		(120)	
Fully connected		(84)	
Fully connected		(2)	

Table 2 The best achieved MAPE of the hyperparameter study. The numbers in parenthesis describe the used parameters and random seed

Model	Heat pump	Cement
LSTM	0.77% (32;16; $Seed_2$)	2.14% (64;4; $Seed_1$)
CNN	1.15% (64;16; $Seed_1$)	2.27% (16;32; $Seed_3$)

cement dataset consists of only 5500 and 605 target data points but with 19 input parameters.

To train the weights in neural networks, initial values must be defined at the beginning, which are usually set 'randomly'. The random seed defines the random status to produce reproducible results. In addition, different seeds can be tried out to identify a particularly bad or good initialization (Table 2).

4.1 Hyperparameter Grid Search

In this section the results of the parameter optimization of both datasets will be presented. The selection of the hyperparameters to be optimized remains the same for both datasets, while only the random seed changes, since the random selection of the weight initialization can have a strong influence on the result. The hyperparameters for these experiments are shown in the Table 3.

Initial results with the heat pump dataset in Fig. 4 showed that the CNN (orange) mostly produced better and more robust results. However, it also shows that different random seeds are important. The best MAPE according to Table 3 from the LSTM in the first seed is 2.23% and in the second 0.77%. With the cement dataset the LSTM

Table 3 Hyperparameter for the grid search. *Grid A* are the different sizes for the first layer and *Grid B* for the second

	CNN/LSTM			
Grid A	8	16	32	64
Grid B	8	16	32	64

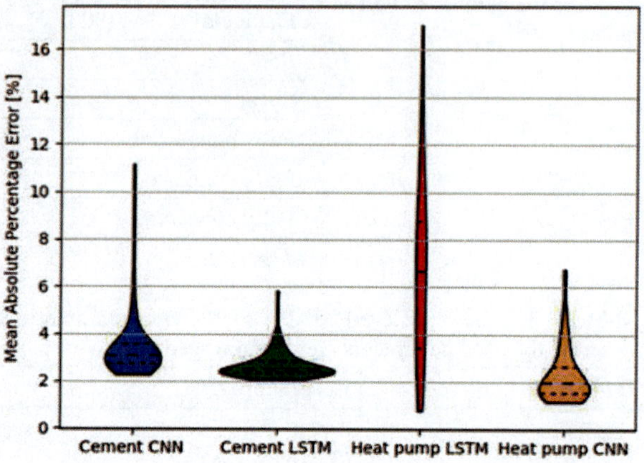

Fig. 5 Violin plot from MAPE of both models and datasets over all seeds

seems to be more stable but with a worse best MAPE then the CNN. Figure 5 shows a violin plot for both network architectures and both data branches. It visualizes the results summarized for the 3 random seeds.

The peaks of the violins denote the highest and lowest MAPE. The width indicates the distribution of the values and the 3 lines the 3 quartiles. This diagram suggests that in the case of the cement dataset, the number of LSTM cells and the selection of the random seed does not have a large influence on the result. The distribution for the CNN-based model looks similar on both datasets, only shifted in height. While the results from the LSTM for the heat pump scatter strongly regardless of the parameter choice, the cement dataset shows more consistent and better results than the CNN. The Table 3 provides two interesting insights: The best MAPE is very similar for the cement dataset for both architectures, while for the heat pump dataset the LSTM gives a much better result (Fig. 6).

4.2 *Reaction to Outlier*

The other series of experiments investigate the behaviour of the models to outliers in the data. For this, we manipulated the dataset and added artificial outliers by replacing 50,000 contiguous data points (12% of input data) from 5 features with the max value of the respective feature. Before the forecast, an attempt is made to detect the outliers

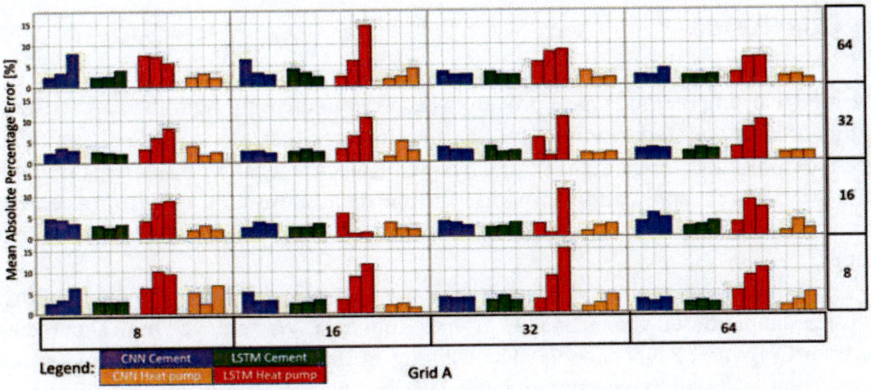

Fig. 6 Results of the hyperparameter study for both networks (CNN/RNN) and datasets (cement/heat pump) in MAPE for 3 random seeds

and replace them with the median of the last 3 days. For this purpose, an autoencoder (AE) is used. The Fig. 7 shows the deviation from the ground truth for the RRSB_D prediction, where the grey area indicates data manipulation. It is noticeable that the LSTM shows significantly better results, especially in the manipulated area beside one outlier.

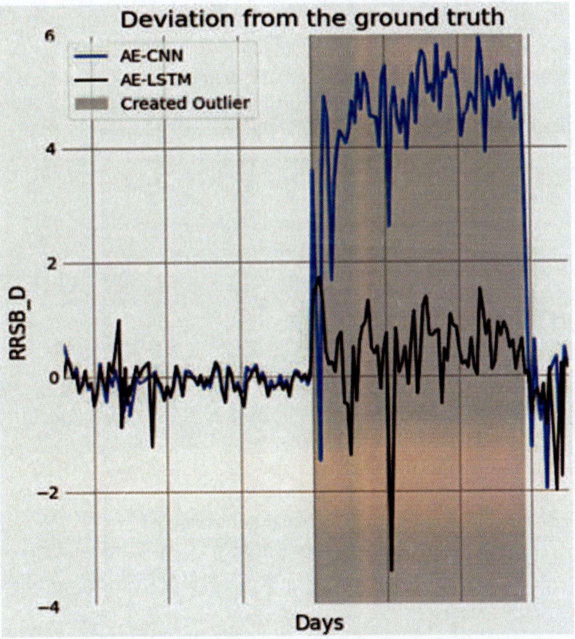

Fig. 7 Reaction of the methods to outlier. Two curves showing the deviation from the ground truth for the target value while a data manipulation is taking place (dark grey area)

5 Conclusion

Machine learning methods have proven to be very effective for building soft sensors. In this article, we study two neural net architectures that are specifically suitable for modelling time series data. Both CNN and RNN are used with real sensory data from two different application domains. The one dataset was collected from a cement production process. The other dataset came from the operation of a gas-fired heat pump.

The trained models were assessed by the mean absolute percentage error (MAPE) between the predicted values and the ground truth data. We achieved high accuracies for both CNN and RNN models. The training of the models was conducted many times using different hyperparameters and various numerical training initialisations. The respective accuracies of the trained models is not the only relevant criterium. We also looked at the robustness of the models in the presence of outliers and anomalies, i.e., how good are the predictions when the time sequence contains abnormal values. The results on the cement dataset can be considered more robust with the CNN models and the RNN models show better robustness on the heat pump dataset. In practical applications, a soft sensor should be operating in concert with a detector for outliers and anomalies. Beside the comparison of CNN and RNN, we presented some preliminary work on the integration of such detectors. For the goal of robust and sustainable applications of soft sensors in complex industrial processes we see a large potential for a fusion of different machines learning methods that may operate in an ensemble or supportively act as decision aids for handling special situations.

Acknowledgements The authors acknowledge the financial support by the German Federal Ministry for Economic Affairs and Energy (BMWi) in the framework of Industrial Collective Research (IGF), Project 21152N/2. We are thankful to Philipp Fleiger who took the initiative for this project. We would like to thank Jens Lippel, Carsten Gallus, and Martin Becker for valuable discussions and their suggestions for the improvement of this article.

References

1. Rao, M., Corbin, J., Wang, Q.: Soft sensors for quality prediction in batch chemical pulping processes. In: Proceedings of 8th IEEE International Symposium on Intelligent Control, pp. 150–155. IEEE (1993)
2. Kadlec, P., Gabrys, B., Strandt, S.: Data-driven Soft Sensors in the process industry. Comput. Chem. Eng. **33**(4), 795–814 (2009)
3. Sevilla, J., Pulido, C.: Virtual industrial sensors trough neural networks. Demonstration examples in nuclear power plants. In: IMTC/98 Conference Proceedings. IEEE Instrumentation and Measurement Technology Conference. Where Instrumentation is Going (Cat. No.98CH36222), pp. 293–297. IEEE (1998)
4. Casali, A., Gonzalez, G., Vallebuona, G., Perez, C., Vargas, R.: Grindability soft-sensors based on lithological composition and on-line measurements. Miner. Eng. **14**(7), 689–700 (2001)

5. Chella, A., Ciarlini, P., Maniscalco, U.: Neural networks as soft sensors: a comparison in a real world application. In: The 2006 IEEE International Joint Conference on Neural Network Proceedings, pp. 2662–2668. IEEE (2006)
6. Luo, J.X., Shao, H.H.: Developing soft sensors using hybrid soft computing methodology: a neurofuzzy system based on rough set theory and genetic algorithms. Soft Comput. **10**(1), 54–60 (2006)
7. Lippel, J., Becker, M., Zielke, T.: Modeling dynamic processes with deep neural networks: a case study with a gas-fired absorption heat pump. In Proceedings of 9th International Conference on Simulation and Modeling Methodologies, Technologies and Applications, pp. 317–326. SCITEPRESS (2019)
8. Maschler, B., Ganssloser, S., Hablizel, A., Weyrich, M.: Deep learning based soft sensors for industrial machinery. Procedia CIRP **99**, 662–667 (2021)
9. Ke, W., Huang, D., Yang, F., Jiang, Y.: Soft sensor development and applications based on LSTM in deep neural networks. In 2017 IEEE Symposium Series on Computational Intelligence (SSCI), pp. 1–6. IEEE (2017)
10. Jian, H., Lihui, C., Yongfang, X.: Design of soft sensor for industrial antimony flotation based on deep CNN. In: 2020 Chinese Control and Decision Conference (CCDC), pp. 2492–2496. IEEE (2020)
11. Jiang, Y., Yin, S., Dong, J., Kaynak, O.: A review on soft sensors for monitoring, control, and optimization of industrial processes. IEEE Sensors J. **21**(11), 12868–12881 (2021)
12. Schweikardt, F., Spenner, L.: Maschinelles Lernen: Entwicklung eines Softsensors zur Vorhersage der Mahlfeinheit einer Zementmühle, Karlsruhe Institute of Technology (KIT) (2017)
13. Schneider, M., Romer, M., Tschudin, M., Bolio, H.: Sustainable cement production—present and future. Cem. Concr. Res. **41**(7), 642–650 (2011)
14. Andreatta, K., Apóstolo, F., Nunes, R.: Soft sensor for online cement fineness predicting in ball mills. In: Proceedings of the International Seminar of Science and Applied Technology (ISSAT 2020). Atlantis Press (2020)
15. Jiang, X., Yao, L., Huang, G., Qian, J., Shen, B., Xu, L., Ge, Z.: A spatial-information-based semi-supervised soft sensor for f-CaO content prediction in cement industry. In: 2020 IEEE 9th Data Driven Control and Learning Systems Conference (DDCLS), pp. 898–905. IEEE (2020)
16. Zielke, T.: Is artificial intelligence ready for standardization? In: Yilmaz, M., Niemann, J., Clarke, P., Messnarz, R. (eds.) Systems, Software and Services Process Improvement. Communications in Computer and Information Science, pp. 259–274. Springer International Publishing, Cham (2020)
17. Yu, Y., Si, X., Hu, C., Zhang, J.: A review of recurrent neural networks: LSTM cells and network architectures. Neural Comput. **31**(7), 1235–1270 (2019)
18. LeCun, Y., Kavukcuoglu, K., Farabet, C.: Convolutional networks and applications in vision. In: Proceedings of 2010 IEEE International Symposium on Circuits and Systems, pp. 253–256. IEEE (2010)
19. Khair, U., Fahmi, H., Hakim, S.A., Rahim, R.: Forecasting error calculation with mean absolute deviation and mean absolute percentage error. J. Phys.: Conf. Ser. **930**, 12002 (2017)
20. Fortuna, L., Graziani, S., Xibilia, M.G.: Comparison of soft-sensor design methods for industrial plants using small data sets. IEEE Trans. Instrum. Meas. **58**(8), 2444–2451 (2009)
21. Zhou, C., Paffenroth, R.C.: Anomaly Detection with Robust Deep Autoencoders. In: Proceedings of 23rd ACM SIGKDD International Conference on Knowledge Discovery and Data Mining, pp. 665–674. ACM (2017)
22. DIN German Institute for Standardization: 1976-04-00. Graphical representation of particle size distributions; RRSB-grid 19.120. Beuth Verlag GmbH, Berlin 19.120, DIN 66145
23. Gao, P., Zhang, T.S., Wei, J.X., Yu, Q.J.: Evaluation of RRSB distribution and lognormal distribution for describing the particle size distribution of graded cementitious materials. Powder Technol. **331**, 137–145 (2018)
24. Kiranyaz, S., Avci, O., Abdeljaber, O., Ince, T., Gabbouj, M., Inman, D.J.: 1D convolutional neural networks and applications: a survey. Mech. Syst. Signal Process. **151**, 107398 (2021)

Marcel Stöhr holds a Bachelor of Engineering degree in mechanical engineering and is currently working as a research assistant in the development of robust artificial intelligence at the university of applied sciences in Düsseldorf.

Thomas Zielke holds degrees in computer engineering, knowledge based systems and neural computation. He is a professor of computer science at the Düsseldorf University of Applied Sciences, Germany. His current research interests include pattern recognition, machine learning, and artificial intelligence.

Framework for Integrating Intelligent Product Structures into a Flexible Manufacturing System

A. Burkart, G. Bitsch, and I. H. de Kock

Abstract Increasing individualisation of products with a high variety and shorter product lifecycles result in smaller lot sizes, increasing order numbers, and rising data and information processing for manufacturing companies. To cope with these trends, integrated management of the products and manufacturing information is necessary through a "product-driven" manufacturing system. Intelligent products that are integrated as an active element within the controlling and planning of the manufacturing process can represent flexibility advantages for the system. However, there are still challenges regarding system integration and evaluation of product intelligence structures. In light of these trends, this paper proposes a conceptual framework for defining, analysing, and evaluating intelligent products using the example of an assembly system. This paper begins with a classification of the existing problems in the assembly and a definition of the intelligence level. In contrast to previous approaches, the analysis of products is expanded to five dimensions. Based on this, a structured evaluation method for a use case is presented. The structure of solving the assembly problem is provided by the use case-specific ontology model. Results are presented in terms of an assignment of different application areas, linking the problem with the target intelligence class and, depending on the intelligence class of the product, suggesting requirements for implementation. The conceptual framework is evaluated by utilising a case study in a learning factory. Here, the model-mix assembly is controlled actively by the workpiece carrier in terms of transferring the variant-specific work instructions to the operator and the collaborative robot (cobot) at the workstations. The resulting system thus enables better exploitation of the potentials through less frequent errors and shorter search times. Such an implementation

A. Burkart · G. Bitsch
ESB Business School, Reutlingen University, Reutlingen, Germany
e-mail: adrian_helmut.burkart@student.reutlingen-university.de

G. Bitsch
e-mail: guenter.bitsch@reutlingen-university.de

I. H. de Kock (✉)
Department of Industrial Engineering, University of Stellenbosch, Stellenbosch, South Africa
e-mail: imkedk@sun.ac.za

K. von Leipzig et al. (eds.), *Smart, Sustainable Manufacturing in an Ever-Changing World*, Lecture Notes in Production Engineering,
https://doi.org/10.1007/978-3-031-15602-1_47

has demonstrated that the intelligent workpiece carrier represents an additional part for realising a cyber-physical production system (CPPS).

Keywords Intelligent products · Intelligent workpiece carrier · Capability model · Manufacturing control · Flexible manufacturing system

1 Introduction

Current trends in manufacturing show an increasing individualisation of products with a high variety [1]. Besides the great demand for tailored products, product lifecycles have been declined by 30% in the last decades [2]. This results in smaller lot sizes, increasing order numbers, and rising data and information processing for manufacturing companies.

In the area of decentralised autonomy for dynamic manufacturing systems, the potential to integrate intelligent products as an active element has been recognised several times [3]. The ability to increase autonomy level for manufacturing processes and to enable direct interaction with the product lead to a CPPS, according to the definition by Cardin [4]. Therefore, certain use cases of storing product and process data on the product or carrier as well as navigating the product independently through the manufacturing system, are identified [5].

However, in previous research, the focus was not on the holistic analysis and development of use case-specific intelligent products or carries, which is fundamental to provide a structured evaluation method for benefit analysis. In addition, Bertelsmeier et al. [5] highlight further research in the area of overall system integration and mixed product-intelligence structures. Therefore, the focus is on the development of a holistic framework for integrating intelligent product structures into a flexible manufacturing system.

This paper presents the nature of problems in flexible assembly systems and classifies the required intelligent products in four different classes (Sect. 2). Section 3 describes the structure of the conceptual framework, which serves as a guideline for defining, analysing, and designing intelligent products. The application of the framework for a use case at Werk150 is shown in Sect. 4.

2 Literature Review

This section identifies the nature problems in flexible assembly and describes one way to counteract them by applying intelligent product structures.

2.1 The Nature of Problems in Flexible Assembly

To identify which problems are frequently encountered in the field of assembly, a title and abstract analysis of the terms "problem" and "assembly" is carried out. For the research subject, Science Direct, IEE, and Web of Science are used as the main source for the keyword search, as these are the most relevant databases for scientific publications in the areas of engineering and management science. The language is restricted to English, and the timeframe is limited to the period from 2016 to 2020, to investigate the current problems in the assembly. Considering all paper types results in 7618 papers, which are extracted on keywords using a full counting method. After filtering the results by the term problem, a ranking regarding the occurrences is conducted. The most common problems in the assembly are identified in scheduling, balancing, and benchmarking.

After identifying the assembly problems, a scale is required to further understand what kind of problem the assembly deals with. Therefore, the three-part distinction of simple, complicated, and complex problems is used.[1] The problems identified as the most common problems are all categorised into complicated problems because formulae are critical and necessary to solve them. In addition, it requires expertise and methods in the fields of production workflows or production planning. In the case of an interaction of several assembly problems, it can lead to multiple or even competing objectives with a different solution, which represents a complex problem. To solve these complicated or rather complex assembly problems, intelligent product structures represent a solution, which is examined in the following sections.

2.2 Definition of Intelligent Products

Intelligence is the most researched aspect in psychology, with a multitude of definitions.[2] This paper is based on Dörner's definition of operational intelligence as the ability to solve complex problems [8]. The essential characteristics of operational intelligence according to Dörner are:

- Complexity
- Multiple and contradictory objectives
- Lack of transparency
- High connectivity with many interactions
- Momentum of its own.

The complexity of a situation is determined by the number of individual features of the system and the number of possible interventions as well as their mutual influence. In addition, the second component of intelligence consists of an actor, who needs to optimise a section of reality regarding multiple and contradictory objectives. Many

[1] The definitions of each class are investigated by Glouberman and Zimmerman [6].

[2] The different views of the term intelligence are examined in Rost [7].

Table 1 Different classes of intelligent products

Class 1	Class 2	Class 3	Class 4	Functionalities	
			X	Sensor actuation	
		X	X	Decision making	Processing
X	X	X	X	Data processing	
		X	X	Service oriented	Communication
X	X	X	X	Date oriented	
Read	Read/write	Read/write	Read/write	Date	Memorisation
Intelligent data product		Intelligent decisional product		Classification	
			Intelligent product		

features of the situation are not accessible to the actor, which defines the lack of transparency. High connectivity with many interactions means that the influence of a variable does not remain isolated but has side effects. An intervention that is intended to affect a certain part of the system also influences other parts of the system. The last criterion is the momentum of its own, which describes a system that evolves independently of the actor's intervention [8].

2.3 *Classification of Intelligent Products*

Besides Dörner's definition of intelligence, many authors have interlinked the notion of intelligence with the notion of product. However, they often use the terms in different ways. Some researchers have focused on the product manufacturing phase and others on the product use phase. In this paper, the focus is on the manufacturing phase. According to WONG et al., the intelligent product is characterised as a physical and informational product with its unique identification, as well as the capability of communicating effectively with its environment. In addition, it can retain and store data about itself, which need to be displayed by a language. Finally, the capability of participating in or making decisions relevant to its destiny is required [9].[3] In contrast to the predominantly general explanations of the term intelligent product, ZBIB et al. has elaborated classification models in the form of different intelligence levels, which is displayed in Table 1 [11].

Table 1 uses "X" to indicate that the functionality exists in the classes of intelligent products. The complexity of the functionalities increases from the memorisation to sensor actuation. For the memorisation, the functionality of reading indicates that the product needs the ability to read data, for example when the product can read data from a memory while moving through the system. This functionality is expanded in class 2 with writing or rewriting data on a memory. Besides the memorisation of

[3] A survey on intelligent products provides further definitions of intelligent products [10].

data, the communication is separated into a data-oriented and a service-oriented part. For the data-oriented part, the product has data access capacity with the abilities of either reading or reading and writing.

On the other side, service-oriented communication is similar to web services and provides advanced information services. The processing is subdivided into data processing and decision-making. For the reading and storing of data in the first two classes, a form of processing is already required, which requires lower computing unit capabilities than decision-making. For this reason, a distinction is made here. The last functionality describes the ability of exteroceptive interaction with products and devices in the environment, which involves physical interaction or informational interaction. In this case, physical interaction means the usage of sensors and actuators to observe a changing condition of the physical environment according to its target value. Informational interaction considers that the intelligent product communicates with other products or resources that provide services that facilitate its manufacturing [11].

3 Conceptual Framework

This section describes the background and the structure of the conceptual framework for integrating intelligent product structures into a flexible manufacturing system.

3.1 Purpose and Requirement Specification

According to van Aken and Berends [12], the design requirements for the framework are divided into five dimensions. However, this paper focuses on the functional and user requirements to understand the purpose of the development. The functional requirements refer to the framework specifications in terms of the performance demands and represent the central components for the requirements specification. Therefore, the framework should contribute towards defining, analysing, and designing intelligent objects within the context of flexible assembly systems. In addition, the framework should be used to consider product intelligence from a more dimensional perspective and is intended to avoid a narrow perspective, which is limited on the capabilities. The realisation of the solution has the aim to solve a business problem, both from a technical as well as an economic perspective. The major benefit of the conceptual framework for users is the holistic aspect of mapping the capabilities of the current manufacturing resources and deriving productivity improvements.

Regarding the user requirements, the framework should be user-friendly and practicable, which indicates that it is easy to understand and adopt. Besides the practicability, the framework should enable users' input, which is guided by defined actions for the processing. In addition, the framework should facilitate repeated and continuous use.

3.2 Framework Structure

The integration process of intelligent product structures is divided into four sections. Figure 1 presents a holistic method for identifying and evaluating the feasibility for using intelligent products in manufacturing. The first step is about analysing of the existing manufacturing environment. Therefore, the product and additional resources are classified according to their existing functionalities (approach from Sect. 2.3). In addition to the functionality classification, another dimension of two different aggregation levels is considered, as many products are composed of components that can also be independent products. The following distinction is made between the product types [10]:

- Product level: the product only manages information, messages, and decisions about itself; the components of the product cannot be distinguished as individual objects.
- Proxy level: In this case, the product not only manages information about itself, but it also recognises the components of which it is composed and acts as a proxy.

After classifying the product and resources, a classification of the existing assembly problem is examined, following the distinction from Glouberman and Zimmerman [6].

The second step of the framework is divided into two steps: Optional and additive product classification and the feasibility study with the outcome of a desirable solution. Besides the two dimensions of step 1, the additive dimension of autonomy and adaptivity expand the generic analysis. The autonomy is divided into four levels and ranges from non-autonomous to fully autonomous. The intermediate stages are semi-autonomous and autonomous [13].[4] On the other side, the dimension of adaptivity is based on human robot collaboration [14]. The adaptivity levels are divided into four levels: no form of adaptivity, the subordinate model for reactive behaviour, adaptivity by changing the parameters to environmental influences, and at the highest level of modelling one's actions and abilities, while considering the behaviour of other actors.[5] After the initial situation is analysed through the different dimensions, the user is able to show different scenarios by increasing the respective levels. By analysing and evaluating different intelligent product structures, a decision is required whether an increase in functionality represents an improvement of the defined assembly problem. If the increased product functionality does not add value to the use case, the process should be discontinued at this step. However, if the

[4] The report of Durst and Gray provides the level of autonomy for intelligent unmanned systems [13].

[5] Gervasi et al. describe the levels of autonomy in the context of a collaborative robot (cobot) [14].

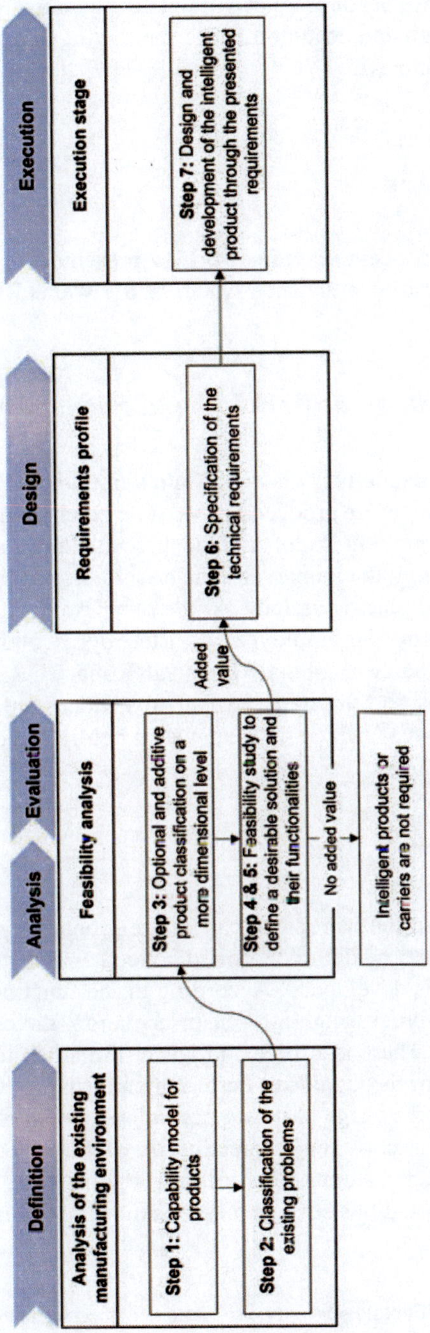

Fig. 1 Conceptual framework

problem can be improved or even eliminated, the third step is about the requirements profile of the intelligent product or product carrier in the manufacturing environment. The final step is the design and development of the intelligent product or carrier in the manufacturing environment.

4 Use Case: Werk150

The application of the conceptual framework is presented by the analysis and development of an information assistance system at the Werk150 in Reutlingen.

4.1 *The Initial Situation at Werk150 and Focus Area*

In Werk150, the scouter assembly is classified into a multi-model assembly line, as three different scooter models are produced in separate batches on a single assembly line. For the use case, the picking station and the three assembly stations are considered. For the picking station, the human and the cobot perform the loading process with different picking and placing actions. At the other two stations, the assembly processes are performed by the employee. Assisted by a cobot, the wedding of the roller is executed. The two cobots, Universal Robot (UR5) and the Franka Emika robot, grip and position the stem. Explanatory videos are provided as work instructions for the employees, which they can watch before and during processing.

4.2 *Definition Phase*

In this step, the cobots and the workpiece carrier are grouped by the classification model of intelligent products, which is described in Sect. 2.3. To group the products, a task-oriented ontology is developed.[6] According to the functionalities of memorisation, communication, processing, and sensor actuation, the cobots at Werk150 are classified into class 4. There is a force and power limitation to detect collisions between robot and operator through drive current monitoring. In this case, physical interaction is provided by the usage of power control sensors to observe a changing condition in electricity[7]. The current workpiece carrier at Werk150 acts as a transport aid which carries components from one processing station to the other. Therefore, it has no additional functionalities such as memorisation, communication, etc., and

[6] A comprehensive overview of ontologies in the production engineering environment is provided in Usman et al. [15].

[7] The occupational safety requirements for collaborative robots are defined in ISO/TS 15,066.

is not classifiable to any of the four capability classes. In this case, it is representing "Class 0" of passive products. It is important to mentation that besides the cobot and workpiece carrier, the generic classification model applies to additional manufacturing resources. For the scenario of classifying the manufacturing resources regarding their functionalities, an ontology is developed. The purpose of the Manufacturing Resource Functionality Classification Ontology (MaRFCO) is to support the listing and classification of the current resources at Werk150. Therefore, the functionality definition element consists of the functionalities described in Sect. 2.3. Each individual of the four intelligence classes is linked with the object property of "has Functionality" to the respective functionalities. In addition, the Intelligence level class has the restriction requiring at least one function to be assigned to it. For this reason, a "Class0" is not assigned to the four intelligence levels, because there are none of the listed requirements provided. It is important to mention, that there should be a special focus on "Class 0" allocations, as there is the potential to improve the functionalities. The classification of the existing problems in Werk150 represents the next sub step of the definition phase. The multi-model assembly line of the scooter products needs to manage the different variants. In managing these, the root cause for assembly errors is caused by human errors, which can be for example assembling the wrong components due to misinterpretation or not reading the instructions [16]. In particularly, newly recruited employees are overloaded by the multitude of work steps and varying product variants with different requirements. In addition, long set-up times are required for the cobots, as it needs to be taught or re-programmed offline when a variant is changed in Werk150.

The mentioned issues represent a typical resource assignment and allocation problem, which appears due to the diversity of the products and the fluctuation of the order stream. The nature of these problems is defined as complicated because it requires expertise and methods in the field of manufacturing to solve the problem. There are many influences on the optimum solution, but using formulae and methods, increases the degree of certainty of outcome.

4.3 Analysis and Evaluation Phase

The second step is about the analysis and comparison of the resources, which are classified according to the five dimensions. The focus is on the workpiece carrier, as it has none of the defined functionalities. The cobots at Werk150 already have a high standard in terms of problem solving and capabilities, which requires no improvement. However, the current workpiece carrier is not able to solve the problem of assigning resources. Therefore, a smart workpiece carrier of class 3 is defined. In addition, the aggregation level for the smart workpiece carrier is classified at proxy level, because it is aware of the transporting components. For the autonomy level, the smart workpiece carrier is categorised at level 3, as its actions are planned independently, for instance, the allocation of the work instructions per working station. However, it is not possible to define this as fully autonomous, as other actors are needed for the decision. In addition, there is a subordinate model for reactive behaviour, which leads to level 2 in adaptivity Depending on the respective product variants and workstation, the required work instruction is selectively transmitted to the machine or operator.

The next step is the analysis and evaluation of the smart workpiece carrier in terms of feasibility. To reduce the long set-up times of robots and the human errors caused by misunderstanding and searching for work instructions, a consensual knowledge model of the variant-specific assembly information is defined as the desirable solution. Therefore, the smart workpiece carrier acts in a central role, as it is responsible for the processing of the components. The intended uses are as follow:

- Step 1: Select a work instruction for the actor (operator and collaborative robot), depending on the station and product variant.
- Step 2: Retrieving the work instructions or production programs for the operator and/or robot.

In addition, the intended end-users are the operator and the cobots, which are searching for the variant-specific work instruction to perform the component assembly. The main benefit of the automatic matchmaking method is the allocation of suitable operations and resources in terms of work instructions for certain product variants. This leads to consequently faster adaptions to changing product variants and fulfils the requirements for a multi-model assembly line. In addition, the allocation of the work instructions occurs tailored based on the actor. In the case of an operator, this means that depending on the operator's qualification level, individual types of work instructions are displayed. For example, an engineer is shown additional technical drawings, whereas an operator is only shown the video form per work instruction.

4.4 Design Phase

Once the use case has been analysed, the next step is to design the structure and functionality for the consensual knowledge model of the variant-specific assembly information. Therefore, an ontology is developed, which is shown in Fig. 2. The logic starts with the class of procedures, which has the subclass work plan. Each work plan creates one product and is linked to a production order. In addition, each work plan consists of several work steps, which are explained by an instruction. To assemble the product or intermediates, each process step is linked to passive and active resources. The requirement of individual work instructions per actor is provided because it is possible to add more than one instruction per work step.

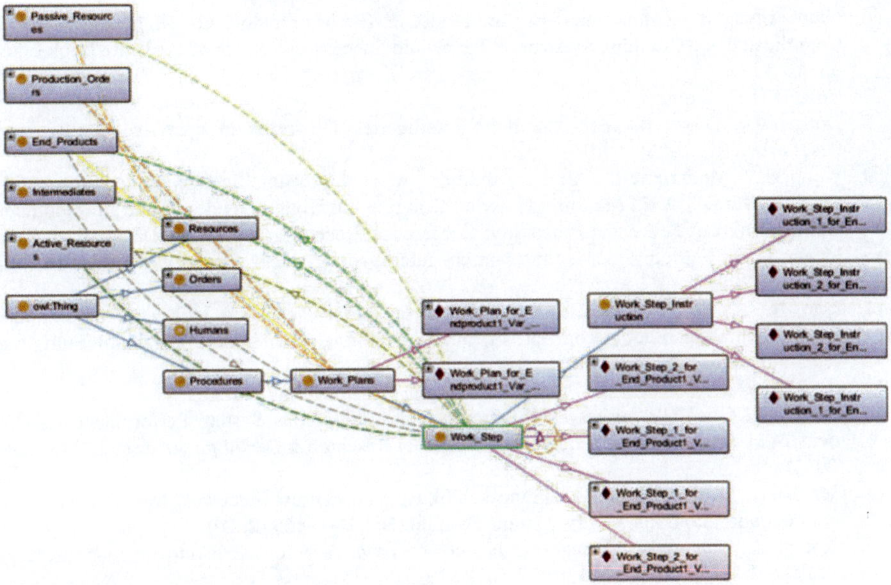

Fig. 2 Work step instruction logic

5 Summary and Outlook

This article presents a holistic analysis and evaluation of intelligent products in process control. Furthermore, the concept was evaluated using a practical use case of the workpiece carrier. In contrast to previous approaches, which usually focus on the four-level classification of intelligence, this method provides five dimensions and their respective benefits. Further research will focus on the IT infrastructure as well as the implementation and integration of the ontology at Werk150 to provide a connection to the current legacy systems.

References

1. Shou, Y., Li, Y., Park, Y.W., et al.: The impact of product complexity and variety on supply chain integration. IJPDLM **47**, 297–317 (2017)
2. Ernst, R., Haar, J.: Globalization, Competitiveness, and Governability (2019)
3. McFarlane, D., Giannikas, V., Wong, A.C., et al.: Product intelligence in industrial control: theory and practice. Ann. Rev. Control. **37**, 69–88 (2013)
4. Cardin, O.: Classification of cyber-physical production systems applications: proposition of an analysis framework. Comput. Ind. **104**, 11–21 (2019)
5. Bertelsmeier F, Schöne S, Trächtler A, et al.: Development and design of intelligent product carriers for flexible networked control of distributed manufacturing processes. In: 2016 24th Mediterranean Conference on Control and Automation (MED), pp. 755–760 (2016)

6. Glouberman, S., Zimmerman, B.: In: Forest, P.-G., Marchildon, G., McIntosh, T. (eds.) Complicated and Complex Systems: What Would Successful Reform of Medicare Look Like? pp. 21–53 (2016)
7. Rost, D.H.: Handbuch Intelligenz (2013)
8. Dörner, D.: Diagnostik der operativen Intelligenz. [Diagnosis of operative intelligence]. Diagnostica **32** (1986)
9. Wong, C.Y., McFarlane, D., Ahmad Zaharudin, A., et al.: The intelligent product driven supply chain. In: Kamel, A.E., Mellouli, K. (eds.) Bridging the Digital Divide: Cyber-Development, Human Progress, Peace and Prosperity: Conference Proceedings, p. 6 (2002)
10. Meyer, G.G., Främling, K., Holmström, J.: Intelligent products: a survey. Comput. Ind. **60**, 137–148 (2009)
11. Zbib, N., Railenanu, S., Sallez, Y., et al.: From passive products to intelligent products: the augmentation module concept. In: 5th International Conference on Digital Enterprise Technology, pp. 243–259 (2008)
12. van Aken, J.E., Berends, H.: Problem Solving in Organizations (2018)
13. Durst, P.J., Gray, W.: Levels of Autonomy and Autonomous System Performance Assessment for Intelligent Unmanned Systems. Engineer Research & Development Center, Vicksburg (2014)
14. Gervasi, R., Mastrogiacomo, L., Franceschini, F.: A conceptual framework to evaluate human-robot collaboration. Int. J. Adv. Manuf. Technol. **108**, 841–865 (2020)
15. Usman, Z., Young, R., Chungoora, N., et al.: Towards a formal manufacturing reference ontology. Int. J. Prod. Res. **51**, 6553–6572 (2013)
16. Personne, R., Matinlassi, V.: Part assurance in a mixed-model assembly line: A case study at Scania Engine Assembly in Södertälje (2014)

Adrian Burkart (B.Eng.) studies Digital Industrial Management and Engineering at ESB Business School. His major field of research is the design and optimization of flexible assembly systems at the ESB Learning Factory.

Prof. Dr. Günter Bitsch is a professor of Digitization in industry at ESB Business School, Reutlingen University. Main research fields are: hybrid decision support systems, (partially) decentralized control systems, scheduling systems and digital shopfloor management systems.

Dr. Imke Hanlu de Kock is a senior lecturer in industrial engineering at Stellenbosch University. Her research focuses on technology management within the context of sustainability transitions. In addition, she has 6 years of experience in consulting.

Manufacturing Technologies I

A Force Controlled Polishing Process Design, Analysis and Simulation Targeted at Selective Laser Melted Ti6Al4V Aero-Engine Components

Quintin de Jongh, Matthew Titus, and Ramesh Kuppuswamy

Abstract Polishing is a manufacturing process used on engineering components, where limitation of friction and wear, is critical. Attempts toward achieving a near net shape (NNS) process has given rightful importance to selective laser melting (SLM) of aeroengine components, followed by a polishing process. This manuscript explores the blast polishing of SLM processed Ti6Al4V alloy components on condition of meeting the ductile regime polishing behaviour. By use of theoretical contact mechanic calculations as well as empirical formulae developed from Finite Element Methods, the machine and algorithm are configured to administer fracture characteristics as well as an appropriate wear mechanism. Higher impinging velocities show greater impact forces, implying faster polishing times. Greater forces cause more surface damage, but this can be reduced largely by increasing the moisture content of the abrasive media. Using the developed analytical model, polishing time was found to exponentially decrease from 91.46 to 0.25 min for impinging velocities varying from 6.28 to 31.4 m/s, while impinging force increased exponentially from 0.0056 to 0.14 N, respectively. The comparative empirical model showed polishing times decrease from 41.18 to 0.615 min and forces increase from 0.11 to 0.761 N for the same range of impinging velocities. Two different media cores for the abrasive particles were considered: a thermoplastic and an organic gelatin. Either core is loosely adhered with a combination of silicon carbide and diamond powder. The thermoplastic grains were fed through a temperature-controlled nozzle arrangement

Q. de Jongh (✉) · M. Titus · R. Kuppuswamy
Advanced Manufacturing Laboratory, Department of Mechanical Engineering, University of Cape Town, Cape Town, South Africa
e-mail: djnqui001@myuct.ac.za; quintindejongh@gmail.com

M. Titus
e-mail: ttsmat002@myuct.ac.za

R. Kuppuswamy
e-mail: Ramesh.kuppuswamy@uct.ac.za

R. Kuppuswamy
A Force Controlled Polishing Process Design, Analysis and Simulation Targeted for the Selective Laser Sintered Aero-Engine Components, 8th International Conference on Competitive Manufacturing, (COMA'22), 9-10 March 2022, Stellenbosch, South Africa

K. von Leipzig et al. (eds.), *Smart, Sustainable Manufacturing in an Ever-Changing World*, Lecture Notes in Production Engineering,
https://doi.org/10.1007/978-3-031-15602-1_48

to enhance their fluidity characteristics while the gelatin was directly applied to the workpiece. Multiple simulations of the process were created to characterize the system at a nanoscale. Varying impinging forces show deformations ranging from 0.056 to 0.548 μm and contact stresses ranging from 4.07 to 15.09 GPa. The simulated results confirm the developed model predictions and thus prove to be an asset in process prediction.

Keywords Polishing · Machine development · Simulation

1 Introduction

In advanced engineering sectors, such as those of the biomedical and aerospace industries, where rapid prototyping and manufacture are required, highly precise components need to be manufactured as efficiently as possible. Due to the high hardness and strength of many of the aero engine components, the polishing process offers unlimited challenges, which must be overcome by meticulous process design. SLM, an additive manufacturing technique, has become popular in these industries due to faster production times and the ability to produce complicated components more easily than traditional manufacturing techniques [1]. However, these parts must undergo additional finishing processes to meet the stringent criteria for use, such as: dimensional accuracy, form accuracy and surface integrity [2]. Nickel powder blast polishing is often used on aero-engine components as it induces repelling charges against the nickel already present on the components, allowing for the control of polishing forces. However, the polishing media cost is high (and hence it is uneconomical) [2]. Electropolishing has also been attempted for aero engineering components. This method induces tensile residual stresses which reduce the fatigue life of the component [3]. Applying traditional lapping and polishing techniques is challenging as most components are featured with special profiles and features [4]. Past attempts using laser polishing of additive manufactured Ti6Al4V has transformed the surface microstructure into a martensitic structure, and hence an additional secondary stress relief becomes a pre-requisite to get the plate-like α phase [5]. Past research on polishing of SLM processed Ti6Al4V parts by centrifugal finishing and shot peening has shown remarkable improvement in surface finish and fatigue strength, but the process involves particularly high cycle times [6]. The blast (Aerolap_polishing process is thus seen as a favourable surface modification technique, as complex shaped surfaces (which are notoriously difficult to process) can be polished efficiently, without damaging the mechanical properties of the workpiece, and without the need for highly trained individuals and/or special conditions to be set up [2]. However, limited research has been done on this process and the relationship between polishing parameters and surface properties have not been fully categorized to date.

This paper describes the design and development of an Aerolap polishing machine, as well as the analysis of its associated process. Fundamental contact mechanics, describing the interactive contact between a sphere and a plane, are well suited to this analysis. Methodology from previous researchers, both analytical and empirical, have been used to model the process and acquire valuable results.

2 Process and Material Classification

Simplified in Fig. 1 is the Aerolap process, which was based off a shot peening machine and involves shooting abrasive media toward a workpiece at carefully designed conditions.

The abrasive media mixture of: gelatin, SiC and diamond, is initially impinged at the work surface at varying velocities. When the abrasive media hits the workpiece, it imparts an impinging force (F). This force creates an indentation in the workpiece of diameter a and deformation δ. This subsequently decreases the surface roughness R_a, and after a certain number of hits, the surface roughness reaches the desired value. Figure 2 supplies a visual aid of the above stated parameters and the composition of the abrasive media. The process is essentially a substantial series of micro indentations on the worksurface created by the impinging abrasives. Peaks of the surface spectrum are plastically deformed to fill the valley, which then improves surface texture.

2.1 Material Properties and Contribution

Just as process classification is important, understanding material properties and parameters (both fixed and fluctuating) is critical to characterizing the interaction that occurs within the polishing process.

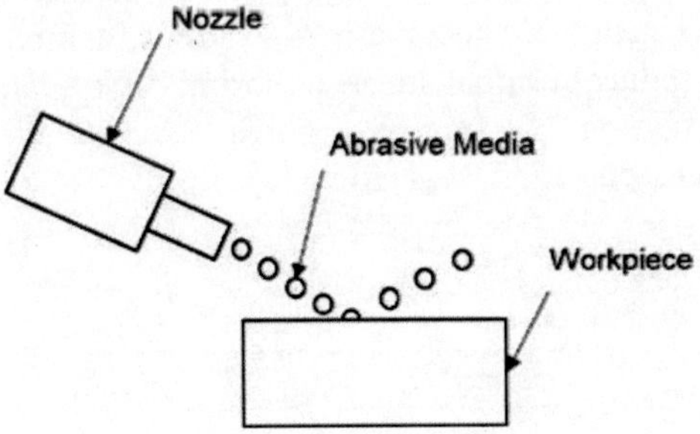

Fig. 1 Simplified aerolap process

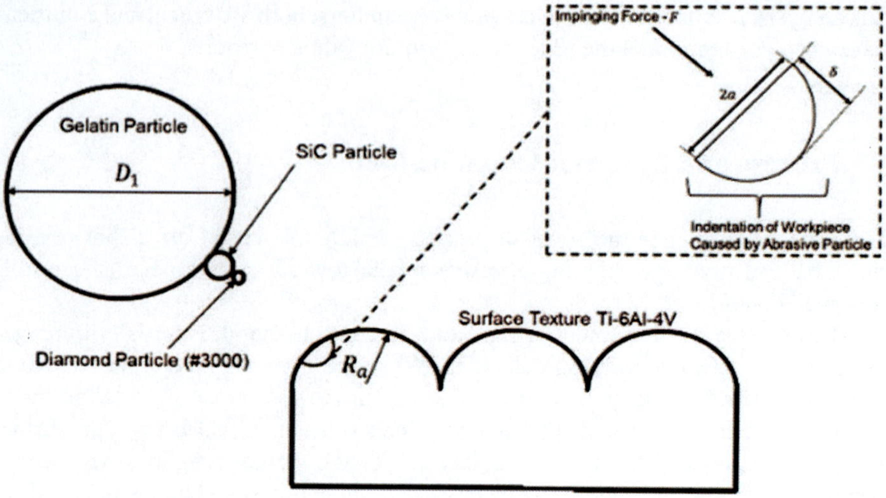

Fig. 2 Abrasive composition and interaction

2.1.1 Workpiece

The materials used to create aero-engine components are typically superalloys (Nickel based) and high strength to weight ratio alloys (notably Ti6Al4V) [7]. Ti6Al4V components are being largely created by additive manufacturing techniques. This paper focuses on the Ti6Al4V alloy, which in addition to use in the aerospace industry, is used extensively in the biomedical/dental industries [1]. 3D printing of titanium has many benefits, including the ability to manipulate strength and density gradient [8], however, multiple problems have become apparent too (including high residual stress accumulation and the agglomeration of pores [8]). SLM processed Ti6Al4V, if not controlled adequately, has a high cooling rate and thus exists in the martensitic β phase (metastable peculiar $\alpha + \beta$ microstructure) [8]. Heat treatments are a viable way to reduce residual stress and bring back a lamellar microstructure while hot isostatic pressing (HIP) can close pores. Table 1 enumerates the properties typical of SLM processed Ti6Al4V [1, 8]:

Table 1 SLM processed Ti6Al4V Properties

Detail	Symbol	Value
Density	$\rho_{Ti6Al4V}$	4410 kg/m^3
Hardness	$H_{Ti6Al4V}$	2270 MPa
Modulus of elasticity	$E_{Ti6Al4V}$	114 GPa
Fracture toughness	$K_{cTi6Al4V}$	$75MPa\sqrt{m}$
Yield strength	$\sigma_{yTi6Al4V}$	900 MPa
Poisson's ratio	$\nu_{Ti6Al4V}$	0.342

2.1.2 Polishing Media/Abrasives

Abrasive choice is of utmost importance when designing for polishing, as the choice of material(s) will directly affect both the process parameters and workpiece properties. The abrasive needs to be harder than the workpiece to effectively facilitate material removal. A combination of SiC (silicon carbide) and #3000 mesh diamond powder was chosen as the outer layer of the abrasive. An inorganic thermoplastic, and an organic material (gelatin) were both investigated as viable materials for the media core. Due to low elastic modulus, both materials deform at the required velocities. The thermoplastic must, however, be heated past its glass transition temperature to deform effectively. A heating nozzle was installed to achieve this. Gelatinous materials can be made from environmentally friendly substances such as seaweed (agar agar) and does not need to be heated. Table 2 shows the material properties of each constituent of the abrasive.

3 Force Control Analysis

To properly design the machine, analysis of the process is required. Evaluations on ductile regime polishing conditions for the workpiece (Ti6Al4V), wear mechanism, and fracture characteristics, were performed.

A study on contact mechanics of media and workpiece interaction allows for vital predictions of stresses between objects that can stop failure from occurring (fracture, fatigue, wear, and yielding are common forms of this) [9]. Both analytical and empirical models were used to analyse the process (with simulations used to verify results).

3.1 Analytical Analysis

This approach is more advantageous when considering the interaction of the combined abrasive (diamond, SiC and gelatin) with the workpiece, and gives more insight into the behaviour of the process from a macroscopic perspective. Where experiments have not been completed, modelling a complex process such as polishing

Table 2 Abrasive consituent properties

Material	Density (ρ) Kg/m^3	Poisson ratio (υ)	Modulus of elasticity (E) Pa	Hardness (H) GPa	Yield strength (S_y) GPa
Gelatin	680	0.5	43,200	–	$1.07e^{-4}$
Diamond	3020	0.148	1180×10^9	105	35
SiC	3500	0.14	330×10^9	28	21

can be quite difficult, especially when multi-layered asperities are used. Previous researchers have used kinetic energy models, and the principle of momentum conservation, [2] as the basis of modelling:

$$F = \frac{m_1 v_1 - m_1 v_2}{t} \tag{1}$$

where F is the impinging force (N); m_1 is the mass of the impinging media (kg); v_1 is the initial velocity of impinging media and v_2 is the final velocity of the impinging media after a stand off distance (d); t is the time taken for the media to cover d. Media mass (m_1) is calculated by multiplying the combined volume of the abrasive by the combined density. The final velocity v_2 is assumed to be zero, because, as the media changes its shape and becomes stagnant, it falls off from the target.

Assuming contact is between a sphere (abrasive) and a plane (workpiece), the following equations for contact radius (a), hertzian contact stress (P_o) and deformation of workpiece (δ) [10] are inferred:

$$a = 0.721\left[F(\eta_1 + n_2)\frac{D_1 D_2}{D_1 + D_2}\right]^{\frac{1}{3}} \tag{2}$$

$$P_o = \frac{1.5F}{\pi a^2} \tag{3}$$

$$\delta = 1.04\left[(\eta_1 + n_2)^2 F^2 \frac{D_1 D_2}{D_1 + D_2}\right]^{1/3} \tag{4}$$

D_1 is abrasive diameter and D_2 is the grain diameter of the workpiece (assumed as 10 μm for Ti6Al4V). η is a material parameter (η_1 for the workpiece and η_2 for the abrasive), computed by:

$$\eta = \frac{1 - \nu}{E} \tag{5}$$

where ν is Poission's ratio and E is the modulus of elasticity of the respective material.

The area of contact (A) is computed as:

$$A = \pi a^2 \tag{6}$$

where a is found by Eq. (2). The deformation of media or workpiece was computed by multiplying the previously acquired deformation by the hardness ratio of the two materials (7) where H_1 is $H_{gelatin}$ and H_2 is $H_{Ti6Al4V}$.

$$\delta_w = H_R \delta\left(H_R = \frac{H_1}{H_2}\right) \tag{7}$$

The polishing time from initial surface texture (R_{a1}) to the finished surface texture (R_{a2}) is computed as:

$$t_{\text{polishing}} = \frac{\text{Total impingments to acheive desired surface roughness}}{\text{Number of impingments per second}}$$

$$t_{polishing} = \frac{\frac{2\pi}{3}\left(R_{a1}^2 - R_{a2}^2\right)}{\frac{\pi n}{6}\left(3a^2 + \delta_{vol}^2\right)\delta_{vol}} \tag{8}$$

where n is the number of media impingements per second. δ_{vol} is the indentation volume [11] on the workpiece:

$$\delta_{vol} = \frac{\pi}{6}\left(3a^2 + \delta^2\right)\delta \tag{9}$$

where a and δ are found by Eqs. (2) and (4) respectively. The process must be run in ductile-regime conditions (processing by plastic deformation and not fracture). Contact stress evaluations were used to ensure that this criterion was met. This was complemented by the use of the Bifano equation [11] for critical depth of cut:

$$d_c = 0.15\frac{E_{Ti6Al4V}}{H_{Ti6Al4V}}\left(\frac{K_{cTi6Al4V}}{H_{Ti6Al4V}}\right)^2 \tag{10}$$

The critical depth of cut for Ti6Al4V was found to be 2.513 mm. All other depths of cuts can be assumed as the depth of indentation (deformation of workpiece) as stated previously.

Theoretical maximum shear stress is given by:

$$\tau_{\max} = \frac{G}{2\pi} \tag{11}$$

where G is the shear modulus (45GPa for Ti6Al4V). $\tau_{\max}$ was found to be 7.2 GPa for Ti6Al4V. Note that $\tau_{\max}$ is the theoretical contact stress to induce ductile regime polishing conditions.

3.1.1 Results

Greater impinging velocities result in an exponential growth in impinging force, from 0.0056 N for 6.28 m/s to 0.1403 N for 31.4 m/s. The depth of cut in the workpiece grows steadily with increasing impinging velocity but is far below the critical depth of cut for all values of impinging velocity (see Fig. 3). The maximum media velocity (31.4 m/s) is sufficient to enable ductile regime conditions (see Fig. 3).

Figure 4 shows the relationship between polishing time and impinging velocity. For a velocity of 6.28 m/s, the polishing time is calculated to be 91.46 min, and for an

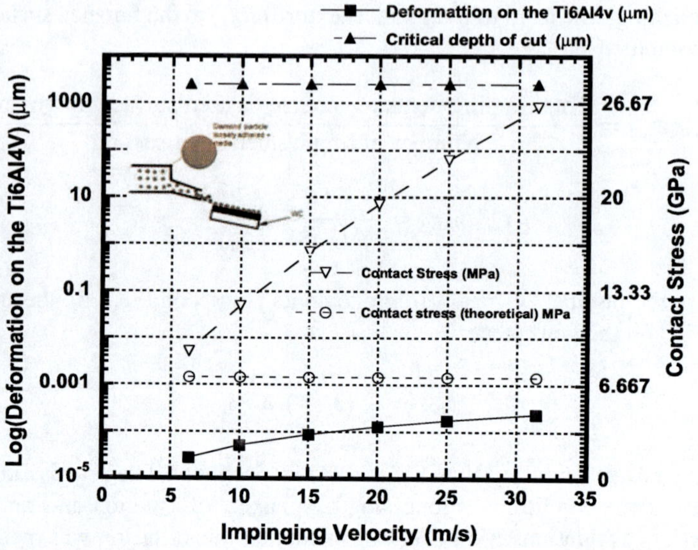

Fig. 3 Workpiece deformations and stresses

impinging velocity of 10 m/s, the polishing time was found to be 16.6 min. Polishing time then exponentially decreases from 3.75 to 0.25 min for impinging velocities of 15 m/s to 31.4 m/s, respectively. This result shows how critical the relationship of impinging velocity is to polishing time (and thereby, polishing efficiency).

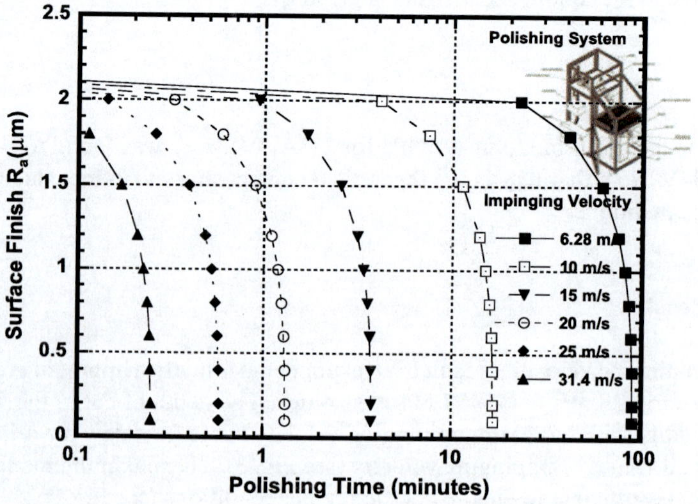

Fig. 4 Polishing times for various impinging velocities

3.2 Empirical Analysis

The following model is based on previously found empirical relations, using the notion of critical values to model the process (based on fundamentally developed hertzian contact mechanics as well as sets of developed and verified empirical formulae). Two concurrent sets of calculations were completed: assuming a single #3000 mesh diamond particle contacts a single asperity of a Ti6Al4V workpiece, and, assuming the entire media (combined gelatin, SiC and diamond) contacts a single Ti6Al4V asperity. Jackson and Green [12] and Kogut and Etsion [13] used FEM to improve on previously developed models that were pioneering for predictions of polishing parameters [14] [15]. Using von Mises yield criteria, they found the following formulae for prediction of critical deformation, ω_c (interference depth), critical area of contact (A_c), and critical contact force (F_c):

$$\omega_C = \left(\frac{\pi C S_y}{2E'}\right)^2 R \tag{12}$$

$$A_C = \pi^3\left(\frac{C S_y R}{2E'}\right)^2 \tag{13}$$

$$F_c = \frac{4}{3}\left(\frac{R}{E'}\right)^2\left(\frac{C}{2}\pi S_y\right)^3 \tag{14}$$

R is known as the equivalent radius of curvature and is determined differently for various contact shapes. For the interaction of a sphere and a plane (the assumption made here), R is equal to the radius of the abrasive. S_y is the abrasive yield strength, E' is the equivalent elastic modulus, found by:

$$\frac{1}{E'} = \frac{1 - v_1^2}{E_1} + \frac{1 - v_2^2}{E_2} \tag{15}$$

where E_1 and E_2 are the elastic moduli of the workpiece and abrasive respectively. ν_1 and ν_2 are the Poisson ratios of the workpiece and abrasive respectively. Also note that C is the critical yield stress coefficient and is found by (16).

$$C = 1.295e^{0.736v} \tag{16}$$

where v is the workpiece Poisson ratio. The critical velocity of abrasive (V_c) before onset of plastic deformation) is found by:

$$V_c = \sqrt{\frac{4\omega_c F_c}{5m_{abr}}} \tag{17}$$

where m_{abr} = mass of the abrasive. By using the previously calculated critical values, we can use the notion of normalization to predict values from empirical relations. Normalized values of spherical deflection, contact area and contact force are given by:

$$\omega^* = \frac{\omega}{\omega_c} \quad A^* = \frac{A}{A_c} \quad F^* = \frac{F}{F_c} \tag{18}$$

For near elastic conditions (where the coefficient of restitution is equal to, or very near, 1) and using their developed FEM empirical model, Jackson and Green [12] found that normalized contact force simplify to:

$$F^* = (\omega^*)^{3/2} \; with \; A^* = \omega^* \tag{19}$$

Maximum output velocity of impinging media is: 31.4 m/s and the normalized velocity (unitless) of diamond particles (at this condition) is 0.028, while the entire media was found to have a normalized velocity of 0.102, both of which are much less than 1, implying that elastic conditions can be assumed [12]. Force at a particular deformation can then be inferred as:

$$F_{imp} = F_c \left(\frac{\omega(V)}{\omega_c} \right)^{3/2} \tag{20}$$

Using normalized force and deformation equations in conjunction with critical velocity and force, the following formula for deformation was inferred:

$$\omega(V) = \left(\frac{5V^2 m_{abr} \omega_c^{3/2}}{4F_c} \right)^{2/5} \tag{21}$$

Force at a particular impinging velocity can then be found by either substituting the inferred (21) into (20) or into the widely used force model [9] show in (22).

$$F_{imp} = \frac{4}{3} E' \sqrt{R} [w(v)]^{3/2} \tag{22}$$

Area of contact can be found quite simply by using the normalized contact area and noting that at elastic conditions, normalized contact area equals normalized deformation. The radius of circular contact area, a can also be found.

$$A = A_C \omega(V), \quad a = \sqrt{\frac{A}{\pi}} \tag{23}$$

Contact pressure/stress is found by (3) in the previous section). Calculation of polishing time is described below.

Firstly, the initial and desired surfaces roughness, R_{a1} and R_{a2} must be determined/chosen. The required volume of grain surface size is calculated by assuming that each surface asperity on the workpiece is a sphere. This determines the volume of required removal from the polishing action. This is found by (24) on the following page.

$$V_{R_a} = \frac{2\pi}{3}\left(R_{a2}^3 - R_{a1}^3\right) \quad (24)$$

The volume of indentation per impinging particle can be found by (25).

$$\mathrm{V_{cut}} = \frac{\pi}{6}\left(\omega(\mathrm{V})^2 + \mathrm{a}^2\right)\omega(\mathrm{V}) \quad (25)$$

where $\omega(V)$ and a are found by (21) and (23) respectively. The number of hits required by the impinging particles is simply found by dividing the difference in volume of grain size by the volume of each particle's impinging cut:

$$Hits = V_{R_a}/V_{cut} \quad (26)$$

Finally, the time for polishing can be inferred by dividing the number of hits required by the number of impinging particles per second:

$$t = \frac{Hits}{particles\ per\ second} \quad (27)$$

3.2.1 Results

By using an output velocity range of between 6.28 m/s and 31.4 m/s, the force per entire media particle was found to be between 0.11 N and 0.761 N (see Fig. 5). Micro analysis using individual diamond particle interaction found impinging forces to be between 0.098 mN and 0.675 mN, with contact stresses between asperity and particle varying between 5.469 GPa and 10.411 GPa. The number of particles impinging upon the workpiece at once varies between 314 and 1570 for impinging velocities between 6.28 m/s and 31.4 m/s (as in the analytical section). Note that diamond particle diameter is assumed to be 2 μm on average. For impinging velocities increasing from 6.28 m/s to 31.4 m/s, polishing times (to polish a Ti6Al4V surface from 2.2 μm to 0.1 μm) decreased exponentially from 41.187 min to 0.615 min.

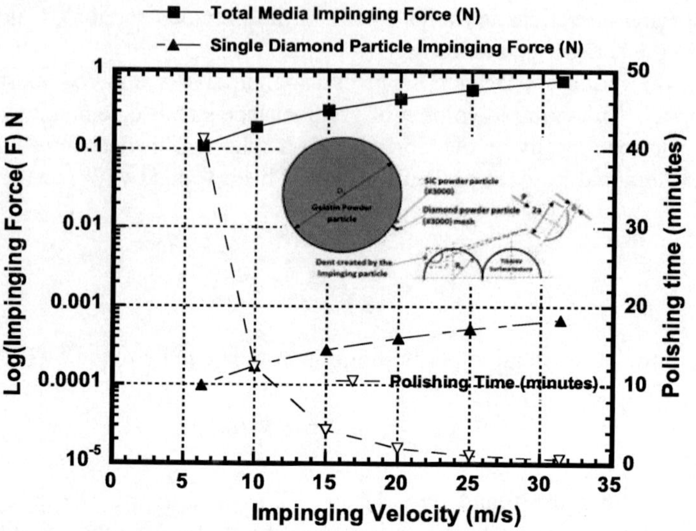

Fig. 5 Empirical method impinging forces and polishing times

4 Machine Design and Development

Figure 6 shows the designed and developed Aerolap flexible media impinging system. A modular design overall, the system consists of many separate yet integrated elements, including an adjustable aluminium frame, a high velocity media impinging unit, a funnelling hopper, a media collection bin with an integrated agitator, and a pneumatic transportation unit. The impinging unit consists of a 5-vane impeller, whose rotating unit is dynamically balanced to a scale of G2.5. The vaned impeller is attached to a spindle capable of a high rate of revolutions per second. This spindle is driven by a flat pulley system with a speed up ratio of 4:1. The driver pulley is attached to an AC induction motor, and the motor speed is controlled by a variable frequency speed drive. Abrasive media is fed to the high velocity impinging unit by the pneumatic transportation system, which is equipped with a feed ejector and a flexible hose system. The media is kept wet by using a Producut Minimum Quantity Lubrication (MQL) system as a mist production unit (supplying mist at between 4 and 8 cc/min). The media, consisting of gelatin, silicon carbide and diamond particles, is mixed with mist using the low-speed agitator (60 to 100 rpm). The media is hydrated to between 30 and 50% water absorption while diamond and SiC powder size is of mesh size #3000. Diamond and SiC particle sizes vary between 1 μm and 3 μm and gelatin particles vary between 1.5 mm and 3 mm. The ratio of gelatin to SiC to diamond is 97:2.9:0.1. The impeller is made from Nylon 66 and is used to propel the flexible media toward the workpiece. Impeller diameter is 120 mm and possesses a speed range of 1000 rpm to 5000 rpm, allowing for impinging speeds of between 6.28 m/s and 31.4 m/s.

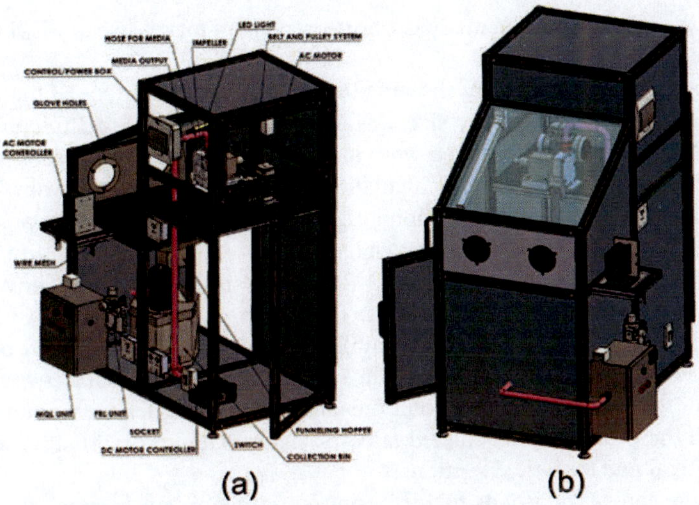

Fig. 6 Labelled assembly with visible inside, assembled view of the completed machine

The pneumatic transportation system transports media from the media mixing unit to the impeller. The feed ejector used is a SCHMALZ SEC-400 and is supplied with air pressure between 2 and 6 bar. A 52 mm inner diameter nitrile rubber hose is used to connect the feed ejector to the impeller. Steady state flow conditions and theoretical flow rates were calculated for the exit of the feed ejector, the inlet and exit of the impeller, and the final exit of media (nozzle outlet). Impeller speed is then suitably adjusted to cater for the system's steady state condition.

A Kistler 9265 dynamometer with a 0.01 N resolution on all three axes is used to measure impinging force. Dewesoft43A DAQ (data acquisition system) is used to capture and evaluate signals. A Taylor Hobson Surtronic S100 profilometer is used before and after polishing to acquire roughness traces. A Leica Cambridge SEM (scanning electron microscope) is used to examine polished surface textures.

5 Simulation

The polishing process in itself is random in nature. Abrasives often impinge at different times, at different angles and are of different size. Process parameters interact with each other in unexpected manners and some polishing effects contrast what is expected [16]. A simulation based model which covers the wide spectrum of process parameters allows for simple verification of models as well as additional insight into the process [16] (particularly for parameters not included in the model, such as temperature increases). The simulation aids in producing the most optimal polishing process possible. Solidworks was used as the simulation program for the process. A static simulation method was chosen and it makes use of FEM to identify the process behaviour.

The process is modelled by showing the asperity on the surface of the workpiece as the impact occurs. Asperity size was assumed as an initial 2.2 micron in diameter while contact areas for force were increased appropriately for each simulation (accordingly to previous section calculations). The simulations were run for various impinging forces (correlated to impinging velocities from the analytical section)—0.0056 N to 0.1404 N for increasing velocities between 6.28 and 31.4 m/s. Table 3 shows the results of the simulation for 6.28, 15, 25 and 31.4 m/s respectively, where the images shown are for displacement. The deformation scale of the images is 10 (with 1 being representitive of true scale). This was done to adequately display the deformations and stresses taking place in a micro simulation. Displacement solutions align with the empirical model solutions and contact stresses were found to agree well with the previously calculated stresses and maximum deformation from both the analytical and empirical sections.

Figure 7 shows the trends of increasing deformation and contact stress for the simulation results. For velocities above 15 m/s the contact stresses are above the theoretical value to induce ductile regime polishing conditions (7.2GPa). The contact stresses never reached the maximum level of 45.2GPa (shear modulus of Ti6Al4V). The results show similiar trends in change of deflection and contact stress with

Table 3 Summarised simulation results

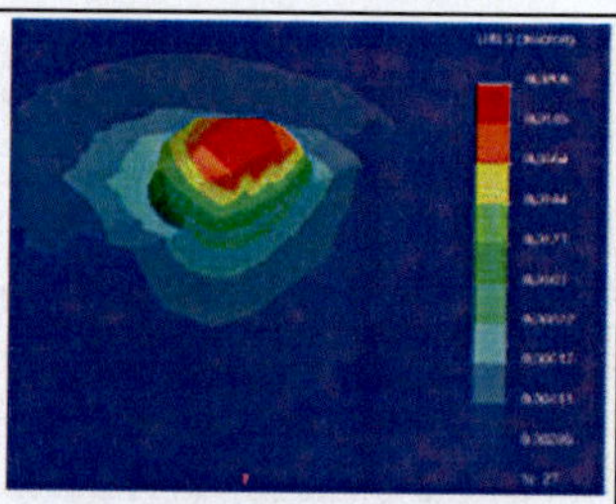 $V_{imp} = 6.28\frac{m}{s}$ $F_{imp} = 0.0056N$ $\delta_{max} = 0.056\mu m$ $\sigma_{cmax} = 4.07GPa$	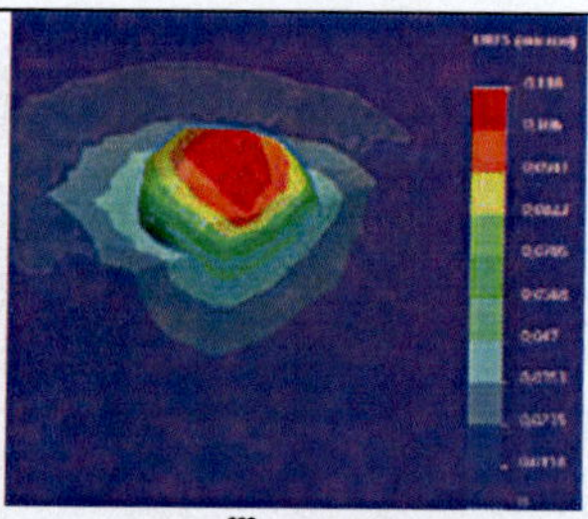 $V_{imp} = 15\frac{m}{s}$ $F_{imp} = 0.0320N$ $\delta_{max} = 0.2\mu m$ $\sigma_{cmax} = 7.781GPa$
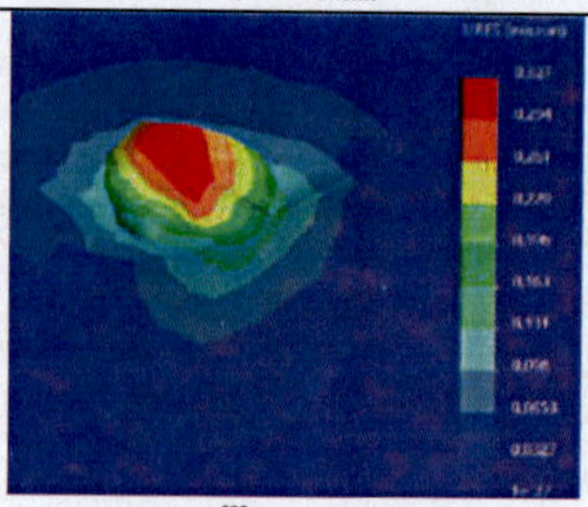 $V_{imp} = 25\frac{m}{s}$ $F_{imp} = 0.0890N$ $\delta_{max} = 0.41\mu m,$ $\sigma_{cmax} = 12.77GPa$	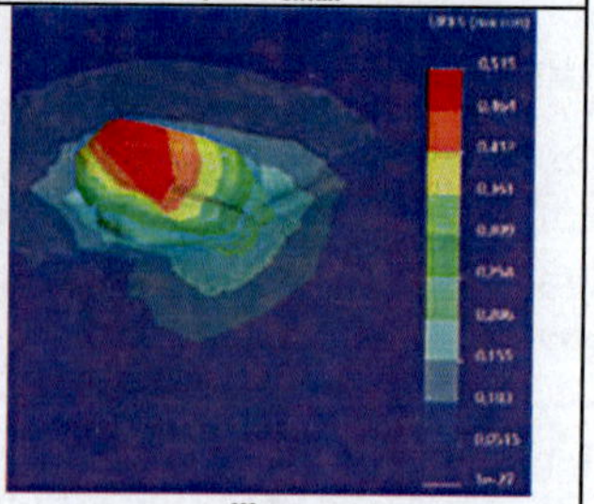 $V_{imp} = 31.4\frac{m}{s}$ $F_{imp} = 0.1404N$ $\delta_{max} = 0.55\mu m$ $\sigma_{cmax} = 15.09GPa$

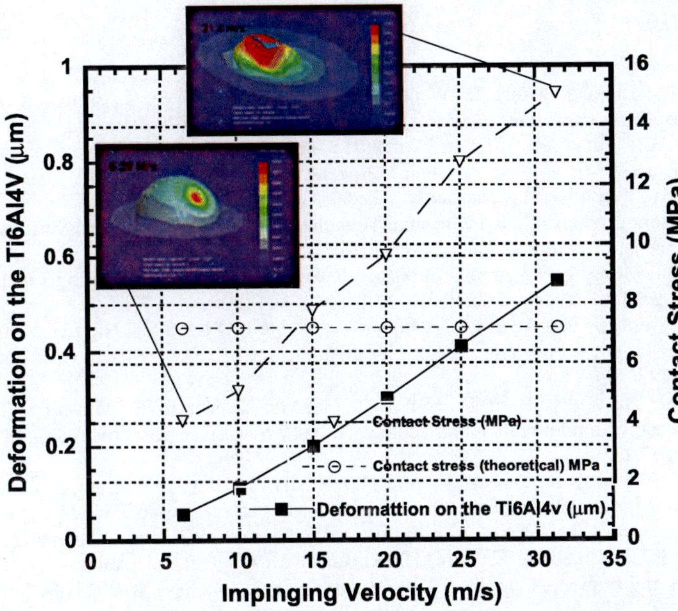

Fig. 7 Simulation trends

increasing impinging velocities (exponential increases to both factors) to that of the analytical and empirical results, thus affirming the previous two models' validities.

6 Conclusions

Analysis of the process proves this to be a viable finishing technique. Materials have been carefully selected and the process has been designed and subsequently proven to meet the ductile regime polishing conditions required for the process to run optimally. Higher impinging speeds prove to polish faster (due to exponential increases in polishing force with increasing polishing velocities) and the negative surface effects of higher speed polishing can be overcome by wetting the abrasive media. Deformation of the workpiece was found to be far below the critical depth of cut of the workpiece and contact stresses proved to be within limits.. Simulations reinforce the analytical and empirical models and solutions. An adjustable design with advanced design features that could easily be implemented in industry and can effectively run the process, has been presented.

Acknowledgements This project was supported by fund NRF GRANT: INCENTIVE FUNDING FOR RATED RESEARCHERS (IPRR)—South Africa through Reference: RA191118492767 and Grant No: 136118. The views expressed and the conclusions drawn in this paper are those of the authors and cannot necessarily be attributed to the references.

References

1. Harada, Y., Ishida, Y., Miura, D., Watanabe, S., Aoki, H., Miyasaka, T., Shinya, A.: Mechanical properties of selective laser sintering. Materials **13**(22), 1–18 (2020)
2. Ramesh, K., Ozbayraktar, S., Saridikmen, H.: Aero-lap polishing of poly crystalline diamond inserts using Multicon media. J. Manuf. Process. **14**(2), 167–173 (2012)
3. Nishimura, Y., Endo, M., Yanase, K., Ikeda, Y., Miyakawa, S., Miyamoto, N.: High cycle fatigue strength of spring steel with small scratches. In: Proceedings of First Structural Integrity Conference and Exhibition. Bangalore (2016)
4. Huang, H., Gong, Z., Chen, X., Zhou, L.: Robotic grinding and polishing for turbine-vane overhaul. J. Mater. Process. Technol. **127**(2), 140–145 (2002)
5. Lee, S., Ahmadi, Z., Pegeus, J., Majouri-Samani, M., Shamsaei, N.: Laser polishing for improving fatigue performance of additive manufactured Ti-6Al-4V parts. Opt. Laser Technol. **134**, 1–12 (2021)
6. Kahlin, M., Ansell, H., Basu, D., Kerwin, A., Newton, L., Smith, B., Moverare, J.: Improved fatigue strength of additively manufactured Ti6Al4V by surface post processing. Int. J. Fatigue **134**, 1–12 (2020)
7. Okura, T.: Materials for Aircraft Engines. University of Colorado Boulder, Boulder (2015)
8. Benedetti, M., Cazzoli, M., Fontanari, V., Leoni, M.: Fatigue limit of Ti6Al4V alloy produced by selective laser sintering. In: 21st European Conference on Fracture. Catania (2016)
9. Jackson, R.L., Ghaednia, H., Lee, H., Rostami, A.: Contact mechanics. In: Tribology for Scientists and Engineers, pp. 93–132. Springer Science+Business Media, New York (2013)
10. Stolarski, T.: Tribology in Machine Design. Butterworth-Heinemann, London (1999)
11. Bifano, T., Dow, T., Scattergood, T.: Ductile-regime grinding: a new technology for machining brittle materials. ASME J. Eng. Ind. **113**(2), 184–189 (1991)
12. Jackson, R., Chusoipin, I., Green, I.: A finite element study of theresidual stress and deformationin hemispherical contacts. J. Tribol. **127**(1), 484–493 (2005)
13. Etsion, I., Kogut, L.: Elastic-plastic contact analysis of a sphere and a rigid flat. J. Appl. Mech. **69**(5), 657–662 (2002)
14. Chang, W., Etsion, I., Bogy, D.: An elastic-plastic model for the contact of rough surfaces. J. Tribol. **110**(1), 57–63 (1987)
15. Zhao, Y., Maletta, D., Chang, L.: An asperity microcontact model incorporating the transition from elastic deformation to fully plastic flow. ASME J. Tribol. 122(1), 86–93 (2000)
16. Maliaris, G., Gakias, C., Malikoutsakis, M., Savaidis, G.: A FEM-based 2D model for simulation and qualitative assessment of shot-peening processes. Materials **14**(2784), 1–14 (2021)

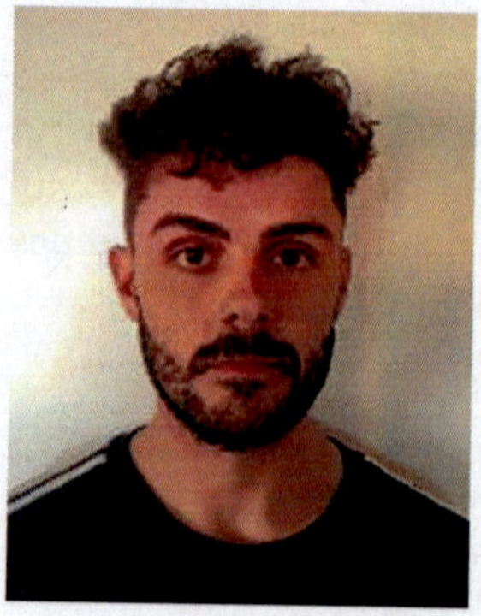

Quintin de Jongh obtained both his B.Sc. Eng and M.Sc. Eng degrees in Mechanical Engineering from the University of Cape Town, focussing on intelligent manufacturing and flexible abrasive assisted polishing.

Matthew Titus acquired both his B.Sc. Eng and M.Sc. Eng degrees in Mechanical Engineering from the University of Cape Town where his focus was on initial design of an Aerolap Polishing Machine and separately using Simulations to characterize the process. He currently works with the CSIR.

Ramesh Kuppuswamy acquired his Ph.D. from the Nanyang Technological University, Singapore and is currently an Associate Professor in the mechanical engineering department at the University of Cape Town.

An Overview of Additive Manufacturing Research Opportunities in Transport Equipment Manufacturing

Rumbidzai Muvunzi, Khumbulani Mpofu, and Ilesanmi Daniyan

Abstract In the South African transport-manufacturing industry, most of the parts are imported from other countries. This has a negative societal and economic impact. Local manufacturing is a viable alternative for creating employment while increasing manufacturing competitiveness. Also, the transport manufacturing industry is highly competitive and often faced with rapid changes in customer taste and needs. This calls for the use of flexible, cost-effective and locally available production methods to meet customer expectations. Additive Manufacturing (AM) is a flexible technology that allows for direct manufacturing of parts from digital models without the need for tooling. This increases opportunities for local manufacturing since the technology does not require tooling. Also, AM is a cost-effective approach for producing spare parts that have become obsolete as a result of the long lifespan of vehicles. This paper aims to provide an overview of AM research opportunities in the transport manufacturing Industry. To achieve this, a thorough literature study was conducted together with an analysis of the industry needs. Based on the study and analysis, a detailed outline of the research opportunities was provided. The study is useful in providing strategies for increasing local production of transport equipment parts using Additive Manufacturing.

Keywords Additive Manufacturing · Flexible technology · Local manufacturing · Transport-manufacturing industry

R. Muvunzi (✉) · K. Mpofu · I. Daniyan
Department of Industrial and Systems Engineering, Cape Peninsula University of Technology, 7535 Cape Town, South Africa
e-mail: muvunzir@cput.ac.za

K. Mpofu
e-mail: mpofuk@tut.ac.za

K. von Leipzig et al. (eds.), *Smart, Sustainable Manufacturing in an Ever-Changing World*, Lecture Notes in Production Engineering,
https://doi.org/10.1007/978-3-031-15602-1_49

1 Introduction

The transport sector in Africa relies on importing most of its parts from other countries [1]. The number of imports increased by 13% in 2020 [2]. As shown in Fig. 1, vehicle parts and accessories were the second largest imports in South Africa last year.

This includes parts from the rail, aerospace and marine industries. The reliance on imported products has a negative societal and economic impact on the country [3]. Local manufacturing is a sustainable move to alleviate poverty by creating employment while increasing manufacturing competitiveness [4]. Rapid technological changes in the nature of vehicles are taking place; these include the development of electric cars, high-speed trains, connected and driverless vehicles. Also, there is increased growth in demand for customized vehicles [5]. Most African countries find it difficult to effectively build sound manufacturing industries because traditional technologies are expensive to set up [1].

Accordingly, it is important to take advantage of flexible, cost-effective and locally available technologies to remain competitive in the Additive Manufacturing (AM) is a group of technologies in which products are made directly from digital models through selective deposition of material without the need for tooling [6]. According to ASTM, AM technologies can be grouped into 7 classes depending on the type of material, method of adding the material and power source used as shown in Fig. 2.

Due to its flexibility, AM creates opportunities for local manufacturing of parts that are costly or difficult to make using traditional methods [7]. AM can accelerate industrial development by helping countries advance industrial competitiveness with less physical investment [8]. The technology is becoming more accessible due to the reduction in the machine and material costs [7]. In South Africa, AM has been

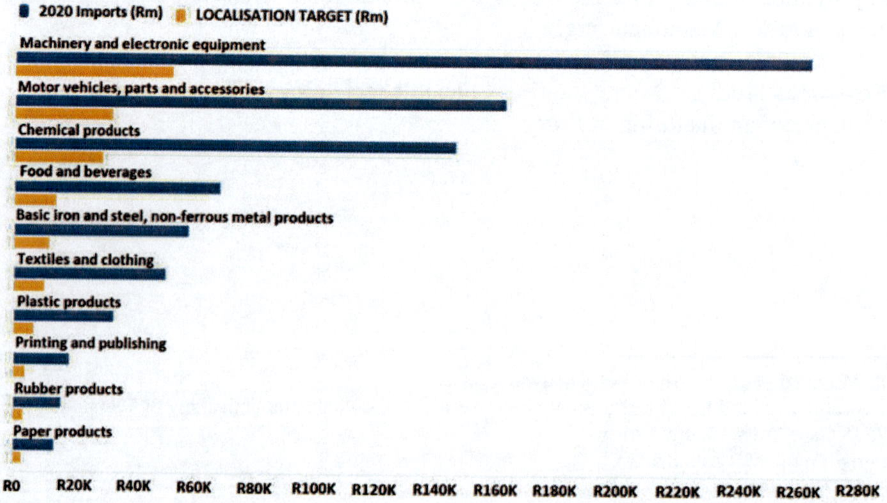

Fig. 1 Localisation targets [2]

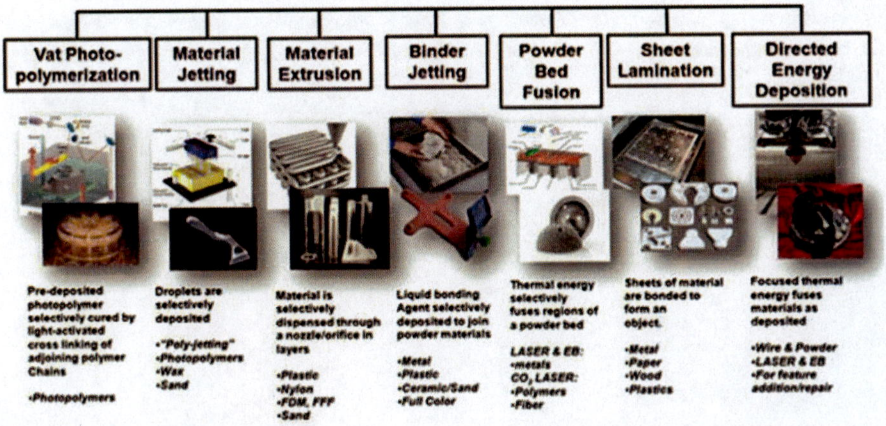

Fig. 2 Classes of AM technologies [6]

mainly explored for the aerospace and medical industries [9, 10]. There is still a gap for other transport sectors such as the rail, automotive and marine industries. Thus, this paper aims to provide an overview of the research opportunities for Additive Manufacturing in the transport manufacturing industry. The structure of the paper is as follows; the second section provides the methodology used in the study. The third section presents the results of the study. The final section gives the discussion and conclusion.

2 Methodology

The methodology employed in this study involves the review of elective literature that are relevant to the subject matter. The articles were searched using various search engines from different academic databases such as Google Scholar, Scopus, and Science direct, ResearchGate amongst others. Some important keywords such as "Additive Manufacturing", "flexible technology","local manufacturing", "transport-manufacturing industry" were employed to aid the search process to ensure that the articles obtained for review are relevant to the subject matter.

Initially, a total of 120 articles were obtained from various academic repositories. This was followed by the elimination of duplicates, unrelated as well as old articles. This brought the number of the articles reviewed to 43. The inclusion criteria include the year of publication (The minimum benchmark was 2013 to ensure that the findings are still relevant to the current situation), the relevance of the title and body of the articles to the subject matter, as well as the nature of the research outputs. In this study, peer-reviewed research outputs were given priority. This is to ensure that the findings of the research outputs are verified and validated based on expert knowledge in the

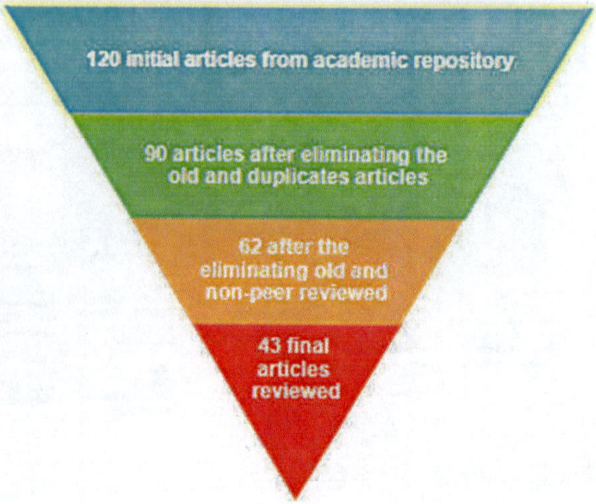

Fig. 3 The framework for the inclusion and exclusion articles

field of study. The framework for the inclusion and exclusion articles is presented in Fig. 3.

3 Results

The following section presents the opportunities for Additive Manufacturing in transport equipment manufacturing.

3.1 Development of Additive Manufacturing Based Hybrid Metal/Polymer Structures

The transport equipment manufacturing industry has a high demand for parts with increased functionality, high strength, lightweight, good fatigue and corrosion-resistant properties [11]. To attain more benefits in a single component, it may be necessary to combine different materials with required properties on different regions of a component [12]. Metals and polymers are the most widely used materials in engineering, if they are both used to produce a component, they can provide added functionality and superior properties [13]. Traditionally, only a few methods such as mechanical fastening and adhesive bonding were explored. However, these methods have limitations such as low geometric design flexibility and lengthy assembly time [14]. Alternatively, AM provides an opportunity for producing multi-material components with customized geometry and tailored material properties at different locations. Consequently, providing tailored functionality.

3.2 Developing New Materials by Combining Ceramic and Metal Powders for Improved Hardness and Toughness

Mixing of standard metal powders (e.g. cold-work steel) with ceramic powders is necessary to improve metallurgical characteristics or to add special features to parts (new high-performance materials) [15]. The combination of ceramic and metal powders (known as metal matrix components MMC) can also improve hardness and toughness to critical parts [16]. Furthermore, the metal matrix components are known for high wear resistance, chemical stability and high thermal performance [15]. However, it is difficult to process these materials using conventional processes due to high costs and energy consumption. Powder-bed AM processes have shown great potential in the processing of these materials due to the high-energy laser or electron beam [17]. However, there is limited information in literature on the development of the matrix components and optimization of the process parameters to produce parts with the required quality.

3.3 Investigating High Strength Alloys for Producing Parts with Superior Properties for Aerospace and Automotive Applications

Some aerospace and automobile applications require materials with high strength to weight ratio, corrosion and fatigue resistance. Materials such as titanium, aluminum and nickel-based alloys (Ti6Al4V, AlSi10Mg, Ni-Alloy 625) provide opportunities for producing components with the above-mentioned characteristics [18–20]. Most of the high-strength alloys are brittle and often prone to residual stresses when build additively, causing the formation of cracks and distortions [18]. Thus, it is necessary to optimize process parameters to produce parts with improved quality, superior properties and improved functionality.

3.4 Additive Manufacturing as an Enabling Technology for Producing Spare Parts for the Transport Equipment Manufacturing Industry

It is difficult to manage the supply of spare parts because of the uncertainty and change in demand. Most companies keep high levels of spare parts inventory, which is costly [21]. On the other hand, AM can be used to optimize the management of spare parts by producing them when needed. This allows for a cost-effective supply of metal and plastic parts when required [22]. Thus, spare parts can be provided fast and reliable, on-site, leading to savings in inventory costs and reducing the lead time for obtaining parts. At the same time, logistics efforts and cost can be reduced by shifting from a make-to-stock to a make-to-order approach while maintaining or even improving lead times [23]. This also leads to a reduced risk of production downtimes.

3.5 *Developing Automated Decision Support Systems for Qualifying Parts for Additive Manufacturing*

Additive manufacturing has provided opportunities to designers and manufacturers such as freedom of design and shorter process chains [24]. Also, the AM technology allows for parts to be produced without the need for tooling. Furthermore, parts produced using AM have been shown to have mechanical properties comparable to those produced conventionally [24]. However, many companies have not yet fully adopted this technology because they do not have the expertise needed to select parts that are suitable for AM application [25]. This is one of the major challenges prohibiting the wide application of AM technology. For a part to be qualified for AM application many issues such as geometry, technical limitations of the AM process, economic and societal implications must be considered [25, 26]. Thus, there is a need for automatic decision support systems for qualifying parts for AM application in the transport industry.

3.6 *Integration of Sensors in Additive Manufactured Components for Smart Manufacturing Applications*

Some transport equipment parts are required to operate under specific conditions such as temperature and moisture [27]. Very often, failure to meet the required conditions can degrade the performance of the system and impose risks of failures and severe consequences. To perform reliably under these harsh conditions, the materials and components need to be properly monitored and the systems need to be optimally controlled. However, most existing sensing technologies are insufficient to work reliably under these harsh conditions. On the other hand, Additive Manufacturing methods such as Selective Laser Melting (SLM) allow for the integration of sensors inside selected production tools and equipment to allow for intelligent, self-adjusting processes [28]. The sensor systems can be integrated in areas where machining and other manufacturing processes are reaching their limits. Sensor systems combined with intelligent feedback control systems in production components will allow for process stability, improved part quality and energy efficiency [29].

3.7 Development of Flexible Modular Tooling Systems Based on Additive Manufacturing

The designs of vehicles are consistently changing and the same applies to the parts. Hence, the production tools should have increased flexibility to adapt to market requirements. Modular tools allow for a variety of parts to be produced faster at reduced costs [30]. Additive manufacturing provides opportunities to produce complex tool modules due to its flexibility [31]. Accordingly, there is a need for further research on the development of modular concepts for moulds and dies using AM as an enabling technology. This includes introducing inserts and modules to different regions on the mould/die with specific properties.

3.8 Challenges Associated with Different AM Technologies

Fused Deposition Modelling (FDM) is the most common AM process because of its cost-effectiveness and ease of use. However, the challenge with the process is the poor surface quality due to the filament striation [32]. Secondly, FDM parts have limited mechanical strength caused by inconsistent adhesion of filament layers [33]. Steriolithography (SLA) is also a common and cost-effective AM process. Some of the challenges of SLA include limited material application and poor mechanical properties [34, 35]. Also, it is challenging to remove resins from parts with fine details [36]. Selective Laser Sintering (SLS) uses a high-powered laser to selectively sinter powdered material. The major challenge with SLS parts is porosity, this compromises the mechanical properties of produced parts [37]. There is a need for further research on new high-performance materials for improved mechanical properties [38]. Selective Laser Melting (SLM) is one of the most common AM metallic processes. However, the cost of equipment is still high and the process is prone to thermal stresses which can lead to dimensional inaccuracy and failure of parts [39, 40]. Also, the use of support material increases the post-processing requirements, making the process costly [41]. Material application is still limited and it is necessary to explore new materials [24]. Regarding polymer jetting, there is a need for further research on post-processing to improve the quality of parts [42]. Also, information on the design of parts for Material Jetting is still limited [43].

4 Conclusions

The paper aimed to reveal AM research opportunities for the transport-manufacturing sector. To achieve this aim a systematic literature review was conducted together with a detailed analysis of industry needs. The first section of the paper explains the need for SA to embrace AM in transport manufacturing. In South Africa, the transport sector mainly relies on importing parts from other countries. Studies have shown that AM is an enabling technology for increasing local production of parts due to its design freedom. The second section of the paper explains the method used in the study. A literature review was conducted to identify potential areas of AM application in accordance with the need of the industry. The third section of the paper provides an outline of the opportunities and challenges associated with AM technologies. Based on the study, the following conclusive remarks are derived.

- There is a need for further research in developing materials with lightweight and high strength properties, which meet the requirements of the transport industry. This will contribute towards the reduction of carbon emissions while improving the efficiency of vehicles. Typical examples of such materials include composites and high-strength alloys.
- There is a need for decision support systems to guide in the selection of suitable parts for AM selection depending on the technical limitation of the technology and product characteristics. Much of the application was mainly in the aerospace sector and there is limited information on the rail and marine industries
- AM also provides opportunities for manufacturing parts with special features to enhance functionality. This includes flexible modular tools to increase part variety in transport manufacturing. Sensors can be incorporated into components to allow the parts to function within the required conditions.

References

1. Black, A., Makundi, B., Mclennan, T.: Africa's Automotive Industry: Potential and Challenges, pp. 1–17 (2017). [Online]. Available: https://www.afdb.org/en/documents/publications/working-paper-series/. Accessed 07 Sept 2021
2. Intellidex: Localisation: what is realistic? (2021). https://www.agbiz.co.za/document/open/intellidex-localisation-what-is-realistic-april-2021. Accessed 5 Jul 2021
3. Barnes, J., Black, A.: State–business bargaining, localisation and supply chain development in the South African auto industry. In: 4th Annual Competition and Economic Development (ACER) Conference, Johannesburg, South Africa, pp. 19–20 (2018)
4. Lopes de Sousa Jabbour, A.B., Ndubisi, N.O., Roman Pais Seles, B.M.: Sustainable development in Asian manufacturing SMEs: progress and directions. Int. J. Prod. Econ. **225**, 107567 (2020). https://doi.org/10.1016/j.ijpe.2019.107567
5. Leal, R., Barreiros, F.M., Alves, L., Romeiro, F., Vasco, J.C., Santos, M., Marto, C.: Additive manufacturing tooling for the automotive industry. Int. J. Adv. Manuf. Technol. 1–7 (2017). https://doi.org/10.1007/s00170-017-0239-8
6. ASTM: ASTM F2792–10 Standard Terminology for Additive Manufacturing Technologies (2010)
7. Thomas, D.: Costs, benefits, and adoption of additive manufacturing: a supply chain perspective. Int. J. Adv. Manuf. Technol. **85**(5–8), 1857–1876 (2016). https://doi.org/10.1007/s00170-015-7973-6
8. Dilberoglu, U.M., Gharehpapagh, B., Yaman, U., Dolen, M.: The role of additive manufacturing in the era of industry 4. 0. Procedia Manuf. **11**, 545–554 (2017). https://doi.org/10.1016/j.promfg.2017.07.148

9. Yadroitsev, I., Krakhmalev, P., Yadroitsava, I.: Selective laser melting of Ti6Al4V alloy for biomedical applications: temperature monitoring and microstructural evolution. J. Alloys Compd. (2014). https://doi.org/10.1016/j.jallcom.2013.08.183
10. du Plessis, A., Yadroitsava, I., Yadroitsev, I.: Ti6Al4V lightweight lattice structures manufactured by laser powder bed fusion for load-bearing applications. Opt. Laser Technol. (2018). https://doi.org/10.1016/j.optlastec.2018.07.050
11. Vasco, J.C.: Additive manufacturing for the automotive industry. Addit. Manuf. 505–530 (2021). https://doi.org/10.1016/B978-0-12-818411-0.00010-0
12. Hertle, S., Kleffel, T., Wörz, A., Drummer, D.: Production of polymer-metal hybrids using extrusion-based additive manufacturing and electrochemically treated aluminum. Addit. Manuf. **33**, 101135 (2020). https://doi.org/10.1016/j.addma.2020.101135
13. Falck, R., Goushegir, S.M., dos Santos, J.F., Amancio-Filho, S.T.: AddJoining: a novel additive manufacturing approach for layered metal-polymer hybrid structures. Mater. Lett. **217**, 211–214 (2018). https://doi.org/10.1016/j.matlet.2018.01.021
14. Matsuzaki, R., Kanatani, T., Todoroki, A.: Multi-material additive manufacturing of polymers and metals using fused filament fabrication and electroforming. Addit. Manuf. **29**, 100812 (2019). https://doi.org/10.1016/j.addma.2019.100812
15. Hu, Y., Cong, W.: A review on laser deposition-additive manufacturing of ceramics and ceramic reinforced metal matrix composites. Ceram. Int. **44**(17), 20599–20612 (2018). https://doi.org/10.1016/j.ceramint.2018.08.083
16. Behera, M.P., Dougherty, T., Singamneni, S.: Conventional and additive manufacturing with metal matrix composites: a perspective. Procedia Manuf. **30**, 159–166 (2019). https://doi.org/10.1016/j.promfg.2019.02.023
17. Li, N., Huang, S., Zhang, G., Qin, R., Liu, W., Xiong, H., Shi, G., Blackburn, J.: Progress in additive manufacturing on new materials: a review. J. Mater. Sci. Technol. **35**(2), 242–269 (2019). https://doi.org/10.1016/j.jmst.2018.09.002
18. Aboulkhair, N.T., Simonelli, M., Parry, L., Ashcroft, I., Tuck, C., Hague, R.: 3D printing of aluminium alloys: additive manufacturing of aluminium alloys using selective laser melting. Prog. Mater. Sci. **106**, 100578 (2019). https://doi.org/10.1016/j.pmatsci.2019.100578
19. Amar, A., Li, J., Xiang, S., Liu, X., Zhou, Y., Le, G., Wang, X., Qu, F., Ma, S., Dong, W., Li, Q.: Additive manufacturing of high strength CrMnFeCoNi-based High Entropy Alloys with TiC addition. Intermetallics **109**, 162–166 (2019). https://doi.org/10.1016/j.intermet.2019.04.005
20. Li, J., Xiang, S., Luan, H., Amar, A., Liu, X., Lu, S., Zeng, Y., Le, G., Wang, X., Qu, F., Jiang, C.: Additive manufacturing of high-strength CrMnFeCoNi high-entropy alloys-based composites with WC addition. J. Mater. Sci. Technol. **35**(11), 2430–2434 (2019). https://doi.org/10.1016/j.jmst.2019.05.062
21. Knofius, N., van der Heijden, M.C., Zijm, W.H.M.: Consolidating spare parts for asset maintenance with additive manufacturing. Int. J. Prod. Econ. **208**, 269–280 (2019). https://doi.org/10.1016/j.ijpe.2018.11.007
22. Eyers, D. (ed.): Managing 3D Printing: Operations Management for Additive Manufacturing. Springer Nature (2020). https://doi.org/10.1007/978-3-030-23323-5
23. Togwe, T., Eveleigh, T.J., Tanju, B.: An additive manufacturing spare parts inventory model for an aviation use case. Eng. Manag. J. **31**(1), 69–80 (2019). https://doi.org/10.1080/10429247.2019.1565618
24. Schmidt, M., Merklein, M., Bourell, D., Dimitrov, D., Hausotte, T., Wegener, K., Overmeyer, L., Vollertsen, F., Levy, G.N.: Laser based additive manufacturing in industry and academia. CIRP Ann. **66**(2), 561–583 (2017). https://doi.org/10.1016/J.CIRP.2017.05.011
25. Yang, S., Page, T., Zhang, Y., Zhao, Y.F.: Towards an automated decision support system for the identification of additive manufacturing part candidates. J. Intell. Manuf. (2020). https://doi.org/10.1007/s10845-020-01545-6
26. Stief, P., Dantan, J.Y., Etienne, A.A., Siadat: A new methodology to analyze the functional and physical architecture of existing products for an assembly oriented product family identification. Procedia CIRP **81**, 1107–1112 (2019). https://doi.org/10.1016/j.procir.2019.03.261

27. Najmon, J.C., Raeisi, S., Tovar, A.: Review of Additive Manufacturing Technologies and Applications In the Aerospace Industry. Elsevier Inc. (2019)
28. Islam, M.N.A., Hossain, M.A., Choudhuri, A., Morton, P., Wicker, R.: Additive manufacturing and hot fire testing of complex injectors with integrated temperature sensors. Am. Soc. Mech. Eng. Power Div. Power **2019** (2019). https://doi.org/10.1115/POWER2019-1938
29. Vasilevitsky, T., Zoran, A.: Steel-Sense. In: Proceedings of the 2016 CHI Conference on Human Factors in Computer Systems, pp. 5731–5742 (2016). https://doi.org/10.1145/2858036.2858309
30. Steffan, K.-E.W.H., Fett, M., Kirchner, E.: Extended approach to optimize modular products through the potentials of additive manufacturing. Proc. Des. Soc. Des. Conf. **1**, 1115–1124 (2020). https://doi.org/10.1017/dsd.2020.172
31. Scholz, S., Mueller, T., Plasch, M., Limbeck, H., Adamietz, R., Iseringhausen, T., Kimmig, D., Dickerhof, M., Woegerer, C.: A modular flexible scalable and reconfigurable system for manufacturing of microsystems based on additive manufacturing and e-printing. Robot. Comput. Integr. Manuf. **40**, 14–23 (2016). https://doi.org/10.1016/j.rcim.2015.12.006
32. Daminabo, S.C., Goel, S., Grammatikos, S.A., Nezhad, H.Y., Thakur, V.K.: Fused deposition modeling-based additive manufacturing (3D printing): techniques for polymer material systems. Mater. Today Chem. **16**, 100248 (2020). https://doi.org/10.1016/j.mtchem.2020.100248
33. Li, Y., Gao, S., Dong, R., Ding, X., Duan, X.: Additive manufacturing of PLA and CF/PLA binding layer specimens via fused deposition modeling. J. Mater. Eng. Perform. **27**(2), 492–500 (2018). https://doi.org/10.1007/s11665-017-3065-0
34. Huang, J., Qin, Q., Wang, J.: A review of stereolithography: processes and systems. Processes **8**(9) (2020). https://doi.org/10.3390/PR8091138
35. Konuray, O., Di Donato, F., Sangermano, M., Bonada, J., Bo, X., Fernández Francos, M.A., Serra Albet, Ramis Juan, X.: Dual-curable stereolithography resins for superior thermomechanical properties. Express. Polym. Lett. **14**(9), 881–894 (2020). https://doi.org/10.3144/expresspolymlett.2020.72
36. Selvaraj, S.B., Jain, S., Day, N.: The evaluation of effect of post processing process on parts printed using photopolymer resin by stereolithography additive manufacturing. J. Multidiscip. Eng. Sci. Stud. **7**(6), 3929–3934 (2021). [Online]. Available: http://www.jmess.org/wp-content/uploads/2021/06/JMESSP13420760.pdf. Accessed 28 Sept 2021
37. Singh, S., Sharma, V.S., Sachdeva, A.: Progress in selective laser sintering using metallic powders: a review. Mater. Sci. Technol. (United Kingdom) **32**(8), 760–772 (2016). https://doi.org/10.1179/1743284715Y.0000000136
38. Chen, A.N., Wu, J.M., Liu, K., Chen, J.Y., Huan, X., Chen, P., Li, C.H., Shi, Y.S.: High-performance ceramic parts with complex shape prepared by selective laser sintering : a review. Adv. Appl. Ceram. **117**(2), 100–117 (2018). https://doi.org/10.1080/17436753.2017.1379586
39. Yap, C.Y., Chua, C.K., Dong, Z.L., Liu, Z.H., Zhang, D.Q., Loh, L.E.: Review of selective laser melting: materials and applications. Appl. Phys. Rev. 041101 (2015). https://doi.org/10.1063/1.4935926
40. Fang, Z., Wu, Z., Huang, C., Wu, C.: Review on residual stress in selective laser melting additive manufacturing of alloy parts. Opt. Laser Technol. **129**(15), 106283 (2020). https://doi.org/10.1016/j.optlastec.2020.106283
41. Gan, M.X., Wong, C.H.: Practical support structures for selective laser melting. J. Mater. Process. Technol. **238**, 474–484 (2016). https://doi.org/10.1016/J.JMATPROTEC.2016.08.006
42. Ziaee, M., Crane, N.B.: Binder jetting: a review of process, materials, and methods. Addit. Manuf. **28**, 781–801 (2019). https://doi.org/10.1016/J.ADDMA.2019.05.031
43. Gülcan, O., Günaydın, K., Tamer, A.: The state of the art of material jetting—a critical review. Polymers (Basel) **13**(16) (2021). https://doi.org/10.3390/polym13162829

Rumbidzai Muvunzi is a Lecturer in the Department of Industrial and Systems Engineering at Cape Peninsula University of Technology. Her research interests are on the application of Additive Manufacturing in the transport industry.

Khumbulani Mpofu is the Research Chair in Manufacturing and Skills Development, Department of Industrial Engineering at Tshwane University of Technology. His research interest include industrial engineering and advanced manufacturing.

Ilesanmi Afolabi Daniyan is a Postdoctoral Research Fellow at the Department of Industrial Engineering, Tshwane University of Technology, Pretoria, South Africa. His research interest include production engineering and advanced manufacturing.

Dimensional Stability of Mineral Cast for Precision Machinery

Eduard Relea, Lukas Weiss, and Konrad Wegener

Abstract Precision machinery employs a variety of materials for their structures: cast iron, welded steel, natural stone, ultra-high performance concrete (UHPC), and mineral cast (MC). Mineral cast is a composite material with an epoxy matrix and various aggregates as a filler, and has been successfully employed for precision machinery structures for over 30 years. For this work, mineral cast specimens were tested by completely immersing them in fluids, so that the conditions to which mineral cast is exposed during the operating conditions of the precision machinery can be simulated. Test specimens with standard dimensions of 80 × 80 × 400 mm were first immersed in demineralized water or grinding oil for 900 h and then dried for 800 h in a climate test chamber at constant ambient conditions of 20 °C and 50% air relative humidity. The change in length of the mineral cast specimens during the drying period was recorded. The shrinkage of the samples was measured by the difference in the length of the samples before and after the drying process. The water and grinding oil soaking tests show that their influence must be considered in case of high-precision applications, as their effects could be directly comparable to minor thermal variations.

Keywords Mineral cast · Precision machinery · Dimensional stability · Material coating

E. Relea (✉) · L. Weiss
Inspire AG, Transfer Institute for Mechatronic Systems and Manufacturing Technology, Technoparkstrasse 1, 8005 Zurich, Switzerland
e-mail: relea@inspire.ethz.ch

L. Weiss
e-mail: weiss@inspire.ethz.ch

K. Wegener
Institute for Machine Tools and Manufacturing (IWF), ETH Zurich, Tannenstrasse 3, 8092 Zurich, Switzerland
e-mail: wegener@iwf.mavt.ethz.ch

K. von Leipzig et al. (eds.), *Smart, Sustainable Manufacturing in an Ever-Changing World*, Lecture Notes in Production Engineering,
https://doi.org/10.1007/978-3-031-15602-1_50

1 Introduction

The structures of precision machinery (PMs), typically embodied as columns and beds, as shown in Fig. 1, have the task to guarantee the correct geometry of the machine elements under static, dynamic, and thermo-mechanical loads, as well as media exposure, hence they are fundamental components for the accuracy of the whole machine. The long-term stability of the employed materials for PM is a crucial matter in machinery design, as precision machinery is intended to be operated over an extensive period of time in the regime of some tens of thousands hours [2]. Materials commonly employed for precision machinery structures are welded steel, cast iron, mineral cast (MC), and, occasionally granite or ultra-high performance concrete (UHPC). The main physical properties of these materials are given in Table 1. Each of these materials offers a number of distinct properties, each with their own set of properties, some beneficial with others presenting drawbacks, which makes them desirable for a diverse range of applications. PM demands high stiffness and moderate strength, which leads to design for stiffness [3]. This means that on one hand, stresses are low, far below the strength of the materials, but on the other hand precision machinery requires stability in the micrometer range under loads and for a time span of decades, therefore even minor deformation, for instance due to creep can cause a problem [4]. The materials must be able to maintain their properties for the entire duration that the PM is intended to be employed for, so any type of interference must be minimized. The machine tool industry requires high control on the geometrical and dimensional stability of the machine structure in order to ensure the manufacture of precise and accurate parts in the micrometer range. This means that even the smallest deviation from the nominal values of the precision machinery can result in unacceptable part deviations [5].

Fig. 1 Machine tool structure of a 5-axis-machining center [1]

Table 1 Material properties of typically employed materials for precision machinery structures

Property [Unit]	Mineral cast	Welded steel	Cast iron	Granite	UHPC
Density [kg/m^3]	2400	7800	7200	2900	2450
Elastic modulus [GPa]	41	210	110	55	45
Tensile strength [MPa]	14	400	200	39	8
Compressive strength [MPa]	105	450	640	150	130
Thermal conductivity [W/m·K]	3	50	45	3	2
Specific heat capacity [kJ/kg·K]	0.73	0.5	0.46	0.79	0.75
Thermal expansion coefficient [μm/m K]	15	12	10	6	11
Poisson ratio	0.25	0.3	0.26	0.25	0.21

2 State of the Art

Mineral cast (MC) is a material that consists of washed and dried selected rock aggregates of selected sizes, fillers, sand, called the filler, and a resin binder as matrix. The most commonly employed resins are epoxies. Compared to methacrylate resins and unsaturated polyester resins, they show less volume shrinkage, a longer pot life, and high long-term stability. Moreover, according to [6], the binder between the mineral components also acts as a lubricant during processing. The greater the proportion of binding agent, the better the flowability of the mineral casting becomes, and the easier it is to cast. However, due to the aforementioned stiffness-driven design, a high volume ratio of the filler to the binder of 9:1 or above is targeted, since the Young's modulus of the resin is one to two orders of magnitude lower than that of the filler, and it is also the most expensive component. Since the mineral cast properties are determined by the properties of the fillers and the recipe, i.e. the proportions of fillers and resin, there are specific requirements for their density, the coefficient of thermal expansion and thermal conductivity, the tensile and compressive strength, and the modulus of elasticity (Fig. 2).

Mineral cast is a cold cast that takes place in molds that can be made of wood, plastic, or steel, depending on the desired final part tolerances, as well as the total production numbers. During the casting process, the mold is shaken to aggregate, compact and vent the mixture, so that the air inclusions can be reduced to less than 1% in volume. The cold casting process allows the direct integration of anchor bolts, tubing for hydraulics, wiring, tanks, and hollow bodies [7]. Mineral cast relies upon short lead-time, flexibility and industrially controlled and repeatable material properties. Thermally stable MC machine structures can be manufactured, as this material offers high heat capacity, combined with low heat conductivity. Furthermore, it possesses a moderate thermal expansion coefficient, comparatively to alternatively employed PM materials, as shown in Table 1. Erbe et al. [8] performed experimental tests to determine the tensile and compressive strength of mineral cast, while the Young's Modulus and Poisson coefficient were investigated by Kepczak et al. [9].

Fig. 2 Broken mineral cast specimen cross section

Schulz et al. [10] concluded that the ultimate fatigue strength of mineral cast with a filler material of quartz with a particle size of 0–8 mm and 7 wt% epoxy resin is 5 MPa. The fracture and shape change of mineral cast was experimentally studied by Dey et al. [11] with three different resins, who obtained the best results with an epoxy resin binder. With regard to creep, when mineral cast is subject to compressive loads greater than 20% of its nominal compressive strength, creep in the sub micrometer-range may occur at room temperature, as shown by [12]. The damping properties of mineral cast were studied by [13] showing that the logarithmic decrement of mineral cast is ten times higher than that of cast iron. With regard to the dimensional stability of PM materials, according to [14], when natural stone and granite are exposed to water or high air relative humidity, and successively dried, the contraction due to water soaking is greater than after exposure to high relative humidity air. However, with regard to studies of mineral cast exposure to media commonly encountered in precision machinery, like water or grinding oil, the literature shows a gap.

3 Experimental Study

For this work, comparative experiments were set up to investigate and evaluate the influence of water and grinding oil on the dimensional stability behavior of mineral cast. As shown in Fig. 4, the specimens were immersed in demineralized water or grinding oil for 900 h, then dried in a climate chamber for 800 h at 50% relative humidity air and constant temperature of 20 °C. The employed

SintoGrind TTS grinding oil is a high-performance cooling lubricant formulated from synthetically produced polyalphaolefin (PAO) based fluid of the manufacturer Oelheld. This product is free of chlorine and heavy metals, possesses no hazardous elements, and exhibits stable viscosity at different temperatures. The precision machinery structures are typically primed, then painted; therefore, the influence of the primer and paint acting as a barrier to water absorption was also investigated. The EFDEDUR UR1916M primer and the EFDEDUR GS9141H paint from the manufacturer FreiLacke both employ an isocyanate crosslinkable polyacrylate resin, which is compatible with the utilized epoxy resin of the MC specimens. The change in length for two specimens for each sample type were measured during drying, as shown in Fig. 3, in order to cross check the results between them and the shrinking results were recorded and compared.

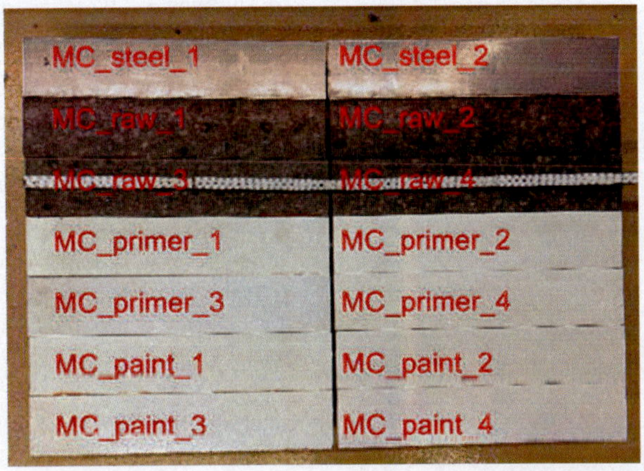

Fig. 3 Mineral cast specimens

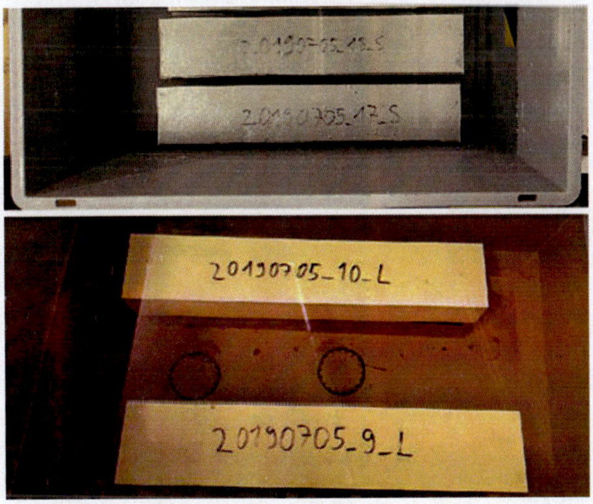

Fig. 4 Instance of a specimen water and grinding oil exposure

Table 2 List of tested specimens

Name of the specimen	Coating type	Water testing	Grinding oil testing
MC_steel_1	Stainless steel sheets	X	
MC_steel_2	Stainless steel sheets	X	
MC_raw_1	No coating	X	
MC_raw_2	No coating	X	
MC_raw_3	No coating		X
MC_raw_4	No coating		X
MC_primer_1	Paint primer	X	
MC_primer_2	Paint primer	X	
MC_primer_3	Paint primer		X
MC_primer_4	Paint primer		X
MC_paint_1	Paint	X	
MC_paint_2	Paint	X	
MC_paint_3	Paint		X
MC_paint_4	Paint		X

3.1 Specimen Description

The tests were performed on specimens measuring 400 mm in length, with a cross section of 80 × 80 mm. The guided minimum wall thickness was fivefold the size of the largest aggregate used in the mineral cast mixture, which is 16 mm, as shown in Fig. 2. This size relation of the cross section was chosen based on the mineral cast manufacturers' recommendation for the minimum wall thickness of mineral cast for precision machinery structures that still fully guarantees the physical properties of the specification sheet. Despite its price disadvantage, epoxy resin was employed as binder for the mineral cast mixture, as literature [12] shows that it offers the best physical properties, so inferior technical performance would not be acceptable in the case of precision machinery structures.

Besides the raw mineral cast specimens, three other specimens with different coatings were tested, as listed in Table 2, and shown in Fig. 3. Two stainless steel shielded specimens were used as reference, as stainless steel is inert and impenetrable to water and grinding oil.

3.2 Test Rig

The goal of the experiment is the change of dimension of the different specimens, when exposed to the fluids. From the experimentation point of view, it is easier

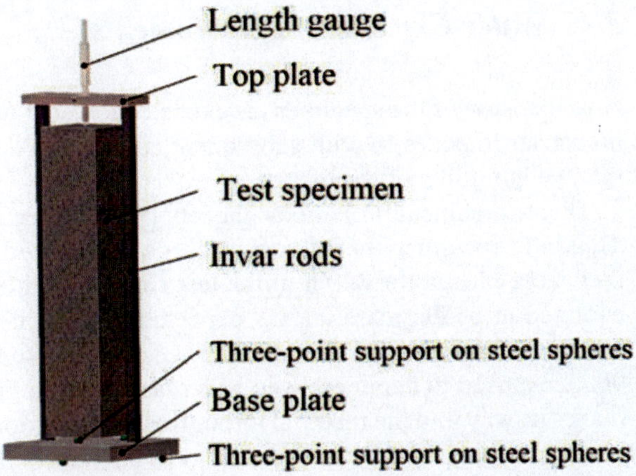

Fig. 5 Invar test stand for length variation measurements

to measure the shrinkage upon re-drying than the expansion due to fluid soaking. Therefore, after 900 h exposure to demineralized water or grinding oil, the shrinking measurements were performed in a specifically designed test bench, shown in Fig. 5, over the course of 800 h drying.

The specimens were positioned statically determined on three spheres, and their length variation was recorded by a length gauge positioned on Invar rods in order to exclude any potential thermal influences.

3.3 *Sensors*

The employed linear sensors to record the length variation of the specimens over time were the ST 1280 from Heidenhain Germany, shown in Fig. 6a. These sensors have a range uncertainty of <0.3 μm, and a repeatability of 0.25 μm. Due to the Invar construction the sensors can guarantee their small measuring uncertainty over a relatively wide temperature range of 10–40 °C. Finally, the temperature during the test was recorded with an analog temperature sensor with an uncertainty of ± 0.1 °C, shown in Fig. 6b.

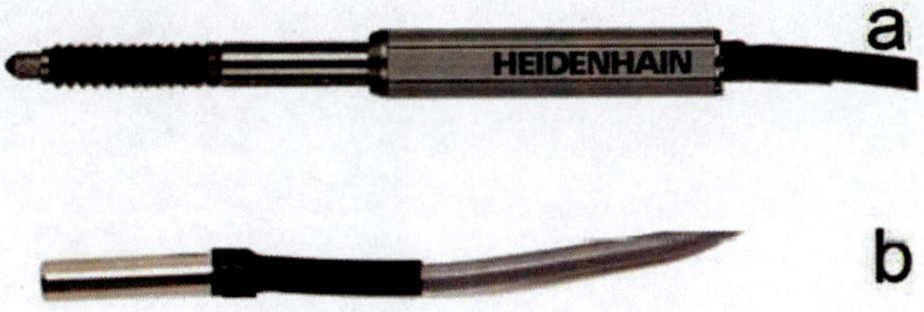

Fig. 6 Displacement (**a**) and temperature (**b**) sensors

3.4 Water Exposure Measurements

An expansion of the specimens is expected to occur when they are soaked in water if no barrier impedes it, with subsequent contraction during drying at 20 °C and 50% relative humidity air exposure.

Due to technical limitations, the length expansion could not be recorded in water. Thus, the specimens were soaked in demineralized water for 900 h, after which they were placed for 800 h in the test rig visible in Fig. 7, in order to quantify the contraction of the material. As expected, the stainless steel shielded mineral cast specimens utilized as references showed almost zero length variation after spending 900 h exposed to demineralized water, as shown in Fig. 8. According to [15], water makes its way into the material through capillary action, and, if the pore walls impede water from expanding, pressures up to 70 bars can build up and exert against the pore walls.

The raw mineral cast specimens showed a considerable length variation of 7 μm during the drying, as shown in Fig. 9.

The primed MC specimens showed a quarter of the length variation compared to the raw MC specimens, while it resulted that the paint on top of the primer made

Fig. 7 Drying of the water exposed specimens in the climate chamber

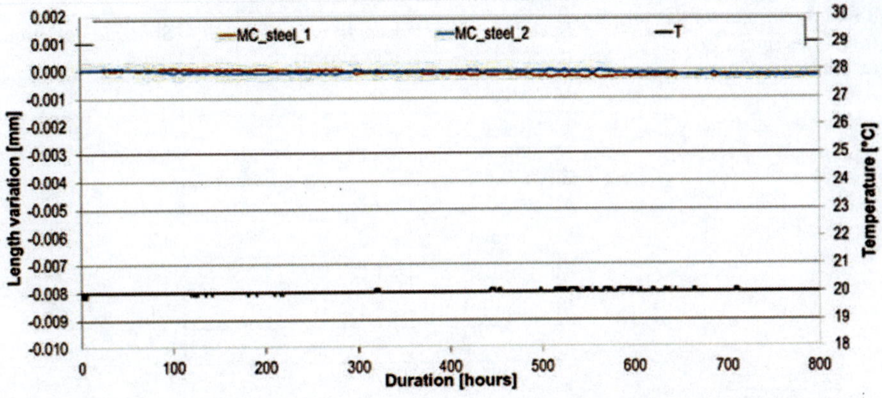

Fig. 8 Linear shrinkage during drying of stainless steel clad MC specimens exposed to water

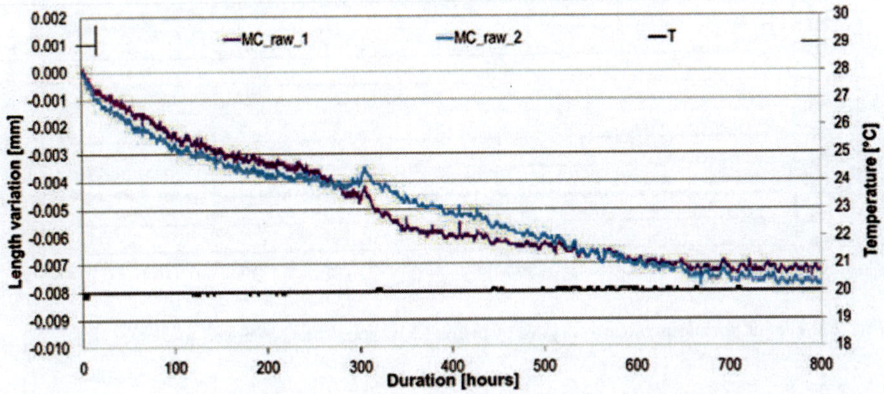

Fig. 9 Linear shrinkage during drying of raw MC specimens exposed to water

matters worse by trapping the water inside the MC, as listed in Table 3 (Figs. 10 and 11).

3.5 *Grinding Oil Exposure Measurements*

Just like for the water tests, the same procedure was followed also for the grinding oil exposed specimens (Fig. 12).

The raw specimens, when exposed to grinding oil, showed a length variation of 2 μm, which is comparable to that of the primed specimens, as shown in Figs. 13 and 14.

Table 3 Results for the tested specimens in demineralized water

Specimen name	Length variation after 800 h [μm]	Standard deviation [μm]
MC_steel_1	−0.2	0
MC_steel_2	−0.2	
MC_raw_1	−7.3	0.25
MC_raw_2	−7.8	
MC_primer_1	−1.3	0.2
MC_primer_2	−1.7	
MC_paint_1	−5.1	0.1
MC_paint_2	−4.9	

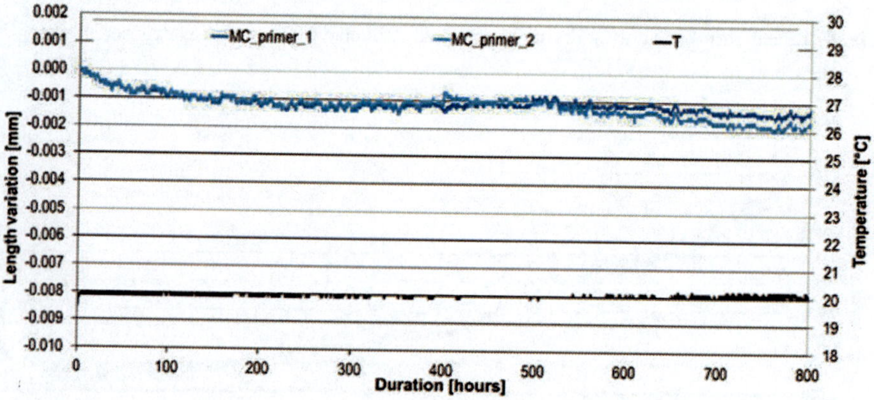

Fig. 10 Linear shrinkage during drying of primed MC specimens exposed to water

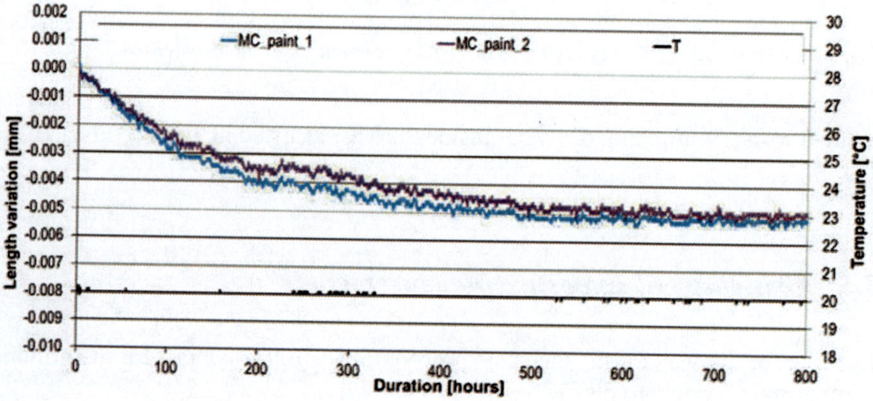

Fig. 11 Linear shrinkage during drying of painted MC specimens exposed to water

Fig. 12 Drying of the grinding oil exposed specimens in the climate chamber

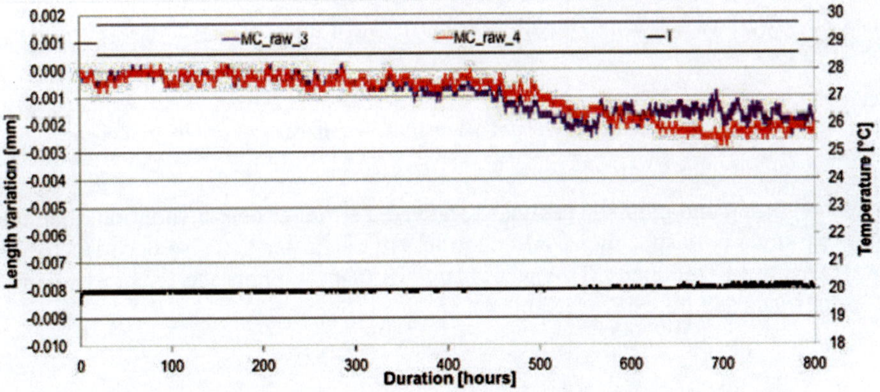

Fig. 13 Linear shrinkage during drying of raw MC specimens exposed to grinding oil

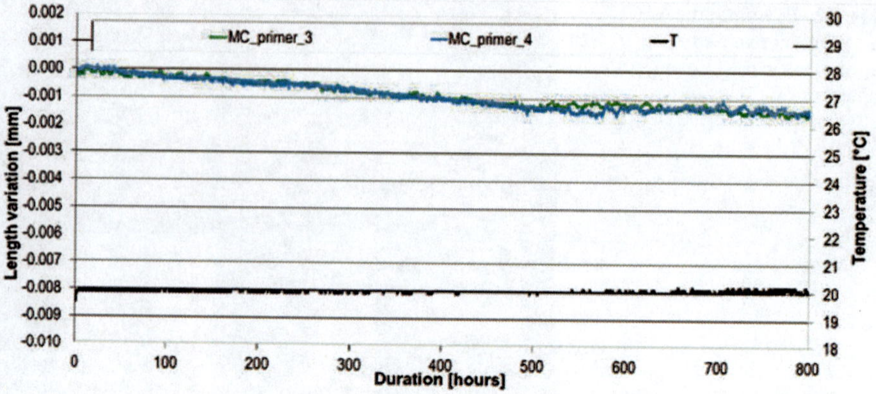

Fig. 14 Linear shrinkage during drying of primed MC specimens exposed to grinding oil

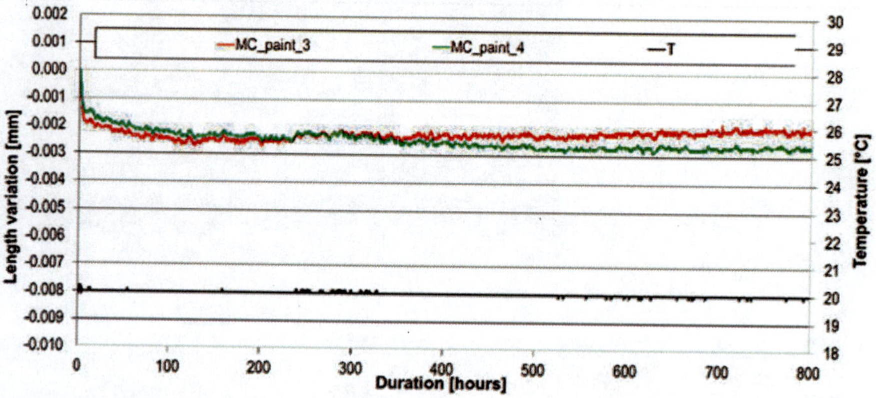

Fig. 15 Linear shrinkage during drying of painted MC specimens exposed to grinding oil

The primed and painted specimens showed a stronger length variation, than that of the primed only specimens. Although less marked, this had also occurred for the water exposed specimens. This would suggest that the paint acts as an exit barrier for the water and oil (Fig. 15, Table 4).

4 Conclusion and Outlook

This paper reports on the dimensional changes of raw mineral cast, and three alternative coatings, when exposed to common MT media conditions, like water or grinding oil. This allowed carrying out a direct comparison between these materials, as the experiment was performed concurrently under the exact same conditions. The

Table 4 Results for the tested specimens in grinding oil

Specimen name	Length variation after 800 h [μm]	Standard deviation [μm]
MC_raw_3	−1.8	0.2
MC_raw_4	−2.2	
MC_primer_3	−1.6	0.1
MC_primer_4	−1.4	
MC_paint_3	−2.0	0.3
MC_paint_4	−2.6	

employed test bench was outlined, as well as each of the examined specimens. The water and grinding oil soaking tests carried out in Sects. 3.4 and 3.5 show that their influence must be considered in case of high-precision applications, as their effects could be directly comparable to minor thermal variations. Raw mineral cast showed large length variations due to the missing entry barrier to water or grinding oil. The difference between length variations caused by water and grinding oil in raw MC specimens, however, is three times larger, while in the case of painted specimens this was reduced to two fold, suggesting from both cases that water has strong capillary effects. For the primer-coated specimens this difference was negligible, while the paint coating had a worsening impact in both cases. Although the performed tests and analysis were only done at macroscopic level, this allows for suggesting appropriate countermeasures to reduce this effect in precision machinery. The measurement of length variation during the exposure to water and grinding oil in an apt measurement setup would allow quantifying the direct impact of media exposure in the first place, and compare consequently to the drying in the climate chamber. This would give an insight on the reversibility of the process.

References

1. https://www.hurco.eu/products/5-axis-machining-centers/swivel-head-and-rotary-table/vmx-60-srti/
2. Neugebauer, R.: Werkzeugmaschinen – Aufbau, Funktion und Anwendung von spanenden und abtragenden Werkzeugmaschinen. Springer Vieweg (2012)
3. López de Lacalle, L.N., Lamikiz, A.: Machine Tools for High Performance Machining. Springer-Verlag (2009)
4. Relea, E., Pfyffer, B., Weiss, L., Wegener, K.: Experimental comparative investigation on creep behavior of mineral cast, ultra-high-performance concrete, and natural stone for precision machinery structures. IJATM (2021)
5. Weck, M., Brecher, C.: Werkzeugmaschinen 2 – Konstruktion und Berechnung. VDI (2017)
6. Chandra, S., Ohama, Y.: Polymers in Concrete. CRC Press (1994)
7. Möhring, H.C., Brecher, C., Abele, E., Fleischer, J., Bleicher, F.: Materials in machine tool structures. CIRP Ann. Manuf. Technol. **64**, 725–748 (2015)
8. Ergun, A., Gerger, M.N.: Usability of polymer concrete as a machine-making material regarding fatigue strength. World J. Eng. Technol. **1**, 59–64 (2013)

9. Kepczak, N., Pawlowski, W.: Mechanical properties of the mineral cast material at the macro and micro level. Mech. Mech. Eng. **20**(3), 249–254 (2016)
10. Schulz, H.: Reaktionsharzbeton im Werkzeugmaschinenbau. Industrie Anzeiger **14**(21), 41–42 (1986)
11. Dey, H.J.: Das Verformungs und Bruchverhalten von Reaktionsharzbeton und die Auswirkungen auf Maschi- nenbauteile. Hanser (1991)
12. Erbe, T., Krol, J., Theska, R.: Mineral casting as material for machine base frames of precision machines. In: Twenty Third Annual Meeting of the American Society for Precision Engineering and Twelfth ICPE (2008)
13. Kepczak, N., Pawlowski, W., Kaczmarek, L.: Cast iron and mineral cast applied for machine tool bed - dynamic behavior analysis, Arch. Metall. Mater. **60**(2),1023–1029 (2015). https://doi.org/10.1515/amm-2015-0254
14. Relea, E., Weiss, L., Wegener, K.: Experimental Study on the Geometrical and Dimensional Stability of Natural Stone Sorts for Precision Machinery. Elsevier LCE (2019)
15. Winkle, E.M.: Stone: Properties, Durability in Man's Environment, Springer-Verlag (1973)

Eduard Relea obtained his M.Sc. Eng. degree in Mechanical Engineering from the Polytechnic University of Turin. His specific focus and area of research interest at inspire AG are composite materials and machine tool structures.

Lukas Weiss holds M.Sc. Eng. degree in Electrical Engineering from the ETH Zurich. He is currently Head of the Machine Group at inspire AG, and his research interest is in the Technology and Innovation Management.

Konrad Wegener holds M.Sc. Eng. and Ph.D. degrees in Mechanical Engineering from the TU Braunschweig. In 2003 he was appointed Professor and Head of the Institute for Machine Tools and Manufacturing (IWF) at the ETH Zurich, Switzerland.

Effect of Mercerization on Coconut Fiber Surface Condition for Use in Natural Fiber-Reinforced Polymer Composites

S. P. Simelane and D. M. Madyira

Abstract The usage of natural fibers requires that they should be pretreated in preparation for their application as reinforcement in Natural Fiber-Reinforced polymer composites. The treatment modifies the fiber surface to improve bonding between the fibers and the polymer matrix. This paper reports on the effects of sodium hydroxide (NaOH) treatment on the surface of coconut fibers. The fibers were subjected to 5, 10, 15 and 20% NaOH concentrations and soaked for 4 hours. They were then thoroughly rinsed and allowed to dry in open air for 7 days after which they were dried in an oven for 30 min. Untreated and treated coconut fibers were observed under the Scanning Electron Microscope. It was noted that each of the NaOH concentrations modified the surface morphology of the fibers differently. The resultant colour of the treated fibers was seen to get darker as the solution concentration increased. The increase in alkali concentration striped the surface of more constituents, thus exposing not only the microfibrils, but also "pits" and other surface components, rendering the fiber surface coarse. This would promote improved bonding within a composite material.

Keywords Coconut fiber · Surface treatment · Sodium hydroxide

1 Introduction

The recent pronouncements on global warming and climate change and the need to pursue sustainability in human development can no longer be ignored. The use of environmentally-friendly materials, products and processes will not only help in reducing the negative impact of non-degradable materials and pollution generating

S. P. Simelane (✉)
Department of Mechanical and Industrial Engineering Technology, University of Johannesburg, Johannesburg, South Africa
e-mail: psimelane@uj.ac.za

D. M. Madyira
Department of Mechanical Engineering Science, University of Johannesburg, Johannesburg, South Africa
e-mail: dmadyira@uj.ac.za

K. von Leipzig et al. (eds.), *Smart, Sustainable Manufacturing in an Ever-Changing World*, Lecture Notes in Production Engineering,
https://doi.org/10.1007/978-3-031-15602-1_51

activities associated with petroleum-based products. It may well contribute to job creation and a better use of the natural resources [1]. Conventional materials such as metals were replaced by polymers, not because the polymers were cleaner or 'greener', but because they could be easily processed. Productivity of polymer products was also high with reduced costs [2]. To enhance the performance of these polymers, fibers were used as reinforcements, thus meeting the demand for high strength-to-weight ratio. Other desirable properties were that the material could be lightweight, low density, and of high specific stiffness, high tensile strength, as well as high modulus [2–4]. However, the continued use of synthetic materials such as carbon, glass and other reinforcements for polymers, with their inherent problems of waste disposal, led to significant interest in natural biodegradable materials. This interest was enhanced by environmental consciousness and the need to recycle and reuse. The result was the incorporation of natural fibers as reinforcement in composite materials.

Current and potential applications of coir and other natural fibers were explored and reported on by various researchers. Hasan et al. [5] presented a comprehensive review of the latest developments in coir fiber biocomposites. They also conducted a SWOT (Strength, Weaknesses, Opportunities and Threats) analysis of coir fibers and their composites. Among other conclusions, they contend that the challenging problem of voids in the fabrication of NFRC could be reduced by the pretreatment of natural fibers, and this study was also motivated by this contention. Akampumuza et al. [6] gave an account of how the automobile industry has become part of the biomaterial industry [6]. Apart from the natural fiber-based products that are depicted in Fig. 1, sporting goods, furniture and aerospace items such as wings and tails can also benefit from coconut fibers [5]. This research intends to be part of these developments.

Fig. 1 Application of NFRC on Mercedes Benz E-class components, ranging from floor mats to door panels [2]

Much has been reported on the subject of natural fiber reinforced composites (NFRC) [2–4], and the fibers are fairly well classified in terms of origin, morphological structure, chemical composition as well as physical and mechanical properties [7–9]. The mechanical behaviour of fiber-reinforced composites and the interaction between the fibers and the matrix into which they are embedded, have also been extensively studied [10]. However, there is still a need to improve the performance of natural fibers and the adhesion between the hydrophobic (water rejecting) polymer and the hydrophilic (water absorbing) fibres. Tajvidi et al. [11] conceded that 'the development of a definitive theory for the mechanism of bonding by coupling agents in composites is a complex problem'. They also point out that it is necessary to consider the 'morphology of the interphase, the acid-based reaction at the interface, surface energy, and the wetting phenomenon'.

Among the natural fibers that have shown significant potential and continue to do so, are coconut fibers or coir fibers. They are readily available and have unique properties and unusually high lignin content, up to about 46% [7, 8]. They also have high durability, and can withstand environmental degradation such as rotting, moth and fungal attack [7]. For these reasons, coconut by-products are amongst the preferred natural resources for various conventional and new applications, including as reinforcement in polymer composites [5].

Coconut fibers, which are extracted from coconut husks (Fig. 2) by retting and/or mechanical extraction such as decortication, are said to contain mainly lignin, which is a complex organic polymer that is deposited in the fibre walls. This makes them rigid and woody. They also contain hemicellulose, which is any of a group of complex carbohydrates that surround the cellulosic fibres of plant cells. Other examples of carbohydrates that also surround the cellulose are pectins. So coconut fibers are said to be made of cellulose. Cellulose consists of very long unbranched fibrils composed exclusively of glucose, held together by hydrogen bonding [12]. These and other substances are the target of this study on natural fiber treatment.

Figure 2 depicts a coconut (a) from which the fibers are extracted from the outer husk and (b) the extracted fibres. The outer husk is found between the hard internal shell and the outer skin.

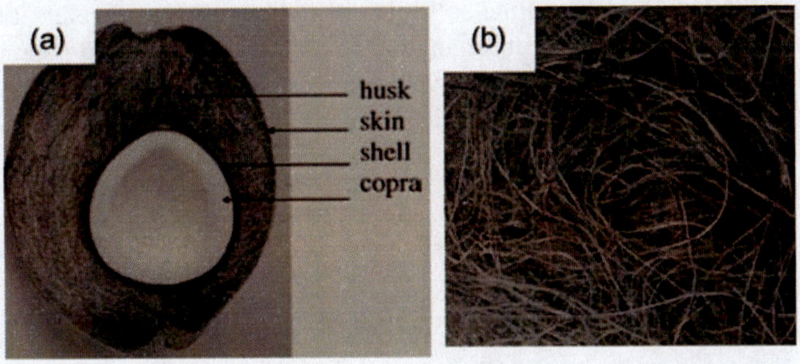

Fig. 2 Images of the parts of a coconut (**a**) and extracted coir fibers (**b**) [12]

Table 1 Chemical composition of coir fibers [13]

Item	Percentage
Water soluble	5.25
Pectin and related compounds	3.00
Hemicellulose	0.25
Lignin	45.84
Cellulose	43.44
Ash	2.22

Table 1 lists the chemical components of a typical coconut fiber, and it is noted that these values vary according to the conditions under which the coconuts were grown and how the tests were conducted [13].

Like most if not all natural fibers, coconut fibers have inherent drawbacks such as low or poor resistance to moisture. There is also poor interfacial interaction between the fibers and the matrix. To overcome these challenges, various methods of modification have been practiced and are still the subject of most research in NFRC [12, 14]. Ahmad et al.[9] gave a comprehensible account of the two main methods of modifying the surface of the plant fibres for better interaction with a matrix substance. These are physical treatments which improve the mechanical properties such as strength, modulus and elongation of the fibers, and chemical treatments which are meant to result in improved contact between the fiber and the matrix. A third method, the physicochemical treatment, is a combination of physical and chemical treatment. It produces clean and fine natural fibres or fibrils with high cellulose content [9].

Alkaline treatment or mercerization is one of the popular chemical treatment methods and it was the subject of this study. It was chosen because of the ease with which it can be applied and more understanding of the method has been gained over time. This treatment uses NaOH to remove lignin, wax and other impurities, and also shatters the hydrogen bonds, thus changing the topology of the fiber surface and promoting roughness [10].

NaOH, with which the coir fibers were treated, is a waxy, opaque and corrosive inorganic compound that contains sodium cations (Na+) and hydroxide anions (OH-). It is a colourless crystalline solid that is highly soluble in water. It is used on a day-to-day basis to manufacture many products such as wood and paper products, aluminium, commercial drain and oven cleaners, as well as soap and detergents. Nam et al. [15] describe the reaction of sodium hydroxide with coir as:

$$\text{Coir} - \text{OH} + \text{NaOH} \rightarrow \text{Coir} - \text{O}^{-} - \text{Na}^{+} + \text{H}_2\text{O}$$

This reaction means that sodium hydroxide promotes the formation of ions or the ionization of the hydroxyl group to the alkoxide, thus disrupting the hydrogen bonds in the network structure of the fiber and the surface becomes rougher.

To understand the interaction between the natural fibers and the polymer matrix, it is important first of all to understand the morphology of untreated fibers and later that of treated fibers. The aim is to influence the performance of natural fibers and thus the interface between the fibers and the polymer. The objective of this study was to investigate the surface structure of coir fiber before and after NaOH treatment and the aim is to contribute to the understanding of the effects of surface treatment on natural fibers.

2 Materials and Method

Coconut fibers were obtained from the Coir Institute of South Africa, in Sandton, Johannesburg. In this study, coir fibers were subjected to varying concentration levels of NaOH while maintaining the same duration of exposure, so as to obtain the optimum treatment regime in the coir fibers. The aim was to observe the morphological and/or structural changes that alkaline-based immersion induce on the coir fiber surface. Only visual observations and SEM analysis were made in terms of colour changes and surface topology. A comparison was made with published literature and untreated fibers.

The coconut fibres were washed thoroughly with deionised water to remove any dust and loose particles from the surface and dried in open air for 7 days. A small sample was then placed on the SEM stub, and a graphite rod was sharpened into a smaller diameter to allow it to obtain high current density with high enough temperature to achieve evaporation. The sample was then placed inside a Q150Tϵ Quorum sputter coater and coated with graphite for 10 min and the result was a sputter film in the range of 2–20 nm. This was done to provide an electrically conductive thin film which could also represent the surface topology of the coconut fibre.

The untreated coir sample was then placed inside the Vega 3 Tescan XMU Scanning Electron Microscope at about 10 kV. The working distances (WD) ranged from 13.52 mm at 55 × magnification and 1 μm scale, to 13.03 mm at 2000 × magnification and 20 μm scale respectively. However, only three of the images were discussed in this report together with the treated fiber images.

Table 2 outlines the masses of both the fibers and the NaOH (in brackets) as measured using the OHAUS Pioneer™ Balance. The values in brackets are precisely what was used in 500 mL of deionised water because the removal or addition of a single pellet could not give the exact values of 25 g, or 50 g, and so on. The fibers were immersed in 500 mL beakers for 4 h in the 5%, 10%, 15%, 20% (w/v) NaOH respectively and washed several times with water. The final wash was with deionised water to remove any NaOH sticking to the fibre surface. The fibers were then dried in open air for 7 days.

The treated fibers were taken to another laboratory where a Quorum Q300T ES sputter coater was used. The whole process from creating a vacuum in the chamber

Table 2 Sodium hydroxide solution preparation

Sodium hydroxide (NaOH)	Mass of NaOH
Untreated	
5%	25 g (25.094 g)
10%	50 g (50.089 g)
15%	75 g (75.001 g)
20%	100 g (100.026)

until coating took about 10 min, and the thickness of the film was about 200 nm. Again a TESCAN VEGA 3 Scanning Electron Microscope was used.

The geometric features that were studied on the fiber surfaces, including the effects of treatment on the fiber diameter, were, for reasonable accuracy, measured using the open source ImageJ software. For the untreated and treated fibers, the known distance of 100 μm as found on the SEM image sheet was used as the basis for the conversion of pixels into metric measurements.

3 Results and Discussion

3.1 Surface Topology of Untreated Fibers

Figure 3 shows images of the Scanning Electron Microscope at 503 × magnification, 1000 × magnification and 2000 × magnification of the untreated fibers, respectively. The measured mean diameter using ImageJ was found to be 216.314 μm. The unit cells that make up the coir surface are seen to be running almost in a parallel orientation. They have an average diameter of 8.2 μm. There are also small protrusions in-between these longitudinal unit cells. Higher magnification of 1000 × reveals that the overall surface is fairly smooth and the spaces between the cells is filled by lignin and the fatty substances that constitute the untreated coir surface. This was also observed by Bismarck et al. [16].

Figure 3c at even higher magnification of 2000 × shows globular protrusions or patches and cuticles [17]. Mulinari et al. [18] attribute these patches to the high amounts of debris that are stuck to the surface of the fibers and result from non-cellulosic material coating [19]. Carvalho et al. [20] also observed a "layer of oils, waxes and extractives", part of what constitutes the lignocellulosic fiber. Surface treatment is meant to deal with these layers by removing wax, pectin, lignin, hemicellulose as well as other impurities [9, 18, 21].

What follows is a discussion on the observations made. The high concentration of NaOH resulting in the removal of more lignin and non-cellulosic material does

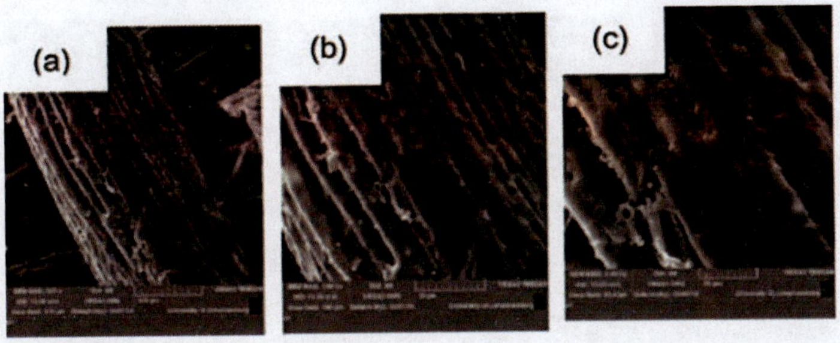

Fig. 3 SEM micrograph of untreated coir fiber: **a** 503 × magnification; **b** 1000 × magnification and **c** 2000 × magnification

not mean a lighter fiber upon rinsing. The colour of solution that is left in the beaker is directly proportional to the fiber that is taken out and washed.

3.2 *Visual Appearance Observation*

The changes in colour experienced by the coir fiber can easily be visualised as the NaOH concentration levels increase. The fiber that is treated with 5% NaOH is not too different from the untreated fiber, but the texture has changed such that the fibers start assuming a somewhat curly and springy posture. The treatment solution has nonetheless started to change colour. It was also reported by Asasutjarit et al. [22] that NaOH treatment of between 2 and 6% does not have a significant effect on the colour changes in fibers. As the percentage is increased to 10%, 15% and finally to 20%, the colour darkens, becoming reddish brown, then brown colour dominates until it is dark brown/deep red and shiny as shown in Fig. 4. This is attributed to the removal of lignin and other non-cellulosic material. While this is true for coir fiber, other researchers have reported the opposite effect as the NaOH concentration is increased beyond 5% for other natural fibers. In this instance the fibers become lighter and rougher [22].

3.3 *Surface Topology SEM*

The inspection of the treated coir fiber under the SEM reveals the structure and morphology of the fibers. The extent to which the lignin and the hemicellulose have been removed by the various concentrations of NaOH is also revealed.

Figure 4 shows the effect of the different NaOH concentrations on the surface of the fibers at around 500 × and 1000 × magnification. At 5% treatment the surface still shows the remnants of the impurities and debris. It is fairly smooth compared to higher alkali treatments, but the pits are starting to show. At 10% the cellulose

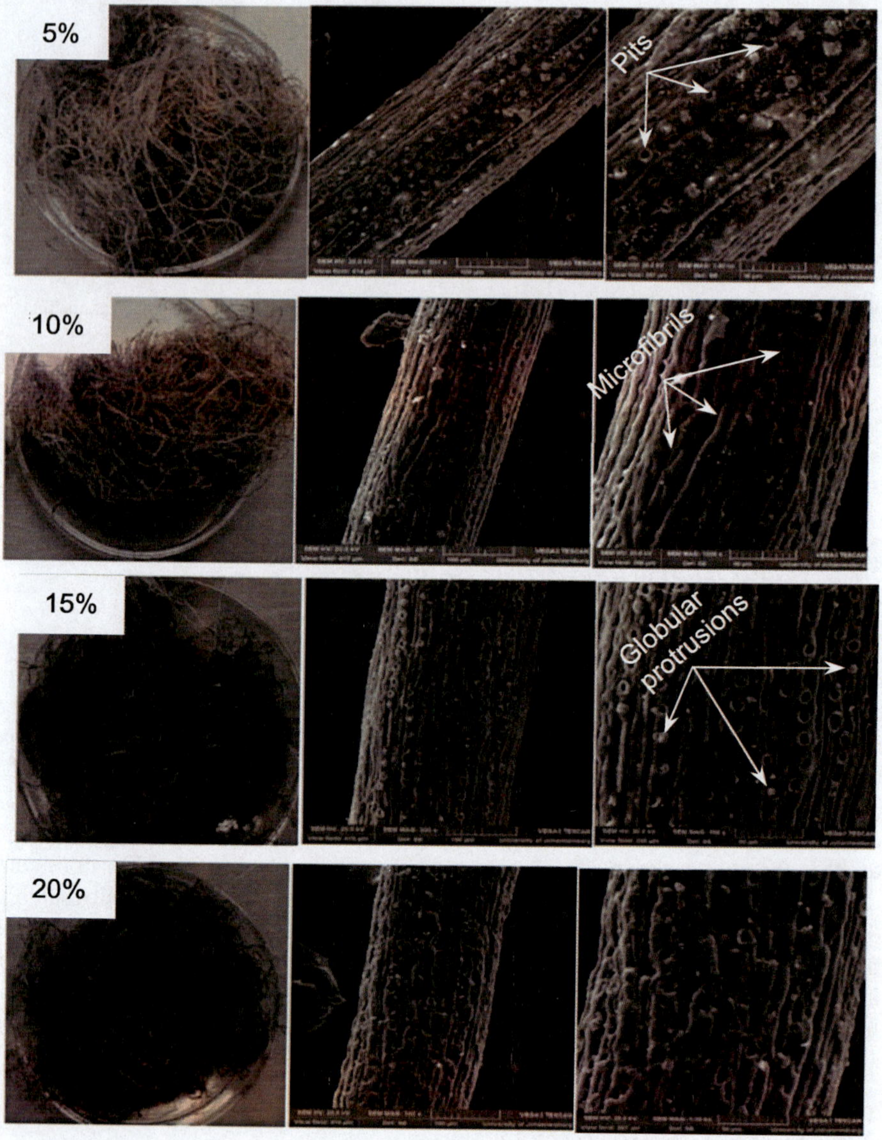

Fig. 4 Surface morphology of coconut fibers at different NaOH concentrations

microfibrils or Parenchyma cells, which inherently constitute lignocellulose coconut fibers, together with globular protrusions [20], are already well-defined. The grooves between them are much more pronounced than in 5% treatment. However, some of the unit cells are still held together by residual lignin. This indicates that further treatment may be needed to allow for more separation of the microfibrils, with the possibility of better matrix penetration and better interaction between the fibers and the polymer.

As the NaOH concentration increases to 15%, not only are the microfibrils well and truly pronounced, but globular protrusions and "pits" are much more evident. The "pits" diameter average about 9.8 μm, slightly beyond the diameter of the microfibrils. These "pits" or circular holes are observed to run longitudinally along the parenchyma cell. They are exposed as a result of the removal of wax and other extractives. The parenchyma cells are used to carry water and nutrients to the different points in the plant such as leaves and roots [20]. Rout et al. [23] contend that at 15% the structure of the lignin is damaged and the carbon chains of the cellulose disconnect. We argue that the exposure of "pits" and globular protrusions at this concentration means an increase in effective surface area and roughness, and better matrix penetration into the fiber. The concequence of this is better bonding between the fiber and the polymer matrix. The advantage or otherwise of this observation lies in the eventual application of the treated fiber as a reinforcement. This is a subject for further studies.

The alkali that is absorbed by the cellulose in the fiber and the washing and drying of the fibers does not change the chemical properties but rather the physical properties of the cellulose. However the waxy cuticle layer which is formed by the fatty acids and their condensation products gets dissolved and leached out by the alkali.

Further increase in NaOH concentration to 20% and subsequent removal of the lignin and hemicellulose reaches a point where the integrity of the fiber may be compromised. No regular pattern on the surface morphology can be seen. The wall structure of the fiber seems to disintegrate. However, the increased roughness may mean an increase in the effective surface area and improved interaction between the fiber and the polymer matrix for which the surface is being prepared.

For purposes of visual comparison, Fig. 5 shows the progression of the effect of alkaline treatment from 5 to 20%.

Changes in diameter as a result of alkaline treatment, taken at various points along the length of the fibers and averaged, are presented in Table 3. It is pointed out that, because nature never provides perfectly geometrically similar natural fibers, every effort was made to take measurements of nearly similar fibers along corresponding points. For each of the treatment regimes, and the untreated fibers, seven samples were used. These samples correspond to the number of mechanical tests that were conducted in a separate study to determine the effect of mercerization on the mechanical properties of the coir fibers.

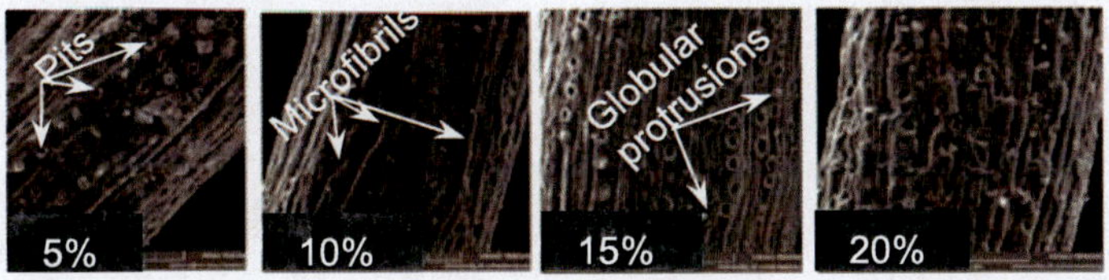

Fig. 5 A side by side comparison of the effects of alkalinisation on the coconut fibre surface

Table 3 Changes in fiber diameter as a result of mercerization

Untreated	216.314 μm
5% NaOH	209.6 μm
10% NaOH	184.7 μm
15% NaOH	230.8 μm
20% NaOH	193.5 μm

The observation in Table 3 is that the coir fibers as received from nature are never homogeneous, even from the adjacent fibers of the same husk. Also the mercerization process does not remove the extractants in a predetermined and regular pattern like in mechanical and other industrial processes such as machining.

An increase in NaOH concentration from 10 to 15% may be expected to result in the reduction of the diameter. A diameter of 184.7 μm at 10% and that of 193.5 μm at 20% respectively seem odd. This may be the result of the difficulty in manually picking geometrically identical fibers. It is argued, and indeed observed under SEM, that even a single fiber is not perfectly round along its length. Not all the geometric features that constitute the fiber surface occur at regular intervals and in a regular pattern. Whilst many researchers reported on selecting natural fibers of the same diameter, it was observed that a fairly wide range was put on these sizes, especially in procedures that require manual hand-laying. Perhaps this is meant to accommodate the irregular topology of these natural fibers. A process and a method of modifying and selecting fibers of identical topology may be attempted. One of various physical treatment methods referred to in the introduction may be used, but care should be taken that they do not offset the cost and time reduction endeavours of using natural fibers. Perhaps it should be at the composite fabrication stage that caution is exercised in controlling the geometries of the material to suit the intended application, as long as the natural fibers are properly and optimally treated.

4 Conclusion

It is concluded that:

- The observation of treated and untreaded fibers under SEM does help in identifying the topology of the coir fibers that is affected by NaOH during mercerization.
- The treatment of coconut fiber with NaOH was able to remove waxes and debris from the surface of the fiber, thus exposing microfibrils, "pits" and globular protrusions.
- The alkali-treated coconut fibers displayed springiness and a difficulty to be kept straight after treatment, as they tend to maintain a twist. This should be borne in mind during the fabrication of the composites because there may be issues with the desired alignment of the fibers.

- The resultant colour of the fiber as well as the solution darkened further with increased alkaline concentration, which may be an indication of the opposite of the bleaching effect in which the removal of a substance from a surface is supposed to 'clean' that surface.
- The increases in sodium hydroxide concentration beyond 15% may result in disintegration of the fiber surface. Although the surface roughness is increased, this may be detrimental to the strength and integrity of the fiber.
- Based on the results of the investigation of the effects of alkali treatment on the fiber surface from this and other studies from published literature, the optimum concentration of NaOH can be determined at the time when the treated fibers are used as reinforcement. This can also include the determination of the best possible treatment duration, the drying conditions and so on.
- The removal of the extractants from the coir fiber surface during mercerization does not occur with predetermined regularity, even along the length of a single fiber.

5 Recommendations

It is recommended that:

- The amount of NaOH concentration be carefully controlled to preserve the structural integrity of the coconut fiber.
- The constituents of the coconut surface fiber be thoroughly examined to determine which of the ingredients is dominant in improving the surface roughness during treatment and the mechanics thereof, so that a targeted approach can be adopted.

References

1. Ravindranath, D.A., S. Radhakrishnan, S.: Coir pith: wealth from waste. India International Coir Fair 2016, Coimbatore (2016)
2. Saheb, D.N., Jog, J.P.: Natural fiber polymer composites: a review. Adv. Polym. Technol. **18**(4), 351–363 (1999)
3. Zhang, M. Q., Rong, M. Z., Lu, X.: All-plant fiber composites. In: Composites Technologies for 2020: Proceedings of the Fourth Asian-Australasian Conference on Composite Materials (ACCM-4), University of Sydney, Australia, 6–9 July 2004, p. 3. Woodhead Publishing (2004)
4. Taj, S., Munawar, M.A., Khan, S.: Natural fiber-reinforced poplymer composites. Proc. Pakistan Acad. Sci. **44**(2), 129–144 (2007)
5. Hasan, K., Horvath, P., Bak, M., Alpar, T.: A state-of-the-art review on coir fiber-reinforced biocomposites. Royal Soc. Chem. J. **11**, 10548–10571 (2021)
6. Akampumuza, O., Wambua, P.M., Ahmed, A., Li, W., Qin, X.: Review of the applications of biocomposites in the automotive industry. Polym. Compos. (2017). https://doi.org/10.1002/pc.,23847,2553-2569
7. Adeniyi, A.G., Onifade, D.V., Ighalo, J.O., Adeoye, A.S.: A review of coir fiber reinforced polymer composites. Composites: Part B, **176**, 107305 (2019)

8. Yan, L., Chouw, N., Huang, L., Kasal, B.: Effect of Alkali treatment on microstructure and mechanical properties of coir fibres, coir fibre reinforced-polymer composites and reinforced-cementitious composites. Constr. Build. Mater. **112**, 168–182 (2016)
9. Ahmad, R., Hamid, R., Osman, S.A.: Physical and chemical modification of plant fibres for reinforcement in cementitious composites. Adv. Civ. Eng., Article ID 5185806 (2019)
10. John, M.J., Anandjiwala, R.D.: Recent developments in chemical modification and characterization of natural fiber-reinforced composites. Polym. Compos **29**, 187–207 (2008)
11. Tajvidi, M., Falk, R.H., Hermanson, J.C.: Time-temperature superposition principle applied to a kenaf-fiber/high-density polyethylene composite. J. Appl. Polym. Sci. **97**, 1995–2004 (2005)
12. Faruk, O., Bledzki, A.K., Fink, H., Sain, M.: Biocomposites reinforced with natural fibers: 2000–2010. Prog. Polym. Sci. **37**, 1552–1596 (2012)
13. Bledzki, A.K., Gassan, J.: Composites Reinforced with Cellulose-based Fibers. Prog. Polym. Sci. **24**, 221–274 (1999)
14. Biagiotti, J., Puglia, D., Kenny, J.M.: A review on natural fibre-based composites-part I. J. Natl. Fibers **1**(2), 37–68 (2004)
15. Nam, T.H., Ogihara, S., Tung, N.H., Kobayashi, S.: Effect of Alkali treatment on interfacial and mechanical properties of coir fiber-reinforced poly(butylene succinate) biodegradable composites. Composites: Part B **42**, 1648–1656 (2011)
16. Bismarck, A., Mohanty, A.K., Aranberri-Askargorta, I., Czapla, S., Misra, M., Hinrichsen, G., Springer, J.: Surface characterization of natural fibers surface properties and the water up-take behavior of modified sisal and coir fibers. Green Chem. **3**, 100–107 (2001)
17. Arsyad, M., Wardana, I.N.G., Pratikto, P., Irawan, Y.S.: The morphology of coconut fiber surface under chemical treatment. Revista Matéria **20**(1), 169–177 (2015)
18. Mulinari, D.R., Baptista, C.A.R.P., Souza, J.V.C., Voorwald, H.J.C.: Mechanical properties of coconut fibers reinforced polyester composites. Procedia Eng. **10**, 2074–2079 (2011)
19. Reddy, N.: Sustainable Applications of Coir and Other Coconut By-products. Springer Nature, Switzerland (2019)
20. Carvalho, K.C.C., Mulinari, D.R., Voorwald, H.J.C., Coiffi, M.O.H.: Chemical modification effect on the mechanical properties of HIPS/Coconut fiber composites. BioResources **5**, 1143–1155 (2010)
21. Karthikeyan, A., Balamurugan, K.: Effect of Alkali treatment and fiber length on impact behavior of coir fiber reinforced epoxy composites. J. Sci. Ind. Res. **71**, 627–631 (2012)
22. Asasutjarit, C., Charoenvai, S., Hirunlabh, J., Joseph Khedari, J.: Materials and mechanical properties of pretreated coir-based green composites. Composites: Part B, **40**, 633–637 (2009)
23. Rout, J., Tripathy, S.S., Nayak, S.K., Misra, M., Mohanty, A.K.: Scanning electron microscopy study of chemically modified coir fibers. J. Appl. Polym. Sci. **79**, 1169–1177 (2000)

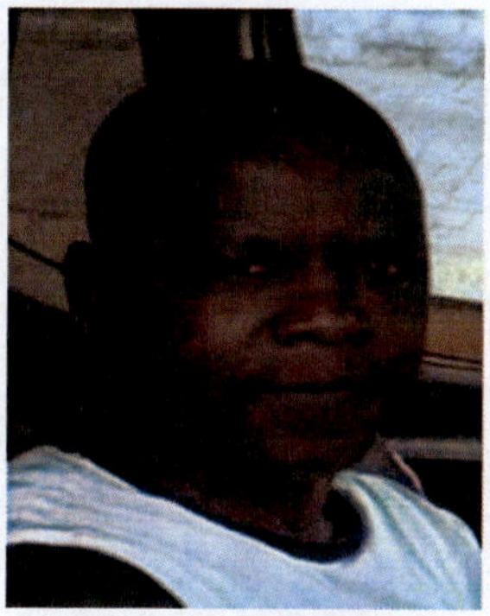

Mr. Sphiwe P. Simelane lectures Fluid Mechanics and Turbomachines, and is a Doctoral candidate in the area of Natural Fiber-Reinforced Composites in the Department of Mechanical and Industrial Engineering Technology, Faculty of Engineering and the Built Environment at the University of Johannesburg.

Prof. Daniel M. Madyira is an Associate Professor in the Department of Mechanical Engineering Science, Faculty of Engineering and the Built Environment at the University of Johannesburg. He is a registered professional mechanical engineer with Engineering Council of South Africa (ECSA). His wide research interests include biomass value chain development for renewable energy applications.

Manufacturing Technologies II

Understanding the Structural Integrity and Post-processing of L-PBF As-Built Ti-6Al-4V Parts: A Literature Review

W. M. I. Makhetha, G. M. Ter Haar, N. Sacks, and T. H. Becker

Abstract Despite the exponential increase in research and industrial application of laser powder bed fusion (L-PBF), Ti-6Al-4V parts produced still fail to meet the structural integrity of most functional requirements in the as-built conditions. The major drawbacks are high surface roughness, high residual stresses, martensitic microstructure, high porosity and anisotropic material properties. The drawbacks are a result of the inherent nature of a typical L-PBF process characterised by rapid heating and cooling and line-by-line and layer-by-layer processes. To provide more insight and confidence in the L-PBF technology, the causes of these drawbacks must be carefully investigated. This involves the integration of three factors: technology, skills and industry requirements. Currently, such understanding is not widely available. The focus is usually separated (on materials and processes or standardisation and qualification capabilities) instead of being integrated. This paper is a two-fold literature review of firstly, the L-PBF output parameters (resultant part characteristics), which are critical to the structural integrity of Ti-6Al-4V produced by L-PBF and secondly, the potential problem-solving strategies for the martensitic

W. M. I. Makhetha (✉) · N. Sacks
Department of Industrial Engineering, Stellenbosch University, Stellenbosch, South Africa
e-mail: Makhetham@cput.ac.za

N. Sacks
e-mail: natashasacks@sun.ac.za

W. M. I. Makhetha · G. M. Ter Haar · T. H. Becker
Department of Mechanical and Mechatronic Engineering, Stellenbosch University, Stellenbosch, South Africa
e-mail: gterhaar@sun.ac.za

T. H. Becker
e-mail: tbecker@sun.ac.za

W. M. I. Makhetha
Department of Industrial Design, Cape Peninsula University of Technology, Cape Town, South Africa

T. H. Becker
Centre for Materials Engeerineering, Department of Mechanical Engineering, University of Cape Town, Cape Town, South Africa

K. von Leipzig et al. (eds.), *Smart, Sustainable Manufacturing in an Ever-Changing World*, Lecture Notes in Production Engineering,
https://doi.org/10.1007/978-3-031-15602-1_52

microstructure, porosity, residual stresses and surface roughness. The methodology used involved a triangulation of the literature followed by an inclusion–exclusion criterion of the relevant literature. The outcome of this work was a contribution towards a better understanding of the technology to boost the industry confidence in additive manufacturing.

Keywords Laser powder bed fusion · Ti-6Al-4V as-built · Structural integrity

1 Introduction

Laser powder bed fusion (L-PBF) is one of many techniques within the canopy of additive manufacturing (AM) technologies [1–3]. Broadly defined, these technologies are used to join material layer-by-layer to make three-dimensional (3D) products from computer-aided design (CAD) models [4–7]. In this way, the AM techniques remove traditional manufacturing constraints, thereby providing almost unchallenged and sustainable manufacturing capabilities to improve manufacturing efficiency drastically. The interest in optimising metal additive manufacturing (MAM) technologies has grown exponentially in recent years [8–17]. A widely reported method of classifying MAM techniques puts these techniques into two categories as either powder bed fusion (PBF) or direct energy deposition (DED) [18–23]. While various materials are being explored for MAM, more research has been conducted on titanium alloys. Ti-6Al-4V is popular in medical and aerospace industries due to its high specific strength, corrosion resistance and low density (about 4.5 g/cm^3), among other properties [12, 24, 25]. These combined properties make titanium alloys an excellent choice for structural parts in applications such as airframes, aero-engines and bio-medical devices. Hence, this paper focuses on the laser PBF (L-PBF) processing of Ti-6Al-4V material.

Although parts produced with L-PBF are finding applications in aircraft components and medical implants [13, 22, 26–29], there are concerns over the integrity of the as-built L-PBF parts [13, 30–35]. Such concerns result from many L-PBF process parameters, which can be grouped as powder, laser, and machine-specific parameters. The powder parameters are often characterised in terms of particle size, shape and flowability [28, 36], which are a direct result of the technique used to produce the powder. Some of the common ones include gas atomisation (GA), rotary atomisation (RA), plasma atomisation (PA), water atomisation (WA) and plasma rotating electrode process (PREP) [37–40]. Nonetheless, metal powders used in L-PBF are assumed to be nominally spherical and have a particle size distribution designed to facilitate good packing behaviour, such that the final manufactured part has good mechanical properties and is optimally dense [28]. The main trade-off in terms of powder size distribution is the cost of powder production. Smaller powder particles may cost more as a feedstock (than a larger size range) due to the cost of producing such particles. The laser parameters are often characterised by laser velocity, laser power, exposure strategy and spot size [41]. The machine-specific parameters are often characterised by building path, layer thickness and system atmosphere [42]. Often, the as-built L-PBF parts do not consistently meet final application requirements in aerospace, for instance [31, 32].

There have been several approaches towards optimising L-PBF process parameters to improve the as-built attributes of Ti-6Al-4V parts. For instance, better Ti-6Al-4V part quality has been reported following optimisation of process parameters such as layer thickness [43], laser scanning strategies [33], built direction and part orientation [44, 45]. A major research focus has been towards determining the optimal laser-power and laser scan speed combination since these two process parameters are critical in achieving a stable melt-pool and thereby a dense part. One such combination was explored in a study by Majumdar et al. [46], who reported that the combination of high laser scan speed (>1000 mm/s) with low laser power (<120 W) led to poor powder consolidation. The authors attributed this to the insufficient supply of thermal energy per unit volume. As a result, the authors reported high porosity levels (>15%), high roughness, low hardness, and increased susceptibility to corrosion of as-built parts. On the other hand, the authors in the same article reported that the combination of low laser scan speed (<200 mm/s) and high laser power (>100 W) resulted in the supply of excessive thermal energy per unit volume, which in turn resulted in increased layer-wise laser penetration depth and keyholing. The authors concluded that the combination of intermediate ("balanced") laser power and laser scan speed is key to producing high-density L-PBF as-built Ti-6Al-4V with low amounts of evenly distributed pores, low roughness and hardness that is similar to conventionally produced Ti-6Al-4V. Other attempts have been made to minimise the development of residual stresses in L-PBF parts. For example, Kempen et al. [46] used a combination of base-plate heating and optimal process parameters to relieve residual stress to a certain degree. Ali et al. [47] also demonstrated a reduction in residual stresses development in a study that attributed this effect to the increase in the pre-heat temperature of the powder bed, which effectively meant reducing the thermal gradients. A different attempt to reduce residual stresses proposed by Masoomi et al. [48] in a simulation study showed that a reduction in the build-up of residual stresses could be achieved by employing optimised scanning strategies.

Nonetheless, the improvement of part quality by process parameter optimisation alone is not enough for parts to meet the functional requirements of industries such as aerospace. For instance, process parameters alone cannot eliminate the characteristic rough surface of L-PBF parts [49]. As such, the current work aims to summarise key attributes of as-built L-PBF Ti-6Al-4V, which determine structural integrity and solutions to common drawbacks of the process.

2 Approach

This literature review extracts existing knowledge from the literature to find key areas of consensus. The aim is to understand better the links between the process and the integrity of the parts produced.

2.1 Criterion for Selecting the Literature

The first part of the study was conducted to determine the criterion for selecting the literature. This was done through a search from multiple sources, as shown in Table 1.

Following the triangulation, the second part of the methodology was to narrow and focus the literature into specific functional attributes. The inclusion/exclusion criterion used is shown schematically in Fig. 1. The criterion follows a standard systematic literature inclusion/exclusion criterion which, according to Tranfield et al. [50], has been used over three decades to provide evidence-based practices which ensure quality and scientific reproducibility in all types of literature survey, irrespective of an academic discipline or industry practice.

The chosen categories in the exclusion/inclusion criterion in Fig. 1 were intended to identify correlations in terms of the main structural aspects of the material, such as microstructure, porosity residual stress and surface roughness. This was done to understand what is known and established from the core contributions, highlight the extent to which consensus is shared, and provide a detailed audit trail back to the core contributions to justify the links between the correlations. This required the use of different databases to include different contributions (published and unpublished

Table 1 Summary of the combination of literature search terms used in the study

Method of triangulation	Examples	Comment
Research giants	Journals, authors, research institutions and industries	The primary contributors were journal publications, and the measure to determine the selected journals as giants were based on the number of citations, impact factor and number of publications
Publication type	Journal articles, reviews, conference papers, conference review papers, books, Master's and PhD theses	Original journal articles formed a significant part of this work
Data source	Science Direct, Google Scholar and Scopus	These were used as primary data sources according to the systematic inclusion and exclusion criteria shown in Fig. 1
Search terms	Microstructure, porosity, residual stresses and surface roughness	The search terms were used in combination with the theme "L-PBF (SLM) Ti-6Al-4V" and screened by title match, which resulted in a total of 802 articles, of which 150 were selected as eligible after full-text assessments and 87 included in the current paper

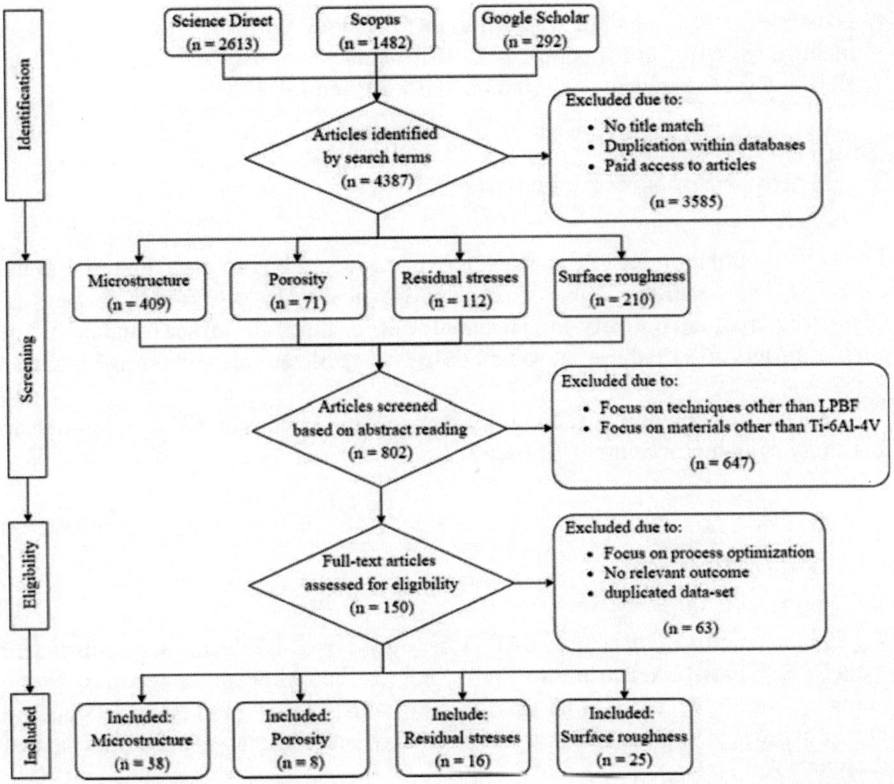

Fig. 1 The inclusion–exclusion criteria of documents used in this study, where "n" stands for the number of articles selected

original journal articles, conference proceedings, industry reports, Master's and Ph.D. theses, and academic web portals such as the Stellenbosch University e-library).

2.2 Limitations of the Methodology

Following the exclusion/inclusion criterion shown in Fig. 1, the following complications which let to overlap and repetition of information were identified:

- Limited results came up under porosity. This was due to porosity being one of many defects of the L-PBF process.
- Limited results came up under surface roughness. This was because the literature primarily reports surface roughness under surface morphology, surface chemistry or surface effects.

- Database search outcome was different for L-PBF compared to selective laser melting (SLM). This was due to L-PBF being previously referred to as SLM. Hence, a more extensive search outcome was seen for SLM.

3 Summary of Key Literature Findings

The L-PBF outputs refer to the characteristic aspects of Ti-6Al-4V material in the L-PBF as-built condition. The as-built state is typically characterised by martensitic microstructure, high porosity, high residual stresses and high surface roughness. The relative proportion of these attributes based on publication numbers is shown in Fig. 2.

The data are shown in Fig. 2a only highlights the major contributing journals to this study from a minimum of 5 articles.

3.1 Microstructure

The as-built microstructure of Ti-6Al-4V produced by L-PBF has been investigated widely [32, 51–54]. According to Fig. 2, microstructure accounts for most of the published literature on Ti-6Al-4V produced by L-PBF. This is because understanding microstructure is fundamental in order to determine the associated mechanical behaviour.

The consensus of more than 90% of the selected studies shows that the microstructure of L-PBF as-built Ti-6Al-4V is predominantly martensitic, with columnar prior-β grains, as shown in Fig. 3. The development of this microstructure is commonly attributed to a diffusionless transformation due to the fast cooling of the melt pool in a typical L-PBF process [32, 55, 56]. The α' laths of the L-PBF as-built Ti-6Al-4V microstructure are generally acicular with thicknesses typically ranging between 300 and 500 nm and are usually within wide and long columnar prior-β grains with the mid-length average width of about 103 ± 32 μm [32, 53–55, 57].

As shown in Table 2, the L-PBF as-built Ti-6Al-4V generally has high UTS (>1000 MPa), high YS (>900 MPa) and a low ε (total elongation <8%). The values are evidently well above the standard requirements, except for the values of elongation and strain to failure. In terms of the influential features, both the α' laths and the prior-β grains are important attributes in this type of microstructure. The acicular α' structure is responsible for retarding the movement of dislocations and cracks, thereby influencing strength associated with this microstructure. At the same time, the prior-β grains are responsible for the anisotropic behaviour (in both tensile strength and ductility) usually associated with this type of microstructure.

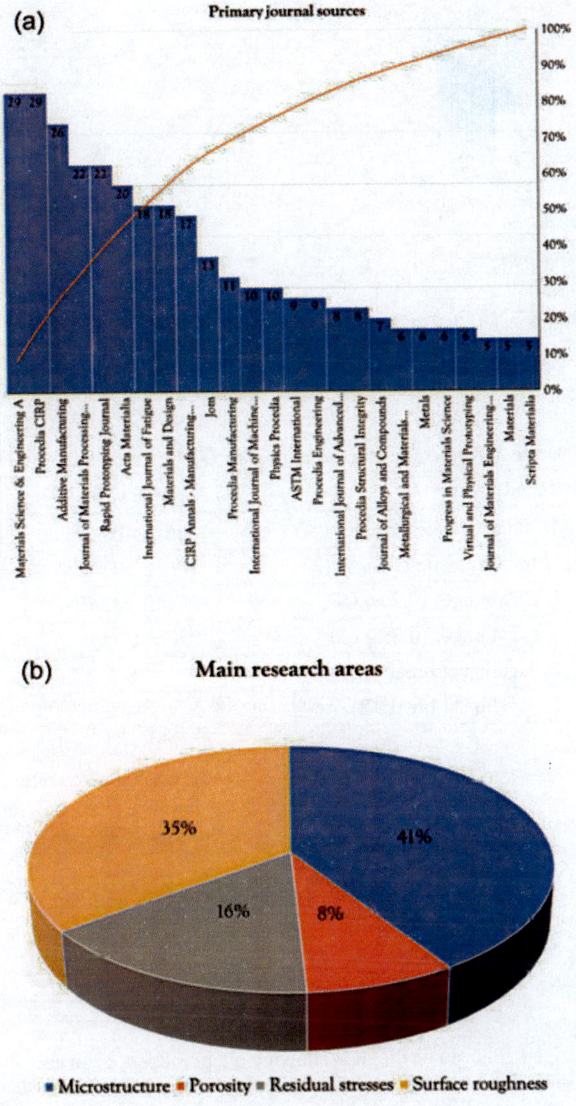

Fig. 2 Results of search terms combinations of (**a**) primary journal sources and (**b**) proportion of publications on selected attributes for structural integrity

3.2 *Porosity*

The L-PBF as-built Ti-6Al-4V is characterised by many types of defects such as pores, microcracks and delamination. The current work focuses on pores (porosity) because they are inherent to the L-PBF processing of Ti-6Al-4V and are the most widely investigated type of defects in parts produced by the technology. Figure 4 shows the significant influences on porosity and the parameters mostly affected following the triangulation of the literature.

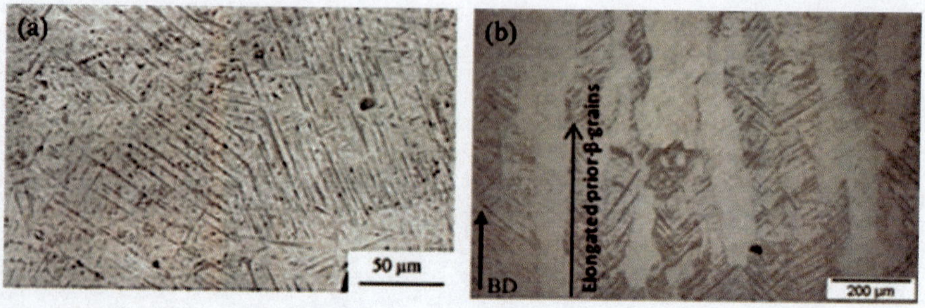

Fig. 3 Micrographs of typical L-PBF as-built Ti-6Al-4V, **a** showing acicular α′ phase (XY –plane) and **b** showing the morphology of elongated prior-β grains (ZX/ZY plane BD—Build direction) [58]

Table 2 Typical tensile properties of L-PBF as-built Ti-6Al-4V [42, 54, 59–64]

Tensile properties	As-built Ti-6Al-4V	Ti-6Al-4V specifications
Ultimate tensile strength (XY)	1206 ± 40 MPa	≥890 MPa
Ultimate tensile strength (Z)	1166 ± 50 MPa	≥855 MPa
Yield strength (0.2%) (XY)	1125 ± 60 MPa	≥800 MPa
Yield strength (0.2%) (Z)	970 ± 70 MPa	≥758 MPa
Elongation at break (XY)	6 ± 3%	≥10%
Elongation at break (Z)	4 ± 3%	≥10%

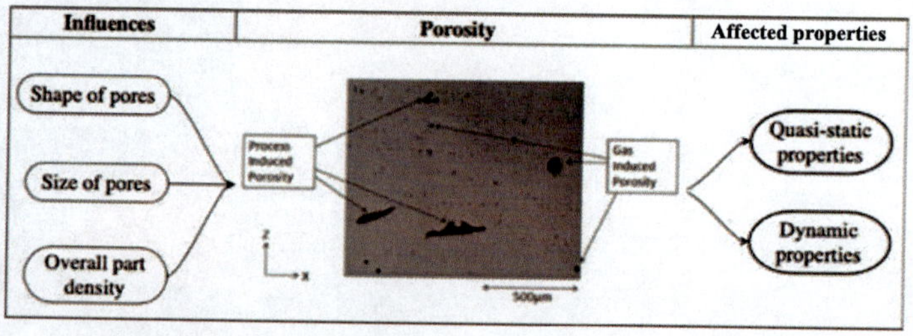

Fig. 4 Schematic diagram showing major influences on porosity and the properties mainly affected [67]. The image of two common types of pores in L-PBF is adapted from [38] with permission from Taylor and Francis

Porosity in L-PBF as-built Ti-6Al-4V parts is process-induced and forms either due to local overheating or insufficient process energy for complete melting [65]. Two types of pores common in L-PBF as-built Ti-6Al-4V are lack-of-fusion pores and gas-entrapped pores (shown in Fig. 4). The lack-of-fusion pores (typically 100–150 μm long [66]) are usually seen at the boundaries of adjacent layers.

On the other hand, the gas-entrapped pores (typically 10–100 μm [32, 66]) are believed to be determined by the quality of the feedstock material (which in turn is determined by the gas atomisation of powder which could contain entrapped microporosity of argon gas bubbles) as opposed to being process-induced [18, 61].

While porosity can adversely affect both the quasi-static and dynamic properties, as shown in Fig. 4, it is particularly detrimental to dynamic properties resulting in poor high cycle fatigue (HCF) resistance of Ti-6Al-4V. The effect of porosity on both dynamic and quasi-static properties is influenced by three major characteristics of pores, namely shape, size and overall part density, as shown in Fig. 4. The shape often seen in as-built L-PBF Ti-6Al-4V parts is either irregular or spherical. The lack-of-fusion pores (often irregularly shaped) have an aspect ratio larger than one, while the gas pores (often spherically shaped) can be considered to have an aspect ratio near one due to their round shape [31]. As such, linear and irregularly shaped pores are more detrimental to mechanical properties than spherically shaped pores. This is because the sharp tips of the long axis of lack-of-fusion pores are known to incur higher and more concentrated stresses, which will initiate cracks more easily [58, 68]. In terms of size, larger pores are more detrimental to mechanical properties than smaller pores because fatigue cracks initiate first from the largest defect if the shape factor is similar [58]. In addition, larger pores mean the microstructure is discontinuous and less ductile. The overall density of the part is determined by the overall number of pores in the part and is often identified as either macroscopic or microscopic pores. The smaller number of pores present in the part (microscopic) means the denser a part is, which in turn means better properties [69]. The reverse is true for macroscopic pores.

3.3 *Residual Stresses*

Residual stresses can occur in parts either as compressive or tensile residual stresses, and in some cases, a part can have both compressive and tensile stresses in different areas. Compressive residual stresses are generally beneficial, while tensile residual stresses are detrimental. The stresses referred to in this paper are the tensile residual stresses, as these are inevitable due to the temperature gradient mechanism (TGM) inherent to the L-PBF process. As shown in Fig. 5 the mechanism of residual stress development occurs through the introduction of high thermal gradients occurring around the laser spot, followed by shrinkage due to thermal contraction as the molten top layers cool down. This takes place between the base plate and the first layer of the L-PBF process and continues between accumulating layers as the process continues.

As shown in Fig. 5, residual stresses affect quasi-static, dynamic and geometrical tolerances. Generally, residual stresses are classified according to the scale at which they occur, either microscopic or macroscopic [71]. The microscopic residual stresses are usually more localised with minimal effect on mechanical properties. On the other hand, the macroscopic residual stresses typically vary over a considerable distance (across the dimensions of the part) and are typically associated with detrimental effects on material properties.

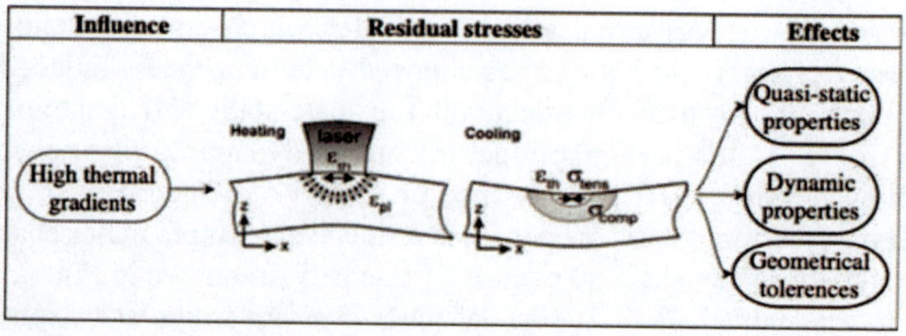

Fig. 5 Schematic representation of the main influence on residual stresses, temperature gradient mechanism (TGM) and the affected properties [67]. TGM was adopted from [70] with permission from Elsevier

3.4 *Surface Roughness*

The inherent average surface roughness (R_a) for L-PBF as-built Ti-6Al-4V is usually reported between 5 and 40 μm [29, 72–77]. This is too rough for applications such as aerospace. Figure 6 shows the major influences on surface roughness and the parameters mostly affected following the triangulation of the literature. The poor surface quality of L-PBF parts is attributed to many building process parameters such as powder particles, build direction, recoater blade, scanning speed, etc. [15, 78].

Even though there is no consensus about the parameters which influence surface roughness the most, the poor surface quality is predominantly linked to three factors, summarised in Fig. 6, as open-pores and other defects on the surface, partially melted powder adhered onto the surface and the staircase effect [79, 80].

The open pores and defects on the surface are dependent on the process parameters such as laser power. According to a review paper of AM of Ti-6Al-4V by Liu and Shin [31], the adhered partially melted powder is attributed to the fact that L-PBF processes keep the unused powders in the powder-bed, which subsequently gets attached to the surface. According to Kruth et al. [70], the balling phenomenon occurs when the molten material does not wet the underlying substrate due to surface tension, which tends to spheroidize the molten metal, causing a rough and bead-shaped surface that

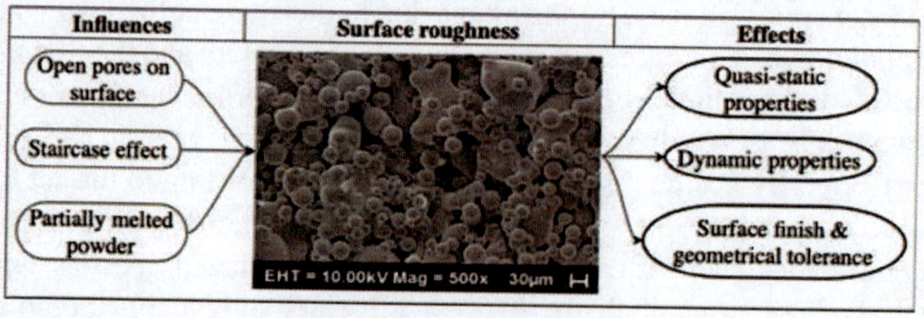

Fig. 6 Schematic representation of the main influences on surface roughness and the affected properties [67]. (Image showing as-built R_a was reproduced from ref. [72], under the terms of the Creative Commons CC BY license)

impedes a smooth subsequent layer deposition. This effect increases with an increase in the number of deposition layers and is made worse by poor laser parameters.

The detrimental effect of high surface roughness is that it serves as direct initiation of microcracks. In addition, surface finish quality is much more detrimental as compared to internal defects (porosity) on fatigue properties [81]. This is because surface roughness acts as multiple stress concentrators. This was confirmed by Chastand et al. [81], after they demonstrated that shorter fatigue lives were associated with the surface defects, while the internal defects required a higher number of cycles to failure.

4 Problem-Solving Strategies of L-PBF Outputs

Despite the ongoing research to optimise the L-PBF process parameters, current commercially produced parts do not meet the requirements in as-built conditions. Hence, a definitive need for the following post-processing strategies as problem-solvers.

4.1 Microstructure

The problem-solver for L-PBF as-built Ti-6Al-4V martensitic microstructure is post-processing through heat-treatments. These heat-treatments are mainly carried out for two reasons; to decompose the α' martensitic microstructure into a dual-phase $\alpha + \beta$ matrix and to change the size and morphology of the prior-β grains. The heat-treatments are usually carried out below or above β-transus temperature and are accordingly referred to as sub-transus or super-transus heat-treatments. In terms of crystallography, sub transus and super-transus mean Ti-6Al-4V exist in two crystallographic allotropes: hexagonal close-packed (hcp) α-phase and body-centred cubic (bcc) β-phase. Generally, the major change in microstructure as a result of sub-transus heat-treatments is the decomposition of α' into α and the formation of β, and the subsequent coarsening of α lamellae [61, 82]. The effect of α lamellae coarsening is a moderate increase in ductility at the expense of strength. This trend continues with an increase in heat-treatment temperature from about 780 °C up to the β-transus temperature at a holding time of 1–4 h followed by either furnace or air cooling [61, 73, 82].

On the other hand, the major change in microstructure resulting from super-transus heat-treatments is a significant growth of the columnar prior-β grains, which, in addition to the increase in ductility at the expense of strength, also means the apparent anisotropic deformation behaviour is partially mitigated [66]. Besides the sub- and super-transus heat-treatment, a better balance of quasi-static and dynamic properties of Ti-6Al-4V has been achieved through duplex anneal post-processing heat-treatments [83], which results in a bimodal microstructure. The duplex anneal process involves two annealing temperature stages combined with specific holding times followed by two-stage cooling methods. An example of a typical duplex anneal approach for L-PBF Ti-6Al-4V by Ter Haar and Becker [83] involved annealing at 910 °C/8 h and 750 °C/4 h holding temperatures followed by water quenching and furnace cooling, respectively. The high temperature allowed for fragmentation of the α phase, while the second low-temperature treatment allowed for the decomposition of the α′ phase to a lamellar α + β microstructure.

4.2 *Surface Roughness*

The problem-solvers for the high surface roughness include machining and polishing techniques. This is usually achieved through standardised conventional techniques. Overall, machining treatments work best on flat surfaces. For complex, high-quality near-net parts, the polishing and chemical milling treatments become the ideal post-processing solutions to achieve desired surface finishes and geometrical tolerances. The improvement of both quasi-static and dynamic mechanical properties following these methods is predominantly 40–50% higher than as-built [77, 84–87].

4.3 *Residual Stresses*

Despite the efforts to reduce the residual stresses by base-plate pre-heating and re-scanning strategies, the residual stresses seen in the as-built parts still fall in the range of 100–500 MPa [88], which are still higher than the maximum allowable stresses for most parts' application. The problem-solvers are stress-relief heat-treatments. Several reports have shown an effective reduction in residual stresses in Ti-6Al-4V produced by L-PBF after heat-treatments ranging between 480 and 650 °C for 1–4 h, followed by either furnace or air cooling [37, 89, 90]. Lately, a reduction in residual stresses of about 90% was reported by Ter Haar and Becker [91] following a heat-treatment at 560 °C. They also showed that the stress relief rate below 560 °C is extremely slow but increases with an increase in temperature from 560 to 610 °C. This was accompanied by a noticeable increase in hardness. Hence, concluded that it is better to stress relieve at 610 °C.

4.4 *Porosity*

Generally, if porosity is below the minimum specification, there is no detrimental effect on mechanical properties. This is because the smaller number of pores present in the part (microscopic) the denser a part is, which in turn means better quasi-static properties [69]. However, if porosity is above the minimum specifications, the potential problem-solver is hot isostatic pressing (HIPing). This has a combination effect, capable of improving both quasi-static and dynamic mechanical properties. A typical HIPing process for Ti-6Al-4V is carried out under an inert gas atmosphere at a 100 MPa minimum pressure within the temperature range of 895 to 955 °C. The pressure enables the closure of internal pores and cracks to increase material density, resulting in better fatigue life. The inherent heat-treatment effect influences microstructure refinement, which leads to an improvement of quasi-static properties.

5 Conclusion

Even though complex parameters characterise the L-PBF process, three conclusions can be drawn from this review. Firstly, the review identified four attributes of the as-built condition which are key to the structural integrity of L-PBF Ti-6Al-4V: microstructure, porosity, residual stresses and surface roughness. Secondly, the paper summarises a body of knowledge through a technical review of the L-PBF processing of Ti-6Al-4V and highlights a definitive need for post-processing solutions to address the as-built issues predominantly seen in these parts. Thirdly, the problem-solving strategies to ensure that these attributes are improved were provided.

Acknowledgements The authors would like to acknowledge the Stellenbosch Technology Centre, Laboratory for Advanced Manufacturing (STC-LAM) and Materials Engineering group at Stellenbosch University for the resources and equipment used.

References

1. Doubrovski, Z., Verlinden, J.C., Geraedts, J.M.P.: Optimal design for additive manufacturing: opportunities and challenges, 1–12 (2011)
2. Bourell, D.L.D., Beaman, J.J., Leu, M.C., Rosen, D.W.: A brief history of additive manufacturing and the 2009 roadmap for additive manufacturing: looking back and looking ahead US-Turkey Workshop on … 2005–2005 (2009)
3. Atzeni, E., Salmi, A.: Economics of additive manufacturing for end-usable metal parts. Int. J. Adv. Manuf. Technol., (2012)
4. Sreenivasan, R., Goel, A., Bourell, D.L.: Sustainability issues in laser-based additive manufacturing. Phys. Procedia **5**, 81–90 (2010)
5. Wohlers, T., Gornet, T.: History of additive manufacturing wohlers report 2014—3D Printing and Additive Manufacturing State of the Industry, 1–34 (2014)
6. Gibson, I., Rosen, D., Stucker, B.: Additive Manufacturing Technologies (2015)
7. ASTM International 2013 Standard Terminology for Additive Manufacturing Technologies Rapid Manufacturing Association, F2792–12a 10–2.
8. Gao, W., Zhang, Y., Ramanujan, D., Ramani, K., Chen, Y., Williams, C.B., Wang, C.C.L., Shin, Y.C., Zhang, S., Zavattieri, P.D.: The status, challenges, and future of additive manufacturing in engineering. Comput. Aided Des. **69**, 65–89 (2015)
9. Romano, S., Brandão, A., Gumpinger, J., Gschweitl, M., Beretta, S.: Qualification of AM parts: extreme value statistics applied to tomographic measurements. Mater. Des. **131**, 32–48 (2017)

10. Attaran, M.: The rise of 3-D printing: the advantages of additive manufacturing over traditional manufacturing. Bus. Horizons **60**, 677–688 (2017)
11. Bartolomeu, F., Faria, S., Carvalho, O., Pinto, E., Alves, N., Silva, F.S., Miranda, G.: Predictive models for physical and mechanical properties of Ti6Al4V produced by Selective Laser Melting. Mater Sci Eng: A **663**, 181–192 (2016)
12. Chastand, V., Tezenas, A., Cadoret, Y., Quaegebeur, P., Maia, W., Charkaluk, E.: Fatigue characterization of Titanium Ti-6Al-4V samples produced by Additive Manufacturing. Procedia Struct. Integrity **2**, 3168–3176 (2016)
13. Guo, N., Leu, M.C.: Additive manufacturing: technology, applications and research needs. Frontiers Mech. Eng. **8**, 215–243 (2013)
14. Seabra, M., Azevedo, J., Araújo, A., Reis, L., Pinto, E., Alves, N., Santos, R., Mortágua, J.P.: Selective laser melting (SLM) and topology optimization for lighter aerospace components. Procedia Strucct. Integrity **1**, 289–296 (2016)
15. Vandenbroucke, B., Kruth, J.: Selective laser melting of biocompatible metals for rapid manufacturing of medical parts. Rapid Prototyping J. **13**, 196–203 (2007)
16. Osakada, K., Shiomi, M.: Flexible manufacturing of metallic products by selective laser melting of powder. Int. J. Mach. Tools Manuf **46**, 1188–1193 (2006)
17. Yadroitsev, I., Bertrand, P., Smurov, I.: Parametric analysis of the selective laser melting process. Appl. Surf. Sci. **253**, 8064–8069 (2007)
18. Kim FH, Moylan SP (2018) Literature Review of Metal Additive Manufacturing Defects NIST Advanced Manufacturing Series, pp 100–16
19. Frazier, W.E.: Metal additive manufacturing: a review. J. Mater. Eng. Perform. **23**, 1917–1928 (2014)
20. Kruth, J.-P., Leu, M.C.C., Nakagawa, T.: Progress in additive manufacturing and rapid prototyping. CIRP Ann. Manuf. Technol. **47**, 525–540 (1998)
21. Beese, A.M., Carroll, B.E.: Review of mechanical properties of Ti-6Al-4V made by laser-based additive manufacturing using powder feedstock. Jom **68**, 724–734 (2016)
22. Harun, W.S.W., Kamariah, M.S.I.N., Muhamad, N., Ghani, S.A.C., Ahmad, F., Mohamed, Z.: A review of powder additive manufacturing processes for metallic biomaterials. Powder Technol. **327**, 128–151 (2018)
23. Dutta, B., Froes(Sam), F.H.: The Additive Manufacturing (AM) of titanium alloys. Metal Powder Report **72**, 96–106 (2017)
24. Guo, R., Xu, L., Wu, J., Yang, R., Zong, B.Y.: Microstructural evolution and mechanical properties of powder metallurgy Ti–6Al–4V alloy based on heat response. Mater. Sci. Eng.: A **639**, 327–334 (2015)
25. Peters, M., Kumpfert, J., Ward, C.H., Leyens, C.: Titanium alloys for aerospace applications. Adv. Eng. Mater. **5**, 419–427 (2003)
26. Uriondo, A., Esperon-Miguez, M., Perinpanayagam, S.: The present and future of additive manufacturing in the aerospace sector: a review of important aspects. Proc. Inst. Mech. Eng., Part G: J. Aerosp. Eng. **229**, 2132–2147 (2015)
27. Simpson, T.W., Williams, C.B., Hripko, M.: Preparing industry for additive manufacturing and its applications: Summary & recommendations from a National Science Foundation workshop Additive Manufacturing (2016)
28. Slotwinski, J.A., Garboczi, E.J., Stutzman, P.E., Ferraris, C.F., Watson, S.S., Peltz, M.A.: Characterization of metal powders used for additive manufacturing. J. Res. Nat. Inst. Stand. Technol. **119**, 460–493 (2014)
29. Townsend, A., Senin, N., Blunt, L., Leach, R.K., Taylor, J.S.: Surface texture metrology for metal additive manufacturing: a review. Precis. Eng. **46**, 34–47 (2016)
30. Parry, L., Ashcroft, I.A., Wildman, R.D.: Understanding the effect of laser scan strategy on residual stress in selective laser melting through thermo-mechanical simulation. Additive Manuf. **12**, 1–15 (2016)
31. Liu, S., Shin, Y.C.: Additive manufacturing of Ti6Al4V alloy: a review. Mater. Design **164**, 107552 (2019)

32. Agius, D., Kourousis, K., Wallbrink, C.: A review of the as-built SLM Ti-6Al-4V mechanical properties towards achieving fatigue resistant designs. Metals **8**, 75 (2018)
33. Leuders, S., Thöne, M., Riemer, A., Niendorf, T., Tröster, T., Richard, H.A., Maier, H.J.: On the mechanical behaviour of titanium alloy TiAl6V4 manufactured by selective laser melting: fatigue resistance and crack growth performance. Int. J. Fatigue **48**, 300–307 (2013)
34. Kruth, J.-P.P., Mercelis, P., Van, V.J., Froyen, L., Rombouts, M., Van Vaerenbergh, J., Froyen, L., Rombouts, M., Van, V.J., Froyen, L., Rombouts, M.: Binding mechanisms in selective laser sintering and selective laser melting. Rapid Prototyping J. **11**, 26–36 (2005)
35. Shiomi, M., Osakada, K., Nakamura, K., Yamashita, T., Abe, F.: Residual stress within metallic model made by selective laser melting process. CIRP Ann. Manuf. Technol. **53**, 195–198 (2004)
36. Sutton, A.T., Kriewall, C.S., Leu, M.C., Newkirk, J.W.: Powder characterisation techniques and effects of powder characteristics on part properties in powder-bed fusion processes. Virtual Phys. Prototyping **12**, 3–29 (2017)
37. DebRoy, T., Wei, H.L., Zuback, J.S., Mukherjee, T., Elmer, J.W., Milewski, J.O., Beese, A.M., Wilson-Heid, A., De, A., Zhang, W.: Additive manufacturing of metallic components—process, structure and properties. Progress Mater. Sci. **92**, 112–224 (2018)
38. Sames, W.J., List, F.A., Pannala, S., Dehoff, R.R., Babu, S.S.: The metallurgy and processing science of metal additive manufacturing. Int. Mater. Rev. **61**, 315–360 (2016)
39. Anderson, I.E., Figliola, R.S., Morton, H.: Flow mechanisms in high pressure gas atomization. Mater. Sci. Eng., A **148**, 101–114 (1991)
40. Seki, Y., Okamoto, S., Takigawa, H., Kawai, N.: Effect of atomization variables on powder characteristics in the high-pressured water atomization process. Met. Powder Rep. **45**, 38–40 (1990)
41. Shi, Q., Gu, D., Xia, M., Cao, S., Rong, T.: Effects of laser processing parameters on thermal behavior and melting/solidification mechanism during selective laser melting of TiC/Inconel 718 composites. Optics Laser Technol. **84**, 9–22 (2016)
42. Shipley, H., McDonnell, D., Culleton, M., Coull, R., Lupoi, R., O'Donnell, G., Trimble, D.: Optimisation of process parameters to address fundamental challenges during selective laser melting of Ti-6Al-4V: a review. Int. J. Mach. Tools Manuf. **128**, 1–20 (2018)
43. Sufiiarov, V.S., Popovich, A.A., Borisov, E.V., Polozov, I.A., Masaylo, D.V., Orlov, A.V.: The effect of layer thickness at selective laser melting. Procedia Eng. **174**, 126–134 (2017)
44. Leutenecker-Twelsiek, B., Klahn, C., Meboldt, M.: Considering part orientation in design for additive manufacturing. Procedia CIRP **50**, 408–413 (2016)
45. Das, P., Chandran, R., Samant, R., Anand, S.: Optimum part build orientation in additive manufacturing for minimizing part errors and support structures. Procedia Manuf. **1**, 343–354 (2015)
46. Kempen, K., Vrancken, B., Buls, S., Thijs, L., Van Humbeeck, J., Kruth, J.P.: Selective laser melting of crack-free high density M2 high speed steel parts by baseplate preheating. J. Manuf. Sci. Eng. Trans. ASME., **136** (2014)
47. Ali, H., Ma, L., Ghadbeigi, H., Mumtaz, K.: In-situ residual stress reduction, martensitic decomposition and mechanical properties enhancement through high temperature powder bed pre-heating of Selective Laser Melted Ti6Al4V. Mater Sci Eng A **695**, 211–220 (2017)
48. Masoomi, M., Thompson, S.M., Shamsaei, N.: Laser powder bed fusion of Ti-6Al-4V parts: Thermal modeling and mechanical implications. Int J Mach Tools Manuf **118–119**, 73–90 (2017)
49. Kumbhar, N.N., Mulay, A.V.: Post processing methods used to improve surface finish of products which are manufactured by additive manufacturing technologies: a review. J. Inst. Eng. (India): Series C **99**, 481–487 (2018)
50. Tranfield, D., Denyer, D., Smart, P.: Towards a methodology for developing evidence-informed management knowledge by means of systematic review* introduction: the need for an evidence-informed approach. British J. Manage. **14**, 207–222 (2003)
51. Neikter, M., Huang, A., Wu, X.: Microstructural characterization of binary microstructure pattern in selective laser-melted Ti-6Al-4V. Int. J. Adv. Manuf. Technol. **104**, 1381–1391 (2019)

52. Zhao, X., Li, S., Zhang, M., Liu, Y., Sercombe, T.B., Wang, S., Hao, Y., Yang, R., Murr, L.E.: Comparison of the microstructures and mechanical properties of Ti-6Al-4V fabricated by selective laser melting and electron beam melting. Mater. Des. **95**, 21–31 (2016)
53. Murr, L.E., Quinones, S.A., Gaytan, S.M., Lopez, M.I., Rodela, A., Martinez, E.Y., Hernandez, D.H., Martinez, E., Medina, F., Wicker, R.B.: Microstructure and mechanical behavior of Ti-6Al-4V produced by rapid-layer manufacturing, for biomedical applications. J. Mech. Behav. Biomed. Mater. **2**, 20–32 (2009)
54. Vilaro, T., Colin, C., Bartout, J.D.: As-fabricated and heat-treated microstructures of the Ti-6Al-4V alloy processed by selective laser melting metallurgical and materials transactions a: physical. Metall. Mater. Sci. **42**, 3190–3199 (2011)
55. Kumar, P., Ramamurty, U.: Microstructural optimization through heat treatment for enhancing the fracture toughness and fatigue crack growth resistance of selective laser melted Ti–6Al–4V alloy. Acta Materialia **169**, 45–59 (2019)
56. Yang, J., Yu, H., Yin, J., Gao, M., Wang, Z., Zeng, X.: Formation and control of martensite in Ti-6Al-4V alloy produced by selective laser melting. Mater. Des. **108**, 308–318 (2016)
57. Simonelli, M., Tse, Y.Y., Tuck, C.: Effect of the build orientation on the mechanical properties and fracture modes of SLM Ti-6Al-4V. Mater. Sci. Eng. A **616**, 1–11 (2014)
58. Liu, S., Shin, Y.C.: Additive manufacturing of Ti6Al4V alloy: a review. Mater. Des. **164**, 8–12 (2019)
59. Lewandowski, J.J., Seifi, M.: Metal additive manufacturing: a review of mechanical properties. Annu. Rev. Mater. Res. **46**, 151–186 (2016)
60. Rafi, H.K., Karthik, N.V., Gong, H., Starr, T.L., Stucker, B.E.: Microstructures and mechanical properties of Ti6Al4V parts fabricated by selective laser melting and electron beam melting. J. Mater. Eng. Perform. **22**, 3872–3883 (2013)
61. Qian, M., Xu, W., Brandt, M., Tang, H.P.: Additive manufacturing and postprocessing of Ti-6Al-4V for superior mechanical properties. MRS Bull. **41**, 775–784 (2016)
62. Facchini, L., Magalini, E., Robotti, P., Molinari, A., Höges, S., Wissenbach, K.: Ductility of a Ti-6Al-4V alloy produced by selective laser melting of prealloyed powders. Rapid Prototyping J. **16**, 450–459 (2010)
63. ASTM International: Standard Specification for Additive Manufacturing Titanium-6 Aluminum-4 Vanadium ELI (Extra Low Interstitial) with Powder Bed Fusion ASTM International, F3001 − 14 i 1–8 (2014)
64. Gorsse, S., Hutchinson, C., Gouné, M., Banerjee, R.: Additive manufacturing of metals: a brief review of the characteristic microstructures and properties of steels, Ti-6Al-4V and high-entropy alloys. Sci. Technol. Adv. Mater. **18**, 584–610 (2017)
65. Wits, W.W., Carmignato, S., Zanini, F., Vaneker, T.H.J.: Porosity testing methods for the quality assessment of selective laser melted parts. CIRP Ann. Manuf. Technol. **65**, 201–204 (2016)
66. Bourell, D., Kruth, J.P., Leu, M., Levy, G., Rosen, D., Beese, A.M., Clare, A.: Materials for additive manufacturing. CIRP Ann. Manuf. Technol. **66**, 659–681 (2017)
67. Makhetha, W.M.I., Becker, T.H., Sacks, N.: Post-Processing Framework for As-Built LPBF Ti-6Al-4V Parts Towards Meeting Industry Functional Requirements JOM (2022)
68. Masuo, H., Tanaka, Y., Morokoshi, S., Yagura, H., Uchida, T., Yamamoto, Y., Murakami, Y.: Influence of defects, surface roughness and HIP on the fatigue strength of Ti-6Al-4V manufactured by additive manufacturing. Int. J. Fatigue **117**, 163–179 (2018)
69. Cunningham, R., Narra, S.P., Ozturk, T., Beuth, J., Rollett, A.D.: Evaluating the effect of processing parameters on porosity in electron beam melted Ti-6Al-4V via synchrotron X-ray microtomography. Jom **68**, 765–771 (2016)
70. Kruth, J.P., Froyen, L., Van Vaerenbergh, J., Mercelis, P., Rombouts, M., Lauwers, B.: Selective laser melting of iron-based powder. J. Mater. Process. Technol. **149**, 616–622 (2004)
71. Kandil, F.A., Lord, J.D., Fry, A.T., Grant, P.V.: A review of residual stress measurement methods—a guide to technical selection NPL. Materials Centre, 1–42 (2001)
72. Vaithilingam, J., Prina, E., Goodridge, R.D., Hague, R.J.M., Edmondson, S., Rose, F.R.A.J., Christie, S.D.R.: Surface chemistry of Ti6Al4V components fabricated using selective laser melting for biomedical applications. Mater. Sci. Eng. C **67**, 294–303 (2016)

73. de Formanoir, C., Michotte, S., Rigo, O., Germain, L., Godet, S.: Electron beam melted Ti-6Al-4V: Microstructure, texture and mechanical behavior of the as-built and heat-treated material. Mater. Sci. Eng. A. **652**, 105–119 (2016)
74. Kasperovich, G., Hausmann, J.: Improvement of fatigue resistance and ductility of TiAl6V4 processed by selective laser melting. J. Mater. Process. Technol. **220**, 202–214 (2015)
75. Alcisto, J., Enriquez, A., Garcia, H., Hinkson, S., Steelman, T., Silverman, E., Valdovino, P., Gigerenzer, H., Foyos, J., Ogren, J., Dorey, J., Karg, K., McDonald, T., Es-Said, O.S.: Tensile properties and microstructures of laser-formed Ti-6Al-4V. J. Mater. Eng. Perform. **20**, 203–212 (2011)
76. Palanivel, S., Dutt, A.K., Faierson, E.J., Mishra, R.S.: Spatially dependent properties in a laser additive manufactured Ti-6Al-4V component. Mater. Sci. Eng. A **654**, 39–52 (2016)
77. Sun, Y.Y., Gulizia, S., Oh, C.H., Fraser, D., Leary, M., Yang, Y.F., Qian, M.: The influence of as-built surface conditions on mechanical properties of Ti-6Al-4V additively manufactured by selective electron beam melting. Jom **68**, 791–798 (2016)
78. Yadroitsev, I., Smurov, I.: Surface morphology in selective laser melting of metal powders. Phys. Procedia **12**, 264–270 (2011)
79. Strano, G., Hao, L., Everson, R.M., Evans, K.E.: Surface roughness analysis, modelling and prediction in selective laser melting. J. Mater. Process. Technol. **213**, 589–597 (2013)
80. Li, Y., Yang, H., Lin, X., Huang, W., Li, J., Zhou, Y.: The influences of processing parameters on forming characterizations during laser rapid forming. Mater. Sci. Eng., A **360**, 18–25 (2003)
81. Chastand, V., Quaegebeur, P., Maia, W., Charkaluk, E.: Comparative study of fatigue properties of Ti-6Al-4V specimens built by electron beam melting (EBM) and selective laser melting (SLM). Mater. Charact. **143**, 76–81 (2018)
82. Lütjering, G., Williams, J.C.: Titanium : Engineering Materials and Processes Edition SPRINGER second edi 1–442 (2007)
83. Ter Haar, G.M., Becker, T.H.: Selective laser melting produced Ti-6Al-4V: post-process heat treatments to achieve superior tensile properties. Materials **11** (2018)
84. Gong, H., Rafi, K., Starr, T., Stucker, B.: Effect of defects on fatigue tests of as-built TI-6AL-4V parts fabricated by selective laser melting. In 23rd Annual International Solid Freeform Fabrication Symposium—An Additive Manufacturing Conference, SFF, pp. 499–506 (2012)
85. Wycisk, E., Solbach, A., Siddique, S., Herzog, D., Walther, F., Emmelmann, C.: Effects of defects in laser additive manufactured Ti-6Al-4V on fatigue properties. Phys. Procedia **56**, 371–378 (2014)
86. Mower, T.M., Long, M.J.: Mechanical behavior of additive manufactured, powder-bed laser-fused materials. Mater. Sci. Eng. A **651**, 198–213 (2016)
87. Greitemeier, D., Palm, F., Syassen, F., Melz, T.: Fatigue performance of additive manufactured TiAl6V4 using electron and laser beam melting. Int. J. Fatigue **94**, 211–217 (2017)
88. Vayssette, B., Saintier, N., Brugger, C., Elmay, M., Pessard, E.: Surface roughness of Ti-6Al-4V parts obtained by SLM and EBM: effect on the high cycle fatigue life. Procedia Eng. **213**, 89–97 (2018)
89. Ali, H., Ghadbeigi, H., Mumtaz, K.: Effect of scanning strategies on residual stress and mechanical properties of Selective Laser Melted Ti6Al4V. Mater. Sci. Eng. A **712**, 175–187 (2018)
90. Tong, J., Bowen, C.R., Persson, J., Plummer, A.: Mechanical properties of titanium-based Ti–6Al–4V alloys manufactured by powder bed additive manufacture Materials. Sci. Technol. **33**, 138–148 (2017)
91. Ter Haar, G.M., Becker, T.H.: Low temperature stress relief and martensitic decomposition in selective laser melting produced Ti6Al4V. Mater. Des. Process. Commun., 2–7 (2020)

William Makhetha holds M.Sc. Eng degree in Materials Engineering from the University of Cape Town. He is currently a Ph.D. candidate in Industrial Engineering at the Stellenbosch University. His area of research interest is in metal additive manufacturing and its application in industry.

Gerrit Ter Haar completed his master's degree cum laude on the topic of additive manufacturing of titanium alloys. He obtained his Ph.D. in the same field which he completed in 2021. Dr. Ter Haar currently holds the position of lecturer at the engineering faculty of Stellenbosch University.

Natasha Sacks is a Professor of Advanced Manufacturing at Stellenbosch University. Prof. Sack's research interests include materials science, and additive and subtractive manufacturing.

Thorsten Becker is the acting Director for the Centre of Materials Engineering at the University of Cape Town. Prof. Becker's research interest is in structural integrity: fatigue, fracture and creep. His work closely collaborates with local and international institutions to better understand the process inherent attributes and their link to the material's structural performance.

Towards a Virtual Optical Coordinate Measurement Machine

Z. Luthuli, K. Schreve, and O. A. Kurger

Abstract Optical coordinate measuring machines offer portability and the ability to capture vast quantities of data points quickly. But traceability of measurements in these systems is difficult since they are fraught with many challenges affecting the uncertainty of measurements, e.g. image noise, variable setups, point cloud based measurements. The VDI (*Verein Deutscher Ingenieure*) German standard is used for verification of these systems using non-complex artefacts that are easy to measure and have already been calibrated with other traceable measuring machines. ISO 10360–13 were only released in 2020. However, neither the VDI nor ISO standard cater for freeform surfaces. The Virtual CMM (Coordinate Measurement Machine) technique solves the problem of traceability of freeform surface measurements. This technique has only been applied to tactile CMMs and articulated arm optical CMMs, also known as discrete point systems, since they measure one point at a time. This research paper reports on the development and application of the Virtual CMM technology to a stereovision scanning optical coordinate measuring system.

Keywords Optical CMM · Uncertainty of measurement · Freeform surfaces

1 Introduction

A complete measurement result must consist of the measured value and its uncertainty [1]. Obtaining task specific uncertainty in any optical coordinate measurement system

Z. Luthuli · K. Schreve (✉)
Department of Mechanical and Mechatronic Engineering, Stellenbosch University, Stellenbosch, South Africa
e-mail: kschreve@sun.ac.za

Z. Luthuli
e-mail: zamokuhle.luthuli@alumni.uct.ac.za

O. A. Kurger
National Metrology Institution of South Africa, Pretoria, South Africa

K. von Leipzig et al. (eds.), *Smart, Sustainable Manufacturing in an Ever-Changing World*, Lecture Notes in Production Engineering,
https://doi.org/10.1007/978-3-031-15602-1_53

is difficult. This is due to varied uncertainty contributors and challenges associated with such systems [2].

Monte Carlo simulations can be used to determine measurement uncertainties. This idea was first used for estimating CMM uncertainties by PTB researchers [3]. They called their concept the Virtual Coordinate Measuring Machine (VCMM). The technique has also been applied in Articulated Arm CMM technology and other discrete point measurement systems [4]. Beraldin [4] proposed the SOP-36 performance verification procedure for non-contact measurement systems. It is based on the VDI standard [5], but requires fewer measurements then the VDI standard. Neither the SOP-26, VDI or recently released ISO 10360–13 standard procedure cover performance verification of freeform surface measuring systems.

This paper is concerned with the application of the Virtual CMM technology to scanning optical CMMs. To the authors' knowledge, no virtual CMM exists in the open domain to estimate the measurement uncertainty of optical CMMs.

2 Literature Review

Early attempts to express the measurement uncertainty were simply based on using the reported accuracy of the device [6] or the device's maximum permissible error (MPE) [6]. These methods obviously ignore the complexities of optical scanning systems and task specific conditions. A slightly better approach is the multiple measurements method [7] but is time consuming and require suitable reference standards. Simulative methods, essentially virtual measurement machines, appear to be the best approach currently available. The methods model the kinematic chain of the measurement system and then typically uses a Monte Carlo simulation to find the measurement uncertainty for any task.

Sladek and Gaska [6] successfully tested a Virtual CMM model was by performing common metrological tasks such as measurements of point to point distances, plane to plane distances, diameter of a sphere, form error of a sphere and distance between the centres of two spheres. For each measurement, the standard uncertainty was determined according to the methodology used in the classical methods of determining the uncertainty of measurements. This was compared to the uncertainty obtained by the virtual CMM [3]. The virtual CMM method proved to be consistent with the essence of the coordinate measuring technique which is a measurement of a single point. This is why it is currently considered as the most effective method among the known methods of uncertainty measurements [8].

A Virtual Articulated Arm Coordinate Measuring Machine (VAACMM) allows for a near real time determination of single measurement results and its uncertainty [8]. A VAACMM was used in the automotive industry for quality control purposes [8].

Optical systems present new challenges to finding the measurement uncertainty. The system setup is intrinsically variable since the cameras' intrinsic and extrinsic parameters may change with the setup. It is impossible to determine uncertainty results only based on measured error components of the CMM [9]. Most 3D imaging systems cannot measure a single point and can only scan a region and produce a point cloud, this brings up two issues which are sources of error [10]: The fact that the points are inherently noisy and the question of which point in the point cloud to use.

3 Optical VCMM Development and Testing

The following is a description of steps the VCMM will follow. The description is separated into the task related calibration steps and the actual measurement steps. The performance verification of the stereovision system used in this research and shown in Fig. 1 was successfully performed using part 3 of the VDI/VDE 2634 standard [5].

3.1 Task Related Calibration

Step 1: A calibration plate was manufactured by machining grooves into the plate, as shown in Fig. 2. The coordinates of the corners of each rectangle on the plate were then measured using a tactile CMM at NMISA (National Metrology Institute of South Africa). Their uncertainties were calculated based on the measurement system and measurement conditions. A total of 128 points were measured in the calibration plate and will be called world points. The tactile CMM is traceable to the laser standard at NMISA, thus the point measurements are traceable.

Step 2: The edges of the rectangles are detected using OpenCV. Lines are fitted to the detected edges and the intersection of the lines give the image coordinates of the

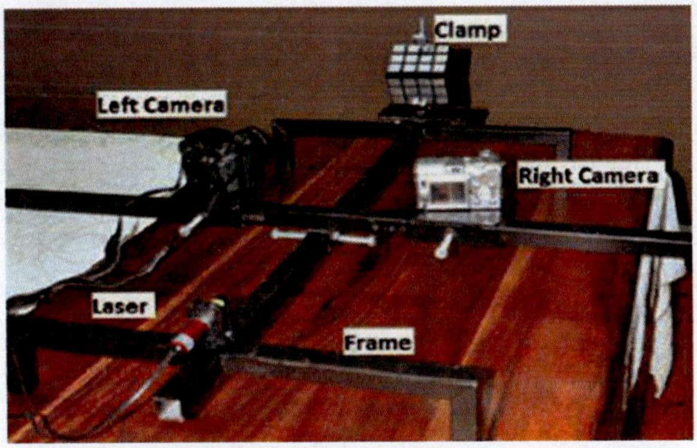

Fig. 1 Stereovision system [11] Cameras used: Canon EOS 4000D and Nikon D750

Fig. 2 Calibration plate (350 × 350 × 350 mm)

corners of the rectangles. The detected corners are matched to the world points. In this description, "points" and "corners" are interchangeable.

Step 3: The camera matrix [12] is then calculated as well as lens distortion parameters. Undistorted image points are then calculated for both cameras of the stereovision system.

Step 4: A Monte Carlo simulation is applied to repeat steps 1–3. Assuming a normal distribution of the measurements, artificial noise is added to both the detected image points and world points. All the standard deviations on the VCMM programme used for the image points and world points need to be a representation of real measurements.

Step 5: Each output of the parameters calculated by the programme is stored as a representative sample of potential results for that specific parameter. Then the average and standard deviations of the distortion parameters and camera matrices are calculated.

Step 6: The calibration image points are triangulated using the data from step 5 to get world points which are referred to as triangulated world points. The deviations between the triangulated world points and the original world points from the tactile CMM are also calculated.

Step 7: The simulation is done several times and the triangulated world points and the deviations from step 6 are stored as a representative sample of results for each world point and later their averages are calculated and used as final results for each point, i.e. the average of the sample for each triangulated world point is the measurement for that world point and the average deviation is the uncertainty for that point.

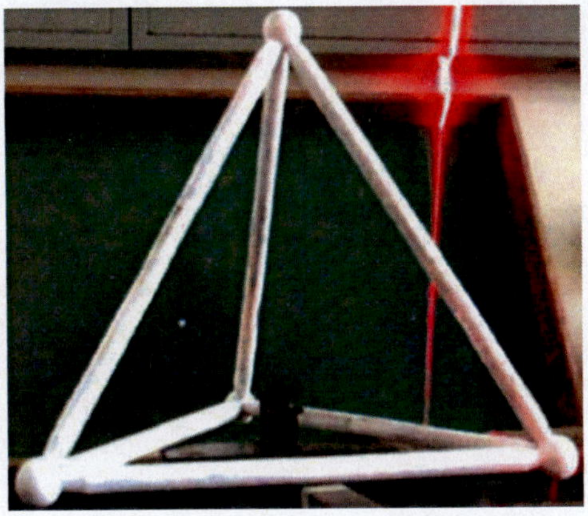

Fig. 3 Tetrahedron artefact

3.2 Actual Measurements

The last procedure after the performance verification and task related calibration in a traceability chain is the performance of the actual measurements. The tetrahedron artefact (Fig. 3) was measured using the VCMM. The advantage of the VCMM is that it gives an uncertainty of a point in 3D space.

In the experiments, a laser line is projected on a sphere of the tetrahedron artefact. Both cameras simultaneously take pictures of the artefact. This was repeated for all the spheres of the tetrahedron artefact.

The following steps describe the measurement steps of the VCMM.

Step 1: As described in [11], OpenCV and Python are used to detect and match the points on the laser line.

Step 2: Many image points of the tetrahedron artefact that lie on the intended tetrahedron spheres were calculated. These image points and their standard deviation were added to the VCMM programme using the Monte Carlo simulation function.

Step 3: The sphere image points and camera matrices from the calibration phase were used to triangulate the sphere surface points.

Step 4: Fit a sphere to the sphere surface points and calculate distances between the spheres.

Step 5: The VCMM calculate the uncertainty of the x y and z coordinates of each point. Using the law of error propagation [12], the following functions for the uncertainties of the sphere radius (U_r) and the distance between two sphere centres (U_D) can be derived:

$$U_r = k\sqrt{\frac{\left(U_x^2 + U_{x_i}^2\right)(x - x_i)^2 + \left(U_y^2 + U_{y_i}^2\right)(y - y_i)^2 + \left(U_z^2 + U_{z_i}^2\right)(z - z_i)^2}{\left(\sqrt{(x - x_i)^2 + (y - y_i)^2 + (z - z_i)^2}\right)^2}} \tag{1}$$

$$U_D = k\sqrt{\frac{(U_{x_1}^2 + U_{x_2}^2)(x_1 - x_2)^2 + (U_{y_1}^2 + U_{y_2}^2)(y_1 - y_2)^2 + (U_{z_1}^2 + U_{z_2}^2)(z_1 - z_2)^2}{\left(\sqrt{(x_1 - x_2)^2 + (y_1 - y_2)^2 + (z_1 - z_2)^2}\right)^2}} \tag{2}$$

where x_i, y_i and z_i are centre points of the sphere i and x, y, and z are points that lie on the surface of the sphere. U_i is an expanded uncertainty of variable i, e.g. U_{x_2} is the expanded uncertainty of x_2. k is the coverage factor [13]. After the surface has been described mathematically, the law of error propagation can be applied similarly to equations for U_r and U_D to get the function uncertainties which is an uncertainty of the freeform surface.

Some freeform surfaces are difficult to describe or estimate as a function. Since the VCMM calculates a point in 3D space with its uncertainty, a lot of points that lie on the surface of the freeform shape can be measured, using the VCMM, with their uncertainty, in that way measuring the form of that complex artefact. Therefore, full traceability of freeform surfaces can be achieved with optical CMMs using the VCMM technique.

Step 8: All the above steps were repeated at least 1000 times and the sphere radius and the distance between two spheres were stored as a representative sample of results for the tetrahedron artefact measurements. Later the sample averages were calculated and used as final measurement results.

4 Experimental Testing of the VCMM

Performance verification of the stereovision system was performed based on VDI/VDE 2634 Part 3 [5]. Based on this procedure, a measurement volume of 350 × 350 × 350 mm was selected, this volume is enough to fit the tetrahedron artefact.

When these performance verification tests were performed, a maximum possible error of 2.737 mm was calculated for the distance between two sphere centres and a maximum possible error of 0.538 mm was calculated for the sphere diameters. See [2] for verification of these results. These calculated maximum possible errors include all the significant uncertainty contributors to the measurement procedure when the performance verification was performed.

The second procedure after the performance verification in a traceability chain is the task related calibration of the stereovision system [4]. The task related calibration was done using the VCMM technique as explained above. Figure 4 below shows the system setup during the task related calibration procedure.

Many tests were performed during calibration of the stereovision system. The system was first tested for algorithm error. Here the VCMM programme described in Section "Task Related Calibration" runs without any noise added. Figure 5 shows the results of this test.

Fig. 4 Stereovision setup

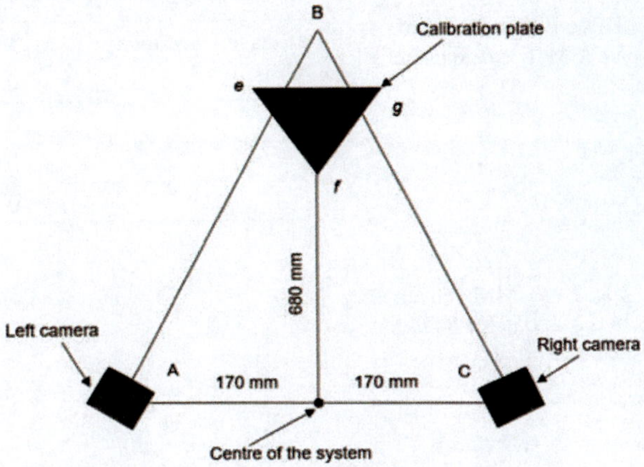

Fig. 5 Algorithm error in the VCMM at zero noise

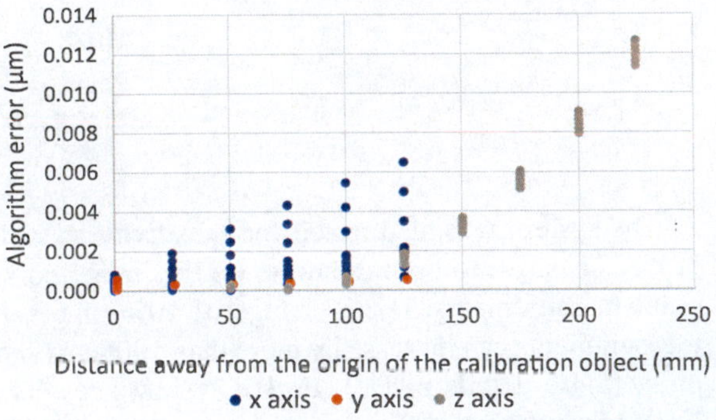

This error is inherently included in the final results on the VCMM programme. More tests were performed on the VCMM with image measurement noise and world point measurement noise included. The relationship between the camera focal length and the uncertainty of measurements was tested (Fig. 6). The test is for a world point at (0,26,50).

Fig. 6 Effect of changing the focal length

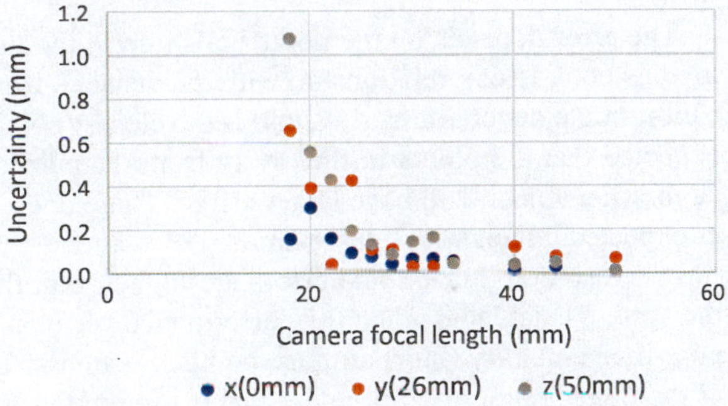

Table 1 Sensitivity of VCMM to image noise

Image Uncertainty	Measurement uncertainty [mm]	
	0.10 pixels	0.11 pixels
X	0.077	0.104
Y	0.012	0.031
Z	0.189	0.268

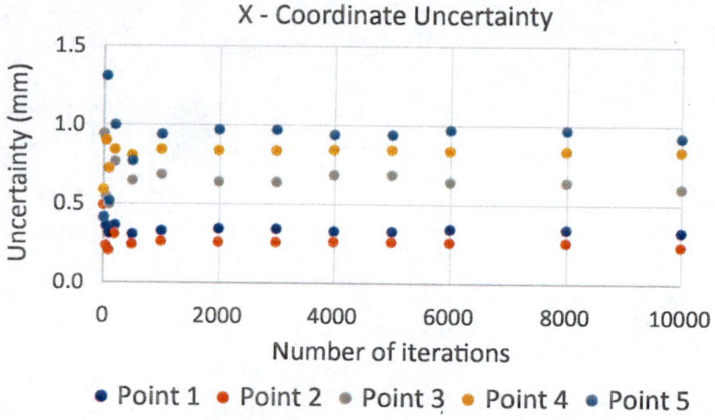

Fig. 7 VCMM convergence in the X axis for 5 randomly selected points

The system was also tested for sensitivity as well as convergence. The system proved sensitive to both the noise on the image point coordinates as well as world point measurements. For a change of 10% of pixel value, the uncertainty of the measurements can increase by more than 100%, as shown in Table 1 for one example point (more details in [2]). Figure 7 shows results on the convergence test. The VCMM results typically converged after 1000 iterations.

The measurement volume was calibrated ensuring that the image point measurement noise as well as world point noise used on the VCMM was a representation of the real-world measurements. More than 1000 iterations were done with the VCMM. Figure 8 shows the measurement uncertainty results from the VCMM for points measured on the calibration object (Fig. 2). In this graph, the expected error is:

$$\in = \sqrt{U_x^2 + U_y^2 + U_z^2} \tag{3}$$

The error depends on the point's position in the measurement volume, therefore there is not a strong relationship with the distance from the origin. The origin of the calibration plate coordinate system is the closest point to the cameras; therefore it is expected that the points further away from the camera, and thus to the back of the calibration object, will have larger errors. There does seem to be an increasing trend as expected, albeit with a large amount of scatter.

It is also noted that some errors are bigger than the maximum possible error of the system calculated when the performance verification was done. It can be due to poor image quality (blur) at those points. It can also be due to a poor measurement of the image point or world point that is used in the VCMM.

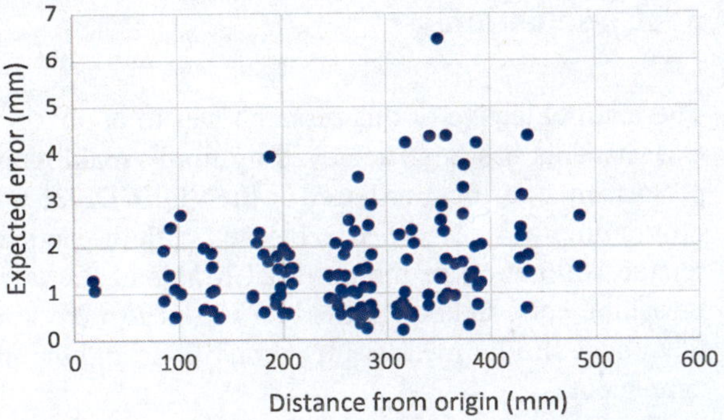

Fig. 8 General error for points measured on the calibration object (Fig. 3)

The VCMM calculates the expanded uncertainty (with coverage factor of 2) of measurements and gives it as an output of the simulation. The system also inherently adds the algorithm approximation errors on the measurement uncertainty of the calculated measurements. It was ensured that there is more than a 95% level of confidence in both the world point measurements and the image point measurements. Therefore, based on the above-mentioned reasons and the GUM guide on the combination of uncertainty contributors [13], it is understandable to have more than a 95% confidence level on the measurement results from the developed VCMM.

The last procedure in a traceability chain after the successful performance verification and task related calibration using the VCMM on the stereovision system is to perform the actual measurements, also using the VCMM. An additional advantage of using the VCMM is that it calculates the 3D uncertainty of a point in the selected volume, this allows the user to be able to see how the uncertainties are distributed within the selected volume and place the workpiece on the part of the work volume where there are smaller uncertainties. Measurements of the tetrahedron artefact were done using the VCMM and Table 2 shows the measurement result of the artefact from the VCMM. The same tetrahedron artefact was independently measured using two other CMMs at similar measurement conditions and results are also included in Table 2. In this table, D is the sphere diameter, L is a distance between two sphere centres, M is the measurement and U is the expanded uncertainty at a 95% confidence level (k = 2).

Table 2 VCMM simulation measurement results of a tetrahedron artefact

Item		D (mm)	L (mm)
Tactile CMM	M	25.006	300.257
	U	0.002	0.003
HP 3D Scanner	M	23.985	299.089
	U	1.173	2.149
VCMM	M	25.276	299.145
	U	0.687	1.475

5 Conclustion

The main objective of this research was to apply the VCMM technology in optical CMMs. This has been achieved by firstly making an optical CMM and doing the performance verification based on the VDI/VDE 2634–3 standard [5]. Then a VCMM programme was developed to be used with the optical CMM for the successful task related calibration of the optical CMM and the taking of the tetrahedron artefact measurements. Other CMMs were used to take measurements of the same artefact under similar measurement conditions and the measurement results were all in agreement.

The developed optical CMM using the VCMM measured an uncertainty of 1.475 mm for the measured distance of 299.145 mm and an uncertainty of 0.687 mm for a diameter of 25.276 mm. These measurements were more accurate than the HP 3D scanner (an optical area scanning system used for comparison).

In the future, the uncertainty achieved from the optical VCMM system can be improved significantly by investing more time in developing a more robust stereovision system including the calibration plate, taking more accurate tactile CMM measurements and image points of the calibration plate and calculating the noise of image points as well as include all the possible uncertainty contributors on the VCMM.

References

1. Sladek, J., Gaska, A., Olszewska, M., Kupiec, R., Krawczyk, M.: Virtual Coordinate measuring machine built using laser tracer system and spherical standard. Metrol. Measure. Syst. **20**(1), 77–86 (2013)
2. Luthuli, Z.: Traceability of Measurements in Optical Coordinate Measuring Machines, Master Thesis, Stellenbosch University, South Africa (2020)
3. Trenk, M., Franke, M., Schwenke, H.: The "Virtual CMM" a Software Tool for Uncertainty Evaluation—Practical Application in an Accredited Calibration Lab, Department of Coordinate Metrology, 38116 Braunschweig, Germany (2009)
4. Beraldin, J.: Basic Theory on Surface Measurement Uncertainty of 3D Imaging Systems, Institute for Information Technology, Canada (2009)
5. VDE/VDI 2634—1: German Standard, Optical 3D measuring systems in Imaging systems with point by point probing (2002)
6. Sladek, J., Gaska, A.: Evaluation of coordinate measurement uncertainty with the use of virtual machine model based on monte carlo method. Measurement **45**(6), 1564–1575 (2012)
7. BS EN ISO 15530—3, Geometric Product Specifications—Coordinate Measuring Machines, Technique for determining the uncertainty of measurements: Use of calibrated workpieces or measurement standards (2011)
8. Sladek, J., Gaska, A., Olzewska, M., Kupiec, R., Krawczyk, M.: Virtual coordinate measuring machine built using laser tracer system and spherical standards. Metrol. Measure. Syst. **20**(1), 77–86 (2013)
9. Busch, K., Kunzmann, H., Waldele, F.: Calibration of coordinate measuring machines. Precis. Eng. **7**(3), 139–144 (1985)

10. Muralikrishnan, B., Rachakonda, P., Shilling, M., Lee, V., Blackburn, C., Sawyer, D., Cheok G., Cournoyer, L.: ASTM E57.02—1: Instrument run off at NIST, in Background Information and Key Findings, US Department of Commerce (2016)
11. Henning B., Schreve, K.: Laser based stereovision measurement of aspherical mirrors, Test and Measurement Conference, Muldersdrift, South Africa (2013)
12. Hartley, R., Zisserman, A.: Multiple View Geometry in Computer Vision. Cambridge University Press, Cambridge (2003)
13. BIPM JCGM 101: Guide to the Expression of Uncertainty in Measurements (2008)

Zamokuhle Luthuli holds a B.Sc. Mechanical Engineering degree from the University of Cape Town and M.Sc. Mechanical Engineering degree from Stellenbosch University. He is currently working at the National Metrology Institute of South Africa as an Engineer/Metrologist in the Length department. His specific focus and area of work and research interest is in coordinate measuring machines.

Kristiaan Schreve holds a Ph.D. in mechanical engineering from Stellenbosch University. He is currently a professor in the Department of Mechanical and Mechatronic Engineering at the same university. His research interest is in machine vision and robotics.

Oelof Kruger has a Masters Degree in Physics from University of KwaZulu-Natal. He has been involved with Dimensional Metrology over the last 30 years working for the National Metrology Institute of South Africa.

Turn-Milled High-Friction Surfaces—Investigations on the Influence of Nominal Surface Pressure and Load Direction

R. Funke and A. Schubert

Abstract An increase of the coefficient of static friction (COF) leads to higher transmissible forces in friction-locked connections. Experimental studies on turn-milling are carried out, focusing on the generation of friction-increasing surface microstructures. Single-edged TiAlN coated cemented carbide end milling cutters are applied to microstructure end faces of cylindrical specimens made of the steel 1.7225 (42CrMo4) in quenched and tempered (+QT) heat treatment condition. Face turned 1.0503 (C45) steel specimens are used as counter bodies. Two different tool types with sharp and chamfered (0.2 mm × 45°) corner are utilised to achieve profile tip angles α_{pt} of 88° and 135°. The investigations focus on the influence of the nominal surface pressure p_{nom} on the COF. Furthermore, the frictional behaviour of sawtooth-shaped microstructures as a function of the loading direction is studied. Accordingly, friction tests with microstructures produced with a sharp corner and tool inclination angles of $\beta = 15°$, 30° and 46° are conducted. The coefficient of static friction is determined using a torsion test bench. The microstructured surface causes a significant change in the torque-slip response. While the slipping curves of the reference test with face turned specimens exhibit a gradual increase, the curves resulting from tests with microstructured specimens show a local maximum, which is already achieved at low relative displacements. This indicates a micro positive locking of the structures of the harder friction partner with the softer surface of the counter body. The smaller profile tip angle leads to higher values of the COF, varying between $\mu = 0.45$ at $p_{nom} = 100$ MPa and $\mu = 0.54$ at $p_{nom} = 300$ MPa. Sawtooth-shaped asymmetric roughness profiles exhibit anisotropic friction properties.

Keywords Turn-milling · Static friction · Surface microstructuring

R. Funke (✉) · A. Schubert
Professorship Micromanufacturing Technology, Chemnitz University of Technology, 09107 Chemnitz, Germany
e-mail: roman.funke@mb.tu-chemnitz.de

A. Schubert
e-mail: andreas.schubert@mb.tu-chemnitz.de

K. von Leipzig et al. (eds.), *Smart, Sustainable Manufacturing in an Ever-Changing World*, Lecture Notes in Production Engineering,
https://doi.org/10.1007/978-3-031-15602-1_54

1 Introduction

In industrial applications such as the wind power or the automotive industry, flange connections or centrally bolted friction connections are used for torque transmission. The trend towards higher power densities sets increasing demands on the performance of such assemblies. According to Coulomb's law, the friction force results from the product of the applied normal force and the COF. Increasing the normal force is often not desired or possible, for example in the wind power industry. On the one hand, the dimensions of the assemblies should not be increased further. On the other hand, the effective surface pressure is already close to the yield point of the friction partners, which means that a further increase in the normal force is not possible. Thus, an increase of the static friction offers the potential to decrease the necessary normal force or rather to realise smaller and lighter assemblies. The properties of the surfaces in contact play a central role in the tribological properties of a tribological system. Coatings and special shims are used to increase the friction in industrial applications [1]. Another approach is the microstructuring of the surface using suitable manufacturing processes. Laser texturing of powertrain components is already applied in the automotive industry [2, 3]. Laser ablation is used to create molten and re-solidified bulged microstructures with a high hardness that penetrate the surface of the counter body and lead to an increase of friction. A similar approach is described in [4]. The maximum coefficient of static friction achieved was $\mu = 0.53$ at a surface pressure of 30 MPa. A further increase of the pressure resulted in decreased values of the COF. A combination of powerful ultrashort pulsed lasers and ultrafast polygon mirror based scan systems can increase the productivity of laser texturing significantly [5]. However, these methods increase the complexity of the process chain. They limit the user's flexibility or require expensive process equipment. A promising approach is the microstructuring by an adapted final machining process.

In previous investigations it was shown that an increase in the COF can be achieved with turn-milled surfaces [6]. At a nominal surface pressure of 100 MPa values up to $\mu = 0.69$ were gained for microstructures with a profile tip angle $\alpha_{pt} = 88°$ and $\mu = 0.44$ for microstructures with $\alpha_{pt} = 135°$. The main objective of our investigations described in this paper was to determine the influence of the surface pressure on the COF. Furthermore, we examined whether asymmetric roughness profiles result in an anisotropy of the frictional properties.

2 Methodology

2.1 *Microstructuring by Turn-Milling*

The kinematics of the turn-milling process is shown in Fig. 1. It can be described as facing using a rotating end mill. Thus, in contrast to facing, the cutting movement results from the rotation of the tool n_t. The feed motion consists of two components:

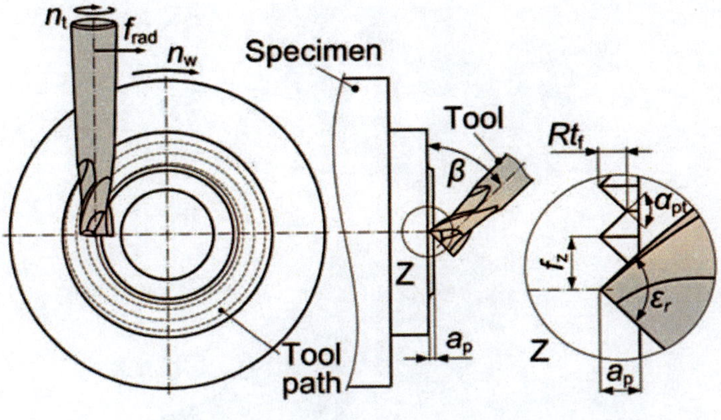

Fig. 1 Kinematics of turn-milling and resulting roughness profile using the example of a sharp corner geometry

The rotation of the specimen n_w and the radial feed f_{rad} in the direction of the specimen axis. This creates a spiral tool path on the annular end face of the specimen. The tool is inclined by an inclination angle β. Based on the arrangement of the tool and the direction of rotation of the specimen shown in Fig. 1, it is a drill cut. Due to the interrupted cut a characteristic roughness profile is generated along the tool path (Fig. 1, Detail Z). The profile tip angle α_{pt} is determined by the corner geometry. The profile height in the direction of feed motion Rt_f is additionally defined by the feed per tooth f_z and β. The feed velocity v_f along the tool path is kept constant to achieve a consistent profile height. The experiments were conducted on a 5-axis milling centre KERN Pyramid Nano, Fig. 2. Minimum quantity lubrication with polyol ester was applied with a flow rate of 27 ml h^{-1}. The annular end faces with an outer diameter of $D_o = 30$ mm and an inner diameter of $D_i = 15$ mm of cylindrical specimens were microstructured. The material used is the steel 1.7225 in quenched and tempered (+QT) heat treatment condition possessing a hardness of 340 HV10.

To create different profile tip angles of the microstructures, two types of TiAlN coated cemented carbide end milling cutters with a diameter of 6 mm were used, Fig. 3. The selection of the corner geometries was based on the results of previous investigations [6]. The microstructures with $\alpha_{pt} = 88°$ (sharp corner) resulted in the highest values of the COF. The values achieved with microstructures exhibiting a profile tip angle $\alpha_{pt} = 135°$ (chamfered corner) were slightly lower, but exhibited a less scattering. These two variants of microstructuring were used to investigate the influence of the nominal surface pressure on the COF as described in Sect. 2.2. Due to the drill cut carried out, the cutting edge on the face side of the sharp tool is the major cutting edge in the turn-milling process. This tool type features a nominal sharp corner, a tool cutting edge angle of the major cutting edge of $\kappa = 2°$ and a tool included angle $\varepsilon_r = 88°$. The resulting profile tip angle of the microstructures is $\alpha_{pt} = 88°$. The tool inclination angle β was set to 46° to achieve a symmetrical profile in the direction of feed motion. The other examined corner geometry was

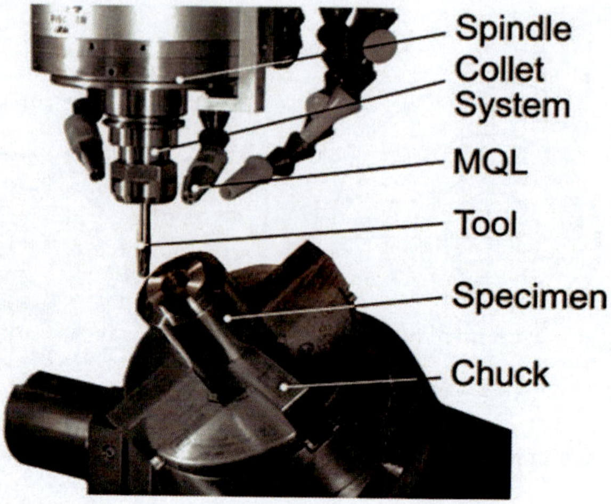

Fig. 2 Experimental setup

a chamfer with the dimensions 0.2 mm x 45°. The tool inclination angle was set to 22.5° to achieve symmetrical profile tips with $\alpha_{pt} = 135°$. Using this inclination angle affects the engagement conditions of the tool in such a way, that only the parts of the corner marked in Fig. 3 (bottom right) create the final surface topography. Both tools are standard tools with two cutting edges. To avoid a possible influence of a radial deviation of the corners on the surface microstructure, for each tool one cutting edge was removed by grinding before the experiments. The process parameters were kept constant using a cutting speed $v_c = 100$ m/min, a depth of cut $a_p = 0.2$ mm, $f_z = 0.125$ mm and $f_{rad} = 0.2$ mm in all experiments. To determine the direction-dependent friction properties of asymmetric sawtooth-shaped profiles, experiments were conducted using the sharp corner and the above-mentioned cutting data. The tool inclination angle was set to 15°, 30° and 46°.

2.2 *Tribological Evaluation*

A test bench of the Professorship Machine Elements and Product Development at Chemnitz University of Technology was used to determine the COF, Fig. 4a.

During the bench test, two specimens were coaxially fastened with a normal force F_N resulting in a corresponding nominal surface pressure p_{nom} in the annular contact area. Subsequently, they were twisted against each other until reaching a defined angle $\varphi = 5°$. This twisting angle φ was converted into a relative displacement s_r along the mean friction diameter D_m as follows:

$$s_r = \varphi \cdot \frac{D_m}{2}, \tag{1}$$

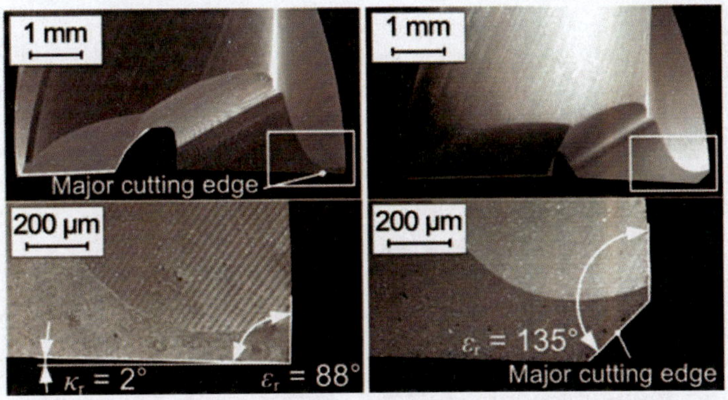

Fig. 3 SEM micrographs of the tools used for turn-milling of microstructures

$$D_{\mathrm{m}} = \frac{2}{3}\left(\frac{D_o^3 - D_i^3}{D_o^2 - D_i^2}\right) = 23.33\mathrm{mm}. \tag{2}$$

In principle three different types of slipping curves can be identified, Fig. 4b. Since all curves determined within the experiments showed a characteristic of type A, the COF μ was calculated on the basis of the maximum torque $T_{\max}$ determined within the entire measuring range, F_{N} and D_{m} as described in [1]:

$$\mu = \frac{2 \cdot T_{\max}}{F_N \cdot D_m}. \tag{3}$$

In each test a new microstructured specimen was mated with a face turned counter body without any additional microstructuring. The counter bodies are made from the medium carbon steel 1.0503 (C45) and possess a hardness of around 200 HV10. They were machined using an indexable insert of the specification CCMT 09T304 WS with a wiper corner, a cutting speed $v_{\mathrm{c}} = 200$ m min^{-1}, a feed $f = 0.1$ mm and a depth of cut $a_{\mathrm{p}} = 0.3$ mm. The specimens featured a mean roughness depth $Rz \approx 1$ µm. All tests were conducted dry. The specimens were cleaned in ultrasonic bath before the bench tests. The nominal surface pressure was varied in steps of 50 MPa between 100 MPa and 300 MPa.

2.3 *Surface Evaluation*

Using a Keyence VK-X1000 microscope, large-scale overview images were taken based on focus variation. Furthermore, each specimen machined was measured using a 3D laser scanning microscope Keyence type VK-9700. The data was analysed

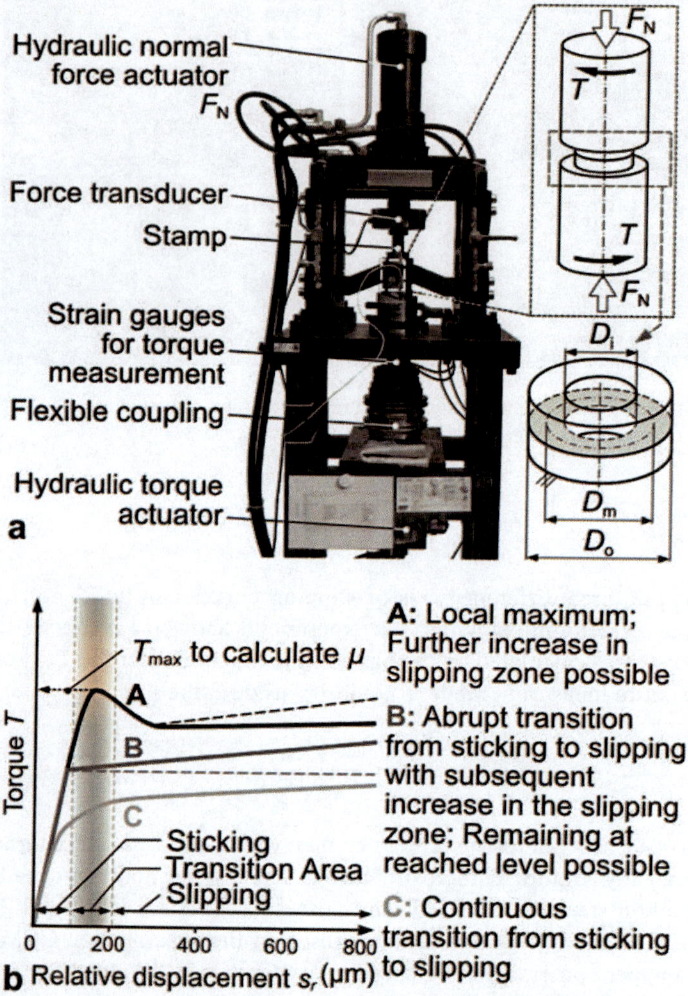

Fig. 4 Determination of the COF: **a** Test bench, **b** three types of typical slipping curves (based on [7] and [1]) and the principle of the evaluation method of μ on the basis of curve A

with the surface analysis software MountainsMap. The raw data was filtered with a smoothing median filter and levelled with the method of the least squares plane.

3 Results and Discussion

3.1 *Influence of Profile Tip Angle and Nominal Surface Pressure*

The influence of the tool geometry on the surface topography is shown in Fig. 5. As described in [6, 8] the intersection of adjacent single microstructures results in regular patterns on the surface. Although the surfaces were produced using identical cutting data they differ significantly. In both variants, the microstructures are aligned transversely to the load direction realised in the test bench tests. The surface generated with the chamfered corner exhibits more or less uniform patterns of parallel ridges. On the surface produced with the sharp corner such ridges only occur locally (cf. Figure 8).

Sa	3.82 μm	8.27 μm
Vmp	0.167 ml m^{-2}	0.795 ml m^{-2}

As expected, the surface with the smaller profile tip angle possesses a higher profile height or arithmetic mean height *Sa*. Since a good correlation between the peak material volume *Vmp* and the COF was shown in previous investigations [6], this parameter is also used for evaluation. This relationship is also confirmed by the results of the bench test shown in Fig. 6. The bars represent the mean COF values μ

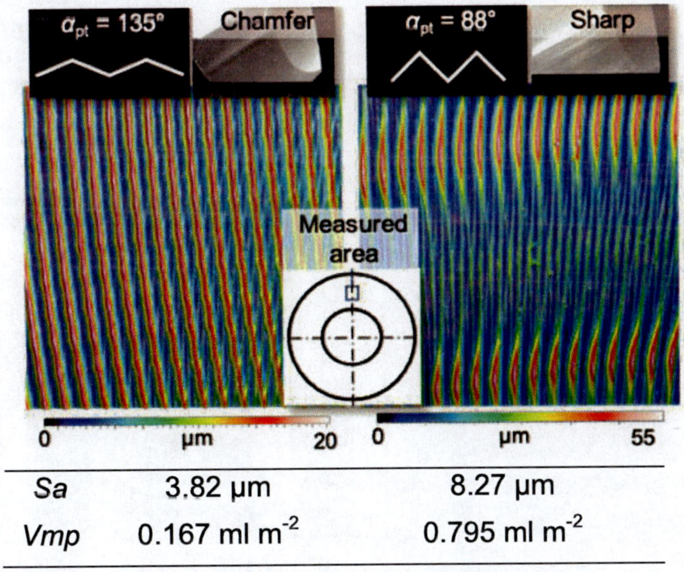

Fig. 5 Surface details of the machined specimens. (Size: (2 × 2) mm^2)

of three individual bench tests. The error bars denote the minimum and maximum values. Microstructures with the smaller profile tip angle lead to a higher COF. The values tend to increase with growing nominal surface pressure and range between $\mu = 0.45$ at $p_{\text{nom}} = 100$ MPa and $\mu = 0.54$ at $p_{\text{nom}} = 300$ MPa, a difference of about 20%. The bench tests with microstructures with $\alpha_{\text{pt}} = 135°$ reveal values between 0.34 and 0.4 without a discernible trend. The friction characteristic determined in the bench tests with microstructured specimens differ significantly from those obtained with two unstructured, face turned specimens with roughness values of $Rz \approx 1$ µm of the reference test, Fig. 7. The bench tests with microstructured specimens lead to slipping curves of the type A (cf. Figure 4b). This indicates a micro positive locking of the microstructures which penetrate the surface of the counter body. It is assumed that the increase in the COF compared to the reference test with unstructured specimens of the same material pairing results mainly from an increase in the deformative friction component. The torque curve of the reference test shows the characteristic of type C. This is typical for tribological systems in which the adhesive component of the friction is predominant. The differences between the torque curves can be explained on the basis of the surfaces of the friction partners after the bench test, Fig. 8. The profil peaks of the microstructures penetrate the surface of the softer counter body under the effect of the normal force. It is assumed that this interlocking withstand an increasing tangential force until one of the two failure mechanisms occurs: either the peaks are sheared off or they plough in the surface of the counter body. The analysis of the surfaces shows that apparently both effects take place for the experiments using the microstructures with the smaller profile tip angle. This effect already occurs with the smallest of the nominal surface pressures examined. After the tests, residues of the material from the counter body are present between the partially sheared off microstructures. Even at $p_{\text{nom}} = 300$ MPa, contact between the friction partners solely occurs in the area of the surface patterns. In the case of the microstructures with the larger profile tip angle, only local indentations of the profile tips on the surface of the counter body are present after the bench test at $p_{\text{nom}} = 100$ MPa.

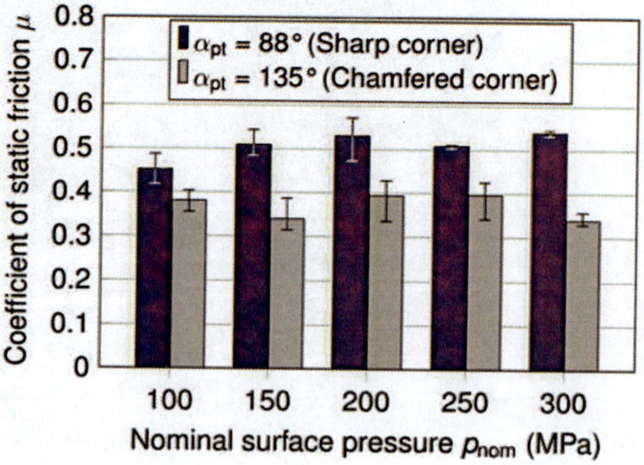

Fig. 6 Influence of the profile tip angle and the nominal surface pressure on the COF

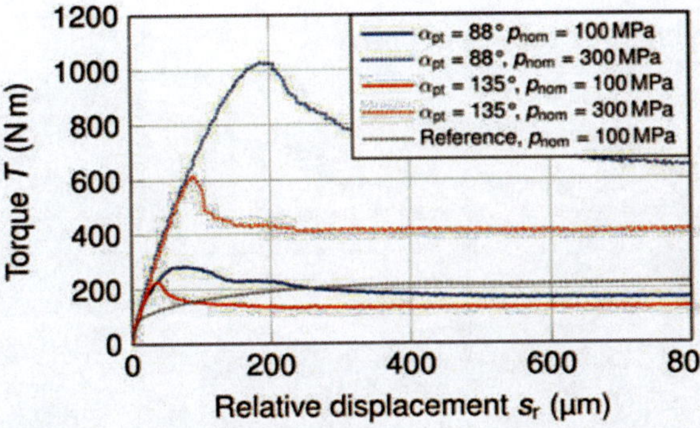

Fig. 7 Torque curves against the relative displacement for microstructures of different profile tip angles and nominal surface pressures

At surface pressures in excess of approx. 250 MPa microploughing also occurs. However, no material is removed from the counter body even at $p_{\mathrm{nom}} = 300$ MPa.

3.2 *Direction-Depending Friction Properties*

The influence of the tool inclination angle β on the surface topography is shown in Fig. 9. The images on the left represent the measured topography of the machined surface. The surface detail on the right illustrate the ideal topography generated with a Matlab-based dexel model. The comparison shows a good overall agreement, although there are differences in the height of the profiles and the shape of the patterns. As stated in [8], there is a high sensitivity of the surface patterns with regard to a change of the feed per tooth. According to this, varying f_z in the micrometre range leads to a significant modification of the patterns. The deviations between the real and the simulated surfaces suggest that there are small differences of f_z from the nominal value in the machining process. The values of *Vmp* for the measured surfaces are significantly lower than the values of the simulated surfaces.

The profile sections extracted from the topography data are shown in Fig. 10 a. The microstructures produced with tool inclination angles of 15° and 30° possess a sawtooth-shaped roughness profile, while using $\beta = 46°$ results in an almost symmetrical profile. The load directions *against* and *along* were realised by changing the direction of rotation in the bench test. With the load orientation *against*, all three variants lead to a nearly identical COF $\mu = 0.44$ ($\beta = 15°$ and 46°) and $\mu = 0.45$ ($\beta = 30°$). The load direction *along* leads to a significant reduction in the COF by 40% to $\mu = 0.27$ for microstructures machined with $\beta = 15°$ and by around 30% to $\mu = 0.31$ for $\beta = 30°$.

The microstructures with symmetrical profiles even lead to a slightly increased COF of $\mu = 0.47$. However, for this variant the slipping curves exhibit two

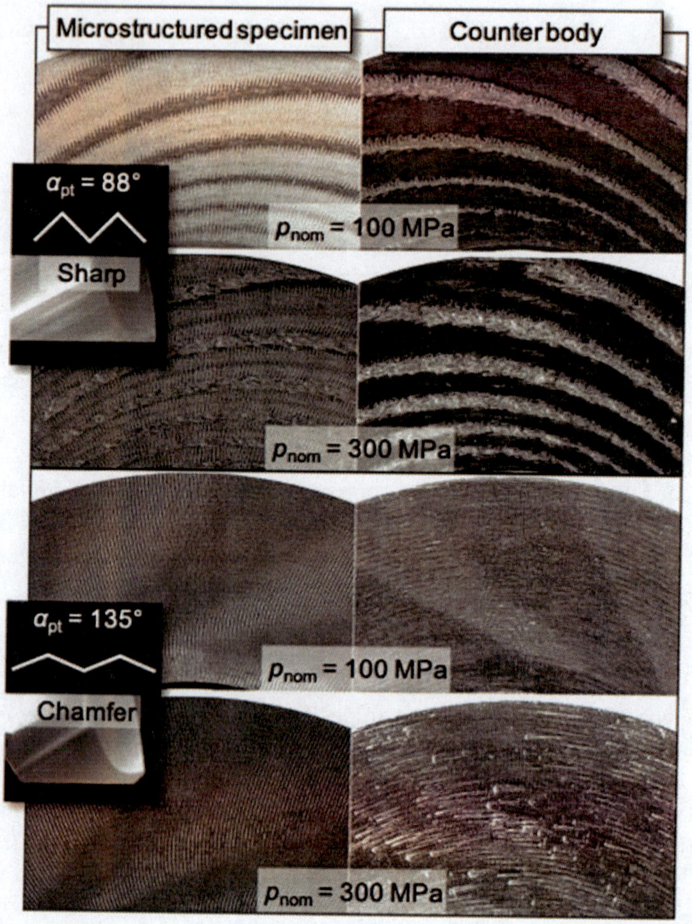

Fig. 8 Surfaces of different friction partners after the bench test

pronounced maxima during the bench test, Fig. 11. The torque initially drops after a first peak, but rises again with increasing relative displacement. This effect is only evident for this structural variant and can also be detected in a somewhat weakened form for the load direction against. The cause is still unclear. It is assumed that during ploughing material increasingly builds up in front of the microstructures until they finally shear off. The results show that the correlation of *Vmp* and the COF, that was determined in previous investigations for symmetrical profiles does not longer apply for asymmetrical profiles.

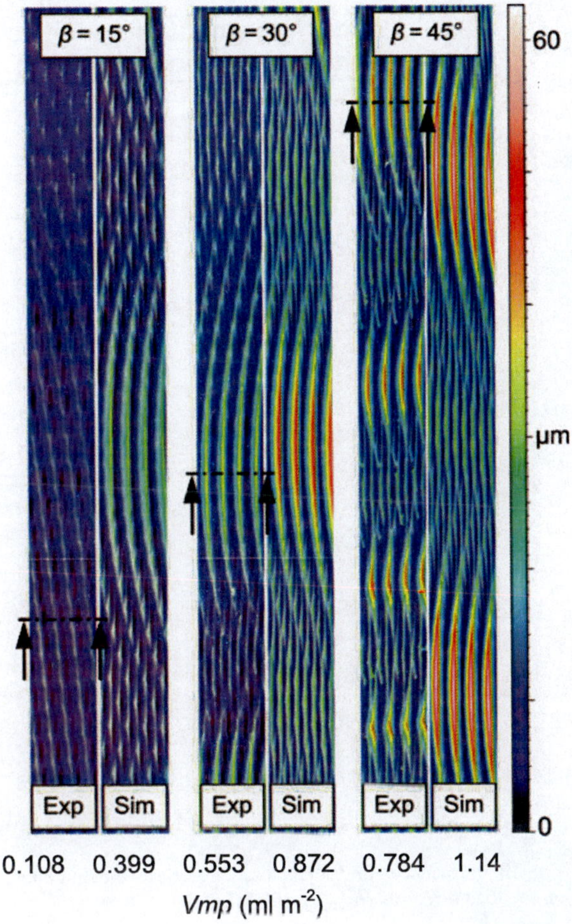

Fig. 9 Surface topography of machined and simulated surfaces using different tool inclination angles. (Tool: sharp corner. Size of the surface sections: (0.48 × 6) mm^2)

4 Conclusions

From the experimental investigations the following conclusions can be drawn:

- The microstructures generated by turn-milling penetrate the surface of the counter body. This leads to a micro positive locking and an increase of the COF compared to the reference test with unstructured specimens. The highest COF $\mu = 0.54$ is achieved using microstructures with $\alpha_{pt} = 88°$ at $p_{nom} = 300$ MPa.
- Microstructures with the smaller profile tip angle $\alpha_{pt} = 88°$ consistently lead to higher values of the COF compared to microstructures with $\alpha_{pt} = 135°$.
- The nominal surface pressure shows no clear influence on the COF. The values of μ vary between 0.45 at $p_{nom} = 100$ MPa and 0.54 at $p_{nom} = 300$ MPa for the smaller profile tip angle and 0.34 at $p_{nom} = 150$ MPa and 300 MPa and 0.4 $p_{nom} = 250$ MPa for $\alpha_{pt} = 135°$.

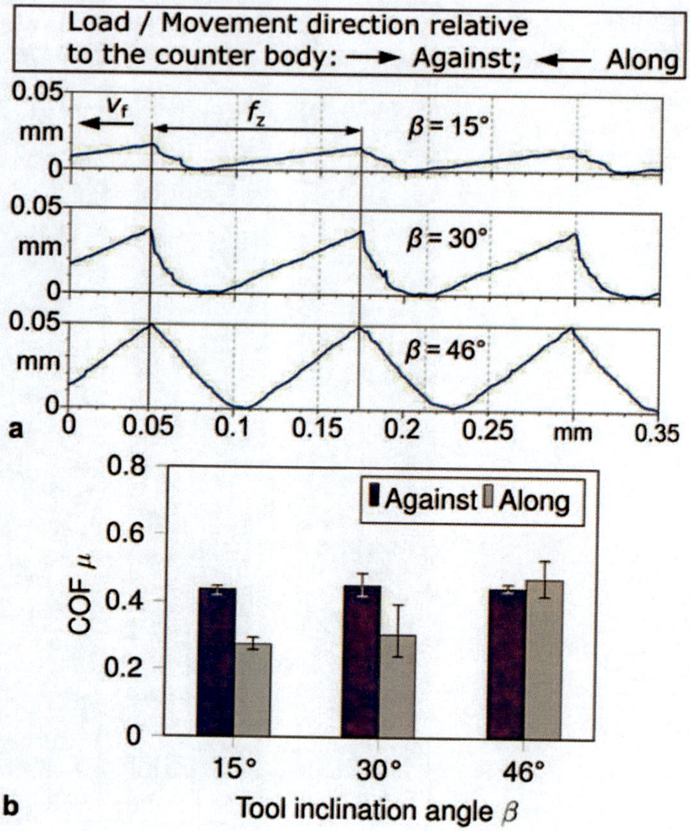

Fig. 10 Influence of the tool inclination angle on the roughness profile (**a**) and the COF μ depending on the load direction ($p_{nom} = 100$ MPa) (**b**)

Fig. 11 Influence of the tool inclination angle and the load direction on the slipping curve

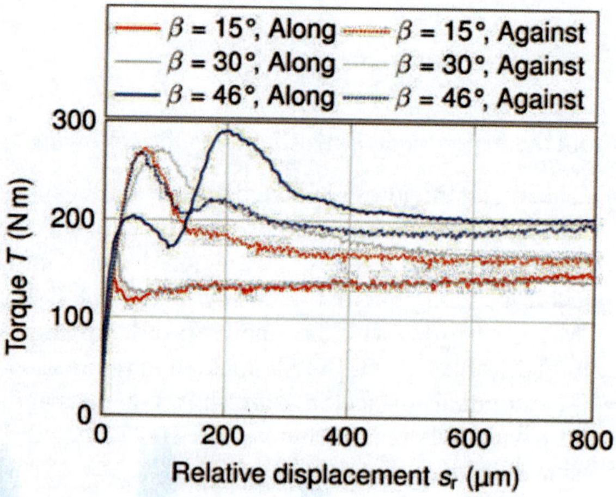

- Sawtooth-shaped asymmetrical roughness profiles exhibit directional friction properties. The COF decreases for the loading direction *along* the sawtooth profile. Microstructures with a symmetrical roughness profile, on the other hand, does not show any pronounced anisotropy of the friction properties.

Acknowledgements The authors gratefully acknowledge the funding of this research by the Deutsche Forschungs-gemeinschaft (German Research Foundation, DFG) within the project "Aimed setting of coefficient of static friction via finishing by turn-milling" (Project-ID: 411688125).

References

1. Leidich, E., Reiß, F.: Transferability of model-based static coefficients of friction to real components. MTZ Worldwide **79**, 62–67 (2018)
2. Flores, G., Wiens, A.: Herstellung von Funktionsoberflächen mit hoher Haftreibung durch Laserstrukturieren (written in German). VDI-Z Integrierte Produktion **158**, 62–64 (2016)
3. Flores, G., Wiens, A.: Haftreibung im Verbrennungsmotor. Laserbearbeitungsver-fahren erzeugen Oberflächen mit Haft-reibungsfunktionalität (written in German). VDI- Berichte **2230**, 109–117 (2014)
4. Schille, J., Ullmann, F., Schneider, L., Gräfensteiner, M., Schiefer, S., Gerlach, M., Leidich, E., Exner, H.: Experimental study on laser surface texturing for friction coefficient enhancement. J Laser Micro Nanoeng **10**, 245–253 (2015)
5. Schille, J., Schneider, L., Mauersberger, S., Szokup, S., Höhn, S., Pötschke, J., Reiß, F., Leidich, E., Löschner, U.: High-Rate laser surface texturing for advanced tribological functionality. Lubricants **8**, 1–20 (2020)
6. Funke, R., Schubert, A.: Effects of process parameters during turn-milling of microstructured surfaces on the coefficient of static friction. Procedia CIRP **77**, 255–258 (2018)
7. N. N.: Boltcouncil – Research Council on Structural Connections, Specification for Structural Joints Using High-Strength Bolds, www.boltcouncil.org (2009)
8. Funke, R., Börner, R., Schubert, A.: Geometrische Oberflächencharakterisierung von durch spanende Verfahren hergestellten Mikrostrukturen (written in German). Sächsisches Geometriesymposium **2020**, 143–153 (2020)

Roman Funke holds his Dipl.-Ing. degree in Mechanical Engineering from Chemnitz University of Technology. He is research associate at the Professorship Micromanufacturing Technology at Chemnitz University of Technology.

Andreas Schubert obtained his Ph.D. degree in Mechanical Engineering from Technische Universität Dresden. In 2003 he was appointed Professor for Micromanufacturing Technology at Chemnitz University of Technology. Furthermore, he is head of the Competence Center Micromanufacturing and Surface Technologies – KoMOT at Fraunhofer IWU in Chemnitz.

The Development and Inverse Kinematics of a 5 DOF Parallel Kinematic Architecture Machining System

W. Dharmalingum, J. Padayachee, J. Collins, and G. Bright

Abstract Innovative solutions to South Africa's economic challenges are required in order provide a better future for the country. One area of challenge, lies in the manufacturing industry. Traditional machining centres are typically expensive, bulky and heavy. This paper presents the research and development of a novel architecture parallel kinematic machine (PKM) for machining purposes. Parallel architectures offer several promising advantages over the more traditional serial machines. The presented machine offers five degrees of freedom and is suitable for drilling and milling tasks. It has been designed to overcome the challenges of weight and cost of traditional machines. The development of the mechanical design is discussed as well as a study of the inverse kinematics of the machine.

Keywords Manufacturing · PKM · Inverse Kinematics

1 Introduction

South Africa faces several challenges in its economic development. One of the areas of significant challenge and opportunity lies is the manufacturing sector. The generation of employment opportunities is a key factor in producing economic growth. The manufacturing industry can be a significant source of job opportunities since it is often labour intensive [1].

W. Dharmalingum (✉) · J. Padayachee · J. Collins · G. Bright
Mechatronics and Robotics Research Group, Department of Mechanical Engineering, University of KwaZulu-Natal, Durban, South Africa
e-mail: DharmalingumW@ukzn.ac.za

J. Padayachee
e-mail: padayacheej@ukzn.ac.za

J. Collins
e-mail: CollinsJ@ukzn.ac.za

G. Bright
e-mail: Brightg@ukzn.ac.za

K. von Leipzig et al. (eds.), *Smart, Sustainable Manufacturing in an Ever-Changing World*, Lecture Notes in Production Engineering,
https://doi.org/10.1007/978-3-031-15602-1_55

One of the major inhibitors to the growth of the manufacturing sector and entrepreneurship is the high initial capital investment to start a business [2, 3]. Machines often need to be imported and this adds significantly to the cost. Manufacturing equipment, especially modern CNC production machines are large and bulky further adding to the capital outlay.

Not only is equipment sourced from overseas but in many instances the technical 'know-how' to maintain and repair these machines also has to be imported. In instances where a machine needs to be repaired or even serviced, personnel have to be flown from another country. This further adds to the monetary outlay required to run the business. And, if the event is an emergency breakdown, valuable production time could be lost waiting for the expert to arrive from overseas.

Another cause for concern with many machining centres is that they are based on the use of proprietary software [4]. Each machine producer typically has its own software specific to that make and, in some cases, even at a machine level. This requires a unique knowledge of how to programme different machines. It also means that data cannot be transferred between machines or the different processes often necessary to produce a part.

This paper showcases the research undertaken by the Mechatronics and Robotics Research Group (MR^2G) to try and address some of these obstacles to the progress of manufacturing technical ability and capacity in developing nations such as South Africa.

Can precision manufacturing equipment be developed that is?

(a) Not prohibitively expensive to the new investor
(b) Not imported
(c) Portable enough to allow for a 'garage workshop' to use it.
(d) Developed locally so that expertise to maintain it is readily available
(e) Software neutral and in-line with modern programming practices which allows for portability of data between machines and processes.

Research was conducted into the area of Parallel Kinematic Machines (PKMs). Traditionally, machining centres have been built on a serial or linear architecture design. Serial architectures have the distinct disadvantage that they may only have one point of contact with the base of the machine and that errors are compounded along the length of the serial connections. In order to provide the necessary rigidity along the serial arm, significant bulk is added to the machine. The bulk, of course adds to the cost of the machine and reduces its portability.

Parallel machines are based on the idea that there are several connections to the base of the machine. This offers the promise of increased rigidity for reduced weight [5]. The reduction in weight can also lead to faster motion within the machine's workspace. This has been used to great advantage by pick-and-place robots such as the ABB IRB 340 Flex Picker [6].

Parallel architectures, while offering several advantages, do have their drawbacks. One significant drawback is that relative to their size, their workspace may be quite small [7].

The paper presents the development of novel PKM designed for machining purposes. It walks through the progress of the research from the successful build of a prototype machine to the development of a full-scale machining centre with added features such as a mounted spindle and unique gripper accessory. A study of the inverse kinematics of the unique architecture is also presented.

2 Desktop Prototype

The initial phase of developing the machining centre involved the production of a desktop model. The architecture along with the kinematics of the desktop model were presented at COMA 19 [8]. It is briefly described here to show the progression of the research.

The structural type classification of PKMs presented by Koseki et al. [9] was used as a basis for determining the structure of the design. Further inspiration was drawn from the very well-known Stewart–Gough hexapod platform. The hexapod platform offers good strength and precision, all six degrees of freedom but a limited workspace relative to its size [10].

The design that the research group settled on incorporated a novel configuration of the six powered prismatic joints. All the rotational joints in the assembly were passive joints. The six legs were coupled together in three pairs which resulted in a machine that could produce five degrees of freedom. There were three translational degrees of freedom, two independent rotational degrees of freedom about the X and Y axes and a parasitic rotation about the Z-axis. A photograph of the desktop prototype is shown in Fig. 1.

3 Full-Scale Design

The prototype shown in Fig. 1 is inverted from what the final machining centre would be. This was done to make for ease of concept trial. A suspended frame would not be required to be built and different concept iterations could readily be switched out during the development process of the final architecture.

3.1 The Machine Frame

In order to build the full-scale machine, an appropriate frame was designed to rigidly suspend the three pairs of legs in an arrangement suitable for machining purposes. The frame is shown in Fig. 2 and was 2 m tall by 1 m deep by 1.5 m wide.

As can be seen from Fig. 2, the majority of the framework was constructed from square tubing. The roof structure consisted of 5 mm steel welded to the frame. The

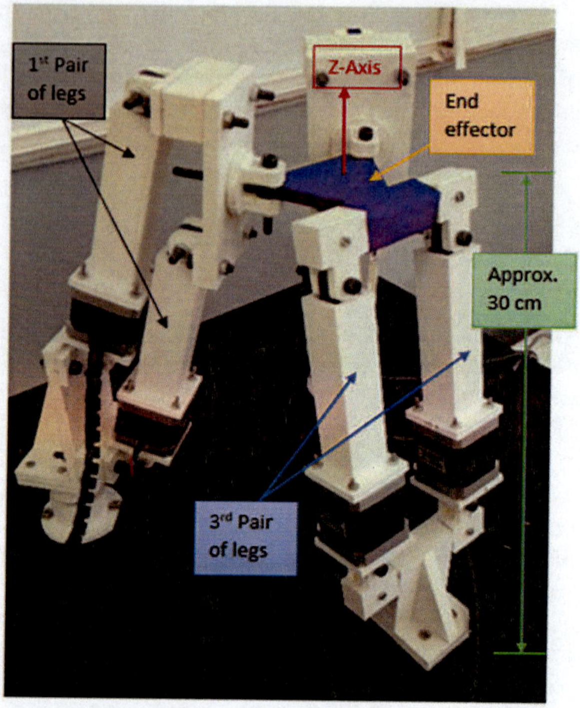

Fig. 1 The desktop prototype

figure also shows that the roof frame structure has extra members to increase rigidity and provide a stable platform from which to hang the actuator arms.

3.2 *Linear Actuator Design and Assembly*

The linear actuator design was based on the use of electric motors rather than hydraulic or pneumatic actuators. Even though electric actuators may be heavier with a slower dynamic ability, controlling them would be easier. Precision is paramount when considering machining applications and using electric actuators would facilitate precise behaviour from the machining system. A single linear actuator assembly is shown in Fig. 3. Each linear actuator provided a stroke length of 250 mm.

The assembly of the linear actuators together with their coupling brackets and attachment to the maching spindle is shown in Fig. 4. The mounting points for the pairs of legs were arranged 120° apart to provide stability and strength to the structure. This arrangement also maximised the workspace of the machine. The arrangment of the legs and the spindle is shown in Fig. 4.

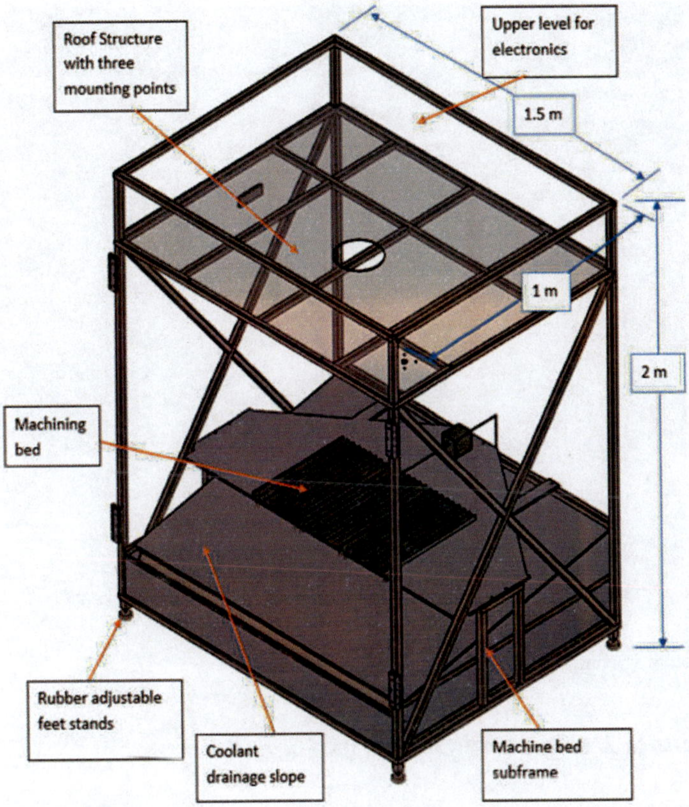

Fig. 2 The machine frame

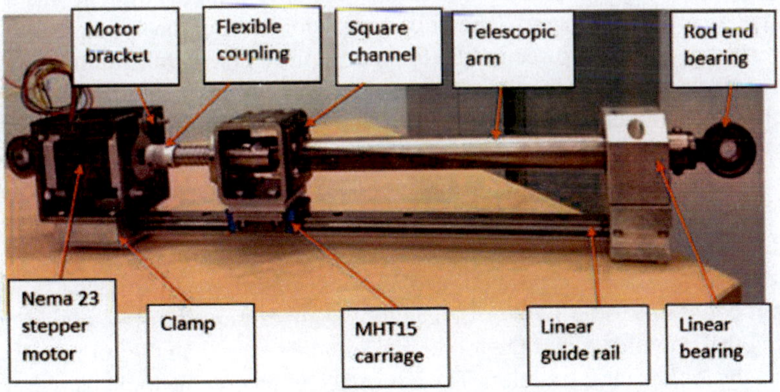

Fig. 3 Linear actuator

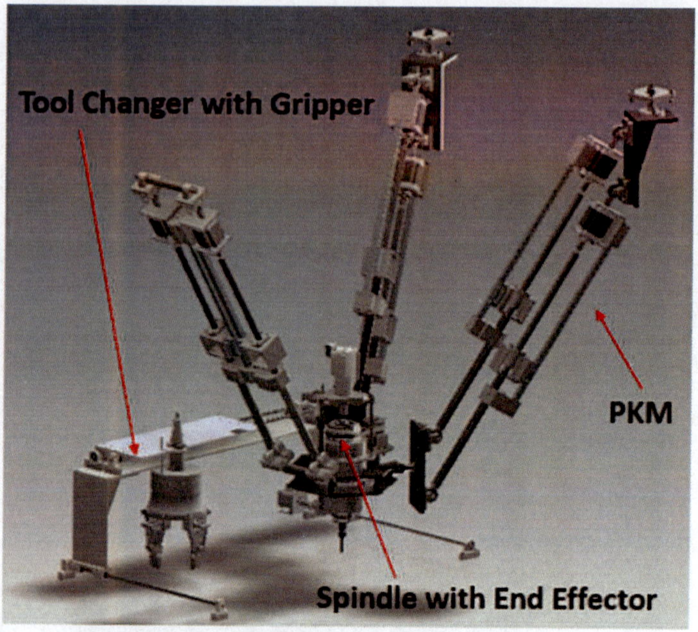

Fig. 4 Machining configuration of the actuators and spindle

3.3 Spindle Design and Other Auxiliary Equipment

The spindle was designed to allow for the intended milling and drilling application of the machine. It incorporated a BT30 tool taper and the drive was able to deliver a maximum operating speed of 10 000 rpm. A pneumatic clutch system was used for the drawbar actuation system. The pull clamp tool locking mechnism could exhert 10 kN of force in order to secure the tooling in place during machining operations. The spindle was driven using a timing belt design with a 1.5:1 pulley ratio between the pulleys. The reduction ratio was able to increase the torque of the spindle but at the cost of revolutional speed. The spindle assembly is shown in Fig. 5.

To enhance the capability of the machine a gripper mechanism was designed. The additional feature of having a gripper would allow the machine to work with parts without the need for human intervention.

The machine also incorporated a tool changing station. In this iteration of the design it could only hold two tools. The tool changer moved in two planar degrees meaning that it could deliver the tool to the spindle rather than the spindle having to move to the tool changing station. The advantage of having the tool changer move is that the tools could be stored away from the workspace of the machine. Since PKMs have a limited workspace for their size, this would allow for all of the workspace to be used for machining and not sacrifice any of it for tool changing. The developed machine required a budget of about R100,000 whereas a typical used CNC milling

Fig. 5 The spindle assembly

would cost in the region of a million rand [11]. In terms of weight, the developed machine came in at a few hundred kilograms whereas CNC machines often start in the tons of kilograms.

4 Inverse Kinematic Analysis

In order to develop a control system for a machining system, one has to ascertain the inverse kinematic relationship between the end effector and the actuator positions. Knowing what the required placement of each actuator needs to be in order to position machine head in the desired location is paramount to developing the control system for the machine. This section describes the methodology used to determine the inverse kinematics of the unique architecture.

The novel architecture was created using revolute and prismatic joints which only permit 1-DOF each. The use of 1-DOF joints to improve tighter tolerances and obtain higher machine accuracies has not been widely investigated for PKMs. As a substitute to universal joints, two individual revolute joints were implemented in series where applicable. The PKM is capable of three translations, independent rotations about the x and y axes and possesses a parasitic rotation about the z axis. The parasitic rotation is induced when rotations about the x and y axes occur sequentially or simultaneously. A linear actuator was used in each leg to achieve prismatic actuation. The PKM topology is shown in Fig. 6.

The use of revolute joints in series resulted nested kinematic loops that are formed by a pair of coplanar legs. The legs of the PKM are therefore grouped as leg pair

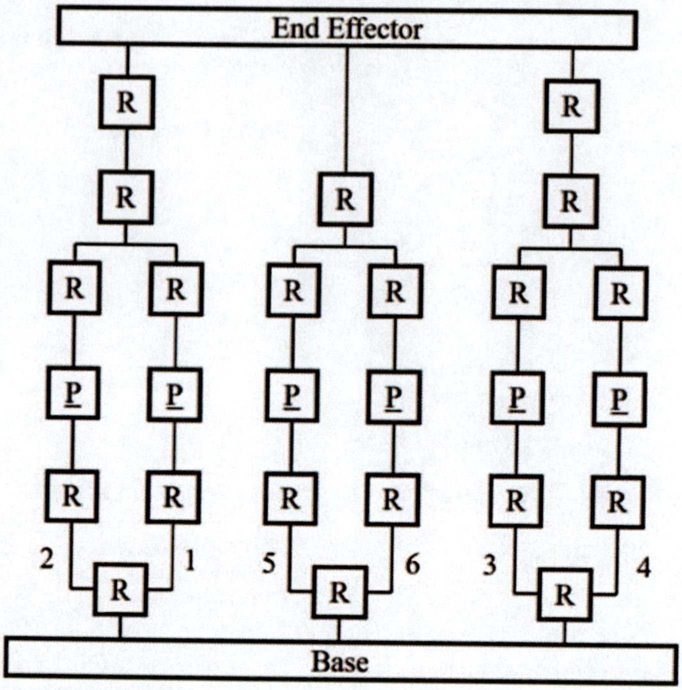

Fig. 6 PKM topology

1–2, leg pair 3–4 and leg-pair 5–6. When the end effector undergoes translation, the nested kinematic loop exhibits a Parallelogram (Pa) structure. When the end effector undergoes rotational motion, the nested kinematic loop exhibits an Irregular Quadrilateral (IQ) structure. Hence the PKM was named the 2-R(Pa-IQ)RR, R(Pa-IQ)R Parallel Manipulator.

Additional variables were created by joint offsets due to the serial arrangement of some revolute joints [12, 13]. An extension of the geometric (vector) method was employed to address the additional variables. An outer and inner vector loop calculation was employed. The outer vector loop is illustrated in Fig. 7. The outer vector loop was constructed using points O, A, D and P. The global and local coordinate systems were placed at point O and point P respectively. The vector $\overrightarrow{AD}$ (Point D relative to Point A) needed to be solved. The parasitic rotation was not investigated hence the rotation matrix was simplified as shown in Eq. 2.

Using leg pair comprised of leg 1 and 2, Eq. 1 was developed.

$$\overrightarrow{OA_{1,2}} + \overrightarrow{A_{1,2}D_{1,2}} = \overrightarrow{OP} + R(\beta, \alpha)\overrightarrow{PD_{1,2}} \tag{1}$$

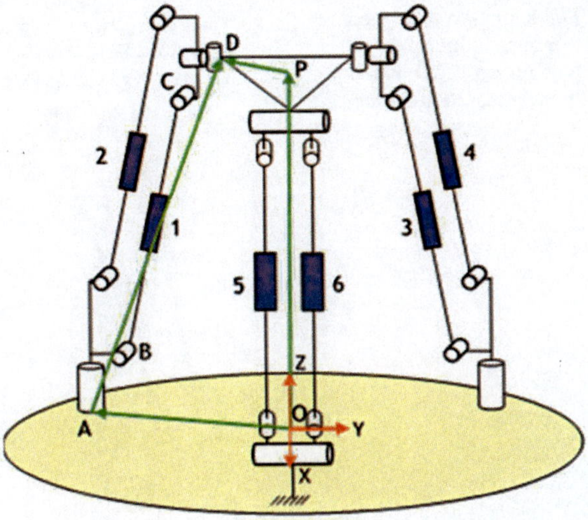

Fig. 7 Outer vector loop diagram

$$\begin{bmatrix} \left(\overrightarrow{OA_{1,2}}\right)_x \\ \left(\overrightarrow{OA_{1,2}}\right)_y \\ \left(\overrightarrow{OA_{1,2}}\right)_z \end{bmatrix} + \begin{bmatrix} \left(\overrightarrow{A_{1,2}D_{1,2}}\right)_x \\ \left(\overrightarrow{A_{1,2}D_{1,2}}\right)_y \\ \left(\overrightarrow{A_{1,2}D_{1,2}}\right)_z \end{bmatrix} = \begin{bmatrix} \left(\overrightarrow{OP}\right)_x \\ \left(\overrightarrow{OP}\right)_y \\ \left(\overrightarrow{OP}\right)_z \end{bmatrix} + \begin{bmatrix} c\beta & s\beta\, s\alpha & s\beta\, c\alpha \\ 0 & c\alpha & -s\alpha \\ -s\beta & c\beta\, s\alpha & c\beta\, c\alpha \end{bmatrix} \begin{bmatrix} \left(\overrightarrow{PD_{1,2}}\right)_x \\ \left(\overrightarrow{PD_{1,2}}\right)_y \\ \left(\overrightarrow{PD_{1,2}}\right)_z \end{bmatrix} \tag{2}$$

where c = cosine *and* s = sine

If L, M and N represent the first, second and third rows of the expanded form of Eq. 2, the magnitude of vector $\overrightarrow{A_{1,2}D_{1,2}}$ is represented by Eq. 3:

$$\left|\overrightarrow{A_{1,2}D_{1,2}}\right| = \left| \begin{matrix} \left(\overrightarrow{A_{1,2}D_{1,2}}\right)_x \\ \left(\overrightarrow{A_{1,2}D_{1,2}}\right)_y \\ \left(\overrightarrow{A_{1,2}D_{1,2}}\right)_z \end{matrix} \right| = \sqrt{(L)^2 + (M)^2 + (N)^2} \tag{3}$$

The inner vector loop diagrams are shown in Fig. 8. Each pair of legs move along a common plane due to the mechanical arrangement of the joints. Vector $\overrightarrow{AD}$ was used in the inner vector loop equation to solve the length of the actuators 1 and 2 which are represented by vector $\overrightarrow{B_1C_1}$ and $\overrightarrow{B_2C_2}$ respectively. Vector $\overrightarrow{AB}$ and vector $\overrightarrow{CD}$ are machine parameters. This method was applied to solve the leg lengths of all other leg pairs.

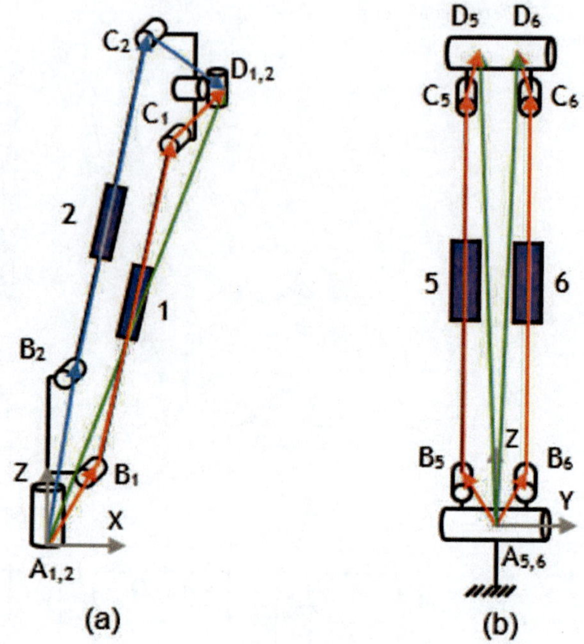

Fig. 8 Inner vector loop diagrams. **a** Inner vector loop for leg pair 1 and 2. **b** Inner vector loop for leg pair 5 and 6

The inner vector loop concerning leg pair 1–2 is illustrated by Eq. 4:

$$\overrightarrow{B_iC_i} = \overrightarrow{A_{1,2}D_{1,2}} - \overrightarrow{A_{1,2}B_i} - \overrightarrow{C_iD_{1,2}} \tag{4}$$

where $i = 1or2$

Due to the coplanar nature of the leg pairs, the vector $\overrightarrow{AD}$ was reduced from a 3-dimensional vector to a 2-dimensional vector. The y component of vector $\overrightarrow{AD}$ was set to zero without losing vector integrity. The x value of vector $\overrightarrow{AD}$ could be calculated using the Theorem of Pythagoras since the z value was determined from the outer vector loop calculation. Leg pair 3–4 is identical to leg pair 1–2 and the same method was applied to leg pair 3–4. Leg pair 5–6 employed the same method except that the y value of vector $\overrightarrow{AD}$ was retained and the x value was set to zero. The z value was calculated using the Theorem of Pythagoras.

The inner vector loop calculation in Eq. 5 represents actuator 1 and 2.

$$\begin{bmatrix} \left(\overrightarrow{B_iC_i}\right)_x \\ \left(\overrightarrow{B_iC_i}\right)_y \\ \left(\overrightarrow{B_iC_i}\right)_z \end{bmatrix} = \begin{bmatrix} \left(\overrightarrow{A_{1,2}D_{1,2}}\right)_x \\ \left(\overrightarrow{A_{1,2}D_{1,2}}\right)_y \\ \left(\overrightarrow{A_{1,2}D_{1,2}}\right)_z \end{bmatrix} - \begin{bmatrix} \left(\overrightarrow{A_{1,2}B_i}\right)_x \\ \left(\overrightarrow{A_{1,2}B_i}\right)_y \\ \left(\overrightarrow{A_{1,2}B_i}\right)_z \end{bmatrix} - \begin{bmatrix} \left(\overrightarrow{C_iD_{1,2}}\right)_x \\ \left(\overrightarrow{C_iD_{1,2}}\right)_y \\ \left(\overrightarrow{C_iD_{1,2}}\right)_z \end{bmatrix} \tag{5}$$

Equation 5 is in terms of the x–z plane and all y matrix entries are zero.

$$\begin{bmatrix} \left(\overrightarrow{B_iC_i}\right)_x \\ 0 \\ \left(\overrightarrow{B_iC_i}\right)_z \end{bmatrix} = \begin{bmatrix} \sqrt{\left|\overrightarrow{A_{1,2}D_{1,2}}\right|^2 - \left(\overrightarrow{A_{1,2}D_{1,2}}\right)_z^2} - \left(\overrightarrow{A_{1,2}B_i}\right)_x - \left(\overrightarrow{C_iD_{1,2}}\right)_x \\ 0 \\ \left(\overrightarrow{A_{1,2}D_{1,2}}\right)_z - \left(\overrightarrow{A_{1,2}B_i}\right)_z - \left(\overrightarrow{C_iD_{1,2}}\right)_z \end{bmatrix} \tag{6}$$

Q and R represent the first and third rows of the right-hand side of Eq. 6. The magnitude of vector $\overrightarrow{B_iC_i}$ is calculated as shown in Eq. 7.

$$\left|\overrightarrow{B_iC_i}\right| = \left|\begin{matrix} \left(\overrightarrow{B_iC_i}\right)_x \\ 0 \\ \left(\overrightarrow{B_iC_i}\right)_z \end{matrix}\right| = \sqrt{(Q)^2 + (R)^2} \tag{7}$$

When the end effector exhibits rotation, the vector $\overrightarrow{C_iD_{1,2}}$ vector changes to accommodate the rotation. Figure 9 illustrates that when the mounting bracket leans forward, the x and z vector components of vector $\overrightarrow{C_iD_{1,2}}$ changes. Additional variables such as angle ψ_1 and positions of point E and F were included to calculate $\left(\overrightarrow{C_iD_{1,2}}\right)_x'$ and $\left(\overrightarrow{C_iD_{1,2}}\right)_z'$. The same methodology was applied to leg pair 3–4 and also when the mounting brackets leaned backward.

The altered $\overrightarrow{C_1D_{1,2}}$ vector for leg length 1 is shown in Eq. 8 and 9.

$$\left(\overrightarrow{C_1D_{1,2}}\right)_x' = \left|\overrightarrow{C_1D_{1,2}}\right| \cos\psi_1 \tag{8}$$

$$\left(\overrightarrow{C_1D_{1,2}}\right)_z' = \left|\overrightarrow{C_1D_{1,2}}\right| \sin\psi_1 \tag{9}$$

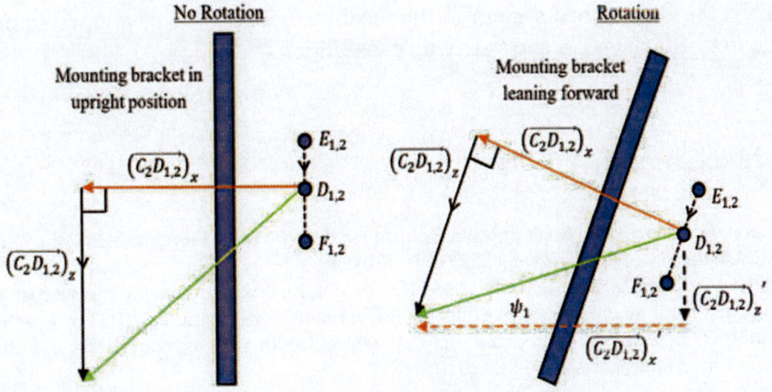

Fig. 9 An example of vector CD being altered to accommodate rotation

When rotation occurs $\left(\overrightarrow{C_i D_{1,2}}\right)'_x$ and $\left(\overrightarrow{C_i D_{1,2}}\right)'_z$ replaces $\left(\overrightarrow{C_i D_{1,2}}\right)_x$ and $\left(\overrightarrow{C_i D_{1,2}}\right)_z$ respectively in Eq. 5. Concerning leg pair 1–2 and leg pair 3–4, mounting bracket can rotate forward or backward or can remain upright. The common revolute joint that connects leg 5, leg 6 and the end effector can perform a positive or negative rotation or remain horizontal. Thirteen different combinations can be achieved due to the positions of the mounting brackets and leg pair 5–6 common revolute joint with the end effector. The inverse kinematic solution accommodated for all combinations.

5 Conclusions

In an effort to overcome some of the challenges facing entrepreneurs in the manufacturing sector in developing nations such as South Africa, research was conducted into the development of a unique parallel kinematic machining centre. The machining centre could offer the possibility of a more financially accessible and portable machine compared to existing CNC machines.

Initially, a desktop prototype was built and tested. A full-scale machine including a spindle was then designed and built with additional features included such as a tool changing system and a gripper attachment.

The inverse kinematics of the unique architecture of the machine were presented as the initial step in developing a control system for the machining centre. The inverse kinematics were solved using a vector loop (geometric) method.

Future work will address the issues of tolerances and precision of the developed machine. These areas may present significant challenges for such a lightweight machine. At this stage in the development, no testing has been done regarding these aspects.

A study of the workspace of the full-scale design is yet to be conducted. It has been completed on the desktop model and that work is still to be published.

Lastly, the full control system of the machine needs to be developed in order to be able to test the system and verify its capabilities.

References

1. Bhorat, H., Rooney, C.: State of Manufacturing in South Africa. Development Policy Research Unit University of Cape Town, Cape Town (2017)
2. Lekhanya, L.M., Mason, R.B.: Selected Key external factors influencing the success of rural small and medium enterprises in South Africa. J. Enterprising Culture **22**(03), 331–348 (2014)
3. Fatoki, O.O.: Graduate entrepreneurial intention in South Africa: motivations and obstacles. Int. J. Bus. Manage. **5**(9), 87–98 (2010)
4. Liu, L., Yao, Y., Li, J.: A review of the application of component-based software development in open CNC systems. Int. J. Adv. Manuf. Technol. **107**(9), 3727–3753 (2020)

5. He, J., Gao, F., Meng, X., Guo, W.: Type synthesis for 4-DOF parallel press mechanism using GF set theory. Chin. J. Mech. Eng. **28**(4), 851–859 (2015)
6. "ABB IRB Packaging Robot Series," [Online]. Available: https://www.robots.com/robots/abb-irb-340. [Accessed 19 September 2021]
7. Fassi, I., Wiens, G.J.: Multiaxis machining: PKMs and traditional machining centers. J. Manuf. Process. **2**(1), 1–14 (2000)
8. Dharmalingum, W., Padayachee, J., Bright, G.: The design of a 5 degree of freedom parallel kinematic manipulator for machining applications. In *2019 International Conference on Competitive Manufacturing (COMA'19)*, Stellenbosch (2019)
9. Koseki, Y., Arai, T., Sugimoto, K., Takatuji, T., Goto, M.: Design and accuracy evaluation of high-speed and high precision parallel mechanism. In *IEEE International Conference on Robotics and Automation Proceedings (Cat. No.98CH36146)*, vol. 2, pp. 1340–1345 (1998)
10. Dasgupta, B., Mruthyunjaya, T.: The Stewart platform manipulator: a review. Mech. Mach. Theory **35**(1), 15–40 (2000)
11. CNC Machines: 2018 Used CNC Machine Price: Average CNC Machine Price. 8 January 2019. [Online]. Available: https://www.manufacturingtomorrow.com/article/2018/12/2018-used-cnc-machine-price-average-cnc-machine-price/12676/. [Acedido em 8 December 2021]
12. Yu, Y., Xu, Z., Wu, Q., Yu, P., He, S., Wang, G.: Kinematic analysis and testing of a 6-RRRPRR parallel manipulator. Proc. Inst. Mech. Eng. C J. Mech. Eng. Sci. **231**(13), 2515–2527 (2017)
13. Dalvand, M.M., Shirinzadeh, B., Nahavandi, S.: Inverse kinematics analysis of 6-RRCRR parallel manipulators in, 9–12 July 2013 2013. In *2013 IEEE/ASME International Conference on Advanced Intelligent Mechatronics* (2013). https://doi.org/10.1109/AIM.2013.65841

Wesley Dharmalingum obtained his MSc. Eng degree in Mechanical Engineering from the University of KwaZulu-Natal in 2020. He is currently pursuing a PhD degree at the University of KwaZulu Natal, South Africa, in the research area of mechanism theory and parallel kinematic manipulators.

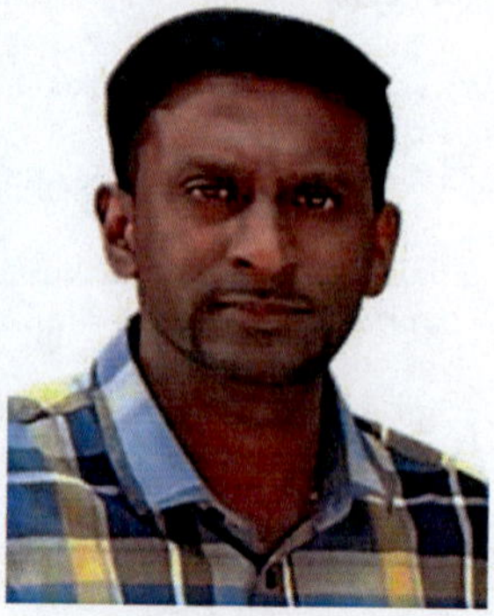

Jared Padayachee holds MSc. Eng and PhD degrees in Mechanical Engineering from the University of KwaZulu-Natal. He is currently a senior lecturer in the Department of Mechanical Engineering at the University of KwaZulu-Natal. His area of research interest is Optimisation in Robotics and Manufacturing.

James Collins holds an MSc. Eng degree in Mechanical Engineering from the University of KwaZulu-Natal. He is currently pursuing a Ph.D. degree at the University of KwaZulu-Natal, South Africa. His specific focus and area of research are robotic manipulators and Step-NC machining.

Glen Bright possesses MSc.Eng and Ph.D. degrees in in Mechanical Engineering from the University of KwaZulu-Natal. He is currently the Dean of Engineering at the University of KwaZulu-Natal. His area of research interest is in Mechatronics and Robotics.

Smart Data

Driving Big Data Capabilities and Sustainable Innovation in Organisations

Tanja von Leipzig, Jacques du Toit, and Frank Ortmann

Abstract Big data can not only provide a glimpse into the current state of a business, but may also provide a foundation for discovering new business opportunities, driving process improvement and innovation and ultimately improving the bottom line. However, realising this explicit value from big data is not without challenges. It is estimated that few big data endeavours succeed, with only a small portion of analytic insights actually delivering measurable business value. The challenges are multifaceted, including factors such as a lack of an overall big data strategy, insufficient buy-in from executive management, resistance to technology adoption, inadequate technical and soft skills and team structures, and poorly-directed investments. Without an understanding of the current landscape or state of the art as far as technology and advanced analytics are concerned, along with a clear roadmap to guide their big data efforts, organisations will find it more difficult to realise the value that big data promises. In this paper some of the uncertainties and challenges faced by organisations with respect to big data are addressed, by presenting a model which evaluates an organisation's capabilities with regard to data centricity and provide an actionable roadmap for the implementation and improvement of big data endeavours. This enables organisations to focus their efforts on creating value from big data, where the model informs continuous efforts in improving organisational efficiency and effectiveness, and driving sustainable innovation.

Keywords Big data · Data-centric · Maturity model

T. von Leipzig (✉)
Industrial Engineering, Stellenbosch University, Stellenbosch, South Africa
e-mail: tanja.vonleipzig@gmail.com

J. du Toit
Physical and Mathematical Analysis (PMA), Stellenbosch University, Stellenbosch, South Africa
e-mail: jacques@spatialedge.co.za

F. Ortmann
Applied Mathematics, Stellenbosch University, Stellenbosch, South Africa
e-mail: frank@spatialedge.co.za

K. von Leipzig et al. (eds.), *Smart, Sustainable Manufacturing in an Ever-Changing World*, Lecture Notes in Production Engineering,
https://doi.org/10.1007/978-3-031-15602-1_56

1 Introduction

Big data is an important, strategic business asset across many industries [1–3]. It provides a glimpse into the current state of a business, provides a foundation for discovering new business opportunities, and drives process improvement and innovation. Big data typically enables improved decision-making, better risk management, increased transparency and more efficient processes [1, 2, 4, 5]. Furthermore, it enables the identification of new market segments and offerings, as well as the discovery of new business models, products and services [1, 2]. Consequently, it provides opportunities to improve revenue and reduce costs.

Unfortunately, many organisations are struggling to realise the value that big data promises [1, 2, 6]. It has been estimated that 85% of big data projects fail [6], whilst 87% of data science projects never make it to production [7]—a critical capability in realising value through advanced analytics. It is further estimated that by 2022, 90% of corporate strategies across all industries will explicitly mention data and analytics as critical business assets [8], but only 20% of all analytics insights will deliver real business value [6]. Organisations are becoming increasingly aware of the importance of big data and advanced analytics in driving organisational effectiveness and efficiency, but face numerous challenges in realising the value.

These challenges are multifaceted, but managerial, cultural and people-related challenges far outweigh the technological ones [2, 9]. Commonly experienced challenges include the lack of an effective overall strategy for big data, insufficient buy-in and explicit support from executive management, resistance to the adoption of new technology, inadequate technical and soft skills, team structures which are unconducive to productivity, and poorly-directed investments [1, 2, 10–12].

To address these challenges, organisations wishing to be data-centric need to have a good understanding of big data, modern technologies and the advanced analytics landscape. To this end, maturity models are useful tools to understand how best to implement and derive value from new technologies or capabilities within an organisation [13]. They provide guidance on how to focus organisational efforts to drive value and enable organisations to effectively balance divergent objectives [2, 10].

2 A Review of Existing Maturity Models

Existing models in big data and related fields were evaluated. Various limitations, omissions and opportunities were identified that, if addressed, would provide more effective guidance for organisations to realise the value of big data.

2.1 Existing Models

Building on the works of [2, 14], 16 maturity models in the fields of big data and analytics were evaluated in terms of the following categories:

1. completeness of the model structure;
2. quality of the model development and evaluation;
3. ease of application; and
4. big data value creation.

In the first category, the model completeness is evaluated by considering aspects such as the model purpose, composition and structure. In the second category, quality is evaluated in terms of the trustworthiness and stability of the model. In this category the design approach, model evolution and level of verification and validation are evaluated. The third category (*ease of application*) considers the model application as well as comprehensibility and practicality of results, while aspects such as model actuality, relevance and performance are considered in the fourth category (*big data value creation*).

The results of the evaluation of the maturity models are depicted in Fig. 1, with the TDWI [15], Fraunhofer Industry 4.0 Checkup [16] and IDC [17] models ranking as the top three models across the four categories.

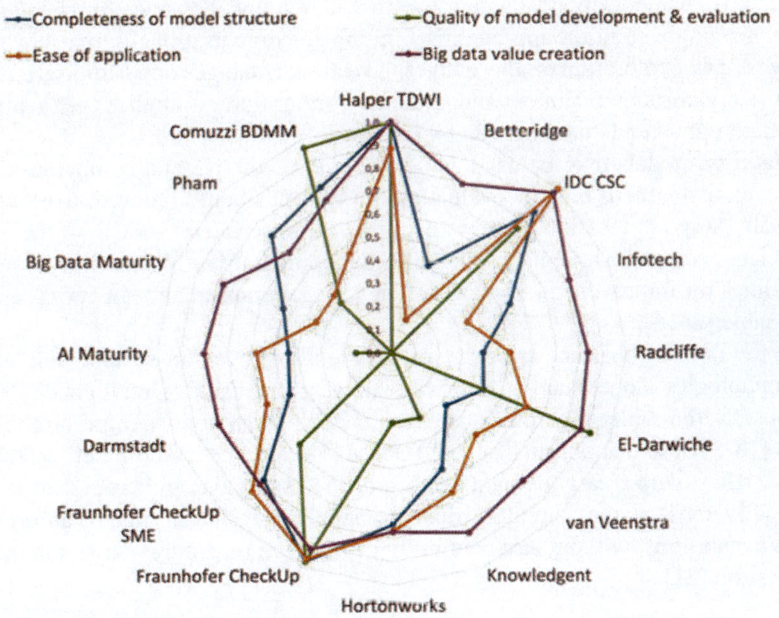

Fig. 1 Benchmarking scores of maturity models

2.2 Limitations and Shortcomings

Whilst maturity models are a useful guidance tool for organisations, they are also frequently criticised. Maturity models may be classified into two distinct groups, namely, those developed by academics (e.g., [5, 18]) and those developed in industry (e.g., [17, 19, 20]). Where models developed by academia are generally well-documented in terms of research and theoretical foundation, they often lack practicality and relevance [2, 5, 18]. In some cases, the models are developed purely based on other existing models, without questioning the appropriateness or validity thereof. In contrast, those developed in industry are usually developed as consulting tools, where one would assume that the practical relevance is high. However, the development process and compilation of the model is typically not well-documented or publicised [2, 14, 16]. This makes it difficult to validate the models and more importantly, exposes them to the biases of the vendor or consulting partner involved in the development [2, 18].

Further criticism includes the static, inflexible nature of most maturity models [2, 21]. Although some are developed in an iterative manner, most are treated as complete and static once they have been applied. They are developed based on current best practices but quickly become outdated as technology, and the field in which they are applied, evolves.

Furthermore, the models themselves are often linear and provide only a single perspective, namely, an ideal solution path without any allowance for changes [21–24]. This single solution approach is limiting for organisations in that it does not allow for any exploration of alternatives, directional changes or customisation to the specific organisational context and strategy. Furthermore, it implies that maturity is a predetermined end-state that can be achieved.

Maturity models may be used as descriptive, predictive and comparative tools. However, many focus only on evaluating the current situation (descriptive), without providing any explanation of how to improve (prescriptive), nor how the assessment score compares to industry benchmarks (comparative). Where they do provide guidelines for improvement, they are often very generalised and not very practical / actionable [2, 16].

In the field of big data, maturity models tend to focus on specific subdomains or technologies alone, such as data warehousing or business intelligence, but fail to consider the wider organisational aspects such as strategy, culture and relevant skills [18]. Those that do, such as the BDMM [18] are intended for well-established organisations with clearly defined and established strategies and processes.

Finally, most maturity models provide a single snapshot in time of an organisation's capabilities, without any continuity or measure of progress after the model's application [21].

2.3 *Opportunities*

The big data, analytics and technology landscapes are constantly changing, and subsequently require a more flexible, dynamic approach to improvement. For a maturity model to be effective in this environment, it should be developed continuously, to incorporate new and/or improved tools, technologies and best practices.

Realising value from big data is often context-dependent and organisations may approach this differently. As such, a maturity model should be nonlinear to allow for, and encourage, directional changes and shifts. Furthermore, it should provide a mechanism for continuous evaluation and improvement.

Considering the myriad of challenges in leveraging big data, a holistic perspective which encompasses technical, organisational, process- and people-related challenges may lead to better outcomes.

3 Methodology and Design Approach

To effectively design a flexible, dynamic maturity model for big data and analytics, an iterative design methodology that drives continuous improvement of the model is proposed.

3.1 *Methodology*

Braun [2] expands on the works of [14, 25, 26] and proposes a generic framework for maturity model development. This framework provides a foundation for iteratively developing maturity models and addresses one of the primary shortcomings (the lack of documentation) by emphasising this throughout the development process. However, even this framework focuses predominantly on the research and development phases of the model and does not include model application/implementation. Furthermore, it does not explicitly cater for iterative and continuous improvement of the model after its creation. To ensure the development of a dynamic, flexible model, the framework was extended to include a model implementation phase as well as a design iteration stage.

Given that the field of application is ever-changing, a landscape scouting phase was added. Here, the big data and analytics landscape is continuously monitored to anticipate relevant changes. This stage serves as an additional trigger of subsequent design iterations, ensuring that the model remains relevant by incorporating the latest tools and best practices. The resulting methodology, depicted in Fig. 2, was used to design the maturity model of this paper.

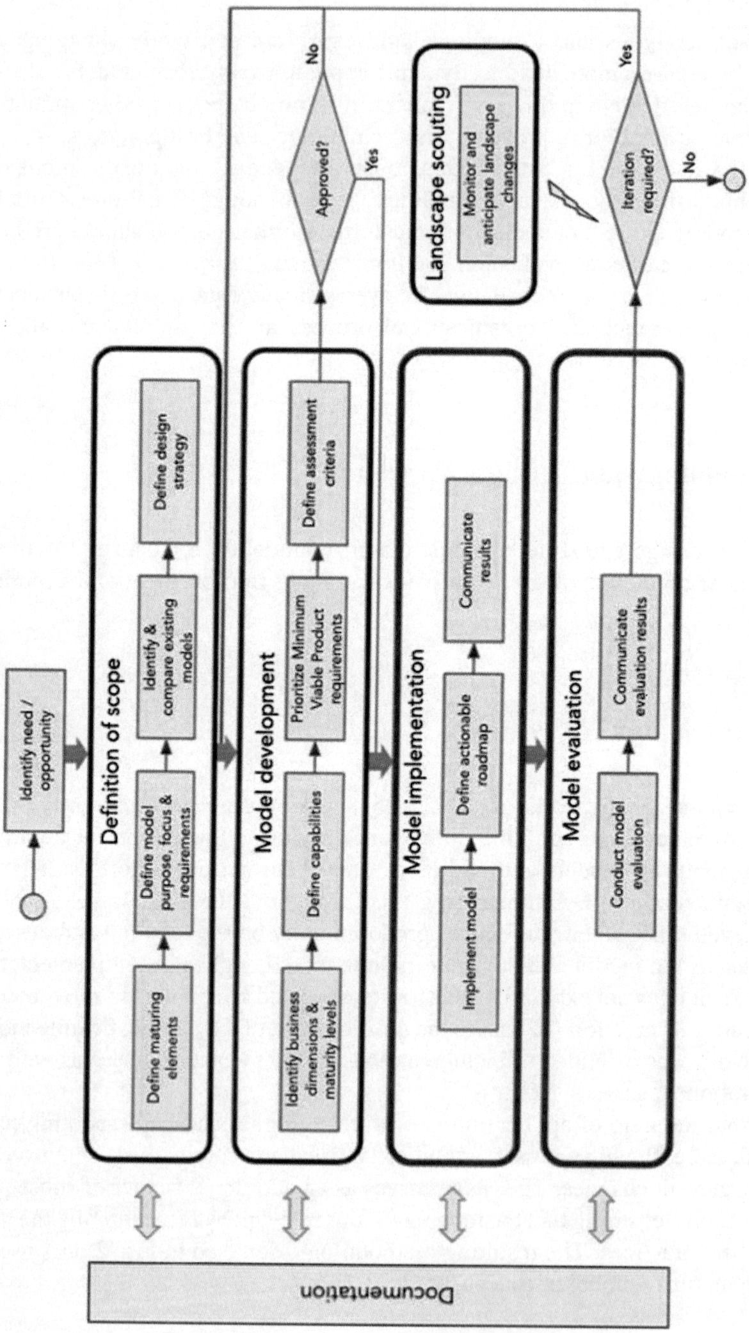

Fig. 2 Iterative design methodology, adapted from [2]

3.2 Design Iterations

An initial set of dimensions and sub-dimensions was defined through the analysis of existing research and models. In particular, different aspects of the highest-ranked models were considered (see §2.1) and refined in consultation with experts in the field of big data analytics. Four rounds of focus group sessions were conducted with six executives, all with direct experience in software engineering, data science and/or machine learning engineering within big data environments.

A comparative analysis was conducted in which each sub-dimension was evaluated in combination with every other sub-dimension to capture dependencies, correlations and/or possible trade-offs. If correlations or overlaps were found, sub-dimensions were consolidated to reduce the model complexity. Dependencies and trade-offs between sub-dimensions were noted for later consideration during the model implementation phase.

For each sub-dimension, experts were asked to define the best case and worst case scenarios to establish the range of the maturity scale. These scenarios were used to formulate initial questions, along with a set of five possible responses (ranging from the worst to the best case scenarios), as shown by the example in Table 1.

A multipronged approach was used to define the questions for each sub-dimension. This was achieved by considering questions from the top-ranked models, conducting research into pertinent industry standards and best practices, as well as by considering the input received from industry experts during the focus group sessions. In the initial iterations, experts were tasked with individually defining questions for each dimension. Considering the immense time effort required, this approach was adapted to conducting these focus group sessions with experts.

After a number of design iterations, an additional perspective was added in the form of key factors. Based on research and the input from experts, the factors critical to the success of big data initiatives in organisations were identified. In addition to

Table 1 Sample question from maturity model

Question	Option 1	Option 2	Option 3	Option 4	Option 5
How influential are outputs of your data and analytics products on decisions made across the various business areas and levels?	Results are delivered at meetings and decisions are predominantly made based on expertise	Insights are delivered at meetings and may contribute somewhat, but decisions are still largely made without data	Some business areas are starting to make decisions based on data, but do not completely trust the analytics products yet	Analytics products are self-service for the decision-makers and used directly for decision making	Data/analytics products provide augmented intelligence and decision support, directly driving decision-making throughout the organisation. No human-in-the-loop is applied where possible

occupying a sub-dimension, questions were assigned relevant key factors (discussed in §4.1). In three additional focus group sessions with the same experts, key factors were consolidated, refined and ranked in terms of importance (i.e. how critical they are for the success of big data initiatives) following a pairwise comparison. These ranked key factors were considered in combination with those prioritised in literature to avoid expert bias. Further revisions were made to the model to ensure that the most important key factors had sufficient coverage in the model.

4 Big Data Analytics Maturity Model

Following the iterative design methodology (see §3), a maturity model for big data was developed. An overview of the model and the continuous improvement thereof, as well as the application scenario and differentiating factors, are presented.

4.1 Model Overview

The maturity model measures an organisation's capabilities across seven dimensions, categorised into 33 sub-dimensions, shown in Table 2. The model comprises 43 questions across the seven dimensions. Each question has five answer options ranging from low to high maturity.

A set of 25 key factors that are central to the success of big data and analytics organisations were defined. The top six key factors (both in terms of importance and model coverage) are:

- Data-driven decisions;
- Enterprise-wide strategy;
- Return-on-investment (ROI)-driven;
- Analytics capabilities;
- Evangelist leadership; and
- Scalable architecture.

Most key factors span multiple dimensions and provide an additional perspective and scoring mechanism to assess an organisation's maturity.

4.2 Continuous Model Improvement

As discussed in §3, continuous improvement is an important component of the model to ensure continued relevance in a dynamic environment, where the model may be

Table 2 Model dimensions and subdimensions

Dimension	Description	Sub-dimensions
Strategy and alignment	This dimension considers the extent to which the organisational vision, strategy and funding support the big data initiatives of the organisation. It considers the level of advocacy and alignment within the organisational structure for the initiative(s) and evaluates the business case for big data	Vision and strategy; funding; advocacy; alignment; business case
Process management	The way data-centric and big data initiatives are planned, funded, managed and evaluated are considered in this dimension	Operations; program measurement; investment focus
Data management	This dimension considers how data are managed in support of big data and analytics efforts. Specifically, how data are collected, stored, and processed, as well as the quality, accessibility and availability thereof	Data collection; data volume, variety, velocity and veracity; data storage; data processing standards; data processing; accessibility and availability
Analytics	In this dimension the utilisation and delivery of data analytics results are evaluated, along with the visualisation, reporting, data-driven decision-making capabilities of the organisation	Analytic techniques; delivery methods; reporting and visualisation; data-driven decision making
Governance	The coherency of, and compliance to, the organisation's data governance strategy in support of its big data initiatives are considered in this dimension, alongside technical standards, data protection and security	Policies; structure; compliance; stewardship; security and privacy
Organisation and people	This dimension considers how organisational culture impacts big data initiatives, as well as the levels of enterprise-wide capabilities and collaboration efforts	Skills and roles; leadership model; culture; performance management; cross-functional practices

(continued)

Table 2 (continued)

Dimension	Description	Sub-dimensions
Technology and infrastructure	This dimension considers the extent to which the infrastructure supports the initiatives and its potential users, as well as the technologies and integrations in place	Information digitised; architecture; functionality; analytic / big data tools; integration

updated with the latest best practices and innovations, and user feedback is incorporated. Consequently, an automated pipeline was produced using open source, cloud-based scripts and tools. Specifically, Google tools and services including Google Sheets, Google Slides, and Google Apps Scripts were used to create the automated pipeline. Changes or additions to the model dimensions, key factors and/or questions may be made directly via the input mechanism of the model, and are automatically propagated throughout the model (e.g. recomputing key factor coverage).

4.3 Model Application

The assessment is conducted by an industry expert that facilitates a workshop with key stakeholders at the organisation being assessed. A group approach is taken to ensure that the various perspectives within the organisation are accurately captured in the answers. This ensures a more realistic snapshot of the organisation, and a higher accuracy of the results. Furthermore, discussions between stakeholders may lead to valuable insights that would otherwise not be uncovered. Once the assessment questions have been submitted, the model outputs indicate the organisation's strengths and areas of improvement. These results are calculated in terms of the model dimensions and subdimensions, as well as the key factors. These scores are interpreted by external experts based on the current context, processes, goals and constraints of the organisation.

The results and personalised recommendations are then presented back to the organisation in another group session. Here, the experts provide insights into the recommendations, and help the organisation understand the trade-offs between the various options available to them in terms of technologies and improvements. In collaboration with the organisation's key stakeholders, and under consideration of the specific context and capabilities, the experts formulate an actionable roadmap for the organisation. Where applicable, practical resources are shared with the organisation to facilitate the recommendations and build internal capacity. These resources include curated or widely available best practices (e.g. machine learning operations guidelines), specialised subject documents (e.g. model feature stores — referred to as *monographs*), or recommendations for targeted training.

A critical component of the model application is the re-assessment and feedback loop. After the recommendations and roadmap are presented, a reasonable time period for re-assessment is determined. After this period, the assessment is repeated to measure improvements made. Here, the differences in scores are used to inform any directional changes (i.e. where recommended actions may not be as successful as anticipated) and to reprioritise and refocus efforts in response to any contextual nuances. This feedback loop is critical to the continuous improvement of the model itself and forms part of the iteration loop depicted in Fig. 2.

4.4 Model Differentiators

A primary differentiator of the model is the iterative design methodology followed in its development. Rather than trying to develop an exhaustive, complete model from the start, an iterative approach is adopted to allow for the continuous improvement of the model. This is especially valuable in the field of big data where the application environment is highly dynamic.

Where existing models are developed either from an academic or an industry perspective, the model presented was developed using a combined approach (see §3). In this case, the model development leveraged existing models and industry standards, current best practices in industry, as well as the knowledge and experience of various experts.

The key factor approach ensures that the critical factors in the big data analytics landscape are sufficiently covered by the model. It also provides an additional avenue for interpretation of the results and recommended actions.

The customised model application allows for increased flexibility of the model. This is based on the assumption that there is no single, optimal solution path for all organisations. Rather, the results of each application are interpreted and prioritised for the specific organisation. Depending on the current context and goals of organisations, two organisations with the same score may receive different proposals to improve their outcomes.

The repeated application of the model provides a mechanism for regular progress reflection, course correction (where required) and continuous improvement within the organisation. Additionally, it enables and encourages the continuous improvement of the model itself [16, 18, 25].

5 Validation

The model was validated throughout its development (see §4), evaluated against the benchmark models (see §2) and subsequently applied to two relevant enterprises. The validation was conducted by the authors, together with the six big data experts included in the initial focus groups.

5.1 Evaluation Against Established Benchmark

As a first point of validation, the model was evaluated against the benchmark established in §2.1. The overall model score, with respect to the evaluation categories defined in §2.1, surpasses that of the benchmark models, with a score of 19.00 (compared to the 16.52, 16.40 and 15.40 scores of the TDWI, Fraunhofer and IDC models, respectively). This is not a surprising result given that the model's starting point was a combination of these three models, with a focus on expanding the successful traits of each model.

The model scores consistently well across the four categories of §2.1, only scoring lower than one of the benchmark models in category 3 (*ease of application*). In this case, a model applied in a workshop setting with the input of industry experts is not as simple to apply as an online self-assessment. Moreover, the complexity is increased due to the incorporation and consideration of the enterprise's context in the interpretation of results and prioritisation of recommendations. Although this increases the application complexity, it increases the model accuracy and relevance for the organisation and is considered a differentiating feature of the model.

One aspect within the model completeness category in which the model can be improved, is the purpose of use. The model currently classifies an organisation into a maturity level (descriptive) and provides recommendations for improvement (prescriptive). It does not currently provide a comparison to benchmark organisations, a feature that the benchmark models do provide. The comparative level of the model is planned for future design iterations and the versioning of models should make comparisons more applicable when the model is regularly updated.

The model scores highest in the big data value creation category. This score may be attributed to the involvement of industry experts in the model development.

5.2 Case Studies

The model was applied to two relevant enterprises. As part of the initial validation, the assessments were completed by experts with direct and continuing interaction with these enterprises. The comparative results of the two enterprises are depicted in Fig. 3. The results were interpreted considering the context of each organisation and used to formulate prioritised recommendations. These results and recommendations were shared and discussed with the experts to ensure that they reflect their expectations and experiences. These discussions led to further improvements to the model.

In the case of Enterprise A, the model was applied at different points by considering the maturity of the organisation currently and retrospectively, as shown in Fig. 4. For each area in which the maturity level increased, experts reviewed the actions and activities that led to these changes. In areas where improvements were not as significant (e.g., in the process management dimension), the reasoning thereof and

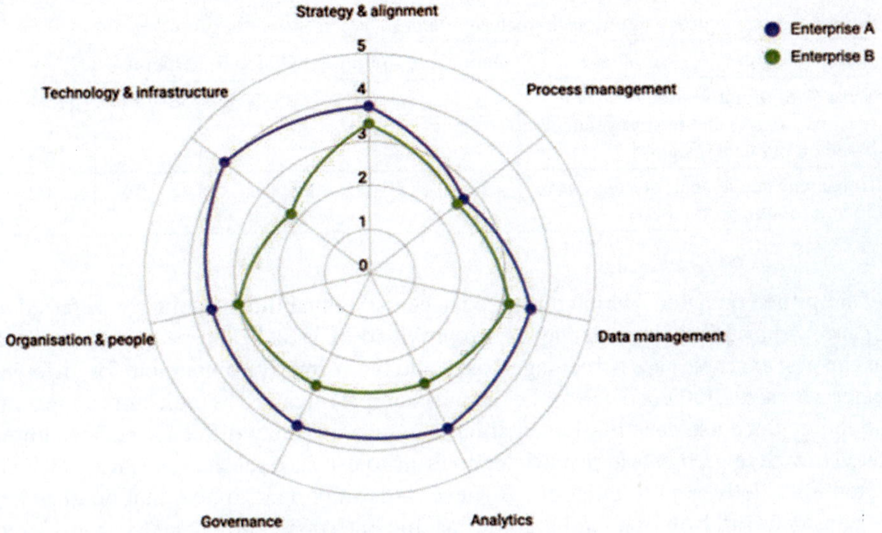

Fig. 3 Comparative results of two organisations

difficulties experienced were discussed. These discussions of the comparative results led to further refinements of the model and the reprioritisation of recommendations.

Enterprise B was only assessed at one point in time, but was also used to validate an approach which emerged in more recent design iterations, namely that of a set

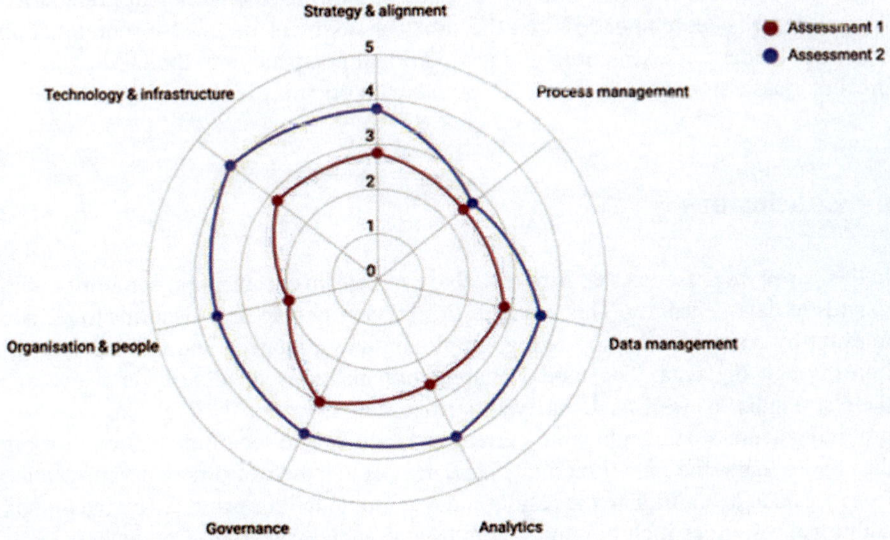

Fig. 4 Results of subsequent assessments

Table 3 Example proxy questions in maturity model

Proxy question	Option 1	Option 2	Option 3	Option 4	Option 5
What % of initiatives focus on new revenue streams and business models based on big data insights?	<10%	10–25%	25–50%	50–75%	>75%
How many active analytics use cases do you have in production?	<10	10–50	50–75	75–100	> 100

of simplified questions which measured necessary capabilities explicitly (referred to as *proxy* questions). Two examples are provided in Table 3. Some questions (such as the first example) are percentage-based and thus directly comparable for different enterprise sizes. Others (such as the second example) are used as indicators of general adoption since use case implementation in enterprises may differ for reasons unrelated to size (e.g., banks apply stricter controls to use case releases or startups which oftentimes deliver more quickly). However, it is important to note that no question is considered in isolation, and that the specific enterprise and context is considered when interpreting the results and prioritising recommendations.

The complete, standard assessment was completed, as well as the smaller set of proxy questions. The results of these two assessments were considered comparatively to determine whether the combined approach provides a better understanding of the organisation's landscape. The experts found the proxy questions easier and faster to answer, and the proxy questions resulted in maturity scores across the dimensions that were similar to those provided by the detailed questionnaire. They provide insights into very practical and actionable aspects of the organisation. However, without the additional scores of the full assessment, they do not provide sufficient granularity. Therefore, it is recommended that the proxy assessment be used in combination with the standard assessment to add an additional perspective to the model, or as a precursor assessment to share initial insights with organisations.

6 Conclusions

In this paper we present a maturity model to evaluate an organisation's maturity with regards to data-centricity. This model is designed to enable organisations to remain relevant by continuously improving, employing key practices and driving innovation through big data. The model enables organisations to identify and leverage their strengths as well as identify, prioritise and improve their weaknesses. By providing actionable insights into current capabilities and recommendations for big data endeavours, the model facilitates and accelerates their efforts in creating value from big data. In addition to the results and recommendations provided by the model, additional resources including monographs and knowledge guides on key concepts and current best practices are shared to build internal capabilities and understanding.

Targeted training is provided to ensure that big data capabilities and innovation are sustained.

Furthermore, this model is designed to provide a holistic understanding of an organisation's capabilities and identify critical areas for current big data endeavours.

The model was validated through its application in two case studies in which the results and prioritised recommendations were validated through discussions with experts. Future work includes applying the model to more organisations. Once sufficient data points are available, the model will be extended to include the comparative aspects.

In contrast to other models, one of the main contributions of this research effort is the design approach followed and the transparency thereof. Many models, especially those developed by industry, lack documentation and fail to provide insights into the development and composition of the models. A further contribution is the extensive review and evaluation of existing maturity models, building on the work of [2]. In the field of big data specifically, most maturity models focus on the technology or data aspects, but none were found that also address the advanced analytics and machine learning aspects. This model evaluates an organisation's capabilities in data, technology and analytics areas, in combination with organisational and people-related capabilities.

The maturity model presented is not intended as a recipe for innovation and success in the realm of big data. Rather, it is intended as a tool to better understand an organisation's capabilities and the current landscape in terms of technology and advanced analytics, and to provide continuous guidance on where (and how) efforts should be focused to realise explicit value from big data.

References

1. Barham, H.: Achieving competitive advantage through big data: a literature review. In *2017 Portland International Conference on Management of Engineering and Technology (PICMET)*, Portland, Oregon, USA (2017)
2. Braun, H.: Evaluation of big data maturity models—a benchmarking study to support big data maturity assessment in organizations. Tampere University of Technology, Tampere (2015)
3. Motau, M.: Assessment of Big Data Analytics Readiness in South African Governmental Parastatals. Tshwane (2016)
4. Coleman, S., Göb, R., Manco, G., Pievatolo, A., Tort-Martorell, X., Reis, M.S.: How can SMEs benefit from big data? challenges and a path forward. Qual. Reliab. Eng. Int. **32**(6), 2151–2164 (2016)
5. Pham, C.M.: Building a maturity framework for big data cybersecurity analytics. In *Applying Business Intelligence Initiatives in Healthcare and Organizational Settings*, IGI Global, pp. 164–183 (2019)
6. Herschel, G., Brethenoux, E., Idoine, C., Kronz, A., Hunter, E., Horvath, M.: Predicts 2019: Analytics and BI Strategy. Gartner (2018)
7. VentureBeat: Why do 87% of data science projects never make it to production?. 19 July 2019. [Online]. Available: https://venturebeat.com/2019/07/19/why-do-87-of-data-science-projects-never-make-it-into-production/. [Accessed 25 Sept 2021]

8. Schulte, W.R., Buytendijk, F., Duncan, A.D., Laney, D., White, A., Waller, G., Howson, C., Logan, V.: Predicts 2019: Data and Analytics Strategy. Gartner (2018)
9. S. L. E. S. R. H. M., LaValle, K.N.: Big data, analytics and the path from insights to value. MIT Sloan Manage. Rev. **52**(2), 21–31 (2011)
10. Farah, B.: A value based big data maturity model. J. Manage. Policy Practice **18**(1), 11–18 (2017)
11. Angrave, D., Charlwood, A., Kirkpatrick, I., Lawrence, M., Stuart, M.: HR and analytics: why HR is set to fail the big data challenge. Hum. Resour. Manag. J. **26**(1), 1–11 (2016)
12. Lukianoff, M.: The Success and Resounding Failure of "Big Data"," 4 February 2020. [Online]. Available: https://towardsdatascience.com/the-success-resounding-failure-of-big-data-50b3f17756f1. [Accessed 25 Sept 2021]
13. Hüner, K., Ofner, M., Otto, B.: Towards a maturity model for corporate data quality management. In *24th Annual ACM Symposium on Applied Computing (ACM SAC 2009)* (2009)
14. T. R. P., Mettler, W.R.: Towards a classification of maturity models in information systems. In: D'Atri, A., De Marco, M., Braccini, A.M., Cabiddu, F. (eds.) Management of the Interconnected World, Berlin, pp. 333–340. Germany, Springer (2010)
15. Halper, F., Krishnan, K.: TDWI big data maturity model guide (2013). [Online]. Available: https://tdwi.org/~/media/3BF039A2F7E1464B8290D8A9880FEC22.pdfma. [Accessed 05 Oct 2021]
16. Häberer, S., Lau, L.K., Behrendt, F.: Development of an Industrie 4.0 Maturity Index for Small and Medium-Sized Enterprises. In: 7th IESM Conference, Saarbrücken, Germany (2017)
17. IDC: IDC Future IT Maturity Assessment," [Online]. Available: https://www.idc.com/itexecutive/planning-guides/maturity-assessment. [Accessed 16 Aug 2020]
18. M. A. P. A. Comuzzi: How organisations leverage: big data: a maturity model. Indus. Manage. Data Syst., **116**(8):1468–1492 (2016)
19. N., Betteridge, N.C.: A maturity model for big data and analytics. IBM (2014). [Online]. Available: https://whitehallmedia.co.uk/blog/2015/12/29/a-maturity-model-for-big-data-and-analytics/. [Accessed 14 Aug 2020]
20. Infotech, "Big Data Maturity Assessment Tool," 2013. [Online]. Available: http://www.infotech.com/research/it-big-data-maturity-assessment-tool. [Accessed 10 10 2021].
21. O'Reilly, B.: Why Maturity Models Don't Work. 4 December 2019. [Online]. Available: https://barryoreilly.com/explore/blog/why-maturity-models-dont-work/#:~:text=Maturity%20models%20are%20not%20scientific,easily%20become%20tomorrow's%20obsolete%20ones. [Accessed 25 Sept 2021]
22. Eckert, A.: A 2-dimensional maturity model: de- and recomposing traditional maturity models for improved management consultancy. Structuremine, Walldorf, Germany
23. Lindemulder, M.: Development of a Continuous Improvement Maturity Model Assessment Instrument. In: *5th IBA Bachelor Thesis Conference*, Twente, Netherlands (2015)
24. Fowler, M.: Maturity Models. 26 August 2014. [Online]. Available: https://martinfowler.com/bliki/MaturityModel.html. [Accessed 18 Sept 2021]
25. De Bruin, T., Freeze, R., Kaulkarni, U., Rosemann, M.: Understanding the main phases of developing a maturity assessment model. In: *Proceedings of the 13th European Conference on Information Systems*, Regensburg (2005)
26. J. K. R., Becker, P.J.: Developing maturity models for IT management. Bus. Inf. Syst. Eng., **1**(3), 213–222 (2009)

Tanja von Leipzig obtained her B.Eng degree in Industrial Engineering from Stellenbosch University (SU) in 2014, and went on to complete her M.Sc. degree in Operations Management from Reutlingen University, Germany in 2017. She completed her Ph.D. in Industrial Engineering at SU in 2022, with her specific research area being adaptive games for bidirectional learning in education and training. She has designed and developed various learning games for both university modules and industry training. She is currently working at a big data analytics company, where she is involved in the development of a big data maturity model as well as training development for various big data engineering related disciplines.

Jacques du Toit obtained his bachelor's degree in physical and mathematical analysis (PMA) from Stellenbosch University (SU) in 2003, majoring in mathematics and computer science. He went on to complete his honours degree in PMA, and received his master's degree cum laude in Applied Mathematics in 2009, also from SU. He then followed master's courses in Artificial Intelligence at the University of Amsterdam (UVA) and received his Ph.D. in Operations Research in 2014 from SU. He is currently working at a big data analytics company where his responsibilities include industry training in big data methodologies and realising value from data using the latest technologies. He has worked on various data science, machine learning and big data problems in industry, as well as big data science projects like the SKA.

Frank Ortmann obtained his B.Sc. degree in physics from University of Natal in 2001, majoring in Physics and Computational Physics. After completing his honours degree in Physics at the same university, he moved to Stellenbosch University (SU) to study for an M.Sc. degree in Applied Mathematics. By the end of 2010 he has completed his Ph.D. in Operations Research at SU.

He is co-founder and executive director a big data analytics company based in Stellenbosch where he is focussing on ensuring models are able to realise their value for business. He has previously worked in the fields of superconducting circuit design and optimisation, high-frequency data capture and storage, and leveraging geospatial information for planning purposes.

Indoor Positioning Using a Single PTZ Camera

J. Hermann, A. H. Basson, K. H. von Leipzig, and V. Hummel

Abstract The market for indoor positioning systems for a variety of applications has grown strongly in recent years. A wide range of systems is available, varying considerably in terms of accuracy, price and technology used. The suitability of the systems is highly dependent on the intended application. This paper presents a concept to use a single low-cost PTZ camera in combination with fiducial markers for indoor position and orientation determination. The intended use case is to capture a plant layout consisting of position, orientation and unique identity of individual facilities. Important factors to consider for the selection of a camera have been identified and the transformation of the marker pose in camera coordinates into a selectable plant coordinate system is described. The concept is illustrated by an exemplary practical implementation and its results.

Keywords Optical indoor localisation system · PTZ camera · Plant layout

1 Introduction

This paper presents the concept of an indoor positioning system based on the use of fiducial markers and a pan/tilt/zoom (PTZ) camera. The intended use case is to capture the actual factory layout, especially in modular and flexible factories. Specifically,

J. Hermann (✉) · A. H. Basson · K. H. von Leipzig · V. Hummel
University of Stellenbosch, Stellenbosch, South Africa
e-mail: julian.hermann@reutlingen-university.de

A. H. Basson
e-mail: ahb@sun.ac.za

K. H. von Leipzig
e-mail: kvl@sun.ac.za

V. Hummel
e-mail: vera.hummel@reutlingen-university.de

J. Hermann · V. Hummel
ESB Business School, Reutlingen University, Reutlingen, Germany

K. von Leipzig et al. (eds.), *Smart, Sustainable Manufacturing in an Ever-Changing World*, Lecture Notes in Production Engineering,
https://doi.org/10.1007/978-3-031-15602-1_57

the pose of the facilities involved in the production process, such as workstations, conveyor belts, machines, etc., must be determined. The pose includes not only the position but also the orientation of the facilities. The determined layout can then be used for simulation purposes based on the real conditions and for the creation of dynamic maps which, for example, can make source and destination locations for automated guided vehicles (AGVs) more flexible. The reason for developing this method is that current indoor positioning systems do not meet the high requirements regarding the determination of the layout. This primarily concerns the accuracy and lack of orientation detection. Typically, most non-optical procedures do not have an orientation determination. In addition, the position accuracies are lower than 15 cm but a very high scanning frequency is usually possible [1]. Since a layout change always involves a great effort, it can be assumed that it will not be changed very often, so the scanning frequency is negligible. In addition, other indoor positioning systems are based on active marker devices and complex infrastructure, which increases their cost. Production system planning should in future not be treated only as an initial planning project [2]. A continuous production system planning and adaptation will become more and more important. The practice also shows that 45% of the average factory planner's working time is spent on manual data acquisition [3] (p. 357). Besides the use of PTZ cameras to capture the layout, there are several other studies which use other measurement techniques. For example, Mütze et al. identified the automatic detection of the factory layout as a possible use case for an ultrawideband based real time localisation system by attaching four tags to each workstation [4]. Another approach is described in [5], where data acquisition is performed with a 3D laser scanner and the resulting point cloud is used to classify factory objects using machine learning. The following describes a concept that aims to fulfil the requirements for the factory layout recording, according to [6].

2 Concept Description

The basic idea is that a PTZ camera is mounted on the ceiling or another elevated point in a factory building overlooking the existing facilities (Fig. 1).

In order to identify the factory layout, the facilities and end-devices are equipped with visual markers with a unique ID. The installation of these markers and their pose within the respective facility coordinate system must be noted. To avoid mutual shadowing, they should be placed at similar heights. Additionally, at least 3 reference markers (subsequently termed as landmarks) must be placed at known positions within the plant/world coordinate system. In order to detect the markers, the first step is to perform a rough scan in which the camera rotates at a low zoom level and determines the rough direction of the markers by recording the respective angular position. Subsequently, the previously determined rough positions are moved at a higher zoom level and the angles are readjusted until the center of the camera image corresponds to the center of the marker. These more accurately determined angular

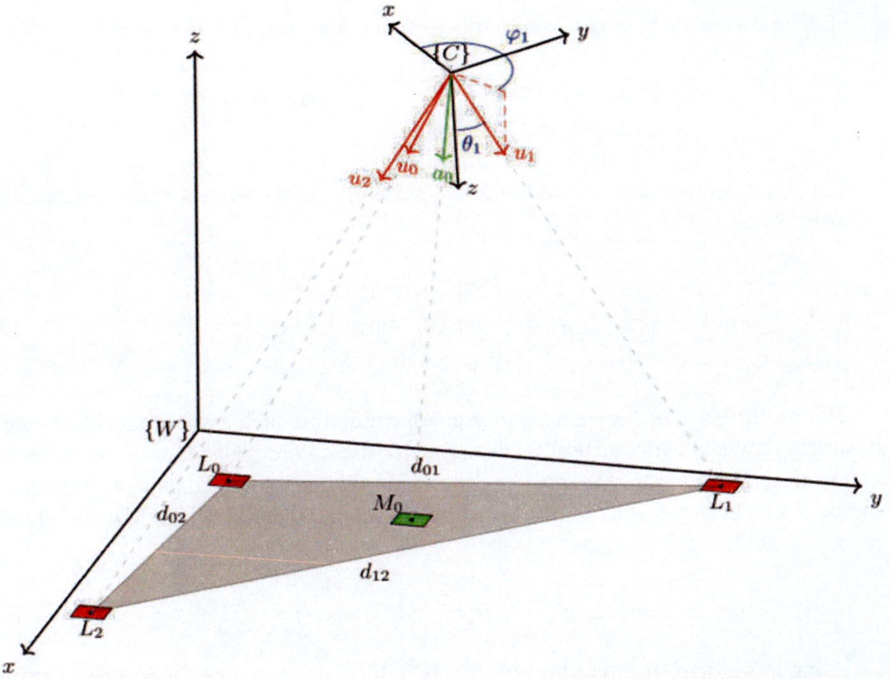

Fig. 1 Overview of geometries for pose estimation using a PTZ camera

positions are then used to determine the required position parameters. The calculation steps are explained in the following subsections.

2.1 *Determination of the Camera Position*

The determination of the position and orientation of a scanner device (in this case the PTZ camera) by using the directional measurements is known as resection in the field of geodesy. An overview of state-of-the-art methods on this topic is described by Awange and Paláncz [7] and Haralick [8]. The described methods usually solve the fourth-degree polynomial equations for the calculation of the spatial resection based on Grunert [9] with different algorithms. As these methods provide multiple solutions, they are rather unsuitable for this concept as they are difficult to automate. For this reason, Donner's method is used for this concept which can be automated nearly free of errors [10]. In our application it is assumed that the pose of the camera coordinate system within the plant coordinate system is not exactly known. In order to calculate these on the basis of [10], the 3D coordinates of n landmarks ($n \geq 3$) inside the world/plant coordinate system (${}^{W}L_i$) are required. Also, the measured azimuth and elevation angle (φ_i, θ_i) of the camera to each landmark is needed. The

Euclidean distances between landmarks in space are initially calculated (1).

$$d_{ij}^2 = (x_i - x_j)^2 + (y_i - y_j)^2 + (z_i - z_j)^2$$
$$\text{for } i \in \mathbb{N} \text{ and } j \in \mathbb{N} \text{ and } i \neq j \tag{1}$$

The unit vectors for the individual landmark points are calculated from the measured angles (2).

$$u_i = \begin{bmatrix} \sin(\varphi_i)\sin(\theta_i) \\ \cos(\varphi_i)\sin(\theta_i) \\ \cos(\theta_i) \end{bmatrix} \tag{2}$$

The multiples t_i of the vectors u_i are determined in such a way that the squared distances between their end points correspond to the squared distances d_{ij}^2 between the connection points (1). From applying the cosine theorem for each pair of connection directions it follows that $u_i \cdot u_j$ is the cosine of the spatial angle between the landmark points L_i and L_j (3).

$$d_{ij}^2 = t_i^2 + t_j^2 - 2t_i t_j u_i \cdot u_j \tag{3}$$

Using more than three calibration points will produce an overdetermined system of equations. To solve the system of equations, a least-square solve routine (in this case as predefined Python function) is used. After the determination of the t_i, the landmark points L_i in the camera coordinate system are obtained (4).

$$^C L_i = t_i u_i \tag{4}$$

2.2 Transformation Parameter Calculation

To perform the 3 dimensional coordinate transformation from the camera coordinate system to the world coordinate system, the Helmert transformation is used. As the transformation parameters are initially unknown, they are calculated using the landmarks (identical points) with known positions in both coordinate systems.

$$^{\mathrm{W}}\mathbf{L_i} = \mathbf{\Delta} + \mathrm{m} \cdot \mathbf{R} \cdot {}^{\mathrm{C}}\mathbf{L_i} \tag{5}$$

$$^{\mathrm{W}}\mathrm{L_i} = \begin{bmatrix} x_i \\ y_i \\ z_i \end{bmatrix}, {}^{C}\mathrm{L_i} = \begin{bmatrix} x_i \\ y_i \\ z_i \end{bmatrix}, \Delta = \begin{bmatrix} \Delta x_i \\ \Delta y_i \\ \Delta z_i \end{bmatrix}, R = R_z R_y R_x$$

$$R_x = \begin{bmatrix} 1 & 0 & 0 \\ 0 & \cos\varepsilon_x & \sin\varepsilon_x \\ 0 & -\sin\varepsilon_x & \cos\varepsilon_x \end{bmatrix}$$

$$R_y = \begin{bmatrix} \cos\varepsilon_y & 0 & -\sin\varepsilon_y \\ 0 & 1 & 0 \\ \sin\varepsilon_y & 0 & \cos\varepsilon_y \end{bmatrix}$$

$$R_z = \begin{bmatrix} \cos\varepsilon_z & \sin\varepsilon_z & 0 \\ -\sin\varepsilon_z & \cos\varepsilon_z & 0 \\ 0 & 0 & 1 \end{bmatrix}$$

By inserting the landmark coordinates of the camera ($^{\mathbf{C}}\mathbf{L_i}$) and the world coordinate system ($^{\mathbf{W}}\mathbf{L_i}$) into Eq. (4), an overdetermined system of equations is created, which can be solved with a least square routine. This results in the rotation matrix **R**, the scaling factor m and the translation vector $\mathbf{\Delta}$. Thus, the relation between the two coordinate systems is clearly determined.

2.3 *Determination of the Object Markers Inside the Camera Coordinate System*

To determine the positions of the measuring points based on the measured angles, the subsequent approach is followed. A surface is constructed by the landmark points within the camera coordinate system. If more than three points are used, the plane must be balanced. This is calculated according to [11], on the basis of the points $^{C}L_i$, to obtain a plane in point-normal form:

$$B : n \cdot \left(\vec{r} - \vec{r_0}\right) = 0 \tag{6}$$

In order to determine the positions of the measuring points, the intersection of the direction vector $\vec{a_i}$ with the previously constructed plane B is calculated. If the measuring points are not at the same height as the landmarks, the plane B must be shifted along its normal vector according to the height difference from the initially known z-position of the marker, resulting in a plane B_i for each point.

$$^{C}M_i = \frac{\vec{n} \cdot \vec{r_0}}{\vec{n} \cdot \vec{a_i}} \cdot \vec{a_i} \tag{7}$$

The direction vectors to the measuring points result as the unit vectors of the landmarks from the measured azimuth and elevation angles (7).

$$^{C}M_i = \lambda \overrightarrow{a_i} = \begin{bmatrix} \sin(\varphi_i)\sin(\theta_i) \\ \cos(\varphi_i)\sin(\theta_i) \\ \cos(\theta_i) \end{bmatrix} \quad (8)$$

2.4 Determination of the Rotation Angle

To calculate the rotation of the markers, the homography matrix is used. The 3 × 3 homography matrix describes the projection from a tag with homogeneous coordinates. The output of the homography matrix is already available as a standard function in the analysis library for many fiducial markers, such as AprilTags [13]. To calculate the orientation and the translation in space, the intrinstic parameters of the camera are needed. For the intended application only the rotation around the Z-axis is needed. According to [14], the parallel lines in the planar surface of AprilTags also appear parallel in the image. This implies that the new image can be interpreted as a frontal view and thus an affine transformation can be applied. Therefore, for the entries of the homography matrix (9), $h_{00} = cos\theta$ and $h_{10} = sin\theta$ can be assumed. The angle θ can then be determined by applying (10). To correct the image skew resulting from the pan angle of the camera, the pan angle must be added to the theta angle to compensate for it.

$$H = \begin{bmatrix} h_{00} & h_{01} & h_{02} \\ h_{10} & h_{11} & h_{12} \\ h_{20} & h_{21} & h_{22} \end{bmatrix} \quad (9)$$

$$\theta = \arctan 2(h_{10}, h_{00}) \quad (10)$$

3 Experimental Setup

In order to apply the theory of the concept, it was implemented in a test setup and initial measurements were taken to determine the accuracy and speed. For this purpose, 28 AprilTags were attached to the floor of a factory building as shown in Fig. 2. The tags were each placed at a distance of 1 m, the tags with IDs from 0 to 15 were placed in a line and the tags with IDs from 16 to 27 were placed in a line right-angled to the first line. The camera was placed at a height of about 4 m, without considering the orientation, and placed in an arbitrary location so that all tags are visible. To test the concept, tags 0, 15 and 27 are used as landmarks and the origin of the world coordinate system is placed in the center of tag 15.

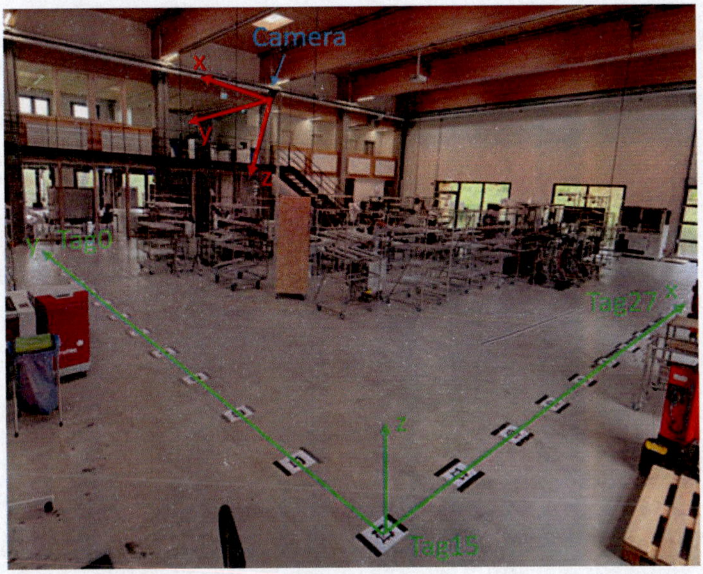

Fig. 2 Test setup using a PTZ camera and AprilTags

As the positions of the tags in the world coordinate system are on the X or Y axis and their distances were measured with a laser measuring device, these are known and can later be compared with the results of the camera measurement. The individual components are described in the following subsections.

3.1 The PTZ Camera

The core element is the camera. A surveillance camera of the type M5525 from Axis with a price of about 1000 € is used. The pan range is 360° with endless rotation and the tilt angle can be moved up to 90°. The angles can be set and read out with a resolution of 0.01°. The image resolution is 1920 × 1080 pixels and the camera has a 10 × zoom. The control of the camera is enabled by an integrated web service (Vapix API), which allows to control and read out practically all relevant parameters by simple http requests. As a preparatory measure, the error of the tilt angle was determined in order to compensate it for the calculations. As seen in Fig. 3 a linear error can be assumed.

The error is highest when the angle is set to 0° (horizontal orientation of the camera). The error decreases as the camera is moved downwards. By applying a linear regression, this error can be compensated.

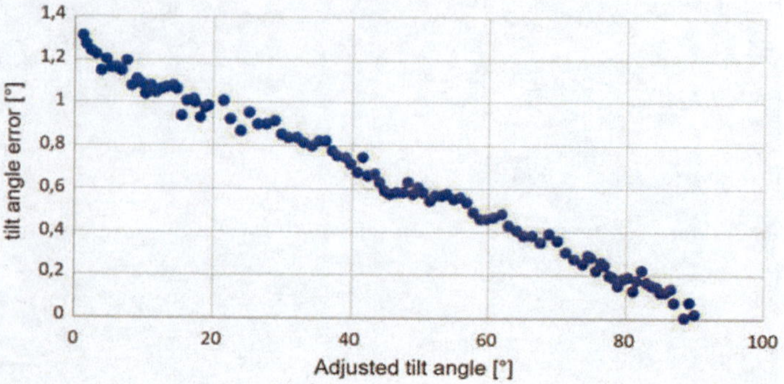

Fig. 3 Tilt angle error depending on adjusted angle

3.2 *Fiducial Markers*

There are countless different fiducial markers for the purpose of pose estimation. A good overview and comparison between some markers can be found under [12]. The AprilTags [15] are used for the test implementation, but other markers can also be used if they provide orientation estimation. One great advantage of the AprilTags is the availability of different tag families, which differ in the number of unique IDs. The standard family "TagStandard52h13" offers a maximum number of 48,714 unique IDs. An example of a tag of the family "TagStandard52h13" is shown in Fig. 4. Flexible AprilTag layouts can also be created for special requirements.

Basically, most fiducials are designed to measure distances based on camera parameters, but since only angles are measured in the concept described above, this feature is not needed.

Fig. 4 AprilTag TagStandard52h13 with ID 0

Fig. 5 Image analysis of an AprilTag at 10 × zoom

3.3 Measuring Procedure

First, the rough scanning was performed in a 2 × zoom state. The rough angle positions were saved. The subsequent fine scanning was performed with the maximum 10 × zoom. For image analysis, the implementation of the AprilTag algorithm from AprilRobotics on github was used (https://github.com/AprilRobotics/apriltag). The image analysis provides, for example, the image coordinates of the corner points, the center point and the corresponding ID of an AprilTag. In Fig. 5, the parameters are displayed graphically and superimposed on the image. The single images of the video stream are analysed in a loop and the pan and tilt angle of the camera are adjusted until the image center is on the center of the AprilTag.

The single images of the video stream are analysed in a loop and the pan and tilt angle of the camera are adjusted until the image center is on the center of the AprilTag. At the end of each fine scan, the angles and homography matrix of the last image are saved for subsequent calculation. These calculations were described in the subsections 2.1–2.4.

3.4 Measurement Results

The measurement was repeated 25 times for each AprilTag and in each case the position errors in the XY plane were calculated. These are shown in Fig. 6 per tag ID as boxplot.

The highest total error for all tags is 52 mm. The reason that the Landmark position errors are not 0 is because the least square method is used to calculate the

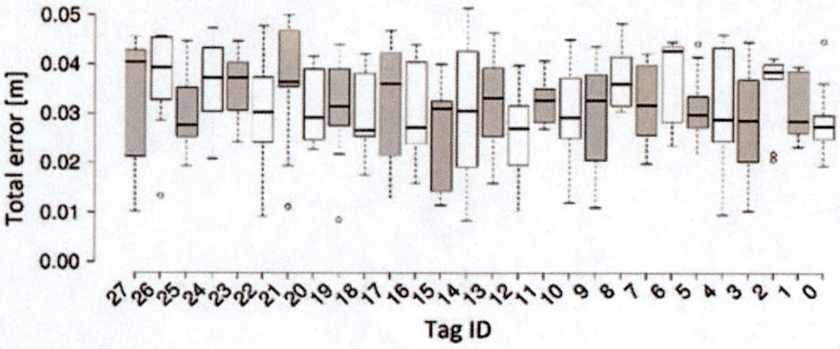

Fig. 6 Total error in the XY plane per AprilTag

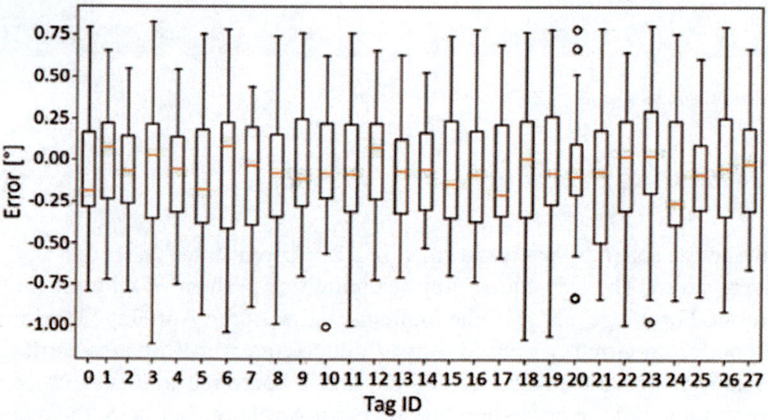

Fig. 7 Rotation error in XY plane per AprilTag

transformation parameters. Figure 7 shows the rotation error in the XY plane. For each tag the measurement was repeated 25 times. The highest rotation error of all tags is $-1.15°$.

4 Conclusion and Future Work

The first test setup confirmed the basic functionality of the concept. A reasonably high positional and rotational accuracy could be achieved. Of course, the accuracy depends on the accuracy of the pan-tilt mechanism and the image resolution and will differ with other camera models. In the presented test setup, all tags were placed on the same level and the camera was temporarily mounted. In future work, the camera will be mounted on the ceiling of the factory floor and a larger AprilTag test area will be set up with different heights to confirm the results. Subsequently, the AprilTags will be attached to the facilities and their allocation to the coordinate system is carried out to determine the pose of the facilities from the measured pose of the AprilTags.

References

1. Brena, Ramon, F., et al.: Evolution of indoor positioning technologies: a survey. J. Sens., (2017)
2. Uhlemann, T.H.J., Lehmann, C., Steinhilper, R.: The digital twin: realizing the cyber-physical production system for industry 4.0. Procedia Cirp **61**, 335–340 (2017)
3. Pawellek, G.: Ganzheitliche Fabrikplanung: Grundlagen, Vorgehensweise. Springer, EDV-Unterstützung (2014)
4. Mütze, A., Hingst, L., Rochow, N., Miebach, T., Nyhuis, P.: Use Cases of Real-Time Locating Systems for Factory Planning and Production Monitoring (2021)
5. Sommer, M., Stjepandić, J., Stobrawa, S., von Soden, M.: Automatic generation of digital twin based on scanning and object recognition. In Transdisciplinary Engineering for Complex Socio-technical Systems, pp. 645–654. IOS Press (2019)
6. Hermann, J., Leipzig, K. V., Hummel, V., Basson, A.: Requirements analysis for digital shadows of production plant layouts. In *Towards Sustainable Customization: Bridging Smart Products and Manufacturing Systems*, pp. 347–355. Springer, Cham (2021)
7. Awange, J.L., Paláncz, B.: Geospatial algebraic computations. Theory and applications, 3rd edn. Springer, Berlin (2016)
8. Haralick, B.M., Lee, C.N., Ottenberg, K., Nölle, M.: Review and analysis of solutions of the three point perspective pose estimation problem. Int. J. Comput. Vision **13**(3), 331–356 (1994)
9. Grunert, J. A.: Das Pothenotsche Problem in erweiterter Gestalt; nebst Bemerkungen über seine Anwendungen in der Geodäsie. Grunerts Archiv für Mathematik und Physik **1**: 238–241 (1841) (cited in: [1], p.335).
10. Donner, R.U.: Eine Lösung für den räumlichen Rückwärtsschnitt in der Vermessungstechnik. *ZfV-Zeitschrift für Geodäsie, Geoinformation und Landmanagement*, (zfv 1/2020) (2020)
11. Hanson, R.J., Norris, M.J.: Analysis of measurements based on the singular value decomposition. SIAM J. Sci. Stat. Comput. **2**(3), 363–373 (1981)
12. Kalaitzakis, M., Cain, B., Carroll, S., Ambrosi, A., Whitehead, C., Vitzilaios, N.: Fiducial markers for pose estimation. J. Intell. Rob. Syst. **101**(4), 1–26 (2021)
13. Krogius, M., Haggenmiller, A., Olson, E.: Flexible layouts for fiducial tags. In *2019 IEEE/RSJ International Conference on Intelligent Robots and Systems (IROS)*, pp. 1898–1903. IEEE (2019)
14. Collins, R.T., Beveridge, J.R.: Matching perspective views of coplanar structures using projective unwarping and similarity matching. In *Proceedings of IEEE Conference on Computer Vision and Pattern Recognition*, pp. 240–245. IEEE (1993)
15. Olson, E.: AprilTag: A robust and flexible visual fiducial system. In *2011 IEEE International Conference on Robotics and Automation*, pp. 3400–3407. IEEE (2011)

Julian Hermann obtained a M.Sc. degree in Mechatronic Engineering from Reutlingen University. He holds a position as researcher at ESB Business School, Reutlingen University, Germany and is currently pursuing a Ph.D. at Stellenbosch University (Industrial Eng. Department).

Anton Basson obtained his Ph.D. in Aerospace Engineering at Penn State University in 1992. He is a Professor in Mechanical Engineering and co-leader of the MAD research group.

Konrad von Leipzig obtained both his B.Eng and M.Eng degrees in Industrial Engineering. He later also completed a B.Com degree. He has been a lecturer at the Department of Industrial Engineering at Stellenbosch University since 1987. He forms part of the team undertaking the realisation of the Learning Factory. His research areas include process, supply chain, logistics and financial management.

Vera Hummel Prof. Dr.-Ing., Dipl.-Ing., has been a professor at the ESB Business School, Reutlingen University since 2010, vice-dean research of the ESB and held the position of an associated Professor at the Stellenbosch University, Industrial Engineering department. She is a founding member of the "Initiative of European Learning Factories" and the initiator and head of the "Werk150" the factory of the ESB for research, education and industry training at the Reutlingen University.

Data Analytics in Industrial Engineering for Economic Sustainability: A Use Case on Planning and Controlling of Rework

Ralph Hensel, Thomas Mayr, and Mathias Keil

Abstract Currently, Industrial Engineering (IE) is predominantly focusing on planning and rationalization of production processes. Studies in the automotive industry are proving that about 30% of the productivity are hidden in additional manufacturing times and not value-adding, such as rework. In order to improve the company's overall productivity, early recognition and correction of quality defects becomes increasingly important to reduce defects caused by the product or the process. With this, the determined planning, optimization, and control of the rework processes moves into the scope of Industrial Engineering. Though, rework activities are characterized by its unpredictable occurrence probability, especially regarding quality defect causes and consequential rework tasks as well as respective execution times. The present article will introduce a methodological approach using Data Mining-methods to analyze recorded product- and process-related quality data in the automotive industry. Besides the descriptive statistics used for diagnostic analysis of effort drivers for rework, regression trees were built for a vehicle-specific rework time forecast, which will be presented, too. These prediction-models were fed with different pre-shaped datasets to gather information about the performance of the forecast.

Keywords Rework · Industrial data science · Data mining · Productivity

1 Introduction

By taking over the responsibility for company's overall productivity management, Industrial Engineering (IE) plays a vital role to secure economic sustainability in a highly competitive world [1]. Unfortunately, in its everyday business IE is predominantly focusing on planning and rationalization of production processes. Experiences in the automotive industry are showing that about 30% of the productivity are hidden in additional manufacturing times, such as rework. These activities are not value-adding and should be avoided to realize the ideal of a zero-defect production but

R. Hensel (✉) · T. Mayr · M. Keil
Department of Industrial Engineering Methods and Standardization, AUDI AG, Ingolstadt, Germany

K. von Leipzig et al. (eds.), *Smart, Sustainable Manufacturing in an Ever-Changing World*, Lecture Notes in Production Engineering,
https://doi.org/10.1007/978-3-031-15602-1_58

are still contributing to the quality-equitable completion of the product [2]. With the aim to improve the company's productivity holistically, early defect recognition and correction becomes increasingly important to reduce quality defects caused by the product, either the constructional design or the supplier itself, and the process, such as technological disruptions or human errors. With this, the planning, optimization, and control of rework processes moves into the scope of Industrial Engineering. Rework activities are characterized by its unpredictable occurrence probability, especially regarding quality defect causes and consequential rework tasks as well as respective execution times. Due to a lack of methods, for Industrial Engineering it is almost impossible to fulfil these ambitious responsibilities, though [3, 4].

The utilization of Data Mining methods in product development and production planning offers considerable potential for the development and implementation of strategies for the optimization of both, products, and processes [5]. Hence, Data Mining contributes to the transition of Industrial Engineering from a reactive to a predictive approach in productivity management. The deployment of supervised and unsupervised machine learning methods allows to discover patterns from structured or unstructured data, to identify regularities and with this to discover new knowledge [6, 7]. With regards to rework planning, Data mining methods could help to investigate the relationship between different characteristics that are influencing product quality and rework with the aim to develop classification models for data-based and computational forecasting of future occurrence of quality defects to derive measures for improvements thereof.

In the following, a methodological approach specifically developed for the automotive industry will be introduced using Data Mining-methods to analyse product- and process-related quality data gathered from MES systems, firstly to determine optimization potential for improvement of production processes and the product itself. Secondly, the presented approach not only helps to predict expectable rework efforts, but also to target the number of employees required for rework in early planning phase of car development projects as well as in series production.

2 Methodological Approach

For the systematic execution of Data Mining projects, the Cross-industry standard process for data mining (CRISP-DM) widely gained acceptance amongst data mining experts, since it is industry, tool, and application neutral [8]. CRISP-DM breaks the process of data mining into six major phases: (1) Business Understanding, (2) Data Understanding, (3) Data Preparation, (4) Modelling, (5) Evaluation, and (6) Deployment. It solves existing problems in data mining by involving professional experts of the respective problem area into the phases of process and data understanding to consider the industry-specific framework conditions for data preparation and modelling accordingly. Figure 1 shows the concept of the presented use case on planning and controlling of rework by help of data mining, which is based on the CRISP-DM model described before.

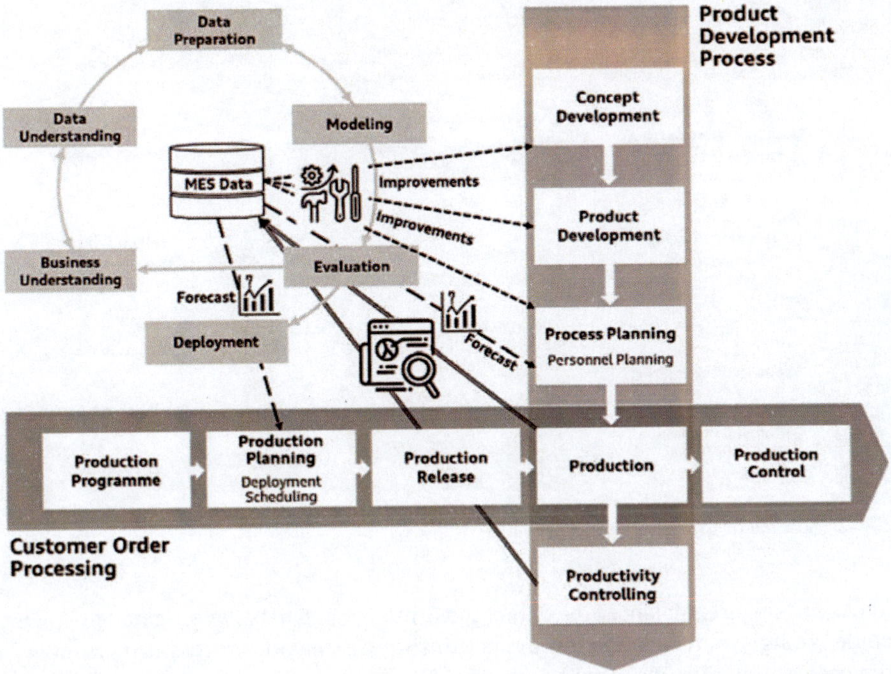

Fig. 1 Concept of data-based planning and controlling of rework by help of data mining

3 Concept Application

Since the concept is based on the CRISP-DM model, the structure of the present paper conforms to the six process steps explained above.

3.1 *Business Understanding*

In the car assembly defects on each individual vehicle are documented by help of mobile devices and recorded in a centralized quality data base. The correction of the defects could be either executed at the assembly line or at specific rework stations subsequent to the final assembly, what strongly depends on the severity as well as the complexity of the defects and with this on the expenditure of rework time expected. Although quality data is recorded in the database, information about expectable defects and required rework operations as well as their duration are currently not taken into consideration for personnel planning and deployment scheduling.

In a broader sense, the rework process starts with the discharge of defective vehicles, which are transported to the rework area, either specific rework stations if available, or a buffer area first. The defects are described in the quality data base, but

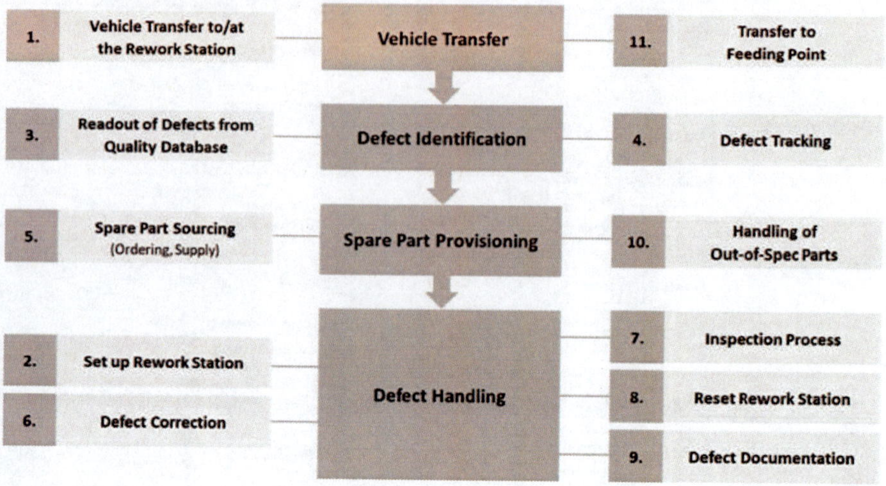

Fig. 2 Scheme of the rework process (s.l.)

under certain circumstances the defect cause might need to be investigated first, what could strongly influence the throughput time of the rework process, for example in the case of complex electronic problems. Furthermore, exchange parts might need to be sourced or reject parts disposed. These material handling tasks are usually not carried out by the reworker himself, who is mainly responsible for correcting the defect by repair work. After a final inspection the vehicle is discharged from the rework area and the correction of the defect being documented in the quality database. Figure 2 gives an overview of the rework process described before.

3.2 *Data Understanding*

In general, the required information for the forecast models is provided by two data sources. The first data set, the production program, contains the produced cars with their respective specifications, such as vehicle-ID, vehicle model, car body style, manufacturing date and additional features, for instance engine-transmission-combination. Figure 3 shows the structure of the specifications described.

The second data source contains the rework data of the produced vehicles, that was collected and documented in a subsystem of the MES. This data set contains information about the vehicle-ID, description of the defects that occurred, NOK assembly groups, NOK subassemblies, NOK components as well as data and time, when the defect was firstly and when it was corrected. Furthermore, the execution time for correcting the defects is documented by the worker in the rework area. Figure 4 displays the deviation of the rework time (in minutes) for the NOK higher-level group of components, clearly illustrating the broad distribution.

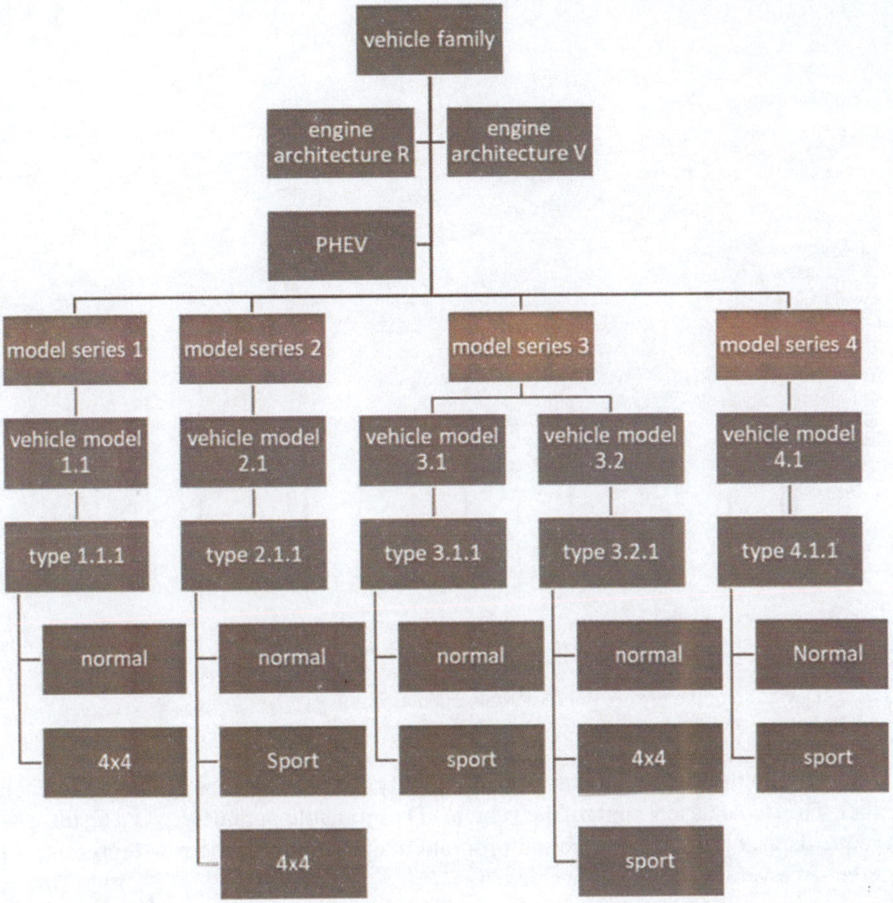

Fig. 3 Hierarchical structure of specifications

3.3 *Data Preparation*

Prior to developing a forecast model that is supposed to determine rework time, the data need to be preprocessed. Due to the fact, that the data set includes a continuous vehicle production plan, it needs to be cut back to a specific timeframe, which represents one full year. However, the historical data from 2020 contains the lockdown during the Covid-19 pandemic.

Furthermore, pre-series vehicles running through the assembly line for testing future vehicle projects, also need to be eliminated in the dataset. Additionally, the dataset required the elimination of missing values for further processing.

Because of the widespread rework time per defect and specific model, mean average rework time per vehicle was considered as a relevant attribute to determine plausible results.

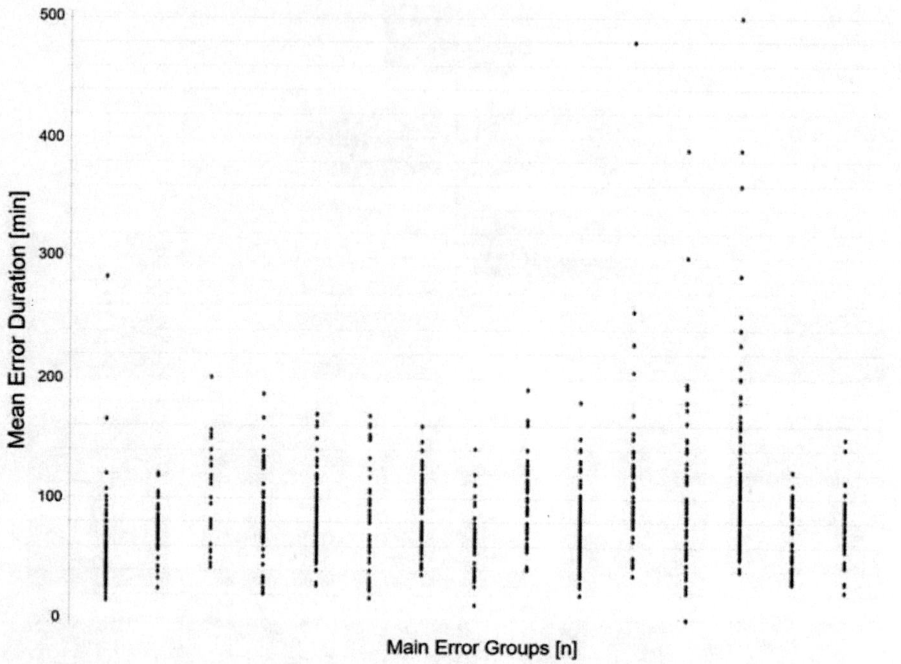

Fig. 4 Deviation of the rework time per NOK main assemblage

Finally, both production data set and quality data set need to be linked to each other. The two datasets contain the vehicle-ID as joinable parameter. As a result, one unique dataset with the production program and a column if the row represents an defect-free vehicle or a NOK vehicle.

3.4 Modelling

Once the raw data set is cleaned up, joined and relevant specifications were selected, the forecasting models were developed. A supervised learning method is used in this case. Based on a random sample selection of vehicles a training data set is defined, containing the target attribute—in this case a Boolean variable representing a zero-defect vehicle or not. This model is trained to identify non-observed NOK vehicles in the test data set. Key aspect in developing a forecast model is the selection of a suitable forecasting type and the subsequent optimization of the procedure parameters to reach a high forecast quality. In this case two different learning procedures are discussed: firstly, a decision tree model and secondly, a random forest model. The decision tree model creates a hierarchical set of rules, checking the feature space and gradually splits the characteristic values in the sense of axis parallelism [6]. The set of rules is structured as a tree—starting from the root along nodes and branches to a single leaf. The training data is split at the root and nodes are representing their characteristic values, to minimize the heterogeneity of the target value in the training data in every outgoing branch [10]. The majority class of the respectively assigned training objects in the leaves determines the target value. Decision tree methods are applied in as classification models as well as in regression models [9, 11].

The forecast quality of data-mining procedures is strongly related to the characteristics of the used data set. In this case, two different models are used, which additionally provide various settings, like splitting criteria and pruning methods. In order to adapt the data set to the forecasting model a splitting ratio of the dataset with a ratio of 4:1 is necessary: 80% of the data set being used for training the model and 20% set served for testing and validating the trained model. The data set is split with random selection. Every branch of the tree in a decision tree model represents a decision, an occurrence or a response. The difference between random forest models and decision tree models is that, random forest models use an ensemble learning approach, which generates a variety of sampled decision trees and finally labels the output through dependency of the single trees. The advantage of decision tree models in comparison to random forest models is the traceability, as these models use a branching method that represents every possible outcome of a single decision.

In the present case study, the data analytics toolset of the software KNIME Analytics Platform was used for the analysis. It is an open-source platform for data analytics, reporting and integration, which contains various components for machine learning and data mining. For the random forest model, KNIME contains tree options Gini Index, Information Gain and Information Gain Ratio as splitting criteria. For the following evaluation, the performance of the model was tested with different splitting criteria. In contrast, the decision tree model is adjustable with specifications like the root of the tree and with or without pruning. It is possible to configure the root as the default value, for internal determination through KNIME, for reaching best results in forecasting. Figure 5 gives a brief insight into the generation of both models that were used in the case study.

3.5 Evaluation

Both forecasting methods are fed with the same data, as it can be seen in the picture above. The training data run through the respective learner of the model. The learner forwards the trained function to the respective predictor. In the learner node of the respective model, some adjustments of parameters like tree depth, partition criteria and pruning options can be made [12]. Finally, the test data are used for validating the models. The scorer nodes output shows a 2 by 2 table with false-true results. Remarkable is the fact that the random forest model with the best spotted configuration reaches the same performance as the decision tree model with a default root vision. With a forecasting accuracy of ca. 62% the model identifies if it is a zero-defect vehicle or a NOK vehicle.

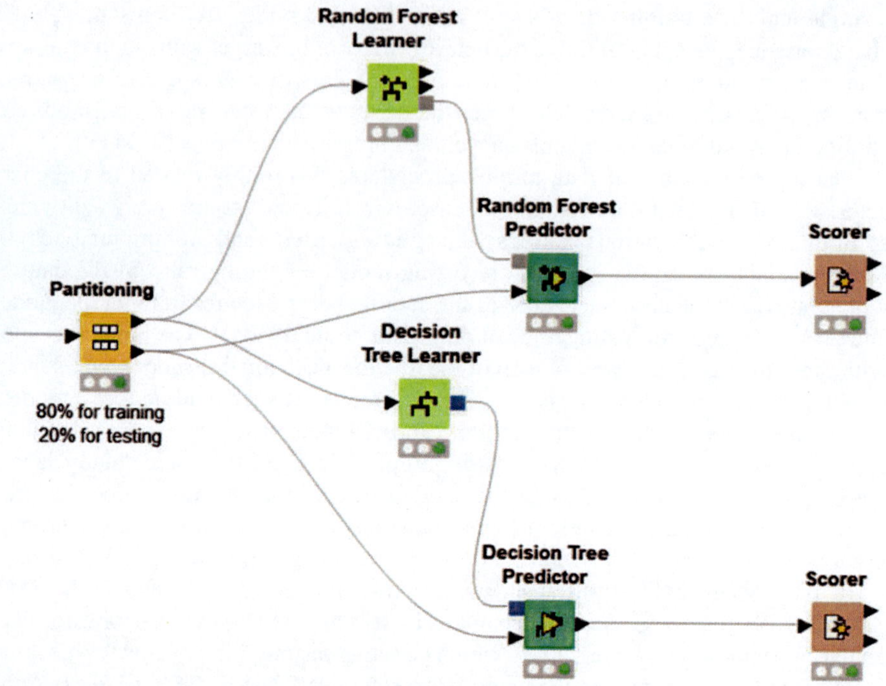

Fig. 5 Modelling random forest and decision tree in KNIME

The input data only contains nine distinctive features, which leads to difficulties in identifying NOK vehicles. For improving the forecasting performance, it is necessary to add more distinctive features to the data set. Alone, adding one distinction between straight-type engine and v-type engine leads to a performance gain of about 5%. This gain describes the information content of each specification. Figure 6 shows the performance gain per specification, which is an indication for the amount of information added to the random forest model.

Based on the vehicle family and the vehicle model, these two specifications include the specification model series. Therefore, no performance gain with model series occurs. Remarkable is the low performance gain of the random forest model concerning the specification sport and PHEV. This means it is not necessary for the rework planning, to differentiate between these two specifications. The impact of each specification to the performance should be clear for further development of the model. It is essential to preprocess the data set in a specific way concerning the distinction to reach better forecasting accuracy. After the classification in NOK vehicles and OK vehicles, the mean rework time of each specific vehicle is calculated. With the information of the mean rework time and the production program the personnel planning now is capable to utilize the rework crew to capacity.

Confusion matrices and several statistical quality criteria are suitable for evaluating the classification quality of the methods used. A confusion matrix consists of

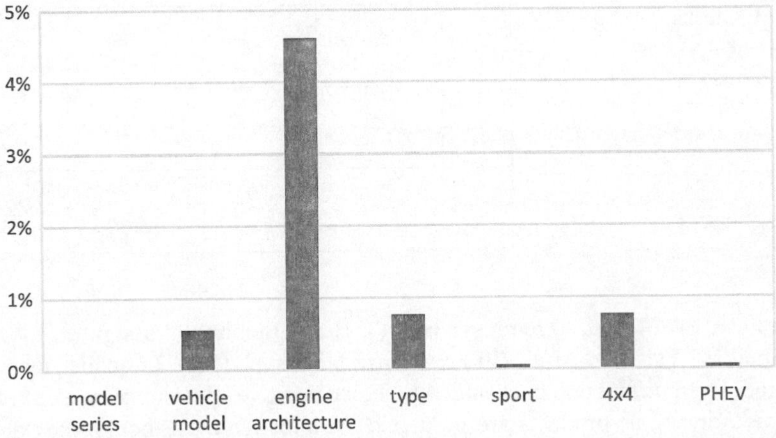

Fig. 6 Performance gain per specification

two dimensions, one dimension is indexed by the actual class of an object, the other is indexed by the class that the classifier predicts [13] shown in Fig. 7, the confusion matrix in this case consists of two rows and two columns, therefore two represents the number of classes to be analyzed.

The correctly classified objects are on the diagonal of the matrix and the other entries of the matrix are the objects that have been assigned to the wrong class. In the present case of binary classification, correctly assigned objects are classified as

		True Label		
		Positive (TP+FN)	Negative (FP+TN)	
Prediction	Positive (TP+FP)	True Positive (TP)	False Positive (FP) pseudo-error	Precision $= \frac{TP}{TP + FP}$
	Negative (FN+TN)	False Negative (FN) slip	True Negative (TN)	Negative Predictive Value (NPV) $= \frac{TN}{FN + TN}$
		Recall $= \frac{TP}{TP + FN}$	Specificity $= \frac{TN}{FP + TN}$	Accuracy $= \frac{TP + TN}{TP + TN + FP + FN}$

Fig. 7 Confusion matrix

Fig. 8 Confusion matrix—random forest model

Prediction	Positive	12524	8296	Precision 0.739
	Negative	4415	7970	NPV 0.49
		Recall 0.602	Specificity 0.644	Accuracy 0.617

"True Positive" (TP) or "True Negative" (TN) and incorrectly assigned objects are described as "False Positive" (FP) or "False Negative" (FN). In quality inspection in an industrial production environment, "False Positive" decisions are also referred to pseudo-errors, as products are predicted as defective even though they are free of defects. "False Negative" decisions are also referred to slippage, since defective products are not recognized as such due to a wrong classification. Based on the information of the confusion matrix, various statistical quality criteria can be determined as conditional probabilities, which enable the quality of the classification to be assessed. When predicting defective cars, it must be ensured that faulty cars are identified with a high degree of confidence, meaning that the slip is as low as possible. To evaluate the method in the present use case, the recall is particularly suitable, which sets the number of correctly classified as false classified vehicles in relation to the population of all actual false vehicles. The precision indicates which proportion of the vehicles predicted to be defective was false. However, since a minimization of the slip is assigned much greater importance than a reduction of pseudo-errors, a lower precision compared to a better recall is accepted. Accuracy is also a good indicator of the quality of the classification of the model, but in this case, it is not as important as the recall.

As the two confusion matrices in Figs. 8 and 9 are indicating, both models show a good recall, but also a poor accuracy. The confusion matrix of the random forest model (Fig. 8) shows a recall of 0.602 and the decision tree model (Fig. 9) shows a recall of 0.599. In this case the random forest model has a slightly higher recall at the same accuracy. Therefore, considering the random forest model as more suitable for this classification application.

3.6 Deployment

For the sustainable implementation of the developed concept for data-based planning and controlling of rework into industrial practice, it is necessary to create a process control logic on a conceptual level, which allows to determine the amount of employees required for rework in early planning phase of car development projects as well as in series production. Studies are showing that only 3% of data science project are implemented productively, since the models developed can often not be

Fig. 9 Confusion matrix—decision tree model

		True Label		
		Positive	Negative	
Prediction	Positive	12705	8491	Precision 0.75
	Negative	4234	7775	NPV 0.478
		Recall 0.599	Specificity 0.647	Accuracy 0.617

connected to the companies' IT systems and the data sources required [14]. Therefore, further conceptual effort needs to be put firstly, into setting up the infrastructure in order to connect the production program and the quality data to the ERP system. Secondly, the accuracy of the forecast needs to be increased, by updating the quality data and reviewing the developed model continuously.

4 Conclusion

The present paper introduced a methodological approach that allows to make predictions about expectable quality defects in the automotive industry based on statistical data of NOK vehicles. By using the mean average rework time per vehicle even the rework time resulting thereof can be predicted, too.

However, rework times that could be derived from the classification models cannot be compared with the requirements on the determination of target times as usually stipulated in Industrial Engineering. Against the backdrop of the presented use case the forecast allows to determine both, the capacity utilization of the rework area and the number of workers needed with acceptable inaccuracy, though. Currently, for the determination of rework times there is no concept available that allows their prediction with reasonable effort. By saying so, the concept presented in this paper can be considered as a cutting edge for IE.

The results of the use case are demonstrating that data sets with an utterly uneven distribution of the characteristics of the respective classes are requiring a certain data preparation to identify specifications that are determining the behavior of the data set. Hence, it is possible that certain NOK groups of components can be classified better the others. Therefore, in the next step, the recorded defects (NOK groups of components and types of defects) shall be classified by help of Clustering methods into groups to increase the quality of the forecast.

References

1. Hensel-Unger, R.: Entwicklung einer Gestaltungssystematik für das Industrial Engineering (IE). Universitätsverlag, Chemnitz (2011)
2. DIN EN ISO 9000:2015–11, Quality Management Systems—Fundamentals and Vocabulary [ISO 9000:2015]
3. Flapper, S.D., Fransoo, J.C., Broekmeulen, R.A.C., Inderfurth, K.: Planning and control of rework in the process industries: a review. Prod. Plann. Control **13**(1), 26–34 (2002)
4. Longard, L., Brungs, F., Hertle, C., Roeth, J., Metternich, J.: Reduced rework through data analytics and machine learning—a three level development approach. SSRN Electron. J., (2021)
5. Eickelmann, M., Wiegand, M., Konrad, B., Deuse, J.: Die Bedeutung von Data-Mining im Kontext von Industrie 4.0. Zeitschrift für wirtschaftlichen Fabrikbetrieb. **110**(11), 738–743 (2015)
6. Fayyad, U., Piatetsky-Shapiro, G., Smyth, P.: From data mining to knowledge discovery in databases. AI Mag. **17**(3), 37–54 (1996)
7. Hastie, T., Tibshiriani, R., Friedman, J.: The Elements of Statistical Learning. Data Mining, Inference, and Prediction, 2nd edn. Springer, New York (2009)
8. Quinlan, J.R.: Induction of decision trees. Mach. Learning **1**(1), 81–106 (1986)
9. Witten, I.H., Frank, E., Hall, M.: Data Mining: Practical Machine Learing Tools and Techniques, 3rd edn. Morgan Kaufman, Boston/USA (2011)
10. Cleve, J., Lämmel, U.: Data Mining, 2nd edn. Oldenburg Verlag, München (2014)
11. Olson, D.L., Desheng, W.: Predictive Data Mining Models (Computational Risk Management). Springer, Heidelberg (2017)
12. Shafer, J., Agrawal, R., Mehta, M.: SPRINT: A Scalable Parallel Classifier for Data Mining. In: Proc. 1996 Int. Conf. Very Large Data Bases, S., pp. 544–555 (1996)
13. Deng, X., Liu, Q., Deng, Y., Mahadevan, S.: An improved method to construct basic probability assignment based on the confusion matrix for classification problem. In: Elsevier, Information Sciences, S. 250–261 (2015)
14. Mierswa, I.: Manifesto for Better Data Science. Industrial Data Science Conference 2020. Dortmund, 21./ 22 (2020)

Ralph Hensel obtained his Ph.D. degree in Work Science (Industrial Engineering) from the Chemnitz University of Technology. Since 2016 he is Senior Expert for Ergonomics and Industrial Engineering at Audi's Industrial Engineering Department.

Thomas Mayr is a Ph.D. student at Audi's Industrial Engineering Department in Ingolstadt and does research on Industrial Data Science. He holds a Master's degree in Industrial Engineering from Technical University of Vienna/ Austria.

Mathias Keil is currently the Head of the Department Industrial Engineering Methods and Standardization at AUDI AG in Ingolstadt. He holds a Ph.D. degree in Work Science (Industrial Engineering) from Chemnitz University of Technology.

Finite Element Analysis of the Stress Distribution in a Novel Brake Beam of a Railcar

Ilesanmi Daniyan, Khumbulani Mpofu, Felix Ale, and Rumbidzai Muvunzi

Abstract The brake beam is an important component of the brake beam assembly and railcar suspension system. A properly designed brake beam will enable effective transmission of the braking force via the brake shoes to the the outer tread of the rail wheels. In this study, the Finite Element (FE) method was employed for investigating the performance of a novel brake beam design. The Finite Element Analysis (FEA) of the component is to ensure that the component possess satisfactory strength and rigidity vis-à-vis the functional requirements. This was done in the Solidworks 2018 environment using the von Mises stress and failure criterion for the evaluation of stress induced in the component member. The results obtained from the manual calculations and simulations of the brake torque and angular velocity of the motor agree significantly, thus, indicating that the suitability of the design data for implementation. The FEA of the brake beam indicates that the material selected for the brake beam possesses adequate satisfactory strength and rigidity to withstand the stress induced without yielding to failure or undergoing permanent deformation. This study provides some information relating to the FEA of brake beam which can assist manufcturers during the manufacturing phase of the component.

Keywords Brake beam · Brake torque · Finite element analysis · Railcar · Stress

I. Daniyan (✉) · K. Mpofu · R. Muvunzi
Department of Industrial Engineering, Tshwane University of Technology, Pretoria 0001, South Africa
e-mail: afolabiilesanmi@yahoo.com

K. Mpofu
e-mail: mpofuk@tut.ac.za

R. Muvunzi
e-mail: MuvunziR@tut.ac.za

F. Ale
Department of Engineering and Space Systems, National Space Research and Development Agency, Abuja, Nigeria

K. von Leipzig et al. (eds.), *Smart, Sustainable Manufacturing in an Ever-Changing World*, Lecture Notes in Production Engineering,
https://doi.org/10.1007/978-3-031-15602-1_59

1 Introduction

The brake beam is a subcomponent of the brake beam assembly usually mounted between opposite side frames [1]. Every railcar system usually has two brake beams for the transmission of the braking force via the brake shoes to the the outer tread of the rail wheels [1]. During the railcar operation, the displacement of the brake beam may occur due to induced stresses or railcar maneuverings along curved paths. The displacement of the brake beam from the normal position may cause ineffective transmission of the braking force thereby causing the rail wheels to wear.

Furthermore, the displacement of the brake beam can also affect the control of the railcar during movement which may cause accident. Conventionally, the brake beam is usually mounted between the opposite side frames with opposite right and left hand ends [1]. A strut is usually attached to the brake beam while the brake heads are attached to the brake beam and struts subassembly to hold the brake shoe designed for the engagement of the railcar wheel. For the conventional brake beam, the continuous application of braking force to the brake beam can trigger lateral or longitudinal deflections. It can also cause vibration or unwanted movement of the side frames. This can generate friction and wear in the brake beam sub-assemblies such as the brake shoe, brake lever etc. The resultant effect of the friction can also cause a gradual wear of the rail car wheel. Usually a clearance is usually created between the brake beam and the accessories for conventional design, however, the continuous movement of the railcar beam and its accessories can cause an extension in the clearance thus keeping the brake beam out of alignment. This implies that in order to improve the current design of the brake beam of railcars, there is a need for reduction in the friction and reciprocating movement of the brake beam accessories during operation. In a bid to address the current challenge of brake beam displacement during operation, this work seek to unveil a novel design in which on end of the brake beam is fixed while the other is free.

In order to promote the rigidity of the brake beam, the beam is designed as a cantilever such that the free end will be exposed to vertical loads. First, the choice of the cantilever beam will promote greater rigidity, lesser flexibility of the brake beam with significant reduction in the number of additional supports and bracing. The Finite Element Analysis (FEA) employs the finite element method for the investigation the behavior of a component part under certain conditions. The method has been widely used for investigating the behavior of some railcar components. For instance, Daniyan et al. [2] employs the FEA approach for the investigation of the behaviour of the traction link of a railcar. The results obtained provided an insight into the magnitude of the stress, displacement and strain distributions during the normal operating condition of the traction link. Nejad [3] studied the variation of residual stresses in railway wheels using the FEA. Under varying loads, the the results obtained from the FEA was validated via physical experimentations. It was reported that the FEA results correlated significantly with the ones obtained during the physical experimentations. In addition, Kukulski et al. [4] used the FEA technique for the investigation of the strength of a railway surface. The results obtained captured the behaviour of the railway surface under varying loads.

The use of the FEA technique involving the isotropic and kinematic hardening models for AISI 1035 for weld stress prediction during assembly operations of the lower and upper bracket of railcars have been reported [5, 6]. Zulkilifi [7] employed the FEA approach to study the interaction between the rail track and wheel. The findings indicate that the rolling contact contributes significantly to the rail surface fracture compared to braking force. It can be inferred from this study that the braking characteristics of the rail car which is partly a function the braking assembly can cause rail surface fracture. Song et al. [8] carried out a 3-D FEA of high-speed train–bridge interactions. The work featured the modelling of the rails and sleepers of the track structure. Sysyn et al. [9] traced frequent track failures to unsupported sleepers or void zones in ballasted tracks. The authors employed the FEA approach to investigate the sleeper–ballast dynamic impact in the void zone. The result obtained showed significant impact acceleration and rail deflection at the void zone.

Zeng et al. [10] investigated the load bearing capacity of the rail track using the FEA approach validated via physical experimentations.

The findings from these studies indicate the appropriateness of the finite element method for the simulation and analysis of rail components. The novelty of this work lies in the design and FEA of a cantilever brake beam as opposed to the conventional freely supported brake beam. Although, the cantilever brake beam boasts of easy design and fabrication without the need for support at the fixed end, however it is susceptible to large deflections. Hence, the need for the use of the FE method for the investigation of the load carrying capacity of the components as well as its behaviour under the required loading or functional conditions. This study can assist manufacturers with design data during the manufacturing phase of the brake beam. It can also provide an insight into the performance of the brake beam vis-à-vis the service requirements.

2 Methodology

The span of the cantilever beam is a function of the depth of the beam, nature and magnitude of the load to be imposed as well as the nature of material to be employed [11].

For this novel design the span of the cantilever beam was selected as 2.5 m while the maximum load was calculated as 2,206,496.25 N. The calculation was informed from the work of Orlova et al. [12]

The overall load (W) is given as 2,206,496.25 N. Assuming that the beam is uniformly loaded thoughout its length, the magnitude of the uniformly distributed load (w) is expressed as Eq. 1 [13].

$$W = wL \tag{1}$$

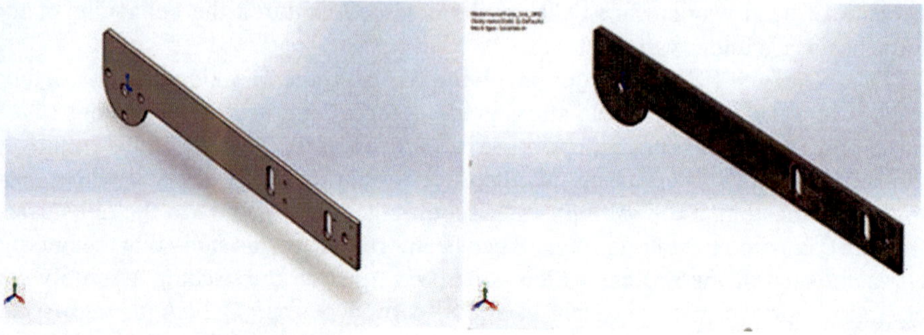

Fig. 1 The model of the brake beam

where L is the length of the beam (m).

The magnitude of the uniformly distributed load throughout the beam length is calculated as 882.5 kN/m. This value equals the maximum resultant force at the fixed end of the beam.

The magnitude of the bending moment (M) of the brake beam varies from zero at the free end to a maximum value of 1103 kNM at the fixed end. This is obtained form Eq. 2 [13]. Hence, the main reinfrcement should be properly provided at the upper layer of the beam to withstand the tensile stress.

$$M = w\left(\frac{L}{2}\right) \tag{2}$$

Figure 1 presents the model of the brake beam which determine the degree of rigidity of the system.

The FEA of the brake beam component was carried out to ensure that the components possess satisfactory strength and rigidity vis-à-vis the functional requirements. This was done in the Solidworks 2018 environment using the von Mises stress and failure criterion for the evaluation of stress induced in the component member.

The choice of alloy steel for the development of the brake beam stem from the fact alloy steel boasts of high strength and high corossion resistance. It is suitable for use in parts which requires high strength, heat and corossion resistance [14]. The brake beam requires high strength to withstand both the rail and load disturbances. The load disturbances is a function of the load carried by the brake sub assembly while the rail disturbances can result from irregular rail profiles [15–17]. The right selection of material of the brake beam can enhance the braking characteristics of the railcar, minimise friction due to the wheel-rail interaction, minimise wheel and rail wear and enhance the control and safety of the railcar.

The material's properties including the meshing information are presented in Tables 1 and 2.

The FEA was carried out to gain an insight into the nature of stress and deformation induced as well as the dynamic response of the materials under stated loading

Table 1 The material properties of the brake beam

Properties	Specification and value
Material	Alloy steel
Model type	Linear elastic isotropic
Failure criterion	Von Mises stress analysis
Yield strength	6.220422E08 N/m^2
Tensile strength	7.23826 E08 N/m^2
Elastic modulus	2.10 E11 N/m^2
Poison's ratio	0.28
Mass density	7700 kg/m^3
Shear modulus	7.9 E10 N/m^2
Thermal expansion coefficient	1.30E-05/K

Table 2 Meshing information

Mesh type	Solid mesh
Mesher used	Standard mesh
Jacobian points	4 Points
Element size (mm)	3.10249
Tolerance (mm)	0.155124
Total nodes	51,496
Total elements	29,924
Maximum aspect ratio	6.0018
% of elements with aspect Ratio <3	99.5
% of elements with aspect Ratio >10	0
% of distorted elements (Jacobian)	0

conditions. FEA permits the investigation of the performance of the brake beam to be developed in details at the design stage vis-a-vis its performance real-world scenario conditions. This is to prevent failure and wear of the railcar wheel. The FEA will also permits the understanding of the working conditions of the component as well as the identification of the feasible range of the process parameters that is required for the cmponent development. Hence, it minimises the number of the prototypes to be developed, and experimental trials to be performed while also minimising the chances for redesign or redevelopment. Thus, the FEA approach is process and time effective.

A geometric Computer Aided Design (CAD) model of the brake beam was carried out in the Solidworks 2018 environment. This is followed by the discretization of the model with a standard mesh having 4 Jacobian points. The mesh selected balances the efficiency with computational cost. The mesh convergence produced mesh size of 0.2 mm as the optimal element size of 3.10249 to mesh the parts into finite elements thus providing the needed level of accuracy. This mesh size ensures that the outcome of the FEA is unaffected by the changing the size of the mesh. The outcome of the FEA is a function of the selection of the right mesh sizw and ease of convergence.

In-situ mechanical and thermal loading profiles were imposed in the analyses to determine the feasibility of the material in meeting the required functional requirements. Non-rotational mechanical boundary conditions were also imposed on the free end of the brake beam in the static general analysis. This was carried out to ensure the rigidity of the brake beam against rotation at the free end. At the free end, the brake beam is allowed to move only in the X–Y–Z axes during the required operation. A total load of $2.20 \times 10^6 N$ was imposed on the designed beam of span 2.5 m and the behaviour of the brake beam was studied under varying loading conditions.

The Von Misses stress employed as the failure criterion to investigate the stress induced and the respnse of the material to the stresses developed. Thereafeter a comparative analysis of the magnitude of the Von Mises stress was carried out in relation to the yield strength of the material. Once the yield strength is not exceeded, it implies that the material will meet the functional requirement under the stated conditions and vice versa. strength of the materials.

Equation 3 presents the expression for the Von Misses stress analysis [18].

$$(\sigma t_1)^2 + (\sigma t_2)^2 - 2\sigma t_1 \times \sigma t_2 = \left(\frac{\sigma y_t}{F.S}\right)^2 \tag{3}$$

where σ is the stress induced in the material (Pa), σy_t is the yield strength of the material (Pa) and $F.S$ is the factor of safety.

The stress (σ) is expressed as Eq. 4 [19].

$$\sigma = \frac{F}{A} \tag{4}$$

where F is the force applied on the brake beam (N) and A is the brake beam crossectional area (mm^2). The elastic and plastic strains were determined using using Eqs. 5 and 6 [19] while Eq. 7 etimates the magnitude of the maximum deflection (δ) on the brake beam with the assumption that the beam is loaded uniformly throughout its length [12].

$$\varepsilon_e = \frac{\sigma}{E} \tag{5}$$

$$\varepsilon_p = \varepsilon - \varepsilon_e \tag{6}$$

$$\delta = \frac{WL^4}{8EI} \tag{7}$$

where ε_e and the ε_p are the elastic and plastic strains respectively, E is the modulus of elasticity (Pa) while $\varepsilon - \varepsilon_e$ denotes the change in length (mm), P is the applied load (N), L is the length of the beam (mm) and I is the mment of inertia (mm^4).

The investigation of the braking or traction torque applied to the wheel was carried out via simulation in the MATLAB 2018b environment. The braking or traction torque can cause damage the wheelset, the railhead and offset the rail track offset thereby affecting the railcar operations [20]. This is because the braking torque can affect the wheel–rail contact characteristics with the potential to offset the balance of the railcar. Furthermore, the that the speed profile of the railcar is usually affected by the wheel-rail interaction and braking characteristics [20]. Hence, the need to investigate the maximum torque required to drive the brake beam in order to balance speed with safety. The computation of the braking torque was obtained from Eq. 8 [21].

$$B_T = (F_{T1} + F_{T2})r_w \tag{8}$$

where B_T is the braking force in Nm, F_{T1} *and* F_{T2} are the tangential forces in hangers 1 and 2 respectively, while r_w is the nominal radis of the wheel (mm).

3 Results and Discussion

Figures 2 and 3 present the applied braking torque as well as the angular velocity of the motor respectively. From the plot obtained from the simulation results in the MATLAB 2020b environment, the applied torque was found to be slightly above 20 N-m. This agrees with the designed calculations carried out manually which computed the applied torque as 25 Nm. Also, the angular velocity of the motor from the simulation results was observed to be slightly above 50 rad/s. This also agrees significantly with the designed calculations done manually which computed the magnitude of the angular velocity of the motor as 52.366 rad/s. These results implies that the brake beam can achieve substantial speed and stability as the railcar responds to the braking torque.

Table 3 presents the reaction forces, which act on the brake beam. The magnitude of the total resultant force was 45.4711 N.

Figures 4, 5 and 6 present the distribution stress induced in the brake beam as well as the deformation due to stress and the corresponding strain. The stress induced ranges from a minimum value of 21.05 Pa to a maximum value of 8.553 MPa (Fig. 4). Considering the maximum load imposed on the brake beam (2,206,496.25 N) as well as the yield strength of the alloy steel selected for the brake beam 622 MPa, the break beam is unlikely to yield to the stresses induced or fail in service under the stated functional requirements. This is due to the fact that the value of the maximum stress induced were found to be lower than the yield strength of the material. The displacement due to the stress induced was also found to be small ranging from a minimum value of 1.0×10^{-30} mm to a maximum value of 4.388×10^{-2} mm (Fig. 5). This minimal displacement is unlikely to affect the performance of the robotic system under the stated functional requirements. The strain developed due to the stress induced ranges from a minimum value of 3.27257×10^{-10} mm to a

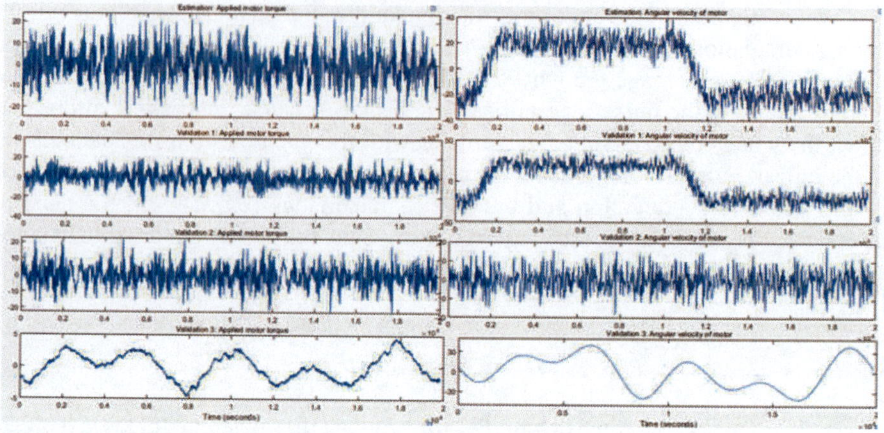

Fig. 2 The simulated magnitude of the applied motor torque

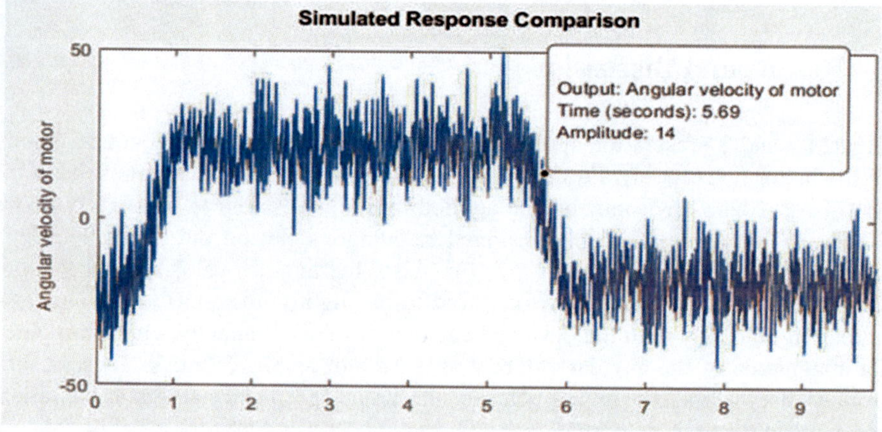

Fig. 3 The simulated magnitude of the angular velocity of motor

Table 3 The reaction forces

Reaction forces	Value (N)
Sum X	0.000218391
Sum Y	45.4711
Sum Z	−0.00046742
Resultant	45.4711

maximum value of 3.021×10^{-5} mm (Fig. 6). This implies that the material selected for the brake beam possesses adequate satisfactory strength and rigidity to withstand the stress induced without yielding to failure or undergoing permanent deformation. The higher the stress imposed, the greater the deformation and the strain developed and vice versa.

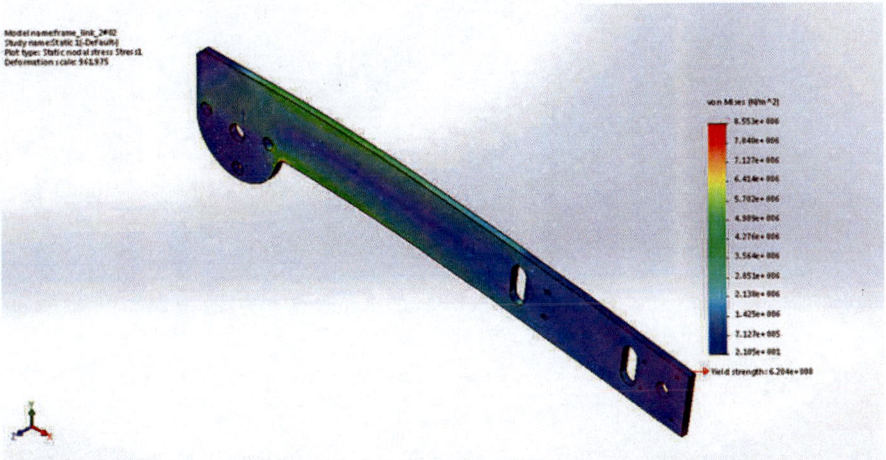

Fig. 4 The stress induced in the brake beam

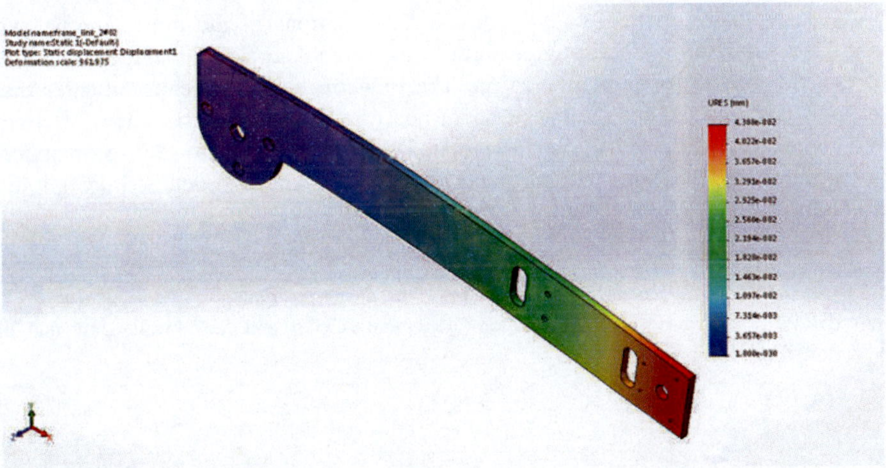

Fig. 5 The deformation due to stress induced in the brake beam

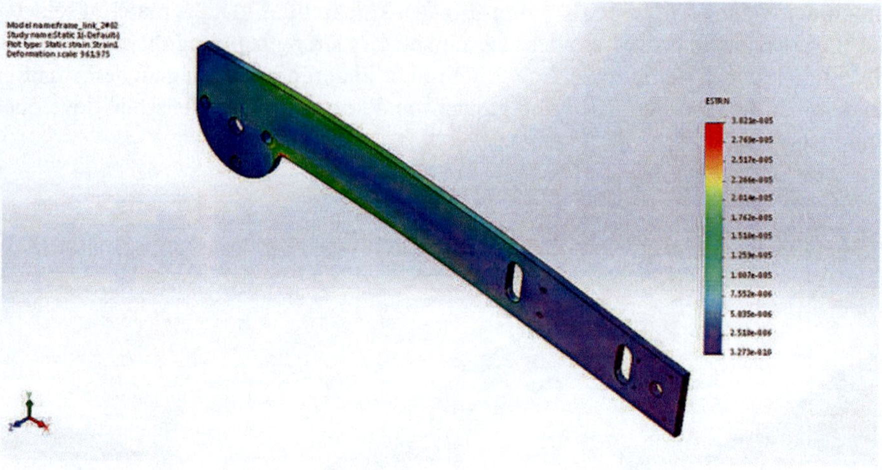

Fig. 6 The strain due to the stress induced in the brake beam

4 Conclusions

The aim of this work was to investigate the stress distribution in the novel brake beam of a railcar using the FEA approach.

This was achieved using the Solidworks and MATLAB software for the computer aided design and modeling as well as simulation of the rake beam. The results obtained from the manual calculations and simulations of the motor torque and angular velocity of the motor agree significantly, thus, indicating that the suitability of the design data for implementation. The FEA of the brake beam indicates that the material selected for the brake beam possesses adequate satisfactory strength and rigidity to withstand the stress induced without yielding to failure or undergoing permanent deformation. Future works can consider the detailed design and development of the brake beam.

Acknowledgements Funding. The authors disclosed receipt of the following financial support for the research: Technology Innovation Agency (TIA) South Africa, Gibela Rail Transport Consortium (GRTC), National Research Foundation (NRF grant 123575) and the Tshwane University of Technology (TUT)."

References

1. Reese, N., Myers, B., Halford, J.: Brake beam assembly for a railway car truck. US Patent no. US 8,925,466 B2, pp. 1–14 (2015)
2. Daniyan, I.A., Mpofu, K., Muvunzi, R., Fameso, F., Ramatsetse, B.I.: Model design and finite element analysis of the traction link of a railcar. Procedia CIRP **100**, 37–42 (2021)

3. Nejad, R.M.: Using three-dimensional finite element analysis for simulation of residual stresses in railway wheels. Eng. Failure Anal. **45**, 449–455 (2014)
4. Kukulski, J., Jacyna, M., Gołebiowski, P.: Finite element method in assessing strength properties of a railway surface and its elements. Symmetry **11**(1014), 1–29 (2019)
5. Daniyan, I.A., Mpofu, K., Fameso, F.O., Adeodu, A.O., Bello, K.A.: Development and simulation of isotropic hardening for AISI 1035 for weld stress prediction during design and welding assembly of the lower bracket of rail cars. Procedia CIRP **84**, 916–922 (2019)
6. Daniyan, I.A., Mpofu, K., Fameso, F., Ale, F.: Simulation of kinematic hardening model for carbon steel AISI 1035 weld stress prediction during the welding assembly of a railcar. Procedia CIRP **93**, 520–526 (2020)
7. Zulkifli, M.A., Basaruddin, K.S., Abdul Rahim, Y., Afendi, M., Gurubaran, P., Ibrahim, I.: Three dimensional finite element analysis on railway rail. IOP Conf. Series: Mater. Sci. Eng. **429**(012010), 1–6 (2018)
8. Song, M.K., Noh, H.C., Choi, C.K.: A new three-dimensional finite element analysis model of high-speed train–bridge interactions. Eng. Struct. **2003**(25), 1611–1626 (2003)
9. Sysyn, M., Przybylowicz, M., Nabochenko, O., Liu, J.: Mechanism of sleeper–ballast dynamic impact and residual settlements accumulation in zones with unsupported sleepers. Sustainability **13**(7740), 1–25 (2021)
10. Zeng, Z., Peng, G., Guo, W., Huang, X., Wang, W., Hu, J., Li, S., Shuaibu, A.A., Yuan, Y., Du, X.: Research on mechanical performance of improved low vibration track and its feasibility analysis for heavy-haul railway applications. Appl. Sci. **11**(10232), 1–27 (2021)
11. The Contructor. What you should know about cantilever beam [Online] Available at https://theconstructor.org/structural-engg/cantilever-beams/167474/. [Accessed 23 Sept 2021]
12. Orlova, A., Savushkin, R., Boronenko, I.Y., Kyakk, K., Rudakova, E., Gusev, A., Fedorova, V., Tanicheva, N.: Design of Unpowered Railway Vehicles from: Handbook of Railway Vehicle Dynamics, pp. 43–114. CRC Press, US (2019)
13. Bansal, R.K.: Strength of Materials, 4th edn. Laxmi Publication, New Delhi, India (2009)
14. Carvil, J.: Engineering Materials: in Mechanical Engineer's Data Handbook, pp. 218–266 (1993)
15. Daniyan, I. A., Mpofu, K., Adeodu, A.O., Daniyan, O.L.: Model design, simulation and control of rail car suspension system. Published Proceedings of the 2019 IEEE 10th International Conference on Mechanical and Intelligent Manufacturing Technologies (ICMIMT 2019), South Africa, pp. 37–42 (2019). IEEE:978-1-5386-7972-2/19
16. Daniyan, I.A., Mpofu, K., Daniyan, O.L., Adeodu, A.O.: Dynamic modelling and simulation of rail car suspension systems using classic controls. Cogent Eng. **6**(1602927), 1–20 (2019)
17. Daniyan, I.A., Mpofu, K., Adeodu, A.O., Bello, K.A.: The use of adaptive fuzzy-PID for vibration control in the suspension system of a railcar. 2020 IEEE 11th International Conference on Mechanical and Intelligent Manufacturing Technologies, South Africa. Added IEEE Xplore, pp. 130–134 (2020)
18. Dupen, B.: Applied strength of materials for engineering technology. Appl. Strength Mater. Eng. Technol. **6**, 152 (2014)
19. Khurmi, R.S., Khurmi, N.: A Textbook of Strength of Materials. S. Chand and Company, India (2019)
20. Dhanasekar, M., Cole, C., Handoko, Y.: Experimental evaluation of the effect of braking torque on bogie dynamics. Int. J. Heavy Vehicle Syst. **14**(3), 308–330 (2007)
21. Zhang, Z., Dhanasekar, M.: Dynamics of railway wagons subjected to braking/traction torque. Veh. Syst. Dyn. **47**(3), 285–307 (2009)

Ilesanmi Afolabi Daniyan is a Postdoctoral Research Fellow at the Department of Industrial Engineering, Tshwane University of Technology, Pretoria, South Africa. His research interest include production engineering and advanced manufacturing.

Khumbulani Mpofu is the Research Chair in Manufacturing and Skills Development, Department of Industrial Engineering, Tshwane University of Technology, Pretoria, South Africa. His research interest include industrial engineering and advanced manufacturing.

Felix Ale is an Assistant Professor with the Institute of Space Science and Engineering, Abuja, and Assistant Director with the National Space Research and Development Agency, Abuja, Nigeria. His research interest include Modelling and simulation, cybersecurity, artificial intelligence.

Rumbidzai Muvunzi is a Postdoctoral Research Fellow at the Department of Industrial Engineering, Tshwane University of Technology, Pretoria, South Africa. Her research interest include advanced and additive manufacturing.

Manufacturing Technology and Materials

A Kinematics Study on a Ni-D Electroplating Process for Enhancing the SuperAbrasive Grinding Wheel Quality

Sofian Eljozoli and Ramesh Kuppuswamy

Abstract The new age development on material processing creates an acute need to establish innovations on the electroplating process particularly towards enhancing the process efficiency, uniformity and improving the adhesion strength between the substrate and the plating. The Nickel-Diamond (Ni-D) plating process for processing a super-abrasive grinding wheel uses a plating bath contain; suspended diamond particles, grinding wheel(cathode), Nickel(anode) and electrolytic solution. The applied electrical current dissolves the nickel anode ions and deposit them on the cathode surface, simultaneously, the suspended diamond particles are dragged by the movement of nickel ions and deposited on the grinding wheel surface. The low quality of electroplated grinding wheels is often a result of non-uniform coating or poor adhesion between the abrasive particles and the substrate. Often, this results in scrap of the Ni-D electroplated component and thus enhances the wastage of resources such as: material, time, and labour. Also, in the Ni-D plating process the particles are suspended in the electrolyte using: bath agitation and electrophoresis process. Past research investigations have found that increased agitation generally enhances the number of particles in the metal deposit, but excessive agitation leads to a turbulent flow which destabilizes the Ni-D plating process and lower agitation reduces the suspension of Ni-D particles. This manuscript unveils the kinematics of the diamond abrasives and nickel ions that facilitate a better bonding to the grinding wheel substrate while applying a newly devised propeller. The kinematics study includes the forces acting on diamond particle and its trajectory while using multiple nickel anodes and a single cathode (grinding wheel substrate). The effect of process

S. Eljozoli · R. Kuppuswamy (✉)
Advanced Manufacturing Laboratory, Department of Mechanical Engineering, University of Cape Town, Cape Town, South Africa
e-mail: Ramesh.kuppuswamy@uct.ac.za

S. Eljozoli
e-mail: eljsof001@myuct.ac.za

R. Kuppuswamy
A Kinematics Study of Diamond Abrasives and Nickel-Ions on an Electroplating Process towards Enhancing the Quality of a Super-abrasive Grinding Wheel, 8th International Conference on Competitive Manufacturing, (COMA'22), 9-10 March 2022, Stellenbosch, South Africa

K. von Leipzig et al. (eds.), *Smart, Sustainable Manufacturing in an Ever-Changing World*, Lecture Notes in Production Engineering,
https://doi.org/10.1007/978-3-031-15602-1_60

parameters: agitation speed, agitation power, size and concentration of diamond particles in the electrolyte towards achieving a uniform adhesion along with an enhanced process efficiency were discussed.

Keywords Electroplating · Grinding wheel · Diamond · Suspension

1 Introduction

Grinding efficiency is one of the most important considerations in selection of grinding operation conditions because it has a significant impact on productivity, quality, energy consumption and overall cost [1]. Electroplated grinding wheels offer many distinct advantages particularly on profile grinding, high speed grinding and grinding of superalloys and ceramics [2]. A significant number of reports have shown that a uniform distribution of abrasive particles into a metal matrix can lead to improvement of mechanical, tribological, and electrochemical properties [3]. Past research have attempted on improving the abrasive deposition for the electroplated super abrasive grinding wheel through use of etching to create small grooves or cavities in the substrate before the deposition and metallization of the abrasive particles prior to electroplating. Also, past findings suggest several techniques to improve the uniformity of coating which are: changing the voltage between each anode and the cathode product; and changing the current density between the cathode–anode combination. However, these approaches enhance the cost as it requires use of multiple power supplies, and the associated complexity towards realizing the precise positioning of the anodes in the bath space. A theorical research suggests the use of optimal shape, multiple anodes along with the rotation of the grinding wheel around its axis during the electroplating process seem to be a promising direction for improving the uniformity of electroplating coatings on products [4]. Past attempts were also seen on electroless nickel -diamond plating in which the substrate is immersed in a chemical charged electrolyte solution and then the pre-heated Ni-D powder are dispensed over the substrate. This method avoids the usage of current distribution, throwing power, and anode–cathode orientation within the bath. After deposition the substate undergoes a post heat treatment process for a better adhesion and hardness enhancement [5]. It was also observed that use of azobenzene surfactant in Ni-D electroplating process exhibits a higher co-deposition especially for the smaller Diamond particles [6]. Attempts were also done on heating the electrolyte (watt bath) solution to around 50 °C so that both Nickel ions and the suspended diamond particles remain charged and promotes an enhanced adhesion [7]. A zeta potential measurement study on the mechanism of Ni-Diamond plating reveals that diamond particles attracted to the cathode electrophoretically after adsorption of positive nickel ions. Therefore, maintaining an optimum potential and stirring speed is critical for facilitating the desired migration of the charged and suspended diamond particles [8]. Research evidence also unveils the use of electrolytic degreasing for a period 5 min to achieve a better adhesion of Ni-D particles on the substrate [9]. Use

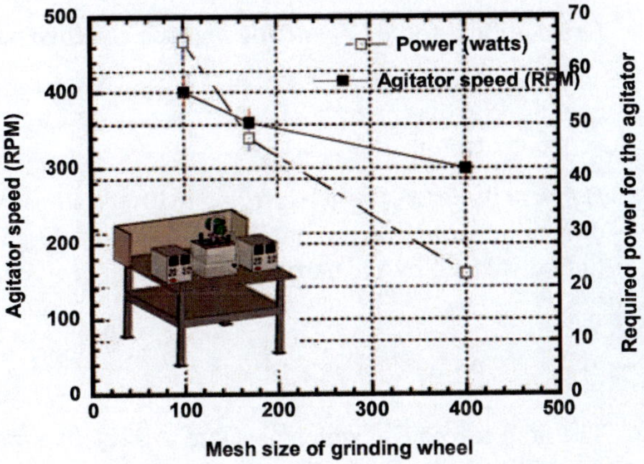

Fig. 2 Agitator speed and powder for the electroplating the Ni-D grinding wheel

$$N_{js} = \frac{S.v^{0.1} d_p^{0.2} (\frac{g(\rho_p - \rho)}{\rho})^{0.45} X^{0.13}}{D^{0.85}} \tag{1}$$

where ν is kinematic viscosity of electrolyte 0.00000167 m^2/s; ρ_p is Diamond density 3510 kg/m^3; ρ density of electrolyte 1200 kg/m^3; D is the impeller maximum diameter 0.1 m and d_p is the diamond particle size (0.00004 m for #400; 0.0001 m for #170; and 0.00017 m for #100. The electrolyte used is a mixture of several chemical (see Table 1) and hence the equiavelt density id caculated as;

$$\rho = \frac{100}{\frac{65.4}{1} + \frac{29.44}{1.913} + \frac{1.963}{3.55} + \frac{1.963}{1.435} + \frac{0.524}{1.206} + \frac{0.0065}{1.01} + \frac{0.00098}{1.015} + \frac{0.654}{3.51}}$$

$$\rho = \frac{1.2\,\text{gm}}{\text{cm}^3} = 1200\,\text{Kg/m}^3$$

N_{js} is computed after substituting the properties of electrolyte, diamond, impeller diameter and the results are shown in Fig. 2. The power (P) required to run the agiataor is computed using Eq. 2 as:

$$\text{P(in Watts)} = \frac{n}{\eta} N_P \rho N^3 D^5 \tag{2}$$

where N_p is the power constant, n is the factor of safety and η is the efficiency of the motor [13].

Both power and speed required for effective electroplating for the various mesh size of grinding wheel is given in Fig. 2. The findings suggests that fine mesh abrasives (#400) require less speed and power than the coarse mesh mesh (#100) mesh abrasives.

To understand the trajectory of diamond abarasive particle a force analysis acting on each abrasive particle was studied. The forces acting on the abrasive are [13, 14];

- Gravitational force (F_g) acting on each abrasive particle is given in Eqs. 3 as:

$$F_g = \frac{1}{6}\pi d_p^3 \rho_p g \tag{3}$$

- Buoyant force (F_b) which is trying to lift the abrasive particle towards the opposite direction of the gravitational force. Buoyant force is the product of mass of the fluid displaced by the particle. The buoyant force is given in Eq. 4 as:

$$F_b = \frac{m_p}{\rho_p}\rho \tag{4}$$

- The drag force (F_d) which acts due to a relative motion between the particle and fluid. The drag force is given in Eq. 5 as:

$$F_d = \frac{C_D u^2 \rho A_P}{2} \tag{5}$$

where C_D is the drag coefficient (assumed $C_D = 12$ using the graph [14]; U is the terminal velocity of the abrasive particle (m/s); A_p is the projected area of the abrasive particle (m^2).

The terminal velocity of the spherical particle was computed using Eq. 6 as:

$$u_t = \sqrt{\frac{4g(\rho_p - \rho)D_p}{3C_D\rho}} \tag{6}$$

Substituting the properties of electrolyte, abrasive particles the net force acting on the abrasive particle is shown in Fig. 3. The net force acting on the abrasive particle was overcome by the applied force imparted by the propeller at a given speed and torque. Also, the applied force imparts acceleration to the abrasive particle and hence follows a certain path so that particle stay within the grinding wheel zone. A four-blade propeller of diameter 100 mm was used for this study. The propeller was powered through a high torque motor of specification # 0IK1GN-AW3U / 0GN9K, with a gear ratio 9:1. The blade is 2 × 12 × 45 mm and perpendicular to the axis. Shown in Fig. 4 is the position of propeller and grinding wheel when both are immersed in the electrolyte. The radial torque and the radial force applied to the propeller motor are converted into axial force as shown in the sub-set of Fig. 4.

Similar to the axial force which is a summation of ($F_{v1} + F_{v2}$) forces the axial velocity was also computed from the radial velocity. The acceleration of the abrasive particle is computed using the applied force and mass of the abrasive particle. Using the computed acceleration, the path of the particle was plotted treating as a projectile submerged in the electrolyte. The path of the abrasive particle at a radius of 5, 25 and 50 mm was computed for mesh size #400 and #170 and the results are shown in Fig. 5.

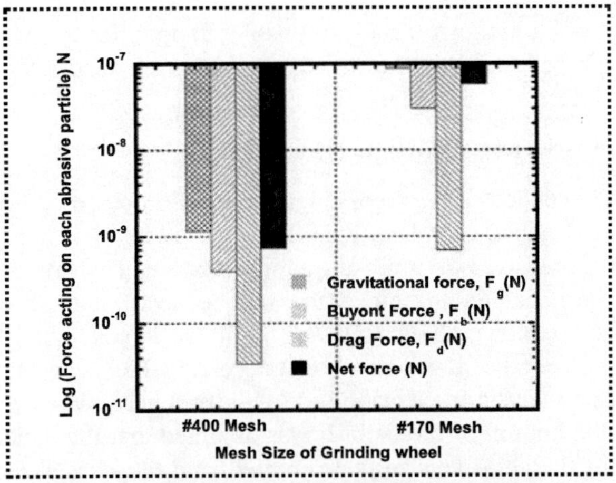

Fig. 3 Forces acting on the abrasive particle

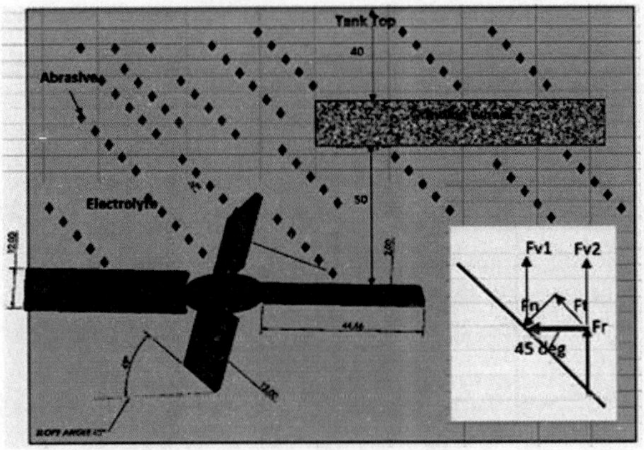

Fig. 4 Propeller and grinding wheel position in the electrolyte tank

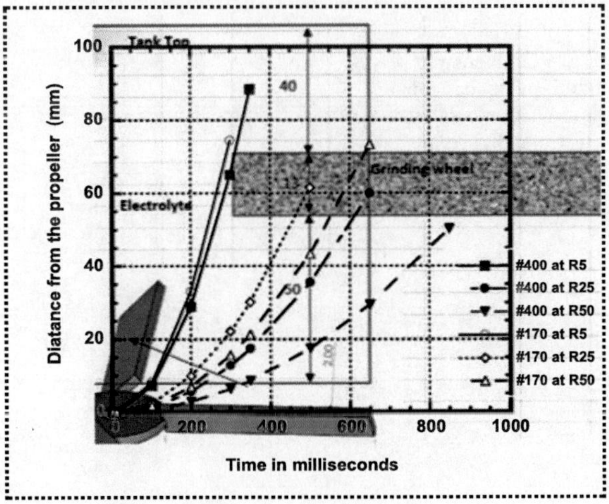

Fig. 5 Path of abrasive particle for mesh size #400 and #170 at a speed 100 RPM

The kinematic analysis enables to optimize the position of propeller and speed of propeller so that the abrasive particles are suspended particle are near the grinding wheel.

- Electroplating time calculation

The expected electroplating time was computed with administration of multi-anodes and cathode combinations. The computation was done for single layer electroplating of abrasives on a nickel-plating mixture. The first subsection is about an estimate for the amount of nickel that will be electroplated onto the substrate. The following subsection is about analysing the suitable cycle time. Then the subsequent subsection is about the minimum amount of coating thickness. Moreover, the fourth subsection is the analysis about magnitude of the optimal flow rate of slurry as it is redeployed from the bottom of the bath. It was assumed that the nickel plating is done so that 30% of the abrasive size is protruded from the nickel coating. Also, the nickel coating is expected to fill the diamond grooves provided at the substrate. As a result, the coating thickness is expected to be 0.12 mm and 0.15 mm for mesh size #400 and #170 respectively. The electroplating time was computed using the Faraday law as given in Eq. 7.

$$t = \frac{1}{3600}\frac{1}{\eta n_a}\frac{wnF}{A_{wt}I} \quad (7)$$

where w = mass of nickel plating; I = current (3 to 20 A); t = time (hrs); η = current efficiency (0.95); n = valency of nickel ion (2); F = Faraday's constant (96,500); A_{wt} = molar mass of nickel (58.69 gm); and n_a = number of anode–cathode combination [14].

Shown in Fig. 6 is the influence of multi anodes on the electroplating time for mesh size #400 and #170 grinding wheel.

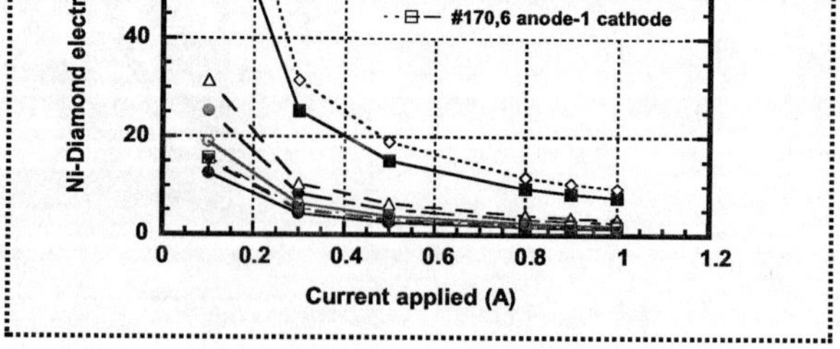

Fig. 6 Influence of multi-anode–cathode combination on Ni-Diamond electroplating time

4 Simulation

The Ni-Diamond deposition with administration of multi-anode was simulated using COMSOL software. Table 2 list down the input parameters used in the COMSOL simulation software.

The simulation study unveils the expected Ni-D electroplated results for the grinding wheel when run at slow RPM of value 10~20 RPM. The finite element solver software COMSOL have also simulated and predicted the plating thickness under varied process conditions such as changing number of anodes, temperature, and current density. The result shows that the variation of thickness is relatively small between the thinnest and thickest part when using six anodes around the rotating grinding. The simulation was carried by selecting a secondary current distribution and time dependent Ni-D plating parameters as the inputs to the Software. The COMSOL software works on the larger principle of conservation of charges and current in the electrolyte and along with the mass conservation between anode and cathode. It uses the Gauss's law and Nernst–Planck equations to determine the Nickel ions charge density and transport of nickel ions towards the cathode (grinding wheel). Also, the Nernet_plank equation includes the transport of diamond particles that are suspended in the electrolyte. The electrolyte used is a mixture of chemicals as mentioned in Table 1, the equivalent conductivity (kappa value) was included in this analysis. Based on the inputs of anode, cathode and electrolyte concentration, the current density was computed using the Butler-Volmer equation and the computed values were used in the analysis. To simulate the grinding wheel (cathode) rotation,

Table 2 Inputs used in the COMSOL simulation

Detail	Value
Density of diamond	3510 kg/m^3
Density of nickel	8910 kg/m^3
Density of electrolyte	1200 kg/m^3
Conductivity of the electrolyte	10 S/m
Average cathode current density	Range 2 ~ 26 A/dm^2; 5 A/dm^2 used
Molar mass of nickel	58.69 gm
Valence of nickel	2
No of anodes to cathode	1 –1 ~ 6–1
Equilibrium potential, nickel reaction	–0.26 V
Exchange current density, nickel reaction	0.1 A/m^2
Exchange current density, hydrogen reaction	2E-5 A/m^2
Grinding wheel size	D180 × 13 mm

the rotating and deforming domain were defined, and a suitable mesh was created to get the simulated results. The COMSOL allows mesh size and element-to-element expansion rate at the cathode and the mesh was devised to conform the geometric features such as faces and edges. As the cathode was allowed to rotate at 10 RPM, a complete revolution of cathode will take 6 s. Shown in Fig. 7 is the effect of multi-anode on the Ni-Diamond electroplating process while applying an average current density of value 5 A/dm^2 for 6 s. The growth rate of Nickel-Diamond deposition was studied at every 0.25 secs and the main observations are given below:

- After 0.25 s of rotation of cathode, Ni-Diamond deposition was found to be 0.035, 0.018, 0.009 and 0.006 μm for 1, 2, 4 and 6 anodes respectively.
- During the same period (0.25 s) the spread of Ni-Diamond deposition was found to be more uniform for multi-anodes than the single anode.
- After 3 s of rotation of cathode, Ni-Diamond deposition was found to be 0.05, 0.035, 0.032 and 0.032 μm for 1, 2, 4 and 6 anodes respectively.
- During the same period (3 s) the spread of Ni-Diamond deposition was found to be more uniform for multi-anodes than the single anode.
- After 6 s of rotation of cathode, Ni-Diamond deposition was found to be 0.07, 0.06, 0.056 and 0.056 μm for 1, 2, 4 and 6 anodes respectively (See Fig. 7).
- During the same period (6 s) the spread of Ni-Diamond deposition was found to be more uniform for multi-anodes than the single anode. After 6 s of rotation of cathode, the variation Ni-Diamond deposition was found to be within 0.07 ~ 0.030, 0.06 ~ 0.04, 0.056 ~ 0.044, and 0.056 ~ 0.044 μm for 1, 2, 4 and 6 anodes respectively (see Fig. 7).

The electroplating thickness was computed using the application of Faraday's law alongwith the remodification of the Eq. 7 and given as;

$$\delta = \frac{t\eta n_a A_{wt} I}{\pi D w \rho_{ni} n F} \quad (8)$$

where D = diamater of grinding wheel (180 mm), w = width of grinding wheel (13 mm) and ρ_{ni} = density of nickel (8910 kg/m^3).

Subsituting the applied ni-D electroplating conditions, the computed ni-D thickness shows an increasing trend from 0.2 to 8.8 μm for a electroplating time of 6 s. However the combined effects: (i) rotation of the grinding wheel and (ii) increase of the number of anodes, influences the uniformity of the Ni-D depostion as shown in Fig. 7.

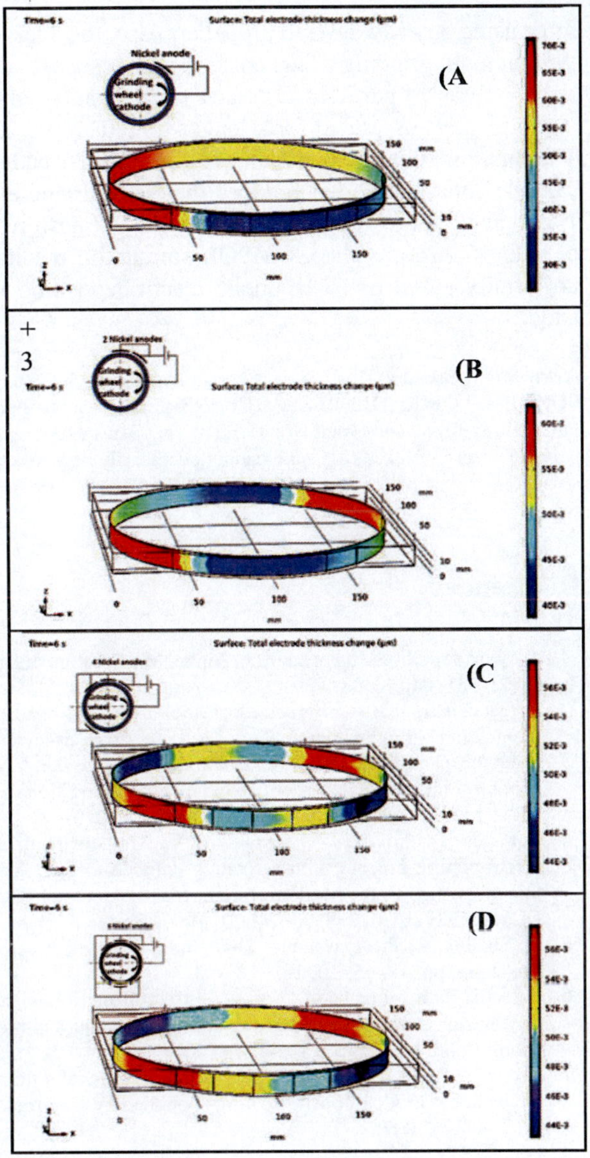

Fig. 7 Effect of multi-anode on the Ni-Diamond electroplating process while applying an average current density of value 5 A/dm^2 for 6 s, with rotation of the grinding wheel. **A** 1 anode-1 cathode, **B** 2 anode-1 cathode, **C** 4 anode-1 cathode, and **D** 6 anode-1 cathode conditions

5 Conclusions

The parameters that affect uniformity and enhanced nickel-diamond deposition on the electroplated grinding wheels were confirmed through analysis and simulation methods. The analysis study demonstrates the kinematic path of diamond abrasive, particularly how the forces acting on the abrasive were overcome with the applied force using a newly devised propeller. Also, the analysis suggests the importance of the (cathode) grinding wheel position with respect to the propeller for different mesh size of abrasive particles. The analytical results confirms that fine mesh abrasive particles stay suspended for a longer time than the coarse mesh abrasive particles. The findings suggests that increase of 1 anode-1 cathode combination to 6 anode-1 cathode combination has reduced the Ni-Diamond electroplating time from 7.5 to 1.55 h and 9.44 to 1.57 h for #400 and #170 mesh respectively at a current density of value 5 A/dm^2. The COMSOL simulation results also confirms the enhanced uniformity effect of Ni-Diamond electrodeposition with administration of multi-anodes.

Acknowledgements This project was supported by fund NRF GRANT: INCENTIVE FUNDING FOR RATED RESEARCHERS (IPRR) –South Africa through Reference: Reference: RA191118492767 and Grant No: 136118. The views expressed and the conclusions drawn in this paper are those of the authors and cannot necessarily be attributed to the references.

References

1. Karpuschewski, B., Knoche, H.J., Hipke, M.: Gear finishing by abrasive processes. Ann. CIRP **57**(2), 621–640 (2008)
2. de Medeiros, R.B.D., Figueiredo Filho, J.M., Mashhadikarimi, M., Gomes, U.: Diamond integrated coating by electroplating process-An overview. Atena publishers, Finland (2020)
3. Solovjev, D.S., Solovjeva, I.A., Konkina, V.V., Litovka, Y.V.: Improving the uniformity of the coating thickness distribution during electroplating treatment of products using multi anode baths. Mater. Today: Proc. **19**(5), 1895–1898 (2019)
4. Solovjev, D., Solovjeva, I., Konkina, V.: Mathematical modelling and optimization of the electroplating process with a rotating cathode to reduce the non-uniformity of the coating thickness. Int. Conf. Modern Trends Manuf. Technol. Equipment: Mech. Eng. Mater. Sci. (ICMTMTE 2019), **298**, 00014 (2019)
5. G, Shella., M, Pushpavanam.: Diamond-Dispensed Electroless and Nickel Coatings, Metal Finishing, pp. 42–46 (2002)
6. Shrestha, N.K., Takebe, T., Saji, T.: Effect of particle size on the co-deposition of diamond with nickel in presence of a redox-active surfactant and mechanical property of the coatings. Diam. Relat. Mater. **15**, 1570–1575 (2006)
7. Lee, H.K., Lee, H.Y., Jeon, J.M.: Co-deposition of micro-and nano sized SiC particles in the nickel matrix composite coatings obtained by electroplating. Surf. Coat. Technol. **201**(8), 4711–4717 (2007)
8. Lee, E., Choi, J.W.: A study on the mechanism of formation of electrodeposited Ni–diamond coatings. Surf. Coat. Technol. **148**, 234–240 (2001)
9. Hiromichi, O., Ryuuichi, I., Okuno, K., Osamu, O.: Fabrication of electroplated micro grinding wheels and manufacturing of microstructures with ultrasonic vibration. Key Eng. Mater. **238**(239), 9–14 (2003)
10. Denis, S.S., Solovjeva, I.A., Victoria, V.K., Yuri, V.L.: Improving the uniformity of the coating thickness distribution during electroplating treatment of products using multi anode baths. Mater. Today: Proc. **19**(5), 1895–1898 (2019)

11. Lelevic, A., Walsh, F.: Electrodeposition of Ni P composite coatings: a review. Surf. Coat. Technol. **378**(10), 1016 (2019)
12. Qin, J., Zhang, X., Xue, Y., Kumar Das, M., Thueploy, A., Limpanart, S., Liu, R.: The high concentration and uniform distribution of diamond particles in Ni-diamond composite coatings by sediment co-deposition. Surf. Interface Anal. **47**(3), 331–339 (2015)
13. Ayranci, I., Machado, M.B., Madej, A.M., Derksen, J.J., Nobes, D.S., Kresta, S.M.: Effect of geometry on the mechanisms for off-bottom solids suspension in a stirred tank. Chem. Eng. Sci. **79**, 163–176 (2012)
14. Schlesinger, M., Paunovic, M.: Modern Electroplating 5th Edition, Wiley (2014)

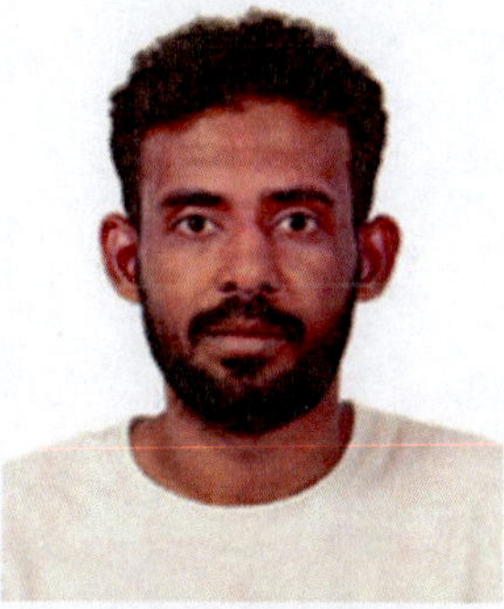

Sofian Eljozoli obtained his bachelor's degree in Mechanical Engineering from the University of Khartoum. He is currently studying towards achieving his M.Sc. Eng. in Mechanical Engineering, focusing on mechanical design and manufacturing.

Ramesh Kuppuswamy acquired his Ph.D. from the Nanyang Technological University, Singapore and is currently an Associate Professor in the mechanical engineering department at the University of Cape Town.

The Effect of Minimum Quantity Lubrication on Selected Surface Integrity Attributes When Machining Grade 4 Titanium Alloy

Alpheus N. Maponya and Rudolph F. Laubscher

Abstract This paper presents an experimental investigation on the influence of Minimum Quantity Lubrication (MQL) process parameters on selected aspects of surface integrity when turning grade 4 titanium alloy. MQL process parameters that were varied include the oil flow rate (50–90 ml/h), nozzle distance (20–40 mm), and airflow rate (25–37 l/h). Surface roughness, residual stress state, and cutting forces were evaluated for dry, flood, and MQL turning conditions using the smaller is better grey relation criterion. Grey relation analysis of the results demonstrated that when utilizing MQL, the best surface finish, lowest cutting forces, and largest compressive residual stresses can be obtained for selected MQL parameter sets but not all at once. The results demonstrate that MQL is a viable alternative to either dry or flood cooling during outside turning of grade 4 titanium alloy, and may result in an improved surface integrity state.

Keywords Grade 4 titanium · Minimum quantity lubrication (MQL) · Residual stresses · Average surface roughness (Ra) · Grey relation analysis (GRA) technique

1 Introduction

Titanium has a relatively low density when compared to steel and other high-performance alloys [1]. In addition, it is a preferred high-performance alloy in many industries due to its good mechanical properties, its excellent corrosion resistance at elevated temperatures, elevated strength to weight ratio, excellent erosion resistance, biomedical capabilities, and high fatigue strength [2]. Typical industry use of titanium may require extensive machining that will affect the surface integrity [3]. However, apart from all the positive characteristics of this material, titanium alloys

A. N. Maponya (✉) · R. F. Laubscher
Faculty of Engineering and the Built Environment, Mechanical Engineering Department, University of Johannesburg, Auckland Park, Johannesburg, South Africa
e-mail: 201214296@student.uj.ac.za

R. F. Laubscher
e-mail: Rflaubscher@uj.ac.za

K. von Leipzig et al. (eds.), *Smart, Sustainable Manufacturing in an Ever-Changing World*, Lecture Notes in Production Engineering,
https://doi.org/10.1007/978-3-031-15602-1_61

generally have low machinability when compared to other metals. This characteristic is due to their low thermal conductivity, high chemical reactivity, and low elastic modulus [4, 5]. Titanium tends to react with various tool materials that may lead to increased tool wear [6, 7]. Tool wear negatively affects surface finish hence; various lubricants are utilized when machining titanium to keep tool wear under control.

Conventional flood cooling with oil-based lubricants consumes significant quantities of lubricants when cutting titanium alloys and their disposal adds to pollution [8, 9]. Increasing environmental concerns and regulations on contamination and pollution have forced the manufacturing industry to look for alternative cutting solutions that are environmentally friendly, sustainable, and have a low economic impact [9]. This solution must offer equivalent if not better results than the conventional flood cooling method regarding various attributes of surface integrity. Current literature showed that MQL assisted machining might be a viable alternative for cutting various alloys including titanium. MQL is a low-cost, near dry, and environmentally friendly alternative to flood cooling. MQL system consists of a lubricant reservoir, a pressurized air tank, a control system, flow tubes with a spray outlet. It works by passing high-pressure air and cutting fluid mixture through a fluid-flow control module where atomization takes place before being injected directly into the cutting zone [1, 10, 11]. The most significant MQL process parameters typically include the nozzle from the cutting interface, the lubricant, and the airflow rate (Directly related to air pressure). MQL typically uses a minimal amount of cutting fluid (50–500 ml/h) three to four orders of magnitude lower than typical flood cooling or less when compared to flood cooling that uses 1–10 L per minute of cutting fluid [12, 13]. Surface integrity attributes typically most affected by the cutting strategy include surface roughness, surface hardness, and residual stress state [14, 15]. The residual stress and surface finish may significantly affect the fatigue performance of the finished part. Residual stresses are induced either mechanically through machining processes or thermally through heat treatment processes. The surface residual stress state plays a critical role in corrosion resistance and fatigue life of the part depending on their nature (tensile or compressive) [16, 17]. Previous work done on MQL demonstrated that the optimal MQL parameters for machining grade 4 titanium are MQL flow rate of 70 ml/h, nozzle distance of 30 mm, and air pressure of 4 bar [18]. Their work shares a similar scope of work, workpiece material, and methodology as the current investigation.

Hence, this study investigates the influence of varying minimum quantity lubrication (MQL) process parameters on selected surface integrity descriptors when turning Grade 4 titanium alloy. The analysis includes surface roughness, residual stress, and cutting forces using the grey relation analysis. This method is used for analyzing relationships between data sets and multiple attribute solution decision-making purposes. It works by normalizing the measured values within a range of 0–1 and the high values favor the output response depending on the selected GRA criterion [19]. The MQL results are compared to dry cutting and flood cooling conditions.

2 Experimental Program

2.1 *Specimen Preparation*

The workpiece is a 75 mm (dia) by 500 mm long commercially available Grade 4 titanium alloy bar. Table 3 shows the material properties for the grade 4 titanium bar (Fig. 1c) as per the material certificate. The titanium bar was cut into two 250 mm equal parts to fit in the CNC machine. For stability and stiffness to be maintained during cutting, a center drill was used to make a hole on one side of the bar where the center spindle fits.

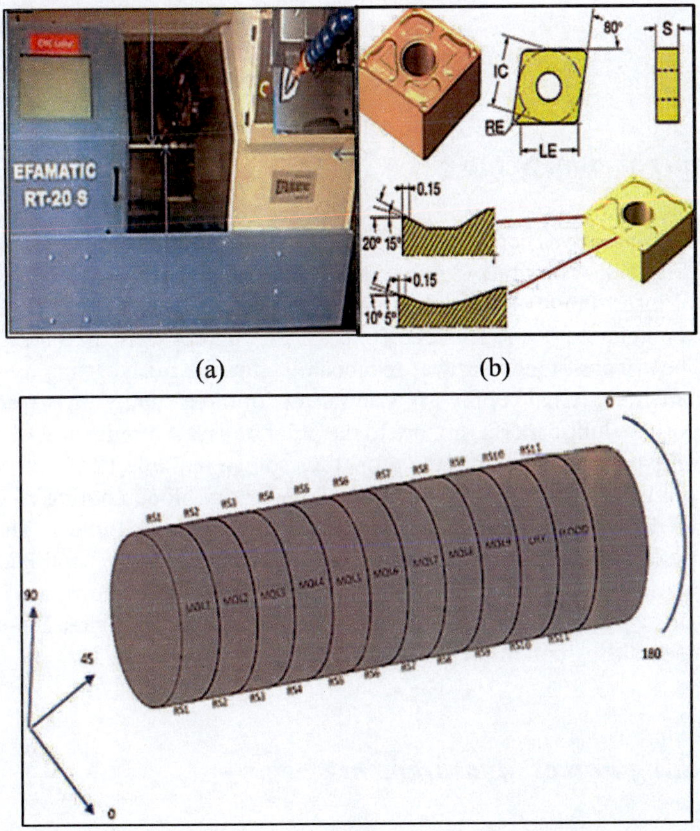

(a) (b)

(c)

Fig. 1 **a** Experimental setup with MQL and flood cooling nozzle configuration; **b** cutting insert with chip breaker technology; **c** machined titanium bar

Table 1 Producut 9 EP properties

Pour point	8° C
Flashpoint	290° C
Kinematic viscosity	39.11 mm^2/s at 40° C
Density	0.9199 g/cm^3at 20 °C

Table 2 MQL parameters values used for the experiments

Parameters	Symbols	Units	Values
MQL-Flow rates	MFR	ml/h	50, 70, 90
Nozzle Distance	ND	mm	20, 30, 40
Airflow rate	A_{FR}	l/h	25, 31, 37
Fixed parameters	Cutting speed 125 m/min; feed rate: 0.2 mm/rev; depth of cut: 1 mm (diameter), cut distance: 20 mm, injection angle 45°		

2.2 *Experimental Setup*

The turning process was conducted on an Efamatic RT-20 slant bed CNC production lathe (see Fig. 1a). A Sandvik Coromant CNMG 12 04 08-XF GC15 coated carbide cutting tool insert (an 80° kite with a flank clearance angle of 0° (see Fig. 1b) with chip breaker technology and DCLNL 2020 K insert holder were used for the cutting process. The effect of tool wear was reduced by allowing fresh cutting tool for each test. A Producut external supply MQL system with three supply lines (see Fig. 1a) that transport the lubricant oil mixture to the point of use. A Producut 9 EP lubricant oil used for experiments and has the properties shown in Table 1.

The titanium bar (see Fig. 1c) was turned under dry, flood cooling, and varying MQL (MQL1 to MQL9 see Fig. 1c) conditions are shown in Table 2. The varying MQL parameters were used to determine the optimal parameter combinations that yield favorable outputs including the center points. Also, the airflow rates in Table 2 corresponds to varying air pressures of 3, 4 (Optimized), and 5 bars. 3 points were measured per cutting condition (Table 3).

2.3 *Cutting Forces Measurement*

A 9265B Kistler dynamometer and a Kistler type 5011B charge amplifier were used to measure the output signals during machining. The signals from the dynamometer were converted into corresponding cutting forces using a build-in calibrated constant. The average cutting forces were measured for a 20 mm length per cutting condition (see Fig. 1c).

Table 3 Material properties for the titanium bar

Chemical Composition (%)	C	Fe	N	O	H	Others		Ti
Requirements	Max 0.08	Max 0.50	Max 0.05	Max 0.40	Max 0.015	Each	Total	Balance
Actual	0.02	0.24	0.02	0.25	0.001	≤0.1	≤0.4	
Mechanical test	Tensile (MPa)		Yield 0.20% (MPa)			Elongation In 4D (%)	Reduction of area (%)	Ultrasonic test
Requirements	Min 550		Min 483			Min 15	Min 25	100%
Actual	610		532			24	39	Acceptable

2.4 *Surface Roughness Measurements*

A Hommel T500 measurement instrument was used to measure surface roughness (Ra, Rz, Rmax). These measurements were taken for every 20 mm section that was machined on the bar. The measurements were taken at the start of the cut (0 mm), in the middle (10 mm), at the end of the cut (20 mm). The average of the three values was used for the analysis.

2.5 *Residual Stress Measurements*

The XRD measurements were conducted at the Nelson Mandela University (NMU). The residual stresses were measured with a Proto iXRD Portable Residual Stress & Retained Austenite Measurement System. The X-ray diffraction system measures the residual stresses by exposing the titanium workpiece to a specific X-ray beam. Refracted beam angles are measured and the $\sin2\psi$ technique is used to estimate the residual stress. The system MGR40P Goniometer was used to measure two values of the near-surface residual stresses. The values were measured in the middle of each 20 mm cut length per cut condition at points 0° and 180° (see Fig. 1c). The average of the two values was used for the analysis.

3 Results and Discussion

This section presents the grey relation analysis (GR) and comparison study for the cutting forces, surface roughness, and near-surface residual stresses. The mean grey relation responses suggest that the optimal overall parameters are level 1 (MQL flow rate: 50 ml/h), level 2 (Nozzle distance: 30 mm), and level 2 (Airflow rate: 31 l/h). The MQL flow rate deviated from the centre point optimal parameters. These values correspond to the highest grey value where the actual data values are small.

The nozzle distance also showed a high impact at intermediate values compared to the lubricant flow rate, suggesting an optimal nozzle distance. Compared to each other, the lubrication stream seems to be the least significant parameter compared to the airflow rate and nozzle distance, which exhibited a very similar influence. Since the optimal MQL parameters are not achievable at the same time, each of these three optimal results are compared to dry cutting and flood cooling results. This section gives a more detailed breakdown of how each output reacts to the change in cutting conditions.

3.1 Effect of MQL on Cutting Forces

3.1.1 Grey Relation Analysis of the Resultant Cutting Force

The Grey relation responses for the cutting force as a function of MQL parameters are presented in Fig. 2. The smaller is better grey relation criterion was used since we require minimum cutting forces. The results indicate that the lowest cutting forces (corresponding to the highest GRG values) were obtained at the highest lubricant and airflow rates. This is due to the optimum lubrication in the cutting zone and improved cooling process. This is due to the larger flow rate that effectively reduces friction and therefore cutting force. The nozzle distance demonstrated the lowest cutting force at an intermediate distance where sufficient heat is generated, softening the workpiece material, and plastic deformation occurs much more easily under these conditions. This result suggests that the machining or shear zone becomes more adiabatic, and the generated heat cannot be dispersed rapidly as the material flows. This heat increases the material's ductility, which aids grain boundary dislocation motion and requires low cutting forces during machining.

The lubrication is less effective if the nozzle is too close or too far from the cutting zone where the cutting forces were observed to be appreciably larger. For both situations, optimal lubricant quantity does not reach the cutting zone leading to an increase resistance due to friction, heat, and tool wear. Thus leading to high cutting forces similar to low airflow rates were fewer oil droplets reach the cutting zone.

Consequently, this causes a high heat generation between the cutting tool and workpiece, which causes evaporation of the oil mist in the cutting zone or lose of its

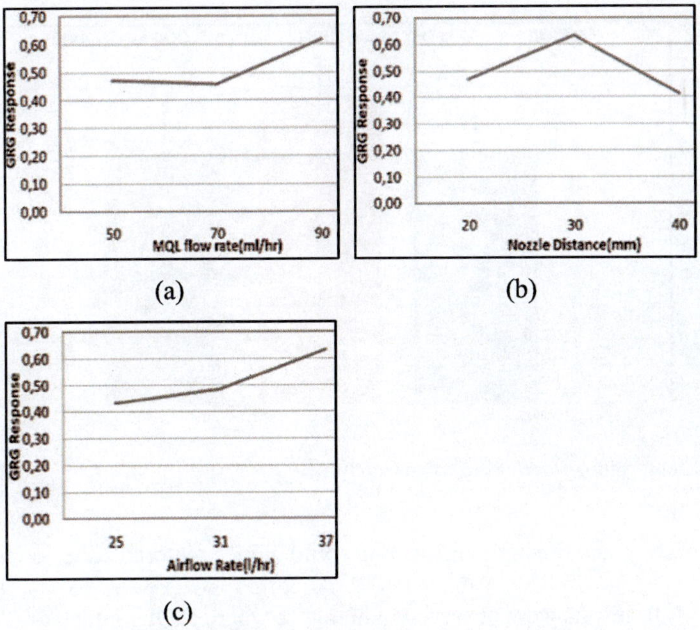

Fig. 2 Variation of GR response of the cutting force as a function of **a** oil flow rate, **b** nozzle distance, **c** airflow rate

thermal conductivity. Thus, also leading to an increase in cutting forces. These results are similar to those highlighted in previous work [20], where a small amount of oil mist vaporizes before effectively removing the heat. A high airflow rate atomizes the oil droplets more effectively, improving their penetration and the conditions that lead to a decrease in the cutting forces due to low friction. The results show that an increase in the airflow rate decreases the cutting forces. Therefore, the optimal MQL parameters that yield the lowest overall resultant cutting forces, are a maximum lubricant flow rate of 90 ml/h, an intermediate nozzle distance of 30 mm, and a maximum airflow rate of 37 l/h (see Fig. 2).

3.1.2 Comparison Study

The lowest overall resultant cutting force was demonstrated for MQL assisted machining when compared to both flood and dry cutting (see Fig. 3). Machining using the obtained optimal MQL parameters displayed a 19.7% lower resultant cutting force when compared to dry cutting and a smaller 6.5% for flood cooling. Flood cooling showed 14.2% reduction when compared to dry cutting. This is due to the effective lubrication accomplished by the micro-droplets of the aerosol. The high pressure and high velocity droplets from the MQL strategy sufficiently penetrate the cutting zone

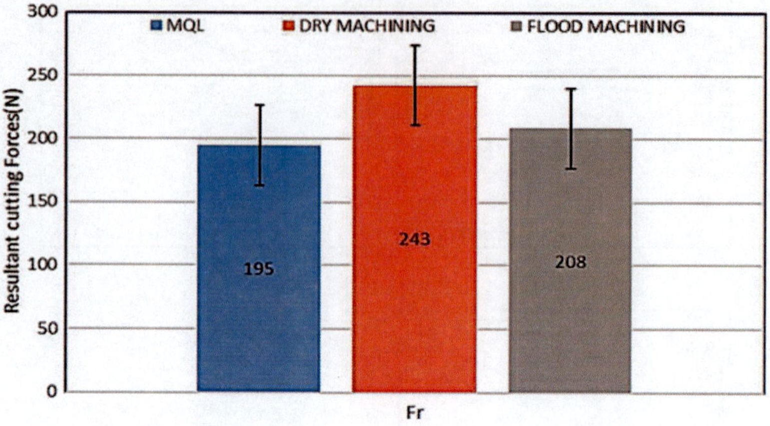

Fig. 3 Resultant cutting force's comparison study

reducing friction and the high airflow rate could reduce material adhesion to the tool face.

Better MQL results were observed in literature where MQL yielded a 40% cutting force reduction, while flood cooling showed a 26% higher cutting force reduction than MQL and 19% less cutting forces than dry conditions. Furthermore, MQL also showed a 30% surface roughness improvement and a 36% reduction in cutting temperature compared with flood cooling and dry machining conditions [21].

The cutting forces obtained for dry cutting are noticeably lower than those obtained in literature [22], when machining C45 steel grade. A similar result was observed by [23] where the lowest cutting forces were obtained under MQL conditions. Also, high MQL parameters were good for cutting force similar to those observed for surface roughness.

3.2 *Effect of MQL on Surface Roughness*

3.2.1 Grey Relation Analysis of the Surface Roughness

The "smaller is better" grey relation criterion was also applicable for the average surface roughness (Ra) since the smoothest surface is generally beneficial and required, which corresponds to the largest grey relation value. The results indicate that increasing the lubrication flow rate improved the average surface roughness A similar result was observed in the available literature [24], where a high lubricant flow rate yielded the lowest surface roughness.

At a low MQL flow rate, the machining zone becomes near dry and friction increases due to insufficient lubrication. The increase in friction leads to an increase in cutting forces to overcome the increased frictional resistance and tool wear. Consequently, this leads to an increase in chatter due to a combination of high cutting forces, insufficient lubrication, and a low elastic modulus of the titanium alloy. Thus, leading to a coarser surface finish. Increasing nozzle distance reduced the efficiency of the lubricant in dispersing the heat generated and increased workpiece deformation. Similar to high airflow rate and low flow rates that cause significant lubricant dispersion or low reach, limiting the concentration of oil droplets that reach the cutting zone at higher nozzle distances. This leads to titanium adhering to the tool face leading to tool failure and yielding poor surface finish due to the near dry conditions [25]. This also significantly influence the resulting surface groove size and again, leads to higher surface roughness.

A low airflow rate gives poor results due to inadequate lubricant reaching the machining zone as a result of the low flow of the transport medium. This result suggests that 31 ml/h is an optimum airflow, which leads to gradual heat dispersion in the cutting zone due to improved cooling. This means that the material is relatively easier to machine, and low cutting forces are required, which also results in low chatter, tool wear; thus, leading to low surface roughness.

The coated cutting tools play a role in providing lower surface roughness but this depends on additional factors such as substrate material, tool coating combinations, and the thermos-physical conditions of both tool and workpiece [26]. Hence, the best overall surface finish results were achieved at the highest MQL lubricant flow rate of 90 ml/h, at a minimum nozzle distance of 20 mm, and an intermediate airflow rate of 31 l/h (see Fig. 4).

3.2.2 Comparison Study

Using these obtained optimal MQL parameters for surface roughness. The evaluated surface roughness attributes (Ra, Rz, and Rmax) all indicated a good surface finish for all MQL cutting experiments. MQL assisted machining displayed the best surface finish when compared to flood cooling and dry cutting. All thee surface roughness attributes (Ra, Rz, and Rmax) show the same trend for all thee cutting conditions (MQL, Dry cutting, and Flood cooling). Ra, Rz, and Rmax results were found to be below 1 μm, 4 μm, and 4 μm respectively for each MQL cutting condition, indicating a good surface finish that helps minimize crack initiation and inhibit failure due to fatigue [27].

When considering the average (Ra) surface roughness component, MQL improved the surface roughness by 42.5% (−0.4 μm) while flood cooling exhibited a 24% (−0.21 μm) improvement when compared to dry cutting results (see Fig. 5). A 30% reduction in surface roughness was observed under MQL conditions when compared to dry and flood cooling conditions [28]. Also, the inferior surface finish in dry machining is a consequence of machining a difficult to machine alloy such as titanium due to the material adhesion to the tool face leading to tool failure and resulting in a poor surface finish. A low 50 ml/h MQL flow rate yielded the best result when compared to dry cutting in literature [29]. It was also found that the

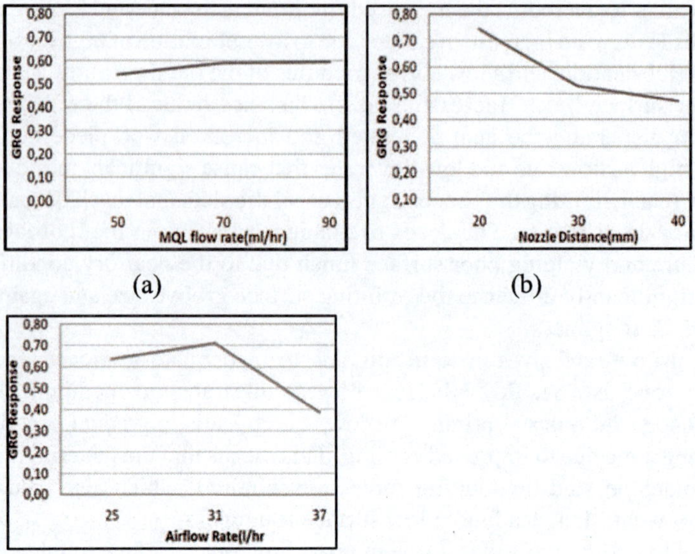

Fig. 4 Variation of GR response of the surface roughness as a function of **a** oil flow rate, **b** nozzle distance, **c** airflow rate

Nano fluid-based MQL yielded better surface finishes than dry cutting and produces surface finishes similar to those observed under wet cutting [30].

MQL offered enhanced performance and exhibited a better surface finish when compared to dry and flood cooling conditions [31].

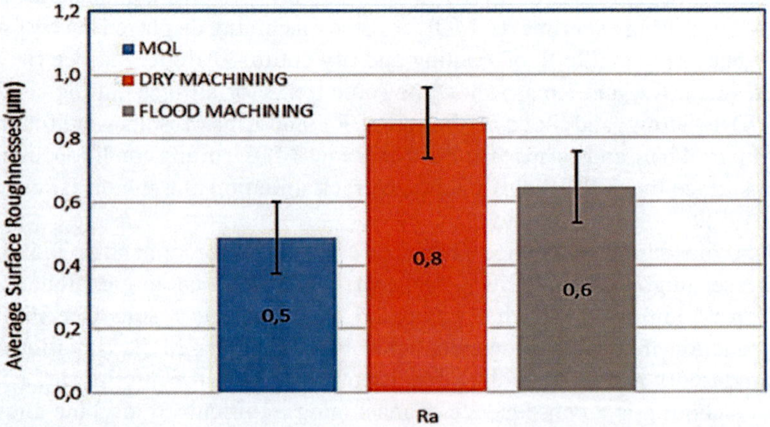

Fig. 5 Surface roughness comparison study for MQL6, Dry cutting, and flood cooling

3.3 *Effect of MQL on Residual Stress*

3.3.1 Grey Relation Analysis of the Primary Residual Stress

Residual stresses are defined as stresses that remain in the material once thermal or mechanical loads are removed. The results demonstrate that there are intermediate values for airflow and nozzle distance that produced optimum results. The intermediate nozzle distance seems to be a localized optimum where sufficient heat is produced and effectively cooled to produce an optimum (highest) compressive residual stress state. An increase in the MQL oil flow rate had an adverse effect on the residual stresses. The highest residual result is obtained for the lowest lubrication flow rate, where plastic deformation, cutting forces, and heat generation due to friction are most significant. The uneven plastic deformation due to mechanical loads, changes in material volume, and thermal effect as a result of solid-phase transformation. This is due to transmission of high temperature from the tool to the workpiece during machining and this change in phase leads to grain size raising gradually results in compressive residual stresses followed by cooling of the material [32]. The opposite is observed for high lubricant and low airflow rates. Similarly, by increasing the airflow rate beyond the optimum, lubrication delivery becomes more efficient resulting in less heat being produced even though cooling is more efficient, which leads to a lower residual stress.

Therefore, the best residual stress state is obtained at a low lubrication flow rate of 50 ml/h, an intermediate nozzle distance of 30 mm from the cutting zone, and an intermediate airflow rate of 31 l/h optimal MQL parameters (see Fig. 6).

3.3.2 Comparison Study

The optimal MQL machining resulted in 20.6% (+91.4 MPa) and 12.3% (+54.5 MPa) higher residual stresses when compared to dry and flood cooling respectively. The highest compressive residual stresses were observed in all cases along the main cutting direction (see Fig. 7). It was revealed that if mechanical loads are more dominant, compressive residual stresses profiles are more probable. While tensile residual stresses result when thermal loads govern [33]. This may be the reason for a low residual stress was observed in dry cutting where high heat is generated increasing the tensile stresses that offsets the compressive residual stress balance. A high MQL flow rate had a positive impact on the surface roughness similar to the cutting force. Best residual stresses were obtained at an intermediate MQL flow rate value. An intermediate nozzle distance gave the best cutting forces and residual stresses, while minimum nozzle distance favours surface roughness. However, a high airflow rate was good for cutting forces and an intermediate airflow rate was good for both the surface roughness and residual stresses.

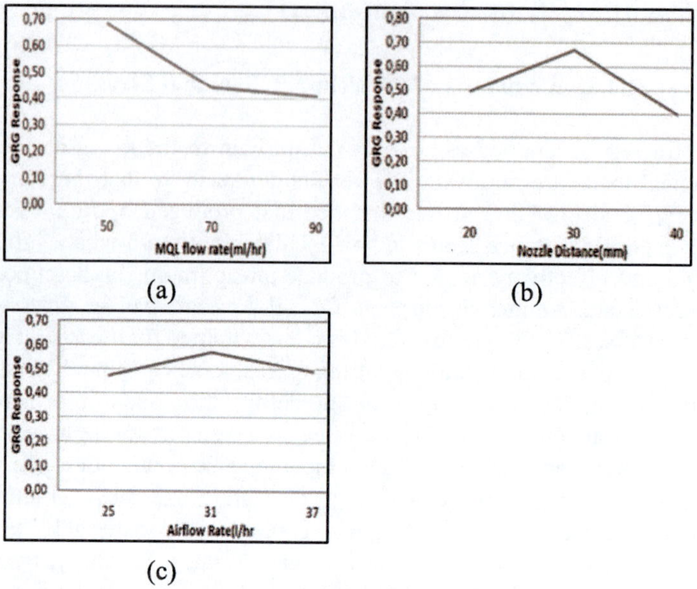

Fig. 6 Variation of GR response of the residual stress as a function of **a** oil flow rate, **b** nozzle distance, **c** airflow rate

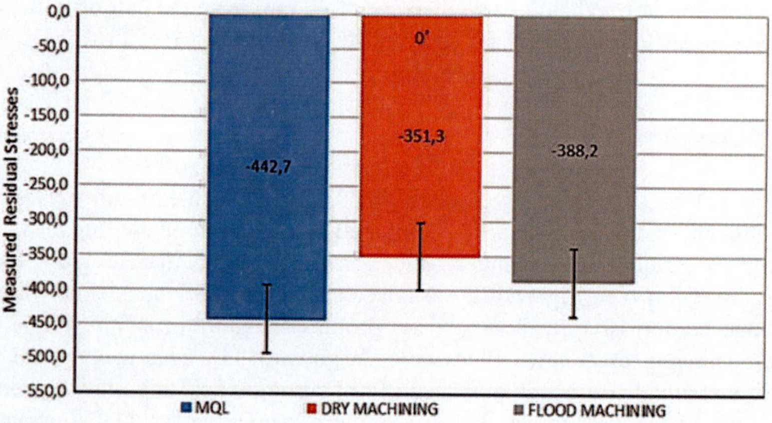

Fig. 7 Residual stresses comparison study MQL2, Dry cutting, and flood cooling

4 Conclusions

In this paper, the effects of MQL parameters on selected attributes of surface integrity during the turning of grade 4 titanium alloy were investigated experimentally. MQL parameters such as MQL oil flow rate, nozzle distance, and airflow rate were varied while cutting speed, feed rate, cutting length, and depth of cut were kept constant. The three surface integrity descriptors reacted differently to varying MQL parameters. In comparison, the optimal MQL parameter that yielded the lowest cutting forces had an adverse effect on surface roughness. While optimal parameters that yielded the lowest surface roughness results had the second-lowest cutting force but worst residual stress results, and optimal parameters that yielded the highest compressive residual stress results also yielded the worst cutting forces and surface roughness.

MQL exhibited a 30% improvement in the resultant cutting forces and 42.5% lower average surface roughness when compared to dry cutting and flood cooling. This result is due to the improved lubricant penetration into the cutting zone and effective cooling. Varying the MQL parameters impacted the resultant residual stress state significantly and yielded mainly compressive residual stresses in all directions. In general, a 20.6% higher residual stress was obtained for the lowest lubrication flow rate, where plastic deformation, cutting forces, and heat generation due to friction are most significant. The cutting force analysis exhibited a variation of 31.6 N. The results for the surface roughness had a variation of 0.11 μm. While the residual stress results had a variation of 49.3 MPa. Which indicates a significant influence of the MQL parameters on the output results. The combination of low surface roughness and compressive residual stresses is known to minimize crack initiation and improve the fatigue strength of the part. This positive result further demonstrates the potential for MQL as a viable alternative for dry cutting and flood cooling especially with the ability to utilize specific MQL process parameter sets to engineer a specific outcome.

Acknowledgements We acknowledge the support from Nelson Mandela University (NMU) and the mechanical engineering department at the university of Johannesburg (UJ).

References

1. Gupta, K., Laubscher, R.: Sustainable Machining of Titanium Alloys: A Critical Review. ResearchGate, Johannesburg (2016)
2. Ezugwu, E., Bonney, J., Yamane, Y.: An overview of the machinability of aeroengine alloys. J. Mater. Process. Technol. **134**(2), 233–253 (2003)
3. Hatt, O., Crawforth, P., Jackson, M.: On the mechanism of tool crater wear during titanium alloy machining. Wear **374–375**, 15–20 (2017)
4. Rahim, E., Sasahara, H.: A study of the effect of palm oil as MQL lubricant on high speed drilling of titanium alloys. Tribol. Int. **44**, 309–317 (2011)
5. Khettabi, R., Fatmi, L., Masounave, J., Songmene, V.: On the micro and nanoparticle emission during machining of titanium and aluminum alloys. CIRP J. Manuf. Sci. Technol. **6**, 175–180 (2013)
6. Niknam, S.A., Khettabi, R., Songmene, V.: Machinability and machining of titanium alloys: a review. Mater. Form., Mach. Tribol. (2014)
7. Madyira, D., Laubscher, R., van Rensburg, N.J., Henning, P.: High speed machining induced residual stresses in Grade 5 titanium alloy. J. Mater. Des. Appl. **3**(227), 208–215 (2012)
8. Liu, Z., An, Q., Xu, J., Chen, M., Han, S.: Wear performance of (nc-AlTiN)/(a-Si3N4) coating and (nc-AlCrN)/(aSi3N4) coating in high-speed machining of titanium alloys under dry and minimum quantity lubrication (MQL) conditions. Wear **305**, 249–259 (2013)

9. Benjamin, D.M., Sabarish, V.N., Hariharan, M., Raj, D.S.: On the benefits of sub-zero air supplemented minimum quantity lubrication systems: An experimental and mechanistic investigation on end milling of Ti-6-Al-4-V alloy. Tribol. Int. **119**, 464–473 (2018)
10. Sartori, S., Ghiotti, A., Bruschi, S.: Solid lubricant-assisted minimum quantity lubrication and cooling strategies to improve Ti6Al4V machinability in finishing turning **118**, 287–294 (2018)
11. DGUV: Minimum Quantity Lubrication for Machining Operations. Deutsche Gesetzliche Unfallversicherung (DGUV), Berlin (2010)
12. Kuo, C., Hsu, Y., Chung, C., Chen, C.-C.A.: Multiple criteria optimisation in coated abrasive grinding of titanium alloy using minimum quantity lubrication. Int. J. Mach. Tools Manuf. **115**, 47–59 (2017)
13. Cai, X.J., Liu, Z.Q., Chen, M., An, Q.L.: An experimental investigation on effects of minimum quantity lubrication oil supply rate in high-speed end milling of Ti-6AI-AV. Proc. Inst. Mech. Eng., Part B: J. Eng. Manuf. **226**(11), 1784–1792 (2012)
14. Ulutan, D., Ozel, T.: Machining induced surface integrity in titanium and nickel alloys: a review. Int. J. Mach. Tools Manuf. **51**, 250–280 (2011)
15. Pawade, R., Joshi, S.S., Brahmankar, P.: Effect of machining parameters and cutting edge geometry on surface integrity of high-speed turned Inconel 718. Int. J. Mach. Tools Manuf. **48**(1), 15–28 (2008)
16. Pan, Z., Feng, Y., Liang, T.Y.: Material microstructure affected machining: a review. Manuf. Rev. **4**(5) (2017)
17. John, J.W.H., Pricea, W.H.: Comparison of experimental and theoretical residual stresses in welds: the issue of gauge volume. Int. J. Mech. Sci. **50**, 513–521 (2008)
18. Mathonsi, T.N.: Outside Turning of Commercially Pure Titanium Alloys by Minimum Quantity Lubrication. The University of Johannesburg, Mechanical Department, Johannesburg (2018)
19. Wei, G., Yi, W.: Grey relation analysis method for multiple attributes decision making with incomplete information in intuitionistic fuzzy setting. In: Chinese Control and Decision Conference, 2008
20. Jeet, S., Kar, S.: Review on application of minimum quantity lubrication (MQL) in metal turning operations using conventional and Nano-lubricants based cutting fluids. Int. J. Adv. Mech. Eng. **8**(1), 63–70 (2018)
21. Lohar, D., Nanavaty, C.: Performance evaluation of minimum quantity lubrication (MQL) using CBN tool during hard turning of AISI 4340 and its comparison with dry and wet turning. Bonfring Int. J. Ind. Eng. Manag. Sci. **3**(3) (2013)
22. Korka, Z.I., Micloşină, C.O., Cojocaru, V.: An experimental study of the cutting forces in metal turning. ANALELE UNIVERSITĂłII "EFTIMIE MURGU", REŞITA (2013)
23. Hadad, M., Sadeghi, B.: Minimum quantity lubrication-MQL turning of AISI 4140 steel alloy. J. Clean. Prod. **54**, 332–343 (2013)
24. Uysal, A., Demiren, F., Altan, E.: Applying minimum quantity lubrication (MQL) method on milling of martensitic stainless steel by using Nano Mos2 reinforced vegetable cutting fluid. Procedia Soc. Behav. Sci. **195**, 2742–2747 (2015)
25. Dureja, J., Singh, R., Singh, T., Singh, P., Dogra, M., Bhatti, M.: Performance evaluation of coated carbide tool in machining of stainless steel (AISI 202) under minimum quantity lubrication (MQL). Int. J. Precis. Eng. Manuf. **2**, 123–129 (2015)
26. Hamdan, A., Sarhan, A.A.D., Hamdi, M.: An optimization method of the machining parameters in high-speed machining of stainless steel using coated carbide tool for best surface finish. Int. J. Adv. Manuf. Technol. **58**, 81–91 (2012)
27. Dhar, N., Kamruzzaman, M., Ahmed, M.: Effect of minimum quantity lubrication (MQL) on tool wear and surface roughness in turning AISI-4340 steel. J. Mater. Process. Technol. **172**, 299–304 (2006)
28. Amrita, M., Srikant, R.R., Sitaramaraju, A.V.: Performance evaluation of nanographite-based cutting fluid in machining process. Mater. Manuf. Processes **29**(5), 600–605 (2014)
29. Yazid, M.Z.A., CheHaron, C., Ghani, J., Ibrahim, G., Said, A.: Surface integrity of Inconel 718 when finish turning with PVD coated carbide tool under MQL. Procedia Eng. **19**, 396–401 (2011)

30. Sharma, A.K., Tiwari, A.K., Dixit, A.R.: Effects of minimum quantity lubrication (MQL) in machining processes using conventional and nanofluid based cutting fluids: a comprehensive review. J. Clean. Prod. **127**, 1–18 (2016)
31. Ramana, M.V.: Optimization and influence of process parameters on surface roughness in turning of titanium alloy under different lubricant conditions. Mater. Today: Proc. **4**, 8328–8335 (2017)
32. Totten, G.E., Mackenzie, D.S.: Handbook of Aluminum. Marcel Dekker Inc., New York (2003)
33. Jacobson, M.: Surface integrity of hard-turned M50 steel. Proc. Inst. Mech. Eng., Part B: J. Eng. Manuf. **1**(216), 47–54 (2002)

Alpheus Ngwako Maponya holds a B.Eng. degree in Mechanical Engineering from the University of Johannesburg. He is currently a Master's student in the Department of Mechanical Engineering. His specific focus and area of research interest is in the effect of MQL in titanium machining.

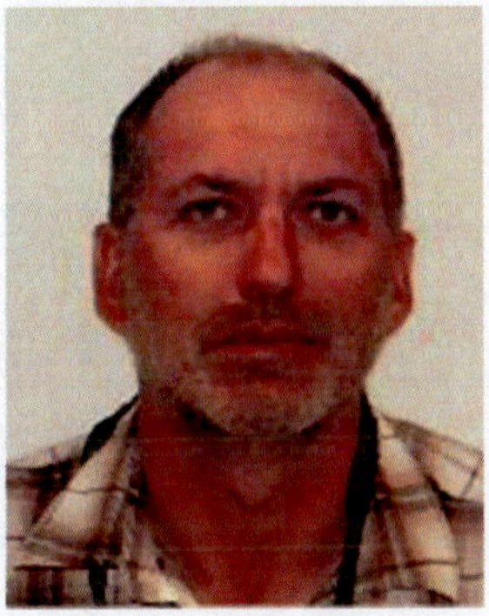

Rudolph Laubscher holds a D.Ing from RAU. He is currently an associate professor at the University of Johannesburg. His research interests are associated with physical metallurgy, manufacturing and FEA.

Laser Shock Peening: A NbC Based Cermet Enhancement Alternative for Improved GCI Interrupted Face-Milling

R. M. Genga, D. Glaser, P. Rokebrand, L. A. Cornish, M. Woydt, T. Gradt, A. Janse van Vuuren, and C. Polese

Abstract Attempts were made to improve the properties and machining performance of NbC based cermet cutting inserts for face milling of grey cast iron (GCI) (BS 1452/GG35) by spark plasma sintering (SPS), use of sub-stoichiometric NbC

R. M. Genga (✉)
Academic Development Unit (ADU), University of the Witwatersrand, Johannesburg, South Africa
e-mail: Rodney.Genga@wits.ac.za

R. M. Genga · L. A. Cornish · C. Polese
DSI-NRF Centre of Excellence in Strong Materials, University of the Witwatersrand, Johannesburg, South Africa
e-mail: Lesley.Cornish@wits.ac.za

C. Polese
e-mail: claudia.polese@wits.ac.za

R. M. Genga · C. Polese
School of Mechanical, Industrial and Aeronautical Engineering, University of the Witwatersrand, Johannesburg, South Africa

R. M. Genga · L. A. Cornish · C. Polese
Africa Research Universities Alliance (ARUA) Centre of Excellence in Materials, Energy & Nanotechnology, University of the Witwatersrand, Johannesburg, South Africa

D. Glaser
CSIR National Laser Centre, Pretoria 0001, South Africa
e-mail: DGlaser@csir.co.za

Mechanical Engineering, Nelson Mandela University, Gqeberha, South Africa

P. Rokebrand
LLC, C3 Analytics, Redwood City, USA

L. A. Cornish
School of Chemical and Metallurgical Engineering, University of the Witwatersrand, Johannesburg, South Africa

M. Woydt
MATRILUB (Materials, Tribology, Lubrication), Berlin, Germany
e-mail: m.woydt@matrilub.de

K. von Leipzig et al. (eds.), *Smart, Sustainable Manufacturing in an Ever-Changing World*, Lecture Notes in Production Engineering,
https://doi.org/10.1007/978-3-031-15602-1_62

($NbC_{0.88}$), Cr_3C_2 and Mo_2C additives and laser shock peening (LSP). The microstructure, hardness, fracture toughness and Young's modulus of WC–Co, $NbC_{0.88}$-Co and $NbC_{0.88}$-Ni cermet-based inserts were investigated for machining performance. Additions of Mo_2C and manufacture by SPS significantly refined the $NbC_{0.88}$-Ni/Co cermet's carbide grain size from ~5.0 μm to <0.9 μm, which increased the hardness (by ~4GPa) and wear resistance. Laser shock peening (LSP) improved the fracture toughness of all the inserts, from 10% (in the SPS WC based in cermets) to ~100% (in the LPS NbC based cermets). Uncoated cutting inserts were manufactured from the sintered cermets in an SNMA (double sided square insert without a chipformer on the cutting edge) shape (12.7 × 12.7 × 4.3 mm^3, 1.6 mm nose radius). The inserts were used for interrupted face-milling of GCI at cutting speeds (v_c) ranging from 100 m/minute (400 rpm)–500 m/minute (2000 rpm) and depths of cut (a_p) of 0.2–1.0 mm. The insert wear was measured after every pass and analyzed by annular dark field scanning transmission electron microscopy (ADF-STEM). Cutting temperatures were measured with a high-speed thermal camera and forces were measured using a Kistler dynamometer. During machining at v_c = 200 m/minute and a_p = 1.0 mm, WC–Co inserts had the lowest flank wear rate (FWR), although LSP significantly improved the tool life of the $NbC_{0.88}$-Ni inserts, significantly reducing the FWR from 148.63 μm/minute to 99.79 μm/minute (by ~33%) and the average resultant force from 1257 ± 15 N to 535 ± 15 N. During machining at v_c = 500 m/minute and a_p = 0.2 mm, LSP significantly lowered the FWRs of the $NbC_{0.88}$ inserts, giving better tool life than all the WC based inserts. Generally, LSP improved the $NbC_{0.88}$ inserts' tool life, reducing the flank wear rate by up to 33% during roughing and 63% during finishing.

Keywords Niobium carbide · Laser shock peening · Face milling · Tool wear

1 Introduction

Machining has been the most used metal shaping process for decades, which is primarily due to the adaptability and versatility of the turning and milling processes in response to industrial and market needs [1]. Milling is a machining operation which cuts by rotation of the cutting inserts in the opposite direction to that of the feed of the work piece [1]. Face milling is a process where the cutting action is achieved by cutting insert rotation about an axis perpendicular to work piece [1]. Due to the rotation motion, the inserts undergo a periodic engagement (cutting) and disengagement (cooling) with the work piece per revolution [1, 2]. This results in rapid heating and cooling cycles per revolution, leading to thermal shock at the insert cutting edge

T. Gradt
BAM Federal Institute for Materials Research and Testing, Berlin, Germany

A. Janse van Vuuren
CHRTEM, Nelson Mandela University, Gqeberha, South Africa

[2]. Additionally, depending on the work piece, the inserts are subjected to abrasion, adhesion, diffusion and mechanical wear mechanisms [2, 3]. These factors make face-milling one of the most challenging tribological processes, requiring inserts with good combinations of hardness, strength, toughness and chemical stability at elevated temperatures, such as provided by cermets [3].

In general, cermets consist of ceramics (e.g. tungsten carbide) embedded in a ductile binder matrix (e.g. cobalt) [3]. Tungsten carbide (WC)—cobalt (Co) based cermets (also cemented carbides) are commercially the most successful carbides since their inception in 1923 [3], due to the good combination of microstructure, mechanical, and wear properties [2, 3]. However, due to recent supply constraints and increasing cost [4], as well as poor chemical stability, particularly for machining of steels and cast irons [5], niobium carbide (NbC) has been investigated as a potential alternative. Niobium carbide has better high temperature properties than WC, such as retention of hot hardness at elevated temperatures, particularly when machining steels and cast irons [4], as well as comparable hardness and significantly lower density [5]. Generally, NbC-Co cements produced by conventional liquid phase sintering (LPS) have lower hardness and fracture toughness than WC–Co cemented carbides, due to excessive NbC grain growth [5]. This grain growth occurs at the high sintering temperatures and long sintering times during LPS [2, 5]. The excessive NbC grain coarsening can be prevented by rapid sintering techniques such as spark plasma sintering (SPS) as well as the addition of grain growth inhibitors such as Cr_3C_2 and Mo_2C [5, 6]. The mechanical properties of NbC cermets can also be improved by using sub-stoichiometric $NbC_{0.88}$ that has been reported to have higher hot hardness than WC at temperatures above 700 °C [7]. Furthermore, substitution of Co by Ni was beneficial, as Ni has better oxidation and thermal cracking resistance, as well as a stable ductile face centred cubic (fcc) structure at all temperatures [3, 8].

Surface engineering processes such as peening have been reported to improve the resistance to crack-based phenomena such as fracture, fatigue and stress corrosion cracking [9, 10]. Through peening, beneficial residual stresses are introduced which can improve fracture toughness (K_{IC}) [9–11]. The use of peening for ceramics and cermets is problematic since brittle materials may not exhibit significant plastic deformation [12, 13], and hence the development of the residual stresses required for improved fracture toughness may not be possible. Laser shock peening (LSP) is an advanced form of peening [9] that enables precise control and repeatability of peening, making it suitable for brittle materials, such as cermets, which have a very limited range of deformation and residual stress generation. The LSP process uses high-powered laser pulses which are forced onto the target to generate rapid plasma expansions [10]. An inertial confinement medium is used to confine and enhance the pressure of the plasma to achieve high pressure of several GPa [10]. The high pressure acting over a short time interval drives a shock wave through the solid target with sufficient strength to exceed the dynamic yield strength, generating beneficial compressive residual stresses (CRs) [9, 10]. The CRs inhibit crack propagation under static and cyclic loading, thus improving the K_{IC} and fatigue life [14].

In this work, an attempt to improve the properties and machining performance of NbC based cermet cutting inserts for face milling BS 1452, grade 17 grey cast iron

(GCI) (DIN 1692 GG35) by spark plasma sintering (SPS), use of sub-stoichiometric NbC ($NbC_{0.88}$), Cr_3C_2 and Mo_2C additives and laser shock peening (LSP) was made. The microstructural and mechanical properties, as well as the wear performance, after face-milling of the developed NbC based inserts were compared to WC–Co based inserts.

2 Experimental Procedure

2.1 Sintering

The characteristics of the starting powder are given in Table 1. Powders of different compositions were consolidated by spark plasma sintering (SPS) and conventional liquid phase sintering (LPS). Spark plasma sintering was done in a HP D5, FCT Systeme sintering furnace. The powders were poured into cylindrical graphite dies with inner and outer diameters of 20.9 mm and 40 mm respectively, and 48 mm height. Horizontal and vertical graphite foils were used to separate the powders from the die and punch set-up. Hexagonal boron nitride was placed on the graphite foil to prevent carbon diffusion from the graphite foil to the powders during sintering. The composite powder assemblies were heated in a vacuum (2 Pa) in two steps, for example WC-10Co (wt%) powders were first heated to 1000 °C at a rate of 200 °C/min and subsequently to 1220 °C at a heating rate of 100 °C/min, and the temperature was held at 1220 °C for 5 min during sintering.

A cooling rate of 200 °C/min was used for all samples. The applied pressure was adjusted from 16 to 30 MPa at 1000 °C, and from 30 to 50 MPa at 1220 °C within 30 s. The pressure was then held constant at 50 MPa throughout the rapid sintering cycle. Different sintering profiles, depending on the powder compositions were used (Table 2) to achieve good densification.

Liquid phase sintering (HIP, Ultra Temp, USA) was done by heating the different powder compositions in a vacuum (0.04 MPa) at an initial heating rate of 2.4 °C/min up to 1200 °C. At 1200 °C, cobalt loss protection (CLP) was carried out by the

Table 1 Sintering powders

Materials	Particle size (μm)	Crystal structure	Purity (wt%)	Source
WC	0.8	Hexagonal	>99.00	H.C. Starck, Germany
Co	0.9	Hcp	>99.80	OMG Americas, USA
Ni	5.9	Fcc	>99.00	Atlantic Equipment Engineers, Micron Metals Inc., USA
$NbC_{0.88}$	1.57	Cubic	>99.00	Höganäs AB, Sweden
Cr_3C_2	0.8	Hexagonal	>99.00	H.C. Starck, Germany

Table 2 Sintering conditions

Composition (wt%)	Abbreviation	Temperature (°C) and dwell time	Pressure (MPa)
$WC\text{-}0.8Cr_3C_2\text{-}8Co$	WC-8Co-L	1430 °C for 75 min	4.4
$WC\text{-}0.8Cr_3C_2\text{-}8Co$	WC-8Co-S	1220 °C for 5 min	50
$NbC_{0.88}\text{-}8Co$	NbC-8Co-L	1430 °C for 75 min	4.4
$NbC_{0.88}\text{-}8Co$	NbC-8Co-S	1260 °C for 4 min	50
$NbC_{0.88}\text{-}8Ni$	NbC-8Ni-L	1430 °C for 75 min	4.4
$NbC_{0.88}\text{-}8Ni$	NbC-8Ni-S	1280 °C for 4 min	50
$NbC_{0.88}\text{-}4Mo_2C\text{-}8Ni$	$NbC\text{-}Mo_2C\text{-}8Ni\text{-}L$	1430 °C for 75 min	4.4
$NbC_{0.88}\text{-}4Mo_2C\text{-}8Ni$	$NbC\text{-}Mo_2C\text{-}8Ni\text{-}S$	1280 °C for 4 min	60

addition of argon gas at a pressure of 0.37 MPa, and a heating rate 3.5 °C/min up to 1430 °C. The temperature was held constant for 75 min, and for the last 20 min, hot isostatic pressing (HIP) was done at 4.4 MPa to eliminate all the surface porosity [3]. The sample was then water cooled at a rate of 3.5 °C/min.

2.2 *Characterization*

Microstructures of the sintered cemented carbides were examined in a field emission scanning electron microscope (SIGMA, Karl Zeiss, Germany) (SEM) and annular dark field scanning transmission electron microscopy (ADF-STEM) (JEOL 2100, with a LaB_6 filament, JEOL, Japan). Image analysis was done on the SEM images using ImagJ software to obtain the mean/aveage carbide grain size of the sintered samples. Vickers hardness (HV_{30}) was measured on polished specimens after standard metallographic preparation, using a load of 30 kg (VHT 003 MTA, Vickers Limited, United Kingdom), calculating an average from five indentations at different regions on each sample. The criteria for the accurate derivation of fracture toughness (K_{1c}) using Shetty's equation [15] were satisfied: $1.25 \leq c/a \leq 2.25$ and $0.25 \leq I/a \leq 2.5$, where c is the crack length from the centre of the indentation to the crack tip, a is half the diagonal of the indentation and I is the difference between c and a. An ultrasonic thickness gauge (45MG, Olympus, USA) was used to measure the velocity of sound in the transverse (v_T) and longitudinal (v_L) directions which were used to calculate the Young's modulus of the samples. Laser shock peening of the samples was carried out using a Spectra-Physics Quanta-Ray Pro 270 ND: YAG Laser. A power intesity of 12GW and energy of 1080 J was applied, and a spot size of 1 mm used. Water was used as the inertia confinement medium.

Cutting tool inserts were manufactured from the sintered compositions and used to conduct face-milling tests on BS 1452, grade 17 (DIN 1691 GG35) grey cast iron. The inserts were square, 12.7 by 12.7 mm^2 with a thickness of 4.3 mm, and a nose radius of 1.6 mm (SNMA insert shape which is a double sided square insert

Table 3 Face milling parameters

Test No.	Cutting speed (m/minute)	Depth of cut (mm)	Spindle speed (rpm)	Feed rate (mm/minute)
1	200	1	400	40
2	400	0.5	800	80
3	500	0.2	2000	200

without a chipformer on the cutting edge). Square cutting inserts were manufactured from the cylindrical 20 mm diameter, 5 mm height sintered WC and NbC cemented carbides by EDM wire cutting (A422S, CHMER, Taiwan). Chamfering (chamfer angle of 20° and a chamfer width of 0.2 mm) was done on the edges of the square inserts by grinding (40LR, Tacchella, Italy) using a D46 diamond grinding wheel. During face milling, the inserts were secured on a Pilot F75SN12080 cutting tool holder attached to a Bridgeport GX1000 CNC milling machine with an Oi-MC Fanuc series controller. Milling inserts were designed and manufactured from the sintered samples. The inserts were squares of the same dimensions already descibed (SNMA shaped insert). The work piece was 150 × 150 × 50 mm^3. Laser shock peening of the inserts' flank cutting edges was done and the original and laser shock peened inserts were compared during face milling. The face milling conditions are shown in Table 3, and the feed of 0.1 mm/tooth and radial depth of cut of 80 mm were held constant at all cutting conditions. Maximum temperatures and thermal variation per second were measured using a Flir thermal imaging camera (D5000, Nikon, Japan). Force measurements were carried out using a Kistler force gauge (9366CC0, Kistler multicomponent force link, Switzerland). The flank wear and crater wear were evaluated using an optical microscope (DM6000 M, LEICA, Germany) with a digital camera (DFC490, LEICA, Germany) and high angle annular dark field scanning transmission electron microscopy (HAADF-STEM) (JEOL 2100, with a LaB_6 filament, JEOL, Japan) was used to deduce the phase compositions after wear.

3 Results and Discussion

3.1 *Mechanical Properties*

The Vickers hardness (HV_{30}) and fracture toughness (K_{IC}) of the sintered samples are shown in Fig. 1. The SPS samples had higher hardness values than those produced by LPS, with the WC-0.8Cr_3C_2-8Co (wt%) sample (WC-8Co-s) having the highest hardness and $NbC_{0.88}$-8Co (wt%) (NbC-8Co-L) and $NbC_{0.88}$-8Ni (wt%) (NbC-8Ni-L) having the lowest hardness values. Generally, the lower sintering temperatures and shorter sintering dwell times during SPS compared to LPS (Table 2) inhibited continuous Ostwald ripping, reducing the carbide grain growth [2, 6] and increasing hardness [2, 3, 6]. Both LPS and SPS WC-0.8Cr_3C_2-8Co (wt%) samples had higher hardness values than all the NbC based samples (Fig. 1). The higher hardness of the WC based samples was attributed to three main reasons: (i) WC (22.5 GPa) has a higher hardness than NbC (19.4 GPa) [5, 6], (ii) finer grain size (Figs. 2 and 3) due to good growth inhibition effect of Cr_3C_2 in the WC-10Co samples [2, 3], and (iii) better wetting behaviour between WC and Co than NbC and Co/Ni, giving a more homogeneous microstructure (Fig. 3), thus, improving the mechanical properties [2, 3]. Additions of Mo_2C to $NbC_{0.88}$-8Ni (wt%) reduced the NbC grain size (Figs. 2 and 4) increasing the hardness by >2 GPa. Further combinations of Mo_2C addition and SPS grain size of <1.0 μm (Figs. 2 and 4) and hardness increase of >3 GPa (compared to NbC-8Ni-L).

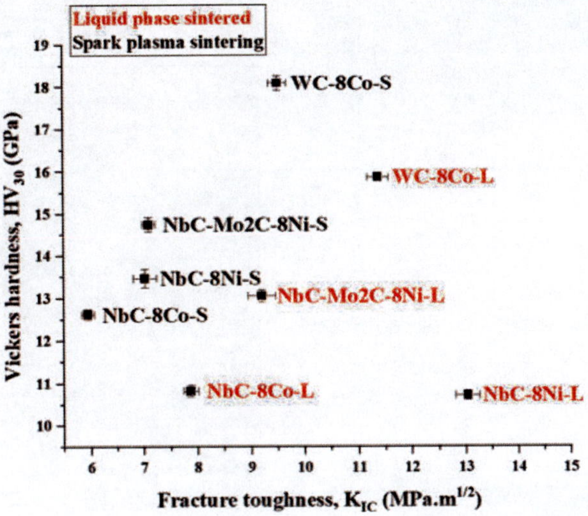

Fig. 1 Relationship between Vickers hardness and fracture toughness of sintered samples

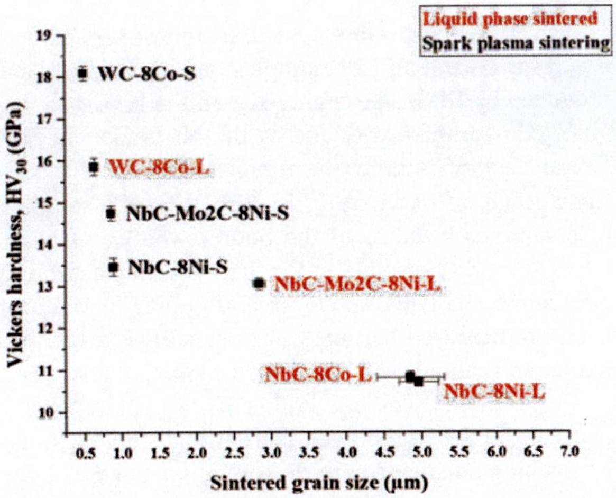

Fig. 2 Effect of grain size on Vickers hardness of sintered samples

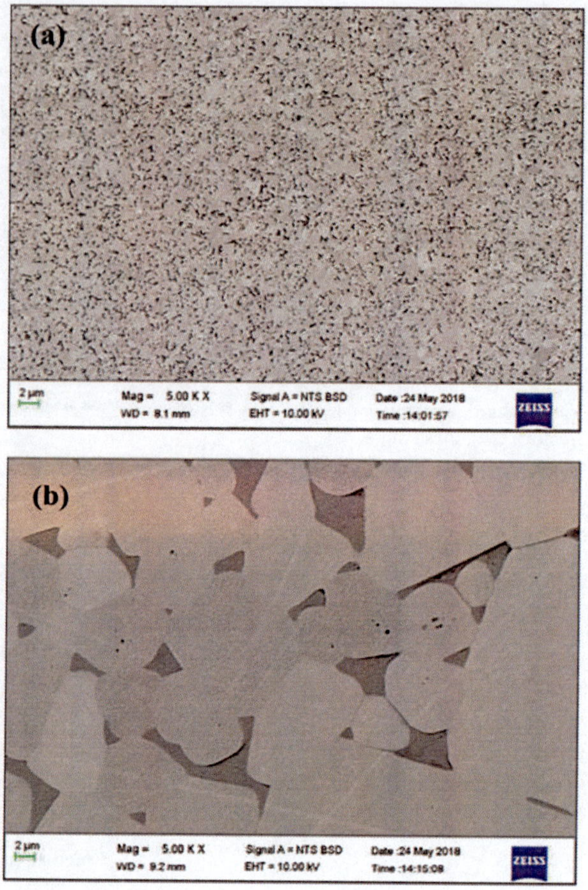

Fig. 3 SEM-BSE images of **a** LPS WC-8Co (wt%), showing WC (light) and Co (dark), and LPS, **b** NbC-8Ni (wt%), showing NbC (light), Ni (dark) using a scale of 2 μm.

The NbC-8Ni-L sample had the highest K_{IC}, while the NbC-8Co-S had the lowest K_{IC} (Fig. 1), and all LPS samples had higher K_{IC} than similar sample compositions produced by SPS. The higher K_{IC} in the NbC-8Ni-L sample than the WC-8Co and NbC-8Co samples was due to the Ni binder. Nickel has a higher plasticity than Co and always retains the more ductile fcc structure (does not undergo a phase transformation like Co) [3, 8]. The WC-8Co samples had high K_{IC} than the NbC-8Co samples because of the poorer wetting of NbC by Co than of WC, and the lower solubility of NbC in Co than WC in Co, which promotes formation of brittle interconnected NbC networks and poorer distribution of Co [6]. Although Mo_2C addition increased hardness, it reduced $K_{IC,}$ which was due to the formation of the harder and more brittle (Nb, Mo)C solid solution [16, 17].

The LPS samples had higher K_{IC} than similar compositions produced by SPS, because of the better binder distribution from the transient liquid binder phase during LPS which enhanced carbide solubility, as well as the capillary action of the liquid phase in the pores during the secondary rearrangement stage of sintering [3].

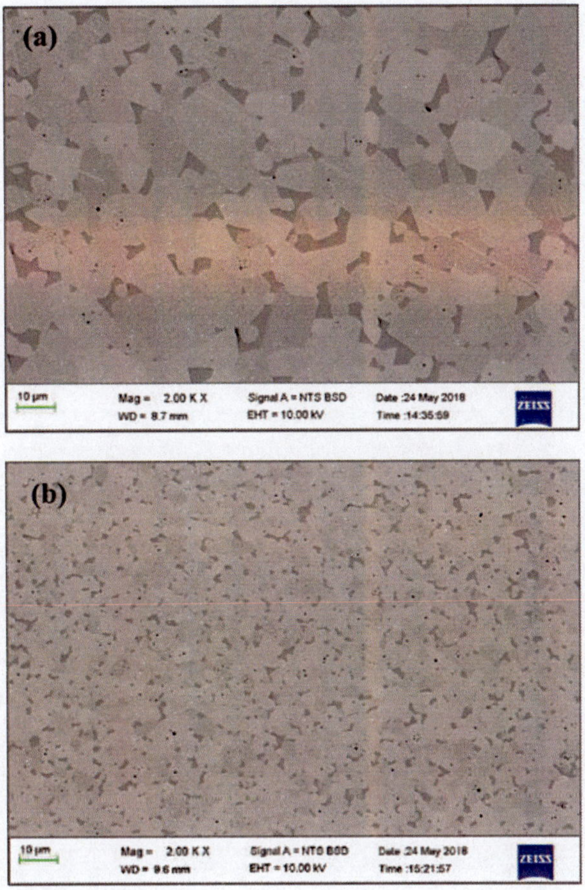

Fig. 4 SEM-BSE images of **a** LPS WC-8Co (wt%), showing WC (light) and Co (dark), and LPS, **b** NbC-8Ni (wt%), showing NbC (light), Ni (dark) using a scale of 10 µm.

Generally, laser shock peening had negligible effect on the samples' hardness, but increased K_{IC} in all the samples (Figs. 5 and 6). The increased K_{IC} was indicated by reduced crack length after Vickers hardness (HV_{30}) tests. For example, Fig. 7 shows reduced crack lengths in the NbC-Mo_2C-8Ni-L sample after laser shock peening. The indentation crack lengths were used to calculate the Shetty's fracture toughness [15], where longer crack lengths indicate lower fracture toughnesses [15].

All the samples, irrespective of the sintering technique had increased K_{IC} and the LPS samples had higher increases in K_{IC} than the corresponding SPS samples (Figs. 5 and 6). Additionally, the LPS samples with originally low K_{IC} values (~7 MPa.m$^{1/2}$) and low hardness, e.g. NbC-8Co-L, had higher increases in K_{IC} after laser shock peening than LPS samples with higher original K_{IC} and hardness values.

There was no change in crack propagation mode before and after laser shock peening (LSP). Transgranular crack propagation (which is the most critical fracture mode in cemented carbides [18]) was the main mode of fracture in all the samples. Reduced crack lengths after LSP without a changed crack propagation mode was

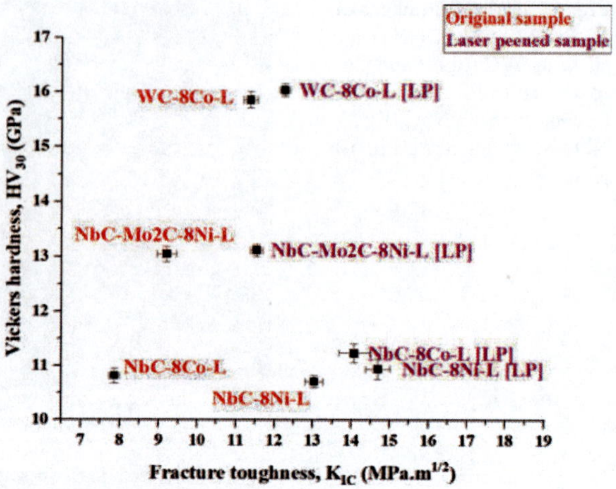

Fig. 5 Effect of laser shock peening on the relationship between Vickers hardness and fracture toughness of the liquid phase sintered samples

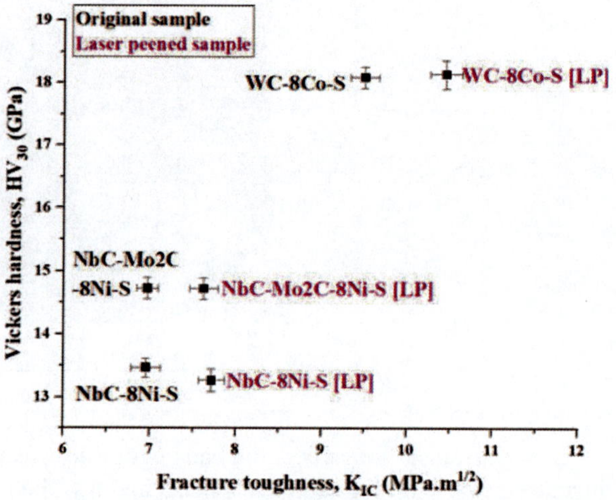

Fig. 6 Effect of laser shock peening on the relationship between Vickers hardness and fracture toughness of the spark plasma sintered samples

attributed to the induced compressive residual stresses which limited of transgranular crack propagation [11, 14]. Laser shock peening had negligible effect on the surface roughness.

3.2 *Young's Modulus*

The WC samples had higher Young's moduli (E) (by >140 GPa) than all the NbC samples, irrespective of the sintering method (Fig. 8) due to WC having a higher E than NbC [19]. Sintering techniques had negligible effect on the samples' Young's moduli (Fig. 8). In the NbC samples, Co substitution by Ni had negligible effect

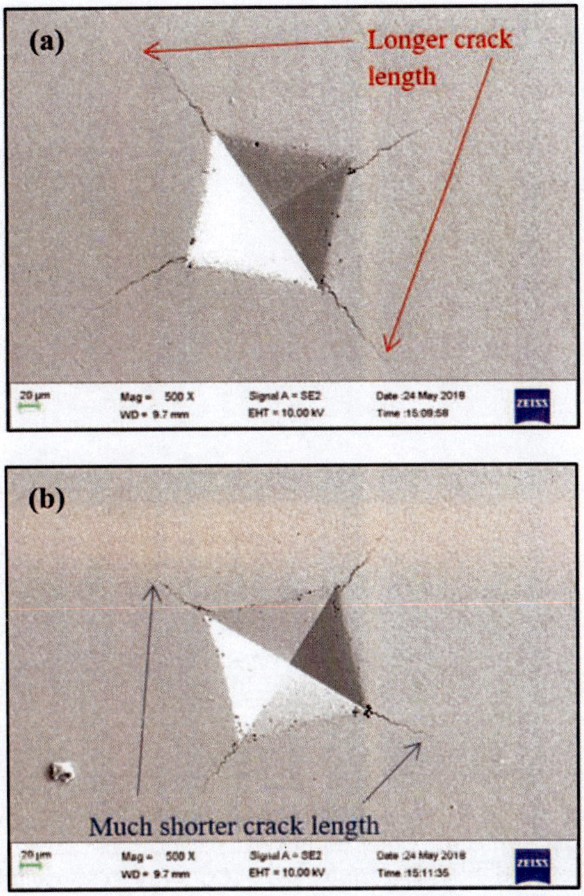

Fig. 7 SEM-SE images of NbC-Mo_2C-8Ni-L, showing: **a** long radial cracks before LSP, and **b** much shorter radial cracks after LSP

on the E, because of similar solubility of Co and Ni with NbC [20], resulting in similar bonding at the NbC/Co and NbC/Ni interfaces because the Young's modulus is proportional to the bonding strength in a material [21]. Slightly reduced E occurred in the LPS NbC-8Ni and NbC-Mo2C-8Ni samples than SPS samples (Fig. 8). The slight reduction in E could be due to slightly lower material bonding strength [22] achieved in the samples from LPS than SPS.

3.3 Face Milling

Cutting tool inserts were produced from selected samples and used for roughing, semi-finishing and finishing machining of BS 1452, grade 17 (DIN 1691 GG35) grey cast iron. The average resultant force (F_R), average cutting temperature (T_a) and flank wear rates (FWRs) of the inserts during face-milling at a cutting speed

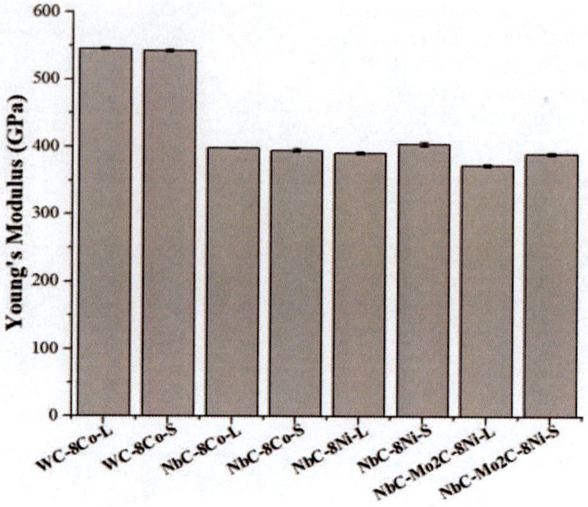

Fig. 8 Comparison of samples' Young's moduli

(v_c) of 200 m/min and depth of cut (a_p) of 1 mm are shown in Table 4. The average resultant force (F_R) had a fixed error of ± 15 N from the Kistler force gauge. The WC-8Co-L insert had the lowest FWR followed by the NbC-Mo_2C-8Ni-S, while the NbC-8Ni-S insert had the highest FWR (Table 4). The lower FWR in the WC-8Co-L insert than WC-8Co-S and all the NbC inserts during roughening was due to the combination good hardness (>15 GPa), high K_{IC} (>11 MPa.m$^{1/2}$) and high E, that improved the abrasion and circular impact resistance during roughing [5, 17]. Addition of Mo_2C to NbC-8Ni inserts reduced the FWRs due to refinement of the microstructure which improved the abrasion wear resistance [3]. Laser shock peening reduced the FWRs for the WC-8Co-S (by ~18%) and NbC-8Ni-S (by ~33%) (Table 4) due to increased K_{IC}. However, LSP had a converse effect on the WC-8Co-L and NbC-Mo_2C-8Ni-S inserts, which could have been due to slight damage/fracturing of the cutting edges in processing. During LSP processing, a protective coating was used to protect the insert cutting edge from direct contact with the laser, and rarely, the coating would experience slight damage, exposing small areas of the cutting edge to the laser, which led to slight cracking and chipping.

During semi-finishing at $v_c = 400$ m/minute and $a_p = 0.5$ mm, the NbC based inserts generally had lower FWRs than the WC based inserts, and the NbC-Mo_2C-8Ni-L insert had the lowest FWR (Table 5). There are no error bars in any of the FWR results because one cutting test per insert was done for each cutting parameter. This meant that only a single set of maximum flank wear values were attained and used to calculate the FWRs. Hence, there are no error bars. However, during semi-finishing, 3 out of 5 NbC based inserts had FWRs below 200 μm/min. While 3 of the 4 WC based inserts had FWR above 200 μm/min, indicating better NbC insert performance during semi-finishing.

The lowest FWR of the NbC-Mo_2C-8Ni-L insert of all the NbC inserts was caused by the combination of good hardness (use of sub-stoichiometric NbC and Mo_2C

Table 4 Insert machining properties during roughing face milling at $v_c = 200$ m/min and a_p = 1 mm

Insert	F_R (N)	Ta (°C)	FWR (μm/min)
WC-8Co-L	845	173.14 ± 1.69	66.53
WC-8Co-L [LP]	354	231.00 ± 1.80	79.58
WC-8Co-S	854	182.60 ± 1.20	130.69
WC-8Co-S [LP]	381	265.53 ± 3.48	107.79
NbC-8Ni-S	1257	246.27 ± 2.34	148.63
NbC-8Ni-S [LP]	536	348.38 ± 3.67	99.79
NbC-Mo_2C-8Ni-L	656	294.46 ± 3.13	125.47
NbC-Mo_2C-8Ni-S	546	233.88 ± 2.53	94.74
NbC-Mo_2C-8Ni-S [LP]	822	337.74 ± 3.56	159.28

Table 5 Insert machining properties during semi-finishing face milling at $v_c = 400$ m/min and a_p = 0.5 mm

Insert	F_R (N)	Ta (°C)	FWR (μm/min)
WC-8Co-L	824	169.90 ± 1.81	189.47
WC-8Co-L [LP]	964	199.15 ± 2.27	204.63
WC-8Co-S	888	197.64 ± 2.15	287.16
WC-8Co-S [LP]	926	175.17 ± 2.29	269.50
NbC-8Ni-S	910	281.20 ± 1.59	287.16
NbC-8Ni-S [LP]	903	240.63 ± 2.43	245.90
NbC-Mo_2C-8Ni-L	1041	277.21 ± 2.53	184.42
NbC-Mo_2C-8Ni-S	1059	257.82 ± 2.88	189.47
NbC-Mo_2C-8Ni-S [LP]	1125	277.81 ± 2.96	174.32

addition) which allowed good abrasion wear [5] and good fracture toughness for cyclic impact at the high cutting speed [17]. The lower FWRs of the NbC inserts than the WC-8Co inserts was due to the higher chemical stability of NbC than WC during machining of ferrous alloys [5, 17]. This was confirmed by HAADF-STEM mapping of the worn WC-8Co-L insert cutting edge interface (Fig. 9) which revealed smoothly worn and rounded WC grains, which indicated diffusion between the insert and the work piece [23]. However, HAADF-STEM mapping images of the worn NbC-Mo_2C-8Ni-L insert cutting edge interface revealed angular-shaped NbC grains (Fig. 10), demonstrating the good chemical stability of NbC during machining of grey cast iron [5, 17]. The higher FWR of WC inserts was due to chemical wear caused by diffusion of carbon from the insert into the work piece because of iron's higher affinity for carbon than tungsten [3]. Laser shock peening had negligible effect on the inserts' wear rate during semi-finishing, with the NbC-8Ni-S showing slight reduction in FWR.

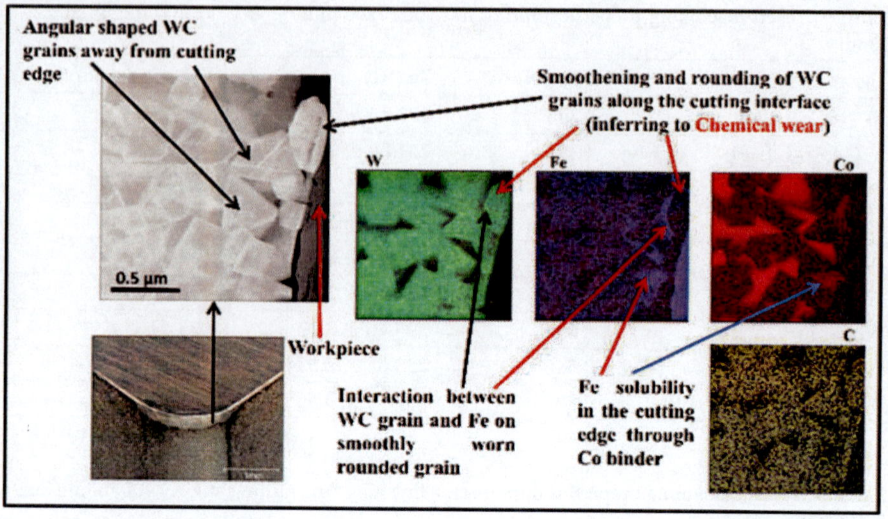

Fig. 9 HAADF-STEM mapping images of the worn cutting interface WC-8Co-L, showing W (green), Fe (blue), Co (red) and C (yellow)

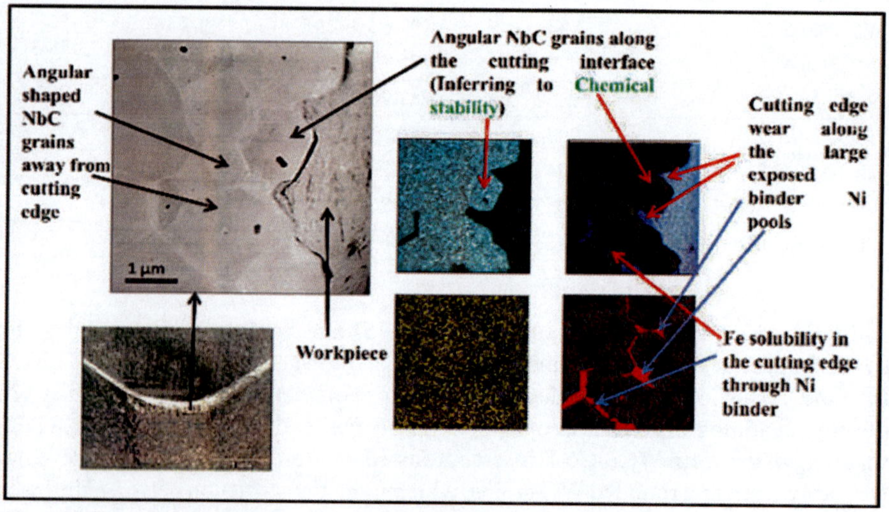

Fig. 10 HAADF-STEM mapping images of the worn cutting interface NbC-Mo_2C-8Ni-L, showing Nb (light green), Fe (blue), Ni (red) and C (yellow)

Increasing v_c from 400 to 500 m/min and reducing the depth of cut a_p from 0.5 to 0.2 mm increased F_R and FWR, and usually reduced the cutting temperature (Table 6). The inserts had different cutting temperatures, and this was attributed to different types of insert flank wear and flank wear rates [23]. The increased F_R and FWR were caused by increased impact (increased number of collisions with work piece

Table 6 Insert machining properties during finishing face milling at v_c = 500 m/min and a_p = 0.2 mm

Insert	F_R (N)	Ta (°C)	FWR (μm/min)
WC-8Co-L	1160	111.86 ± 1.22	282.11
WC-8Co-L [LP]	1233	121.14 ± 1.33	371.58
WC-8Co-S	1204	155.31 ± 1.69	243.16
WC-8Co-S [LP]	1222	122.97 ± 1.43	313.68
NbC-8Ni-S	1217	151.97 ± 1.66	493.68
NbC-8Ni-S [LP]	1311	143.84 ± 1.87	230.53
NbC-Mo_2C-8Ni-L	1319	173.81 ± 1.89	340.00
NbC-Mo_2C-8Ni-S	1258	161.45 ± 1.76	365.26
NbC-Mo_2C-8Ni-S [LP]	1237	142.25 ± 1.55	134.74

per second), increasing friction as well as blunting and abrasion of the cutting edge [2]. Considering the inserts that were not laser shock peened, the WC-8Co-S insert had the lowest FWR because of the high hardness, which is critical for abrasive wear resistance during finishing [3], and good K_{IC}. Although the WC based inserts had the lowest FWRs during finishing, they still experienced considerable wear rates (>200 μm/min) from chemical wear.

Laser shock peening of the inserts significantly reduced FWRs for the NbC inserts so that the NbC-Mo_2C-8Ni-S insert had the lowest FWR of 134 μm/min. This low FWR for NbC-Mo_2C-8Ni-S was due to the combination of good hardness, increased K_{IC} due to LSP and chemical stability. Laser shock peening had negligible effect on the WC based inserts FWR values at the high cutting speed, since chemical wear which still took place due to iron's higher affinity for carbon than WC, resulting in loss of C from WC, weakening the cutting edge [3].

4 Conclusions

Spark plasma sintering (SPS) of WC and NbC cermets resulted in higher hardness and lower fracture toughness (K_{IC}) than liquid phase sintering (LPS). The WC based samples had the best combination of hardness, K_{IC} and Young's modulus (E) compared to all the NbC based samples. Addition of Mo_2C improved the hardness of the NbC-8Ni samples. Laser shock peening (LSP) increased the K_{IC} of all the samples. During roughing face milling, the WC-10Co-L insert had the lowest flank wear rate (FWR) because of the good combination of hardness, high K_{IC} and high Young's modulus (E), which are critical properties for roughing. Laser shock peening significantly reduced the FWR for WC-8Co-S and NbC 8Ni-S inserts due to increased K_{IC}. During semi-finishing, the NbC inserts had lower FWRs than the WC inserts, with the NbC-Mo_2C-8Ni insert having the lowest FWR values due to the better chemical stability of NbC than WC in machining Fe alloys. During finishing, although the WC inserts experienced considerable flank wear (chemical wear), they had the lowest FWR than NbC inserts for inserts that did not have LSP. Laser shock peening significantly reduced the FWR during finishing of the NbC inserts giving much lower FWRs than the WC inserts. Overall, during roughing, the WC inserts had the lowest FWRs, whereas during semi-finishing and finishing, the NbC inserts had the lowest FWRs due mainly due to better chemical stability and improved K_{IC} due to laser shock peening.

Acknowledgements This work was financially supported by Department of Science and Innovation and the National Research Foundation, South Africa. The authors are grateful to Dr Chris Freemantle and Mr Joseph Oupa Mukwevho from Pilot Tools (Pty) Ltd for provision of powders and liquid phase sintering facilities, Mr Lufuno Mbauand Mr Ntsakisi Baloyi are thanked for assistance with the machining and lab work, as well as Mr Gerrard Peters for discussions and advice. The authors are also grateful to Companhia Brasileira de Metallurgia e Mineração (CBMM), São Paulo, Brazil, for supplying the NbC powder. Finally, I would like thank God for making this possible.

References

1. Childs, T., Maekawa, K., Obikawa, T., Yamane, Y.: Metal Machining—Theory and Application. Wiley & Sons Inc., New York, USA (2000)
2. Genga, R.M., Akdogan, G., Polese, C., Garrett, J.C., Cornish, L.A.: Abrasion wear, thermal shock and impact resistance of WC-cemented carbides produced by PECS and LPS. Int. J. Refract. Met. Hard Mater. **29**, 133–142 (2015)
3. Upahyaya, G.S.: Cemented Tungsten Carbide Production, Properties and Testing. Noyes Publications, New Jersey, USA (1998)
4. Woydt, M., Mohrbacher, H.: The use of niobium carbide (NbC) as cutting tools and for wear resistance tribosystems. Int. J. Refract. Met. Hard Mater. **49**, 212–218 (2015)
5. Genga, R.M., Cornish, L.A., Woydt, M., Janse van Vuuren, A., Polese, C.: Microstructure, mechanical and machining properties of LPS and SPS NbC cemented carbides for face-milling of grey cast iron. Int. J. Refract. Met. Hard Mater. **73**, 111–120 (2018)
6. Huang, S.G., Liu, R.L., Li, L., Van der Biest, O., Vleugels, J.: NbC as a grain growth inhibitor and carbide in WC–Co hardmetals. Int. J. Refract. Met. Hard Mater. **26**, 389–395 (2008)
7. Woydt, M., Mohrbacher, H.: The background for the use of hardmetals and MMCs based on niobium carbide (NbC) as cutting tools and for wear resistant tribosystems. In: Conference Proceedings of 3rd International Conference on Stone and Concrete Machining, Bochum, ISBN 987-3-943063-20-2
8. Penrice, T.W.: Alternative binders for hard metals. J. Mater. Shaping Technol. **5**, 35–39 (1987)
9. Gao, Y.K.: Improvement of fatigue property in 7050–T7451 Aluminium Alloy by laser peening and shot peening. Mater. Sci. Eng. **528**, 3823–3828 (2011)
10. Montross, C.S., Wei, T., Ye, L., Clark, G., Mai, Y.: Laser shock processeing and its effect on microstructure and properties of metal alloys: a review. Int. J. Fat. **24**, 1021–1036 (2002)
11. Rubio-Gonzalez, C., Felix-Martinez, C., Gomez-Rosas, G., Ocaña, J.L., Morales, M., Porro, J.A.: Effect of laser shock processing on fatigue crack growth of duplex stainless steel. Mat. Sci. Eng. **528**, 914–919 (2011)
12. Wulf, P., Frey, T.: Shot peening of ceramics: damage or benefit. Cer. Forum Int. **79**, E25–E28 (2002)
13. Wulf, P., Frey, T.: Strengthening of ceramics by shot peening. J. Euro Ceram. Soc. **26**, 2639–2645 (2006)

14. Wang, C., Jiang, C., Cai, F., Zhao, Y., Zhu, K., Chai, Z.: Effect of shot peening on the residual stresses and microstructure of tungsten cemented carbide. Mat. Des. **95**, 159–164 (2016)
15. Shetty, D.K., Wright, I.G., Mincer, P.N., Clauer, A.H.: Indentation fracture of WC-Co cermets. J. Mater. Sci. **20**, 1873–1882 (1985)
16. Genga, R.M., Rokebrand, P., Cornish, L.A., Brandt, G., Kelling, N., Woydt, M., Janse van Vuuren, A., Polese, C.: High-temperature sliding wear, elastic modulus and transverse rupture strength of Ni bonded NbC and WC cermets. Int. J. Refract. Met. Hard Mater. **87**, 105143 (2019)
17. Genga, R.M., Rokebrand, P., Cornish, L.A., Zeman, P., Brajer, J., Woydt, M., Janse van Vuuren, A., Polese, C.: Roughing, semi-finishing and finishing of laser surface modified nickel bonded NbC and WC inserts for grey cast iron (GCI) face-milling. Int. J. Refract. Met. Hard Mater. **86**, 105128 (2019)
18. Chermant, J.L., Osterstock, F.: Fracture toughness and toughness of WC-Co composites. J Mat. Sci. **11**, 1939–1951 (1976)
19. Mohrbacher, H., Woydt, M., Huang, S., Vleugels, J.: Niobium carbide—An innovative and sustainable high-performance material for tooling, friction and wear applications. Adv. Mater. Sci. Environ. Energy Tech. **260**, 67–81 (2016)
20. Tracey, V.A.: Nickel in hardmetals. Int. J. Refract. Met. Hard Mater **11**, 137–149 (1992)
21. Askeland, D.R., Fulay, P.P., 2010, Essentials of Material Science and Engineering, 2nd Edition, Cengage Learning, Stanford, UK (2010)
22. Banerjee, D., Lal, G.K., Upadhaya, S.G.: Effect of binder-phase modification and the Cr_3C_2 addition on the properties of WC-10Co cemented carbide. JMEPEG **4**, 563–572 (1995)
23. Trent, E.M., Wright, P.K.: Metal Cutting, 4th edn. UK, Butterworth-Heinemann, London (2001)

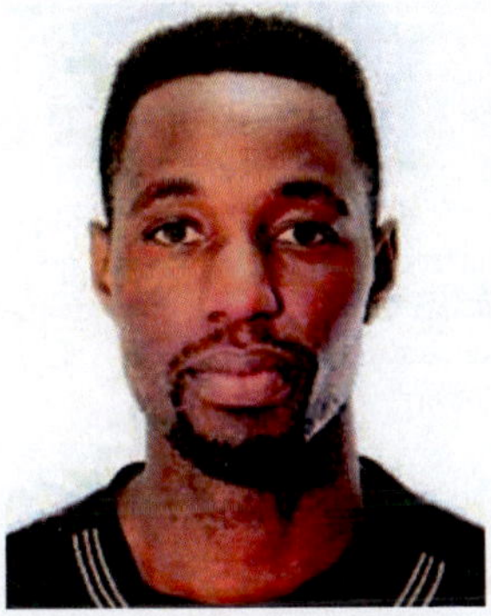

Rodney Genga Associate Professor, Ph.D. (Eng.), Director: Academic Development Unit Priciple Researcher: School of Mechanical, Industrial and Aeronautical Engineering, Faculty of Engineering & the Built-Environment University of the Witwatersrand.

Daniel Graser Dr. (Ph.D. Eng.), Senior engineer at the CSIR, Research associate: Mechanical Engineering, Nelson Mandela University.

Patrick Rokebrand Dr. (Ph.D. Eng.), Senior Data Consultant at C3 Analytics, LLC VP: Kimble's Aviation and Logistical Services, Inc. VP: Rock Harbor, LLC.

Lesley Cornish School of Chemical & Metallurgical Engineering, University of the Witwatersrand, Director: DSI-NRF Centre of Excellence in Strong Materials, hosted by the University of the Witwatersrand, Johannesburg Director: ARUA CoE Materials, Energy and Nanotechnology.

Mathias Woydt is managing partner of MATRILUB Materials Tribology Lubrication, with more than 35 years of experience in R&D and with more than 350 publications and 51 priority patents filed. He is elected STLE Fellow and recipient of ASTM award of excellence and board member of the German Society for Tribology.

Claudia Polese Associate Professor School of Mechanical, Industrial and Aeronautical Engineering. Polese is Associate Professor in Aerospace Manufacturing and Design Head of Aeronautical Engineering Stream.

Evaluating the Relationship Between Powder Characteristics, Defects, and Final Build Properties for L-PBF WC–Co

P. Govender, D. Hagedorn-Hansen, D. C. Blaine, and N. Sacks

Abstract With rising interest in additive manufacturing (AM) techniques, there is increased focus on research that evaluates critical parameters that guide the selection of powders suitable for AM. This study focuses on two spray-dried WC–Co powders (12 and 17 wt% Co) and evaluates the relationship between typical powder characteristics, defects occurring post printing and the final build parameters. The precursor powders and parts produced using a laser powder bed fusion (L-PBF) process were characterized. Suitable WC–Co printing parameters were used for building cutting tool inserts. The as-built parts were analysed for density and defect formation due to printing. The final part properties were evaluated and related back to the precursor powder properties.

Keywords Tungsten carbide-cobalt · Powder characteristics · Laser powder bed fusion · cutting tools

P. Govender (✉) · D. C. Blaine
Department of Mechanical and Mechatronic Engineering, University of Stellenbosch, Stellenbosch, South Africa
e-mail: 18375391@sun.ac.za

D. C. Blaine
e-mail: dcblaine@sun.ac.za

P. Govender · D. Hagedorn-Hansen · D. C. Blaine · N. Sacks
DSI-NRF Centre of Excellence in Strong Materials, Johannesburg, South Africa
e-mail: devonh@sun.ac.za

N. Sacks
e-mail: natashasacks@sun.ac.za

D. Hagedorn-Hansen · N. Sacks
Department of Industrial Engineering, University of Stellenbosch, Stellenbosch, South Africa

K. von Leipzig et al. (eds.), *Smart, Sustainable Manufacturing in an Ever-Changing World*, Lecture Notes in Production Engineering,
https://doi.org/10.1007/978-3-031-15602-1_63

1 Introduction

Cemented carbides are considered a workhorse material for a wide range of manufacturing industries. They are metal matrix composites (MMC) where metallic carbides, such as WC or TiC, act as reinforcing particles inside a metallic alloy binder phase, typically Co- or Ni-based, which acts as the metal matrix [1, 2]. The carbides contribute to the hardness and wear resistance, while the binder provides toughness and counters the brittle nature of the carbides [1, 3]. Cemented carbides are traditionally produced through press and sinter techniques; typical products are cutting tool inserts and high wear forming tools, such as dies. They are well-suited to these applications as they possess an excellent combination of thermal stability, hardness, and toughness [1, 4].

Laser powder bed fusion (L-PBF), an additive manufacturing (AM) technology, provides a potential alternative to conventional subtractive manufacturing processes, by offering advantages in the reduction of production time and feasibility of manufacturing complex geometries [2, 3, 5]. Producing high quality L-PBF parts depends on the use of powder that has is suited for AM.

The aim of this study was to determine the link between standard powder metallurgy (PM) powder characteristics, novel powder spreadability parameters, and the as-built properties of L-PBF WC–Co ISO 504:1975 designated CNMA 120404 cutting tool inserts. For this study specifically, spray-dried WC–Co powders, with Co contents of 12 and 17 wt%, respectively, were evaluated.

2 Materials and Methods

2.1 Raw Materials

Two different commercially available, agglomerated, and sintered WC–Co powders, that are typically used for high velocity oxygen fuel (HVOF) coatings, were used in this study: Praxair 1342 VM (WC-12Co) and Kennametal JetKote 117 (WC-17Co). As L-PBF of WC–Co is still under development, there are no AM powders available commercially. For ease of reference the respective powders are referred to as WC-12Co and WC-17Co, respectively. Table 1 shows the chemical composition of these powders, as reported by the suppliers.

Table 1 Powder composition, as per supplier datasheet

Powder description	WC (wt%)	Co (wt%)	Other (wt%)
WC-12Co	82.5	11.72	5.77
WC-17Co	77.8	17	5.2

2.2 *Methodology*

Standard powder characterisation methods were used to evaluate the raw powders. These methods included apparent density and flow rate, measured according to ASTM standards B212 and B213. Particle morphology and size was determined through image analysis using a scanning electron microscope (SEM), with concomitant energy dispersive spectroscopy (EDS) used for elemental analysis. The particle size distribution was measured using the light scattering technique with a Micromeritics Saturn DigiSizer.

Spreadability metrics, powder bed spread density (PBD) and percentage coverage, were determined with the use of a custom designed powder spreadability rig [6] that replicates the powder spreading process of L-PBF. Both the WC-12Co and WC-17Co powders were used to print the CNMA cutting tool inserts on a Concept Laser M2 L-PBF machine. Two sets of inserts were produced from each powder keeping the laser power (190 W) and scan speed (550 mm/s) constant while varying the hatch spacing at two different levels (80 and 100 μm). The layer thickness and laser spot size were kept constant at 30 μm and ~50 μm, respectively. The parts were removed from the base plate with an Agie Charmilles CA20 wire electrical discharge machine (W-EDM).

The as-built inserts were prepared for analysis using standard metallographic procedures for WC–Co. Density was measured using the Archimedes principle. Porosity as well as grain size and microstructure were investigated using optical microscopy and SEM. The hardness of the inserts was measured using the Rockwell hardness scale A (HRA) testing with a load of 60 kgf for a hold time of 3 s. The cobalt content was measured with energy dispersive spectroscopy (EDS) on the SEM. The presence of visible defects was determined for each set of build parameters. Powder X-ray diffraction (XRD) was performed on a Bruker D2 Phaser instrument. The influence of the precursor powders was then related to the as-built material characteristics.

3 Results and Discussions

3.1 *Initial Characterization*

3.1.1 Particle Morphology and Size

Figure 1 shows SEM images of the WC-12Co and WC-17Co powders at two levels of magnification. Both powders show spherical particles that are the sintered agglomerates of WC grains fused together by Co; this is the typical morphology that is expected for HVOF WC–Co powders. The particles of the WC-17Co powder are slightly more elongated than those of the WC-12Co powder. Table 2 reports the powder particle size distribution (PSD), as determined through light scattering technique. The PSDs

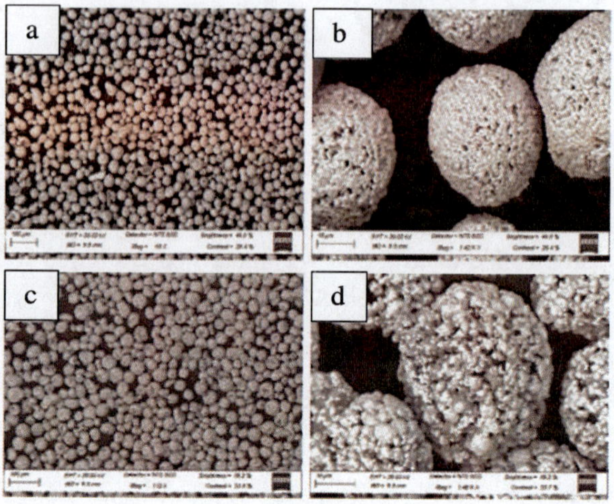

Fig. 1 SEM images of WC-12Co (**a, b**) and WC-17Co (**c, d**) with scale bars of 100 μm and 10 μm, respectively for (**a, b**) and (**c, d**)

Table 2 Powder particle size distributions

Powder designation	D10 (μm)	D50 (μm)	D90 (μm)
WC-12Co	21.50	31.94	43.05
WC-17Co	24.96	36.96	50.59

of both powders are similar, ranging from D10 around 20 μm to D90 around 50 μm. The WC-12Co powder is slightly smaller on the whole (around 13–14% across D10, D50 and D90) with a slightly narrower spread.

Elemental composition was determined by EDS spot analysis to identify WC grains and the Co regions that connect the WC grains. These features are indicated by the labels in Fig. 2, with WC showing lighter in colour than the darker grey Co regions. This visualizes the Co is the binder matrix while the WC grains are the reinforcing phase in the composite particles.

3.1.2 Powder Characteristics

The apparent density and flow rate for each powder are listed in Table 3. Each test was repeated three times, with the mean values along with the standard deviations reported. It is misleading to compare the apparent density values as the Co content of the two powders, and therefore the theoretical full density of the powder particles, differs. According to general relationships for full density WC–Co, a 12 wt% Co WC–Co material has a full density of 14.3 g/cm^3 while the density when the Co content is 17 wt% is 13.7 g/cm^3 [7]. By normalising the apparent density with the theoretical full density of each powder, the relative or fractional apparent density was calculated and is also reported in Table 3. As the relative density reflects the volume

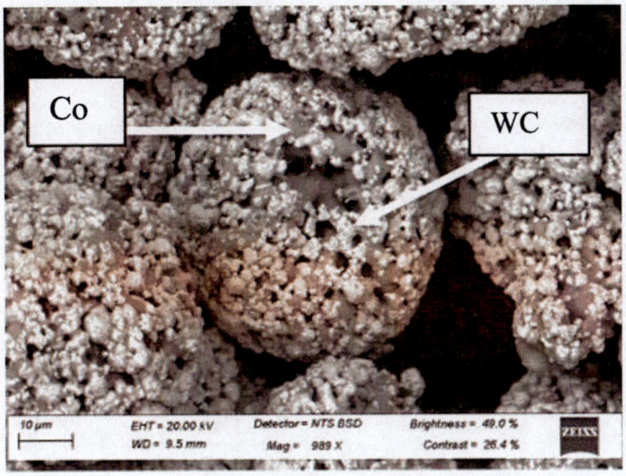

Fig. 2 SEM image of WC-17Co agglomerated powder particles with EDS spot analysis verification of WC and Co particles inside the agglomerates

Table 3 Powder properties—apparent density and flow rate

Powder	Apparent density (g/cm^3) {relative density}	Flow rate (s/50 g) {Volume powder/50 g (cm^3)}
WC-12Co	5.39 ± 0.01 {37.7%}	14.18 ± 0.98 {3.50}
WC-17Co	3.98 ± 0.00 {29.0%}	21.73 ± 0.53 {3.65}

fraction of actual material mass in a porous volume, the fractional porosity (volume % of air) is estimated as (1—relative density).

The WC-12Co powder has a higher relative apparent density than the WC-17Co powder; this indicates that the WC-12Co powder packs more densely, has less air gaps between powder particles, than the WC-17Co powder.

The implication of the higher full density of the WC-12Co on the flow rate can be partially explained by considering the volume of actual material that is occupied by the 50 g sample mass of each powder, also reported in Table 3. WC-12Co is denser, so the individual powder particles (without air gaps) occupy a smaller volume per unit mass. As the two powders have similar PSDs, the number of particles per unit mass of the WC-12Co is smaller than that of the WC-17Co powder and so the WC-12Co powder will take a shorter time to flow through the funnel. Nevertheless, the difference in volume per 50 g of powder is only 4%, while the difference in flow rate is 53%. Considering these two factors simultaneously indicates that the WC-17Co is more cohesive (more resistant to flow) than the WC-12Co powder.

3.1.3 Spreadability Metrics

The spreadability metrics, PBD and percentage coverage, are reported in Tables 4 and 5, respectively. The PBD was determined at layer heights of 60, 80 and 100 μm, while the percentage coverage was determined at layer heights of 40, 60 and 80 μm.

Table 4 Powder bed spread density (PBD)

Powder	Layer height (μm)		
	60	80	100
	PBD (g/cm^3) {relative density}		
WC-12Co	4.28 {29.9%}	4.51 {31.5%}	4.64 {32.5%}
WC-17Co	2.96 {21.6%}	2.52 {18.4%}	2.79 {20.4%}

Table 5 Percentage coverage (by area)

	Layer height (μm)		
	40	60	80
	Percentage coverage		
WC-12Co (%)	52	81	88
WC-17Co (%)	56	82	56

Both powders have a PBD lower than their apparent density (Table 3); however, the difference between the relative PBD and relative apparent density is larger, 7.4–10.6%, for the WC-17Co as compared to the WC-12Co powder, 5.2–7.8%. The implication of this result is that the reduction in packing density due to the action of spreading the powder across the baseplate is higher for the WC-17Co powder than the WC-12Co. This correlates with the flow rate result that indicates that the WC-17Co is more cohesive, more resistant to flow and spreading, than the WC-12Co powder. The elongated particle shape may have increased interparticle friction that may have affected the flowability. Similar results are presented by Snow et al. [8]. The larger particle sizes in WC-17Co would hinder spreadability at each layer height. This was observed by Cordova et al. [9] who also noted that large powder particles or agglomerates block the path for further powder spread. This creates lines or tracks across the base plate where little too no powder is present; these areas of low powder result in less material being spread across the base plate, resulting in a lower PBD.

For WC-12Co, there is a directly proportional relationship between percentage coverage and layer height. In contrast, the percentage coverage for WC-17Co increases as layer height increases from 40 to 60 μm, but then decreases substantially at a layer height of 80 μm. At least 10% of the particles in both powders are larger than the first layer height, 40 μm. This explains the low percentage coverage results at a layer height of 40 μm, as many particles would not fit under the gap between the base plate and re-coater blade and would be pushed across the plate instead of being deposited as a powder layer. Additionally, the more cohesive WC-17Co powder may form clumps or agglomerates of the composite particles, that are not broken up by the shear force of the re-coater blade moving across the baseplate when a higher layer height is used, such as is the case for the 80 μm layer height result. The result is that the percentage coverage drops after an optimal layer height is reached.

3.2 *Post Printing Characterization*

3.2.1 Prints and Defects

The prints were successfully performed. There were, however, issues with the WC-12Co inserts as many of them curled and warped off the base plate during the L-PBF process. This may be due to insufficient laser power, yet the parameters chosen were found to be the most beneficial to produce near defect-free components [10]. The final samples are shown in Fig. 3.

The physical properties of the inserts were measured and are tabulated in Table 6 [10]. It should be noted that even though WC-17Co powder had lower relative apparent density and relative PBD than WC-12Co, the as-built density of the final L-PBF inserts built of the WC-17Co was higher than the WC-12Co inserts. Another way of viewing this result is by considering that even though the loose WC-12Co powder packs more densely and spreads more densely than the WC-17Co powder, with less air gaps between powder particles, the residual level of porosity in the as-built inserts was higher for the WC-12Co inserts. This is a counter-intuitive result but it can be explained by considering the influence of the higher Co content. This is discussed in more detail after considering the implications of the hardness testing results.

The hardness measured for WC-17Co inserts was higher than that of the WC-12Co inserts. Two factors influence the hardness: the Co content and the residual porosity in the as-built material. Typically, a cemented carbide with a higher WC content and thus lower Co content, relates to a higher hardness [7]. A higher level of residual porosity

Fig. 3 Printed CNMA samples from (Top) WC-12Co and (Bottom) WC-17Co [10]

Table 6 CNMA insert properties

	WC-12Co		WC-17Co	
Hatch Spacing (μm)	80	100	80	100
Density [g/cm^3] (%)	12.48 (86.1)	12.32 (85.0)	12.84 (93.0)	12.59 (91.2)
Hardness [HRA]	53.2 ± 10.7	54.2 ± 9.6	57.3 ± 8.1	60.3 ± 3.2
wt% Co	9.45	9.99	16.32	16.67

correlates to a lower hardness. While the difference in Co content between WC-12Co and WC-17Co is 5 wt%, the corresponding difference in fractional Co volume is (26.5–19.3) = 7.2 vol.%. Additionally, EDS analysis of the inserts indicated that the final Co content of the WC-12Co inserts was <10 wt%, indicating that around 2 wt% Co evaporated during printing. In contrast, <1 wt% of Co evaporated during printing of the WC-17Co, as shown in Table 6. While the WC-17Co samples had a higher vol.% of Co, which should correspond to a lower hardness, their as-built residual porosity (1—relative as-built density) was lower. The additional Co allows for a more stable melt since a higher fractional volume of the powder forms a liquid phase (Co) when exposed to the laser; this results in less evaporation, and better wetting and densification of the microstructure during printing. Thus the influence of the higher Co content in WC-17Co has a significant effect on both the density and residual porosity. The effect of the residual porosity on the hardness of the final as-built material was more significant than the Co content, with the denser WC-17Co displaying a higher hardness, despite it having a higher Co content.

3.2.2 Phase Identification

Figures 4 and 5 show comparisons of XRD results for each powder, as well as their respective printed inserts. XRD identifies the crystalline phases present in a material. The results reveal that W_2C and other η-phases, although absent in the precursor powders, are observed in all the as-built inserts. The formation of the W_2C phase results from the heat generated by the L-PBF process. A greater quantity of W_2C was measured in the inserts with a lower hatch spacing (80 μm). These correlates to increased laser track overlap at a lower hatch spacing, which results in a higher temperature at the intersection point due to the higher energy input per unit area. It is known that W_2C and η-phases are brittle in nature and affect the mechanical properties of the final samples [5].

4 Conclusion

Two commercially available WC–Co powders with Co contents of 12 and 17 wt%, respectively, were evaluated for this investigation. The powders were evaluated

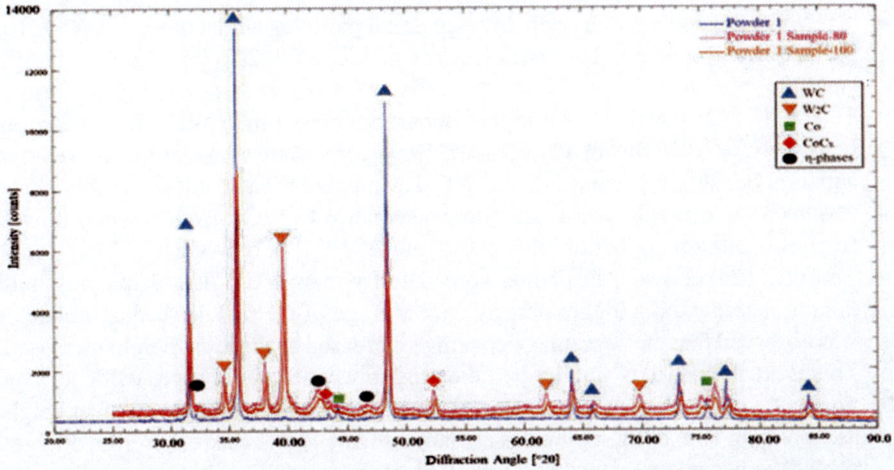

Fig. 4 X-Ray Diffraction phase analysis results for WC-12Co and printed insert samples from each hatch spacing

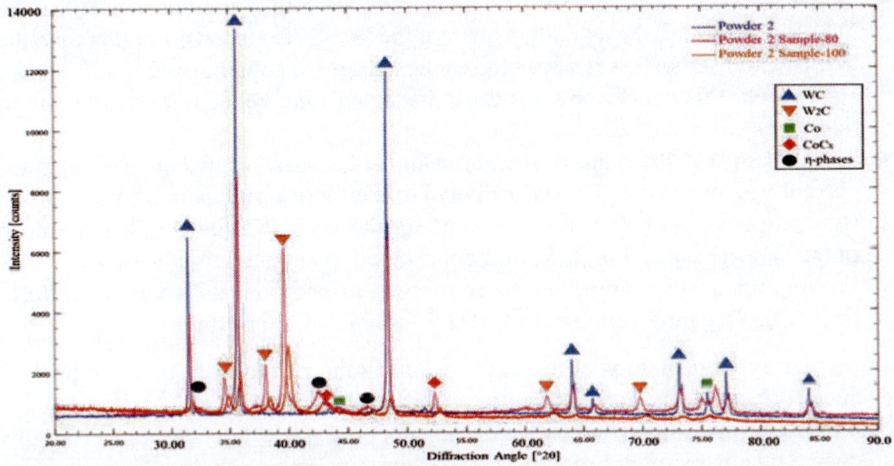

Fig. 5 X-Ray Diffraction phase analysis results for WC-17Co and printed insert samples from each hatch spacing

according to their powder characteristics: particle size and shape, apparent density, and flow rate, as well as the spreadability performance, as indicated by powder bed spread density and percentage coverage. The powders were then used to print CNMA inserts using L-PBF, which were then further characterised. The following conclusions are drawn from the results:

- The WC-12Co powder has primarily spherical particles whilst those of WC-17Co are slightly elongated. The particle sizes of the WC-12Co powder are slightly smaller.
- The WC-12Co powder has a higher apparent density than WC-17Co, which is partially due to its lower Co content. However, when considering the relative apparent density, it is clear that the WC-12Co powder packs more densely.
- The WC-12Co powder has a faster flow rate than WC-17Co. Both powders flowed freely, despite the more cohesive nature of the WC-17Co powder.
- The WC-12Co powder has better spreadability than WC-17Co, displaying both denser packing and a higher percentage coverage of the powder bed baseplate. It was observed that the percentage coverage increased as the layer height increased. This is attributed to its smaller particles and low cohesive forces as evident in the flow rate test. Both powders displayed poor spreadability at a 40 μm layer height as more than 10% of their powders are larger than 40 μm and are therefore dragged across the baseplate instead of being deposited.
- Inserts built from the WC-12Co powder warped off the base plate. Insufficient laser power, inadequate powder spreading in the case of WC-17Co, and the development of residual stress in the inserts may have contributed to this result.
- While the WC-17Co powder does not spread as well or pack as densely as the WC-12Co powder, the as-built density of the WC-17Co inserts is higher than the WC-12Co inserts. This is attributed to the higher Co content resulting in a more stable melt; this conceivably results in better printing behaviour contributing to denser parts.
- The as-built WC-17Co inserts measured higher hardness, even though a higher Co content was measured. This is attributed to the more significant influence of the residual porosity, that has a lower value for the WC-17Co inserts. The presence of the W_2C phase and η-phase also contributed to an increase in hardness.
- Printing parameters were seen to be optimal as components were successfully built with very little variation in Co wt% before and after printing.

Further experimentation is needed to quantify the relationship between powder characteristics, defects, and final build properties for L-PBF WC–Co. Various alloys could also be investigated to see similarities between powders investigated in this study and whether similar properties/defects are found post printing.

Acknowledgements The authors wish to acknowledge the financial support received from the Department of Science and Innovation and the National Research Foundation in South Africa (Grant Nos: 41292 and 129313). The opinions expressed and conclusions arrived at are those of the author/s and are not necessarily to be attributed to the Centre of Excellence in Strong Materials.

References

1. Schubert, W.D., Lasser, E., Bohlke, W.: Cemented Carbides—A Success Story, p. 2. International Tungsten Industry Association (2010)

2. Petersson, A.: Cemented Carbide Sintering: Constitutive Relations and Microstructural Evolution. KTH Materials Science and Engineering, Stockholm (2004)
3. Da Silva, A.G.P., Schubert, W.D., Lux, B.: The role of the binder phase in the WC-Co sintering. Mater. Res. **4**(2), 59–62 (2001)
4. Kumar, S.: Manufacturing of WC-Co moulds using SLS machine. J. Mat. Proc. Tech. **209**, 3840–3848 (2009)
5. Bricin, D., Kriz, A.: Processability of WC-Co powder mixtures using SLM additive technology. MM Sci. J., Czech Republic, pp. 2939–2945 (2019). https://doi.org/10.17973/MMSJ.2019_06_2018115
6. Parker, B.S., Blaine, D.C.: Blending of powders for in-situ alloying of Ti-6Al-4V laser powder bed fusion, MEng thesis, Stellenbosch University, South Africa, [unpublished] (2021)
7. Landolt-Börnstein, Properties of Hardmetals and Cermets, Group VIII. Springer Materials (2002)
8. Snow, Z., Martukanitz, R., Joshi, S.: On the development of powder spreadability metrics and feedstock requirements for powder bed fusion additive manufacturing. Addit. Manuf. **28**, 78–86 (2019)
9. Cordova, L., Bor, T., de Smith, M., Campos, M., Tinga, T.: Measuring the spreadability of pre-treated and moisturized powders for laser powder bed fusion. Addit. Manuf. **32**, 101082 (2020)
10. Hagedorn-Hansen, D.: Laser powder bed fusion of tungsten carbide cobalt cutting tools, Ph.D. thesis, Stellenbosch University, South Africa, [unpublished] (2021)

Preyin Govender obtained his Beng and M.Eng. degree in Mechanical Engineering from the Stellenbosch University. Currently, he is persuing his Ph.D. in Mechanical Engineering.

Devon Hagedorn-Hansen obtained his M.Eng. degree in Industrial Engineering from Stellenbosch University and has recently concluded his defence for his Ph.D. in Industrial Engineering. Devon is the founder and Managing Director of HH Industries.

Deborah Blaine is an Associate Professor of Mechanical and Mechatronics Engineering at Stellenbosch University. She is also the Ceramics Focus Area coordinator for the DSI-NRF CoE for Strong Materials.

Natasha Sacks is a Professor of Advanced Manufacturing at Stellenbosch University. Her research interests include materials science, and additive and subtractive manufacturing.

Manufacturing Systems

An Overview of the Manufacturing Systems: A Literature Survey

Nokulunga Zamahlubi Dlamini, Khumbulani Mpofu, Ilesanmi Daniyan, and Boitumelo Ramatsetse

Abstract To date manufacturing industries aims at achieving a growing variation of tailored, superior, high excellence and quality products in flexible sets. The transition from traditional machine systems to current reconfigurable machine (RM) requires consistency in achieving the requirements brought by the changes on the market demand, product life cycle and flexibility. This manuscript presents a literature review about the manufacturing system. The paper highlights the concepts of RM, dedicated machine (DM) and flexible machine (FM). It also highlights the application areas as well as the methodology and tools, by existing works. The search of the articles was conducted by inserting search strings in scientific search engines and academic databases to find relevant contributions on the analysed topic. The trend of the literature shows a gradual shift from dedicated machines to flexible machines and now reconfigurable machines. The findings of this work provide an insight into the requirements for the development of sustainable and reconfigurable manufacturing systems.

Keywords Dedicated machines · Flexible machines · Manufacturing systems · Reconfigurable machines

---

N. Z. Dlamini (✉) · K. Mpofu · I. Daniyan
Department of Industrial Engineering, Tshwane University of Technology, Pretoria 0001, South Africa
e-mail: 221820916@tut4life.ac.za; lungad213@gmail.com

K. Mpofu
e-mail: MpofuK@tut.ac.za

B. Ramatsetse
Educational Information and Engineering Technology, Wits School of Education, Johannesburg 2193, South Africa
e-mail: boitumelo.ramatsetse@wits.ac.za

K. von Leipzig et al. (eds.), *Smart, Sustainable Manufacturing in an Ever-Changing World*, Lecture Notes in Production Engineering,
https://doi.org/10.1007/978-3-031-15602-1_64

1 Introduction

The increasing market competitiveness, recent advances in manufacturing technologies as well as the dynamic nature of products and service requirements necessitate the reconfiguration of machine tools. In order to rapidly and efficiently respond to the various changes that characterise the manufacturing processes and new product variations, cost-effective reconfigurable systems are developed. In this digital era, manufacturing industries are faced with the challenges of improving the designs, manufacturing process and standard of their products and services. Dedicated manufacturing systems (DMSs), flexible manufacturing systems (FMSs) and reconfigurable machine systems (RMS) represent the evolution trends of manufacturing systems in the quest for improved profitability and competitiveness. Koren et al. [1] states that Reconfigurable Machine (RM) is a machine whose modules can be altered to provide either alternative functionality or incremental increase in its production rate to meet changing demand. The original position of a RM can continuously change or be modified to create new operations and functionality to meet production capacity. Koren et al. [1] further, states that RM has two objectives. First, to adapt the machine functionality to fit a new member of a family parts and the second is to increase the machine production rate by adding resources.

Aboufazeli [2] categorised the different manufacturing systems into three namely: Dedicated Manufacturing System (DMS), Flexible Manufacturing System (FMS) and Reconfigurable Manufacturing System (RMS). The DMS are usually designed for a single product with different simultaneous operations. It boasts of high volume production but with fixed capacity, few product variants and low flexibility.

Dedicated manufacturing systems (DMS) as defined by Katz [3] finds application when part production volumes are high and constant, with the parts unchanged. Dedicated machine (DM) used in a (DMS) is designed around a specific part that is mass produced. DM are created to perform single operations with high reliability and repeatability, and high productivity and therefore is relatively simple and less expensive.

Katz [3] states that flexible manufacturing systems (FMS) are used when the required quantities are relatively lower and many modifications in the part design are foreseen, or more than one type of product is produced on the same line simultaneously. Flexible machines (FM) that are used in (FMS) are created to perform most operations in flexible manner. The system can produce many different parts, alter, and undertake different positions or states in response to changing requirements with little consequence in performance, cost, time and effort.

The particular features of these machine systems will be elucidated in this paper.

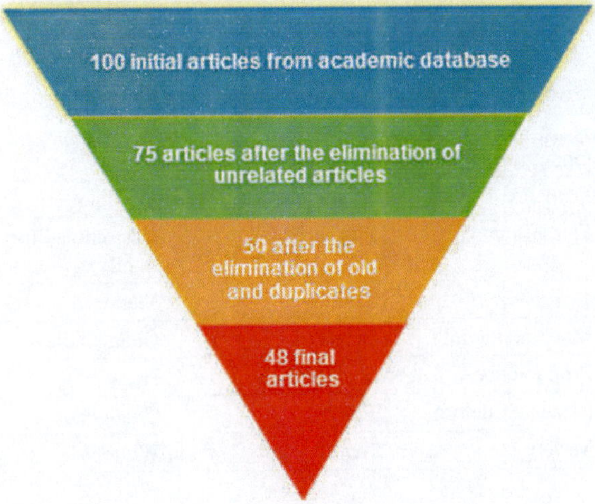

Fig. 1 The framework for the inclusion and exclusion articles

2 Methodology

The search of the articles is conducted by inserting search strings in scientific search engines, Google Scholar (scholar.google.com), Scopus (scopus.com) and Science direct (sciencedirect.com) mainly, to find relevant contributions on the analysed topic. The analysis includes the most relevant literature contributions published. In the phase of article screening, search strings include 'reconfigurable machine', 'reconfigurable manufacturing system', and 'reconfiguration' as basic terms. Initially a total of 100 articles were selected from various academic database. The final articles reviewed were 48 following the elimination of old, duplicate and unrelated articles. The framework for the inclusion and exclusion articles is presented in Fig. 1.

3 Literature Review

The quest for the localisation of certain components in order to promote industrialisation and the development of Gross Domestic Product (GDP) requires a reconfigurable system capable of handling such manufacturing processes [4]. Furthermore, many manufacturers seek to gain a competitive edge by delivering high quality products which are cost effective with reduction in the manufacturing lead time. Nowadays, manufacturing processes continue to witness evolution with numerous change coupled with the introduction of new designs and complex components. In order to solve these challenges there is a need to develop a system which integrate both the ending and cutting functions.

Table 1 Differences between dedicated, reconfigurable and flexible machines [5, 6]

Machine type	Dedicated	Reconfigurable	Flexible
Given part geometry	Fixed part geometry	Part geometry fits a part family	Any part geometry
Operation speed	Very fast	Fast	Slow
Demand	Stable	Variable	Variable
Flexibility	Not flexible	Customized flexibility for part family	Full flexibility
Cost per part	Low	Medium	Reasonable
Machine module	Fixed	Changeable	Fixed
Productivity	Very high	High	Low
System structure	Fixed	Changeable	Changeable
Variety	No	Wide	High

Machine types differ in their characteristics, RM are designed around the common characteristics of part families, and this feature differentiates them from dedicated or flexible machine.

Table 1 differentiates the characteristics of machines.

On the other hand, the FMS is designed for varieties of products, volume, mix and operations, hence, it boast of high flexibility and scalability but with a lower throughput compared to the DMS, hence, the system is expensive compared to the DMS.

The RMS combines six principles of modularity, scalability, integrability, convertibility, customisation or flexibility, and diagnosability to enable a rapid change in the capacity and functionality of a system in respond to the dynamic demand and product requirements [5].

The modularity principle deals with the compartmentalisation of the entire system into various parts or sub-systems. The combination of the separate parts or sub-systems changes the overall structure and functionality of the system based on the need. The scalability features of the RMS deals with the ability of the system to change the production capacity based on the nature of market forces (demand and supply) [5]. This feature is necessary for increasing or decreasing the productivity of the system in response to the forces of demand and supply.

The integrability feature addresses the integration of the different modules or parts accurately and rapidly [5]. It employs mechanical and control interfaces to ensure that the modules fits together properly. The strength of the modules determines the production rate of the system as well as the overall quality and dimensional accuracy of the final products. This feature also reduces the set up and ramp up times thus, increasing the productivity and cost effectiveness of the system.

The convertibility feature of the RMS is a characteristics of the RMS which deals with the ability of the system to change its functionality by changing the system's structure [5]. This is to allow multiple manufacturing operations to be performed on the system by changing certain components of the system. Effective convertibility will aid the development of new product variants without the addition of a new machine or subsystem to the RMS.

Customisation addresses the modification required by the system to adapt to the specialised needs of customers. A particular need of customers can be met by ensuring that the system is flexible enough to allow for the modification of certain components. This will enable the production of different varieties of the same product in a cost effective manner. The higher the ease of system's customisation to meet the variation of products based on customer's demand, the higher the flexibility of the system and vice versa.

The diagnosability feature defines the ability of the system to troubleshoot itself and identify the root cause of defects. This is necessary to reduce manufacturing errors, and to increase the availability, productivity and value added time of the RMS [5]. Gwangwava et al. [7] proposed a methodology for the design and reconfiguration of reconfigurable bending press machine (RBPM). The study presented a function-driven object methodology for the design and reconfiguration of the RBPM. First, the reconfigurability needs were identified followed by the function tree which identifies the primary bending function of the RBPM.

Sibanda et al. [8] highlighted some engineering design which features the life-cycle approach for reconfigurable machines. The study proposed an integrated design method which integrates certain components such design for assembly, design for manufacturing, eco-design, concurrent engineering and reconfigurable design as important elements of the reconfigurable machine.

The aim is to ensure optimal use of resource with effective environmental conservation, cost effectiveness during design and manufacturing, improved product quality with reduction in the manufacturing lead time. Sibanda et al. [9] proposed a class of the reconfigurable machine referred to as the reconfigurable guillotine shear and bending press machine (RGS & BPM). The machine integrates the dual function of bending and cutting of sheet metals. The first step was the development of the part families for the machine followed by the use of the part families for the development of the modules for the machine sizing and part manufacturing. The machine was designed to work on dedicated part families by addition or removal of the developed modules as it applies.

Olabanji and Mpofu [10] presented mathematical models for achieving reconfiguration in a smart assembly work cell. The function of the model is to set the limit of reconfiguration for a Reconfigurable Assembly Fixture (RAF) and too enhance the integration of the RAF operation within the assembly procedures of the systems. The results obtained indicated that the developed mathematical models can be employed for adjusting the dimensions of the RAF and computing its operating parameters.

Olabanji and Mpofu [11] also worked on the design sustainability of reconfigurable machines using four indicators of reconfigurable machines (reconfugurability, manufacturability, functionality and lifecycle analysis) and three indicators of sustainability (economic, social and environment). The study indicated that the performance of the reconfigurable machines can be achieved by improving the design features relating to the sub-indicators without any compromise to other sustainability indicators.

Olabanji et al. [12] presented a distributive approach to mathematical modelling and position control of multiple hydraulic actuators as a clamping system in a reconfigurable assembly fixture. The experimental validation of the electrohydraulic system was carried out to observe the synchronisation of the hydraulic actuators.

The results obtained from the performance evaluation of the system indicate that the controller significantly achieved the position control of all identical actuators in each subsystem. This is evidenced in the marginal settling time, effective transient response and elimination of the overshoot.

Daniyan et al. [13] developed a reconfigurable fixture for holding workpiece during low weight machining operations. The fixture finds application in manufacturing industries as a work holding device for low weight workpiece with a maximum weight of 15 kg during machining operations such as drilling, boring, filing, drilling, cutting or grinding. The fixture employs pneumatic controllers as a novelty which serves an improvement of fixtures. This is due to the fact that the fixture balances operator's ergonomic issues with cost effectiveness, precision, accuracy and smart location. Seloane et al. [14] presented a conceptual design of intelligent reconfigurable welding fixture in order to address some of the challenges relating to assembly operations in the railcar manufacturing industry. The system demonstrated capacity for precise welding assembly, less rate of scrap, improved flexibility and changeability to meet frequent market changes.

4 Results and Discussion

The findings from the literature survey conducted including the method employed for reconfigurable, dedicated and flexible manufacturing systems are presented in Tables 2 and 3 respectively.

The evolution and paradigm changes in reconfigurable manufacturing systems necessitates that the systems are flexible and changeable. Asghar et al. [43] conducted a study where the co-evolution of process planning and machine configurations in which optimal machine capabilities are generated through the application of multi-objective genetic algorithms. Furthermore, based on these capabilities, the system is tested for reconfiguration in case of production changeovers.

Architectural designs require innovative and efficient changes in reconfigurable machine tools as well as reconfigurable manufacturing systems. According to Martin [44] a methodology for designing production system is mainly based on a homogenous matrix for describing the machine structure and manufacturing features for describing parts in terms of geometrical data.

Global competition and the societal implications are rapidly increasing hence the need for variety of good quality products, thus requirements for manufacturing industries to adopt by adjusting its capacity and functionality quickly at low cost. Implementing Lean manufacturing techniques such as Jidoka and Poka-yoke to increase the diagnosability of the system is more useful in the evaluation of a reconfigurable manufacturing system according to Prasad et al.[45]. Functionality and capacity are

Table 2 Summary of findings on reconfigurable systems

Authors	Title	Method	Findings/conclusion
Xu et al. [15]	A method for design of modular reconfigurable machine tools	Method presented in this paper is a method for the design of modular reconfigurable machine tools (MRMTs). The proposed method consists of three steps: module identification, module determination, and layout synthesis	A MATLAB software package has been developed that can design an MRMT for a given part family. A layout synthesis process can then be carried out to produce generalised layouts based on a tree-graph algorithm and reduce them to create specific layouts by examining the selected modules. A layout evaluation process was conducted to find a final design based on a combined index considering three important issues for design of modular machine tool, i.e., reconfigurability, stiffness and error sensitivity
Schlette et al. [16]	3D simulation-based user interfaces for a highly reconfigurable industrial assembly cell	Man machine interaction software used for the development of new highly reconfigurable multi-robot assembly cell	It aimed at addressing needs of SME by (a)providing a highly configurable multi-robot assembly cell, (b)3D simulation-based user interfaces for a fast, easy and safe set up of assembly process along consistent workflow

(continued)

Table 2 (continued)

Authors	Title	Method	Findings/conclusion
Padayachee [17]	Development of a modular reconfigurable machine for reconfigurable manufacturing systems	The mechatronic design approach for the development of reconfigurable machine tools was used in this study. The design perspectives for reconfigurable machines and the principles adopted for Modular Reconfigurable Machines (MRMs) has been developed	Modular reconfigurable machines were proposed in this research as a possible solution to the machining requirements of Reconfigurable Manufacturing Systems (RMSs). The modular nature of these machines permits a change in machining functions and degrees of freedom on a platform, thus enabling adjustable functionality in (RMSs)
Makinde et al. [18]	Review of the status of reconfigurable manufacturing systems (RMS) application in South Africa mining machinery industries	This paper explores on various literature review in relation to reconfigurable manufacturing system concepts, architectural design characteristics, control capabilities and its role in mining machinery industries, and the potential future of this RMS in South Africa mining machinery industries	Reconfigurable manufacturing systems has been used in the mining machinery industries to manufacture Load Haul Dump (LHD) trucks, multiple drilling bits such as the tricone bits, screen mesh of different aperture sizes which has contributed immensely to the 15.3% GDP achieved by the manufacturing sector in South Africa. Global high mineral demand is a positive indicator for great future in this industry if RMS is fully harnessed, utilized and applied in this sector

(continued)

Table 2 (continued)

Authors	Title	Method	Findings/conclusion
Majija et al. [19]	Conceptual development of modular machine tools for reconfigurable manufacturing systems	Presents the conceptual development of modular machine tool, by selecting and compiling a list of modules available commercially off the shelf. In this study, the Morphology Method was used for the concept formulation of a modular machine tool for reconfigurable manufacturing systems. The machine tool is developed in such a way that it can be configured accumulatively in terms of degrees of freedom, to accommodate different confirmations of machine structures	The use of standard components on the development can reduce costs and development time of systems. Flexibility and responsiveness of the RMS was achieved through addition and deletion of modules, which are selected from modules available commercial off the shelf
Ramatsetse et al. [20]	Conceptual design framework for developing a reconfigurable vibrating screen for small and medium mining enterprises	The purpose of this paper was to present a framework that has a structured approach for developing a reconfigurable vibrating screen (RVS) that will be used by small and medium mining enterprises. Design Concepts and design requirements and constraints methodology was adopted to design the framework	The concepts enabling mechanisms for achieving a reconfigurable system were generated. The proposed framework can be utilised at the conceptual stage of the design. The proposed framework was found to be more appropriate to different types of design conditions; therefore, it is recommended that the framework be applied to similar types of reconfigurable designs

(continued)

Table 2 (continued)

Authors	Title	Method	Findings/conclusion
Yoon et al. [21]	Feasibility study on flexibly reconfigurable roll forming process for sheet metal and its implementation	In this study, a new sheet metal forming process called the flexibly reconfigurable roll forming (FRRF) process was proposed as an alternative to both the existing flexible forming process and the conventional die forming process for a multicurved sheet metal surface. The FRRF methodology was employed to manufacture a multicurved sheet metal surface was used	The methods and procedures of the FRRF process for producing three-dimensional curved sheet metal surfaces were described, and finite element analysis (FEA) and simulations of representative multicurved sheet metal surfaces were conducted to evaluate the feasibility of this process

(continued)

Table 2 (continued)

Authors	Title	Method	Findings/conclusion
Yoon et al. [22]	Study on flexibly reconfigurable roll forming process for multi-curved surface of sheet metal	To manufacture three-dimensional sheet metal parts for small-quantity batch production, several flexible forming technologies, including multi-point forming (MPF), have been made as alternatives to the conventional die forming one. However, the existing alternatives cause defects like dimples and wrinkles on the sheet metal during forming, due to their discrete punches. To alleviate these limitations, a new sheet metal forming process, named flexibly reconfigurable roll forming (FRRF) process was proposed. This innovative technology utilises adjustable punches mounted on two reconfigurable rollers	In this investigation, the method, apparatus, and procedure of the FRRF process for sheet metal forming were proposed. It was confirmed that the feasibility has been demonstrated through the numerical simulations of the proposed FRRF process. The FRRF process require uncomplicated components and a comparatively smaller installation space. Thus, this process can reduce the production cost related to the development and maintenance of forming dies. It also requires no additional machining as a result of forming errors such as dimples and wrinkles from the forming process Therefore, this progressive process can be used for sheet metal forming in small-quantity batch production, including applications in the shipbuilding, rapid transit train, automobile, and free-form construction industries

(continued)

Table 2 (continued)

Authors	Title	Method	Findings/conclusion
Carbonari et al. [23]	A new class of reconfigurable parallel kinematic machines	Proposed here was a conceptual design where the spherical joint which connects each leg to the end-effector is realised as a combination of revolute pairs; a locking system allows one to alternatively fix one of the revolute joints, giving the machine different 3-CPU kinematic configurations which correspond to different types of mobility	The article shows that, stemming from the 3-CPS under-actuated kinematics, many different machines can be derived, all characterized by the 3-CPU topology, but possessing different operative capabilities: the switching from one machine to the other can be done when the robot is in the home position by simply actuating cursor that locks one of the revolute joints in the spherical pair of each leg. A sample design of a machine of this class has been 182L

(continued)

Table 2 (continued)

Authors	Title	Method	Findings/conclusion
Mesa et al. [24]	A methodology to define a reconfigurable system architecture for a compact heat exchanger assembly machine	This paper presents a design methodology developed to obtain modular reconfigurable manufacturing systems (RMS). The method integrates the utilisation of modular architecture principles, selection algorithms (analytical hierarchical process), clustering algorithms (average linkage clustering algorithm), family product features and functional system analysis in the classical product design process	A method focused in system reconfiguration at machine level through adding, removing, adjusting, and widening modules from the family product reconfiguration variables was developed. This methodology applies systematic algorithms for selecting the best alternative to reconfigure the particular subfunctions involved in the system reconfiguration. It seeks to establish a platform and reconfigurable modules to develop robust systems that allow easy reconfiguration and adaptation to new market requirements, product launch

(continued)

Table 2 (continued)

Authors	Title	Method	Findings/conclusion
Gupta et al. [25]	A novel approach for part family formation for reconfiguration manufacturing system	A two-phase approach was developed where parts are first grouped into families and then families are sequenced, computing the required machines and modules configuration for each family. In the first phase, parts are grouped into families based on their common features. In the second phase, optimal selection and sequences of the resulted part families was achieved by using a Mixed Integer Linear Programming (MILP) model minimising reconfigurability and under-utilization costs to get the minimum cost solution	This work has presented a novel methodology for grouping parts into families on the basis of operation similarity which is a central issue in the design of reconfigurable manufacturing systems. Principle Component Analysis and Agglomerative Hierarchical K-means algorithm is applied to find the level of similarity in parts. The obtained part families are based on compactness of the family formation on the basis of operational similarity

(continued)

Table 2 (continued)

Authors	Title	Method	Findings/conclusion
Goyal et al. [26]	A novel methodology to measure the responsiveness of reconfigurable machine tool (RMTs) in reconfigurable manufacturing system	In the present research work a novel methodology was developed to assess the responsiveness of an reconfigurable machine tool (RMT) through developing the operational capability and machine reconfigurability metrics. The developed approach has been numerically illustrated and it has been observed that with a marginal increase in the cost, the RMTs offering higher responsiveness are available for the required operations	The result of numerical illustration depicts that with slightly increasing cost, the machines having higher responsiveness are available to cope with the ever changing capacity and functionality requirements. The performance measure developed would help the management to enhance the decision quality in the reconfigurable manufacturing system by providing an apparent trade-off between economy and responsiveness
Olabanji et al. [27]	Design, simulation and experimental investigation of a novel reconfigurable assembly fixture for press brakes	Design was done and experimental investigation was done using FESTO hydraulic test bench	Provision of important information on the accurate positioning and location as well as gripping of press brake frames. The study also provides important information on the design analysis of a reconfigurable assembly fixture for press brakes

(continued)

Table 2 (continued)

Authors	Title	Method	Findings/conclusion
Sibanda et al. [28]	Framework for the development of a new reconfigurable guillotine shear and bending press machine	Design for manufacturing approach	Provision of a structured approach for machine development. A framework that will enhance the development of a reconfigurable guillotine shear and bending press machine tool. This will enable cutting and bending of varying sheet metal products by a single machine with dual functionality
Wang et al. [29]	Formation of part family for reconfigurable manufacturing systems considering bypassing moves and idle machines	Employs the longest common subsequence (LCS) among different part processes identified. Then, based on this LCS, the shortest composite supersequence (SCS) was employed	The algorithm employed improves the accuracy of part family formation and eliminates the ambiguity of part family division. The proposed algorithm has higher practicability
Ashraf and Hasan [30]	Product family formation based on multiple product similarities for a reconfigurable manufacturing system	A novel approach for the product family formation in RMS based on multiple similarities of products linked with modular products comprising of several parts. These similarities include modularity, reusability and commonality	A descriptive mathematical model for the calculation of similarities is demonstrated with the help of a numerical example and the similarity values obtained are depicted in the form of a dendrogram. In the dendrogram, it was found that if the same weights are assigned to all the similarities, eight products will be merged into three families

vital in reconfigurable manufacturing systems. A novel approach based on the module interactions and machine capability can measure the machine reconfigurability and operational capability of a reconfigurable machine tool states Goyal et al. [46]. The is an increase in the usage of Internet of things (IoT) to solve global issues and use of smart systems. Kurniadi et al. [47] states that reconfiguration planning (RP)

Table 3 Summary of findings on flexible manufacturing systems

Authors	Title	Method	Findings/conclusion
Chang and Peterson [31]	Modeling and analysis of flexible manufacturing systems	Computer simulation approach for the investigation of the efficiency of FMS in terms of its lead time, utilisation of resource, queue and inventory levels, throughput analysis, and the number of workstations	The simulation approach aid the understanding of the FMS interms of its responsiveness and as well as changes of demand patterns
Chatterjee and Chakraborty [32]	FMS selection using preference ranking method	Use of multi-criteria decision support models	The multicritera decision models are suitable for making decisions relating to the selection of FMS structure
Taha and Rostam [33]	A hybrid fuzzy analtical hierarchy process (AHP)-PROMETHEE decision support system for machine tool selection in flexible	Use of the multicriteria decision support model (hybrid fuzzy AHP-PROMETHEE decision support system)	Suitability of the hybrid fuzzy AHP-PROMETHEE decision support system for making decisions relating to tool selection in FMS
Buyurgan et al. [34]	Tool allocation in flexible manufacturing system with tool alternatives	Heuristic approach based on life over size ratio	The method boast of effective allocation and use of space in the tool magazines, shorter flow times and reduction in the number if tool changeovers in the FMS
Joseph and Sridharan [35]	Evaluation of routing flexibility of a flexible manufacturing system using simulation modelling and analysis	Discrete even simulation model validated using the fuzzy logic	The proposed approach is suitable for determing the routing efficiency, versatility, variety and flexibility. However, further evaluation which considders the cost of processing and interruption in material flow should be considered

(continued)

Table 3 (continued)

Authors	Title	Method	Findings/conclusion
Singholi [36]	Impact of manufacturing flexibility and pallets on buffer delay in flexible manufacturing systems	Taguchi experimentation and simulation	The number of pallets and routing flexibility are two critical factors which affects the average delay at buffers. There is a feasile comiation of factors which can reduce buffer delay at a certain operating conditions
Jeschke et al. [37]	Industrial Internet of Things and cyber manufacturing systems	Modelling and simulation as well as artificial intelligence and analytics	The use of internet of things devices can make manufacturing systems flexible, smart, adaptable, and reconfigurable
Cheng and Chan [38]	Simulation optimization of part input sequence in a flexible manufacturing system	Simulation of different combinations of parts and input sequences	The simulation–optimization model to the planner for FMS is advantageous
Rybicka et al. [39]	Testing a flexible manufacturing system facility production capacity through discrete event simulation	Discrete event simulation	Selection of optimal parameters for FMS set up
Florescu et al. [40]	Operational parameters estimation for a FMS	Analytical modelling and simulation	The layout and configuration of FMS can influence its production capacity and adaptability
Mahmood et al. [41]	Performance analysis of a FMS	Modelling and simulation approach	The integration of system dynamics reliability functions will enable the identification of the troubled spots in the FMS with effective decision support for improving the performance of the FMS
Daniyan et al. [42]	Design and simulation of a flexible manufacturing system for manufacturing operations of railcar subassemblies	Discrete event modelling and simulation	An inverse relationship between the operating cycle time of the conveyor and the conveyor speed was established

problems can be solved, automated, and controlled with the integration of Internet of things IoT into RMS and the development of mathematical model to solve RP problems to save reconfiguration time, cost, and effort.

The reconfiguration planning problem on sustainability or environment-friendly functions within RMS is sparse. Kurniadi [48] proposes a multi-disciplinary green bill-of-material (MDG-BOM)—an improved Green-BOM concept—with an additional multi-disciplinary feature to minimize emissions and hazardous materials during product development, as well as manage product information across multiple disciplines during the reconfiguration process.

5 Conclusions

The aim of this work was to carry out an overview of the manufacturing systems. This was achieved via the literature survey comprising of 48 peer reviewed articles obtained from different academic databases. The findings show a systematic shift from dedicated machines to flexible machines and reconfigurable machines. The present manufacturing paradigm features the use of internet of things devices to make manufacturing systems smart, adaptable, and reconfigurable. This is in line with the fourth industrial revolution which promises high level of flexibility, responsiveness and productivity. It is envisaged that the findings of this work will provide an insight into the requirements for the development of sustainable and reconfigurable manufacturing systems.

Acknowledgements **Funding** The authors disclosed receipt of the following financial support for the research: Technology Innovation Agency (TIA) South Africa, Gibela Rail Transport Consortium (GRTC), National Research Foundation (NRF grant 123575) and the Tshwane University of Technology (TUT).

References

1. Koren, Y., Shpitalni, M.: Design of reconfigurable manufacturing systems. J. Manuf. Syst. **29**, 130–141 (2010)
2. Aboufazeli, N.: Reconfigurable machine tools design methodologies and measuring reconfigurability for design evaluation. Master's thesis, Royal Institute of Technology, Sweden (2011)
3. Katz, R.: Design principles of reconfigurable machines. Int. J. Adv. Manuf. Technol. **34**, 430–439 (2007)
4. The Report of the Department of Trade and Industry (DTI), South Africa. Select Committee on Trade and International Relations, pp. 1–27 (2018)
5. Koren, Y., Heisel, U., Jovane, F., Moriwaki, T., Pritschow, G., Uslow, G.: Reconfigurable manufacturing systems. CIRP-Ann. Manuf. Technol. **4**(2), 527–540 (1999)
6. Koren Y.: Reconfigurable manufacturing system. In: Chatti, S., Laperrière, L., Reinhart, G., Tolio, T. (eds.) CIRP Encyclopedia of Production Engineering. Springer, Berlin, Heidelberg (2019)

7. Gwangwava, N., Mpofu, K., Tlale, N., Yu, Y.: A methodology for design and reconfiguration of reconfigurable bending press machines (RBPMs). Int. J. Prod. Res. **52**(20), 6019–6032 (2014)
8. Sibanda, V., Mpofu, K., Trimble, J., Kanganga, M.: Engineering design featuring the lifecycle approach for reconfigurable machine tool. Procedia CIRP **84**, 948–953 (2019)
9. Sibanda, V., Mpofu, K., Trimble, J., Kanganga, M.: Development of part families for a reconfigurable machine. J. Eng., Des. Technol. **18**(5), 991–1014 (2020)
10. Olabanji, O.M., Mpofu, K.: A model for achieving reconfiguration in a smart assembly work-cell. Int. J. Adv. Manuf. Technol. **109**, 2777–2793 (2020)
11. Olabanji, O.M., Mpofu, K.: Design Sustainability of Reconfigurable Machines. IEEE Access **8**, 215956–215976 (2020)
12. Olabanji, O.M., Mpofu, K., Battaia, O.: A distributive approach for position control of clamps in a reconfigurable assembly fixture. Int. J. Autom. Control **14**(1), 34–51 (2020)
13. Daniyan, I.A., Adeodu, A.O., Oladapo, B.I., Daniyan, O.L., Ajetomobi, O.R.: Development of a reconfigurable fixture for low weight machining operations. Cogent Eng. **6**(1579455), 1–17 (2019)
14. Seloane, W.T., Mpofu, K., Ramatsetse, B.I., Modungwa, D.: Procedia CIRP **91**, 583–593 (2020)
15. Xu, Z., Xi, F., Liu, L., Chen, L.: A method for design of modular reconfigurable machine tools. Machines **5**, 5 (2017)
16. Schlette, C., Kaigom, E. G., Losch, D., Grinshpun, G., Emde, M., Waspe, R., Wantia, N., Roßmann, J.: 3D simulation-based user interfaces for a highly-reconfigurable industrial assembly cell. In: 2016 IEEE 21st International Conference on Emerging Technologies and Factory Automation (ETFA), 2016, 1–6. IEEE
17. Padayachee, J.: Development of a modular reconfigurable machine for reconfigurable manufacturing systems. Master's thesis, Department of Mechanical Engineering, University of Kwazulu-Nata, South Africa (2010)
18. Makinde, O., Mpofu, K., Popoola, A.: Review of the status of reconfigurable manufacturing systems (RMS) application in South Africa mining machinery industries. Procedia CIRP **17**, 136–141 (2014)
19. Majija, N., Mpofu, K., Modungwa, D.: Conceptual development of modular machine tools for reconfigurable manufacturing systems. In: Advances in Sustainable and Competitive Manufacturing Systems. Springer (2013)
20. Ramatsetse, B. I., Matsebe, O., Mpofu, K., Desai, D.: Conceptual design framework for developing a reconfigurable vibrating screen for small and medium mining enterprises. In: SAIIE25 Proceedings, Stellenbosch, South Africa, 2013
21. Yoon, J.-S., Kim, J., Kim, H.-H., Kang, B.-S.: Feasibility study on flexibly reconfigurable roll forming process for sheet metal and its implementation. Adv. Mech. Eng. **6**, 958925 (2014)
22. Yoon, J.-S., Son, S.-E., Song, W.-J., Kim, J., Kang, B.-S.: Study on flexibly-reconfigurable roll forming process for multi-curved surface of sheet metal. Int. J. Precis. Eng. Manuf. **2014**(15), 1069–1074 (2014)
23. Carbonari, L., Callegari, M., Palmieri, G., Palpacelli, M.-C.: A new class of reconfigurable parallel kinematic machines. Mech. Mach. Theory **79**, 173–183 (2014)
24. Mesa, J., Maury, H., Turizo, J., Bula, A.: A methodology to define a reconfigurable system architecture for a compact heat exchanger assembly machine. Int. J. Adv. Manuf. Technol. **70**, 2199–2210 (2014)
25. Gupta, A., Jain, P., Kumar, D.: A novel approach for part family formation for reconfiguration manufacturing system. Opsearch **51**(1) (2014)
26. Goyal, K.K., Jain, P.K., Jain, M.: A novel methodology to measure the responsiveness of RMTs in reconfigurable manufacturing system. J. Manuf. Syst. **32**, 724–730 (2013)
27. Olabanji, O., Mpofu, K., Battaïa, O.: Design, simulation and experimental investigation of a novel reconfigurable assembly fixture for press brakes. Int. J. Adv. Manuf. Technol. **82**(1/4), 663–679 (2016)
28. Sibanda, V., Mpofu, K., Trimble, J.: Framework for the development of a new reconfigurable guillotine shear and bending press machine. Procedia CIRP **63**, 366–371 (2017)

29. Wang, G.X., Huang, S.H., Shang, X.W., Yan, Y., Du, J.J.: Formation of part family for reconfigurable manufacturing systems considering bypassing moves and idle machines. J. Manuf. Syst. **41**, 120–129 (2016)
30. Ashraf, M., Hasan, F.: Product family formation based on multiple product similarities for a reconfigurable manufacturing system. Int. J. Model. Oper. Manag. **5**(3), 247–265 (2015)
31. Chang, D.A., Peterson, W.R.: Modeling and analysis of flexible manufacturing systems: a simulation study. In: 122nd ASEE Annual Conference and Exposition, 2015, Seattle, WA. pp. 1–19
32. Chatterjee, P., Chakraborty, S.: FMS selection using preference ranking method: a comparative study. Int. J. Ind. Eng. Comput. **5**, 315–338 (2014)
33. Taha, T., Rostam, S.: A hybrid fuzzy AHP-PROMETHEE decision support system for machine tool selection in flexible manufacturing cell. J. Intell. Manuf. **23**, 2137–2149 (2012)
34. Buyurgan, N., Saygin, C., Killic, S.E.: Tool allocation in flexible manufacturing system with tool alternatives. Robot. Comput. Integr. Manuf. **20**, 341–349 (2004)
35. Joseph, O.A., Sridharan, R.: Evaluation of routing flexibility of a flexible manufacturing system using simulation modelling and analysis. Int. J. Adv. Manuf. Technol. **56**, 273–289 (2011)
36. Singholi, A.: Impact of manufacturing flexibility and pallets on buffer delay in flexible manufacturing systems. Int. J. Eng. Manag. Econ. **5**(3–4), 308–330 (2015)
37. Jeschke, S., Brecher, C., Meisen, T., Özdemir, D., Eschert, T.: Industrial Internet of Things and cyber manufacturing systems. In: Jeschke, S., Brecher, C., Song, H., Rawat, D. (eds.) Industrial Internet of Things, pp. 3–19. Springer (2017)
38. Cheng, H.C., Chan, D.Y.K.: Simulation optimization of part input sequence in a flexible manufacturing system. In: Proceedings of the Winter Simulation Conference, Phoenix, AZ, 11–14 Dec 2011, pp. 2374–2382
39. Rybicka, J., Tiwari, A., Enticott, S.: Testing a flexible manufacturing system facility production capacity through discrete event simulation: automotive case study. Int. J. Mech., Aerosp., Ind., Mechatron. Manuf. Eng. **10**(4), 668–672 (2016)
40. Florescu, A., Baarabas, S., Sarbu, F.: Operational parameters estimation for a flexible manufacturing system: a case study. MATEC Web Conf. **112**, 1–6 (2017)
41. Mahmood, K., Karaulova, T., Otto, T., Shevtshenko, E.: "Performance analysis of a flexible manufacturing system performance. Procedia CIRP **2017**(63), 424–429 (2017)
42. Daniyan, I.A., Mpofu, K., Ramatsetse, B.I., Zeferino, E., Monzambe, G., Sekano, E.: Design and simulation of a flexible manufacturing system for manufacturing operations of railcar subassemblies. Procedia Manuf. **54**, 112–117 (2021)
43. Asghar, E., Zaman, U.K.U., Baqai, A.A., Homri, L.: Optimum machine capabilities for reconfigurable manufacturing systems. Int. J. Adv. Manuf. Technol. **95**(9–12), 4397–4417 (2018)
44. Martin, P.: Design of architecture and physical configuration for RMT/RMS: modelling of machines, workpieces, manufacturing. In: Reconfigurable Manufacturing Systems: From Design to Implementation, 2019, p. 57
45. Prasad, D., Jayswal, S.C.: Assessment of a reconfigurable manufacturing system. Benchmarking: An Int. J. **28**(5)1558–1575 (2019)
46. Goyal, K.K., Jain, P.K., Jain, M.: Optimal configuration selection for reconfigurable manufacturing system using NSGA II and TOPSIS. Int. J. Prod. Res. **50**(15), 4175–4191 (2012)
47. Kurniadi, K.A., Ryu, K.: Development of IoT-based reconfigurable manufacturing system to solve reconfiguration planning problem. Procedia Manuf. **11**, 965–972 (2017)
48. Kurniadi, K.A., Ryu, K.: Development of multi-disciplinary green-BOM to maintain sustainability in reconfigurable manufacturing systems. Sustainability **13**(17), 9533 (2021)

Nokulunga Zamahlubi Dlamini A master's degree candidate at Tshwane University of Technology. Her research focuses on virtual reality and reconfigurable machines. Nokulunga holds a B. Tech. in Industrial Engineering from the Vaal University of Technology and has tutoring experience.

Khumbulani Mpofu is the Research Chair in Manufacturing and Skills Development, Department of Industrial Engineering, Tshwane University of Technology, Pretoria, South Africa. His research interest include industrial engineering and advanced manufacturing.

Ilesanmi Daniyan is a Postdoctoral Research Fellow at the Department of Industrial Engineering, Tshwane University of Technology, Pretoria, South Africa. His research interest include production engineering and advanced manufacturing.

Boitumelo Ramatsetse is currently a Lecturer and Researcher at the University of Witwatersrand, South Africa. He has over 9 years of teaching experience in the higher education sector. He is currently an associate member of the Southern African Institute of Industrial Engineers (SAIIE), South African Insitution of Mechanical Engineering (SAIMechE), as well as the Industrial Engineering and Operations Management (IEOM). His research interest include reconfigurable systems, production systems, mineral processing systems, and machine design.

Research Endeavors Towards Predictive Modelling of a Grinding Process

Fungai Jani, Samiksha Naidoo, Quintin de Jongh, and Ramesh Kuppuswamy

Abstract The advent of additive manufacturing has created a paradigm shift in material processing and hence manufacturing processes such as: grinding have gained new momentum towards imparting the finished shape and size to the engineering components. This research unveils the development of an intelligent grinding system for processing a selective laser sintering based Ti6Al4V engineering component and paves the foundation for use on a wide variety of super-alloy materials. The developed system started with a theoretical platform which used a semi-empirical-analytical model for feature-extraction at its core, allowing for prediction of grinding behavior, based on workpiece material, wheel type and wheel-work interface parameters. The developed model was coded into MATLAB; with three notable outputs of the system being: residual stress, workpiece surface temperature and a unitless Coefficient B. An intelligent database using Microsoft Access was made to support the system by allowing for storage and cross-referencing of data. The data in the table include material, wheel, and coolant information, as well as system outputs and operator inputs. This database provided the basis for optimization of parameters. Graphical User Interfaces (GUIs) were created for the intelligent database (Access), system (MATLAB), and for data acquisition (DeweSoft). These three interfaces were connected, allowing for easy navigation and inputs from the operator. Signals were also eliminated through use of discrete wavelet transformations and moving average filters, enabling statistical analysis such as kurtosis coefficients and Werner's force model. A feature correlation engine was built using artificial intelligence: based on a neural network (NN), decision tree and a linear regression model. Through a rigorous

F. Jani · S. Naidoo · Q. de Jongh · R. Kuppuswamy (✉)
University of Cape Town, Cape Town, South Africa
e-mail: Ramesh.kuppuswamy@uct.ac.za

F. Jani
e-mail: jnxfun002@myuct.ac.za

R. Kuppuswamy
Research Endeavors Towards Development of an Intelligent Grinding System, 8th International Conference on Competitive Manufacturing, (COMA'22), 9-10 March 2022, Stellenbosch, South Africa

K. von Leipzig et al. (eds.), *Smart, Sustainable Manufacturing in an Ever-Changing World*, Lecture Notes in Production Engineering,
https://doi.org/10.1007/978-3-031-15602-1_65

designed experimentation process, results showing a great reduction in grinding burn and a presence of residual compressive stress.

Keywords Grinding · Intelligent grinding systems · Grinding performance prediction

1 Introduction

An Intelligent Grinding System (IGS) monitors a grinding process while controlling specific grinding parameters after having analyzed the data in real-time, to achieve the specified grinding requirements. The merits of an IGS are not limited to the achievement of very accurate dimensional characteristics, but also stretch towards the enhancing the grinding productivity along with the minimization of grinding induced surface damages. Residual stress is a combination of mechanical and thermal effects due to the interaction between the surface of the grinding wheel and the workpiece, giving rise to unintended tensile stresses within the part, that affect service life negatively. Various technological advances have enabled a better monitoring and controlling of the grinding process, but the limits of real-time monitoring and control remains as an unfinished agenda [1]. With leveraging the previous research, this research aims to close this gap ever so slightly by developing an IGS that utilizes the merits of artificial intelligence (AI) to monitor, analyze and control the entire grinding process in real-time.

2 Grinding Model Development

Developing an accurate and reliable model is the first and most important step in creating any intelligent grinding system that is capable of monitoring and enabling performance predictions. Many empirical and analytical formulae and equations have been based around the acquisition of grinding force signatures as this parameter either directly or indirectly enable to monitor the grinding induced stresses. Few of the past findings towards grinding induced stress model development are given below. They are,

1. Numerical Modelling

This method makes use of numerical time stepping procedures embedded in mathematical models to acquire the behaviour of the process over time. A past research [2] modelled the surface grinding process focussing on residual stresses and temperatures in the process. Another notable past finding [3] suggests use of FEM to model residual stress formation and draw conclusions regarding tensile residual stresses due to phase transformations. Numerical modelling has the advantages of requiring little data and producing results quickly, however exact solutions are not distinctly clear.

2. Empirical/Experimental Modelling

To use this technique, a large amount of initial data (normally experimental) is required. Such models produce accurate results, but often limited to a pre-set range of grinding conditions and materials. A popular model used in grinding is that of Werner's force model [4], which estimates specific grinding force. The formula is given by:

$$F_n' = K(c_1)^{\gamma}\left(\frac{Q_w}{v_s}\right)^{2\varepsilon-1}(a_e)^{1-\varepsilon}(\mathrm{d}_s)^{1-\varepsilon} \tag{1}$$

The Wener's grinding force model was further adjusted [5] to replace the proportionality factor, cutting edge density and grinding property exponent with a single empirical factor, thus reducing the required empirical factors, along with the maintainence of the desired accuarcies.

3. Analytical Modelling

This modelling technique provides a large amount of insight from both a theoretical and a physical sense. A complete micro-mechanism model (looking at the interaction between grinding grains and asperities of a workpiece) has not been developed yet but models have been created that incorporate features of analytical modelling with aspects from empirical and numerical modelling. Few past findings, [2, 6] and Chen et al. [7], have created models with accurate outputs regarding grinding induced residual stresses. The analytical model is helpful as it produces accurate results over a wide variety of conditions and acts like a building block for the intelligent grinding systems.

3 Experimental Setup and Data Collection

Shown in Fig. 1 is the experimental setup and instrumentation consists of Dynamometer (Kistler 9265), Microphone sensor (GRAS40H) and Piezotron Acoustic Emission Sensor signals (8152B221) for acquisition of process sugnatures during grinding. These sensors are connected to a data acquisition (DAQ) system for processing and analysing the data subsequently. Hardware Dewesift 43 and software Dewesoft X1 were used for data acquisition and analysis.

Ti6Al4V alloy was used for the experimental study. The physical dimensions of the samples were 20 mm × 20 mm × 20 mm cubes. The sample components were manufactured using Selective laser sintering (SLS) processes. SLS processed Ti6Al4V, if not controlled adequately, has a high cooling rate and thus exists in the martensitic β phase (metastable peculiar α + β microstructure). Heat treatments are a viable way to reduce residual stress and bring back a lamellar microstructure while hot isostatic pressing (HIP) can close pores. Grinding of Ti6Al4V was done using a silicon carbide grinding wheel having the specifications 39C100KU9. During the

Fig. 1 Experimental setup for grinding test of Ti6Al4v alloy

grinding operation, the work material was fed into the grinding wheel at a predetrmined table feed, depth of grinding and wheel speed conditions. Grinding tests were conducted until the grinding wheel reaches a condition which requires dressing interruption which is considered an unhealthy situation.

4 Process Parameters and Required Outputs

The developed IGS use the grinding forces/residual stress parameter to determine the grinding wheel failure boundaries. Temperature increase and workpiece burn both directly affect the build-up of residual stress, with the thermal field having some of the greatest influence on magnitude of apparent residual stress. The residual stress parameter is difficult to measure in real time and a past finding suggests using the grinding force parameter for computing the residual stresses [8]. The developed empirical model had the output of a new coefficient called 'coefficient B' and the details are given in Eq. 2.

$$B = P' t_c = \frac{P}{b v_w} \tag{2}$$

where P = total power, b = width of grinding, v_w = workspeed

The total grinding power is given as.

$$P = F_t v_s \tag{3}$$

where Ft is the tangential force and v_s is the grinding wheel speed.

The tangential force in grinding was computed as.

$$F_t = C_p \frac{v_w a_e b}{v_s} + \mu F_n \tag{4}$$

where F_n is the normal force, μ is the wheel-work interface friction, a_e is the depth of grind and C_p is the specific grinding resistance. The normal force F_n is computed as Inada [9];

$$F_n = \frac{\pi v_w a_e b}{2 v_s} \tan \Phi \tag{5}$$

where ϕ is the half of the included angle of the abrasive.

Combining Eqs. 3 and 4, Eq. 2 is rewritten as.

$$B = \frac{(C_p \frac{v_w a_e b}{v_s} + \mu F_n) v_s}{b v_w} \tag{6}$$

Also, the behaviour of grinding force (F_n) has an indirect indication of the abrasive wear of a wheel (V_{abr}) and given as.

$$V_{abr} = \frac{2}{\pi} \frac{\tan \theta}{II} F_n L \tag{7}$$

The grinding process parameter F_n was used to compute the residual stress and enable to determine the type of grinding wheel failue such as wheel loading or abarasive wear. Such proactive findings facilitate to take the necessary corrective actions. Shown in Fig. 2 is the layout of the devleoped IGS system.

The IGS system was created using MATLAB and its associated App Designer with object-oriented programming. The predictive model was configured with the support of a statistical analysis software K-Graph. Also, the theoretical analytical model, was coded into the backend of the system. User inputs such as: workpiece material, wheel type, coolant condition enable the system to use the correct pre-stored grinding constants information. Oncc all inputs have been acquired, the model is run to acquire predicted outputs and recommended grinding conditions/parameters for the system. At this point, the active cycle data is updated, and data is exported into a format applicable for the data acquisition software (DeweSoft). The data (force

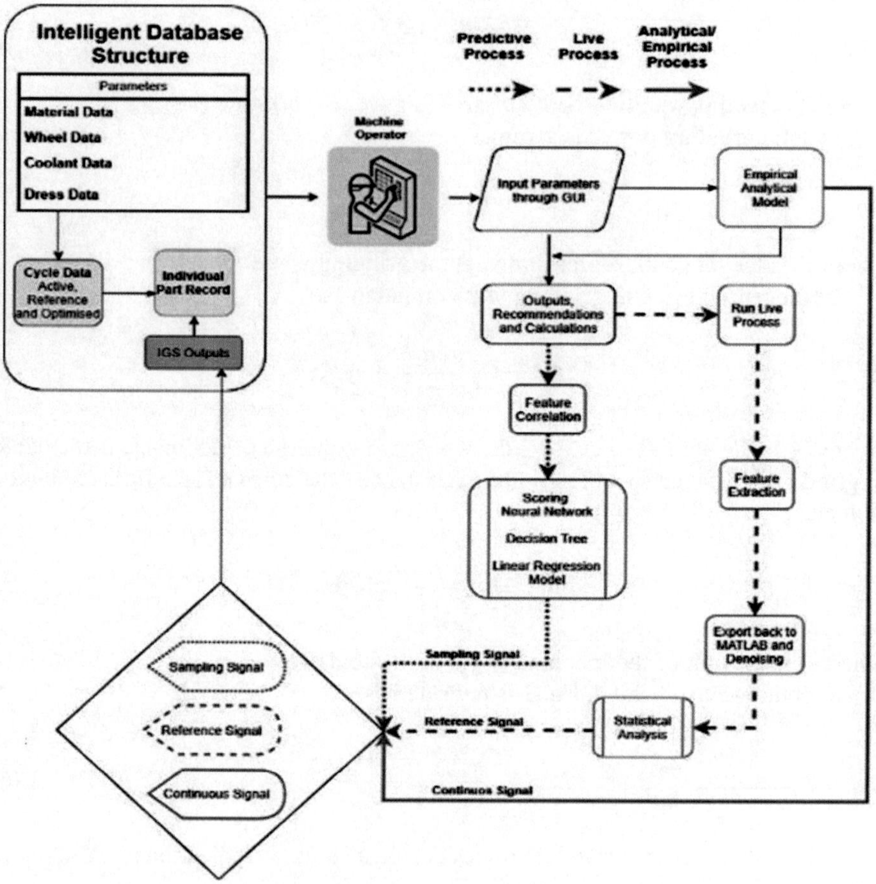

Fig. 2 Layout of the devleoped IGS system

signals) are exported back to MATLAB and denoised appropriately, including a minor statistical procedure that is performed in MATLAB to see if the data collected is representative and is of good quality through the skewness and kurtosis features. In the sampling signal feature correlation and statistical analysis is done followed by a prediction model that predicts target variables. Analysis pages from both MATLAB and Statistical software allows the operator to compare both signals and predictions in terms of grinding forces/residual stress and relates such information to grinding wheel failures such as abrasive wear, and wheel loading. The predictive model trains 3 models, a neural network, decision tree as well as a linear regression model. A model comparison is done within K-graph to determine which predictive model performed the best. Furthermore, we opted to use Support Vector Machine, Naïve Bayes and K-star algorithms as further classification methods which allows us a lot of flexibility in terms of prediction methods, because one prediction model can work better for a specific grinding configuration than the other, and we will be able to analyse that trend.

5 Results and Discussion

Shown in Figs. 3 and 4 are (i) the data acquistion screen image of the devloped IGS and (ii) the behaviour of grinding force, microphone and accelrometer signatures against the volume of material removal for the Ti6Al4V respectively.

In this manuscript, some of the signal processing techniques such as time-domain analysis, spectrum analysis and continuous wavelet transform (CWT) analysis are used to investigate the condition of the grinding wheel based on the force, microphone and accelerometer signatures. Apart from the experimental finding, machine learning techniques were applied to understand and predict the grinding process behaviour. Machine learning technique has three phases, which are; feature extraction, feature selection and feature classification. In feature extraction, statistical features, histogram features, discrete wavelet features and empirical mode decomposition features were extracted from the collected force, microphone and accelerometer signals. In the feature selection technique, a subset of the existing features was selected without any transformation. The decision tree algorithm is a feature selection technique and used in the present study. Machine learning process has two stages in the third phase. In the first stage, the classification algorithms are trained with the help of selected features from the training data of various fault signals. In the second stage, the trained algorithm is tested with the help of selected features from the test data.

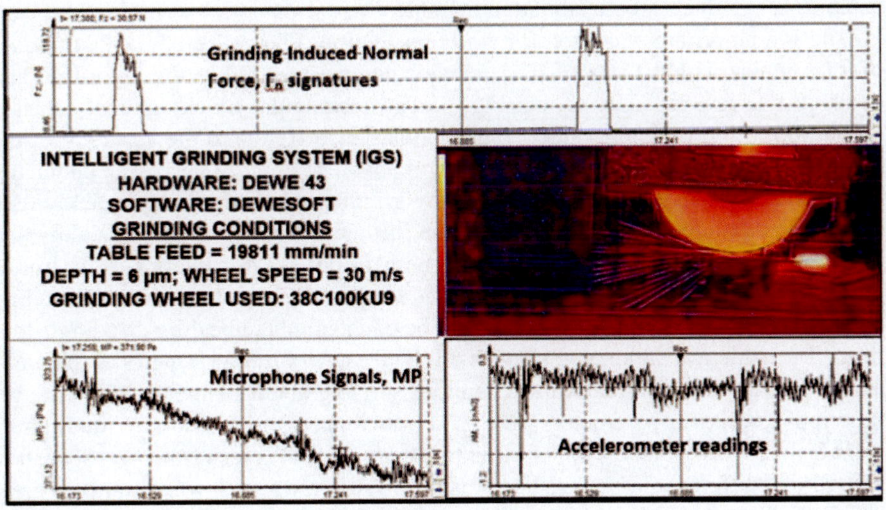

Fig. 3 Data acquistion in the devloped IGS system

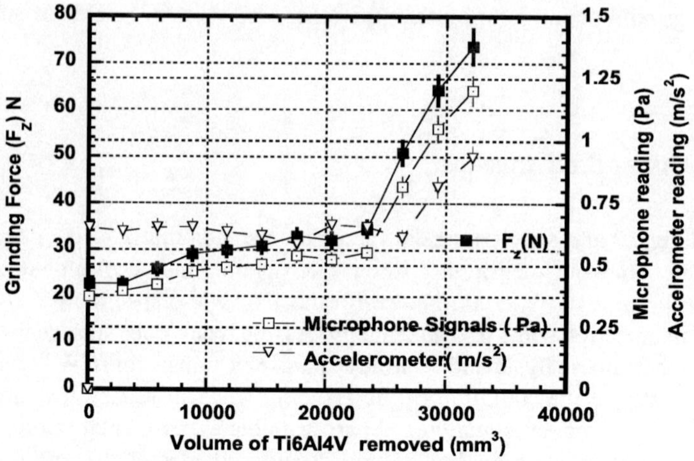

Fig. 4 Behaviour of grinding force, microphone and accelrometer signatures against the volume of material removal for Ti6Al4V

The classification phase identifies the faulty component. The support vector machine (SVM), artificial neural network (ANN), Naïve Bayes, decision tree and K-star algorithms are used as classifiers in this research work. From the force, microphone and acceleration data, descriptive statistical features like skewness, mode, standard error, maximum, minimum, range, sum, mean, standard deviation, median, sample variance and kurtosis are assessed to serve as features. These parameters are called as statistical features. These parameters were treated as an input to the J48 algorithm for feature selection. The computed statistical parameters parameters (10 signals samples per each class) are tabulated in Table 1.

The J48 algorithm was used for feature selection. The collected data comprises of 12 samples with 10 sets of data on each sample was fed to the algorithm and decision tree is formed as shown Fig. 5. The rectangular blocks indicate classes (condition of the grinding wheel). The standard error refers to the accuracy of the sample against the population. Standard error enable to broadly clarrify an unhelathy and healthy grinding wheel as an unhealthy grinding wheel detrioates quickly than an healthier grinding wheel. Kurtosis value further narrow down the unhealthy and healthy grinding perofrmances. It explains how the tail of a distrubution differ from a normal distribution. While the normal distrubution can be equated to a good grinding behavior but the extreme tail values could be equated to a unhealthy grinding situations. The statistical parameter, skewness is a measure of the asymmetry of the probability distribution of a real-valued random variable about its mean. All extracted descriptive statistical parameters were used as an input to the decision tree and significant features were selected out of the extracted features. As seen from Fig. 5, the tree structure of different classes has been formed in such a way that when standard error is greater than 0.161 value it is classified as healthy condition. To further narrow down

Table 1 Extracted statistical features from the grinding force signatures

Class	Sample	Mean	Standard deviation	Variance	Standard error	t factor	Friedman value	*P* value	Kurtosis	Skewness
Healthy Grinding wheel subject to degradation	1	22.7	1.17	1.37	0.371	61.2	76.98 Groups sample 1 to sample 9)		0.373	0.713
	2	22.68	1.1717	1.37	0.37	61.21		0.0232	0.432	0.598
	3	25.78	0.656	0.43	0.21	124.23		0.0216	0.401	0.337
	4	29.08	0.682	0.46	0.215	134.68		0.016	0.412	0.465
	5	29.98	0.52	0.464	0.26	184.51		0.0164	0.39	0.225
	6	30.82	0.59	0.457	0.225	245.71		0.0277	0.29	0.23
	7	32.94	0.59	0.358	0.19	174.04		0.015	0.27	0.25
	8	32.06	0.959	0.92	0.303	105.67		0.025	0.651	0.51
	9	34.5	1.085	1.17	0.343	100.53		0.021	0.547	0.628
Abrasive wear	10	50.835	0.509	0.26	0.161	315.25	87.53 Groups 10 (sample 1 to sample 10)	0.00016	0.247	0.934
Loaded g. wheel	11	64.23	0.41	0.22	0.129	249.2		0.00011	0.213	0.973
	12	73.56	0.36	0.19	0.114	329.8		0.000098	0.189	0.965

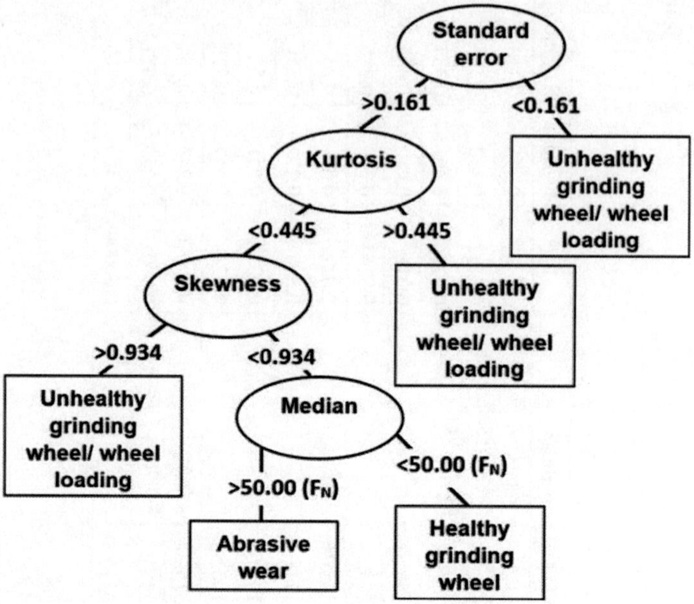

Fig. 5 Decision tree of statistical features of grinding induced force signatures

the accuracy kurtosis, skewness and median were applied, and the figures shows the condition of loaded grinding wheel and grinding wheel features with abrasive wear.

The decision tree has further enabled to perform classification of the grinding wheel behaviour using few selected features. The different classifiers such as ANN, SVM, Naïve Bayes, and K-star algorithms are used to distinguish the different grinding wheel conditions. The performance of each classifier are reported as follows. Totally 120 samples of data from 12 experiments were used to classify the grinding wheel behaviour. The detail findings on each classifier is given below.

1. Support vector machine (SVM)

SVM is an emerging classifier in fault diagnosis. Table 2 shows the confusion matrix of the SVM classifier, and the diagonal elements represent the correctly classified instances. Standard error of each data sample was used as a tool to group the data on each signal grouping. The results are given in Table 2.

From the confusion matrix, out of 120 instances, 109 instances were classified by a SVM algorithm with the overall classification accuracy of about 90.8% for the given force signatures. It should be noted that in all the experiments all the grinding parameters such as table feed, depth of grind, wheel speed were kept constant and hence feature extraction of grinding force/residual stress has largely depended upon the grinding wheel degrading conditions. The classification accuracy could be enhanced with use of a very large data set.

Table 2 SVM based confusion matrix for statistical features of force signatures

Healthy grinding wheel	Grinding wheel featured with abrasive wear	Grinding wheel featured with wheel loading	
84	6	0	Healthy grinding wheel
0	9	1	Grinding wheel featured with abrasive wear
0	4	16	Grinding wheel featured with loading

2. Artificial neural network (ANN)

ANN is one of the classifiers in the area of fault diagnosis/condition monitoring. The salient statistical features were fed to the ANN classifier using the Neural designer software and classification of the grinding wheel behaviour is represented as a confusion matrix as shown in Table 3. In the neural software a gradient descent method was adopted for finding the local minimum of differential function for the grinding induced normal forces (F_N) against time. The results are tabulated in the confusion matrix which suggests that out of 120 instances, 96 instances were correctly classified, and the classification accuracy was found to be 80% which is quite low for fault diagnosis.

3. Naïve Bayes algorithm

The selected features of force signatures were also classified using the Naive Bayes algorthm to classify the grinding wheel peroformance behavior. Bayesian classification uses conditional probability that describes the occurrence of event normal force at pre-defined boundries based on mean value and predefined failure conditions. To increase the accuracy of probability function only force signatures related to upgrinding were taken and the force signatures related to down grinidng were omitted. The results are given in the Table 4.

Table 3 ANN based confusion matrix for statistical features of force signatures

Healthy grinding wheel	Grinding wheel featured with abrasive wear	Grinding wheel featured with wheel loading	
78	8	2	Healthy grinding wheel
0	6	4	Grinding wheel featured with abrasive wear
0	8	12	Grinding wheel featured with wheel loading

Table 4 Naïve Bayes confusion matrix for statistical features of force signatures

Healthy grinding wheel	Grinding wheel featured with abrasive wear	Grinding wheel featured with wheel loading	
90	0	0	Healthy grinding wheel
0	7	3	Grinding wheel featured with abrasive wear
0	5	15	Grinding wheel featured with wheel loading

From the confusion matrix, 8 out of 120 instances were misclassified by the classifier and the overall classification accuracy was about 93.3% which is nearly OK for fault diagnosis. Thus, the combination of Naïve Bayes and statistical features particularly the F_n value enable to establish the fault diagnosis of the grinding wheel. The statistical features were extracted from the acquired force signatures and feature selection was performed using decision tree technique. The performances of the different classifiers were analysed using selected features and the summary of classification efficiencies of different classifiers is as shown in Table 5. From the table, it can be deduced that the classification accuracies are found to be low within a reasonable acceptable limits.

Similar decision tree and classifier methods were adopted for microphone and acclerometer signals and a similar trend were noted. Also an ANOVA analysis is done to validate whether microphone signals is an indirect indiction of force signatures/residual stress. The results are shown in Table 6.

Statics such as P & F factor were used to analyse the data. P & F factor enables to establish the relationship between the different categories: Fn, Mp and accelrometer readings. The smaller the "P" value, the smaller the probability of making mistakes by rejecting the null hypothesis, and consequently, the larger the corresponding coefficient. By evaluating the P-values of the parameters: Fn and microphone values, it is found that the P-values which is smaller for Fn and MP than Fn and acclerometer readings. On a similar note it was observed that there exists different level of satisfaction between Fn, microphone and acclerometer readings. The values of "F" in Table 6 A & B which is >1 suggests that the samples were drawn from a different population.

The t-value measures the size of the difference relative to the variation in your sample data. In other words, t factor suggests the calculated difference represented in

Table 5 Classification accuracies of different clasifiers

Classification accuracy	SVM (%)	ANN (%)	Naïve Bayes (%)
	90.8	80	93.3

Table 6 ANOVA Comparison tests on (A) grinding force and microphione signatures (B) grinding force and accelrometer signatures

Source	DF	SS	MS	F	P
A: Analysis of variance results					
Total	29	25,691,373	885,909.4		
A	1	1,241,754	1,241,754	1.422	0.243
Error	28	24,449,619	873,200.7		
B: Analysis of variance results					
Total	27	37,256,553	1,379,872.3		
A	1	387,383.62	387,383.62	0.2731	0.605
Error	26	36,869,169	1,418,045		

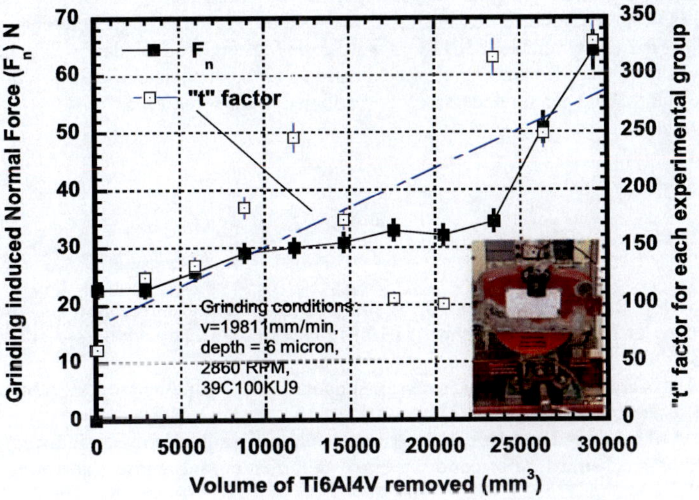

Fig. 6 Grinding force and "t" factor behavior against the volume of grinding

units of standard error. The greater the magnitude of t factor the greater the evidence against the null hypothesis. Shown in Fig. 6 is the behaviour of grinding force and t factor against the volume of material removed. The findings in Fig. 6 enable to predict the behaviour of a new grinding wheel with a small break-in test.

6 Conclusion

An intelligent grinding system was developed using semi-empirical, analytical models and experimental results at its core, to allow for prediction and analysis of grinding conditions based on workpiece material, wheel type, and grinding process parameters. Also, an intelligent database was developed to support the system and to allow for storage and referencing of data, as well as providing the basis for optimisation of the process. The fault diagnosis of the grinding wheel was constructed using grinding force signatures, microphone and accelerometer readings. An extensive analysis using both MATLAB and Statistical software has enabled to construct grinding feature extraction and thus enable to predict the performance of a new grinding wheel after a small break in test. Use of classifiers such as: Support Vector Machine, Artificial Neural Network, Naïve Baynes algorithm and decision tree has clearly defined the boundaries of healthy and unhealthy grinding wheel. The classification accuracies of SVM, ANN and Naïve Bayes were found to be 90.8%, 80% and 93.3% respectively. A "t" factor study along with the grinding force behaviour has further enable to predict the behaviour of a new grinding wheel.

Acknowledgements This project was supported by fund NRF GRANT: INCENTIVE FUNDING FOR RATED RESEARCHERS (IPRR)—South Africa through Reference: Reference: RA191118492767 and Grant No: 136118, as well as the MasterCard Foundation Scholars Program at the University of Cape Town. The views expressed and the conclusions drawn in this paper are those of the authors and cannot necessarily be attributed to the references.

References

1. Kuppuswamy, R., Jani, F., Naidoo, S.: A study on intelligent grinding systems with industrial perspective. Int J Adv Manuf Technol **115**, 3811–3827 (2021). https://doi.org/10.1007/s00170-021-07315-9
2. Hamdi, H., Zahouani, H.: Residual stresses computation in a grinding process. J. Mater. Process. Technol. **147**(3), 277–285 (2004)
3. Zhang, M.M.: Applied mechanics in grinding, part 7: residual stresses induced by the full coupling of mechanical deformation, thermal deformation and phase transformation. Int. J. Mach. Tools Manuf **39**, 1285–1298 (1999)
4. Werner, G.: Influence of work material on grinding forces. Ann. CIRP **27**, 243–248 (1978)
5. Salonitis, V.M.: Empirical estimation of grinding specific forces and energy based on a modified Werner grinding model. In: 14th CIRP Conference on Modelling of Machining Operations. Procedia CIRP **8**, 287–292 (2013)
6. Tonshoff, H.K.: Modelling and simulation of grinding processes. CIRP Ann. **41**(2), 677–688 (1992)
7. X. Chen, W.R.: Analysis of the transitional temperature for tensile residual stress in grinding. J. Mater. Process. Technol. **107**(1–3), 216–220 (2000)
8. Kruszynski, B.W.: Residual stress in grinding. J. Mater. Process. Technol. **109**, 254–257 (2001)
9. Inada, Y.: Studies in ultra high speed grinding, Ph.D. thesis. Sendai: Tohoku University, Japan (1996)

Fungai Jani obtained his B.Sc. Eng. degree in Mechanical Engineering from the University of Cape Town. He is currently studying towards achieving his M.Sc. Eng. in Mechanical Engineering, at the University of Cape Town.

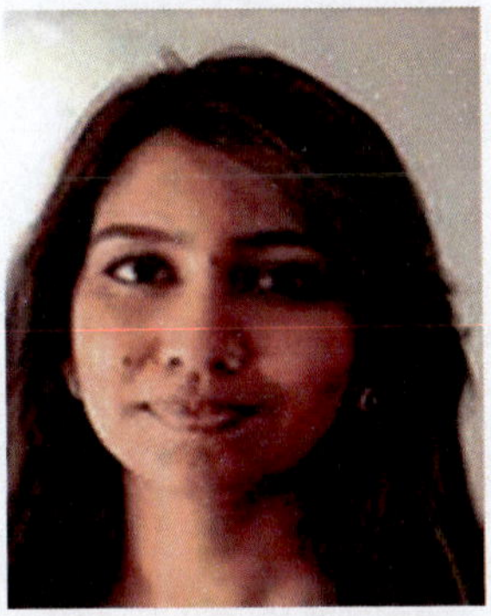

Samiksha Naidoo obtained her B.Sc. Eng. degree in Mechanical Engineering from the University of Cape Town. She is currently studying towards achieving his M.Sc. Eng. in Mechanical Engineering, at the University of Cape Town.

Quintin de Jongh obtained his B.Sc. Eng. degree in Mechanical Engineering from the University of Cape Town. He is currently studying towards achieving his M.Sc. Eng. in Mechanical Engineering, at the University of Cape Town.

Ramesh Kuppuswamy acquired his Ph.D. from the Nanyang Technological University, Singapore and is currently an Associate Professor in the mechanical engineering department at the University of Cape Town.

Role of Grinding Spark Image Recognition on Enhancing the Smart Grinding Technology for Ti6Al4V Alloy

S. Naidoo, F. Jani, and Ramesh Kuppuswamy

Abstract Research has shown that an intelligent grinding system (IGS) improves the reliability and maintenance of grinding operations, as it has the capabilities for real time monitoring and failure prediction. This research addresses the additional feature towards enhancing IGS, through Augmented Reality (AR) techniques. AR will provide a visual monitoring and failure prediction mechanism, as an overlay to the IGS. Thus, allowing the operator to view and compare the grinding process to a simulated model, in real time. This improves the on-line support offered to the operator and has the potential to greatly influence the down time, reliability, quality and inspection of grinding operations. The grinding of the Ti6Al4V alloy results in high temperatures at the grinding wheel-work interfaces and results in a thermal damage. Previous research conducted on IGS, found that operators generally use the acoustic emission (AE) signals and grinding force signatures as a check for characterizing the grinding process. Prediction of the grinding burn in real time offers better monitoring capabilities for mission critical components. In this study the grinding spark characteristics such as: color, spark area and quantity of spark lines was used as a measure for establishing the grinding process behavior. Spark images were used to assess the grinding processes in real time, with the use of a thermal imaging camera. To create the AR environment in the IGS system, a FLIR C5 thermal camera was used to capture the object's depth and dimensions. Conclusively, the research investigations on AR and IGS systems, indicate that it could accurately track the grinding process and extract useful information regarding the failure mechanisms of the grinding wheel and workpiece and the system works in near real time.

S. Naidoo · F. Jani · R. Kuppuswamy (✉)
Department of Mechanical Engineering, University of Cape Town, Cape Town, South Africa
e-mail: Ramesh.kuppuswamy@uct.ac.za

F. Jani
e-mail: jnxfun002@myuct.ac.za

R. Kuppuswamy
Towards configuration and development of an Augment Reality (AR) enhanced Intelligent Grinding System (IGS) for Ti6Al4V alloy, 8th International Conference on Competitive Manufacturing, (COMA'22), 9-10 March 2022, Stellenbosch, South Africa

K. von Leipzig et al. (eds.), *Smart, Sustainable Manufacturing in an Ever-Changing World*, Lecture Notes in Production Engineering,
https://doi.org/10.1007/978-3-031-15602-1_66

Keywords Intelligent grinding system · Image recognition · Augmented reality · Prediction modelling

1 Introduction

Manufacturing processes such as the grinding process, undergoes extensive digitization and the digitized grinding system is often termed as Intelligent Grinding System (IGS). Through technological advancements, IGS has greatly improved its ability to provide process monitoring and fault diagnosis abilities in real time, thus drastically changing the conventional maintenance and predictive reliability techniques. The main aspects contributing to an IGS system is the grinding science, grinding experiment-based signals and the application of algorithms to develop a model that can provide predictive process behaviours. Grinding process monitoring is fundamental to creating a robust IGS system. In order to skilfully capture signals and analyse the grinding processes, a variety of techniques are employed. These include, applying fuzzy logic rules, acoustic sensors, designing sensors into the grinding wheel and developing prediction models that can operate in real time [1].

Augmented Reality (AR) is defined as a system that brings together both the real and virtual environment, it is interactive and operates in real time, and both environments are placed in the 3D space [2]. It is important to distinguish between augmented reality and virtual reality. In the case of virtual reality, the real life environment is completely replaced and one sees and interacts with a new form of an environment that is computer generated [2]. This manuscript explore the aspects required to integrate AR in an IGS system, in creating an enhanced grinding process monitoring system.

2 AR in Manufacturing

Egger and Masood [3] noted that in creating a smart factory, AR is a key technology that allows for a human level of engagement within the manufacturing space. Despite the advancements in AR, there are still areas that require massive improvements in order to create a technological system that is powerful and effective, both for the operator and the manufacturing performance. The authors stress the importance for future AR technologies to be developed with the operator in mind. Even the most advanced technology can fail during implementation, if the user acceptance is poor, thus resulting in failure. The AR tracking system can use a marker based system. In order to improve the marker based system, which is effected by environmental conditions, algorithms can be improved to identify the markers correctly, under a range of conditions. It is noted that another form of improving the AR application in the manufacturing, is for the conditions that affect the operator's experience to be integrated into the software. In addition, a system that adapts to the operation, in real time, can greatly enhance the operators experience [3]. If one had to consider the grinding process, the operator uses the grinding noise as a tell of the state of the grinding wheel [4]. Although this practise is commonly used, it is not a valid indicator, as the result can vary from operator to operator. Lee et al. [4] developed an intelligent system using the data from machining sounds and applying deep learning algorithms to determine the grinding wheel wear. Therefore, by implementing an AR system that takes into account the data from visual and audio signals, processed from the IGS, a more visual and interactive prediction can be achieved in real time.

2.1 Developing a Successful AR Application in Manufacturing

Ong et al. [5] highlight that the AR systems should aim to improve manufacturing by decreasing the lead time, reducing the cost of production and improving the quality. AR has the potential to be greatly beneficial to the grinding of titanium alloy. The accuracy required on hard materials, results in grinding challenges which in turn impacts on the cost and time required to replace the grinding wheel and restart the machining process. There is an opportunity within this space to be explored, through AR, that will reduce time and cost. By decreasing the time to change the grinding wheel, the overall lead time is reduced. If a successful prediction model can be developed to predict the wear on the grinding wheel in real time, this can improve the quality of the workpiece, as it prevents defects from occurring due to a worn grinding wheel. The prediction model will be complimented with an AR system which will provide an interactive visual monitoring system for the operator to perform the grinding process with improved accuracy. Overall, this contributes to a sustainable form of manufacturing as the material wastage of defected components and worn grinding wheels, is reduced. The AR system was devised to align to the following hardware and software requirements, outlined by Ong et al. [5].

- The AR system is reliable, simple to use and cost effective.
- The AR system allows the user to easily interact with both the real world and virtual manufacturing space.
- The AR system is devised to be internet based to allow for collaboration, regardless of geographic location.

2.2 Grinding Spark Analysis

- **Predicting grinding burn**

Grinding burn arises from the high grinding wheel-work interface temperatures and results in a poor surface quality, with changes to the microstructure and an increase in grinding wheel wear which necessitates redressing [6]. Using the grinding burn on the workpiece as a criterion for assessing the wheel life, prior research was able to successfully predict the burn time using an Artificial Neural Network. A back propagation algorithm was used to predict the burn time using various grinding conditions as the inputs, namely the wheel speed, material diameter, in feed and grinding power. In the grinding experiments conducted on C60 steel, the pyrometer measured the spark temperature and found that burn occurred when 900 °C was exceeded. The research concluded that spark temperature was a significant determinant of the grinding zone temperature and that burn time dropped as the grinding power increased. It was also found that the 4-3-1 network, in which the hidden layer consists of 3 nodes, displayed a strong correlation between the predicted burn time and the actual burn time [6].

- **Predicting normal and faulty grinding conditions**

Another study focused on the monitoring the grinding process in real time, to determine whether the condition is normal or faulty [7]. The study characterised a faulty grinding condition by the chatter vibration and burn, which in turn impacts on the surface roughness of the workpiece. Therefore, an online system was established to predict the state. To develop the prediction model, data in the form of signals, thermal images and visual images were acquired. To capture the thermal images, a 16 × 16 IRISYS IRI 1002 thermal camera was used and measured the grinding zone temperature. A Logitech web camera captured the visual images, in order to determine the spark area.

Image processing was conducted using two methods, namely binary thresholding and image processing. The binary thresholding method was applied to the visual images. The method determines the spark area in the grinding zone, in terms of pixels. Each pixel that contains a grinding spark is represented with white, whilst the background (non-spark) pixels is represented by black. The image processing method was applied to thermal images and involved subtracting the image with no sparks (reference image) from an image with sparks that you want to determine the sparks from. The output is an image with only sparks. The results from the experiment found that the spark area was a significant indicator of the grinding zone temperature as an increase in the spark area, corresponded to an increase in the grinding zone temperature. The back propagation neural network was applied to the visual images of the spark area in order to predict a normal or faulty grinding conditions with a 95% accuracy. The radial basis neural network was applied to the thermal images to predict the grinding conditions with close to a 100% accuracy [7].

3 Experiment Setup and Grinding Conditions

Figure 1 is the experimental set up. The grinding of Titanium alloy (Ti6Al4V) uses resin bonded Diamond grinding wheel D400(MD20) N100B. The grinding conditions are; wheel speed of 30 m/s, a feed rate of 3000–30,000 mm/min and a depth of grinding 0–10 μm. The audio signals were captured using a microphone (GRAS40H) and the grinding behaviour with an accelerometer sensor (8152B221-Kistler). The dynamometer was used to acquire the grinding forces, accelerometer to measure the vibrational change and thermal camera to measure the temperature in the grinding zone. The experimental study was focused on thermal analysis of the grinding process in determining the state of the grinding wheel in terms of wear. The temperatures obtained from the grinding zone, was compared against the spark images in the grinding zone obtained from the thermal camera.

The thermal camera selected for the study, is the FLIR C5 model [8, 9]. The camera detects temperatures in the range −20 to 400 °C. The frequency is 8.7 Hz which translates to 8.7 frames per second. The camera is classified as long wave as it has a spectral range of 8–14 μm, which is the range of wavelengths that can be detected. The fixed focus feature of the camera suggests accurate measurements of temperature will be captured as the camera will always be in focus. The noise equivalent temperature difference (NETD), which is the smallest temperature difference between the target and the background, is less than 70 mK. The resolution of the camera is 160 × 120 i.e., 19,200 pixels. The data obtained from the measurement devices was conveyed to Dewesoft Data Acquisition software in real time, where it was further analysed. The

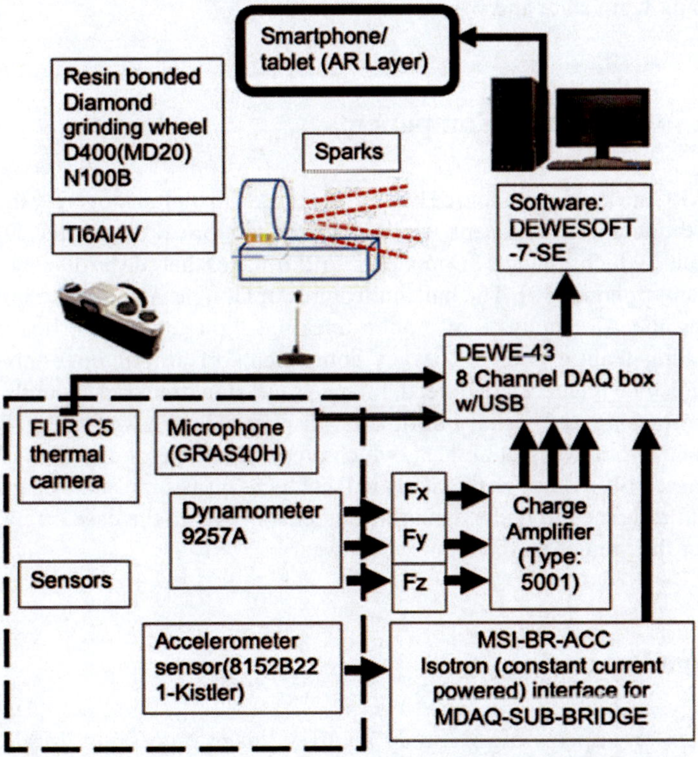

Fig. 1 Grinding experimental set up for feature extraction and correlations

thermal images displayed did undergo image processing on MATLAB and thermal studio software. The prediction model enables implementation in real time, and MATLAB was used along with the Deep Learning toolbox. The augmented reality layer created for the IGS, have used both Google ARCore and Unity softwares for development. The AR monitoring system has enabled to assess the grinding wheel state in real time and provide visual prompts/notifications to the operator.

4 Proposed Framework for Prediction Model of Wheel Wear

The framework developed by Lee et al. [4] in monitoring the grinding wheel wear considered the audio signals from the machining process. The extracted features from the audio signals in each experiment, were applied to a convolution neural network (CNN), in order to classify the extracted features into categories. This was applied in determining the grinding wheel condition. The classification used the back propagation algorithm. The framework (see Fig. 2) was adapted for thermal imaging processing, which would lead to predicting wheel wear in terms of normal, approaching worn state and worn.

5 AR Software and Components

Google ARCore is an open-source library that is used to digitally overlay the information onto the actual environment, whilst Unity the software provides the 3D rendering functionality, which converts 3D models into 2D images that can be displayed onto the tablet or smartphone [10]. The built-in features of Google ARCore are instrumental in creating the AR application: motion tracking, light estimation, image tracking and detecting feature points. The key components of an AR environment is the capturing device (camera), data feed, image processing, tracking capabilities, information processing and visual information display [11]. Shown in Fig. 3, was the development from Wang et al. [11], which presented a diagram of the AR environment for assembly. Although this study will not focus on assembly, the diagram served as a useful reference to understand the elements of AR in the case of a monitoring system for the prediction model of wheel wear.

6 Image Processing

One of the key aspects of this study involves image processing. MATLAB used to conduct the image processing. The process involves the following key stages:

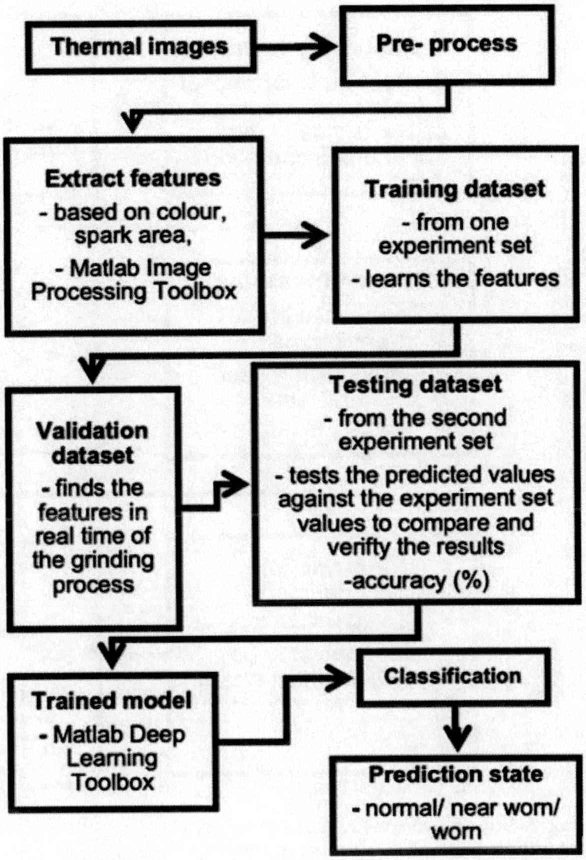

Fig. 2 Proposed framework of prediction model based on thermal imaging analysis

import, pre-process, segment, post-process, and classification [12]. The importing stage involves bringing the image into MATLAB, working with the grey scale and colour images to extract colour planes and intensities and adjusting the contrast. Images in MATLAB are stored as arrays in which every element of the array represents the pixel of the image. The number of array rows is equivalent to the images height, and the number of columns is equivalent to the image's width in pixels. In terms of grey scale images, each element in the array is stored by a number between 0 and 255, whereby the low values are low intensity (dark areas) and the higher values are high intensity (bright areas). For colour images, the elements in the array are similarly stored as numbers between 0 and 255, however, the colour image is stored as a combination of the intensity of 3 colours: red, green and blue (RGB). The colour plane for each of the RGB values is transformed to a grey image, where the higher values represent the brightness or intensity of the specific colour.

The segmenting stage involves separating the image into the characteristics shown on the image, in order to create binary images from the pixel values. The binary image comprised of 1 for the parts of the image you want to keep, and 0 for the parts to be

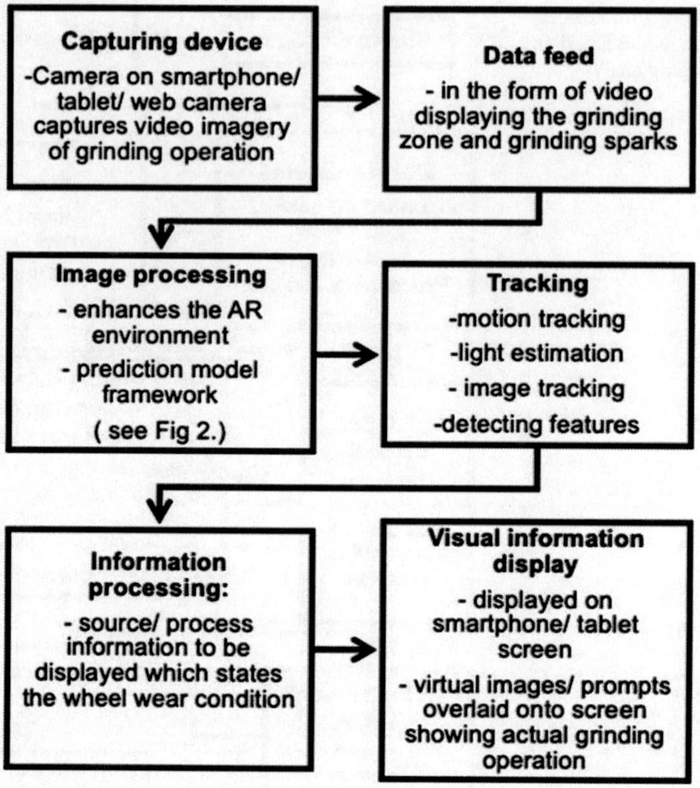

Fig. 3 AR components used for the study [11]

excluded. Using the histogram feature to display the intensities of the grey image, is useful in identifying which values the pixels are concentrated at, and determining whether the image contains more darkness or brightness. Depending upon the image intensities a threshold was introduced and hence the display information contain the necessary features. When dealing with many images, a MATLAB function to identify the best threshold for the image is used.

Pre- and post-processing techniques were performed to improve the segmentation, as well as filtering the noise and removing the background. An averaging filter is used to remove the noise from the image. Once the filter is applied to smooth out the image, a binary image was produced, which is made of two pixels to represent the bright and dark. "Opening" an image was applied to emphasize the dark aspects of the image, and "closing" an image was applied to emphasize the bright regions. This is used to isolate portions of the image.

Classification is developing a metric for the image and then applying this metric across multiple images. In order to process hundreds of images, a datastore is created to access these numerous files. By using a for loop, the program developed classifies the image and applies across all image files in the folder [12].

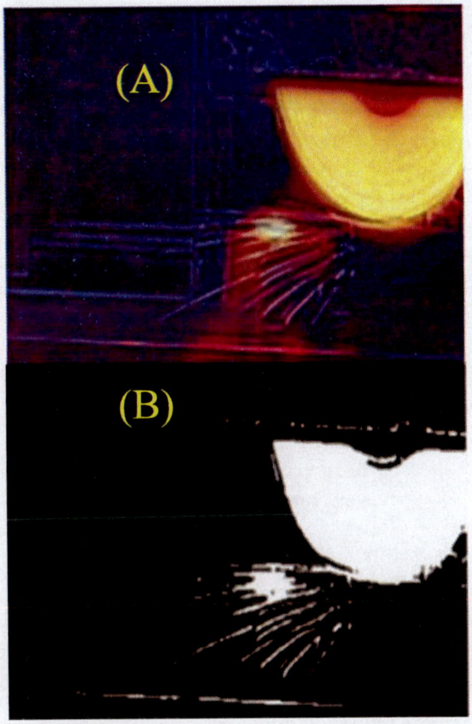

Fig. 4 Thermal image **a** converted to black and **b** white image [11]

The following code sequences in the MATLAB program was developed in order to extract digital information from the image. In the program, a binary image is generated from a thermal image, the spark lines in the area of interest were isolated using the crop feature, lines were identified using the Hough Transform and thereafter plotted and counted [13, 14].

The program in MATLAB converted the initial thermal image to a grey image, thereafter it converted the grey image to black and white (binary image) (Fig. 4).

The area of interest is then cropped from the binary image by stating the size and position of the cropping rectangle.

Thereafter, the Hough Transform was applied. The purpose of the Hough Transform is to identify the lines in the black and white image. It does this by representing a line in terms of ρ and θ.

$$\rho = x \cos\theta + y \sin\theta \quad (1)$$

where

ρ—distance from origin to line, at 90° to the vector
θ—angle from the x-axis to the vector, clockwise from positive x-axis.

```
%% Hough transform

% displays peaks on Hough matrix
[H,T,R] = hough(I_BW_crop);
imshow(H,[],'XData',T,'YData',R,'InitialMagnification','fit');
xlabel('\theta'),ylabel('\rho');
axis on, axis normal, hold on;

% detecting lines in black and white image
P = houghpeaks(H,8,'threshold',ceil(0.5*max(H(:))));
x = T(P(:,2));
y = R(P(:,1));
lines = houghlines(I_BW_crop,T,R,P,'Fillgap',5,'Minlength',5);
figure
imshow(I_BW_crop)
hold on
max_len = 0;
```

Fig. 5 MATLAB code for Hough Transform [11]

Fig. 6 Identified spark lines (represented as yellow lines) with endpoints (green circles) from Hough Transform [11]

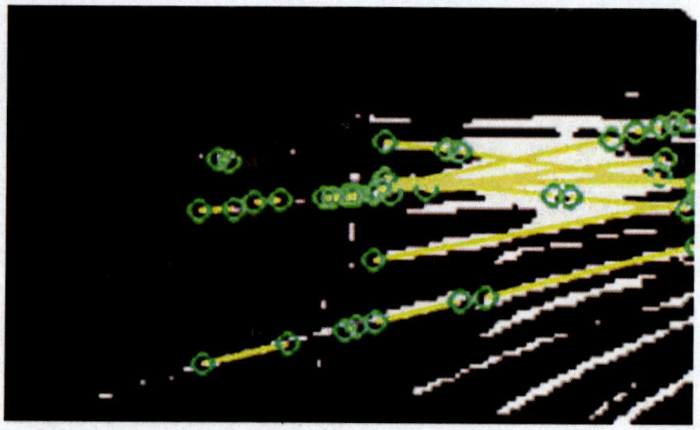

The 'houghpeaks' function in MATLAB returns the row and column co-ordinates of the Hough Transform, which are used to search for lines in the binary image. Thereafter, the identified lines are plotted on the binary mage (Figs. 5 and 6).

The program then determined the quantity of sparks within the cropped spark area. Based on the code, the number of spark lines identified from this image is 22.

7 Conclusions

Previous research found that grinding sparks served as the basis for determining the carbon content in steel and predictions for grinding burn, faulty and normal grinding operations, and the material removal rate. Grinding sparks have been analysed with features such as quantity, spark area, colour, and texture. This study has revealed a method to analyse the spark features: colour, spark area, line features and quantity of lines towards predicting the state of the wheel wear which is used as an early warning system of grinding wheel failure. The prediction model forms part of the image processing component in the AR environment. The influence of AR in IGS allows the prediction model to be translated to a dynamic visual display in the form of text and images. This contributes to sharing knowledge expertise to the operator in real time and improves wheel wear monitoring capabilities, which will in turn reduce down-time and manufacturing costs.

Acknowledgements This project was supported by fund NRF GRANT: INCENTIVE FUNDING FOR RATED RESEARCHERS (IPRR)—South Africa through Reference: Reference: RA191118492767 and Grant No: 136118. The views expressed and the conclusions drawn in this paper are those of the authors and cannot necessarily be attributed to the references.

References

1. Kuppuswamy, R., Jani, F., Naidoo, S., de Jongh, Q.: A study on intelligent grinding systems with industrial perspective. Int. J. Adv. Manuf. Technol. **115**(11), 3811–3827 (2021)
2. Azuma, R.T.: A survey of augmented reality. Presence: Teleoperators Virtual Environ. **6**(4), 355–385 (1997)
3. Egger, J., Masood, T.: Augmented reality in support of intelligent manufacturing—a systematic literature review. Comput. Ind. Eng. **140**, 106195 (2020)
4. Lee, C.-H., Jwo, J.-S., Hsieh, H.-Y., Lin, C.-S.: An intelligent system for grinding wheel condition monitoring based on machining sound and deep learning. IEEE Access **8**, 58279–58289 (2020)
5. Ong, S.K., Yuan, M.L., Nee, A.Y.C.: Augmented reality applications in manufacturing: a survey. Int. J. Prod. Res. **46**(10), 2707–2742 (2008)
6. Deiva Nathan, R., Vijayaraghavan, L., Krishnamurthy, R.: In-process monitoring of grinding burn in the cylindrical grinding of steel. J. Mater. Process. Technol. **91**(1–3), 37–42 (1999)
7. Junejo, F., Amin, I., Hassan, M., Ahmed, A., Hameed, S.: The application of artificial intelligence in grinding operation using sensor fusion. Int. J. GEOMATE **12**(30), 11–18 (2017)
8. Thermal Camera Specs You Should Know Before Buying: Teledyne FLIR (2019). https://www.flir.com/discover/professional-tools/thermal-camera-specs-you-should-know-before-buying/. Accessed 15-8-2021
9. FLIR C5 Data Sheet: (2021). www.flir.com/C5. Accessed 15-8-2021
10. Oufqir, Z., el Abderrahmani, A., Satori, K.: ARKit and ARCore in serve to augmented reality. In: 2020 International Conference on Intelligent Systems and Computer Vision (ISCV) (2020)
11. Wang, X., Ong, S.-K., Nee, A.Y.-C.: A comprehensive survey of augmented reality assembly research. Adv. Manuf. **4**, 1–22 (2016)
12. Image Processing Onramp: (2021). https://matlabacademy.mathworks.com/R2021a/portal.html?course=imageprocessing#chapter=2&lesson=2&section=1. Accessed 6-9-2021
13. Hough Transform—MATLAB Hough: (2021). https://www.mathworks.com/help/images/ref/hough.html. Accessed 8-10-2021
14. Extract line segments based on Hough transform—MATLAB Houghlines: (2021) https://www.mathworks.com/help/images/ref/houghlines.html. Accessed 8-10-2021

Samiksha Naidoo obtained her B.Sc. Eng. degree in Mechanical Engineering from the University of Cape Town. She is currently studying towards achieving her M.Sc. Eng. in Mechanical Engineering, at the University of Cape Town.

Fungai Jani obtained his B.Sc. Eng. degree in Mechanical Engineering from the University of Cape Town. He is currently studying towards achieving his M.Sc. Eng. in Mechanical Engineering, at the University of Cape Town.

Ramesh Kuppuswamy acquired his Ph.D. from the Nanyang Technological University, Singapore and is currently an Associate Professor in the mechanical engineering department at the University of Cape Town.

Hybrid Production Principles: A Framework for the Integration in Aircraft Manufacturing

Alexander Wenzel, Torben Lucht, and Peter Nyhuis

Abstract Steady increasing individual customer requirements characterize production processes in aircraft manufacturing. In addition, these production processes are usually marked by rigid cycle times and a lack of flexibility potential. Due to these constraints, the increasing number of variants in flow production leads to considerable efficiency deficits in the form of high utilization losses. Conventional approaches to handle variant diversity within flow lines (e.g. vertical separation of variant-neutral and variant-specific product components) reach their limits with complex products because the low proportion of standardized, variant-neutral components alleviates the actual advantages of line production. Hybrid production principles (HPP), e.g. a flow-line-based production with bypasses to modular workshops, in combination with a production planning and control (PPC) configuration tailored to these principles, offer high logistics and economic potential for multi-variant aircraft manufacturing. A logistics oriented description of the system behaviour of hybrid production principles does not exist yet. However, this is necessary to investigate possible effects on logistics performance depending on the principles applied. Therefore, this paper identifies and describes factors influencing the selection and design of hybrid production principles concerning production organization and the associated intralogistics and PPC. Based on this, a holistic framework for integrating hybrid production principles in future-oriented aircraft manufacturing is presented. In order to handle the resulting complexity, the framework integrates individual components of strategic and operational planning and thus forms the basis for an integrated digital planning approach. The presented framework in this paper has a conceptual character and needs to be verified due to a transfer to a proper application environment. Therefore individual components of the framework (e.g., logistic models) must be developed and transferred in a consistent structure to enable integrated planning and control of HPP.

A. Wenzel (✉) · T. Lucht · P. Nyhuis
Institute of Production Systems and Logistics, Leibniz University Hannover, An der Universität 2, 30823 Garbsen, Germany
e-mail: wenzel@ifa.uni-hannover.de

K. von Leipzig et al. (eds.), *Smart, Sustainable Manufacturing in an Ever-Changing World*, Lecture Notes in Production Engineering,
https://doi.org/10.1007/978-3-031-15602-1_67

Keywords Hybrid production principles (HPP) · Production planning and control (PPC) · Aircraft manufacturing

1 Introduction

Irrespective of crisis-ridden business years (e.g., due to terrorist attacks or pandemics), the civil aircraft market is characterized by strong growth due to steadily rising traffic volumes [1]. Due to the continuously increasing sales volume in aircraft production, manufacturers increasingly oriented themselves towards measures to decrease throughput times and increase their output [2]. At the same time, the growing number of global competitors, especially in the smaller short-haul aircraft segment, is leading to increasing pressure to innovate and reduce costs. In addition, the globalized demand for customized products at low prices continues to grow [3], with the result that the aircraft market is defined not only by volume growth but also by variant growth [4]. This development mainly affects aircraft assembly since the share of assembly activities exceeds the share of parts production [5]. In addition, assembly, as the last stage of product creation, plays an essential role concerning the efficient satisfaction of customer requirements, like on-time delivery [6]. In this context, the continuous customer influence—even after the start of production—together with highly fluctuating process times due to increasing variety of product derivates as well as the increasing need for scalability caused by fluctuating sales volumes (e.g. in times characterized by crises) require a high degree of flexibility with short reaction times [7]. This confronts aircraft manufactures with significant challenges in planning and operation of their often line-oriented assembly systems. Increasing complexity costs due to additional efforts, e.g. supply planning as well as rising downtimes in the assembly lines, are the consequence [8]. Many efforts to control variant diversity within flow lines, for example, by vertical separation of variant-neutral and variant-specific product components [9] or reducing model-mix losses, e.g. by sequence planning [10], partly reach their limits in the case of highly complex products, since the decreasing number of standardized, variant-neutral components reduces efficiency advantages of line production [11]. For a long-term and efficient control of the increasing diversity of variants, new approaches—so far primarily driven by the automotive industry and hardly scientifically based—envisage a step away from complete line assembly towards hybrid production structures [8]. This paper highlights the potential of these approaches and identifies challenges associated with these approaches in the context of aircraft manufacturing. Based on this, a framework for meeting these challenges is presented, which can serve as the basis for designing future-oriented aircraft assembly systems.

2 A Literature Review: Hybrid Production Principles and Their Potentials

The basic idea behind the idea of hybrid production principles (HPP) can be summarized as follows: Through the development of new forms of organization, the classic principle of exclusive line production, established through the paradigm of lean production, is dissolved in favor of a more workshop-oriented production or expanded in a suitable form to HPP. The degree between the final renunciation and an individual extension of the exclusive line production, for example, with workshop-oriented bypasses, depends on each approach. A common feature of the identified systems is the use of Automated Guided Vehicles (AGVs). The AGVs allow a flexible transport of, e.g. assembly objects or necessary operating resources to different stations [12, 13].

Göppert et al. describe an approach to freely linked assembly systems. Spatial and temporally rigid linkages of the individual assembly resources are completely dissolved, thus enabling an individual and flexible assembly sequence for each order. An order route is defined individually for each order, which means that different order routes can result in several orders of a product type depending on the available assembly resources. This makes it easier to access different resources for different variants, to switch to alternatives if resources are occupied, and to integrate additional assembly resources without considering spatial sequence restrictions. Otherwise, the dynamic order routes increase the complexity of material supply due to changes in locations, sequences, and timing [14]. A similar approach is described by Hüttemann et al. [15]. In this context, these concepts of freely linked or line-less assembly are parallel to workshop-oriented assembly systems' principles because the central line completely disappears [16].

In contrast, approaches such as modular or fluid assembly [17, 18], box assembly [19] or matrix assembly [20], which do not necessarily aim at the complete dissolution of line-based assembly structures, are summarized under the general term of HPP. According to Kampker et al., these hybrid production principles are an approach that enables the assembly of variant-independent product modules within the flow line and variant-specific assembly operations, which are the main drivers of complexity costs, in decoupled assembly segments [8]. Decoupled assembly segments' structures are based on, e.g. a workshop principle, a construction site principle, or even a matrix principle. The respective areas thereby form a kind of bypass to supply or, in the case of competence-changing assembly segments, an extension of the central assembly line [17].

Assembly segments can be used stationary or location-independent, with the result that a rigid linkage between individual workstations in the central assembly line can be dissolved. Higher flexible cycle sequences open up degrees of freedom both within the assembly line and the processes supplying the assembly line. On the one hand, this enables the assembly of product variants with fundamentally different product designs in a highly variable assembly structure. On the other hand, it has the potential to reduce efficiency losses between low and fully equipped variants due to cycle

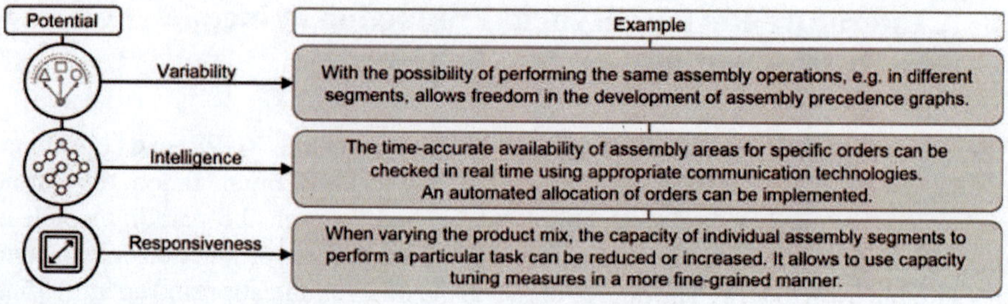

Fig. 1 Potentials of hybrid production principles (based on [15, 18])

time spreading [17, 20]. This leads to an increase in variant flexibility due to partial assembly besides the basic cycle time. At the same time, the additional assembly segments needed to implement HPP causes both capital expenditure and increasing space requirements on the shop floor. Additionally, the partially resulting, unfocused material flow increases the complexity of the assembly processes considerably and thus increases the planning and control effort, for example, concerning material supply [19].

Besides that, three key potentials were identified by moving away from complete line assembly (see Fig. 1). These potentials essentially result from linking location-independent assembly segments and transport processes operated by AGVs [15, 18].

3 Research Methodology

The interdependencies and influencing technical and organizational factors to be considered when integrating HPP into aircraft manufacturing do not allow the research question to be answered based on quantitative analyses. According to the research approach of applied science (AS) by Ulrich [21], design principles can be validly derived based on problem-relevant theories of the basic sciences and procedures of the formal sciences. For the development of this framework, argumentative-deductive reference modeling was used. Hereby, simplified and optimized representations (ideals) of systems are usually created inductively (based on observations) or deductively (e.g. from models) to deepen existing knowledge and generate design options [22]. Weick's 'sensemaking' concept [23] functioned as a guideline to create a valid system understanding. Following this research methodology, the challenges in adopting HPP in aircraft manufacturing are structured below, and the approach developed is presented on this basis.

4 Challenges in the Adoption of Hybrid Production Principles in Aircraft Manufacturing

Fundamentally, the industrial application of highly flexible assembly systems is closely linked to technical and operational/organizational challenges that affect company processes and resources directly related to assembly systems or to accompanying IT systems [15]. To make use of the potentials described in Sect. 2, which are associated with the approaches of HPP, several challenges have to be considered and overcome. These challenges are based on three main problem areas:

The first problem area is the rather practice-driven development of the approach in the automotive industry. Systems, such as box assembly [19] which basic structure is already very close to the approach of HPP, are already described in scientific literature. However, a fundamental understanding of the logistical system behavior of such hybrid production systems is still absent. So far, it is not possible to make any well-founded statements about the performance or cost development resulting from the use of such systems. Production logistic interdependencies resulting from the combination of flow-oriented assembly lines and function-oriented assembly segments (e.g. according to workshop principle) have not been investigated so far. However, this is necessary to enable an objective-oriented and requirements-based design of HPP for aircraft assembly and to evaluate the logistical advantageousness of such structures.

The second problem area relates to the modular assembly segments associated with the HPP, which cause varying space requirements. In aircraft manufacturing, this effect is additionally intensified by the large component dimensions (e.g. fuselage), which already need fundamentally significant floor space requirements within the assembly. Assembly layouts in aircraft manufacturing must consider that as work progresses, the amount of space required increases (e.g., due to the assembly of wings), and individual orders can therefore compete for space on the shop floor [24]. Therefore, the aircraft product itself already imposes limits on hybrid structures in terms of modularity and thus flexibility in aircraft assembly. With the modular assembly areas and AGVs connecting them, which require sufficient travel distances to carry out their individual transport routes, additional participants enter the competitive situation. The traditional boundaries between the strategic design of a factory or unit and operational tasks are becoming increasingly blurred due to the aforementioned influences. The associated complexity is always related to a large amount of information to be processed. Therefore, the use of new information and communication technologies is necessary to implement end-to-end planning processes for digitally controlled large component assembly (see [16]).

The AGV system, which is usually indispensable for the efficient implementation of HPP, as an additional component for aircraft assembly represents the third problem area. In addition to the space requirements of individual AGVs or AGVs operating in a network, which has already been addressed, this problem area also involves technical and organizational challenges in connection with additional components involved in assembly systems. The use of AGVs or Autonomous Mobile Robots (AMR) enables new degrees of freedom in assembly. Still, it is associated with increasing control effort and growing complexity in intralogistics, e.g., material supply [25]. In general, it can be stated that the integration of AGVs and AMRs into the IT landscapes of PPC is a significant challenge. This process is made more difficult by the high safety and quality requirements prevailing in the aircraft industry.

5 Framework for the Integration of Hybrid Production Principles in Aircraft Manufacturing

Hybrid production principles promise great potential for increasing effectiveness and flexibility in multi-variant aircraft manufacturing (see Sect. 2). As outlined in the previous section, combining several basic production principles (e.g. line and workshop production) and a related extension of the production system by AGVS or AMRs results in challenges in the design, description, and evaluation of such production systems. A holistic approach is required to transfer HPP into aircraft manufacturing (focus on final assembly). Figure 2 links the previous observations and transfers them into a macro structure as a basis for the development of the framework. This framework integrates suitable solutions for meeting the identified challenges and their interactions, including necessary information flows and planning-relevant interfaces (see Fig. 3).

The resulting modularity of HPP leads to highly variable layouts of assembly workstations. This must be taken into account in the layout design so that it is possible to react dynamically to necessary layout changes. However, this requires the consideration of operational planning activities (e.g. routing). Therefore, the overriding component within the framework is a consistent digital planning approach

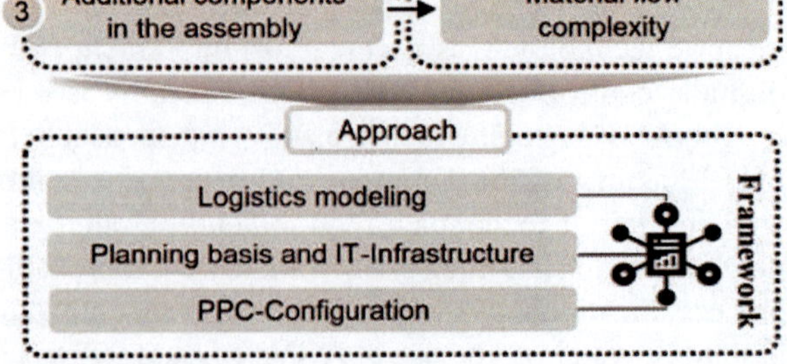

Fig. 2 Macro-structure of the approach

(see Fig. 3), which forms the fundament for the interaction of strategic and operative activities and allows short-term decisions. These decisions can concern, for example, the adjustment of order distributions, assembly sequences, or space requirements. The resulting highly networked processes require the integrated preparation of different information within the planning basis. For this purpose, different digital models are implemented to generate information about products, processes, and the entire factory design during the whole planning process. Model Based System Engineering (MBSE) principles will be forced to combine different digital twins and obtain a complete representation of the aircraft assembly for a potential realization. The Digital Twin model from Lechler [26] could be used as a basis to create the Configured Production System Model (CPSM), which links various information from the digital models and thus provides a gross scope for the HPP-designing part of the framework.

For a detailed design of hybrid production principles and PPC for aircraft manufacturing, it is essential to investigate the logistical system behavior, such as utilization or flexibility resulting from the combination of line and workshop assembly. For this purpose, the approach of logistic modeling is provided as a possible method [27]. First, necessary logistic targets and their interdependencies need to be identified. These are then converted into mathematical models and validated by using simulations. By transferring the findings into a generally applicable form, quantitative cause-effect relationships depending on the selected HPP structure can be analyzed, and design options are derivable. The CPSM provides the information and data required for the logistics modeling. Potentials of HPP in aircraft assembly can be determined concerning the achievement of logistic objectives (e.g. utilization, WIP), and relevant conclusions can be passed on to decision-making instances. In addition, logistics modeling enable the logistics-oriented design and parameterization of planning and control strategies [27, 28]. Therefore, possible factors influencing the

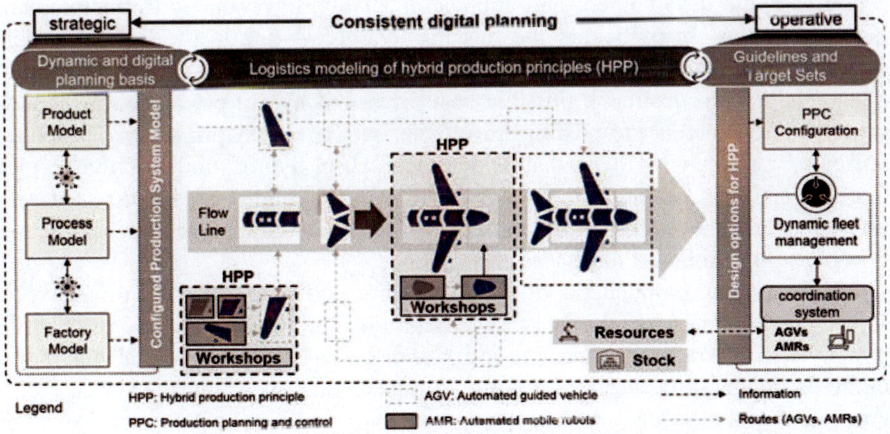

Fig. 3 Framework to enable hybrid production principles in aircraft manufacturing

configuration of a hybrid aircraft assembly concerning system design or intralogistics will be identified (e.g. the number of variants, order time dispersion, transport times). Linked to the development of hybrid assembly structures, suitable design options are derived for an integrated PPC focusing on material flow synchronization, which, e.g. ensures the efficient integration of AGVs. To reduce the complexity and control efforts related to AGVs, an ideal degree of dovetailing between centralized and decentralized planning and control activities is determined in close exchange with information-processing systems. In this context, PPC objectives contain information about possible points of intervention and planning granularity regarding degrees of autonomy, for example, in scheduling or routing specific orders. This information is bundled into guidelines and transferred to a dynamic fleet management system in a suitable form. To ensure a successful realization, the development of an adaptive and integrated coordination procedure, which can harmonize different information streams (e.g. from Real-Time Location Systems), is envisaged at this point. According to guidelines, this system covers all resources (AGVs, AMRs) involved in the aircraft assembly and their degrees of freedom and mutual interactions. Based on the flexibility potentials—resulting from the degrees of autonomy given by the PPC—the system coordinates all relevant resources.

6 Conclusion

The design of hybrid production principles for aircraft manufacturing, combined with a PPC configuration geared to this, holds prospects for more significant increases in efficiency and flexibility in aircraft manufacturing with its many variants. Nevertheless, these approaches—mainly driven by the automotive industry—have hardly found their way into other industry sectors. In part, the complex challenges associated with the development and application of hybrid production principles are reasons for this. In particular, the missing overview—due to the lack of understanding of logistical system behavior in hybrid structures- of company-specific potentials and, for example, possible cost drivers of hybrid production principles often leads to adherence to existing, unsuitable, but transparent principles. This paper presented a framework that transfers the approach of hybrid production principles into aircraft manufacturing, integrates the associated challenges and their interactions, and describes necessary solution components, including essential information flows, planning-relevant interfaces and methods.

The individual components of the framework and their interaction need to be investigated in detail with future work to determine the benefit of hybrid production principles for aircraft manufacturing and provide the basis for the potential integration of these principles into aircraft assembly. It must be examined how far existing mathematical logistic models can be adapted or extended to describe the system behavior of HPP. This includes, for example, the modeling of possible flexibility requirements to assess when and at which assembly stage hybridization would be suitable. In addition, it should be investigating how PPC for hybrid aircraft assembly should be designed to ensure the successful integration of AGVs and AMRs. In cooperation with Helmut Schmidt University and Airbus, these issues are being challenged in the research project "Intelligent Modular Robotics and Integrated Production System Design for Aircraft Manufacturing" (iMoD).

References

1. Biele, A.: Produktionsplanungsansätze In Der Flugzeugmontage: Modellierung, Algorithmen Und Leistungsbewertung. Dissertation—Fernuniversität Hagen, 155p (2018)
2. Hinsch, M. (ed.): Industrielles Luftfahrtmanagemet: Technik Und Organisation Luftfahrttechnischer Betriebe, 2nd edn. Springer-Verlag, Berlin, Heidelberg, 373p
3. Schenk, M., Wirth, S., Müller, E.: Fabrikplanung Und Fabrikbetrieb: Methoden Für Die Wandlungsfähige. Springer-Verlag, Vernetzte Und Ressourceneffiziente Fabrik (2013)
4. Schuh, G.: Produktkomplexität Managen: Strategien; Methoden; Tools, 2nd edn. Carl Hanser Verlag, München, 326p (2005)
5. Sahney, V.N.: Scheduling And Shop Floor Control In Commercial Airplane Manufacturing: Dissertation, Massachusetts Institute Of Technology (2005)
6. Wiendahl, H.P., Gerst, D., Keunecke, L.: Variantenbeherrschung In Der Montage: Konzept Und Praxis Der Flexiblen Produktionsendstufe. Springer, Berlin, Heidelberg, 329p (2004)
7. Mach, F., Zenker, M.: Xxl-Produkte Am Laufenden Band: Einführung Einer Fließfertigung In Die Produktion Von Xxl-Produkten. Zwf - Zeitschrift Für Wirtschaftlichen Fabrikbetrieb **110**(6), 356–359 (2015)
8. Kampker, A., Kawollek, S., Fluchs, S., Marquardt, F.: Einfluss Der Variantenvielfalt Auf Die Automobile Endmontage. Zwf **114**(7–8), 474–479 (2019)
9. Grigutsch, M., Nywlt, J., Schmidt, M., Nyhuis, P.: Highly flexible final production stages—taking advantages of scale effects by reducing internal component variants. Wgp Congress **2012**(907), 127–137 (2014)
10. Koether, R.: Improving productivity in model-mix-assembly. In: Bullinger, H.-J., Warnecke, H.J. (eds.) Toward the Factory of the Future. In: Proceedings of the 8th International Conference on Production Research and 5th Working Conference of the Fraunhofer Institute For Industrial Engineering (Fhg-Iao) at University of Stuttgart, August 20–22, 1985, pp. 761–766. Springer, Berlin, Heidelberg, S.L. (1985)
11. Gans, J.E.: Neu-Und Anpassungsplanung Der Struktur Von Getakteten Fließproduktionssystemen Für Variantenreiche Serienprodukte In Der Montage. Heinz Nixdorf-Institut, Univ, Paderborn (2009)
12. Bubeck, A., Gruhler, M., Reiser, U., Weißhardt, F.: Vom Fahrerlosen Transportsystem Zur Intelligenten Mobilen Automatisierungsplattform, In: Bauernhansl, T., Hompel, M. Ten, Vogel-Heuser, B. (eds.) Industrie 4.0 In Produktion, Automatisierung Und Logistik. Anwendung, Technologien, Migration, pp. 221–233. Springer Vieweg, Wiesbaden (2014)
13. Feldkamp, N., Bergmann, S., Straßburger, S.: Modellierung Und Simulation Von Modularen Produktionssystemen. Simulation In Produktion Und Logistik, pp. 391–401 (2019)
14. Göppert, A., Hüttemann, G., Jung, S., Grunert, D., Schmitt, R.: Frei Verkettete Montagesysteme: Ein Ausblick. Zwf - Zeitschrift Für Wirtschaftlichen Fabrikbetrieb **113**(3), 151–155 (2018)
15. Hüttemann, G., Buckhorst, A.F., Schmitt, R.H.: Modelling and assessing line-less mobile assembly systems. Procedia Cirp **81**, 724–729 (2019)
16. Bornewasser, M., Hinrichsen, S. (eds.): Informatorische Assistenzsysteme In Der Variantenreichen Montage. Springer, Berlin, Heidelberg (2020)
17. Fries, C., Wiendahl, H.-H., Foith-Förster, P.: Planung Zukünftiger Automobilproduktionen. In: Bauernhansl, T., Fechter, M., Dietz, T. (eds.) Entwicklung, Aufbau Und Demonstration Einer

Wandlungsfähigen (Fahrzeug-) Forschungsproduktion, pp. 19–43. Springer, Berlin, Heidelberg (2020)
18. Kern, W., Rusitschka, F., Kopytynski, W., Keckl, S., Bauernhansl, T.: Alternatives to assembly line production in the automotive industry. In: 23rd International Conference on Production Research (2015)
19. Vöhringer, K.-D.: Neue Anforderungen An Die Montagesystemgestaltung Im Automobilbau, In: Milberg, J. (ed.) Wettbewerbsfaktor Zeit In Produktionsunternehmen. Referate Des Münchener Kolloquiums '91 Institut Für Werkzeugmaschinen Und Betriebswissenschaften Technische Universität München 28. Februar / 1. März 1991, pp. 323–336. Springer, Berlin, Heidelberg, S.L. (1991)
20. Greschke, P.: Matrix-Produktion: Konzept Einer Taktunabhängigen Fließfertigung. Bod—Books On Demand, Norderstedt (2016)
21. Ulrich, H.: Skizze Eines Allgemeinen Bezugsrahmens Für Die Managementlehre: Grundlegung Einer Allgemeinen Theorie Der Gestaltung, Lenkung Und Entwicklung Zweckorientierter Sozialer Systeme (1984)
22. Becker, J., Delfmann, P., Knackstedt, R.: Adaptive Reference Modeling: Integrating Configurative and Generic Adaptation Techniques for Information Models, pp. 27–58 (2007)
23. Weick, K.E.: Sensemaking in Organizations. Sage (1995)
24. Prinzhorn, H., Wesebaum, S., Ullmann, G.: Planung Zeitdynamischer Layouts Für Xxl-Produkte. Zwf-Zeitschrift Für Wirtschaftlichen Fabrikbetrieb **1413**(10), 608–611 (2015)
25. Fragapane, G., Koster, R.D., Sgarbossa, F., Strandhagen, J.O.: Planning and control of autonomous mobile robots for intralogistics: literature review and research agenda. Eur. J. Oper. Res. **294**(2), 405–426 (2021)
26. Lechler, T., Fuchs, J., Sjarov, M., Brossog, M., Selmaier, A., Faltus, F., Donhauser, T., Franke, J.: Introduction of a comprehensive structure model for the digital twin in manufacturing. In: 2020 25th IEEE International Conference on Emerging Technologies and Factory Automation (Etfa). IEEE (2020)
27. Nyhuis, P., Wiendahl, H.P.: Fundamentals of Production Logistics: Theory, Tools And Applications, 1st edn. Springer Verlag, Berlin, Heidelberg, 323p (2009)
28. Schmidt, M.: Modellierung Logistischer Prozesse Der Montage. Tewiss Verlag, Garbsen, 160p (2011)

Alexander Wenzel M.Sc. studied industrial engineering with the focus on production and systems engineering at Technical University Braunschweig. Since 2020, he has worked as a research associate at the Institute of Production Systems and Logistics (IFA) at the Leibniz University Hannover.

Torben Lucht M.Sc. studied industrial engineering with the focus on production technology at RWTH Aachen University. Since 2018, he has worked as a research associate at the Institute of Production Systems and Logistics (IFA) at the Leibniz University Hannover.

Peter Nyhuis Prof. Dr.-Ing. habil. studied mechanical engineering at Leibniz University Hannover and subsequently worked as a research assistant at the Institute of Production Systems and Logistics (IFA). He is heading the IFA since 2003. In 2008 he became managing partner of the IPH—Institut für Integrierte Produktion Hannover GmbH.